T0188957

Lecture Notes in Business Information Processing 476

LNBIP reports state-of-the-art results in areas related to business information systems and industrial application software development – timely, at a high level, and in both printed and electronic form.

The type of material published includes

- Proceedings (published in time for the respective event)
- Postproceedings (consisting of thoroughly revised and/or extended final papers)
- Other edited monographs (such as, for example, project reports or invited volumes)
- Tutorials (coherently integrated collections of lectures given at advanced courses, seminars, schools, etc.)
- Award-winning or exceptional theses

LNBIP is abstracted/indexed in DBLP, EI and Scopus. LNBIP volumes are also submitted for the inclusion in ISI Proceedings.

Selmin Nurcan · Andreas L. Opdahl ·
Haralambos Mouratidis · Aggeliki Tsohou
Editors

Research Challenges in Information Science

Information Science and the Connected World

17th International Conference, RCIS 2023
Corfu, Greece, May 23–26, 2023
Proceedings

Editors
Selmin Nurcan
Université Paris 1 Panthéon-Sorbonne
Paris, France

Andreas L. Opdahl
University of Bergen
Bergen, Norway

Haralambos Mouratidis
University of Essex
Colchester, UK

Aggeliki Tsohou
Ionian University
Corfu, Greece

ISSN 1865-1348 ISSN 1865-1356 (electronic)
Lecture Notes in Business Information Processing
ISBN 978-3-031-33079-7 ISBN 978-3-031-33080-3 (eBook)
https://doi.org/10.1007/978-3-031-33080-3

This Springer imprint is published by the registered company Springer Nature Switzerland AG
The registered company address is: Gewerbestrasse 11, 6330 Cham, Switzerland

Preface

We are pleased to present the proceedings of the 17th edition of the International Conference on Research Challenges in Information Science (RCIS 2023), which took place in Corfu, Greece on May 23–26, 2023.

RCIS aims to bring together scientists, researchers, engineers and practitioners from information science in a wide sense. First arranged in 2007, the RCIS conference series aims to provide opportunities for knowledge sharing and dissemination in areas such as information systems and their engineering, user-oriented approaches, data and information management, enterprise management and engineering, domain-specific IS engineering, data science, information infrastructures, and reflective research and practice.

The theme selected for RCIS 2023, "Information Science and the Connected World" reflects the role of information technologies and infrastructures in a world where everyone - and increasingly everything - is permanently connected through technology. In this hyper-connected world, information systems can both be a blessing - as witnessed by the rapid global research collaboration to counteract the Covid crisis - and a curse - as witnessed by the wide and rapid spread of disinformation, for example about Covid and its countermeasures.

The conference program included three engaging keynote talks related to this theme: An opening keynote by Nancy Pouloudi from the Department of Management Science and Technology, Athens University of Economics and Business (AUEB), Greece; "Research Challenges for Large-Scale Interconnected Ecosystems", by Schahram Dustdar, from the Distributed Systems Group, TU Wien, Austria; and "AI and Cybersecurity in the Connected World: A viable partnership?", by Sokratis Katsikas from the Norwegian Center for Cybersecurity in Critical Sectors (NORCICS), Norwegian University of Science and Technology – NTNU, Norway.

A total of 90 abstracts and 87 full papers were submitted to the main track of the conference, from which 28 have been accepted as full papers and included in these proceedings, giving an acceptance rate of 32%. To review and select papers, we relied on a Program Committee composed of 62 members and a Program Board of 11 members. All the submitted papers that were received in an acceptable format and considered in scope underwent a thorough single-blind review process that involved at least three members from the Program Committee and Board. The papers with no clear verdict after the initial round of reviews were discussed online. The discussions were moderated by Program Board members, who concluded each discussion by writing a meta review. The final acceptance decisions were made during a Program Board meeting in Paris on March 15 2023. We are deeply thankful to the Program Board, Program Committee and additional reviewers for their fair, competent and active contribution in selecting papers for publication in the RCIS 2023 conference and proceedings.

In addition to the 28 accepted papers, two journal-first papers were presented in the main conference track after selection by Journal-First Chair Samira Si-Said Cherfi. Besides the main track, RCIS 2023 comprised three other tracks that accepted short papers: the RCIS Forum, the Research Projects track and a Doctoral Consortium.

The Forum track, chaired by Christos Kalloniatis and Maribel Yasmina Santos, invited presentations of research challenges, novel ideas and tool demonstrations. The track received 14 submissions from which 8 were accepted. In addition, 7 more papers first submitted as full papers to the main conference track were invited and accepted as short papers, leading to a total of 15 forum papers that are included in these proceedings.

The Doctoral Consortium track, chaired by Saïd Assar and Tanya E. Vos, invited short descriptions of PhD research projects. The consortium attracted 9 submissions out of which 6 were accepted and included in these proceedings.

Finally, the Research Projects track, chaired by Manos Magkos and Luis Campos, solicited presentations of initial or intermediate project results. The track received 9 submissions, of which 7 were accepted and published in an online volume of the CEUR-WS proceedings.

We wish to express our warm thanks to all the track chairs for their involvement in soliciting and selecting papers for each of these tracks.

The RCIS 2023 Program also included four interesting tutorials: "Blockchain Technology in Information Science: How Blockchains Enable the Connected World", by Pooyan Kazemian of the Weatherhead School of Management, Case Western Reserve University, USA; "Comparing Products using Similarity Matching", by Mike Mannion of the Department of Computing, Glasgow Caledonian University, UK and Hermann Kaindl of the Institute of Computer Technology, TU Wien in Austria; "Getting started with scriptless test automation through the graphical user interface, a hands-on tutorial", by Olivia Rodriguez Valdes, Beatriz Marín, Tanja E. J. Vos, Fernando Pastor Ricos and Lianne V. Hufkens of the Universitat Politècnica de València, Spain and the Open Universiteit in The Netherlands; and "How to Develop and Realize Conceptual Models? The Bee-Up Research & Education Support Tool", by Wilfrid Utz, Patrik Burzynski and Robert Andrei Buchmann of Babeș-Bolyai University, Romania and OMiLAB NPO in Germany. We want to thank the Tutorial Chairs, Raimundas Matulevičius and Natalia Stathakarou, for their great work in attracting and selecting the tutorials, as well as all the tutorial presenters.

We also want to thank Schahram Dustdar for leading a Coaching and Mentoring session for early-career researchers.

Overall, organizing RCIS 2023 was a great pleasure. We want to thank all the members of the organizing committee, chaired by Yannis Karydis, for their engagement and support. We are also grateful to the RCIS 2023 publicity chairs Christoforos Ntantogian, Farhana Sajjad and Hemin Jiang, who ensured the visibility of the conference. Special thanks go to the staff of Springer Nature, who assisted us in the production of these proceedings.

Finally, we thank all the authors for sharing their work and findings in the field of Information Science and all the attendees for their lively participation and constructive feedback to the authors. We hope you will enjoy reading this volume.

April 2023

Selmin Nurcan
Andreas L. Opdahl
Haralambos Mouratidis
Aggeliki Tsohou

Organization

Conference Chairs

General Chairs

Aggeliki Tsohou	Ionian University, Greece
Haralambos Mouratidis	University of Essex, UK

Program Chairs

Selmin Nurcan	Université Paris 1 Panthéon-Sorbonne, France
Andreas Opdahl	University of Bergen, Norway

Organizing Chair

Yannis Karydis	Ionian University, Greece

Doctoral Consortium Chairs

Saïd Assar	Institut Mines-Télécom Business School, France
Tanya E. Vos	Universidad Politécnica de Valencia, Spain

Forum Chairs

Christos Kalloniatis	University of the Aegean, Greece
Maribel Yasmina Santos	University of Minho, Portugal

Workshop Chairs

Shareeful Islam	Anglian Ruskin University, UK
Rebecca Deneckere	Université Paris 1 Panthéon-Sorbonne, France

Tutorial Chairs

Raimundas Matulevičius	University of Tartu, Estonia
Natalia Stathakarou	Karolinska Institutet, Sweden

Research Projects Chair

Manos Magkos	Ionian University, Greece
Luis Campos	PDM&FC, Portugal

Journal-First Chair

Samira Si-Said Cherfi	Conservatoire National des Arts et Métiers – Paris (CNAM), France

Publicity Chairs

Christoforos Ntantogian	Ionian University, Greece
Farhana Sajjad	University of Surrey, UK
Hemin Jiang	University of Science and Technology, China

Program Board

Saïd Assar	Institut Mines-Télécom Business School, France
Marko Bajec	University of Ljubljana, Slovenia
Xavier Franch	Universitat Politècnica de Catalunya, Spain
Renata Guizzardi	Universidade Federal do Espirito Santo, Brazil
Evangelia Kavakli	University of the Aegean, Greece
Pericles Loucopoulos	Institute of Digital Innovation and Research, UK
Oscar Pastor	Universidad Politécnica de Valencia, Spain
Jolita Ralyté	University of Geneva, Switzerland
Colette Rolland	Université Paris 1 Panthéon-Sorbonne, France
Monique Snoeck	Katholieke Universiteit Leuven, Belgium
Jelena Zdravkovic	Stockholm University, Sweden

Program Committee

Rajendra Akerkar	Western Norway Research Institute, Norway
Nour Ali	Brunel University, UK
Raian Ali	Hamad Bin Khalifa University, Qatar
Carina Alves	Universidade Federal de Pernambuco, Brazil
Joao Araujo	Universidade NOVA de Lisboa, Portugal
Fatma Başak Aydemir	Boğaziçi University, Turkey
Dominik Bork	Technische Universität Wien, Austria
Cristina Cabanillas	University of Seville, Spain
Mario Cortes-Cornax	Université Grenoble Alpes, France

Estefanía Serral	Katholieke Universiteit Leuven, Belgium
Ghazaal Sheikhi	University of Bergen, Norway
Samira Si-Said Cherfi	Conservatoire National des Arts et Métiers – Paris (CNAM), France
Anthony Simonofski	University of Namur, Belgium
Pnina Soffer	University of Haifa, Israel
Erick Stattner	University of the French West Indies, France
Angelo Susi	Fondazione Bruno Kessler, Italy
Eric-Oluf Svee	Stockholm University, Sweden
Ernest Teniente	Universitat Politècnica de Catalunya, Spain
Nicolas Travers	Pôle Universitaire Léonard de Vinci, France
Juan Trujillo	University of Alicante, Spain
Jean Vanderdonckt	Université Catholique de Louvain, Belgium
Yves Wautelet	Katholieke Universiteit Leuven, Belgium
Hans Weigand	Tilburg University, The Netherlands
Guohui Xiao	University of Bergen, Norway

Program Committee, Forum

Aikaterini-Georgia Mavroeidi	University of the Aegean, Greece
Alfonso Marquez-Chamorro	University of Seville, Spain
Ana Lavalle	University of Alicante, Spain
Ana León	Universitat Politècnica de València, Spain
António Vieira	University of Évora, Portugal
Carina Andrade	University of Minho, Portugal
Costas Lambrinoudakis	University of Piraeus, Greece
Estrela Ferreira Cruz	Instituto Politécnico de Viana do Castelo, Portugal
Evangelia Kavakli	University of the Aegean, Greece
Hugo A. López	Technical University of Denmark, Denmark
João Moura-Pires	Universidade NOVA de Lisboa, Portugal
Michalis Pavlidis	University of Brighton, UK
Stavros Simou	University of the Aegean, Greece
Vasiliki Diamantopoulou	University of the Aegean, Greece

Program Committee, Doctoral Consortium

Fabiano Dalpiaz	Utrecht University, The Netherlands
Hugo Jonker	Open University, The Netherlands
Manuele Kirsch Pinheiro	Université Paris 1 Panthéon-Sorbonne, France
Beatriz Marín	Universitat Politècnica de València, Spain

Fethi Rabhi	University of New South Wales, Australia
Dalila Tamzalit	University of Nantes, France
Rogier Van de Wetering	Open University, The Netherlands

Additional Reviewers

Agostinelli, Simone
Ali, Syed Juned
Beko, Marko
Benvenuti, Dario
Callegari, Christian
Christopoulou, Eleni
Coelho, Miguel
Curty, Simon
De Luzi, Francesca
Deforce, Boje
Fukuda, Munehiro
Ghanbari, Hadi
Kanavos, Andreas
Karagiannis, Stylianos
Kermanidis, Katia Lida
Langousis, Andreas
Marín, Beatriz
Matos-Carvalho, João Pedro
Mavroeidi, Katerina
Mosquera, David
Mouratidis, Despoina
Muff, Fabian
Nguyen, Khuong
Ringas, Dimitrios
Tsipis, Athanasios
Tsoumanis, Georgios
Weinbach, Bjørn Christian
Yuehgoh, Foutse

Abstracts of Keynote Talks

Research Challenges for Large-Scale Interconnected Ecosystems

Schahram Dustdar

Distributed Systems Group, TU Wien, 1040 Vienna, Austria
dustdar@dsg.tuwien.ac.at

Abstract. In a highly connected world, modern distributed systems deal with uncertain scenarios, where environments, infrastructures, and applications are widely diverse. In the scope of IoT-Edge-Fog-Cloud computing (i.e., the distributed compute continuum), leveraging neuroscience-inspired principles and mechanisms could aid in building more flexible solutions able to generalize over different environments. In this talk we discuss our recent findings and show that highly connected information- and communications systems constitute learning self-adaptive ecosystems.

Keywords: Computing continuum · Edge intelligence · Self-adaptive ecosystems

The Computing Continuum

The transition from Cloud to Edge computing has introduced a new concept known as the Distributed Computing Continuum. This merges the virtually limitless resources of the Cloud with the diverse and proximate nature of the Edge. As a result, the Distributed Computing Continuum integrates the underlying infrastructure of all computing tiers, making infrastructure a priority compared to existing Internet-distributed systems. Current research on Edge computing and the Distributed Computing Continuum concentrates on addressing specific issues, leading to specialized solutions with limited applicability. Several examples include, which is tailored for an ultra-dense network, which presents a solution for a static system description, and, which proposes an Edge-Cloud orchestration that necessitates Cloud centralization. In this article, we aim to present guidelines for more generalized solutions by emphasizing four crucial aspects that need thorough analysis and consensus among researchers, stakeholders, and the scientific community to advance the development of the Distributed Computing Continuum.

Firstly, the Distributed Computing Continuum demands an innovative representation that transcends traditional computer system architectures. Systems within the Continuum consist of a vast array of diverse devices and networks. The functional requirements of such systems can naturally evolve, dynamically alter the running services, or experience unexpected events. These changes will impact the underlying infrastructure configuration, rendering previous architectural representations obsolete. For example, Edge infrastructure necessitates dynamic adaptation to new devices and network connections, resulting in an entirely new system from a design standpoint. This differs from Cloud computing, where changes in infrastructure can be updated without affecting the application. Another challenge for Distributed Computing Continuum Systems is their modeling. It is essential to distinguish between representation and modeling, with representation describing the system, its components, relationships, and features, while modeling addresses the system's dynamic behavior and component interactions. Both concepts require a degree of compatibility for an ideal system representation and model. The complexity and openness of these systems make it difficult to ensure the accuracy of adaptation strategies; in Cloud computing, this is typically resolved by considering a single elasticity strategy per component. Additionally, assessing the impact of adaptation on the entire system is complex, as using a different set of Edge devices may necessitate data transmission through alternate networks, potentially affecting privacy and security constraints.

The third essential element is a lifelong learning framework. The dynamic environment, diverse user behaviors, evolving functional requirements, and long-term infrastructure usage necessitate the development of a learning framework to maintain high-quality standards throughout the system's life cycle. This aligns with the concept of lifelong learning for self-adaptive systems presented in. The final aspect requiring consensus is the business model. A successful business model has been a significant driver for Cloud computing. However, the multi-tenant and multi-proprietary nature of the underlying infrastructure presents a more complex set of stakeholders for Distributed Computing Continuum Systems. To attract the necessary collaborations and investments for the development of such an ambitious computing tier, establishing agreements to enable the optimal business model is crucial.

The primary goal of this keynote is to emphasize the need for a comprehensive perspective on the emerging Distributed Computing Continuum Systems. These systems are far from being fully realized, and we believe that shared foundations are necessary for their advancement. We present what we consider to be critical research challenges and potential research roadmaps for each aspect, aiming to inspire discussion and innovation within the research community.

In conclusion, Distributed Computing Continuum Systems require widespread agreement on representation, modeling, a lifelong learning framework, and a business model to facilitate their development. This article offers initial ideas for constructing these essential building blocks. Following this, we present a technique or concept for each aspect to outline our vision for Distributed Computing Continuum Systems. Finally, we discuss the overall integration of these concepts and techniques before concluding and suggesting future work.

AI and Cybersecurity in the Connected World: A Viable Partnership?

Sokratis Katsikas

Department of Information Security and Communication Technology, Norwegian University of Science and Technology, 2815 Gjøvik, Norway
sokratis.katsikas@ntnu.no
https://www.ntnu.no/iik

Abstract. Artificial Intelligence (AI) and Cybersecurity are domains of intense research activity. However, how one domain influences the other is less widely known and has been insufficiently researched. In this talk we provide an overview of the interrelationships between AI and cybersecurity.

Keywords: Artificial intelligence · Cybersecurity

A tri-dimensional relationship between AI and cybersecurity exists: (i) cybersecurity of AI; (ii) AI to support cybersecurity; (iii) malicious use of AI [1]. Machine learning (ML) is the major tool in today's AI systems. In the following we focus on ML.

The majority of ML methodologies operate with the assumption that their environment is benign. However, this assumption does not always hold. Attackers may wish to either reduce the confidence in the result or cause some form of misclassification, i.e, to alter the output classification of an input example. The latter may take the form of a *targeted misclassification* or that of a *source/target misclassification* [3]. An *adversarial attack* is defined as *the purposeful manipulation of an AI system with the end goal of causing it to malfunction.* Adversaries can maliciously modify the training *(poisoning attacks)* or the test data *(evasion attacks)* [2] or they can launch an *exploratory attack* to attempt to gain as much as possible about the learning algorithm of the victim system [3]. Several defensive strategies exist, but adversarial examples are hard to defend [3]. General purpose technical and organisational standards (such as ISO-IEC 27001 and ISO-IEC 9001) can contribute to mitigating some of the risks faced by AI with the help of specific guidance on how they can be applied in an AI context [4].

AI can be and has been used also as a tool/means to support cybersecurity (e.g., by developing more effective security controls) and to facilitate the efforts of law enforcement and other public authorities to better respond to cybercrime [1]. Examples of such

---

This work has been partially funded by the Research Council of Norway under grant nr. 310105 - Norwegian Centre for Cybersecurity in Critical Sectors (NORCICS).

use include Intrusion detection; Detection of malicious objects (e.g., documents, websites); Detection of anomalous or malicious activities; Digital forensics support; Identification of fake and compromised Online Social Network and user/employee accounts; Threat intelligence; Decision support systems and recommenders; Authentication; and Vulnerability detection. Nevertheless, challenges continue to exist, such as the high precision requirements in many cybersecurity applications; challenges of practically meaningful evaluation and comparison of ML models; and the challenge of designing ML-based IDSs that will perform better than "conventional" techniques under real-world conditions in certain environments, e.g., industrial control systems [5].

AI can be leveraged by adversaries to create more sophisticated types of attacks. AI can make cyber attacks more effective; more targeted; cheaper; faster; more autonomous; more difficult to attribute; and more interactive. Such use cases include the use of AI for evading authentication controls (e.g. CAPTCHA); autonomous/automatic movement in cyber attacks; enhanced social engineering and identity theft; highly targeted massive spear phishing attacks; highly interactive and scalable social engineering attacks [6].

In conclusion, AI systems are being increasingly used in everyday life, including in mission critical systems. Such systems can improve cybersecurity, but can also be misused and they are vulnerable to cyberattacks whose impact can be catastrophic. Therefore, we need to improve our understanding of the interconnections between the two domains; to leverage this understanding to develop methods and tools to address efficiently and effectively the identified challenges; and to validate the methods and tools to be developed in a number of use cases.

References

1. Malatras, A., Agrafiotis, I., Adamczyk, M. (eds.): Securing Machine Learning Algorithms. ENISA, Athens (2021). https://www.enisa.europa.eu/publications/securing-machine-learning-algorithms/@@download/fullReport
2. Pitropakis N., Panaousis, E., Giannetsos, T., Anastasiadis, E., Loukas, G.: A taxonomy and survey of attacks against machine learning. Comput. Sci. Rev. **34** (2019). Article 100199. https://doi.org/10.1016/j.cosrev.2019.100199
3. Chakraborty, A., Alam, M., Dey, V., Chattopadhyay, A., Mukhopadhyay, D.: A survey on adversarial attacks and defences. CAAI Trans. Intell. Technol. **6**(1), 25–45 (2021) 10.1049/cit2.12028
4. Bezombes, P., Brunessaux, S., Cadzow, S.: Cybersecurity of AI and Standardisation. ENISA, Athens (2023). https://www.enisa.europa.eu/publications/cybersecurity-of-ai-and-standardisation/@@download/fullReport
5. Wolsing, K., Thiemt, L., Sloun, C.V., Wagner, E., Wehrle, K., Henze, M.: Can industrial intrusion detection be SIMPLE?. In: Atluri, V., Di Pietro, R., Jensen, C.D., Meng, W. (eds.) ESORICS 2022. LNCS, vol. 13556, pp. 574–594. Springer, Cham (2022). https://doi.org/10.1007/978-3-031-17143-7_28
6. Kaloudi, N., Li, J.: The AI-based cyber threat landscape: a survey. ACM Comput. Surv. **53**(1) (2021). Article 20. https://doi.org/10.1145/3372823

Contents

Forum Papers

Doctoral Consortium Papers

Tutorials

Requirements

Goal Modelling: Design and Manufacturing in Aeronautics

Anouck Chan[1], Anthony Fernandes Pires[1](✉), Thomas Polacsek[1], Stéphanie Roussel[1], François Bouissière[2], Claude Cuiller[2], and Pierre-Eric Dereux[2]

[1] ONERA, Toulouse, France
{anouck.chan,anthony.fernandes_pires,thomas.polacsek, stephanie.roussel}@onera.fr
[2] Airbus, Blagnac, France
{francois.bouissiere,claude.cuiller,pierre-eric.dereux}@airbus.com

Abstract. In aeronautics, the development of a new aircraft is usually organised in sequence. That means the aircraft is designed first, then its industrial system. Therefore, the industrial system may endure stringent constraints due to aircraft design choices. This can result in suboptimal performance with respect to manufacturing. But approaches such as Collaborative Engineering or Concurrent Engineering invite different engineering teams to work simultaneously and together in order to open up new prospects for a product design. In the context of a project that aims at developing methods and tools for co-designing an aircraft and its industrial system, we use Goal-Oriented Requirements Engineering (GORE) to model and to understand their respective expectations but also their dependencies. In this paper, we describe our application of goal modelling based on three iterative attempts. We start from an exploratory stage to have a global picture of the dependencies between the design of an *aircraft nose section* and its *industrial system*. We finish with a focus on a smaller problem in which we understand the key elements of the assembly line performance based on a nose design. For each attempt, we describe our results and feedback, and show how we overcame issues raised at the previous stage. We also highlight the links with known issues about GORE practical application.

Keywords: Collaborative Engineering · Concurrent Engineering · Goal-Oriented Requirements Engineering · Goal Modelling · Aeronautics

1 Introduction

In aeronautics, development is usually a sequential process. Development begins with the definition of the aircraft and the optimisation of its characteristics, and then with the definition of the industrial system and its performance optimisation. By industrial system we mean factories, production processes, supply chain, assembly lines, *etc.* This development cycle leaves very little room for optimising

S. Nurcan et al. (Eds.): RCIS 2023, LNBIP 476, pp. 3–18, 2023.
https://doi.org/10.1007/978-3-031-33080-3_1

or fine-tuning the industrial system's own characteristics. Indeed, the industrial system is quite constrained by the aircraft design. This problem is exacerbated by a silo organisation, with one silo dedicated to aircraft design and another silo dedicated to industrial system design. Such an organisation is almost inevitable in the case of large aircraft manufacturers that have to deal with a large number of stakeholders. However, it reduces the ability to have a global vision and the potential for collaboration across silos.

It is only recently that industrial teams have been involved in the early stages of aircraft development, and aircraft and industrial system designs are seen as part of the same overall system. The idea is that the industrial system own constraints and objectives can also influence the design of the aircraft. This strong relationship between the two designs is typically what leads to Collaborative Engineering, Concurrent Engineering or even Simultaneous Engineering [4,14]. They are well-known approaches that consist in different engineering teams working simultaneously and together in order to obtain an optimum design of several interacting systems[1]. Simultaneously designing two systems together, or what we call "co-designing" in this paper, is particularly challenging when it comes to an aircraft and its production [12]. For instance, a choice of an assembly process, such as drilling instead of welding, can impact aircraft aerodynamics. Conversely, some design considerations, such as the choice of materials, will completely define the manufacturing processes. Optimising the overall system "aircraft/industrial system" design is precisely the purpose of transformation projects in the aerospace industry, such as the *Black Diamond* in Boeing or the *DDMS (Digital Design Manufacturing & Service)* in Airbus [7]. These two companies aim to change their development practices. They want to ensure that, from the early development phases, aircraft and its industrial system are considered together.

In order to meet this challenge, we participate in a project involving academics and industrial actors to define new methods and tools for the co-design of an aircraft and its industrial system. In this respect, it was necessary to have a clear view of each system expectations and how they influence each other. To this end, we decided to apply *Goal Oriented Requirements Engineering (GORE)* approaches. GORE is a part of *Requirements Engineering (RE)* using goals, *i.e.* the objectives the considered system should meet, to express, elicit and negotiate requirements [6]. But in *RE*, the gap between research and practice is a known problem since the 2000s, and studies have been initiated to shed lights on it [3]. Méndez Fernández echoes these findings and proposes the NaPiRE (Naming the Pain in Requirements Engineering) initiative to capture RE practices and problems in industry in order to connect with the research community [10]. As an example of gap, the author points out the predominance of GORE in research but their shortfall in practice [5,9]. Thus, Mavin *et al.* point out the lack of connection between research and industrial practice via a study based on a GORE literature survey and a questionnaire addressed to practitioners [8]. Some recent

[1] For additional information on the specific differences between Collaborative Engineering, Concurrent Engineering or Simultaneous Engineering, please refer to [13].

works have tried to bridge the gap between academic and industry domains by considering more relevant graphical layouts [15] or by investigating how agility and non-functional requirements can be handled together [17].

In a similar spirit, we try to use GORE approaches in an industrial context to identify stakeholders, to clarify both aircraft and industrial system objectives and to elicit inter-dependencies between them. In this paper, we present three iterative attempts to use GORE in collaboration with aeronautics experts. The first attempt focuses on modelling the goals of the aircraft and its industrial system with the experts using a language-free approach. The objective is to softly introduce the notion of goals to the experts and obtain a common vocabulary among their different domains. The second attempt focuses on introducing the iStar GORE language for the goal modelling. The objective is to go a step further in the usage of GORE in order to obtain a more formalised and precised goal model. An additional objective is to use a language close to the goal modelling concepts that could have emerged during the first attempt. The third attempt focuses on a narrower case study in order to push the usage of GORE even further. The objective is to learn from the two previous attempts and re-apply goal modelling from square one on some aspects of the design but in much more details. For each attempt, we present the organisation of the working sessions, the obtained model and the lessons learned from the experience. In addition, we outline the problems we face when operating the sessions.

This article is organised as follows. Section 2 describes the overall case study and its challenges. Section 3 summarises our first attempt to model aircraft and its industrial system following a GORE-like approach without using a specific language. Section 4 presents a second modelling approach for which a GORE language was introduced. Section 5 describes how we overcome some issues we had from the previous approaches by focusing on a smaller case study. Finally, Sect. 6 concludes the article and opens on future work.

2 Case Study Presentation

2.1 Aircraft Nose Section

Our case study focuses on an aircraft nose. We were not interested in any particular aircraft model, but in the general process of designing an aircraft nose and its associated industrial system. A schematic overview of a generic aircraft nose section is given in Fig. 1.

An aircraft is built upon the assembly of cylinder sections which are equipped with all required *aircraft systems* such as electric cables, water pipes, air conditioning, waste treatment *etc.* In addition, the nose section contains the cockpit with its flight instruments (like vertical speed indicator or altimeter) and the landing gear. Because of the diversity and number of elements installed, the nose section is very specific and challenging to design and to build. It faces many constraints from different fields like aerodynamics, electronics, as well as weight or ergonomics. Moreover constraints are closely interlaced. For example, ergonomic

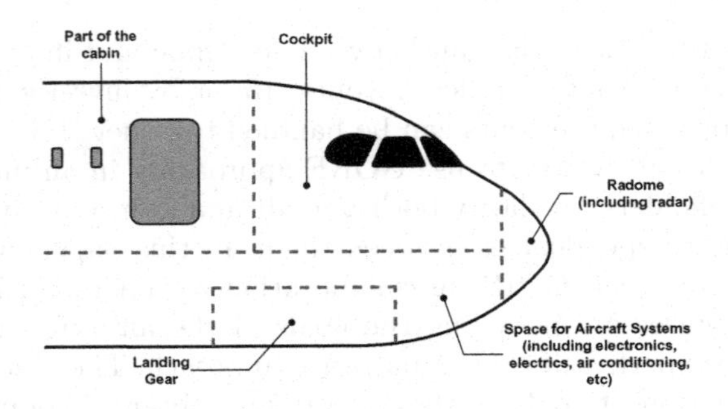

Fig. 1. Generic overview of an aircraft nose section

constraints related to the cockpit (*e.g.* pilots comfort or visibility) are detrimental to the aerodynamic performances. Flying requires a huge number of aircraft systems elements with their own electronics which results in several safety constraints dealing with potential failures. All these elements are heavy, require a large amount of space and must be located as close as possible to the cockpit. Design constraints must also take into account maintenance requirements, particularly between flights. For instance, it is required that faulty equipment can be replaced by functioning one during a turnaround time without interfering with ground operators, crew and passengers.

2.2 Manufacturing Challenges

Design choices of an aircraft have impacts on the way the aircraft is manufactured, and therefore on its industrial system. Moreover, manufacturing and assembling an aircraft nose becomes an even greater challenge when the rate of delivery increase. Thus, the way the nose section assembly is broken down into cluster of tasks is one of the key factors for efficiency. But principles design and associated technologies used to manufacture the primary structure are also essential. For instance, the use of riveting instead of welding to connect parts of the aircraft entails different consequences on industrial system characteristics like machines investment or assembly time. Other elements such as the machining of metal sheets versus cutting them into smaller ones, the number of panels to build the skin or the number of frames and stringers, lead to major differences in terms of cost, assembling time, investment, human resources, *etc.*

On top of these structure assembly challenges, there are ergonomic difficulties for the accessibility to the different areas of the aircraft under construction. For instance, operators generally have to install large and heavy elements (*e.g.* about 70 kg for element of landing gears) in small spaces with limited access (*e.g.* through a small horizontal hatch). Aside from requiring a large amount of time, the installation of these elements may occupy a large area of the aircraft under construction too (*e.g.* the stepladder to access the hatch occupies the space below it, which prevents instruments from being placed at this location). Because no assembly activities can be performed in the occupied areas, this prevents the parallelisation of activities and therefore the overall assembly efficiency.

3 First Application: Modelling the Overall System

In order to build the goal model to understand the dependencies between product and production, we have organised several working sessions with experts. In this section, we first describe these sessions with respect to their participants and their content. Then, we present the resulting goal model and discuss the associated feedback.

3.1 Goal Elicitation: Modus Operandi

Participants. Experts that participated in the working sessions are all members of Airbus (and its various subsidiaries). They all have many years of experience in their specific technical fields but none of them had any prior knowledge of goal-oriented approaches. Some of the experts that we call the core team, attended all the sessions and contributed to the goal model. This team was composed of two architects, specialised in aircraft design and its systems, two industrial experts, specialised in the design and organisation of an assembly line and an expert in the aircraft manufacturer's activities digitalisation. Finally, on top of these experts, there were two academic participants (authors of this paper) that are familiar with GORE approaches and served as facilitators.

Organisation. Six sessions took place over three months and lasted around two hours each. All sessions were organised remotely.

Sessions Content. The first sessions objective was to have a common vocabulary among all the participants. In fact, many professions, competencies, were involved and experts had two very different cultures: the design office and manufacturing. Then, we took a few sessions to introduce the basic concepts of goal modelling approaches, along with the advantage of such approaches. It was only after all the participants had built a shared culture and vocabulary, as well as a clear vision of what a goal is, that a first goal model emerged.

In addition, two more sessions were dedicated to specific technical points. In these two sessions, some experts joined to provide explanations. Firstly, in a half-day teleconference session, two experts came to explain how they had designed an aircraft's nose. Secondly, we had a full day face-to-face session with an assembly processes expert who was able to answer technical questions that arose during the previous working sessions.

Goal Elicitation Support. Experts were free to elicit goals and their links without any formal constraints. It allowed them to confront their different cultures and to share their own vocabulary with others. The working sessions took place on a collaborative online space where each participant could create notes and link notes together. This online space has only one shape in which it is possible to write text. The only way to add semantic, i.e. to distinguish the different concepts, was the use of colours. In practice, the first sessions were dedicated to elicitation and were like *"training wheels"* stages to make the participants start reasoning about goals. We asked them to elicit their different domains goals and the relationships between them, inside their own domains but also intra-domains, in order to design the aircraft nose and its manufacturing. The last sessions were dedicated to consolidate the model by discussing the links and dependencies between the goals of the different domains and by checking the consistency of the modelling. At this point, the participants had a better understanding of goal modelling approaches and a common ground came out in terms of concepts needed for modelling.

3.2 Goal Model

The result of the very free notation approach was a diagram with 123 goals and 120 relationships, distributed almost equally between the design domain and the manufacturing domain. In the end, the participants informally defined and used four types of relationships: *contribution*, *decomposition*, *dependence* and *trade*. The *trade* relationship is unusual in comparison of the other relationships that can be found in other GORE languages. In our diagram, it was used to pinpoint the need for negotiation between the satisfaction of two goals. This model gave us a glance at the concepts manipulated by the participants and their capacity to express goals. An overview of the resulting diagram is shown in Fig. 2. Note that there are 5 dependence relationships between the two domains. The text and some aspects have been removed for confidentiality reasons but it gives a visual insight on the size and shape of the model.

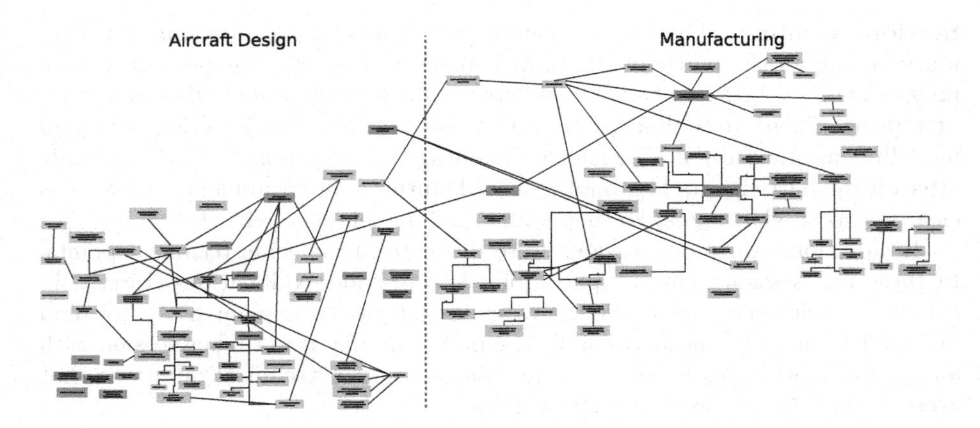

Fig. 2. Overview of the goal elicitation using a online notes-based tool

3.3 Discussion

Positive Aspects. GORE approaches have been well received by the experts. They were particularly satisfied with two aspects.

First, through the GORE approach, the experts could have a global picture of the entire co-design activity. They are traditionally working in silos and it is sometimes not easy to think the activity globally. This view is fundamental, in order to understand the constraints and goals of each domain and to reach an optimal co-design of an aircraft and its factory.

Second, eliciting the goals allowed them to elicit the motivations of their activities. They are missing this capability. Indeed, the motivations are not present in the company processes. Even if experts know that an activity has to be fulfilled, from their experience and from the enterprise process, the reasons for it are not always obvious.

Difficulties Encountered. We encountered various difficulties in this first attempt for applying GORE approaches. Firstly, it was difficult to understand where to start. How do you make industry experts model using GORE when they have never dealt with these concepts? So the beginning of the activity was very confusing. A step-by-step introduction, as we have done, seems to be a good way of addressing the problem in a goal-oriented way.

Secondly, the availability of experts was sometimes a problem. The more we were expanding the model, the more we realised that the core team lacked some very specific technical knowledge. Either the expert domain was not included in the core team or two experts were not present at the same time to discuss a common issue.

Thirdly, we had difficulties to obtain goals and not design solutions or tasks. It was challenging for the participants to reason at goal level. Experts were often inclined to start talking about actions and solutions rather than maintaining a rational goal level. We sometimes had to go through many sessions to be able to express a goal correctly.

Fourthly, this tendency to think in terms of actions and design solutions often led participants to debate very narrow issues, thereby losing the purpose of the activity. The sessions tended to diverge on discussions and debates about technical details and lasted much longer than planned. The sessions required a lot of moderation, it was difficult to find the right balance between letting experts debate in order to find a meaningful goal and stopping them in order to make progress on the model.

Finally, the growing complexity of the model over the sessions was difficult to manage. The activity of co-designing aircraft and its industrial system is, of course, very complex, but the model was becoming equally complex, especially in terms of relationships.

4 Second Application: Introducing a Core Language for Modelling

Following this first attempt to use goal modelling approaches, we decided to take a step back and to assess the ways to proceed. We had obtained a model rich in information for our objective, but informal and with a growing complexity. The experts had also become more familiar with goal modelling. It was time to bring the activity and the experts a step further into the GORE approaches and to introduce a GORE language into the sessions. Based on the emerging types of relationships identified during the free notation approach and inspired by the work of [11], we decided to use iStar [18] to move forward with the sessions. In this section, we describe the sessions with respect to their participants and their content. Then, we present the starting point of the sessions and the resulting goal model. Finally we discuss the feedback of this next step in our application.

4.1 Goal Elicitation: Modus Operandi

Participants. For these sessions, only members of the core team were present. Among them, the two architects, specialised in aircraft design were constantly present. They had some knowledge of the industrial system so they could also help clarify and translate the manufacturing side. In addition, the obtained model was discussed and reviewed with one industrial expert, specialised in the design and organisation of an assembly line, and the expert in the aircraft manufacturer's activities digitalisation. Finally, in addition to these experts, there was one academic participant from the initial two present at all the sessions. The second academic participant took part in discussions on the model between sessions and in the review sessions.

In comparison with the previous attempt, this reduction in the number of participants is due to two factors. First, it was difficult to get all the participants present at every session, their availability being limited. Second, as we already had a lot of information to work with, we were hoping that working in a smaller group of specific experts would make the moderation easier in the sessions without losing too much expertise. Having two aircraft architects with knowledge in the manufacturing was a rare opportunity to take advantage of.

Organisation. Four sessions took place at an average rate of every two weeks during two months and lasted at most two hours each. Two of these sessions were dedicated to modelling and were organised remotely. The two other sessions were a mix of modelling and review sessions. These sessions were organised in an "hybrid" manner: some experts were present in the room, the others were connected remotely due to distance.

Sessions Content. Before starting the attempt, the free notation diagram obtained in the previous section was translated in iStar by the academic participants. During the first session, the language was introduced. Only the concepts actually used for the translation were presented and illustrated to the experts directly on the model. That includes actors, goals, soft goals, tasks, resources, refinement links, contribution links and dependency links. After this introduction, a review was done on the model: goals were reworded or removed if judged too far from our objective, tasks were created and dependencies were clarified. During the second session, the goals and tasks were refined in order to obtain a more detailed view of the aircraft nose design and the manufacturing design domain. In the last sessions, the model was reviewed and amended.

Goal Elicitation Support. During all the sessions, only the academic participants had the hands on the model while the experts had a visual access to the work in progress and were able to guide the modifications. We did not have access to an online and collaborative tool to model with iStar: it was done locally on one academic participant laptop and the screen was shared in live with the experts. Between sessions, the academic participants were discussing the models and were preparing questions or modifications to suggest to the experts at the beginning of the next session. The experts also had access to a visual copy of the model at the end of each session and could review it if necessary but could not modify it.

4.2 Goal Model

In order to structure the free notation diagram, we translated it into iStar and introduced it to the core team. We chose iStar because the dependency relationship is one of its core elements. In fact, as highlighted in [11], identifying and characterising dependency relationships, and more specifically the dependencies between design and production, is a key point in what we call co-design. Moreover, iStar is extensible and could therefore be adapted to meet our needs. For the concepts expressed in the free notation diagram (Fig. 2), the goals were first translated to goals, soft goals or tasks depending on how the academics understood them. Most of the four types of relationships which emerged during the free notation approach were translated in their counterpart in iStar. A *dependence* relationship between domains was translated as a iStar dependency link between the different actors. A *contribution* relationship was translated into a iStar contribution link. The *decomposition* relationship was translated into a iStar refinement link.

The iStar translation of the *trade* relationship was more complicated. In the free notation diagram, the *trade* relationship could be present in two cases: first to pinpoint the need for a negotiation between two solutions to achieve a goal; second, to identify the need for a negotiation about how two goals could be achieved, the way to reach one goal having an impact on the satisfaction of the other one and vice-versa. The iStar translation was realised according to each case. In the first case, goals expressing solutions to the same objective were usually combined together and translated into one iStar task expressed at a higher abstract level, where the solution is yet to decide. In this case, the need for negotiation between those solutions is encompassed in the task and the *trade* relationship is not represented anymore. In the second case, the two goals involved in the *trade* relationship were translated as soft goals or tasks in the iStar model. The *trade* relationship between them was translated in terms of heterogeneous impacts on additional soft goals, using iStar contribution links. These additional soft goals represent the reasons for negotiation.

Starting from this translation of the free notation diagram into a iStar model, the collaborative working sessions were used to iterate over the model and consolidate it. For intellectual property issues, we cannot share the resulting diagram but only describe its main features and give a visual preview of its form in Fig. 3.

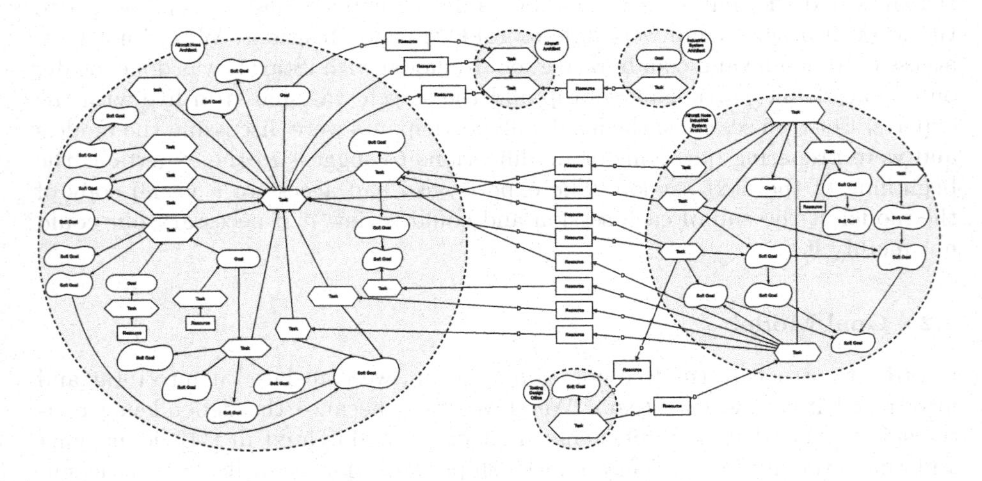

Fig. 3. Overview of the goal elicitation using iStar

The resulting iStar model has 5 actors. Two actors belong to the design domain: the *Aircraft Architect* and the *Aircraft Nose Architect.* Three actors belong to the manufacturing domain: the *Industrial System Architect*, the *Aircraft Nose Industrial System Architect* and the *Tooling Design Office*. The model contains 69 elements, including 30 goals and soft goals, 24 tasks and 15 resources. There are 96 relationships: 14 of them are dependency relationships, the others are mainly refinement and contributions. Among these 14 dependency relationships, 8 are between the *Aircraft Nose Architect* and the *Aircraft Nose Industrial System Architect* (Assembly line).

4.3 Discussion

Positive Aspects. The introduction of the iStar language brought some positive aspects. First, it allows a more goal-oriented modelling. There was an improvement on the expression of goals during sessions and the possibility to define tasks among goals gave a better flexibility in terms of modelling and expressivity. In addition, having a defined language and rules to model helped guide discussions about goals during the sessions. Second, the new version of the goal model brought a better understanding of aircraft design and industrial system dependencies and it was well received by the experts. The dependency description was indeed clearer and more developed than in the free notation diagram.

Difficulties Encountered. We also faced difficulties in this second stage of GORE approaches application. First, even if we introduced a goal-oriented language which helped moderate the discussions, it was still difficult for the experts to think in terms of goals. The modelling still needed a lot of supervision to ensure we did not diverge from the initial objective. Second, despite the fact that the introduction of the iStar language brought some clarity in the modelling, the growing complexity of the model was still an issue. The goal model we obtained after the four sessions was smaller than the free-notation model and more focused on our problematic, but was still incomplete in regards to what is done in reality in terms of aircraft design and manufacturing design. The model highlighted that there was a lot of interactions between the aircraft nose and the industrial system. However, we couldn't manage to take all these dependencies into account in the model. On the other side, if we wanted to focus on a specific part of the model, the associated goals were not refined enough. So, we faced a double problem. On the one hand, we had too many dependencies between the aircraft nose and the industrial system to be able to understand them cognitively. On the other hand, we did not refine the goals enough to be able to understand how to optimise the overall system. Therefore, in order to go deeper into the technical details, we decided to reduce our system of interest.

5 Third Application: Narrower System of Interest

In order to reduce the complexity of the obtained models we decided to focus on a smaller system for this third attempt. In this way we hope to obtain more usable model. Focusing on a smaller system could help to reduce the number of elements while obtaining a more complete model of the system studied. Thus, instead to study the whole industrial system as we did in the previous attempts, we chose to restrict our system of interest to the aircraft nose and its *assembly line*. An assembly line is a set of stations, *i.e.* work spaces, equipped with machines and where operations to assembly an aircraft are done. In addition, compared to the global system studied above, we do not consider here the goals associated with aircraft nose designs. We only focus on understanding the key elements of assembly line performance, and this for a given aircraft nose design.

5.1 Goal Elicitation: Modus Operandi

Participants. In this application, we worked with an intermediate size team compared to the previous approaches. The team was composed of three aircraft and industrial system experts. They were all part of the previous experts team and so had prior experience with goal modelling. They had access to the model of Sects. 3 and 4. We did not make a selection of these experts: those who were available came. On the academic side, there were four researchers. Two of them participated in the previous goal elicitation.

Organisation. We organised eight working sessions over two months. The first session lasted two hours. For the other sessions, there was much more flexibility. Depending on what needed to be discussed, they lasted between 15 min and 2 h. All sessions were done remotely, except the last.

Sessions Content. The first session was dedicated to the problem presentation by the experts. They described the actors, the objectives, what they do and the dependencies in their own words. The purpose was to explain to researchers the main lines of the system to model and the differences with the overall system of interest studied previously. Based on this understanding, a first iStar model was created. Then, the other sessions were dedicated to the consolidation of this model. Between two consecutive sessions, researchers updated the iStar model with their understanding of the problem and prepared some questions on issues that needed to be clarified. On their side, experts worked on answers and completed the model accordingly. Questions were sent by email, but answers were given and discussed collectively in the session. There was no email exchange. Note that for the experts there were no iStar language constraints. So, they decided to use colour and labelled links in order to characterise the impact of tasks on goal satisfaction.

5.2 Goal Model

For this application, we used a more recent version of iStar, iStar 2.0 [2]. We obtained the model presented in Fig. 4. In this model, there are seven actors. Each actor represents a company team with a specific knowledge. Note that a team can sometimes be composed of dozens of persons.

In this model, actors on the industrial system side are: *Assembly line Designers*, *Assembly line Operators*, *Industrial Analyst* and *Industrial Decision-makers.* The function of these actors is to participate in the design of the industrial system. As they share several common goals, we add a high-level actor, namely *Industrial Designers*, that contains these common goals. All industrial system actors are *part of Industrial Designers* actor. Therefore, they not only have their own goals to satisfy but also the *Industrial Designers* goals. On the aircraft design side, we consider a unique actor, *Aircraft Designers*, that designs aircraft. Finally, the *Management Department* actor defines targets for driving economic and strategic company policies.

Actors part of *Industrial Designers* share four soft goals, *i.e.* objectives for which there is no clear definition of when they are achieved. The soft goal *Minimise footprint* expresses the wish to minimise the floor space required by the

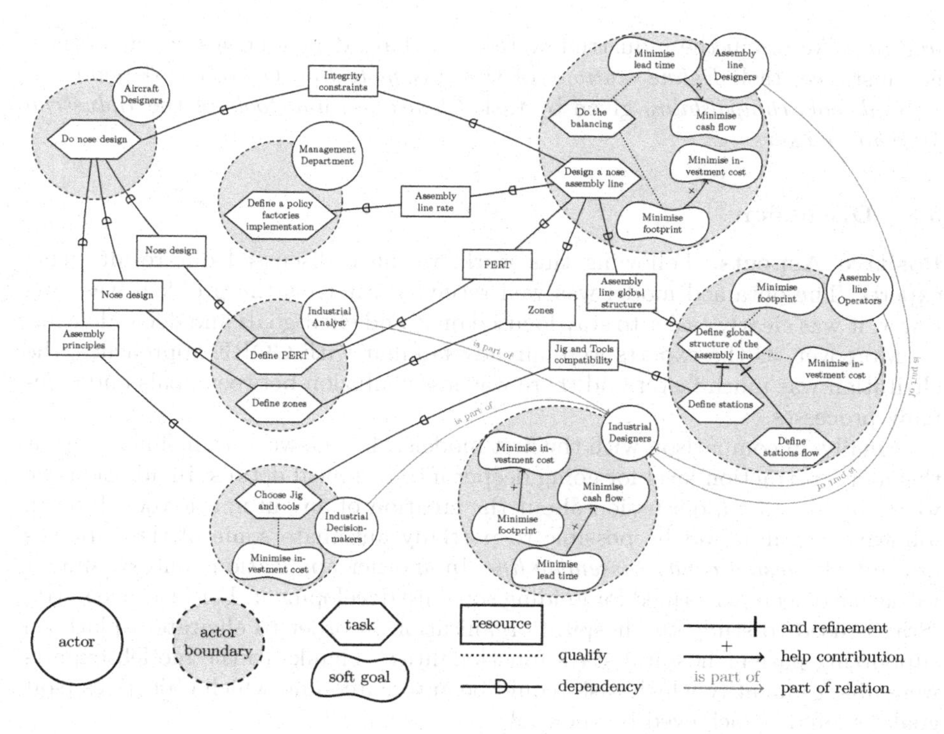

Fig. 4. Aircraft nose assembly line design iStar 2.0 goal model

assembly line. Contributing to its satisfaction also has a positive impact on the satisfaction of the soft goal *Minimise investment cost*. The latter represents the cost implied by the assembly line construction. The soft goal *Minimise cash flow* addresses financial aspects due to the fact that the company must buy elements to build the aircraft (*e.g.* materials or motors) before delivering the aircraft to the airline. Therefore, the soft goal *Minimise lead time*, *i.e.* the total time needed to build a whole aircraft, contributes to the cash flow optimisation.

The activities are closely linked through the *dependency* relationship. For example, the actor *Industrial Decision-makers* has the task *Choose Jig and tools* which has dependence with the task *Do nose design* through the dependum resource *Assembly principles*. This means, in order to choose jig and tools, *i.e.* what type of machines can be used by the operators of the assembly line, *Industrial Decision-makers* need to have *Assembly principles*, *i.e.* how the different parts of the aircraft are joined. The resource *Assembly principles* is provided by the fulfilment of the task *Do nose design*. In order to achieve the task *Design a nose assembly line*, actor *Assembly line Designers* needs *Integrity constraints* that describe the sets of aircraft parts that have to be assembled in the same station. This element is provided by the task *Do nose design*. The actor *Assembly line Designers* also needs to know the number of aircraft built per month. It is given by the actor *Management Department* through the resource *Assembly line rate.* Actors in the industrial system also have dependencies on each other. For instance, task *Define stations* of the *Assembly line Operators* depends on *jig and tools compatibility* given by task *Choose Jig and tools* of the *Industrial Decision-makers.*

5.3 Discussion

Positive Aspects. Following this work, we have discussed our results with experts. The obtained model was well received. First, as the problem was narrower, it was clearly easier to stay focused on it and elicit goals and dependencies.

Moreover, as the experts were already familiar with GORE approaches, the elicitation was much faster and there was less confusion between goals and company processes.

Finally, in comparison with the first models, the one we obtained here was at the good abstraction level for going deeper into technical details. In addition, we were able to start a discussion about the creation of an automatic tool. Indeed, following [1], it would be possible to partially automate some of the subtasks of the task *Design a nose assembly line.* In another application context, similar works have been performed for guiding software development. For instance in [16] iStar is used to study the hospital organisation in order to elicitate a platform supporting patient hospital stays functionalities. Thanks to the model, experts were able to identify which tasks could be automated and which objectives (soft goals) should be achieved by the tool.

Difficulties Encountered. Even if the iStar framework helped us to structure the relationship between actors and elements, we faced some issues that resulted in a model that is more complex than it should. For instance, it is not possible in iStar to have dependencies between two tasks of the same actor. We needed dependencies to value the importance of some resources such as *PERT* or *Zones.* These resources are made available by the dependee task (target of the arrow) and needed by the depender task (source of the arrow). Therefore, to be able to represent these dependencies in iStar, we had to split the industrial system actor into several ones that are all linked to a parent actor in order to maintain the soft goal sharing. Another example of difficulties is the fact that a task of a given actor can only contribute to goals belonging to this actor. So we had to duplicate soft goals several times.

6 Conclusion

We presented applications of GORE approaches in the context of co-designing an aircraft nose section and its industrial system. Our main objective was to understand their respective key goals and dependencies between them. In this industrial context, we attempted to apply goal modelling in three different ways. In the first attempt, we obtained mixed results. Experts hadn't any experience in GORE and had only a few notions in goal modelling. They could freely use an online collaborative graphic tool to model the goals of the whole system of interest. But while GORE allowed experts to have a global picture and structure the elicitation, the resulting goal model was hard to exploit because of the size of the model and the tendency of experts to think in terms of actions rather than goals. In the second attempt, we introduce iStar syntax to a smaller group of experts. It needed a significant moderation by the academic participant to obtain a proper iStar model. But thanks to iStar syntax, this attempt allowed structuring the model and expressing goals at a more appropriate level of description. Nevertheless the model was still difficult to exploit. In the third attempt, the experts had acquired experience in goal elicitation so their supervision was reduced. They were free to use any notation on their diagram while a corresponding iStar model was build by academic researchers. This time, participants focused on a narrower system of interest. The application of GORE was quite successful as the obtained goal model allowed us to go a step further and start discussing the perimeter of an automatic tool.

From these three attempts, our feeling is that organising knowledge in terms of goals is neither easy nor instinctive for industrial practitioners. They are more likely to start by representing a system's goals with actions and design solutions. This may lead to a complex model with an inadequate level of description as we have seen during the first attempt. Using a specific goal language, as iStar, under supervision may help them to think more in terms of goals and improve the modelling. Nevertheless the size of the system of interest may also be an issue as our initial system of interest seemed too broad to obtain an intelligible model. In our specific industrial context, focusing on a smaller system eased the application of the GORE approach and the model obtained in the last attempt was more complete and usable.

For future works, we would like to address the limitations of iStar we had to face, specifically when the goal model is supposed to help the design of methods and tools for supporting decision. In such specific cases, starting from a framework like iStar, it might be possible to adapt it to use Research Operations dedicated vocabulary such as *criteria* instead of *soft goal*. Similarly, it would be possible to use different rules with respect to the dependency relationship that would allow a little more flexibility.

References

1. Chan, A., Fernandes Pires, A., Polacsek, T., Roussel, S.: The aircraft and its manufacturing system: From early requirements to global design. In: Franch, X., Poels, G., Gailly, F., Snoeck, M. (eds.) CAiSE 2022. LNCS, vol. 13295, pp. 164–179. Springer, Cham (2022). https://doi.org/10.1007/978-3-031-07472-1_10
2. Dalpiaz, F., Franch, X., Horkoff, J.: iStar 2.0 language guide. CoRR abs/1605.07767 (2016)
3. Franch, X., et al.: How do practitioners perceive the relevance of requirements engineering research? An ongoing study. In: Proceedings of RE 2017, pp. 382–387. IEEE Computer Society (2017)

4. Göpfert, I., Schulz, M.: Logistics integrated product development in the German automotive industry: current state, trends and challenges. In: Kreowski, H.J., Scholz-Reiter, B., Thoben, K.D. (eds.) Dynamics in Logistics. LNL, pp. 509–519. Springer, Heidelberg (2013). https://doi.org/10.1007/978-3-642-35966-8_43
5. Horkoff, J., et al.: Goal-oriented requirements engineering: a systematic literature map. In: Proceedings of RE 2016, pp. 106–115. IEEE Computer Society (2016)
6. van Lamsweerde, A.: Goal-oriented requirements engineering: a guided tour. In: Proceedings of RE 2001, pp. 249–262. IEEE Computer Society (2001)
7. Mas, F., Arista, R., Oliva, M., Hiebert, B., Gilkerson, I.: An updated review of PLM impact on US and EU aerospace industry. In: Proceedings of ICE/ITMC 2021, pp. 1–5. IEEE (2021)
8. Mavin, A., Wilkinson, P., Teufl, S., Femmer, H., Eckhardt, J., Mund, J.: Does goal-oriented requirements engineering achieve its goal? In: Proceedings of RE 2017, pp. 174–183. IEEE Computer Society (2017)
9. Mylopoulos, J., Chung, L., Yu, E.S.K.: From object-oriented to goal-oriented requirements analysis. Commun. ACM **42**(1), 31–37 (1999)
10. Méndez Fernández, D.: Supporting requirements-engineering research that industry needs: the NaPiRE initiative. IEEE Softw. **35**(1), 112–116 (2018)
11. Nuñez, A.A., Chan, A., Donoso-Arciniega, A., Polacsek, T., Roussel, S.: A collaborative model for connecting product design and assembly line design: an aeronautical case. In: Serral, E., Stirna, J., Ralyté, J., Grabis, J. (eds.) PoEM 2021. LNBIP, vol. 432, pp. 267–280. Springer, Cham (2021). https://doi.org/10.1007/978-3-030-91279-6_19
12. Polacsek, T., Roussel, S., Bouissiere, F., Cuiller, C., Dereux, P.-E., Kersuzan, S.: Towards thinking manufacturing and design together: an aeronautical case study. In: Mayr, H.C., Guizzardi, G., Ma, H., Pastor, O. (eds.) ER 2017. LNCS, vol. 10650, pp. 340–353. Springer, Cham (2017). https://doi.org/10.1007/978-3-319-69904-2_27
13. Putnik, G.D., Putnik, Z.: Defining sequential engineering (SeqE), simultaneous engineering (SE), concurrent engineering (CE) and collaborative engineering (ColE): on similarities and differences. Procedia CIRP **84**, 68–75 (2019)
14. Shenas, D.G., Derakhshan, S.: Organizational approaches to the implementation of simultaneous engineering. Int. J. Oper. Prod. Manag. **14**(10), 30–43 (1994)
15. Wang, Y., Li, T., Zhou, Q., Du, J.: Toward practical adoption of i* framework: an automatic two-level layout approach. Requirements Eng. **26**(3), 301–323 (2021). https://doi.org/10.1007/s00766-021-00346-4
16. Wautelet, Y., Kolp, M., Heng, S., Poelmans, S.: Developing a multi-agent platform supporting patient hospital stays following a socio-technical approach: Management and governance benefits. Telematics Inform. **35**(4), 854–882 (2018)
17. Werner, C., Li, Z.S., Ernst, N., Damian, D.: The lack of shared understanding of non-functional requirements in continuous software engineering: accidental or essential? In: Proceedings of RE 2020, pp. 90–101. IEEE Computer Society (2020)
18. Yu, E.S.K.: Towards modelling and reasoning support for early-phase requirements engineering. In: Proceedings of ISRE 1997, pp. 226–235. IEEE (1997)

Cloud Migration High-Level Requirements

Antoine Aubé[1,2](✉) and Thomas Polacsek[1]

[1] ONERA, Toulouse, France
{antoine.aube,thomas.polacsek}@onera.fr
[2] Stack Labs, Toulouse, France

Abstract. With the increasing adoption of Cloud Computing in the industry, new challenges have emerged for information systems design. In this context, many requirements are met by choosing adequate cloud providers, cloud services, and service configurations. To provide design support, it is necessary to understand what drives the selection of each of these elements. Therefore, it may be interesting to elicitate these high-level requirements, to have an overview of what significantly impacts the cloud environment selection. Here we focus on a particular case of cloud system design: migrations. Through a qualitative study with cloud migration experts, we identify eleven high-level requirements that drive design decisions. We propose an analysis of these results and two classifications to support elicitation and analysis of requirements in cloud migrations.

Keywords: Cloud Computing · System design · Requirements engineering

1 Introduction

Cloud Computing consists of using services to enable computing resources. These services are called *cloud services*, and they may be *Infrastructure as a Service* (*IaaS*) when they deploy virtualized infrastructure elements such as machines or networks, *Platform as a Service* (*PaaS*) when they deploy a framework for executing software, or *Software as a Service* (*SaaS*) when they are software hosted on cloud and provided to their end-users. Cloud services must also ensure rapid elasticity, which means allocating and deallocating resources very quickly, according to users' needs. [12] By configuring cloud services, users enable a *cloud environment* in which applications can be deployed. With Cloud Computing, the notion of infrastructure tends to disappear, as the hardware is hidden behind catalogs of services called *clouds.*

Adopting Cloud Computing can be attractive in terms of cost. Indeed, users pay only for what they consume. Organizations could also expect cost reduction, as they do not need to maintain an on-premise infrastructure with all the inherent costs (e.g., real estate, electricity, Internet access, maintenance workers). However, the reality is not this simple. One does not design information systems

S. Nurcan et al. (Eds.): RCIS 2023, LNBIP 476, pp. 19–34, 2023.
https://doi.org/10.1007/978-3-031-33080-3_2

for the cloud in the same way as traditionally. To be effective, a cloud environment must leverage new concepts brought by Cloud Computing, especially elasticity, to avoid over-consumption. There are also new situations due to the multi-tenancy of the system: it is common to have a part of the system hosted on a cloud and another on-premise. There is a dual objective: to select an environment (i.e., choose the cloud provider, cloud services, and service configurations) and avoid unnecessary costs.

This dual objective is challenging to manage in the early design phases. Indeed, it is not easy to assess the consequences of a cloud service choice on the satisfaction of the requirements in the specification phase. In practice, this lack of foresight leads to overspending and to system designs that do not fulfill the requirements. It is, therefore, a great advantage for cloud adoption to be able to assess these consequences. However, it is necessary to understand beforehand what drives the selection of a cloud provider or a cloud service. Therefore, even if these *drivers*, or *high-level requirements*[1] are various, it may be interesting to elicitate them to have an overview of what high-level requirements have a significant impact on the cloud environment selection. We choose in this study to only focus on migration. Cloud migration means transferring applications to a cloud environment. It involves changes to the system architecture, which depend on the chosen services. We can express our aim in the following research question:

What are the high-level requirements, the drivers, for cloud environment selection (cloud services and service providers) in the migration context?

In order to answer this question, we decided to conduct cloud experts interviews. Our approach is motivated by the lack of existing information in the literature. Indeed, as we will see in Sect. 2, while many works focus on organizations' motivation, the identification of the choice drivers is rarely addressed. Similarly, there is also some work that focuses on the choice of environments, but the requirement elicitation in cloud migration is not addressed. To conduct our interviews, we follow a methodology presented in Sect. 3. Section 4 is dedicated to the interviews, and in Sect. 5, we present the eleven high-level requirements that we extracted from the interviews. In Sect. 6.1, we compare them to the works published in the literature while, in Sect. 6.2, we propose two possible classifications for these requirements and conclude in Sect. 7.

2 Related Work

Since its introduction, many organizations have adopted cloud computing. Some studies have identified motivations that led these organizations to migrate their

[1] While we distinguish requirements (they come from users and stakeholders) and drivers (they are constraints related to the domain), this is not always clear in the early design stage. For instance, security can be as well a requirement and implicitly enforced at some point, as a certain level of security is now considered normal. Therefore, we choose to use both as synonyms in this paper.

systems. For instance, in the early days of cloud computing, Jamshidi *et al.* [10] have led a systematic review of studies about cloud migrations and concludes with six main motivations. In the same year, Nussbaumer *et al.* [13] studied the motivations of Small and Medium-sized Enterprises (*SME*s). Both conclude with a set of motivations that are closely tied to the technical characteristics of cloud services that make them more convenient than on-premise infrastructure. Eight years after, Bremer *et al.* [3] led another systematic literature review of these motivations with a focus on *SME*s. They conclude with a more extensive set of motivations that include most previously found, and many new, that are mainly related to organizational aspects and consider Cloud Computing not simply as an alternative to hosting software but as a means to develop business. These works focus on the motivations for adopting cloud computing while we focus on the drivers for selecting a cloud environment. We notice that some highlighted motivations may be considered as such a driver. For instance, an organization that migrates a system for reliability issues (it is their motivation) will consider new requirements for the system's reliability that will drive the selection of services.

One can distinguish migration strategies by taking perspective from the cloud environment selection. Cloud migration strategies are high-level schemes defining activities to move a system to a cloud. Many works propose a classification of these strategies. For instance, Binz *et al.* [2] distinguish three broad classes of system migration strategies which depend on how the system's components are adapted. In contrast, Andrikopoulos *et al.* [1] focus on the kinds of services leveraged by the migration. In the same spirit, the consulting firm Gartner [8] considers five options, widely called *the 5 R's*. Finally, Zhao and Zhou [17] synthesized several migration strategy classifications and proposed theirs, distinguishing five classes mixing software adaptation and kinds of leveraged services. These classifications describe the possibilities for migration, yet it remains unclear which one to choose to satisfy the requirements best.

In contrast with the strategies, migration methodologies describe processes for achieving migration. They are extensively documented. For example, Hasselbring and Frey propose a model transformation-based methodology, called *CloudMIG* [7], for automating the migration of a legacy system into a *SaaS*. In contrast, Zhang *et al.* [16] propose a generic migration methodology in which a system redesign to *service-oriented architecture* allows to call legacy functionalities from web services hosted on virtual machines. To sum these methodologies up, *MLSAC* [6] is a model-driven framework to express diverse reengineering methods to migrate legacy applications to a cloud. Gholami *et al.* [9] also led an extensive analysis of migration methodologies. While the tailorability of methodologies is one of the survey criteria to adapt the methodology to one migration, the suitability of the methodology for this migration is not addressed. For us, this suitability depends on the migration requirements and needs to be established to determine the best methodology.

3 Methodology

To answer our research question, we conducted a qualitative study to collect feedback from cloud migration experts. Qualitative studies are common in many research fields, such as medical studies [5]. They are often in the form of semi-structured interviews, which are interviews with a pre-established set of open-ended questions. We followed the qualitative surveys empirical standard of *ACM SIGSOFT* [14].

Interviewers. The interviewer is a Ph.D. student who has worked as a cloud developer in the industry for three years.

Sampling. The sampling aimed to interview industry employees who participated in the initial phases of at least one cloud migration. We wanted to filter participants who only have theoretical knowledge of cloud migration. As recommended by McCracken [11], we have looked for diverse profiles with proximity to our research question; in this sense, we have sought participants with various professions who contribute to cloud migrations.

Data Collection. Each interview had a duration of between 45 min and one hour. The interviews were face-to-face or by video conference, one participant at a time. The audio was recorded with the interviewee's agreement for later analysis.

For each interview, we always follow the same pattern. The interview is divided into three parts. Firstly, we explain the purpose of our study and how the interview will be conducted. We also define the specific terminology we are using in order to avoid any ambiguity (e.g., *cloud environment*, *cloud system*, *requirement*). Secondly, interviewees briefly introduce themselves and the context of

Table 1. Interview questions

#	Interview question
1	How did you analyze the components to migrate? What information did you search for?
2	How did you consult the requirements already met by the system to migrate?
3	Do you have metrics or proofs of the satisfaction of those requirements?
4	Did you gather information about how users interact with the system?
5	How did you elicitate the cloud-specific requirements?
6	Were there requirements for the quality of service? If so, what aspects?
7	Were there requirements for the system's environmental performance? If so, what aspects?
8	Was the cost of workforce accounted for in the operating budget?
9	How were expressed the requirements towards the operating budget? Was it for the whole system, one budget per functional unit, etc.? Were these fixed amounts, or was it more complex? Were there tolerance thresholds, etc.?
10	Were there requirements for the software adaptation endeavor?

the migration project. In this step, we collect the company business domain, the role and responsibilities of the interviewee, and their number of years of experience in the field of cloud migration. Then, in the second step, we have a general discussion where the interviewee briefly introduces the context of migration and his role. Our work of identifying the drivers really begins in the third part of the interview. This is the main part, where we collect the interviewee's feedback. To structure this step, we built an interview guide with ten questions which can be found in Table 1. Questions 1 to 4 address the analysis of the existing system. Question 5 focuses on the requirements elicitation method, while the others explore the performance of the system (questions 6 and 7), operating budget (questions 8 and 9), and migration cost (question 10).

4 Corpus

All the interviewees are cloud consultants: they support organizations in migrating to the cloud. They work with a wide range of organizations: small and large companies, government organizations, and international companies. So this allowed us to get feedback on various migration projects (e.g., big or small teams, different business domains). Although they told us about projects in their careers, all the experts in this study were consultants at Stack Labs, a French company specializing in cloud computing consulting, at the time of the interviews.

Table 2. Interviewees' Demographic Information

#	Domain	Role	Exp. (years)
P1	Satellite imagery	Project Manager	18
P2	Satellite imagery	Cloud Architect	10
P3	Satellite imagery	Data Engineer	17
P4	Agriculture	Cloud Developer	19
P5	Nuclear industry	IT Operator	4
P6	Local authorities	Cloud Enabler	16
P7	Consumer goods selling	DevOps	1
P8	Consumer goods selling	Project Manager	8
P9	Audience monetization	Cloud Enabler	15
P10	Ministry	IT Operator	1
P11	Health	DevOps	8

Regarding the migration context, the role of the interviewees, and their number of years of experience in the field of cloud migration, answers are summarized in Table 2. The participants' experience varies from 1 to 19 years, but most had more than ten years of experience in the industry. Among the roles, *DevOps* does not refer properly to the eponym methodology but to a position in charge of both the development and operations; this is how the interviewees named their role.

Note that interviewees encountered difficulties in collecting requirements. For instance, *P1* did not know the maximum budget allocated to operations:

> *"I think they had one, but they never told us about it. I guess they were waiting for our proposal to see if it suited their needs."*

We have two hypotheses to explain this. On the first hand, the interviewees were consultants, so the information may not have been shared to encourage proposing the least expensive architecture possible. On the other hand, the companies might not have been able to estimate such a budget with relevance because they lack expertise in cloud computing.

5 Cloud Migration High-Level Requirements

The following paragraphs present our *Cloud Migration high-level Requirements* (CMR). They are sorted by decreasing occurrences in the interviews.

*Operations Effort (**CMR1**).* Once a cloud system is in production, human operators are charged to monitor and repair it to keep it running smoothly. These operations activities may be automated and assisted by using adequate cloud services. For instance, logging the system's traces is essential for analyzing errors and correct. It requires a great deal of time and expertise to design, configure and operate a reliable logging tooling on *IaaS* because it is up to the operations team to ensure many properties, such as cross-logging between machines (in the case of one of them is not available) or disk space availability over time for traces storage. However, this time and expertise may be spared by the use of some *PaaS*, such as *Cloud Logging* (a *Google Cloud* service), which delivers the tooling for reliably storing traces without operations team involvement.

All interviewees mentioned the need to diminish the operations effort as a cloud environment selection driver. The motivation is not the same in all projects.

In some of them, mostly in small projects without an operations-dedicated team, it is to reduce the workload of teams in place. For instance, the migration of *P4* involved the modernization, via the use of a workflows orchestrator, of many processes which were manually triggered. They choose a *PaaS* delivering *Apache Airflow*[2] over other solutions to avoid the team managing that component by itself. For the same reason, *P2* chose to use *serverless* services over virtual machines for some components:

> *"They had services written in Java and hosted on Tomcat application servers, which perfectly fits [a* PaaS *service], so we used it [...] that costs nothing (i.e., in terms of development) and it is far more practical [...], so there was no point to dismiss this possibility."*

[2] Workflow orchestration platform. https://airflow.apache.org.

In other projects, reducing operations effort eases the organization to scale and postpones hiring new operators. In this regard, *P7* migrated several systems of a company to the same cloud, using the same *PaaS* containers orchestrator to standardize the technologies used in the company and allow the company to merge operations teams, making it easier for them to enable a new system:

> *"They used to have various platforms, some on-premise, others on public or private clouds, and various technologies [...]. That was a pain for the operations teams, so the migration goal was to enforce one platform and one containers manager for all the projects [...]."*

*Cloud Environment' Costs (**CMR2**).* A cloud environment has a recurring cost billed annually to organizations. This cost depends on the resources' configurations and usage quantities.

Many services may fit the same architectural role but are priced differently. For instance, to store files, a virtual disk may be deployed by an *IaaS*, and its cost depends on the size of the disk, its location, and its type. Objects storage services are *PaaS* alternatives priced on the average stored amount of data over the billed period and at each user interaction (e.g., read, write, list). The cheapest solution depends, in fact, on the system usage by its users.

To comply with the performance and cost requirements, *P5* opted for an *IaaS* to host a software forge so that they could configure a bespoke platform. In contrast, the *PaaS* alternative had either undersized or oversized configurations. This way, they lowered the environment's costs while maintaining a high operations effort, which is acceptable when migrating from on-premise infrastructure, as *P9* said:

> *"For each project (i.e., a customer of the company), there needs to deploy a new platform which means that, on-premise, they would deploy one server, or two, or X depending on the load. Cloud computing is a perfect fit for the situation. The on-premise infrastructure made it until the migration, but at an exorbitant cost [...]."*

Minimizing the environment's costs was mentioned in most interviews, but only *P5* made it a priority. In most cases, this goal is less important than other criteria. For instance, *P9* could have used *serverless* services, which would have been a cheaper solution, but members of the team did not want this kind of service:

> *"We (i.e., the DevOps team) faced dogmatic developers who were against using managed services. Using [an IaaS] has already been difficult for them to accept."*

*Cloud System Reliability (**CMR3**).* Organizations put in place a policy called a *Disaster Recovery Plan* (DRP) to respond efficiently in case of system failure. This plan aims to minimize the impact of failures on their customers (e.g., service interruption, data loss). Such failures may be caused by a climatic event (e.g., earthquakes, fire) or malicious operation.

When hosting a system on-premise, it is the organization's responsibility to deploy the whole strategy, which may be costly. This may involve, for example, provisioning a stock of hardware in case some machines shut down or ensuring that there is always an on-site operator to handle the incident. However, the actions are limited. For instance, systems in many organizations are deployed in only one premise. There is, therefore, no backup in the event of a premise-wide failure such as a power outage.

The migrations of *P1*, *P2*, *P3*, *P4* were initiated after a failure that had consequences on their business. Some of the companies had an insufficient *DRP*, and some others did not have one. *P2* told:

> *"The goal [of the migration] for the system's operations was to bring robustness [...] The first objective was to make it reliable because, initially, everything relied on some machines and a single engineer to operate the whole. [...] Their initial setup made it very difficult to recover after a disaster."*

Migration was therefore associated with creating a *DRP*. Indeed, clouds have qualities that facilitate a *DRP* implementation and minimize its cost via the responsibility transfer of a part of the system's operations. It also offers opportunities of making *DRP* more robust. For instance, services are delivered with *Service Level Agreements* (*SLA*) which give insurance about some properties, such as the resources' availability or durability. Resources may also be deployed worldwide to duplicate data and computing units in distant places, so a disaster in one location cannot interrupt the whole service.

The concern of reliability is essential for companies delivering a service with a *SLA* by themselves. The team of *P9* worked with was concerned about migrating for this reason, as the loss of system uptime means customers will demand refunds and, possibly, terminate the contract with the company. As of *P11*, they reported concerns about their team's user experience. Indeed, the system would crash when too many users interacted with it. As a result, they leveraged the elasticity of cloud computing to handle those occasional spikes.

*Latency (**CMR4**).* In many projects, latency experienced by the user was a significant concern because a too-long wait could jeopardize the system's adoption.

It was a network latency issue for the company where *P3* intervened. They used to host their system on-premise in France, which resulted in poor performances for their new customer in South America. To solve this issue, the cloud provider has been chosen so that it can deploy the system both in Western Europe and South America.

In other situations, the company focuses on workload processing speed when they deal with heavy computation tasks. One of the goals of the migration of *P5* was to reduce the software building time (e.g., compilation, GPU-intensive tasks) of a software forge. They used bare metal machine services to configure tailored machines for the task (e.g., with GPU) and to limit the virtualization overhead. They then installed a container orchestration system on top of these machines:

> *"We had two possibilities: either use a managed Kubernetes*[3] *or build it ourselves. We made the second choice for performance reasons. Because the virtual machines behind the managed Kubernetes did not meet the characteristics we were looking for, or, otherwise, it would have been far too costly [...]."*

P4 also had this requirement because the system has a cumbersome workload process twice a year. It must have enough resources to handle it but does not need them for the rest of the year. Elasticity has been the key in this situation, with the tasks orchestration platform leveraging highly scalable serverless computing resources to shorten the periodic workload processing.

A third way to deal with latency was reported by *P9*. Performance was the company's primary concern, so *P9* measured the user-perceived latency before and after the migration.

> *"[...] Before, all the servers were in the same room, now they are in several gigantic data centers, data must pass through local networks, the Internet, etc., so obviously we lose performance. It is something that I measured [...], and they accepted my proposal because I had planned solutions to compensate for the performance loss."*

The mentioned solutions consisted of software adaptation; for instance, they switched the communication protocol to a lighter one.

The interviews highlighted elasticity and the use of tailored *IaaS* as solutions to the latter issue; we expect the resources' collocation to be also one, but no participant mentioned it.

*Need of New Skills (**CMR5**).* Hosting a system on a cloud requires specific technical skills to develop and operate this system. Some of these skills are related to cloud computing specificities, such as elasticity, which must be considered to build an efficient system.

Some others are cloud provider-centered and depend on the chosen services. For instance, a *IaaS*-based environment can host the same software as an on-premise infrastructure, with little or no adaptations. However, it would take a lot of adaptations to host this software on a serverless service, as it needs to use this service's capabilities (e.g., use a supported programming language, use a specific library). Adaptations are specific to a service in particular: if the software is to be hosted on another serverless service, other adaptations will be necessary.

Some interviewees have worked in small teams with developers who are also in charge of the operations. In these cases, the company could not hire new employees. They could not even replace some to avoid the loss of business-related knowledge. Cloud services had, therefore, to be chosen to minimize the need for

[3] Kubernetes, a widespread containers orchestrator: https://kubernetes.io/

formation for the team. For instance, *P2* cited the requirement of keeping *Python* as the development language:

> *"It had to be maintainable by them, in terms of development. [...] For instance, the scripts had been written in Python, and we could not suggest that they should be rewritten in Go because the developers were data scientists. They knew Python because it is widely used in their field, and it is quite difficult to switch languages overnight."*

P6 indicated that the choice of cloud provider was evident in their migration because the operators had experience with one, so they chose it.

*Digital Sovereignty (**CMR6**).* Digital sovereignty does not have a unique definition for organizations, but it is a more and more important theme in the cloud computing industry. Regulations such as the *CLOUD Act* in the United States of America or the European *General Data Protection Regulation* (*GDPR*) show the will of governments to take this theme in charge.

In practice, organizations concerned with digital sovereignty impose constraints on the location of cloud resources and the legislation ruling cloud providers.

This concern was present during the migration of *P5* because the system deals with sensitive data related to French energy sovereignty. It was therefore required that all data and processing remain in France and that the cloud provider follows the French legislation:

> *"The IT direction of [the company] wanted that the system stayed in France. It was about french nuclear plants, so they did not want it to be hosted on [an American cloud provider] [...]."*

Depending on the context, the implementation of digital sovereignty may be less restrictive. For instance, *P6* had to make a trade-off between digital sovereignty and costs:

> *"They wanted their data hosted on a sovereign cloud, so we suggested a trade-off: host the data on [an American cloud provider] and the cipher key on [a French cloud provider]."*

Sometimes, this concern implies not migrating some components to a cloud. For example, *P8* related that some data were stored in China and must remain there, but the chosen cloud provider does not have a data center in this country. So the system is hybrid to meet the digital sovereignty requirements.

*Security (**CMR7**).* It surprises us that only a few interviews mentioned security as a driver of the cloud environment selection.

P6 sliced the migration into phases: the first to migrate the whole system on *IaaS*, then several modernization phases to leverage more interesting services. Many services were put aside later because a network topology was established and validated in the first phase. Using these services would have broken this topology, possibly introducing security issues. It is indeed necessary, as the French state placed some components of the system under the status of *Scientific and technical potential of the Nation*[4], which imposes the implementation of means to protect it from espionage and piracy.

P8 stipulated that security was essential for migration. In particular, their company is divided into multiple sectors, and they wanted the insurance of data tightness between sectors. They leverage their cloud provider's identity and access management to achieve this.

*Adherence to the Cloud Provider (**CMR8**).* Although the amount of research works aimed to reduce cloud vendor lock-in [15], the interviews seem to indicate that the industry has not embraced these solutions. Instead of putting in place measures to limit the vendor lock-in, the team decides whether to accept or not this adherence.

P5 said that the team wanted to minimize the vendor lock-in because it was the company's policy. If the cloud provider changes its conditions or pricing, they want to shift to another provider immediately. To achieve this, they relied on open source standards and widely used software such as *PostgreSQL* delivery as a *PaaS* or *Kubernetes*:

> *"We chose Kubernetes because it is an open industry standard [...] allowing the company to be most independent to other companies."*

On the contrary, some companies require one cloud provider in particular. In the case of *P1*, the company wanted to host the system on *Google Cloud* because they use many other *Google* products and want to take advantage of this ecosystem.

*Cloud Migration' Cost (**CMR9**).* Many requirements involve adaptations to software (e.g., use of *PaaS*), which represent development tasks that increase the cost of migration.

However, the migration budget was very scarce in some projects related to the interviewees. Consequently, the adaptations were confined to the bare minimum, restricting the choice of services. As mentioned before, *P2* explained that some micro-services from the existing system were migrated to *Google App Engine*[5] with very little work because they were written in Java and running on a Tomcat server, which are compatible characteristics with this service. The rest of the system has been hosted on *IaaS* virtual machines, which are less attractive in terms of costs but need no work for the migration.

*Cloud Migration' Duration (**CMR10**).* Some migrations are not constrained by the budget but by time because they have a firm deadline.

[4] French protection program for the Nation's assets: http://www.sgdsn.gouv.fr/missions/protection-du-potentiel-scientifique-et-technique-de-la-nation/.

[5] App Engine, a Google Cloud *PaaS*. https://cloud.google.com/appengine.

This is what *P2* related:

> *"As their activity is seasonal, the migration had to be finalized before the start of the coming season."*

In their case, because the budget was also an issue, they renounced modernization. For instance, they initially intended to change the database to use a *PaaS* database with similar performance at a much lower cost, but the change would have taken too long. If the budget had been less limited, they might have involved more developers and made these adaptations. This driver is, therefore, not the same as the previous one.

*Ecological Efficiency (**CMR11**).* Ecology in information systems has been a popular theme for a while. In 2008, Chen *et al.* [4] published propositions for achieving the ecological efficiency of information systems. However, only a few organizations consider ecology while designing their systems.

Clouds offer opportunities regarding this theme thanks to the pooling of a significant number of resources which, on the first hand, encourages the cloud provider to find solutions to minimize energy consumption and hardware purchases to make substantial savings and, on the other hand, avoids the multiplication of premises and IT resources for each organization.

Some cloud providers integrate features such as low carbon locations or data centers *PUE*[6] dashboards to help their customer design low-impact cloud environments.

In the interviews, only *P8* has worked on a migration where ecological efficiency was considered:

> *"[...] there was another issue, an ecological one. For them, it was the most important to take into account. [...] We were not told why; we suspect that it may be related to their brand image or requirements of their shareholders."*

It affected the cloud environment's selection: resources located in ecological data centers (powered by renewable or nuclear energies, efficient cooling) and resources' collocation to reduce network usage. According to them, cloud providers still provide too few indicators to eco-design a cloud system.

6 Discussion

6.1 Connections with Related Work

Migration Motivations. As mentioned in Sect. 2, some previous works established lists of motivations for migrating to cloud computing. Some may also be understood as drivers for selecting the cloud environment. All drivers from cited works

[6] Power Usage Effectiveness: a standard measure of how efficiently a data center uses energy.

Table 3. Comparison of selection drivers between the study and related works

This study	Jamshidi *et al.* [10]	Nussbaumer *et al.* [13]	Bremer *et al.* [3]
CMR1	*"Maintenability"*		
CMR2	*"Cost saving"*	*"Cost"*	*"Cost reduction"*
CMR3		*"Reliability"*	
CMR4	*"Elasticity to fluctuation"*	*"Performance"*	
CMR5			*"Easiness of use", "Technical understanding of Cloud Computing"*
CMR6			
CMR7		*"Security"*	*"Data security concerns"*
CMR8	*"Interoperability"*		*"Dependency on provider"*
CMR9			
CMR10			*"Reduced set-up time"*
CMR11			
			"Transparency in regard to service"
			"Easy possibility to test services before purchase"
			"Service & Support"

and this study have been listed in Table 3 and annotated whether we consider them as motivations or selection drivers.

The drivers listed as motivations are related to either general characteristics of clouds (e.g., *"Scalability"*, *"Focus on the main business"*) or characteristics of companies that might benefit from cloud computing (e.g., *"Competitive pressure"*, *"Organizational size and structure"*).

As we can see, the interviews highlighted most selection drivers from the literature, plus two new drivers: *Digital sovereignty* (**CMR9**) and *Ecological efficiency* (**CMR11**). In addition, the driver *Adherence to the cloud provider* (**CMR8**) was addressed only in one of its aspects: the related works documented requirements towards reducing this adherence, while the interviews pointed out that some organizations desire to use (or not use) a cloud provider in particular.

Requirements towards *"Service & Support"* do not appear in our study. Two reasons can explain this observation. First, the leading cloud providers' services are now considered well-documented and have a high level of support. This driver, thus, has not appeared in the interviews. Second, the interviewees, being consultants, acted as support during the migration. Support requirements were then satisfied by the qualities of the consultancy delivery rather than those of the cloud services. Their role may also explain why none spoke about *"Easy possibility to test services before purchase"* because, as cloud experts, they have a comprehensive knowledge of the services and the underlying mechanisms. Other types of participants may have highlighted it. The current state of cloud computing in the industry might explain why *"Transparency regarding service"* was not mentioned because, except for some *IaaS*, major cloud providers give little to no clue about the ways resources are delivered.

Migration Methodologies and Cloud Environment Selection Approaches. Most reviewed approaches for migrating a system and selecting a cloud environment do not indicate how to evaluate their suitability, given the migration requirements. While none participant claimed to have followed any of these approaches, we might draw some associations between them and the identified requirements families.

Both *CloudRecommender* and *COCA-PT* aim at minimizing the costs of the environment, which unambiguously matches **CMR2**. Quinton's approach assumes pre-made decisions about software, much like Zhang's methodology. This looks like reported migrations with no software adaptations, so they seem to fit migrations considering **CMR9** or **CMR10**.

6.2 Requirements Classification

After having extracted cloud migration drivers from the interviews, it could be interesting to classify them. Indeed, requirements classifications help engineers and business analysts elicit and analyze requirements.

We first attempted several classifications: one based on the *Goal-Oriented Requirements Engineering* distinction between functional and non-functional requirements, one that distinguishes hard and soft goals, and a last with the kind of services used to meet the requirements (*IaaS*, ...). None was relevant to classifying our *CMR*s.

In the following paragraphs, we present two classifications that we consider relevant.

First, one can consider the *Triple Constraints* of project management. Constraints are: *cost*, *delay*, and *quality*. Of course, these constraints interact with each other, e.g., increasing the quality also increases either the cost or the delay. As shown in Table 4, our high-level requirements can be sorted into three categories: the requirements that minimize costs, those that reduce the delay, and those that increase quality. As in project management, satisfying some requirements of one category can lead to less satisfying other categories. For instance, reducing operation efforts (**CMR1**) is implemented by using *PaaS* and, most of the time, involves software adaptations, which conflicts with the need to minimize the migration duration (**CMR10**).

Table 4. Project management classification

Category	Cloud Migration Drivers
Cost	**CMR1**, **CMR2** **CMR5**, **CMR9**
Delay	**CMR10**
Quality	**CMR3**, **CMR4**, **CMR6**, **CMR7**, **CMR8**, **CMR11**

One can also consider classification based on the DevOps paradigm (development activities and operation activities). If this dichotomy is interesting, we must also consider the characteristics of cloud services that, if well selected, can satisfy some requirements. In Table 5, the requirements are classified according to these three categories. The *need of new skills* (**CMR5**) requirements affect both software development (e.g., knowledge about a programming language) and operations skills (e.g., serverless architecture monitoring). The *adherence to the cloud provider* (**CMR8**) one comes in two versions: either aiming at the system agnosticism to the cloud provider, which results in little to no development at the cost of operations, or accepting vendor lock-in with the desire to use highly managed services, leading to software adaptations and fewer operations.

Table 5. Team's activities classification

Category	Cloud Migration Drivers
Development	**CMR2**, **CMR5**, **CMR8**
Operations	**CMR1**, **CMR3**, **CMR5**, **CMR8**, **CMR9**, **CMR10**
Cloud characteristics	**CMR4**, **CMR6**, **CMR7**, **CMR11**

7 Conclusion

To better understand the drivers of public cloud environment selection during migrations, we conducted a qualitative interviews study with industry cloud experts. The study resulted in eleven cloud migration drivers that impact the choice of cloud providers, cloud services, and cloud services configuration. We have retrieved most drivers known in the scientific literature with this approach and found new ones.

We also provide two classifications for these requirements to support requirements elicitation and analysis in migration projects.

There are several threats to the validity of this study. We followed a standardized approach for leading the interviews, not for their analysis. Moreover, the corpus is tiny. However, we noticed that only one new high-level requirement emerged from discussions after the 6th interview. We, therefore, believe the study to be close to saturation. Finally, the interviewees are similar in their position as consultants, which may lead to missing some requirements.

Future works should address several questions. We reviewed many migration approaches, but there miss a method to determine which one is tailored for a given migration. We think this method would be based on the drivers identified in this paper. In addition, the identified drivers may lead to conflicting requirements, which must be addressed through trade-offs. Yet, it is hard to make the best trade-offs (e.g., what impact on the cloud environment's cost for slightly decreased performance?). A methodological framework should be developed to handle this issue.

References

1. Andrikopoulos, V., Binz, T., Leymann, F., Strauch, S.: How to adapt applications for the cloud environment - challenges and solutions in migrating applications to the cloud. Computing **95**(6), 493–535 (2013)
2. Binz, T., Leymann, F., Schumm, D.: Cmotion: a framework for migration of applications into and between clouds. In: IEEE International Conference on Service-Oriented Computing and Applications, pp. 1–4. IEEE Computer Society (2011)
3. Bremer, M., Walter, T., Fjodorovs, N., Schmid, K.: A systematic literature review on the suitability of cloud migration methods for small and medium-sized enterprises. ESSN: 2701-6277 (2021)
4. Chen, A.J., Boudreau, M., Watson, R.T.: Information systems and ecological sustainability. J. Syst. Inf. Technol. **10**(3), 186–201 (2008)
5. DiCicco-Bloom, B., Crabtree, B.: The qualitative research interview. Med. Educ. **40**, 314–21 (2006). https://doi.org/10.1111/j.1365-2929.2006.02418.x
6. Fahmideh, M., Grundy, J., Beydoun, G., Zowghi, D., Susilo, W., Mougouei, D.: A model-driven approach to reengineering processes in cloud computing. Inf. Softw. Technol. **144**, 106795 (2022)
7. Frey, S., Hasselbring, W.: Model-based migration of legacy software systems into the cloud: the cloudmig approach. Softwaretechnik-Trends **30**(2), 84–85 (2010)
8. Gartner: Migrating applications to the cloud: Rehost, refactor, revise, rebuild, or replace? (2010). https://www.gartner.com/en/documents/1485116. Accessed 07 July 2022
9. Gholami, M.F., Daneshgar, F., Low, G., Beydoun, G.: Cloud migration process - a survey, evaluation framework, and open challenges. J. Syst. Softw. **120**, 31–69 (2016)
10. Jamshidi, P., Ahmad, A., Pahl, C.: Cloud migration research: a systematic review. IEEE Trans. Cloud Comput. **1**(2), 142–157 (2013)
11. McCracken, G.: The Long Interview. Qualitative Research Methods, SAGE Publications (1988). https://books.google.fr/books?id=3N01cl2gtoMC
12. Mell, P., Grance, T.: The NIST Definition of Cloud Computing. Techmical report, 800-145, National Institute of Standards and Technology (NIST), Gaithersburg, MD (2011)
13. Nussbaumer, N., Liu, X.: Cloud migration for SMEs in a service oriented approach. In: IEEE 37th Annual Computer Software and Applications Conference, Workshops, pp. 457–462. IEEE Computer Society (2013)
14. ACM SIGSOFT: Qualitative surveys (interview studies) (2022). https://github.com/acmsigsoft/EmpiricalStandards/blob/master/docs/QualitativeSurveys.md. Accessed 07 July 2022
15. Silva, G.C., Rose, L.M., Calinescu, R.: A systematic review of cloud lock-in solutions. In: IEEE 5th International Conference on Cloud Computing Technology and Science, vol. 2, pp. 363–368. IEEE Computer Society (2013)
16. Zhang, W., Berre, A., Roman, D., Huru, H.: Migrating legacy applications to the service cloud. In: ACM Object-Oriented Programming, Systems, Languages and Applications (OOPSLA) (2009)
17. Zhao, J., Zhou, J.: Strategies and methods for cloud migration. Int. J. Autom. Comput. **11**(2), 143–152 (2014)

Idea Browsing on Digital Participation Platforms: A Mixed-Methods Requirements Study

Antoine Clarinval[1(✉)], Julien Albert[1], Clémentine Schelings[2], Catherine Elsen[2], Bruno Dumas[1], and Annick Castiaux[1]

[1] Namur Digital Institute, University of Namur, Namur, Belgium
antoine.clarinval@unamur.be

[2] Inter'Act Lab, University of Liège, Liège, Belgium

Abstract. Digital participation platforms (DPP) are websites initiated by local governments through which citizens can post and react to ideas for their city. In practice, the majority of DPP users browse the posted ideas without contributing any. This activity, referred to as lurking, has widely recognized positive outcomes, especially in a citizen participation context. However, it has been devoted little attention. In practice, the idea browsing features available on current DPP are limited, and the literature has not evaluated the available approaches nor studied the requirements for idea browsing. In this paper, we report on an evaluation of the filterable list, which is the most common idea browsing approach on DPP. Our findings show that it lacks stimulation hedonic quality and call for a more stimulating approach. Thus, we conducted 11 semi-structured interviews to collect requirements and found that idea browsing on DPP should be supported by the combination of (1) a stimulating interactive representation such as circle packing or thematic trees displayed as entry point and (2) a filterable list for deeper exploration. This article is the first to study requirements for idea browsing features on DPP.

Keywords: Digital participation platform · Content browsing · Lurking · Requirements · AttrakDiff · Mixed-methods

1 Introduction

As part of their daily work, local representatives make decisions on how to allocate the available human and financial resources and design policies that should fit the expectations of the population. However, they often lack knowledge of citizens' needs, and increasingly resort to involving citizens in decision-making processes to capture these missing insights. Such involvement of citizens is not new [3] but it has been further accelerated by the new opportunities offered by information and communication technologies [10,22]. Citizen participation involving these technologies is referred to as digital participation. The United Nations E-Government Survey[1] shows that its adoption has increased rapidly in

[1] https://publicadministration.un.org/en/Research/UN-e-Government-Surveys.

S. Nurcan et al. (Eds.): RCIS 2023, LNBIP 476, pp. 35–50, 2023.
https://doi.org/10.1007/978-3-031-33080-3_3

the last 20 years. One of the most commonly implemented digital participation method is the digital participation platform (DPP), which is an idea generation platform initiated by local governments (e.g., [4,23,25,39]). All citizens can access the platform to browse the posted ideas, contribute ideas of their own, and react to others' ideas.

As usually observed in online communities [35], the large majority of DPP users browses ideas without posting content [14,24]. These users are referred to as "lurkers" [11,36]. While lurking has previously been negatively perceived [46], it is now widely recognized that it has positive outcomes such as the vicarious learning of the community dynamics and the propagation of information in other communities [11,17,48]. The civic nature of DPP gives a special importance to idea browsing, as it is essential for citizens to stay informed [19,38]. The large proportion of DPP users who browse ideas compared to contributors and the importance of this activity in a citizen participation context show that idea browsing is an essential part of DPP. Still, little attention has been devoted to this activity. In practice, the idea browsing features of current DPP are limited to filterable lists and less frequently dot maps. To the best of the authors' knowledge, previous literature has not investigated whether these features satisfactorily support idea browsing on DPP, and the requirements for browsing ideas on these platforms remain unknown.

The goal of this article is to answer these two gaps with a focus on the filterable list, which currently is the most commonly implemented approach on DPP. To achieve this, an evaluation of the filterable list from a selected representative DPP was conducted with 38 respondents who completed the AttrakDiff questionnaire [18]. The results show that while it offers satisfactory pragmatic quality, it lacks stimulation hedonic quality. Based on these results, 11 semi-structured interviews were conducted to (1) explain the results observed in the questionnaire for stimulation hedonic quality and (2) collect requirements for idea browsing on DPP.

2 Background

2.1 Citizen Participation Through Digital Platforms

Citizen participation refers to the involvement of citizens in the decisions taken by their government, excluding participation in the elections and in public life [7]. This process can have several objectives that can be achieved using a wide range of methods [45]. These objectives include informing citizens of the decisions for transparency purposes, collecting their opinion on already defined resource allocation plans, or delegating them the decision power in part or in full [3]. Methods include traditional (i.e., non digital) approaches such as town hall meetings [29] and innovative ones made possible thanks to the new opportunities offered by information and communication technologies [10,22]. Digital participation methods can potentially attract more participants since they make it possible to participate remotely at any time instead of having to physically attend a scheduled event [21]. Citizen participation organized through digital methods is referred to as digital participation, online participation, or e-participation in the literature. It is defined as the use of "information and communication technologies to broaden and deepen political participation by enabling citizens to connect with one another and with their elected representatives" [30].

Digital platforms [4] are among the most common digital participation methods. This paper focuses specifically on digital platforms initiated by local governments allowing citizens to propose ideas of actions, with the possibility to browse and react to others' proposals. They are referred to as digital participation platforms (DPP) in this article. Examples of DPP include "Decide Madrid" (Spain) [39], "Réinventons Liège" (Belgium) [25], and "Better Reykjavik" (Iceland) [23].

2.2 Idea Browsing on Digital Participation Platforms

In group discussions involving many participants such as public hearings [15], it is often observed that only few participants drive the exchanges by sharing their views while the vast majority, although concerned and interested, remains silent and listens. The same phenomenon occurs in online communities. A minority of members, referred to as "actives" or "contributors" [6], feeds the community with content, while the majority, referred to as "lurkers" [35], only consumes the content (i.e., in the context of DPP, consuming content consists in browsing ideas). According to the "90-9-1" participation inequality rule stated by Nielsen [34], it is common for the proportion of lurkers to reach 90% of users, while 9% are occasional contributors and the remaining 1% frequently feeds the community with content. The number of lurkers is argued to be higher in the context of digital citizen participation [12]. After monitoring the use of Regulation Room, which is a participatory rule-making digital platform, it was observed that only approximately 5% of visitors registered as users, and that only a part of them posted comments [14]. In an analysis of the "Réinventons Liège" platform, Lago [24] observed that 17.5% of visitors registered, and that 7% of registered users proposed at least one idea. This represents less than 2% of visitors. The Laugardalur consultation on the Betri Reykjavik platform[2] reports 1,997 users, 125 (6.3%) of which have contributed at least one idea. These numbers show the importance of idea browsing features on DPP, as these features are used by a large majority of users unlike the posting of ideas. In addition, idea browsing also concerns contributors [1,33], as contributing content usually implies browsing the posted content beforehand.

Lurking in online communities has been plagued with negative connotations [11,37,47], lurkers being sometimes considered as selfish free-riders [46] who take advantage of active members' contributions without giving anything in return. However, lurking has now widely recognized positive outcomes and scholars recommended to encourage this behavior [11]. Inside the community, lurking allows newer members to get more familiar with the community and learn its dynamics vicariously [2,26]. This behavior also allows community members to increase their knowledge base [17], and this can have a positive effect beyond the community via information propagation [48]. Lurking has a special importance in the

[2] https://www.betrireykjavik.is/group/3740.

context of DPP. Indeed, ideas potentially impacting the users' environment and daily life are discussed on these platforms, and it is therefore essential that concerned citizens are aware of them. Literature on citizen participation recognizes the importance of accessing information by considering this activity as a form of participation in itself [3] or as a necessary condition for participation [38].

For the reasons explained in the previous paragraphs, idea browsing is of great importance on DPP. However, as generally observed in other types of online communities [31], little attention has been devoted to this activity. In practice, the idea browsing features of current DPP most commonly consist of filterable lists. Although being the most common browsing interface on opinion content platforms, such lists are limited in terms of scalability and engagement [13,42]. Less frequently, dot maps can be found on DPP as well. In the literature, the attention is devoted to the production of content instead of its consumption. Indeed, the information that should be provided when posting ideas [49,50] and the range of opinions that should be provided for idea voting [43] have been studied, among others. However, no previous work has investigated whether the implemented idea browsing approaches perform satisfactorily (**Research Gap 1**), nor what the requirements for idea browsing on DPP are (**Research Gap 2**). These are the two gaps this article aims to answer, with a focus on the filterable list, as it currently stands as the most commonly implemented approach.

3 Methodology

Data was collected following a mixed-methods strategy [20] involving both quantitative and qualitative data. The explanatory mixed-methods design [9] was chosen because it consists in exploring a phenomenon with a larger number of participants using quantitative methods and explaining the findings with richer qualitative data collected from a smaller sample. Research Gaps 1 and 2 were investigated simultaneously, first by means of a questionnaire, and then through semi-structured interviews. The interview guide was designed based on the questionnaire results and aimed to give more depth to the quantitative findings.

3.1 Quantitative Data Collection – Questionnaire

The questionnaire introduces respondents to the goal of the research and to the DPP concept. They are informed that the collected data will be processed anonymously and consent that it can be used for research purposes. The body of the questionnaire is structured into three parts. The first part asks sociodemographic information (i.e., gender, age range, education level, and occupation), previous experience with DPP, and motivators for browsing ideas on a DPP. The second part asks respondents to browse ideas on the DPP of the city of Mons (Belgium) (named "Demain Mons"[3]) and to complete a shortened version of the AttrakDiff questionnaire [27]. The AttrakDiff questionnaire has been

[3] https://mons.citizenlab.co/fr-BE/projects/participez-ici/3.

used in a wide range of application domains and has the advantage of covering both pragmatic and hedonic aspects of interaction [28]. Following [27], the items of the questionnaire were translated into French and half were reverse-coded. The motivation behind the choice of the "Demain Mons" platform is that it relies on CitizenLab, which is a well-established turnkey citizen participation platform implemented in more than 400 local governments worldwide. Therefore, we considered this platform as representative of existing DPP. The ideas on "Demain Mons" are displayed as a filterable list. The third part of the questionnaire aims at gathering respondents' opinion on alternative representations that could be used to browse ideas on a DPP. Respondents are presented with four images illustrating visual idioms and are asked to select the one that relates the most to a set of citizens' ideas or to describe a better fitting representation. The four images were designed by the authors after taking inspiration from existing research and platforms. The list represents the standard approach implemented on DPP. The tree was inspired by previous research on DPP which proposes a representation destined to local representatives [25]. The circle packing (named "bubbles" to avoid technical terms in the questionnaire) is used in the dashboard of Citizenlab[4] destined to local representatives. The light bulbs view is inspired from the way ideas are represented in popular cartoons. The four images are shown in Fig. 1, along with their explanation as included in the questionnaire. The questionnaire was distributed using social media and mailing lists.

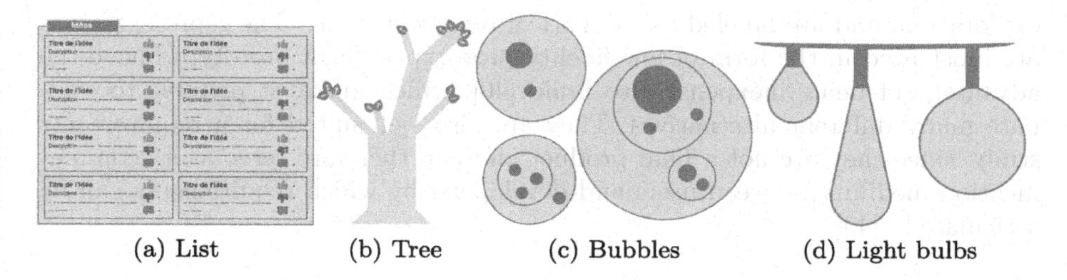

(a) List (b) Tree (c) Bubbles (d) Light bulbs

Fig. 1. Representations of ideas shown in the questionnaire. (a) The *list* includes the title, description, number of likes, dislikes, and comments for each idea. (b) The *tree* groups ideas from the same theme and is divided into branches that correspond to sub-themes. The size of a branch corresponds to the number of ideas and reactions. Each leaf corresponds to a single idea. (c) Each large *bubble* groups the ideas pertaining to a single theme. The inner bubbles correspond to sub-themes. The size of the bubbles varies according to the number of ideas. The ideas are represented by dots which size corresponds to the number of reactions for this idea. (d) On the garland, each *light bulb* groups the ideas of a single theme. The size of the light bulb corresponds to the number of ideas and reactions for that theme.

[4] https://www.citizenlab.co/blog/civic-tech/4-reasons-why-digital-participation-is-easier-than-you-think-2/.

3.2 Qualitative Data Collection – Semi-Structured Interviews

In order to delve deeper into the findings of the questionnaire, semi-structured interviews were conducted. Interviewees were recruited through an open call on a voluntary basis. No incentive was given for participating. The interviews cover three parts in 40–60 minutes. Each part deepens one section of the questionnaire. The first is a discussion on interviewees' motivations for accepting to be interviewed, their general opinion and previous experience with DPP, and their motivators and deterrents for browsing ideas on these platforms, after asking the same socio-demographic information as in the questionnaire. The second part delves deeper into the results of the AttrakDiff evaluation, which revealed that the weakness of the idea browsing list lies in its lack of stimulation hedonic quality (see Sect. 4.2). Interviewees are presented with seven cards showing different representations of a set of ideas. These include the four presented in the questionnaire (i.e., the list, the tree, the bubbles, and the light bulbs) as well as three additional ones suggested by the questionnaire respondents. The vases and the balloons (Fig. 2 (a) and (b)) are direct variations of the light bulbs suggesting to add shape variations. The mindmap (Fig. 2 (c)) has a branch structure representing the hierarchy of themes and subthemes. Apart from the list, six representations of ideas are thus proposed to interviewees. The tree, the bubbles, and the mindmap allow exploring the theme and subtheme hierarchy and constitute the hierarchical representations group. The light bulbs, the balloons, and the vases give information at the level of themes without supporting drill-down exploration, and are labeled as categorical representations. The representations are illustrated in the form of low-fidelity prototypes. Such prototypes have the advantage of being inexpensive to build [40], which makes it possible to evaluate many different alternatives. They are also well-suited for a requirements study, since they are not a final product and can therefore serve as a "communication medium [between users and developers] by which requirements can be articulated" [40].

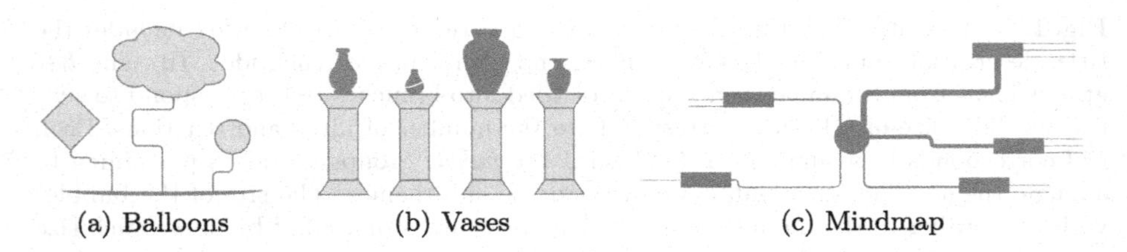

(a) Balloons (b) Vases (c) Mindmap

Fig. 2. Representations of ideas added for the interviews. (a, b) Variations of the light bulb representation. Each theme is represented by a *balloon* or a *vase* of different shape which size varies according to the number of ideas. (c) On the *mindmap*, each node represents a theme which is further divided into subthemes. The thickness of the branches represents the number of ideas in the theme or subtheme they lead to. The leaves of the mindmap represent individual ideas.

After an explanation of the seven representations, interviewees are presented with three pairs of cards on which adjectives are written. Two of the pairs are those used in the shortened AttrakDiff questionnaire to measure stimulation hedonic quality (i.e., Unimaginative – Creative and Dull – Creative). The third one is "Least preferred – Most preferred" and measures general preference. Opposing adjectives are placed on opposed ends of a table and interviewees are asked to rank the seven representations by placing them on the table between the two adjective cards. In order to avoid biasing the ranks by forcing interviewees to choose, ex aequo rankings are permitted. In the third part, interviewees are asked to focus on their few most preferred representations and to imagine what would be their ideal idea browsing approach.

4 Results

4.1 Sample Description

In total, 38 valid completed questionnaires were collected. Regarding socio-demographics, 16 (42%) of the respondents are females. All age groups are represented and the mean age is 40 (approximated from the age intervals). 92% of the respondents hold a higher education degree and 82% are employed or self-employed, the others being unemployed (5%), retired (5%), or studying (8%). In the second phase of the research, 11 participants (3 females) were interviewed. Their average age is approximately 35. 10 are employed and 1 is retired. 10 hold a higher education degree. Their motivation for agreeing to take part in the interviews is that they find that DPP are a "good" and "healthy" initiative from local governments. They believe that it has potential to foster democratic processes, inform citizens, discuss ideas constructively, and help public servants to better understand citizens' needs.

Overall, the respondents and interviewees stated that they would be motivated to browse ideas on such platforms to (1) discover public opinion trends, (2) compare their opinion to others', and (3) consult others' ideas by location, topic of interest or simply out of curiosity. Their main reasons for not browsing ideas on a DPP are (1) the fear that the posted ideas would not effectively be taken into account, making it useless to browse them, (2) the local government being unable to process a high number of ideas, (3) the lack of transparency on the idea selection process, (4) the high number of ideas to browse, and (5) the low usability of the browsing interface.

4.2 Evaluation of the Current Idea Browsing Approach

The results of the evaluation of the idea browsing list show an average of 0.47 for hedonic quality, 1.05 for pragmatic quality, and 0.91 for attractiveness (Fig. 3). Following the official AttrakDiff interpretation guidelines, this indicates that the idea browsing list is "task-oriented." It performs satisfactorily on the attractiveness and pragmatic aspects, although scores around 1 suggest that there are areas of improvement. On the other hand, it is not the case for hedonic quality. Although the Tacky – Stylish score is satisfactory, the Cheap – Premium aspect receives a score of 0.4. The scores are even lower for the stimulation-related aspects. Indeed, the Unimaginative – Creative and Dull – Captivating aspects received an average of 0.3 and 0.1, respectively. This shows that the main issue with the idea browsing approach currently implemented on DPP is its lack of stimulation quality.

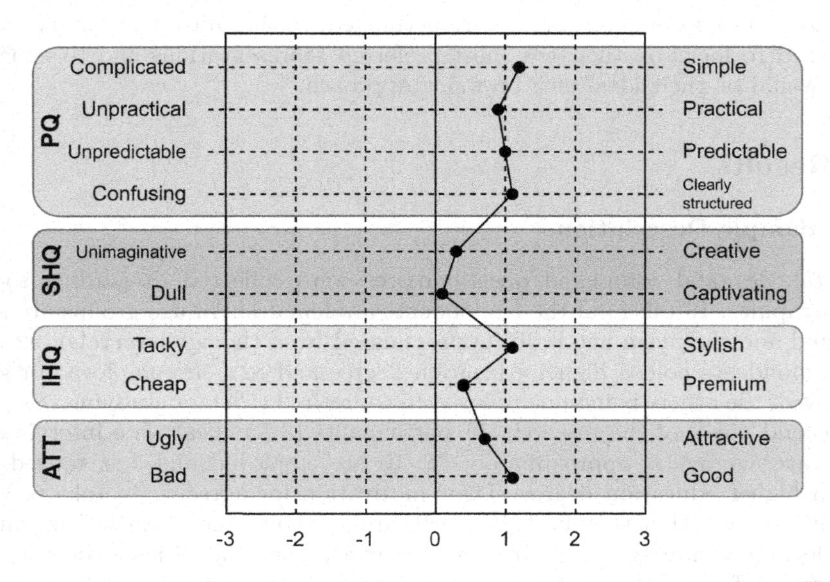

Fig. 3. Average score for each pair of the shortened AttrakDiff questionnaire (**PQ** = Pragmatic Quality, **SHQ** = Stimulation Hedonic Quality, **IHQ** = Identity Hedonic Quality, and **ATT** = ATTractiveness).

In the ranking activity, interviewees were asked to assign a rank from 1 (corresponding to the worst) to 7 to each representation, which in turn received a score equivalent to the rank. When several representations were assigned the same rank, they were each given the same score, corresponding to the average of the positions they would be assigned if there was no equality in the ranks. For example, if the lowest rank is given to three representations, they are each given a score of $(1 + 2 + 3)/3 = 2$. Figure 4 shows the scores derived from the ranks assigned by the eleven interviewees (I1–I11) for the Unimaginative – Creative pair. Nine interviewees ranked the list as the least creative. They explained that lists are very commonplace and that the list is as such the least original representation. The scores given to the list for the Dull – Captivating pair (Fig. 5) are by far the lowest among other representations. Only one interviewee ranked the list higher than second to last. Interviewees explained that a list is "boring" and "painful to browse."

	I1	I2	I3	I4	I5	I6	I7	I8	I9	I10	I11	AVG
List	1	1	1	1	1	1	1	2	1	1.5	1	**1.1**
Tree	2.5	6	4	3	3	7	7	4	4	4	7	**4.7**
Bubbles	4	6	3	5	7	6	3	3	3	3	3	**4.2**
Mindmap	2.5	6	2	2	2	3	5	1	2	1.5	2	**2.6**
Light bulbs	6	3	5	6	5	4	3	6	7	5.5	5	**5.0**
Balloons	5	3	6	7	5	2	3	7	5	7	5	**5.0**
Vases	7	3	7	4	5	5	6	5	6	5.5	5	**5.3**

Fig. 4. Scores derived from the ranks assigned by each interviewee for the **Unimaginative – Creative** pair.

The low ranks given to the list representation were nonetheless nuanced by several interviewees. They noted that although the list was overall the least creative and captivating representation, it was still an efficient approach to browse ideas in details and that it should be provided on a DPP. This is in line with the satisfactory scores observed in the questionnaire for the pragmatic quality.

	I1	I2	I3	I4	I5	I6	I7	I8	I9	I10	I11	AVG
List	1	1	1	2	2	2	1	1.5	1	2	4	**1.7**
Tree	5	6	7	4	3	7	6	4	5	2	7	**5.1**
Bubbles	7	6	2	6	4	5	3	3	6	6.5	5	**4.9**
Mindmap	6	6	6	3	1	6	7	1.5	4	6.5	6	**4.8**
Light bulbs	3	3	5	7	6	3	3	6	7	2	2	**4.3**
Balloons	4	3	4	5	5	1	3	7	3	4	2	**3.7**
Vases	2	3	3	1	7	4	5	5	2	5	2	**3.5**

Fig. 5. Scores derived from the ranks assigned by each interviewee for the **Dull – Captivating** pair.

4.3 Toward a More Stimulating Idea Browsing Approach

The six alternative representations outperform the list on the Unimaginative – Creative, Dull – Captivating, and general preference (Fig. 6) rankings. However, the question of which representation performs the best among these is more divisive. The image association part of the questionnaire revealed that the bubbles view was preferred by 14 respondents, the list by 11, the tree by 8, and the light bulbs by 2. The 3 remaining respondents suggested the three other representations that were added for the interviews. This illustrates the diversity of possible representations, among which no clear preference stands out. The analysis of the interview rankings also shows a large variance in the ratings.

	I1	I2	I3	I4	I5	I6	I7	I8	I9	I10	I11	AVG
List	6	4	1	7	1	1	5	2	1	1	4	**3.0**
Tree	5	6	4	5	3	7	6	6.5	5	3	7	**5.2**
Bubbles	7	7	3	4	5	6	4	4	7	5.5	5.5	**5.3**
Mindmap	4	5	2	6	2	5	7	1	4	5.5	5.5	**4.3**
Light bulbs	2	2	6	3	6	3	2	5	6	3	2	**3.6**
Balloons	3	2	5	2	4	2	1	6.5	3	7	2	**3.4**
Vases	1	2	7	1	7	4	3	3	2	3	2	**3.2**

Fig. 6. Scores derived from the ranks assigned by each interviewee for the **general preference**.

Overall, the hierarchical representations were found less creative but more captivating than the categorical representations. The general preference ranking favors hierarchical representations as well. Although the interviewees reported that they based their preference on several criteria (e.g., practicality, originality), the data suggests that the captivating character is a more important factor than the creativity. Indeed, the scores of the Dull – Captivating ranking are twice more strongly correlated ($r = 0.63$) with general preference scores than those of the Unimaginative – Creative ranking ($r = 0.31$). However, sample size does not allow for a significance analysis of the correlation.

Two representations stand out from the general preference ranking, namely the tree and the bubbles. Interviewees explained that the hierarchical nature of these representations makes them want to explore the deeper levels. Some interviewees explained their preference for the tree metaphor by stating that it is "cool" (2 interviewees) and allows progressing from general themes to specific ideas intuitively (3). The bubbles representation was found eye-catching (3) and has the advantage of changing more dynamically while the shape of the tree would change more slowly (1).

When describing their ideal browsing approach, 4 interviewees centered their idea around the bubbles and 4 others chose the tree, showing divided opinions in line with the image association part of the questionnaire. 2 others expressed a particular attraction toward the vases and 1 toward the balloon. The interviewees who preferred the bubbles or the tree nonetheless explained that another representation should provide more detailed information on a subset of ideas selected after exploring the tree or the bubbles. They all found that the list was well-suited for this purpose, although three interviewees also mentioned the mindmap as an alternative. Thus, although the interviewees did not converge toward a specific idea browsing solution, the majority of them recommend an idea browsing approach that supports the Information Seeking Mantra [44]. The solution they envision shows a stimulating representation – the tree and the bubbles were suggested the most frequently – giving the global picture (**overview**) as entry point. Then, users should be able to identify a subset of ideas by interacting with this representation (**zoom and filter**). Finally, the selected ideas should be displayed as a list to give detailed information (**details-on-demand**).

5 Discussion

5.1 Implications for Research and Practice

The findings presented in this article have direct implications for researchers and practitioners, and in particular for designers of DPP. Regarding Research Gap 1, the results from the AttrakDiff evaluation show that the most commonly implemented approach is not satisfactory in terms of stimulation quality. As for Research Gap 2, all the interviewees suggested an idea browsing approach that differs from the one provided on DPP. Their suggestions did not converge toward a specific solution but highlight a general architecture involving a stimulating representation such as a tree or bubbles as entry point with a list allowing in-depth exploration of a selected subset of ideas. This architecture can serve as basis for designers willing to improve idea browsing on their DPP.

In addition to answering the two research gaps, the interviews hinted motivators and deterrents for browsing ideas on DPP, such as the lack of confidence in the truthfulness of the local government and the perception that the posted ideas could not be processed due to a lack of resources on the local government's end. This extends previous literature [37] with insights specific to DPP, although further research is needed to produce a more complete picture. Furthermore, the findings reported in this article can serve as inspiration for researchers studying similar platforms implemented in different contexts. One example is the digital suggestion box [41]. Lastly, the ranking activity from the interviews was met with a lot of enthusiasm. The mix of a quantitative question and think-aloud for qualitative explanation worked especially well. Interviewees reported that it was playful, encouraged them to be more diligent in the ranking, made the question more "tangible," and made the interview feel less lengthy since it added diversity in its conduct. Based on this positive feedback, we recommend researchers to incorporate similar activities when possible in their data collection instruments.

5.2 Limitations

The research presented in this article has several limitations. The first is the sample size of 38 respondents of the questionnaire. While it gives interesting general tendencies, it is too low to provide any statistically generalizable results. The second relates to the representativeness of the sample. More than 90% of the questionnaire respondents and of the interviewees hold a higher education degree. This indicates that the most educated part of the population is much overrepresented. This is most probably due to the distribution strategy of the questionnaire, which relied exclusively on digital channels including professional social media. The third limitation stems from the illustration of the visual representations used in the questionnaire and the interviews. The representations could have been illustrated in other manners and with different levels of details, which could have impacted the findings. The fourth limitation is that the results could not successfully converge toward a specific idea browsing solution, but rather hinted a general architecture, which reduces the impact of the findings for practitioners. Nonetheless, this general architecture can serve as starting point for a design generation process, such as a design studio [27], able to deliver a solution stemming from a shared vision. The fifth limitation is the focus on stimulation hedonic quality. It is motivated by the results from the questionnaire that show that stimulation is the lowest rated quality, and therefore the most in need for improvement. However, while more stimulating representations were identified, their other qualities (e.g., pragmatic quality) will also have to be assessed when refining the general architecture into specific designs.

5.3 Future Work

Many respondents reported that they had never heard of DPP before, although the largest cities of their region have implemented them recently. This is a very common issue with citizen participation initiatives: they usually attract few new participants and struggle to engage citizens beyond the "usual suspects," which causes representativeness issues. Previous literature on information seeking provides an interesting explanation to this phenomenon. It highlights that individuals can seek information in different ways [32]. The large majority of citizen participation methods, including DPP, only support active encounters with information, meaning that individuals have to make a step forward to encounter the information (i.e., browse ideas on a DPP). Another mode of information seeking consists in serendipitous encounters with information, and has therefore a much greater potential to attract new participating citizens. One way of implementing this mode of information seeking in the context of DPP would be to show a visual representation of the ideas on a public display, which would be deployed in the public space and accessible for browsing to any passerby [51]. Public displays have already showed success in implementing citizen participation initiatives [8] and proved their potential to attract much more citizens than traditional approaches [16]. It would be valuable to research whether showing a representation of the ideas from a DPP in the public space would help to increase the awareness of the platform and in turn attract more lurking and contributing participants, and what would be the impact on the representativeness.

Contributors on DPP should be representative of the population since their input is expected to influence decisions that will affect the whole population. The numbers discussed in Sect. 2 show a very low proportion of contributors among DPP users, let alone among the citizenry, which poses representativity issues.

It is therefore essential to research how lurkers can be encouraged to become contributors. Several leads have been proposed to encourage lurkers to contribute in their online community [5,26,37,47]. Overall, three directions emerge when aggregating these recommendations, along with the cross-cutting concern of ensuring good usability. First, mentoring from elder active participants. Second, implementing mechanisms that echo gamification such as rewards, ranks, and cooperation. Third, offering content browsing mechanisms prompting new contributions. Regarding the latter, it would be valuable to investigate whether implementing an idea browsing approach following the general architecture suggested in this article would actually result in a higher number of contributions. The integration of nudges into representations of ideas would also be interesting to study, as previous research showed that nudges can increase contributions on user-generated content platforms [52].

6 Conclusion

Digital participation platforms (DPP) are online websites put in place by local governments. They are a call for citizens to post and react to ideas of improvement for their city. An important part of the interaction with DPP is to browse the posted ideas. This is necessary for citizens willing to contribute reactions or ideas of their own, but also for those willing to get acquainted with the posted content without contributing. This latter group represents the large majority of users and is referred to as lurkers. However, the idea browsing approaches implemented in current DPP are limited, the most common one being a list filterable by theme. To the best of the authors' knowledge, previous literature has not investigated whether this approach satisfactorily supports idea browsing, and has not studied the requirements for idea browsing on DPP.

In this article, the idea browsing list of a representative DPP was evaluated using the AttrakDiff questionnaire. Results showed improvable but satisfactory pragmatic quality and attractiveness, and insufficient stimulation hedonic quality. Then, interviews were conducted to gain qualitative insights into the questionnaire results and collect requirements for the design of a more stimulating idea browsing approach. While the interviews did not converge toward a specific design, they confirmed that the current idea browsing approach is not satisfactory, and that idea browsing should instead be implemented using a stimulating interactive representation such as circle packing or thematic trees as entry point combined with a list for further exploration. This article makes a step forward in the understanding of the requirements for idea browsing on DPP. It also proposes to investigate in future research how visual representations of ideas could make the content posted on DPP more representative of the population.

Acknowledgements. The research pertaining to these results received financial aid from the Federal Science Policy according to the agreement of subsidy no. [B2/223/P3/

BeCoDigital]. Financial support was also received from the European Regional Development Fund (ERDF) for the Wal-e-Cities project with award number [ETR121200003138] and from the Research Public Service of Wallonia (SPW Recherche) for the project ARIAC by DIGITALWALLONIA4AI with award number [2010235].

References

1. Aristeidou, M., Scanlon, E., Sharples, M.: Profiles of engagement in online communities of citizen science participation. Comput. Hum. Behav. **74**, 246–256 (2017)
2. Arnold, N., Paulus, T.: Using a social networking site for experiential learning: appropriating, lurking, modeling and community building. Internet High. Educ. **13**(4), 188–196 (2010)
3. Arnstein, S.: A ladder of citizen participation. J. Am. Inst. Plann. **35**(4), 216–224 (1969)
4. Berntzen, L., Johannessen, M.R.: The role of citizen participation in municipal smart city projects: lessons learned from Norway. In: Gil-Garcia, J.R., Pardo, T.A., Nam, T. (eds.) Smarter as the New Urban Agenda. PAIT, vol. 11, pp. 299–314. Springer, Cham (2016). https://doi.org/10.1007/978-3-319-17620-8_16
5. Bishop, J.: Increasing participation in online communities: a framework for human-computer interaction. Comput. Hum. Behav. **23**(4), 1881–1893 (2007)
6. Brandtzæg, P.B.: Towards a unified media-user typology (MUT): a meta-analysis and review of the research literature on media-user typologies. Comput. Hum. Behav. **26**(5), 940–956 (2010)
7. Callahan, K.: Citizen participation: models and methods. Int. J. Public Adm. **30**(11), 1179–1196 (2007)
8. Clarinval, A., Simonofski, A., Vanderose, B., Dumas, B.: Public displays and citizen participation: a systematic literature review and research agenda. Transform. Gov. People Process Policy **15**(1), 1–35 (2021)
9. Creswell, J.W., Clark, V.L.P.: Designing and Conducting Mixed Methods Research. Sage, London (2017)
10. Cugurullo, F.: How to build a sandcastle: an analysis of the genesis and development of Masdar City. J. Urban Technol. **20**(1), 23–37 (2013)
11. Edelmann, N.: Reviewing the definitions of "lurkers" and some implications for online research. Cyberpsychol. Behav. Soc. Netw. **16**(9), 645–649 (2013)
12. Edelmann, N., Parycek, P., Schossbock, J.: The unibrennt movement: a successful case of mobilising lurkers in a public sphere. Int. J. Electron. Gov. **4**(1–2), 43–68 (2011)
13. Faridani, S., Bitton, E., Ryokai, K., Goldberg, K.: Opinion space: a scalable tool for browsing online comments. In: Proceedings of the SIGCHI Conference on Human Factors in Computing Systems, pp. 1175–1184. Association for Computing Machinery (2010)
14. Farina, C.R., Epstein, D., Heidt, J.B., Newhart, M.J.: Regulation room: getting "more, better" civic participation in complex government policymaking. Transform. Gov. People Process Policy **7**(4), 501–516 (2013)
15. Fung, A.: Varieties of participation in complex governance. Public Adm. Rev. **66**, 66–75 (2006)
16. Goncalves, J., Hosio, S., Liu, Y., Kostakos, V.: Eliciting situated feedback: a comparison of paper, web forms and public displays. Displays **35**(1), 27–37 (2014)

17. Gray, B.: Informal learning in an online community of practice. J. Dist. Educ. **19**(1), 20–35 (2004)
18. Hassenzahl, M., Burmester, M., Koller, F.: Attrakdiff: Ein fragebogen zur messung wahrgenommener hedonischer und pragmatischer qualität. In: Szwillus, G., Ziegler, J. (eds.) Mensch & Computer 2003, pp. 187–196. Springer, Heidelberg (2003). https://doi.org/10.1007/978-3-322-80058-9_19
19. Irvin, R.A., Stansbury, J.: Citizen participation in decision making: is it worth the effort? Public Adm. Rev. **64**(1), 55–65 (2004)
20. Johnson, R.B., Onwuegbuzie, A.J., Turner, L.A.: Toward a definition of mixed methods research. J. Mixed Methods Res. **1**(2), 112–133 (2007)
21. King, C.S., Feltey, K.M., Susel, B.O.: The question of participation: toward authentic public participation in public administration. Public Adm. Rev. 317–326 (1998)
22. Kitchin, R.: The real-time city? Big data and smart urbanism. GeoJournal **79**(1), 1–14 (2014)
23. Lackaff, D.: Case study: better reykjavik - open municipal policymaking. In: Civic Media: Technology, Design, Practice, p. 229 (2016)
24. Lago, N.: Digital platforms for participation in city plan: typology of citizens' modes of presence. Technical report, University of Mons (2019)
25. Lago, N., Durieux, M., Pouleur, J.-A., Scoubeau, C., Elsen, C., Schelings, C.: Citizen participation through digital platforms: the challenging question of data processing for cities. In: Proceedings of the International Conference on Smart Cities, Systems, Devices and Technologies, pp. 19–25. International Academy, Research, and Industry Association (2019)
26. Lai, H.-M., Chen, T.T.: Knowledge sharing in interest online communities: a comparison of posters and lurkers. Comput. Hum. Behav. **35**, 295–306 (2014)
27. Lallemand, C., Gronier, G.: Méthodes de design UX. Éditions Eyrolles, Paris (2018)
28. Lallemand, C., Koenig, V., Gronier, G., Martin, R.: Création et validation d'une version française du questionnaire attrakdiff pour l'évaluation de l'expérience utilisateur des systèmes interactifs. Eur. Rev. Appl. Psychol. **65**(5), 239–252 (2015)
29. Lukensmeyer, C.J., Brigham, S.: Taking democracy to scale: creating a town hall meeting for the twenty-first century. Natl. Civ. Rev. **91**(4), 351–366 (2002)
30. Macintosh, A.: eParticipation in policy-making: the research and the challenges. In: Cunningham, P., Cunningham, M. (eds.) Exploiting the Knowledge Economy: Issues, Applications and Case Studies, vol. 3, pp. 364–369. IOS Press (2006)
31. Malinen, S.: Understanding user participation in online communities: a systematic literature review of empirical studies. Comput. Hum. Behav. **46**, 228–238 (2015)
32. McKenzie, P.J.: A model of information practices in accounts of everyday-life information seeking. J. Doc. **59**(1), 19–40 (2003)
33. Muller, M.: Lurking as personal trait or situational disposition: lurking and contributing in enterprise social media. In: Proceedings of the ACM Conference on Computer-Supported Cooperative Work, pp. 253–256. Association for Computing Machinery (2012)
34. Nielsen, J.: The 90-9-1 rule for participation inequality in social media and online communities (2006)
35. Nonnecke, B., Preece, J.: Lurker demographics: counting the silent. In: Proceedings of the SIGCHI Conference on Human Factors in Computing Systems, pp. 73–80. Association for Computing Machinery (2000)
36. Nonnecke, B., Preece, J.: Silent participants: getting to know lurkers better. In: Lueg, C., Fisher, D. (eds.) From usenet to CoWebs, pp. 110–132. Springer, London (2003). https://doi.org/10.1007/978-1-4471-0057-7_6

37. Preece, J., Nonnecke, B., Andrews, D.: The top five reasons for lurking: improving community experiences for everyone. Comput. Hum. Behav. **20**(2), 201–223 (2004)
38. Romariz Peixoto, L., Rectem, L., Pouleur, J.-A.: Citizen participation in architecture and urban planning confronted with arnstein's ladder: four experiments into popular neighbourhoods of hainaut demonstrate another hierarchy. Architecture **2**(1), 114–134 (2022)
39. Royo, S., Pina, V., Garcia-Rayado, J.: Decide Madrid: a critical analysis of an award-winning e-participation initiative. Sustainability **12**(4), 1674 (2020)
40. Rudd, J., Stern, K., Isensee, S.: Low vs. high-fidelity prototyping debate. Interactions **3**(1), 76–85 (1996)
41. Sandstrom, C., Bjork, J., et al.: Idea management systems for a changing innovation landscape. Int. J. Prod. Dev. **11**(3–4), 310–324 (2010)
42. Schelings, C.: Renouveau des approches participatives pour la fabrique de la Smart City. PhD thesis, Université de Liège (ULiège) (2021)
43. Serramia, M., et al.: Citizen support aggregation methods for participatory platforms. In: Artificial Intelligence Research and Development, pp. 9–18. IOS Press (2019)
44. Shneiderman, B.: The eyes have it: a task by data type taxonomy for information visualizations. In: Proceedings of the Symposium on Visual Languages, pp. 336–343. Institute of Electrical and Electronics Engineers (1996)
45. Simonofski, A., Snoeck, M., Vanderose, B.: Co-creating e-government services: an empirical analysis of participation methods in Belgium. In: Rodriguez Bolivar, M.P. (ed.) Setting Foundations for the Creation of Public Value in Smart Cities. PAIT, vol. 35, pp. 225–245. Springer, Cham (2019). https://doi.org/10.1007/978-3-319-98953-2_9
46. Smith, M.A., Kollock, P.: Communities in Cyberspace. Routledge, London (1999)
47. Sun, N., Rau, P.P.-L., Ma, L.: Understanding lurkers in online communities: a literature review. Comput. Hum. Behav. **38**, 110–117 (2014)
48. Takahashi, M., Fujimoto, M., Yamasaki, N.: The active lurker: influence of an in-house online community on its outside environment. In: Proceedings of the International SIGGROUP Conference on Supporting Group Work, pp. 1–10. Association for Computing Machinery (2003)
49. Tavanapour, N., Poser, M., Bittner, E.A.: Supporting the idea generation process in citizen participation-toward an interactive system with a conversational agent as facilitator. In: Proceedings of the European Conference on Information Systems, pp. 1–17 (2019)
50. Thiel, S.-K., Lehner, U.: Exploring the effects of game elements in m-participation. In: Proceedings of the British HCI Conference, pp. 65–73 (2015)
51. Moere, A.V., Hill, D.: Designing for the situated and public visualization of urban data. J. Urban Technol. **19**(2), 25–46 (2012)
52. Zeng, Z., et al.: The impact of social nudges on user-generated content for social network platforms. Management Science (2022). https://pubsonline.informs.org/doi/full/10.1287/mnsc.2022.4622

Conceptual Modeling and Ontologies

What Do Users Think About Abstractions of Ontology-Driven Conceptual Models?

Elena Romanenko[1(✉)], Diego Calvanese[1,2], and Giancarlo Guizzardi[3]

[1] Free University of Bozen-Bolzano, 39100 Bolzano, Italy
{eromanenko,diego.calvanese}@unibz.it
[2] Umeå University, 90187 Umeå, Sweden
[3] University of Twente, 7500 Enschede, The Netherlands
g.guizzardi@utwente.nl

Abstract. In a previous paper, we proposed an algorithm for ontology-driven conceptual model abstractions [18]. We have implemented and tested this algorithm over a FAIR Catalog of such models represented in the OntoUML language. This provided evidence for the *correctness* of the algorithm's implementation, i.e., that it correctly implements the model transformation rules prescribed by the algorithm, and its *effectiveness*, i.e., it is able to achieve high compression (summarization) rates over these models. However, in addition to these properties, it is fundamental to test the *validity* of this algorithm, i.e., that it achieves what it is intended to do, namely provide summarizing abstractions over these models whilst preserving the gist of the conceptualization being represented. We performed three user studies to evaluate the usefulness of the resulting abstractions as perceived by modelers. This paper reports on the findings of these user studies and reflects on how they can be exploited to improve the existing algorithm.

Keywords: Conceptual Model Abstraction · Ontology-Driven Conceptual Models · User Study

1 Introduction

The complexity of an (ontology-driven) conceptual model highly correlates with the complexity of the domain and software for which it is designed. According to Guttag [14], one way to reduce complexity is through the abstraction process. Speaking of conceptual modeling, Egyed [7] defined abstraction as "a process that transforms lower-level elements into higher-level elements containing fewer details on a larger granularity". The main idea is to provide the user with a bird's-eye view of the model by filtering out some details.

Our previously suggested algorithm for ontology-driven conceptual model abstraction [18] was implemented and tested over the models from a recently created FAIR catalog for ontology-driven conceptual modeling research [4]. Although the algorithm has shown that it is applicable to a wide range of

S. Nurcan et al. (Eds.): RCIS 2023, LNBIP 476, pp. 53–68, 2023.
https://doi.org/10.1007/978-3-031-33080-3_4

models, the question of the quality of the resulting abstractions was still open. In order to answer it, we conducted three user studies. The results of the studies and our suggestions for algorithm improvement are presented in this paper.

The remainder of the paper is organized as follows: Sect. 2 presents our baseline and background; Sect. 3 describes user studies conducted to find out the ways for improvement of the approach; Sect. 4 elaborates on final considerations and future work.

2 Background

By a conceptual model, one could equally mean a UML Class Diagram as well as a Business Process Model. This is because conceptual models are high-level abstractions used to capture information about the domain and both these languages (among many others) are employed for that. Ontology-driven conceptual models (ODCMs) are usually considered a special class of conceptual models that utilize foundational ontologies to ground modeling elements, modeling languages, and tools [22].

The role of conceptual models in general and ODCMs, in particular, is quite precisely specified in the literature. They are intended to enable clients and analysts to understand one another, to communicate successfully with application programmers, and hence "play a fundamental role in different types of critical semantic interoperability tasks" [11].

However, the complexity of the conceptual model correlates with the complexity of the domain. This sometimes leads to situations where the number of concepts and sub-diagrams goes far beyond the cognitive tractability threshold of those people who are supposed to work with those diagrams. Thus, despite the fact that conceptual models are developed for human communication, one of the most challenging problems is "to understand, comprehend, and work with very large conceptual schemas" [23].

The problem of making conceptual models (and ODCMs) more comprehensible is addressed in the literature by the proposal of different complexity management techniques, and for quite some time this research area has been under active study. For analysis and classification of the existing approaches, one could refer to [23]. Here, it suffices to point out that producing a meaningful but reduced version of the original conceptual model via filtering out the details and keeping the most important notions—also known as *summarizing* or *abstracting*—is one of the most challenging tasks.

Most of the methods for conceptual model summarization are based on classic modeling notations (UML, ER) [23, p.44] and rely on syntactic properties of the model, such as closeness or different types of distances between model elements (see [2]), while in case of ODCMs, there is the possibility to leverage their built-in ontological semantics. The first version of an abstraction algorithm leveraging foundational ontological semantics was introduced in [10], followed by an enhanced version in [18], which was able to abstract more sophisticated models, i.e., models employing a larger number of formal ontological primitives.

For detailed descriptions and justifications of the algorithms, we refer to the previously published papers [10,18]. For the scope of this paper, it is enough to highlight that both algorithms bear a remarkable simplicity in the number of rules, are deterministic and do not require human intervention or *seeding*[1] (as opposed to competing algorithms, e.g., [7]), are computationally efficient and scalable, thus, able to process very large models in a timely manner.

The newest version of the algorithm proposed in [18] defines 11 graph-rewriting rules, which are grouped into three categories, namely, rules for abstracting *(1)* parthood relations (compositions and aggregations), *(2)* different *aspects* of objects (*relators*, *qualities*, and *modes* in terms of Unified Foundational Ontology [13])[2], and *(3)* hierarchies of concepts (generalization relations). Also, one should note that application of the rule does not always imply the complete elimination of the corresponding construct being addressed, e.g., after applying rules from the first group, some of the parthood relations could be kept. It is possible to apply them in a compositional way, so we can abstract both parthood relations and hierarchies. Thus, with three groups of rules, one can receive eight possible models (including the original model and full abstraction, when all rules are applied) [18].

The defined graph-rewriting rules utilised the ontological semantics of UFO. However, since ODCMs in general are not bound to any specific foundational ontology, one can choose the most appropriate to the task at hand. A recent special issue of Applied Ontology journal [5] describes seven of them—BFO, DOLCE, GFO, GUM, TUpper, UFO, and YAMATO. UFO appeared to be the most fruitful for the abstraction algorithm development because of the existence of the UFO-based OntoUML language and corresponding tools. For an in-depth discussion, philosophical justification, and formal characterization of UFO and OntoUML, we refer to [9,12].

The suggested algorithms needed to be properly evaluated from two angles. Although an initial attempt to calculate the compression rate was done in [10], there was a need to assess how much information is reduced on a larger sample of models. Also, we wanted to investigate whether the suggested algorithm provided reasonably good results from the modelers' point of view. Since the rest of the paper is devoted to the latter problem, let us briefly describe the compression results that were obtained.

The ability to assess the algorithm over models becomes possible with the creation of a FAIR model catalog[3] for ontology-driven conceptual modeling research [4] (hereinafter referred to as the Catalog). The Catalog offers a diverse collection of conceptual models, created by modelers with varying modeling

[1] Seeding is the (typically, manual) pre-selection of certain model elements that need to be maintained in the final abstraction.

[2] A characteristic feature of *aspects* in UFO is that they are existentially dependent from the main entities. For example, *quality* `Colour` cannot exist without the object itself. Also, *relator* `Employment` is not possible without `Employer` and `Employee`.

[3] https://w3id.org/ontouml-models.

skills, for a wide range of domains, with different purposes, and currently consists of 135 models.

The problem with the Catalog from the point of view of our research is that it contains all errors that were introduced by the model's authors. Those include not only typos but also modeling mistakes. From the point of view of the Catalog, the decision to keep models as they were created was reasonable, because one of the purposes was an empirical discovery of modeling (anti-)patterns [4]. Unfortunately, this contradicts the goal to assess the quality of the algorithm, since most of the time the original errors would be propagated to the abstraction.

Taking into account the above-mentioned conditions, we selected 41 models for the purpose of the algorithm evaluation. The selected models satisfied the following criteria: *(1)* they contained only those 16 stereotypes, for which the second version of the algorithm was developed (that left us with 71 models out of the original number), and *(2)* they did not contain syntactical modeling errors that could not be easily fixed. The syntactical correctness of models was checked automatically with the OntoUML plugin[4] for Visual Paradigm[5].

A fuller analysis of the algorithm (e.g., in terms of computational complexity) is out of the scope of this paper, so here we report only on the results of evaluating the compression rates produced by the algorithm against the set of selected models. The interested reader may compare these with the results published in [10] for the first version of the algorithm.

As it can be seen from Fig. 1, the algorithm leads to a reduction of the number of concepts as well as the number of relations of about three times for the medium-size models in case of applying all of the proposed rules (the so-called *full abstraction*). The maximum reduction rate happens after removing generalizations relations (abstracting hierarchies of concepts).

Despite achieving large compression rates, these numbers by themselves tell us nothing about the appropriateness of the abstracted models, i.e., to what degree the abstractions are perceived as useful and meaningful by modelers. An initial attempt to compare the abstraction results of the first version of the algorithm to other existing methods was made in [21]. The main hypothesis of the experiment was that the abstraction algorithm (the first version proposed in [10]) produces models capturing the gist of the original model more appropriately than the competing algorithms proposed in [7] and [16].

The experiment was organized as follows. A group of 50 participants with different modeling backgrounds—from students to professionals with years of modeling experience—were presented with the original conceptual model in the car rental domain and with several abstractions. They were asked to rate the models according to their view on the quality of the abstraction and justify their choice.

The suggested algorithm was clearly preferred by practitioners with large modeling experience. However, overall, the experiment did not demonstrate a significant preference for one of the tested abstraction algorithms. This result

[4] https://github.com/OntoUML/ontouml-vp-plugin.

[5] https://www.visual-paradigm.com.

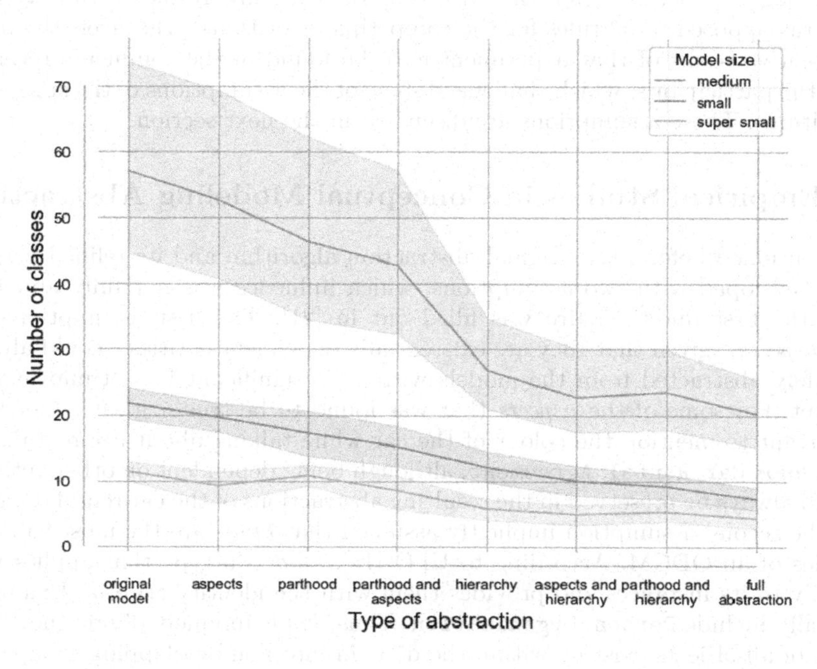

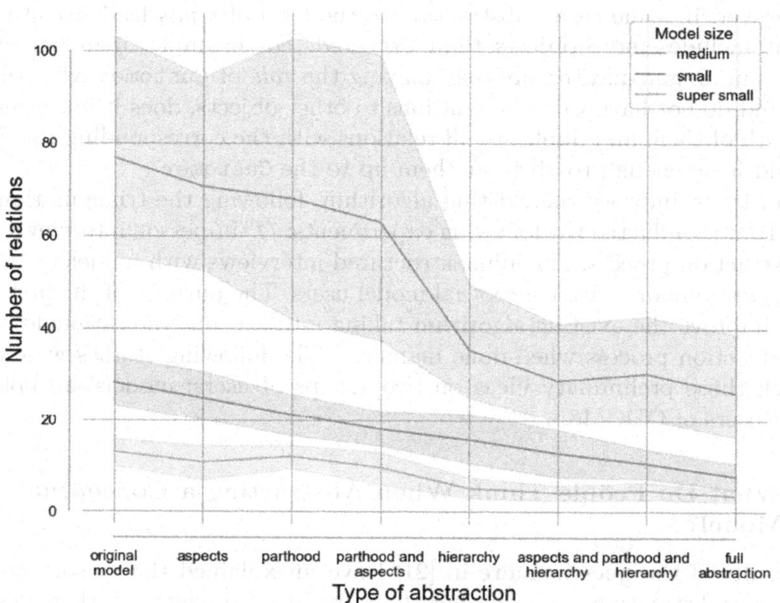

Fig. 1. Compression of the abstracted models.

is not negative per se, given that [10] achieves the same results with only four rules (as opposed to 92 rules for the competing algorithms). However, the main informative result of that experiment is to be found in the comments received from the participants, which challenged some of the assumptions of the suggested algorithms. These assumptions are discussed in the next section.

3 Empirical Studies in Conceptual Modeling Abstraction

As mentioned before, the original abstraction algorithm and its refined version were developed with two assumptions, which influence was not fully apparent until the first questionnaire was filled out in [21]. The first assumption was that *aspects*—given that they are existentially dependent entities—could always be safely abstracted from the models without a significant loss of information content. For some of the *aspects* that was found to be true, e.g., it is not that important to mention the colour of the car while talking about car rental, but the *relator* `Car Rental Agreement`, although being dependent on other entities, should always be preserved in the resulting abstractions of the car rental domain.

The second assumption implicitly assumed that *kinds* are the most valuable entities of an ODCM. According to UFO [13], *kind* is a type that applies necessarily to its instances and provides them with the identity criteria. Examples typically include `Person`, `Organization`. Thus, *kinds* uniquely divide the entire space of all objects existing within the domain into non-overlapping groups.

However, in some cases, abstraction to the level of *kinds* leads to situations when it includes more objects than expected. For example, if in our system `Person` and `Organization` are only playing the *role* of `Customer` who can rent a car, and do not have valuable relations to other objects, does it make sense to keep both of them and duplicate all relations with the corresponding role name, or would it be enough to abstract them up to the `Customer`?

In order to improve the existing algorithm, following the triangulation approach [1], we conducted the following experiments: *(1)* in-person interviews about the abstraction process, *(2)* online structured interviews with modelers, and *(3)* online questionnaire with conceptual model users. The purpose of the first study was to improve the existing algorithm taking into account the rationale behind the abstraction process when done manually. The following studies were aimed to check those preliminary ideas on two groups of users: models' authors and general users of ODCMs.

3.1 What Do People Think When Abstracting a Conceptual Model?

The results of the questionnaire in [21] leave unexplained the reason modelers prefer one abstraction over another. Our first experiment was then designed to find out what people consider when abstracting a conceptual model, i.e., their (often tacit) rationale for choosing the main concepts to be preserved, and how the final abstraction produced by the algorithm corresponds to that target preferred one. Thus, the goal of the study was to answer the following questions: *(1)* What is the rationale for choosing the most valuable concepts of the model (seeding)? *(2)* How does it change depending on the given goal? *(3)* Can abstraction serve as an explanation of the original model?

From our point of view, the major drawback of the algorithm suggested by Egyed [7] is not in its lack of simplicity (i.e., a large number of rules) but in the necessity for the modeler to perform *seeding*—again, an explicit selection of a list of concepts that are considered as the most valuable in that model. The problem with that approach is that, in order to determine those concepts, one needs to be familiar with the domain and with the conceptual model. But if the conceptual model is large and complex, this requires the modeler to deal with the complexity of the model, thus, risking to defeat of the purpose of an abstraction technique. So, on the one hand, the non-determinism of that approach has an advantage in the ability to generate different abstractions according to one's alternative goals, but on the other hand, it requires an expert and cannot be used for supporting users in getting acquainted with a new domain.

The purpose of this first study was two-fold. First, we wanted to understand how conceptual model users abstract from complicated ODCMs and how they perform model seeding. The hypothesis was that by understanding their rationale, we could derive information to (perhaps, partially) automate the seeding process, thus, mitigating the aforementioned problems while preserving some of the advantages of a non-deterministic approach (personalization).

Second, we wanted to preliminarily check the hypothesis that the (simpler) abstracted model is perceived as an explanation of the (complex) original model. We suggested that abstraction could be part of the pragmatic explanation process in the case of ODCMs, in line with what is argued for domain ontologies in [19][6]. Thus, on the one hand, since the abstraction should correspond to the concrete goal, it should be reviewed or even modified in accordance with the given goal. On the other hand, if the already given abstraction contained an error, i.e., a contradiction with the original model, it could pass unnoticed due to overreliance on the given explanation. In other words, if given, the explanations are interpreted as a signal of competence and are simply accepted regardless of their correctness, especially by non-experts (see [3,6] and experiments in eXplainable AI).

For that, we conducted 5 one hour interviews, and to reflect on this, we used the transcripts of think-aloud and retrospective reports of the participants[7].

[6] This view of an abstract conceptual model as a type of explanation is in line with the literature on Design Theory (e.g., [17]). In this community, a conceptual model is taken to be a simplified and useful explanation of how something works from the point of view of an external observer. We come back to this idea of an abstracted model as a sort of pragmatic explanation and its grounds in Sect. 4.

[7] The study has been reviewed and received approval from the Ethics Research Committee at the Free University of Bozen-Bolzano, Italy (Prot. n. 5/2022 from 28/09/2022).

We followed the approach suggested in [8], under the assumption that "cognitive processes are not modified by these verbal reports" [8, p.16].

The experiment was conducted individually and face-to-face with the researcher, using a laptop and a standard well-known UML editor, namely Visual Paradigm. The participants received a pure black-and-white model without any additional notes, also without the OntoUML stereotypes for the classes and associations. The level of expertise was defined as a self-assessment before participation, and the study included two experts in ontology-driven conceptual modeling, two experts in conceptual modeling, and one non-expert but an experienced user of conceptual models. A pilot study with one conceptual modeling expert was conducted to assure the tasks were clear enough and did not raise difficulties.

Each participant was given two tasks to be solved one by one. Both models related to the same domain of a library management system, which was quite general and did not require special knowledge. In the first task, given the ODCM (see Fig. 2), the participant was asked to produce a model abstraction, where the abstraction was defined according to Egyed's algorithm [7]. In order to simplify this process, they were presented with a short narrative telling them why they need to create an abstraction. During the abstracting process, they were asked only to think aloud, without additional comments. After solving the task, they also gave a retrospective reflection on their choices.

In the second task, the participant was asked to change the given abstraction while keeping in mind a concrete goal. The abstraction was produced by the algorithm with some modifications. We introduced a contradiction w.r.t. the original model by making `Person` and `Organization` subtypes of `Client`[8]. In order to make the error even more obvious, we kept some other concepts: `Librarian`, `Employee`, and `Library` (see modifications in pink in Fig. 2). In particular, in this modified abstraction, every `Librarian` is a `Client` of the `Library`, which was not true in the original model. According to the narrative, this abstraction was produced by one of the participant's colleagues.

The results of the protocol analysis were quite interesting. Four out of five interviewees were regularly distracted by the layout of the model (when during the modifications it was deemed "ugly" and "annoying"). Moreover, during the retrospective reports they reflected on "chopping the things that are unconnected to anything else" (from an ontology-driven conceptual modeling expert), and the need to remove all the cardinality constraints ("Of course they are interesting, but if we talk about simplifying and abstracting, I would do that"—from an expert in conceptual modeling).

In other words, languages for (ontology-driven) conceptual modeling are typically visual languages as well. And because of that, it is impossible to completely isolate the assessment of that model from the assessment of the layout of that

[8] For a discussion for why this is an error (in this case, it introduces a logical contradiction in the model), we refer to [9].

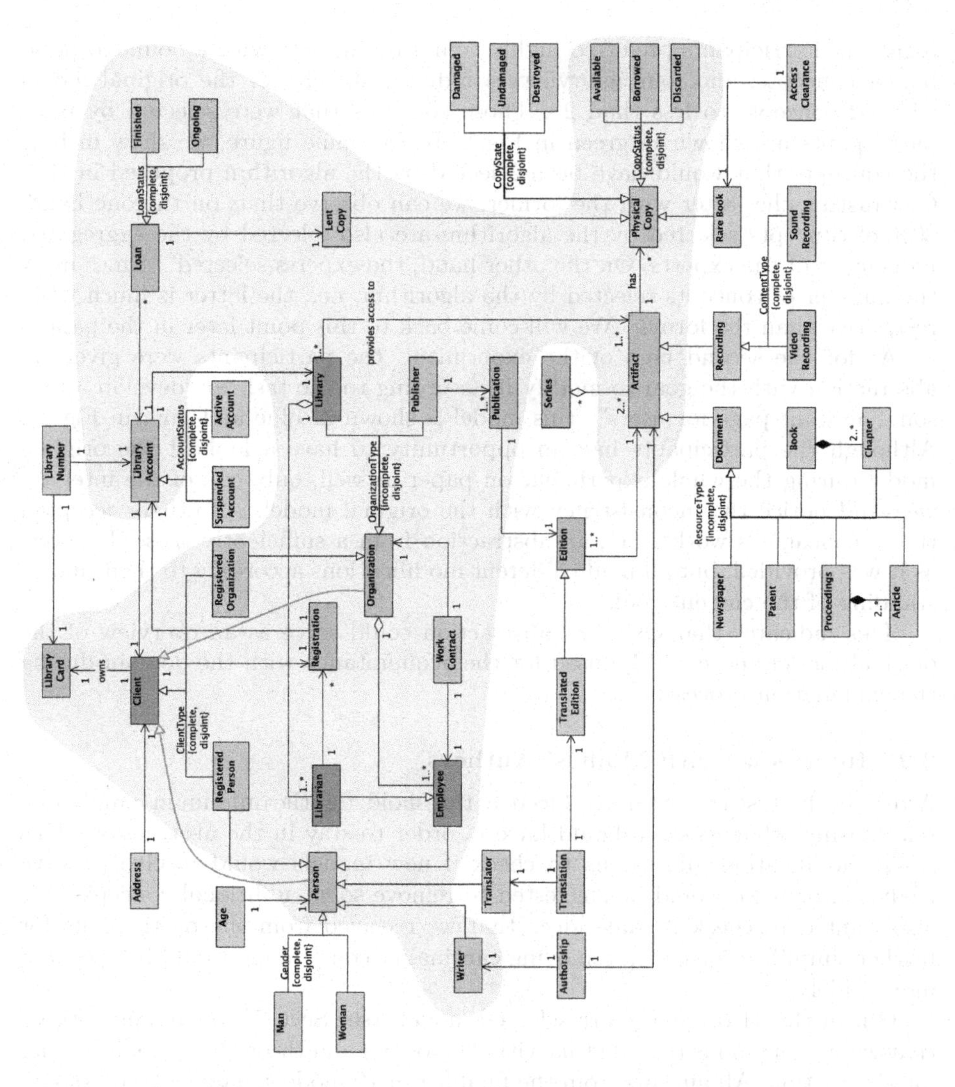

Fig. 2. Library management system model for the first user study. The sub-model selected as the abstraction by most of the participants is in green. The concepts selected by the abstraction algorithm are in blue. In pink, we have an error that was introduced for the second part of the experiment. (Color figure online)

model. The idea to remove cardinality constraints also comes from the desire to have a visually simple model without information that is not needed at that precise moment. Thus, we agree that "the notion of simplicity is essential to characterize abstract representations" [20, p.63].

We also noticed that the idea of a correlation between the number of relations for a given concept and its significance for the domain is surprisingly well-regarded. Participants preferred highly connected models with a bounded number of concepts, and four interviewees reduced the size of the original model with 52 concepts to less than 20. Those concepts that were selected by most participants are shown in green in Fig. 2. In the same figure, we show in blue the concepts that would have been selected by the algorithm proposed in [18]. Contrasting the latter with the former, we can observe that, on the one hand, 80% of concepts selected by the algorithm are also selected by the aggregated judgement of the experts. On the other hand, the experts selected 3 times more the number of concepts selected by the algorithm, i.e., the latter is much more restrictive than the former. We will come back to this point later in the paper.

As for the second part of the experiment, the participants were given an abstraction with the goal to modify it according to the task to "develop a personal account page for users" (this model is shown in blue and pink in Fig. 2). Although the participants had an opportunity to have a look at the original model during the whole experiment on paper as well, only one of the interviewees did notice the inconsistency with the original model. All others accepted their "colleague's work", i.e., an abstraction from a sufficiently trusted source, as it was provided, but all made different modifications according to their understanding of the current goal.

This indicates that ODCM's abstraction could serve as an overview of the original model and could be used for the acquaintance with the domain during the explanation process.

3.2 Interviews with Models' Authors

After the first study, we introduced a threshold for the minimum number of relationships that *aspects* should have in order to stay in the abstraction. This small modification allowed us to check if new models would receive positive feedback, or if we would be suggested to remove some additional concepts. We also wanted to check if some ideas that we received from the participants for further simplifications, e.g., removing cardinality constraints, would be accepted more widely.

Out of the 41 originally pre-selected models (see Sect. 2), we removed those that were anonymous (that left us with 23 models) and those that were too small for abstraction. All authors from the final list of 10 models, namely 26 ontology-driven conceptual modeling experts, received links to the abstractions of their own models and invitations for online structured interviews. The interviews were conducted anonymously, and the questions did not specify which model was being referred to.

After reviewing an abstraction of the original model that was published by them, the authors answered up to 18 questions from three groups (some examples of questions are provided):

1. Questions about the satisfaction with the abstraction:
   - I understand the abstraction of the original model.
   - The abstraction of the original model has sufficient details.
   - The abstraction tells me enough about the domain.
   - The abstraction of the original model contains irrelevant details.
   - The abstraction could be used for the acquaintance with the domain.

2. Questions about the correctness of the abstraction:
   - The abstraction introduced wrong concepts that did not exist in the domain.
   - The abstraction did not introduce any semantically wrong relations.
   - The abstraction reveals some errors that were unintentionally introduced in the original model.
3. Questions about algorithm improvements:
   - Removing cardinalities in the abstraction will make the model clearer.
   - Removing role names in the abstraction will make the model clearer.

Some questions used a 5-point Likert scale, others were left open. In total, we conducted 7 interviews with an average time of about 40 min (including the time for the abstractions' review).

Most of the respondents were able to understand the abstraction of the model quite easily. However, opinions about whether the abstraction contains enough details were divided—3 agreed and 4 disagreed (see Fig. 3). This is even more interesting, taking into account that only two of the authors claimed that the abstraction contained irrelevant details. In other words, authors would prefer to have less restrictive abstractions and were unsatisfied with their conciseness—in line with the judgement of experts, as we have seen before.

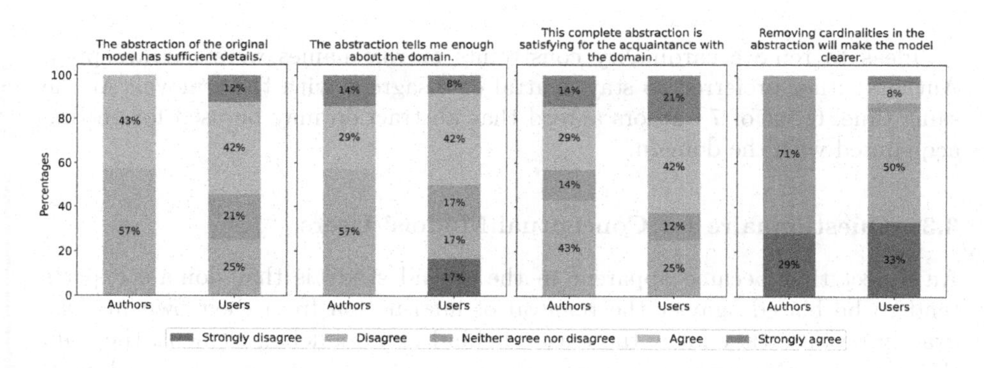

Fig. 3. Comparison of the answers for different groups of users. Percentages less than 8 are not specified.

As for the correctness of the abstraction, we were glad to see an agreement of opinion that the abstraction did not introduce any semantically wrong concepts. The most heated debate happened with the question of whether the abstraction introduced any semantically wrong relations. 5 authors did not notice anything wrong, while two others state that semantically wrong relations have been introduced. In one case the author refused to specify which relation is wrong, but the second case requires additional comments. It involves abstracting parthood.

The algorithm in [18] abstracts a parthood relation by transferring certain properties (including relational ones) from the part to the whole. For example, if the Faculty of Computer Science of UNIBZ has a project with the Government, then UNIBZ has a project with the Government or, more precisely, UNIBZ has the property of "having a faculty that has a project with the Government". However, our study showed that when the part-whole relation is established between two parts of the same whole, the result can seem incorrect to model creators (see the pattern in Fig. 4a and the abstracted result in Fig. 4b. This pattern requires special attention and should be addressed explicitly in the further development of the algorithm.

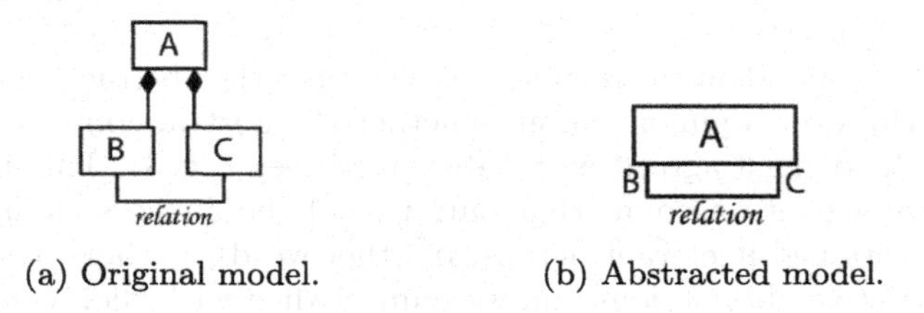

(a) Original model. (b) Abstracted model.

Fig. 4. One of the abstraction patterns for the parthood relation.

Ideas to remove cardinality constraints or role names were not accepted. Authors either preferred to stay neutral or disagreed with the removal. At the same time, 6 out of 7 authors agreed that abstraction may be used for getting acquainted with the domain.

3.3 Questionnaire for Conceptual Models' Users

An aspect that became apparent in the second study is that domain experts tend to be biased against the removal of information from their own models, exactly because they know the reason behind each modeling element, they run the risk of amplifying their importance. In a limit case, they would see all modeling elements as essential, forgetting that, as a *lossy* (as opposed to lossless) technique, abstraction necessarily implies the removal of information content from the model. One of the respondents from the previous study formulated this in the following way:

> *"The idea of conceptual models is to represent the complexity of the entities of a domain and their relationships. When applying an algorithm to generate simpler representations, there is a risk of generating an interpretation bias in the reader. Complex problems often require complex solutions."*

However, the model creators are not the only users of the models they create. In fact, if a conceptual model is successful, the model creator will be just one among a multitude of users. Our hypothesis is that other users of the model would have a different attitude towards abstraction in this respect. With that in mind, we developed a questionnaire for a more general audience. The questionnaire was developed based on two anonymous models from the Catalog and did not require any special knowledge except familiarity with UML Class Diagrams. One of the abstractions, which corresponds to the `bank-model` from the Catalog[9] and was presented in the questionnaire, is shown in Fig. 5.

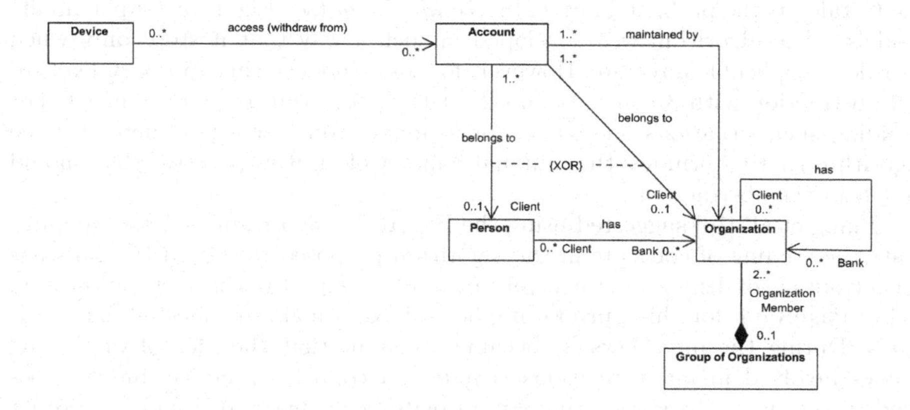

Fig. 5. One of the abstractions used in the questionnaire.

The questionnaire consisted of 20 questions, which were correlated with the questions of the structured online interviews, and it was partially based on the assessment suggested in [15]. In total, we received 24 responses with an average completion time of 15 min.

Again, most of the users (92%) were able to understand the abstraction with ease. For 75% of the respondents, the abstraction had sufficient details (compare this result to the result of the previous study, see Fig. 3), about 62% agreed that the abstraction provided a good overview of the domain, and 29% agreed to reduce an abstraction even more.

When assessing the quality of the abstractions, these conceptual model users were more tepid: about 55% of them were convinced that the abstractions were correct and that they did not introduce any wrong relations and concepts; about 21% suggested that the abstraction could be wrong.

Unexpected consent was reached for the question of whether we should remove cardinality constraints and role names: most of the users (83% in both questions) prefer them to be kept in the abstraction.

We also received interesting feedback on the question of how the participants envisage the abstraction to be useful. Although the question was left open, some of the answers were recurrent, among them "better communication between stakeholders", and "understanding the original model".

[9] https://github.com/OntoUML/ontouml-models/tree/master/models/bank-model.

4 Final Considerations

The abstraction algorithm suggested in [18] lacked a proper evaluation. We conducted several studies with the purpose to gain an understanding of what can be further improved and whether the resulting models are able to keep the semantics of the original model.

The first problem that became visible thanks to the algorithm evaluation over the Catalog is the problem of *excessive compression* (see Fig. 1 for "super small" models). The algorithm was developed in such a way that it stops only when no rule is applicable anymore. However, for some models, this approach leads to full abstraction with 3 concepts and 2 relations, or even only one concept. For avoiding such situations one could, for example, consider a parameter for the algorithm that determines the minimal number of entities (classes) that should be left in the abstraction.

Huang et al. [16] suggested using the PageRank algorithm as a way to automate the seeding of concepts in the algorithm proposed by Egyed [7]. This has the advantage of dispensing the involvement of an expert in the abstraction task. Using PageRank for this purpose implies selecting highly connected classes as seeds. During the interviews, it became apparent that the idea of preserving classes involved in many relations is indeed a common practice: interviewees tended to remove the classes that were isolated or connected to only one other class more often.

In future work, we intend to investigate the use of topological metrics (e.g., the degree of connectivity of a class) in combination with ontological semantics to improve the algorithm in [18]. For example, classes that would otherwise be eliminated by the algorithm could instead be preserved if they are connected over a certain (absolute or relative) threshold. We observed this, especially in the case of *aspects*, in general, and *relators*, in particular (see Sect. 3.1), where the removal of some of these elements led to the model being perceived as incomplete by participants[10].

Unfortunately, the number of participants in the last studies was not very large. However, we suppose that creators of the models and typical users of the same ODCMs have different views on the usefulness of the abstraction, and those views must be taken into account when developing the final system.

Before reusing an existing ODCM one needs to understand it. However, since the number of concepts and diagrams may be large, typical users may face problems in familiarizing themselves with an ODCM. We claim that abstraction could be part of the pragmatic explanation process of an ODCM (as well as sub-models

[10] Formally, the original transformation rules as proposed in [10] only prescribed the abstractions into material relations of those *relators* that were connected to at most two mediation relations. In both [21] and [18], this idea was extrapolated to cover *relators* participating in more than two mediation relations. This experiment confirms a tacit rationale behind the original rule: *relators* participating in more than two mediation relations are exactly those cases that would lead to relation reification in traditional conceptual modeling, i.e., one of those cases in which modelers want to perform model expansion—the exact opposite of abstraction.

for domain ontologies in [19]). In other words, an abstracted model in some cases may serve as an explanation of the original more complex model, and comments from the users received in the last study implicitly support this idea.

Moreover, abstractions, when playing the role of explanations, struggle with the same problems. They are not taken critically (see results from the first study) and should correspond to the current goal. This means that the results of the deterministic algorithm should be used for the first acquaintance with the domain, but for the further explanation process, the most valuable concepts should be perhaps selected explicitly by the user. Also, it is very important to generate a proper abstraction. Otherwise, due to the gap between the abstracted and the original model, a user exposed to the former could have difficulties with understanding the latter.

Acknowledgements. Empirical studies and user experiments are never possible without the generous voluntary participation of several individuals. The authors would like to express their great appreciation for all the people who spent their time answering the questionnaires and participating in the interviews. This research has been partially supported by the Italian Basic Research (PRIN) project HOPE, and by the Province of Bolzano through the project MENS. Diego Calvanese is also supported by the Wallenberg AI, Autonomous Systems and Software Program (WASP), funded by the Knut and Alice Wallenberg Foundation.

References

1. Adams, A., Cox, A.L.: Questionnaires, in-depth interviews and focus groups, pp. 17–34. Cambridge University Press, Cambridge (2008). https://doi.org/10.1017/CBO9780511814570.003
2. Akoka, J., Comyn-Wattiau, I.: Entity-relationship and object-oriented model automatic clustering. Data Knowl. Eng. **20**(2), 87–117 (1996). https://doi.org/10.1016/S0169-023X(96)00007-9
3. Bansal, G., et al.: Does the whole exceed its parts? The effect of AI explanations on complementary team performance. In: Proceedings of the 2021 CHI Conference on Human Factors in Computing Systems (CHI). Association for Computing Machinery (2021). https://doi.org/10.1145/3411764.3445717
4. Barcelos, P.P.F., et al.: A FAIR model catalog for ontology-driven conceptual modeling research. In: Ralyté, J., Chakravarthy, S., Mohania, M., Jeusfeld, M.A., Karlapalem, K. (eds.) ER 2022. LNCS, vol. 13607, pp. 3–17. Springer, Cham (2022). https://doi.org/10.1007/978-3-031-17995-2_1
5. Borgo, S., Galton, A., Kutz, O.: Foundational ontologies in action. Appl. Ontol. **17**, 1–16 (2022). https://doi.org/10.3233/AO-220265
6. Buçinca, Z., Malaya, M.B., Gajos, K.Z.: To trust or to think: cognitive forcing functions can reduce overreliance on AI in AI-assisted decision-making. Proc. ACM Hum.-Comput. Interact. **5**(CSCW1) (2021). https://doi.org/10.1145/3449287
7. Egyed, A.: Automated abstraction of class diagrams. ACM Trans. Softw. Eng. Methodol. **11**(4), 449–491 (2002). https://doi.org/10.1145/606612.606616
8. Ericsson, K.A., Simon, H.A.: Protocol Analysis. The MIT Press, Cambridge (1993)
9. Guizzardi, G.: Ontological foundations for structural conceptual models. CITIT PhD.-thesis series 05-74 Telematica Instituut fundamental research series 015, Centre for Telematics and Information Technology, Enschede (2005)

10. Guizzardi, G., Figueiredo, G., Hedblom, M.M., Poels, G.: Ontology-based model abstraction. In: Proceedings of the 13th International Conference on Research Challenges in Information Science (RCIS), pp. 1–13. IEEE (2019). https://doi.org/10.1109/RCIS.2019.8876971
11. Guizzardi, G., Sales, T.P., Almeida, J.P.A., Poels, G.: Automated conceptual model clustering: a relator-centric approach. Softw. Syst. Model. 1–25 (2021). https://doi.org/10.1007/s10270-021-00919-5
12. Guizzardi, G., et al.: Types and taxonomic structures in conceptual modeling: a novel ontological theory and engineering support. Data Knowl. Eng. **134**, 101891 (2021). https://doi.org/10.1016/j.datak.2021.101891
13. Guizzardi, G., et al.: UFO: unified foundational ontology. Appl. Ontol. **17**(1), 167–210 (2022). https://doi.org/10.3233/AO-210256
14. Guttag, J.: Abstract data types and the development of data structures. Commun. ACM **20**(6), 396–404 (1977)
15. Hoffman, R.R., Mueller, S.T., Klein, G., Litman, J.: Metrics for Explainable AI: Challenges and prospects. arXiv abs/1812.04608 (2018)
16. Huang, L., Duan, Y., Zhou, Z., Shao, L., Sun, X., Hung, P.C.K.: Enhancing UML class diagram abstraction with page rank algorithm and relationship abstraction rules. In: Drira, K., et al. (eds.) ICSOC 2016. LNCS, vol. 10380, pp. 103–116. Springer, Cham (2017). https://doi.org/10.1007/978-3-319-68136-8_10
17. Norman, D.: The Design of Everyday Things: Revised and Expanded Edition. Basic Books (2013)
18. Romanenko, E., Calvanese, D., Guizzardi, G.: Abstracting ontology-driven conceptual models: objects, aspects, events, and their parts. In: Guizzardi, R., Ralyté, J., Franch, X. (eds.) RCIS 2022. LNBIP, vol. 446, pp. 372–388. Springer, Cham (2022). https://doi.org/10.1007/978-3-031-05760-1_22
19. Romanenko, E., Calvanese, D., Guizzardi, G.: Towards pragmatic explanations for domain ontologies. In: Corcho, O., Hollink, L., Kutz, O., Troquard, N., Ekaputra, F.J. (eds.) EKAW 2022. LNAI, vol. 13514, pp. 201–208. Springer, Cham (2022). https://doi.org/10.1007/978-3-031-17105-5_15
20. Saitta, L., Zucker, J.D.: Abstraction in Artificial Intelligence and Complex Systems. Springer, New York (2013). https://doi.org/10.1007/978-1-4614-7052-6
21. de Souza Figueiredo, G.V.: Ontology-based complexity management in conceptual modeling. Ph.D. thesis, Federal University of Espírito Santo (2022)
22. Verdonck, M., Gailly, F.: Insights on the use and application of ontology and conceptual modeling languages in ontology-driven conceptual modeling. In: Comyn-Wattiau, I., Tanaka, K., Song, I.-Y., Yamamoto, S., Saeki, M. (eds.) ER 2016. LNCS, vol. 9974, pp. 83–97. Springer, Cham (2016). https://doi.org/10.1007/978-3-319-46397-1_7
23. Villegas Niño, A.: A filtering engine for large conceptual schemas. Ph.D. thesis, Universitat Politècnica de Catalunya (2013)

On the Semantics of Risk Propagation

Mattia Fumagalli[1(✉)], Gal Engelberg[2], Tiago Prince Sales[3], Ítalo Oliveira[1], Dan Klein[2], Pnina Soffer[4], Riccardo Baratella[1], and Giancarlo Guizzardi[3]

[1] In2Data & Conceptual and Cognitive Modeling Research Group (CORE), Free University of Bozen-Bolzano, Bolzano, Italy
{mattia.fumagalli,idasilvaoliveira,baratellariccardo}@unibz.it

[2] Accenture Israel Cyber R&D Lab, Tel Aviv, Israel
{gal.engelberg,dan.klein}@accenture.com

[3] Semantics, Cybersecurity & Services (SCS), University of Twente, Enschede, The Netherlands
{t.princesales,g.guizzardi}@utwente.nl

[4] University of Haifa, Haifa, Israel
spnina@is.haifa.ac.il

Abstract. *Risk propagation* encompasses a plethora of techniques for analyzing how risk "spreads" in a given system. Albeit commonly used in technical literature, the very notion of risk propagation turns out to be a conceptually imprecise and overloaded one. This might also explain the multitude of modeling solutions that have been proposed in the literature. Having a clear understanding of what exactly risk is, how it be quantified, and in what sense it can be propagated is fundamental for devising high-quality risk assessment and decision-making solutions. In this paper, we exploit a previous well-established work about the nature of risk and related notions with the goal of providing a proper interpretation of the different notions of risk propagation, as well as revealing and harmonizing the alternative semantics for the links used in common risk propagation graphs. Finally, we discuss how these results can be leveraged in practice to model risk propagation scenarios.

Keywords: Risk propagation · risk modeling · ontological analysis

1 Introduction

Our ability to reason about risk is fundamental in our daily lives. In this regard, the increasing enhancement of statistical methods and analytical applications has opened up promising research directions. An exemplary case is the so-called *Risk Propagation* technique [18].

Typically, in risk management, risk propagation provides a model for analyzing how risk "spreads" in a given system–that is, a model for a sort of cascading effect. Risk propagation addresses questions like:

S. Nurcan et al. (Eds.): RCIS 2023, LNBIP 476, pp. 69–86, 2023.
https://doi.org/10.1007/978-3-031-33080-3_5

.i "how does the risk associated with a device in a network 'spreads' through connected devices?";
.ii "how does the risk of my car breaking down affect the risk of me being late for an appointment?";
.iii "how does someone in my office being infected by COVID-19 affect the risk that I get infected as well?".

Risk propagation techniques are often implemented via *probabilistic graphs models* [22], in which a system to be analyzed is encoded as a set of nodes and edges, characterized by correlations and probabilities. Examples include *Bayesian networks* [4,11] and *Fault Trees* [21].

What remains certain is that the work on risk propagation still presents many open challenges from both a theoretical and a technological perspective. For instance, what do people mean when they say that risk propagates? Do they mean that risk propagates physically–like a virus that copies itself and moves through hosts? Is risk something that can be simply encoded as a weight value to be then passed through other nodes in a network? And again, do probabilistic graphs and similar graph models allow us to properly capture all the information about risk and its propagation? Most often what is actually "propagated" are probability values, leveraging some specific measures, like *conditional probability*. So, how should we interpret the notion of risk propagation? Can a further analysis of this notion support current solutions in this domain, and if so, how?

This work stems from the idea that this last question has a positive answer. In particular, we perform what we believe is the first ontological analysis of the notion of risk propagation. Our analysis is guided by the *Common Ontology of Value and Risk (COVER)*, a well-founded ontology of risk from previous research work [24]. As we shall see, our analysis allows us to *.i* explain how the propagation of risk relates to the phenomenon of belief updating; *.ii* explain how talking about the "propagation" of risk can be misleading; and *.iii* identify the concepts and relationships required to capture the cascading effect assumed when talking about risk propagation without incurring in ambiguities and reductions. Our investigation also allows the creation of a unified framework for modeling risk propagation, fostering the clarification of the real-world semantics behind risk propagation graphs, and paving the way for an ontology-based adoption of this technique.

The remainder of this paper is organized as follows. In Sect. 2, we present the research baseline on which we ground our work, namely the ontological foundations provided by COVER. Section 3 illustrates some risk propagation definitions and techniques currently available in the literature. Section 4, provides the core contribution, namely an ontological analysis of the notions of risk propagation and risk propagation graphs. Then, in Sect. 5, we discuss the implications of our findings. Lastly, Sect. 6 presents the final considerations and limitations.

2 Research Baseline

Before delving into the notion of risk propagation, let us introduce the view on the nature of risk formalized in the *Common Ontology of Value and Risk*

(COVER) [24].[1] We will use this ontology as a basis to guide the subsequent analysis of the notion of risk propagation, which, as we shall see, poses entirely new ontological issues w.r.t. the adopted ontology itself. We chose COVER because: *.i* it is based on a foundational ontology; *.ii* it embeds a domain-independent conceptualization of risk; *.iii* it is built upon widespread definitions of risk and shows how the risk is connected to the notion of value. Moreover, COVER has already been connected to different domain ontologies showing its utility in clarifying some related notions (e.g., *trust*, *prevention*, *security*).

2.1 Risk Assumptions in COVER

The first assumption in COVER is that risk is **relative**. An event might be seen as a risk by an observer and as an opportunity by another. To exemplify why this assumption holds, consider the case of a potential robbery. The would-be victim perceives such an event as a risk, i.e., as something she does not want to happen and that would hurt some of her goals. From the would-be robber's perspective, the robbery is a desired event that will help her in achieving some of her goals.

The reason why risk is relative constitutes the second assumption about its nature. Risk is perceived according to **impact on goals** as well as the **importance of these goals** to a given agent, i.e. in order to talk about risk, one needs to account for which goals are "at stake". For instance, if one is concerned with the risk of missing a train, it is because missing a train impacts one's goals, such as arriving on time for a meeting.

The third assumption implied by COVER is that risk is **experiential**. This means that we ultimately ascribe risk to events, not objects. This claim may seem counter-intuitive at first, as many theories refer to entities such as "Object at Risk" and "Asset at Risk" [2]. Here the claim is not that such concepts do not exist. Instead, the assumption is that when assessing the risk an object is exposed to, one aggregates risks ascribed to events that can impact the object. For instance, consider the risks your phone is exposed to. In order to identify and assess them, you will probably need to consider: .i which of your goals depend on your phone (e.g. getting in contact with your friends, being responsive to business e-mails); .ii what can happen to your phone such that it would hinder its capability to achieve your goals (e.g. its screen breaking, it being stolen); and .iii which other events could cause these (e.g. you dropping it on the floor or leaving it unattended in a public space). Then the risk your phone is exposed to is the aggregation of the risk of it falling and breaking, the risk of it being stolen, and so on.

The next assumption is that risk is **contextual**. Thus, the magnitude of the risk an object is exposed to may vary even if all its intrinsic properties (e.g., vulnerabilities) stay the same. To exemplify, let us pick one risk event involving

[1] Note that we took COVER as primitive, which was itself subject to validation and proper comparison to the literature of risk in risk analysis and management at large (e.g., [6,16,17]).

your phone, namely that of dropping it and its screen breaking. Naturally, the properties of the phone influence the magnitude of this risk, such as it having a strengthened glass screen. Still, the properties of the surface on which it was dropped (e.g. its hardness) and of the drop itself (e.g. its height) can significantly increase how risky the drop and breaking event is.

Lastly, another assumption that we derive from COVER is that risk is grounded on **uncertainty** about events and their outcomes. This is a very standard position, as proposed in [16] and extensively discussed in [1], which implies that likelihood is positively correlated with how risky an event is. For instance, the risk of a volcano eruption damaging a city is higher for a city that lies by an active volcano than for a city that lies by a dormant one simply because it is more probable.

2.2 The Ontology of Risk

Figure 1 represents the concepts in COVER that are germane to the objectives of this paper.

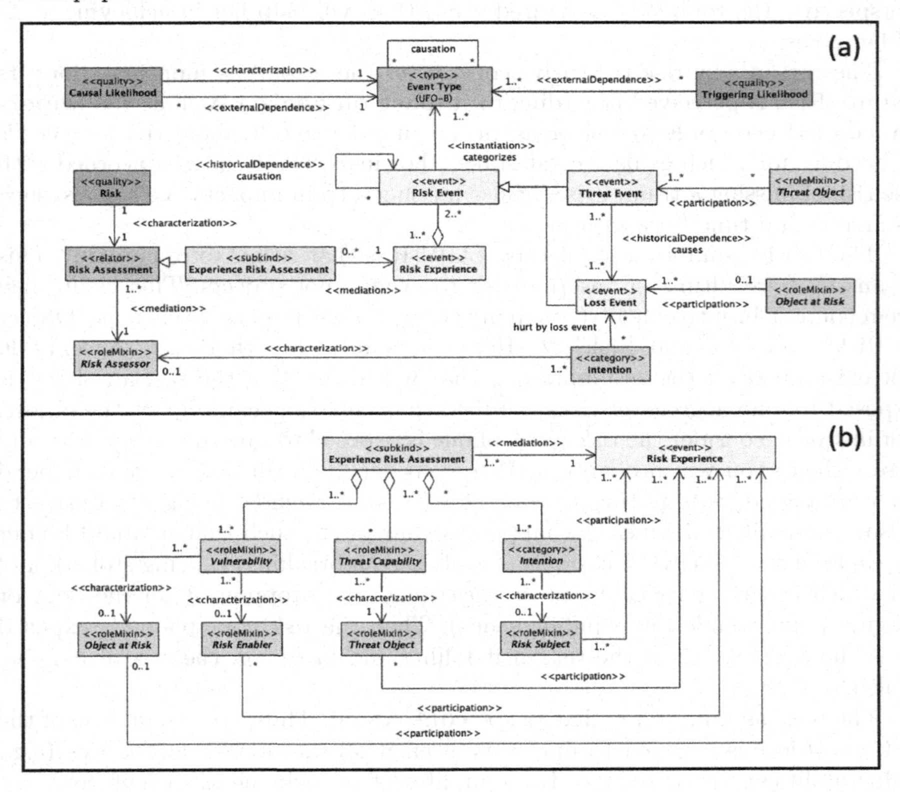

Fig. 1. Two views of the *Common Ontology of Value and Risk (COVER)* [24].

Figure 1 provides two views of COVER (i.e., (a) and (b)), by highlighting the concepts that are key for our analysis. (a) allows understanding the notion of Risk Assessor, namely an agent that makes a Risk Assessment about objects (Object Risk Assessment) and has experiences (Experience Risk Assessment) afforded by these objects. A Risk Experience is a complex event composed of Risk Events, i.e., Threat Events, which may involve the participation of Threat Objects, and Loss Events, which may involve the participation of Objects at Risk. This, as highlighted by view (b), allows us to push the analysis beyond the notions of event and risk assessor's goals, and embed also the concepts of Vulnerability, Threat Capability and Intention, namely *dispositional properties* [3], as aspects that can be involved in risk experiences and propagation.

COVER allows also considering Risk itself as a quantitative measure attributed to a Risk Assessment. The assumption here is that *risk* can be only ascribed to *envisioned experiences* that may (but are not certain to) occur. The ontology addresses this issue by accounting for the existence of future events, as proposed by Guarino [13]. As we will see later on, this aspect will guide us on the analysis and the disambiguation of existing risk propagation models, where event occurrences, event types, and objects are often conflated.

As a final remark, COVER will allow us to further explore how risk is quantified and employed in the propagation process via the concepts of Causal Likelihood and Triggering Likelihood, which are typically expressed by probabilistic measures and, as we will see in the next sections, offers the baseline to understand risk propagation mechanisms. One essential aspect is that considering the ontological grounding of COVER, the likelihood is a quantitative concept that inheres in types of events, not in individuals. Thus, the challenge will be to see how this influences the understanding of current risk propagation approaches and, possibly, the modeling of future ontologically well-founded solutions.

3 On Risk Propagation

What we propose here is an ontological analysis of the notion of risk propagation, for which, as far as we know, there is no related work. The goal of this section is to contextualize that notion by reporting some definitions provided in the literature and giving some representative examples of application. The information below is the result of a review of papers found with criterion *[allintitle: "risk propagation"]* on *Google Scholar*, from 2000 to 2021. The selection of approaches and definitions is not complete but aims to offer a representative view of what is available in the current set of still scarcely generalized and standardized works.

3.1 Some Definitions

The notion of "risk propagation" refers, often rather vaguely, to phenomena in which one can observe that some events *affect* the probability that some other (desirable or undesirable) events happen. Its semantics varies depending on the application context and actual definitions are given in very few papers. Some representative ones we found state that risk propagation is:

.i "the impact on business value spread across operational assets that results from the occurrence of a disruptive event" [12];
.ii "the sequence of inter-dependent risks in the supply network which may or may not lead to a disruption or ripple effect" [10];
.iii "the process by which certain risk units pass certain elements and/or the consequences of risk to other risk units under the influence of necessary external factors" [7];
.iv "how risks originate at one node of the supply chain and create further risks across the supply chain" [5].

A common aspect of these definitions and different senses is their pragmatic orientation. They are always derived from, or highly dependent on, a specific application context or a complementary implementation, namely, the algorithm adopted to perform inferences and take decisions. As an example, in the context of *cyber-security*, risk propagation can be applied to quantify the risk of connected devices, which can be exposed to and compromised by cyberattacks. In this specific scenario, the risk may *originate* (*.i*, *.iii*, *.iv*) from some intervention actions and *propagate* over connected cyber-assets and, eventually, certain events (e.g., processes connected to the cyber assets, such as "vehicle assembly") via certain types of relationship (e.g., correlation, parthood, or causation). The final outcome of the risk propagation, given a certain threat, is then an *assessment* of the potential vulnerabilities of all the selected elements, i.e., cyber assets and related processes.

3.2 Modeling Risk Propagation

Let us consider the following simplified scenario. *"Anna, Bob, and Carl have to make a presentation for a new client. This event is extremely important because it would allow their start-up to gain an important project. On the morning of the presentation, there is heavy traffic congestion on their way to work and the customer only has one 30-minute slot in the early morning. In order to arrive on time and give the presentation, the three must decide whether to take the same means of transportation or each try a different option: subway, car, or bus."*

This example illustrates typical aspects modeled in risk propagation. From it, we can easily understand why the aforementioned risk propagation definitions may arise. As from definition *.i*, we observe how the occurrence of a disruptive event (heavy traffic congestion) impacts business value (potentially losing a customer). Similarly, as from *.ii*, we may talk about a possible *ripple effect* caused by interconnected risk events (e.g., the congestion, the missed presentation, and the loss of a client). As from *.iii*, we may identify multiple risk units, i.e., items for which we may want to calculate the risk, namely the customer, the company, the people involved, and the means of transportation. It is also possible, as from *.iv*, to understand in what sense *risk may originate* at one node (e.g., from the car being stuck in the traffic for a certain amount of minutes). In summary, most of the considerations that emerge from this illustrative scenario suggest a sort of *cascading effect*, which occurs in a network, as if the risk was actually something that could be passed from one node to another.

Risk propagation definitions are often proposed alongside risk propagation techniques, which makes them rather biased by the underlying adopted technology. Figure 2 depicts a risk propagation graph of our scenario, as well as some other techniques proposed in the literature.

Figure 2a is the example we introduced at the beginning of this section. Each node is a risk unit and risk can be spread over the units through edges connecting them. Here we do not stick to any particular assumption about how the risk value is associated with the nodes and how it is propagated. The whole graph could be taken as a *Labeled Property Graph* in which a risk value is associated with each node and propagated through a simple inference mechanism.

Provided in [25], Fig. 2b was designed to measure how risk spreads over a supply chain. Each node in the graph is taken as a *risk unit.* Besides the source node (the leftmost) and the destination node (the rightmost), other nodes representing different transportation steps are provided. Each node is associated with a risk value which then can affect the value of the other nodes. By analyzing the graph one can discover what are the most critical chains in the transportation process and, eventually, adopt mitigation strategies.

Figure 2c depicts a technique for propagating risks in a network based on the *Tropos Goal-Risk Framework*, "a goal-oriented framework for modeling and analyzing risks in the requirement phase of software development" [8]. Here the nodes in the graph may represent agents, tasks, activities, and goals, all of which can be combined to model a risk chain.

Figure 2d also depicts a model for propagating risk over a supply chain [4]. However, differently from Fig. 2b, the model presents the typical structure of a probabilistic network. The nodes being the *target* of an edge are said to be dependent on the corresponding *source* nodes. The nodes that do not present

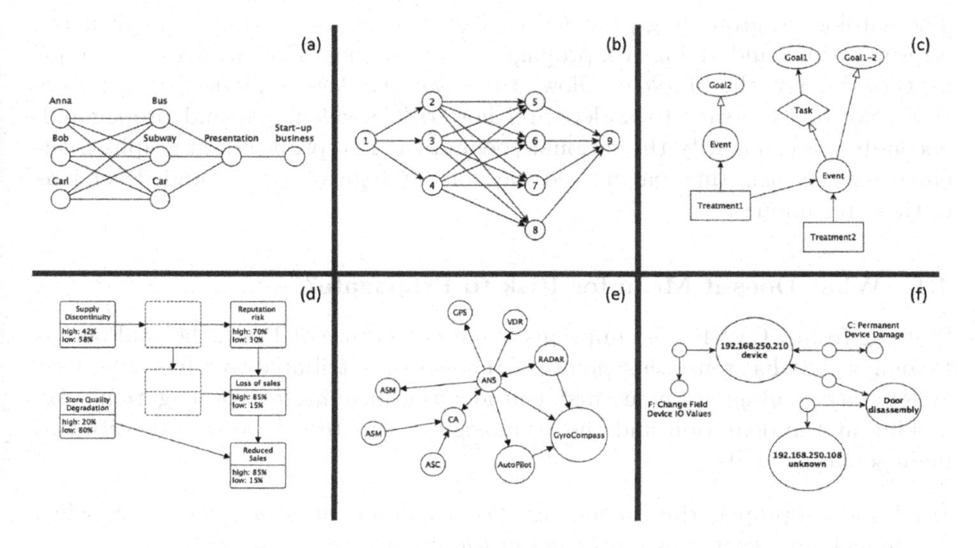

Fig. 2. Different risk propagation graphs. (a) Our example; (b) from [25]; (c) from [8]; (d) from [4]; (e) from [19]; and (f) from [9].

dependencies are said to be independent nodes. In this scenario, the risk propagation mechanism consists of "updating" a risk value ascribed to a certain node according to what *happens* to other nodes. Each node, in this context, clearly represents an event associated with a certain probability value. A user can then query the model to calculate the risk for a corresponding node, assuming that some events in the network occur.

Figure 2e illustrates the technique proposed in [19]. The model is used to perform risk propagation over a cyber-physical system, namely, a cyber-enabled ship. The nodes of the graph represent the different components involved in controlling the navigation of the ship. Similarly to *(b)*, all the edges in the graphs represent a sort of information flow or message-passing mechanism. These edges are then used to calculate how *risk flows* from one node to the others. For instance, the approach allows analyzing how the risk of a radar malfunction can affect other cyber assets (e.g., the collision avoidance system).

Lastly, Fig. 2f is the graph presented in [9], which proposes a knowledge graph-based process-aware risk propagation approach. This work does not provide an ontological analysis of risk propagation but it merely introduces a knowledge graph schema for answering practical queries provided by practitioners. Here risk propagation maps into a kind of *message-passing algorithm* [26]. The nodes in the graph can represent multiple things, such as objects, events, and processes, which are mainly categorized as subclasses of *risk units*. Finally, the edges are used to calculate how the risk values associated with a node affect the values of its neighbor nodes and vice versa.

4 Explaining Risk Propagation

The ontological grounding provided by COVER allows us to make explicit the interpretations underlying risk propagation techniques. This involves two major aspects. Firstly, the ontology allows to *unpack* (*unfold*, *explain*) [15] concepts that may be necessary to understand how risk is calculated and propagated. Secondly, we can clarify the rationale behind the risk propagation graphs modeling assumptions, thus paving the way for ontologically well-founded versions of these techniques.

4.1 What Does it Mean for Risk to Propagate?

Starting from COVER's assumptions about the nature of RISK, the goal here is to analyze in what sense this *quantitative measure* attributed to a RISK ASSESSMENT may *propagate*. Let us first consider the meaning of "propagation". By looking at the definition and the etymology of this term[2] we can identify two main semantic fields.

(a) Firstly, "propagation" concerns *the spreading of something as a belief*, namely an event as a *psychological feature*, a *mental process*.

[2] From https://www.collinsdictionary.com/ and https://www.etymonline.com/.

(b) Secondly, the notion concerns a physical process, namely something that can be observed, which can be analogous to *i. biological reproduction* or *ii.* the *gradual change of an object*, in the sense of extension and enlargement.

Given that RISK is a quantitative measure and that RISK ASSESSMENT is entangled with the RISK EXPERIENCE of a RISK ASSESSOR, definition (a) is naturally suited to the *experiential perspective* fostered by COVER. In this sense, the propagation of risk is *an event that concerns the change or update of a judgment of an agent.* This event, as we are going to discuss below, may occur via *observation*, or via *simulation.* Accordingly, it can be said that risk propagates because the beliefs of a given subject, about a certain situation, change. This supplementary consideration highlights the influence of the notion of *belief propagation* [23] in the definition of risk propagation. Furthermore, this explains why current risk propagation models leverage message-passing algorithms, which represent an implementation of belief propagation as a probability inference technique. However, one key observation is that, in belief propagation, the beliefs updating mechanisms concern only how the LIKELIHOODS associated with some given events are quantified and updated. As we have seen, this is not enough to explain the quantification of risk. *Risk*, indeed, *cannot be mapped one-to-one to a probability value*, and risk assessment is not just a probability quantification. Rather, risk, and then risk propagation, always depend on the subjective judgment of a particular agent about *a given probability of having a certain loss, i.e., a certain event that would have an impact on one of its goals, the latter having a certain measure of importance to that agent.* Looking at the available approaches, this last point is often left implicit, and the propagation of risk ends up being identified as the propagation of probabilities.

Risk Propagation via Observation. Let us take the example provided in Subsect. 3.2. Here risk can be assessed by a RISK ASSESSOR, i.e., Anna, with regard to some possible LOSS EVENTS. Consider, for instance, the event of having a car accident. This could mean Anna losing 500 euros to adjust the damaged car. According to Anna's assessment, the risk of having a car accident can be then calculated as (in a simplified manner) $P(A) \cdot 500$, where $P(A)$ stands for *the probability having a car accident* and 500 euros is the *loss value (i.e., impact) related to the damage.* In this setting, the quantitative measure derived through RISK ASSESSMENT depends on Anna's judgment, as *"the product of the probability that a given (undesired) event happens and the negative value assigned to that given event"*. Now, how the risk of having a car accident can be propagated? Definition (a) of "propagation" suggests that this has to do with some changes in Anna's *beliefs.* Suppose that the event of having an accident with the car is correlated to the possibility of giving a presentation to an important customer. Suppose that not being able of giving the presentation could be considered by Anna as another possible loss event, because her boss will complain and she will not get a bonus of 1.000 euros at the end of the year. In this sense, if the two events are correlated (Anna uses the car the same morning of the presentation), a higher risk of having an accident has an impact, or, "a cascading effect", using the related work terminology, over the risk of failing the presentation. This cascading effect is related to the notion of *conditional probability*, representing the probability that an event occurs *given the occurrence of another event*, and can be taken as the backbone structure of any graph enabling probability inferences. In this setting, keeping fixed the loss values for two events A and F (e.g., 500 and 1.000 euros respectively), if the probability that F occurs depends on the probability that A occurs (i.e., F, given A as $P(F|A)$), the change of risk for A implies a change of risk for F. This depends on an increase of the probabilities associated to A and then of the probabilities associated to F. Again, the loss values remain unaltered. Following the example, then, the propagation occurs when Anna updates her RISK ASSESSMENT according to what happens to some given correlated events composing her RISK EXPERIENCE (e.g., it is raining and there is a lot of traffic, this may increase the risk of having a car accident, then Anna knows that the risk of being late and failing to deliver the presentation just became higher).

Risk Propagation via Simulation. The propagation of risk can only take place because of an existing chain of events in Anna's experience. These events are associated with a corresponding chain of (possible) events at the type level, which are characterized by CAUSAL LIKELIHOOD and TRIGGERING LIKELIHOOD, namely the probability an event occurs and the probability an event causes another event, respectively (see Fig. 1). The fact that something has occurred has an impact on the event-type chain, thus updating the probability values and, accordingly, Anna's quantification of risk. That being said, in the example above, Anna's experience is updated given a new *evidence*, namely by instantiating an EVENT TYPE (e.g., the possibility of having a car accident is realized). The risk related to that event, and other possible correlated loss events, is then updated and such new information is used to update the judgment of the assessor. A different scenario is introduced when the risk is propagated given some simulations run by the assessor. In practice, the new aspect here is that risk propagation is not performed to update the risk assessment *given that something has occurred*, rather it is performed to update the risk assessment *given that something may occur*. Take the presentation example. Anna may want: a) to understand what are the possible transportation actions she can choose when she has to go to the customer; b) to identify possible correlations between those actions (e.g., the probability of arriving on time given the probability of taking the underground); c) to make a ranking of possible loss events (it is better to have a fee for exceeding the speed limits than missing the presentation); given b) and c), d) to select the best option, i.e., the less risky actions in order to not fail the presentation. This involves two main observations. Firstly, the propagation of risk requires the design of an *imaginative risk experience*, which according to COVER naturally maps into a chain of correlated EVENT TYPES. In this sense, risk propagation occurs by simulating the occurrence of some of the (loss) event types taken into consideration, and this can be used to discover new information. For instance, the simulations may allow the discovery of new probabilities and then new risk values. Also, adding new possible events in the causal chain of the simulation may lead to discovering some new objects at risk. Secondly, this simulation could be seen as supplemental to the RISK ASSESSMENT itself.

In this sense, all the models used to implement risk propagation can be taken as a projection of an *imaginative risk experience* that can support a more accurate risk identification process. Concerning this last point, an analysis of the graphical representations used to run risk propagation in the light of COVER plays a pivotal role, since those can be considered as different ways of modeling the *imaginative risk experience* that is necessary to propagate risk via simulations.

4.2 What is in a Risk Propagation Graph?

According to what was discussed in Subsect. 4.1, a risk propagation graph can be seen as a way of supporting the reasoning abilities of a given RISK ASSESSOR. In this respect, multiple elements are involved. For instance, THREAT EVENTS, LOSS EVENTS, THREAT OBJECTS, LOSS OBJECTS, and different possible relationships between them, but also RISK values and the subjects who provide a prior estimation of RISK. In this sense, COVER concepts can be used to discover the multiple interpretations involved in a risk propagation graph.

One first observation is that quite often *types* and *instances* are somehow conflated in these graphs. We may have, indeed, instances of events or event types, where the former is an occurrence of the latter. Similarly, we may have instances of objects or types of objects. This confusion is usually biased by the answering capabilities that should be enabled by the designed graphs. Putting together nodes like "Anna", "Car" and "Failed Presentation", for example, may depend on the queries that the graph should be able to answer. Instance nodes, like "Anna" and "Failed Presentation", usually represent the units that need to be assessed through the propagation process or the things we must decide about (e.g., what is "Anna's risk of losing her job", "given that the presentation failed"). Differently, the nodes representing types are used in the graph to model events having an impact on the final evaluation of risk. That being said, we can straightforwardly divide the types of links in a risk propagation graph as *type-to-type* links and *instance-to-type* links.

Type-to-Type. When two types are related, this generally means that the source node provides a condition for affecting the risk associated with the target node. We have at least the following kinds of uses for the *type-to-type* link in risk propagation graphs:

(1) *event/event (correlation).* This is one of the most common types of links (it usually occurs when the graph adopted for propagating the risk implements a probabilistic network). In COVER the correlation relation is not explicitly represented but can be somehow related to EVENT TYPES characterized by (a weaker version of) CAUSAL-LIKELIHOOD. In this setting, when two events are linked in risk propagation graphs, they are said to be correlated. For instance, the event of the *presentation* is correlated to the event of *transportation.*
(2) *event/event (historicalDependence).* The relations of causality between event types are often assumed as being the backbone link on which a risk propagation graph is built. The problem here is that a clear interpretation is often

left implicit by designers, thus leaving open the possibility of mixing correlation links with causality links, or just simply reducing causality to a kind of strong correlation. To understand how causality is different from correlation and how this is important for risk propagation, think that the risk propagated through causality can be deterministically controlled by modifying or un-linking the cause node. If an event type A is the cause of another event type B, controlling A means controlling the effect on B[3]. This is not the case when we have just a correlation relation. For instance, the risk of arriving late may be correlated to the risk that the presentation performs badly, but it might not be the cause of its failure. Notice that this distinction can be of essential importance for querying and using a risk propagation graph (see for instance the choice of mitigation strategies).

(3) *object/event (object participation).* This other link always denotes the dependency between objects of a certain type and the events these are implicitly involved in. This relation can be also used to discover new correlation or causality links between EVENT TYPES. For instance, suppose we have a graph with the link *Device(Object)* $\rightarrow$ *Computer freezes(Event Type)*. In a simple risk propagation scenario, this link may be used to 'pass' a risk value from an object to a certain event. Following COVER terminology, "Device" can be taken to represent a THREAT OBJECT (TYPE), implicitly involving a certain EVENT TYPE, e.g., as a shorthand proxy for *"Plugging in Device"*. The "Device" THREAT OBJECT thus hides a PARTICIPATION relationship with that RISK EVENT, which, in turn, represents an occurrence of that given EVENT TYPE. This clarification guided by COVER has multiple implications for modeling. As an example, consider the possibility to identify a set of multiple *threat objects* associated with a single type of event. Similarly, this would imply the possibility of making explicit multiple types of OBJECT AT RISK participating in the same event.

(4) *object/object (parthood).* This case usually occurs in graphs representing the connection between digital objects (see the cybersecurity cases like graphs *e* and *f* in Fig. 2). But it is also possible to find examples of it in physical contexts, e.g., supply chain scenarios. Here the interpretation is often one of a parthood link between the objects at hand (e.g., hard disk and laptop) such that VULNERABILITIES of an object can be *activate* or enable the VULNERABILITIES of the other (see [3,24]) when they are connected via these links. Notice that, in this case, we have implicit the *Risk events (Types)* connected to the manifestation of these VULNERABILITIES. So, when writing $(A \rightarrow B) \cdot \alpha$, we have α representing a probability value that refers either to: .i the correlation or causality connecting the RISK EVENT (TYPES) that are the manifestation of these VULNERABILITIES; .ii the probability of a manifestation of the VULNERABILITY of my device (of type) A (which, again, is a RISK EVENT) given that A is part of another device (of type) B; .iii the probability of a manifestation of the VULNERABILITY of my device (of type A) given that it has another device (of type) B as part.

[3] [3] advances and ontology-based discussion on causation and event prevention.

Instance-to-Type. When an instance is related to a type, this generally depends on mixing the semantics of the link in the graph with the possible operations that the link enables.

(1) *individual/event (individual participation)* This possible modeling refers, as in the type-to-type case, to a participation relationship. The different nuance here is that the object in question does not represent a class of objects, but an individual, e.g., a specific IP address, a person like Anna, or a specific product (e.g., "my laptop"). This particular case depends on the need to infer risk values for specific objects in a given system, or the need to identify objects instances that may represent a threat.
(2) *property/event (characterization)* This instance/type relation is less common compared to the previous one. A typical example of this relationship is when one node in the graph is associated with multiple nodes that are considered as "risk characteristics" [20]. For instance, we may have multiple nodes about a delay quantified in minutes (e.g., 15 min), associated with a node representing a transportation step - all these ranges of time have a different impact on the final calculation of risk. We may also have nodes like, "low impact", "high impact", and "medium impact" associated with an event. In these cases, the probability value is an abstract particular (an instance) representing a *quality value* [14] characterizing that event type (e.g., the probability of an event of transportation occurring with a 10' delay).

5 Implications

According to our analysis Risk Propagation is an event that *leverages the causal relations involved in a risk experience network, to make a (possibly more nuanced) calculation of causal and triggering likelihoods and hence of risk values.* Accordingly, risk propagation always involves bundles of different interconnected concepts, representing events and objects, but also *dispositional properties* as *manifestation of events desired by an agent* [24] (e.g., *intentions, vulnerabilities* and *threat capabilities*). Reasoning with the effect of changes in those things can play a key role in the propagation and the analysis of risk. In this respect, we highlight two implications.

Implications on Expressivity. Here we map the *expressivity* of a risk propagation model to the *capability to answer queries about a risk experience.* If we consider the analysis we provided, it turns out that the adoption of graphs with no distinctions, for example, between causation and participation, or between objects and events, is restrictive. To emphasize the implication on expressivity, we gathered feedback from 5 domain security expert analysts, in the domain of cyber-security, who have been involved in risk assessment activities and that have been working on the identification of risk causes and risk dependencies via the application of risk propagation approaches. We ran open-ended interviews and the main open questions we asked were about *relevant queries that lack proper support in existing risk propagation approaches.*

We then selected the following examples as the most representative cases of (currently) not addressable requests. These are used *to infer the risk of*:

(1) *"an object, given the event(s) in which it participates"*;
(2) *"an event, given the event(s) to which it is connected"*;
(3) *"an object, given the object(s) to which it is connected"*;
(4) *"an event, sharing an object with other events"*;
(5) *"an object in an event with another object, which is in another event"*;
(6) *"an event, given different properties characterizing the correlated event"*.

These examples cannot be addressed without the support of an ontology. For example, what if the approach relies only on a probabilistic graph where all the nodes are EVENT TYPES? What if it does not support any distinction between types of relations, like PARTICIPATION or CAUSATION?

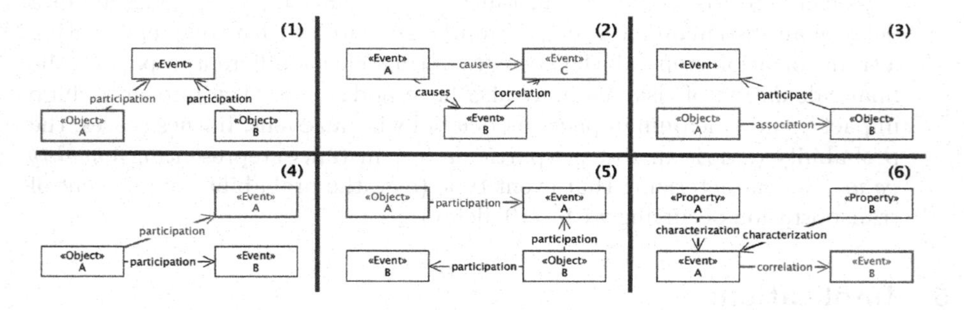

Fig. 3. Patterns that can be used to model the example requests (1–6).

Figure 3 shows of how the above-listed requests/queries may be addressed through graph structures. Multiple nodes having different semantics are involved, e.g., objects, events, and different relations (e.g., associations, participation, correlations, and characterization relations). As we have discussed, all these elements composing a RISK EXPERIENCE can be often conflated and mixed up.

Consider pattern (1). The *mainstream approach* (let us call it MA) is to flatten the structure into a graph where each link can be used to propagate risk values and each node can be associated with a value quantifying risk. In this sense, a person (e.g., Anna) who participates in an event associated with a risk value (e.g., a presentation with a risk of failure), can inherit that risk. Accordingly, all the nodes and relations are involved in the calculation of risk, and no functionality to distinguish objects from events is provided. This is also valid for the semantics of the relations, where no information is provided to understand whether some different links (e.g., "presents" or "look at") are of the same type (e.g., "participation"). Following (MA), all the other patterns highlight similar issues. Consider (2). How can I filter out only chains of events from graphs where I have nodes like "Anna", "Car" and "Presentation", and calculate the risk of one event given only the events to which it is connected? Again, consider (6), given a graph with nodes like "Anna", "Car", "Transportation", "Presentation", "10 mins late", how can we filter out only nodes that are properties characterizing only a specific event?

By leveraging our analysis, instead, we can adopt a new perspective, i.e., an *Ontology-driven Approach (OA)*, where information is represented explicitly. The main suggestion that emerges is that all information regarding the calculation of probability should be represented via a *probabilistic graph* encoding a chain of RISK EVENT TYPES. Similarly, the data used to query and assess the given system of events and objects can be modeled via a *knowledge graph* whose structure concerns a completely different scope. This has three main implications. Firstly, the graph for the probabilities assessment can be fully disambiguated, thus involving a clear understanding of the events making the reference scenario, their causal chains, and a better grounding for the modeling and update of probabilities. Secondly, the knowledge graph, by accounting for (types of) OBJECTS (e.g., OBJECTS AT RISK), their PARTICIPATION in RISK EVENTS and possible different relations between them, enables users to run queries that go beyond the graph structure built upon probabilities inference mechanisms.

Implications on Accuracy. Our analysis can be used also to support the *accuracy* of a risk propagation assessment. Suppose we have a graph including the different concepts we discussed, where nodes and relations may participate in the propagation of risk. This implies a challenge in determining the possible causes of a loss and the possible paths in the graphs that could lead to that loss.

Consider query (3), suppose we have two persons (objects), i.e., A and B, and a loss event that is "presentation" (P). B participates in P, but A does not. Does A inherit the risk of the event by being associated with B, which is involved in P? This depends on *the type of their association.* If A is a "colleague-of" B, A may share the risk of losing the client, if A is just a "friend-of" B, possibly not. Thus, in this case, clarifying the semantics of the graph components allows for discovering critical risk paths. Similarly, suppose that, as from pattern (2) in Fig. 3 we have a chain of events, where event C represents a LOSS EVENT, e.g., the *presentation failure*, event A represents the event "studying till late" and the event B represents the event "arriving late". Some key questions in this simple scenario are: *what leads to a change in the presentation performance?*, *what events are likely to improve the probability that the presentation succeeds?*, and *what one should do to improve the final outcome?*. Distinguishing between *causal* and *correlation* relations, in this case, would be of pivotal importance. For instance, by knowing that what is having a causal impact on the presentation performance is "the fact that I rehearsed a lot", I can consider it much "riskier" to not prepare the presentation than arriving late.

As a final key point, the introduction of dispositional properties as first-class citizens allows for connecting risk propagation with *prevention*, but also to explain some other aspects of the propagation between objects. How, for instance, introducing or blocking threat agents and mitigating vulnerabilities may affect the whole risk propagation process? Moreover, what if, for example, Anna's boss is exposed to financial risk and his "risk can propagate to Anna" because of their contract, i.e., because of the commitments and claims (which are again dispositions) in that contract? Our ontological analysis can be used to explain also that kind of propagation.

6 Conclusion

In this paper, we leveraged the *(COVER)* ontology to run an analysis of *"risk propagation"*. To the extent of our knowledge, ours is the first attempt to investigate this notion through an ontological investigation. The presented analysis allowed us to: explain *.i* how the propagation of risk is somehow concerned with a phenomenon of belief updating; *.ii* explain how talking about the "propagation" of risk can be misleading; *.iii* identify the concepts and relationships required to capture the cascading effect assumed when talking about risk propagation without incurring in ambiguities and reductions. Moreover, the presented work led us to two new important insights. Firstly, the application of ontology-driven conceptual models may play a key role in the explanation of specific applications and techniques. More concretely, the application of COVER played a central role in the explanation of concepts that are necessary to understand how risk is calculated and propagated in the available approaches. Secondly, the analysis suggests that to fully exploit risk propagation and answer queries about how risk is propagated through different kinds of objects and events, the inference facility provided by probabilistic graphs should be integrated with the representation of several other notions involved in the concept of risk, namely, agent, their goals, the importance of these goals, the impact of certain events on their goals, etc.

We have three main plans for future work based on the presented results. Firstly, we aim to explore further how the phenomenon of risk propagation is related to the quantification of *severity*, the activation of *countermeasures*, and the identification of *vulnerabilities*. In this respect, the idea is to exploit some of the results presented in a companion paper [3] and provide an analysis that goes beyond risk and delves into the connected security domain. Secondly, we aim to compare and assess some of the available technologies that are employed to propagate the risk and see how these may be used to revise, extend or confirm our ontological assumptions (see, for instance, the case of models that are able to update probabilities through a cyclic or acyclic chain of events). Finally, we aim to exploit the explanation proposed in this paper and evaluate it over an ontology-driven risk propagation approach, by using real data and showing how this can improve/extend the current state-of-the-art applications.

Acknowledgement. This work was done in collaboration with Accenture Labs, Israel. The research conducted by Mattia Fumagalli is also supported by the *"Dense and Deep Geographic Virtual Knowledge Graphs for Visual Analysis - D2G2"* project, funded by the *Autonomous Province of Bolzano.*

References

1. Aven, T., Renn, O., Rosa, E.A.: On the ontological status of the concept of risk. Saf. Sci. **49**(8), 1074–1079 (2011)

2. Band, I., et al.: Modeling enterprise risk management and security with the archimate language - W172 (2017)
3. Baratella, R., et al.: Understanding and modeling prevention. In: Guizzardi, R., Ralyté, J., Franch, X. (eds.) RCIS 2022, pp. 389–405. Springer, Cham (2022). https://doi.org/10.1007/978-3-031-05760-1_23
4. Cao, S., Bryceson, K., Hine, D.: An ontology-based Bayesian network modelling for supply chain risk propagation. Ind. Manag. Data Syst. (2019)
5. Chaudhuri, A., et al.: Risk propagation and its impact on performance in food processing supply chain: a fuzzy interpretive structural modeling based approach. J. Model. Manag. (2016)
6. Coso, I.: Enterprise risk management-integrated framework. Committee of Sponsoring Organizations of the Treadway Commission, vol. 2 (2004)
7. Deng, X., et al.: Formation mechanism and coping strategy of public emergency for urban sustainability: a perspective of risk propagation in the sociotechnical system. Sustainability **10**(2), 386 (2018)
8. Deng, X., et al.: Risk propagation mechanisms and risk management strategies for a sustainable perishable products supply chain. Comput. Ind. Eng. **135**, 1175–1187 (2019)
9. Engelberg, G., et al.: An ontology-driven approach for process-aware risk propagation. In: 38th ACM/SIGAPP Symposium on Applied Computing (2023)
10. Garvey, M.D., Carnovale, S.: The rippled newsvendor: a new inventory framework for modeling supply chain risk severity in the presence of risk propagation. Int. J. Prod. Econ. **228**, 107752 (2020)
11. Garvey, M.D., et al.: An analytical framework for supply network risk propagation: a Bayesian network approach. Eur. J. Oper. Res. **243**(2), 618–627 (2015)
12. González-Rojas, O., et al.: Quantifying risk propagation within a network of business processes and it services. Bus. Inf. Syst. Eng. **63**(2), 129–143 (2021)
13. Guarino, N.: On the semantics of ongoing and future occurrence identifiers. In: Mayr, H.C., Guizzardi, G., Ma, H., Pastor, O. (eds.) ER 2017. LNCS, vol. 10650, pp. 477–490. Springer, Cham (2017). https://doi.org/10.1007/978-3-319-69904-2_36
14. Guizzardi, G.: Ontological foundations for structural conceptual models (2005)
15. Guizzardi, G., et al.: Ontological unpacking as explanation: the case of the viral conceptual model. In: International Conference on Conceptual Modeling, pp. 356–366 (2021)
16. ISO: Risk Management - Vocabulary, ISO Guide 73:2009 (2009)
17. ISO: ISO 31000:2018 - Risk management - Guidelines (2018)
18. Jiang, J., et al.: Identifying propagation sources in networks: state-of-the-art and comparative studies. IEEE Commun. Surv. Tutor. **19**(1), 465–481 (2016)
19. Kavallieratos, G., et al.: Cyber risk propagation and optimal selection of cybersecurity controls for complex cyberphysical systems. Sensors **21**(5), 1691 (2021)
20. Li, M., et al.: Risk propagation analysis of urban rail transit based on network model. Alex. Eng. J. **59**(3), 1319–1331 (2020)
21. Newman, M.: Networks. Oxford University Press, Oxford (2018)
22. Pearl, J.: Graphical models for probabilistic and causal reasoning. In: Quantified Representation of Uncertainty and Imprecision, pp. 367–389 (1998)
23. Pearl, J.: Reverend bayes on inference engines: a distributed hierarchical approach. In: Probabilistic and Causal Inference: The Works of Judea Pearl, pp. 129–138 (2022)

24. Sales, T.P., Baião, F., Guizzardi, G., Almeida, J.P.A., Guarino, N., Mylopoulos, J.: The common ontology of value and risk. In: Trujillo, J.C., et al. (eds.) ER 2018. LNCS, vol. 11157, pp. 121–135. Springer, Cham (2018). https://doi.org/10.1007/978-3-030-00847-5_11
25. Shin, K., et al.: Risk propagation based dynamic transportation route finding mechanism. Ind. Manag. Data Syst. (2012)
26. Sunil, K., et al.: Message passing algorithm: a tutorial review. IOSR J. Comput. Eng. **2**(3), 12–24 (2012)

The Omnipresent Role of Technology in Social-Ecological Systems

Ontological Discussion and Updated Integrated Framework

Greta Adamo(✉) and Max Willis

BC3 - Basque Centre for Climate Change, Leioa, Spain
greta.adamo@bc3research.org, max@maxwillis.net

Abstract. Technology-driven development is one of the main causes of the triple planetary crises of climate change, biodiversity loss and pollution, yet it is also an important factor in the potential mitigation of and adaptation to these crises. In spite of its omnipresence, technology is often overlooked in the discourses of social and environmental sustainability, while in practice sustainability initiatives often draw criticism for favouring technical solutions or oversimplifying the relationships between society, environment and technology. This article extends our RCIS 2022 publication "Conceptual integration for social-ecological systems: an ontological approach" with an ontological examination of technology in two prominent social-ecological systems paradigms, *social-ecological system framework* (SESF) and *ecosystem services* (ESs) cascade. We ground the ontological analysis of technology on analytical and postphenomenlogical philosophical literature and effect several re-designs to the initially proposed integrated framework. The main aim of this work is to provide a clearer and theoretically founded semantics of technology within SESF and ESs to improve knowledge representation and facilitate comparability of results in support of decision-making for sustainability.

Keywords: Philosophy of technology · Ontological analysis · Social-ecological system framework · Ecosystem services cascade

1 Introduction

Technological development is a major contributor to climate change, biodiversity loss and pollution (e.g. [29]) yet technology is also an essential tool for finding solutions to these tripartite crises [10,15,47]. The fact that very few natural resources are accessible, beneficial or valuable to humans without the existence and availability of some kind of technology to extract them [15,24] reveals a fundamental human-nature relationship as decidedly technological. Technology here refers to resource extraction tools and infrastructures within social-ecological systems (SESs), but also to Information and Communication Technology (ICT), the main epistemological grounds for environmental science and knowledge production via data collection, analysis and for example, climate change predictions [44]. While a few attempts have been made to create over-arching frameworks

S. Nurcan et al. (Eds.): RCIS 2023, LNBIP 476, pp. 87–102, 2023.
https://doi.org/10.1007/978-3-031-33080-3_6

[4,32], the literature on sustainability and related scientific fields often lacks in-depth analyses or assumes an overly simplified view of technology [36]. This tendency, frequently accompanied by ambiguous or unspecified semantics and a lack of theoretical foundations, is reflected in *social-ecological system framework* (SESF) [34] and *ecosystem services* (ESs) cascade [40], two widely applied SES frameworks used to capture common knowledge, define indicators, collect data, visualise results and manage resources for decision-making. The lack of a clear, explicit reification of technology limits the accuracy of those worldviews. In addition there are still challenges of comparing, aligning and integrating SESF and ESs outcomes [39] resulting in scarce information reuse and the compartmentalisation of sustainability knowledge [8].

The aforementioned issues are tackled in this paper using ontological analysis to clarify concepts and define aspects of technology within SESF and ESs cascade that without specification affect the assessments emerging from SESs analysis. Moreover this work proposes a unified conceptual and theoretical base of technology continuing the integration begun in the RCIS 2022 contribution [2]. The broader objectives that guide this extension are: (i) to establish a more explicit semantics of the main SESF and ESs components, (ii) to include technology-related concepts that can improve the quality of knowledge representation, and (iii) integrate the use of both approaches for comparisons of data and results in a unified perspective that will extend the reach of sustainability studies. This analysis of technology is performed by examining the state of the art of SESF and ESs cascade and extracting, where possible, the conceptual components related to technology. These are defined ontologically on the basis of both material and social-critical aspects (see also [18,50]) by incorporating analytical philosophy and applied ontology approaches with postphnomenology [27,49].

The paper begins with a review of some relevant interpretations of technology (Sect. 2), followed by treatments of technology found within SESF and ESs (Sect. 3). It then presents the analysis and inclusion of technology-related concepts in the integrated framework from RCIS 2022 (Sect. 4). Then the new concepts are discussed using the example of aquaculture (Sect. 5) and in conclusion future works are addressed (Sect. 6).

2 Perspectives on Technology

Providing a unified definition of technology is hindered by the prevalence of two contrasting perspectives in the literature, one that focuses primarily on the social dimension of technology called "humanities philosophy of technology", and another that revolves around technology as an engineered product called "engineering philosophy of technology", which is closer to the analytical philosophical heritage [19]. Regardless of their differences, these two are intertwined and both are relevant for a complete and multi-faceted interpretation of technology within sustainability contexts that require social-technological approaches [3]. In particular the "humanities philosophy of technology" discussions concerning "instrumentalism" alongside "critical theory" considerations [17] provide a continuum that enriches the social aspects of the analytical perspective of technology.

Some Philosophical Accounts of Technology. Technology in this work is considered to be an entity created by humans, therefore an artefact. A common way to define artifacts refers to objects created intentionally to achieve a goal [41], thus artifacts are only those that are *intended* (here meant as *prior* to the production stage) [31]. This general definition fits with a widespread interpretation of technology in climate change and sustainability research known as *instrumentalism* [17,36] which assumes that technology is designed to achieve desired results. This vision considers technology as neutral, then not loaded with *specific* social-cultural values (besides perhaps efficiency), and under the control of humans [17]. The neutrality of instrumentalism has nevertheless been challenged [19], for although technology is interpreted as goal-oriented, this does not imply that there are no values embedded in those goals, a point also discussed by Feenberg [17]. Technology emerges from the world of design and engineering and is therefore situated in a wider societal context with constituent needs and values (e.g. the safety or sustainability features of technology). Instrumentalism has been linked to an overabundance of *technology fix* approaches that do not sufficiently consider broader social-ecological systems, which is particularly problematic in the context of sustainability [19].

In contrast with instrumentalism, *critical theory*-derived interpretations regard technology as not only managed by humans but also charged with societal values. In the critical sense technology is more than a neutral tool, it is also a framework that provides for the shaping of society [17]. This idea is reflected in what Feenberg calls, somewhat confusingly, *instrumentalization theory* which encompasses two aspects of technology, its functional-technical properties, also named "primary instrumentalization", and socio-cultural contexts such as power dynamics that influence the design and the designer(s), which is called "secondary instrumentalization" [18]. Among the critical theories of technology described in [36] is *postphenomenology*, that combines (i) phenomenology, in which the object-(artifact) is not only perceived by the human subject but also mediates their experiences through mutual engagement [28], (ii) pragmatism, in which technology is considered only within use contexts [51], and (iii) empirical studies of actual technologies in which philosophical investigations revolve around technology instances [49]. At the heart of postphenomenology is an elaboration of mediating relations (embodiment, hermeneutic, alterity, and background) [27,28,48,49] through which technology mediates human experiences, visions of the world, choices and actions [49].

As discussed in [18], instrumentalization theory presents some analogies with a perspective of technology proposed within the analytical philosophical tradition, that is *the dual nature of technical artifacts* [50]. According to this duality, technical artifacts carry *functions* that cannot be explained only by their *physical characteristics* since functional specifications involve *social-intentional* aspects. Studies of technical artifacts, therefore, entail analysis of their physical properties as well as their intended functions relevant to a *use plan* [50].

In the context of applied ontology, relations between the artefact itself, including its characteristics, and the intentional agents are formalised in [11]. This defines artefact as purposely mentally created by an agent through a process of selection. Those artefacts are modelled by distinguishing their material composition (e.g. physical object and amount of matter) and the artifact itself, i.e. material + intentional aspects. Different from physical artifacts, social artifacts are those that are acknowledged within a community. Rather than grounding the concept of artifact on functions, [11] considers the more general notion of *capacities*, which are useful to describe the qualities of a certain artifact, and *attributed capacities* that are connected with the intentions of its creators. Kassel offers another in-depth formalisation [31], which aligns with ìnstrumentalism to describe artefacts as entities created with prior intentions and successfully produced. Technical artefacts for Kassel are those having properly ascribed functions, considered as the capacity of the artefact to achieve a goal in a context when the properties of the artefact are directed to its intended effects, which can include accidental functions. Those functional aspects can be ascribed privately by an individual or publicly by a social group. A dedicated analytical investigation of *social artifacts* is proposed by Thomasson [46], offering two main theses. The first considers an artifact as something with intended *features* that can range from material properties to normative characteristics and which *could* also include intended functions. Secondly, it observes a less general class of artifact, the *public artifact*, that is recognizable by a certain group or community and is dependent upon *norms*, which guide the behavior of the community's members and contexts. Public artifacts depend upon collectives and are representations of normative practices and shared semantics, an interpretation that is closer to critical theories of technology.

The literature on technology offers many other useful interpretations and definitions. However, we chose the above concepts to begin a discussion using material and social technology perspectives, which ground the analysis in the following sections.

3 Technology in Social-Ecological Systems

SESF is an ontology composed of social and environmental variables organised in multiple conceptual levels, called tiers, aimed at providing a shared understanding to diagnose the sustainability of SESs [34,37,39]. The main SESF tier elements are *resource unit, resource system, actor, and governance system* tailored together in the *action situation* interactive pool. Exogenous elements that might influence the system framework are *social-economic-political settings* and *related ecosystems.* A full list of the second tier variables can be found in [34] Table 1. With its origins in political science and focus on preserving common-pool resources [9], the language of SESF tends towards ideas connected with sustainable resource management. ESs takes a different approach and focuses on the services provided by nature that contribute to human well-being [7,40]. Following the *Common International Classification of Ecosystem Services* (CICES)

[23] the main classes of ecosystem services are *provisioning* (e.g. crop, water), *regulation & maintenance* (e.g. flood protection and pest control) and *cultural* (e.g. scientific and symbolic experiences derived from ESs). The cascade conceptual framework explains ESs based on ecosystem *structure*, *process* and *function* that provide ecosystem *services* that are *beneficial* for human societies and carry *values* [40]. The ESs paradigm is an amalgam of both natural science and ecological-economics [9]; its language and applications lean towards economic valuation and visualization of ecosystem services to orient policy and decision-making [39].

Identifying Technology in SESF and ESs Cascade. The two frameworks adopt different perspectives and scales of analyses even though both SESF and ESs represent an ecological and a social partition [9]. Divergences are also reflected in their approaches to technology, which in SESF is only partially developed and in ESs is notable for its absence. Based on our current knowledge, neither of the two frameworks expresses a direct commitment on how to interpret the role of technology.

SESF includes explicit references to technology and several second tier variables encompass technological artifacts either directly or indirectly. Technology (S7) is attributed to the *social, political and economic settings* that are external factors influencing social-ecological systems at local or global scale [34]; examples comprise relevant existent technology and ICT facilities, such as mobile phones and information systems, yet this variable is the less examined in literature [37]. Technologies available (A9) is a variable of the *actors* element describing the accessibility level of relevant technology, amount of technological artifact available and the extent of physical surface, e.g. land and ocean, in which technology is used [37]. Finally Human-constructed facilities (RS4) are included in SESF's *resource system* and facilitate the realization, maintenance and enhancement of the *resource unit* (stock) [26]. Examples are tourism and transportation infrastructures (e.g. boardwalks and harbours), human-constructed habitats (e.g. artificial reefs or aquaculture ponds) and ICT infrastructures for data processing and management [37]. Implicit references to technology and ICT can be found in several SESF concepts, for example most of the *action situation* second tier variables (e.g. Harvesting (I1), Information sharing (I2), Monitoring activities (I9) and Evaluative activities (I10)). Besides the direct inclusion of technology in SESF and related research of SESs robustness involving for example energy and public infrastructures [5,26,34], technology elements remain second-tier and are not given the same level of prominence as the social and ecological first tier components.

In the ESs cascade artificial entities are not explicitly included, even though most ecosystem services are not immediately available to meet human needs and are implicitly linked to technical artifacts used by humans for extraction or access. Technology then forms part of the *production boundary* [40] where the social and economic systems overlap with the ecological system, and where goods (with their associated benefits and values) are derived from ecosystem services. The ecosystem *services* themselves are considered as desired outputs of the ecological system (derived from ecosystem *structure* and *function*). The CICES nomenclature [23] begins for example with the provisioning (biomass) ecosystem service 1.1.1.1 "Cultivated terrestrial plants (including fungi, algae) grown for nutritional purposes" and gives the example of a wheat field before harvest, which yields goods and benefits that include "Harvested crop; Grain in farmer's store; flour, bread". Harvesting of wheat requires at the very least simple tools, yet modern agriculture typically involves mechanisation and technologies for sorting, cleaning, preparation, storage and transport. The CICES cultural biotic ecosystem services 3.1.2.1 "Intellectual and representative interactions with natural environment" and potentially also 6.1.2.1 "Natural, abiotic characteristics of nature that enable intellectual interactions" both refer to scientific investigation, and by implication the presence and use of technology. Section 4.2 and 4.3 of CICES extended version list material and energy sources that are bound to sophisticated technologies for conversion, transmission and storage.

4 Conceptualising Technology in the Integrated SESs Framework

In this section we provide an informal disambiguation of technology concepts, an extension and re-design of the SESs integrated framework proposed in [2]. Our previous work modelled and integrated several SESF and ESs cascade components. These included the general notion of *resource*, which provides the ground for resource unit and system, *actor* and *governance*, which set the basis to better understand the governance system, *ecosystem structure*, *function* (now *functional role*) and *service*, *value* & *benefit* represented through the *valuation relationship* based on [6]. That investigation followed a simplified ESs version that does not include ecosystem *process*. The ontological analysis of each of those elements was based on relevant applied ontology research especially in the domain of information systems, such as the *Unified Foundational Ontology* (UFO) [22] and the *Descriptive Ontology for Linguistic and Cognitive Engineering* (DOLCE) [33] and related works.

After the review of philosophical literature and the representations of technology in SESF and ESs cascade, this research follows a two-step approach: (i) ontological analysis of technology entities and relations in the SESs frameworks and (ii) re-design of the initially proposed integrated framework. Step one concerns the semantic disambiguation and definition of SESF and ESs technology components, employing "humanities philosophy of technology" and applied ontology as theoretical grounds. Consistency with our previous work is ensured by continuing to reference UFO, DOLCE, and cognate ontological literature, such as the study of resources proposed in [42]. To further analyse the social aspects of technology within SESs, in which technology affects the access to a certain resource or service and thereby shapes the worldviews of actors involved in a certain SESs context, we adopt both analytical works, such as [11,46], and post-phenomenology as a critical perspective of technology. Although we recognise the different theoretical underpinnings of analytical and humanities approaches, we believe that often the challenges of creating a bridge between the two perspectives are related to linguistic and heritage discrepancies and not to absolute divergences. The methodological steps of ontological analysis are provided in [1]. The semantic clarifications are actualised in step two by either re-interpreting existing elements based on the outcomes of step one or by introducing newly defined components and relations. Note that these two steps run parallel.

4.1 Ontological Analysis of Technology and Its Integration

We begin this investigation by considering the general definition of artefact included in Sect. 2 as something intentionally created by agents with certain characteristics that satisfy a purpose and we use this definition as a starting point to extend the notion of *human-made resource* presented in [2]. A *human-made resource* is defined as a role played by an artificial object (*physical object* or *amount of matter*) assigned to *activities* and relevant for a *plan*. This relevance is derived from characteristics or qualities of the artifact that make it suitable to achieve a goal [42]. While the aforementioned descriptions are included in the initial integrated framework, the relation `creates` between the *actor*, which holds *intentional moments* with *propositional content* [22], and the human-made resource is newly introduced (`I re-design`) and refers to the intentional mental creation of the artefact performed by the actor [11]. As in the previous version, resources, including those made by humans, carry *values* because of their functions that are linked to their qualities. Those values were modelled through the contextual relation of value ascription [6], labelled *valuation relationship*, that is the assignment of values from an actor to an entity, more specifically objects and activities. Another update here regards the relation between *informational object* as the provider of propositional and immaterial technical content (e.g. model, code or plan) that is typically `realised by` a support [42] as also formalised in DOLCE Lite Plus (DLP) [20] (`II re-design`).

In the previous work *resources* were classified as *objects* [22], however after further analysis of the literature the concept of resource in the context of SESs has been re-defined as a social-public entity, congruent with definitions provided in [11,46]. Now a resource is classified as a *social object* that plays *social functional roles* and is recognized, formally or informally, within a certain community of people (`III re-design`). One inconsistency arises from this decision, as indeed many human-made and all natural resources also have material properties. To accommodate this we adopt the approach proposed in [42], in which resources play roles (social objects) while retaining their material character. This grounds the framework elements *nr physical object* and *nr amount of matter* together with *hr physical object* and *hr amount of matter*, where "nr" and "hr" stand for "natural resource" and "human-made resource", respectively. These are resources playing roles in a context, yet are also subclasses of the non-role/contextual entity *object*. Inspired by the formalisation of SESF proposed by Hinkel et al. [25] we also extend the "resource" side of the framework by including the concept *resource system* (`IV re-design`) that encompasses both *natural resource* and *human-made resource* (Fig. 1a depicts I, II, III, IV re-designs). Following on with an expansion of *governance system* we explicitly introduced from UFO the entity *social actor* [22][1] which is a kind of actor that subsumes *organisation* and can define *norms* (i.e. information objects), *social commitments* and the newly introduced *social decision*, which is a kind of social commitment (i.e. social moments) [22] `recognised and established by` a social actor. The aforementioned concepts enrich the notion of governance as an *activity* `performed by` the social actor that `regards` *policy*, i.e. a formally agreed plan [2]. Those re-designs (number `V`) are represented in Fig. 1b.

Technological Mediation in SESs. This phase concludes with the `VI re-design` (Fig. 1c); the resulting framework can be found in full here. The `VI re-design` involves the notion of technological mediation inspired by postphenomenology [27,28,48,49] and schematized in [3]. We include this conceptualisation of technology to extend the definition of human-made resources, as social objects recognised within a group of actors, to explicitly capture the mechanism through which technology frames and shapes human actors' *intentional moments* and potentially also their *actions.* Such mediation that the actor experiences occurs due to an engagement between the actor and the technology. Focusing on the four base mediation relations [27,48], (i) *embodiment* between actor and human-made resource occurs when the former adopts the latter to extend their capabilities. For example, the breathing apparatus for underwater exploration mediates and expands the human possibilities to reach into the marine environment. (ii) The *hermeneutic* relations involve the interpretation by the actor of symbolic representations provided by the human-made resources, in the way that a digital coastal map mediates the sailor's knowledge of the route, regulations, and hazards, and then also mediates their navigation of the environment. (iii) The third mediation, *alterity,* occurs when the actor interacts with the human-made resource as a "quasi-other", such as when a sailboat's autopilot is perceived as a member of the crew. Finally (iv) the "invisible" *background* relation occurs when the human-made resource is performing without being noticed by the actor, such as the GPS satellite system upon which mariners depend. These mediations enable human experiences, epistemologies and practices [3,49].

The inclusion of the perspective provided by technological mediation requires a conceptual shift: human-made resources are not only collectively selected and recognized to achieve goals, such as to extract provisioning ESs or use human-constructed facilities, but also ways of engaging with and interpreting situations that have consequences on the actor's worldview(s). For example without sensors and equipment that mediates our information access through hermeneutic interpretation, scientists would not be able to monitor and generate knowledge of certain environmental conditions, nor would relevant information be available for decision-making [3]. These aspects are captured in the re-designed framework through the relation `mediates` between human-made resources and intentional

[1] The paper of Guizzardi at al. [22] refers to *agents*, for the purpose of this work we consider *actor* and *agent* as interchangeable.

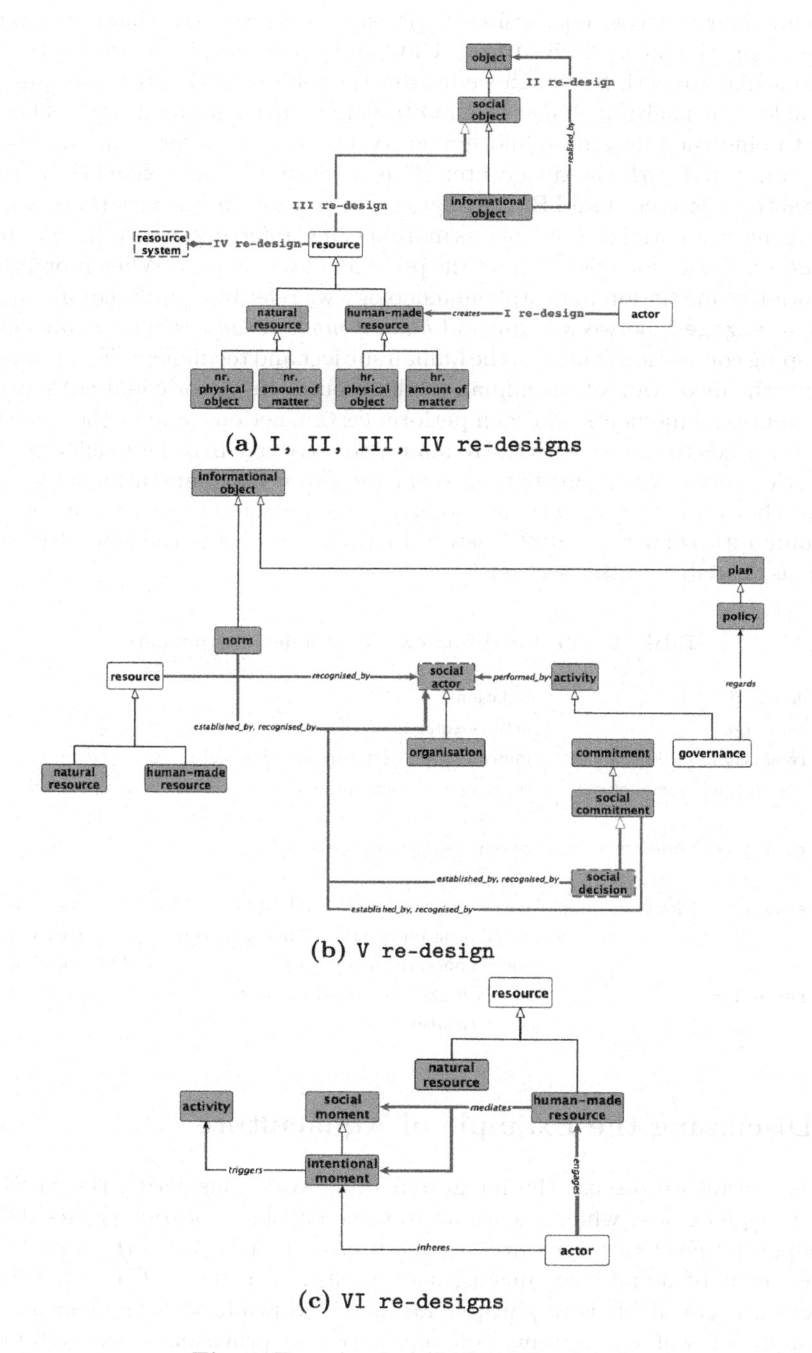

(a) I, II, III, IV re-designs

(b) V re-design

(c) VI re-designs

Fig. 1. Framework re-design snippets.

moments (beliefs, intentions, desires, preferences and decisions) that are `inherent` to the actor, the latter derived from UFO and taking inspiration from the phenomenologist Husserl. Although we borrow the relation `mediates` from postphenomenology, it is also included in UFO to define "material relations". Those, in order to bind the ralata need another entity, called *relator*, for example "town A being connected with the dive center B" is mediated by the existence of a road (the relator) between A and B (the relata) [22]. However in the new framework we present the existential dependence as more specific, referring to certain *intentional moments* that cannot exist without the presence of technology. When representing the notion of mediation in postphenomenology we refer to [3] and require another relation, `engages`, between *actor* and *human-made resources*. This expresses the co-shaping connections between the human subject and technology. To summarise, without the mediation of the human-made resource the actor could not have certain intentional moments and then perform certain actions, and at the same time the actor needs to engage with the human-made resource in order to elicit its characteristics and allow the mediation to emerge. Note that human-made resources mediate both (private) intentional moments and social moments that are inherent to communities of actors (after Kassel). In Table 1 are listed the new entities and relations described in this section.

Table 1. Integrated framework extended components.

Re-design #	Entity	Relation	
`I re-design`	-	actor *creates* hr	
`II re-design`	-	information object *realised by* object	
`III re-design`	*social object*	resource *is a* social object	
		resource *recognised by* social actor	
`IV re-design`	*resource system*	resource system *composed of*	nr
			hr.
`V re-design`	*social decision*	norm *recognised and established by*	social actor
		social decision *recognised and established by*	social actor
		social commitment *recognised and established by*	social actor
`VI re-design`	-	hr. *mediates* intentional moments	
		actor *engages* hr	

5 Discussing the Example of Aquaculture

In this section we discuss the integrated framework considering the SESs scenario of *aquaculture*, which according to the controlled vocabulary AGROVOC developed by *Food and Agricultural Organization* (FAO) [14] is the farming and management of aquatic organisms, such as fish and algae. The expansion of aquaculture worldwide is responsible for increased production and consumption per capita of seafood proteins [16], i.e. a rise in provisioning ecosystem services, but also the transformation of what were once common-pool resources (e.g. free-swimming fish and crustaceans) into private, commercialized goods

[13]. Advances in technology and engineering have fostered intensive production which can have adverse environmental consequences, such as changes in coastal water quality, degradation of coastal habitats, and spread of disease to wild fish populations [13,35], and have introduced social inequalities through loss of access to traditional fishing grounds, livelihoods and scarcity of important dietary proteins that disproportionately affect marginalized communities (see [21,45]). For these reasons aquaculture has become the focus of both SESF [30,38] and ESs [21,35] sustainability investigations, yet important technology-related aspects of this SES remain undisclosed. Although this theoretical examination is not a replacement for a real-world case study, it demonstrates how an attention to technology helps represent authentic SESs settings.

Starting from the initial definition of aquaculture and viewed through the integrated framework, aquaculture is an *activity* (kind of farming) performed by *social actors*, i.e. farm operators and workers, to achieve the planned *goals* of controlling *natural resources*, the aquatic organisms, and restricting access to provisioning *ecosystem services* within the system. This is achieved by performing a series of more specialised activities involving the management of reproduction, growth and protection from predation [13]. In many cases achievement of these goals is dependent on modification of the environment and the employment of dedicated *human-made resources*, such as cages, nets, artificial ponds, aeration equipment and specifically crafted feed. Though dependent on the same *ecosystem services* as wild harvesting, which involve *functional roles* played by *natural resources*, aquaculture systems are often artificial or semi-natural ecosystems.

The management and improvement of production efficiency, particularly in intensive aquaculture, also relies on sophisticated human-made ICT resources that provide knowledge (*informational objects*) derived from marine data [1]. The collection of such data, for example oxygen content, pH, temperature and salinity, is allowed by sensing technologies that mediate (through embodiment, hermeneutic and background relations) the understanding of the state of the aquaculture farm. This can support more efficient production as aquaculture operators monitor bio-chemical conditions affecting the life-cycles of the *natural resources* to determine optimal times for feeding and harvesting, yet can have unintended consequences as operators push the ecological carrying capacity of the *resource system* to it limits, using monitoring technology to maximize harvests. Meanwhile established regulatory frameworks (*governance* and agreed *policies*) often require that same monitoring data to ensure sustainability at local and regional scales [3]. Where *policies* are in place, producers can be required to restrict operations so that biophysical parameters remain within acceptable tolerances. In this case the mediation of technology fosters sustainable ecosystem management yet can have the unintended consequence of raising barriers to participation as prevailing socio-economic conditions particularly in developing countries hinder access to expensive monitoring technologies [43], and can diminish opportunities to exploit the *resource system.*

Aquaculture Examples from SESF and ESs. To complete the discussion we extract two toy examples from real-world SESs case studies, focusing on their

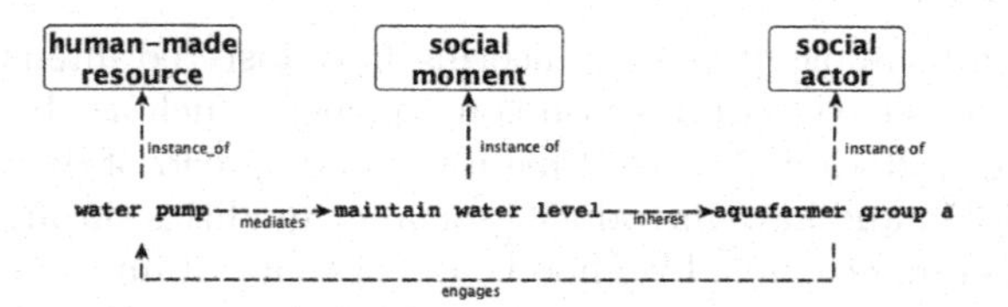

Fig. 2. Example of SESF instances.

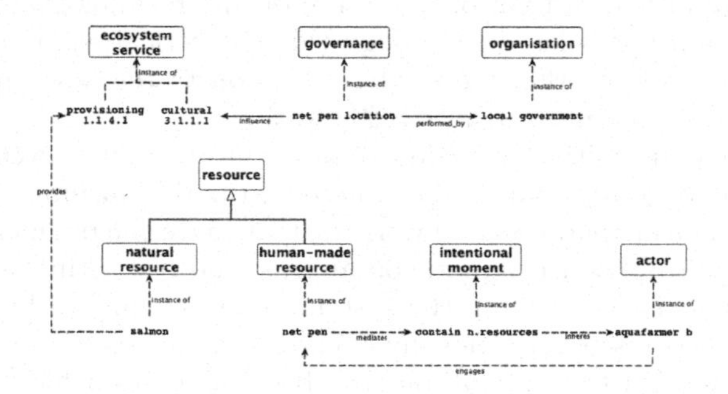

Fig. 3. Example of ESs instances.

technology-related components; the first SESF study regards community-based pond aquaculture in Indonesia [38]; the second examines intensive salmon farming in Chile through ESs [35]. Although the scale of analysis and the methodologies of the two are different, both consider social and ecological parameters. Note that the visual snippets are graphically inspired from examples proposed in [12]. In the SESF study, stakeholder engagement revealed the importance of water pumps for local aquafarmers to maintain pond water levels during the dry season, and a government aid program made a pump available to one group of aquafarmers. Figure 2 represents a snippet of this scenario which depicts "water pump", "aquafarmer group *a*" and "maintain water level" respectively classified as *human-made resource*, *social actor* and *social moment* following the integrated framework. The social moment of the aquafarmer group is mediated by the presence of the water pump that when it is engaged (i.e. adopted) would allow to keep the water in the pond at desired levels throughout the dry season.

The second example, based on ESs, involves an aquafarmer who requires a net pen to control the salmon. Meanwhile government regulations are not in effect to control the placement of net pens which is affecting the seascape and impacting tourism. Figure 3 captures this scenario in which the *intentional moment* "contain natural resource" of the *actor* "aquafarmer b", is mediated by the existence of the *human-made resource* "net pen" which allows for the control of the *natural resource* "salmon" and access to the provisioning *ecosystem service* for nutrition (CICES 1.1.4.1). The aquaculture farms are located in coastal areas which provide valuable cultural *ecosystems services* (e.g. ecotourism), nevertheless the lack of *governance* regulations concerning "net pen location" from the *organisation* "local government" is devaluing the environment in particularly beautiful areas that have high potential for tourism.

The previous discussion of aquaculture is intended to exemplify the use of the same theoretical ground to analise both SESF and ESs investigations. While the discussion of environmental monitoring makes visible unintended consequences of technology, the first case study excerpt shows how the existence of a technology affects group decisions. The second pinpoints trade-offs in SESs impacts and benefits that could be managed by governance activities. Together these illustrate the potential implications of such an integrated framework for sustainable policy development.

Limitations. The possible limitations of this work include (i) the focus only on two SESs frameworks and (ii) the adoption of specific perspectives on technology. From the multitude of SESs theories (see [9]), we focus on SESF and ESs due to their prominence in available literature, and decades of real-world applications. Concerning point two, our discussion is limited by reasons of space and we acknowledge that there are many interesting works that have not been included in the discussion. However we refined our selection to representative works from both the analytical philosophy/applied ontology and critical theory literature that include both material-technical and social-critical examinations of technology.

6 Conclusions and Future Works

In this follow-up to our RCIS 2022 paper we continue the ontological investigation and conceptual integration of SESF and ESs cascade paradigms by exploring technology-related notions. Our hybrid focus melds different perspectives on technology, which prompted several re-designs to the initially proposed integrated framework. This demonstrates how ontological analysis can clarify and facilitate the conceptual integration of different knowledge streams, in this case SESF and ESs cascade. The integrated SESs framework has been updated, yet much remains to be done to achieve useful comparability of data and results from diverse, often extremely localised sustainability research. Potential next steps in this work are to unpack the SESF *action situation*, explore the ESs *production boundary*, and further analyse the value dimension. To evaluate the proposed framework we foresee two interconnected tracks. The first is the development of participatory workshops with in-house domain experts and partners in local SES research to review the framework's concepts and relations, and query the appropriateness and utility of our approach for analysing actual real-world studies. Outcomes from these workshops will support a second track, which involves the implementation of the framework in the *ARtificial Intelligence for Environment & Sustainability* (ARIES) platform [8].

Acknowledgements. We acknowledge the Basque Government IKUR program Supercomputing and Artificial Intelligence (HPC/AI), the María de Maeztu Excellence

Unit 2023-2027 (CEX2021-001201-M) funded by MCIN/AEI /10.13039/501100011033, and the RCIS community for their valuable insights that helped develop this work.

References

1. Adamo, G.: Investigating business process elements: a journey from the field of Business Process Management to ontological analysis, and back. Ph.D. thesis, DIBRIS, Università di Genova, Via Opera Pia, 13 16145 Genova (2020)
2. Adamo, G., Willis, M.: Conceptual integration for social-ecological systems - an ontological approach. In: Guizzardi, R., Ralyté, J., Franch, X. (eds.) RCIS 2022. LNBIP, vol. 446, pp. 321–337. Springer, Cham (2022). https://doi.org/10.1007/978-3-031-05760-1_19
3. Adamo, G., Willis, M.: Technologically mediated practices in sustainability transitions: environmental monitoring and the ocean data buoy. Technol. Forecast. Soc. Chang. **182**, 121841 (2022)
4. Ahlborg, H., Ruiz-Mercado, I., Molander, S., Masera, O.: Bringing technology into social-ecological systems research-motivations for a socio-technical-ecological systems approach. Sustainability **11**(7), 2009 (2019)
5. Anderies, J.M., Janssen, M.A., Ostrom, E.: A framework to analyze the robustness of social-ecological systems from an institutional perspective. Ecol. Soc. **9**(1) (2004)
6. Andersson, B., Guarino, N., Johannesson, P., Livieri, B.: Towards an ontology of value ascription. In: Formal Ontology in Information Systems - Proceedings of the 9th International Conference, FOIS 2016, Annecy, France, 6–9 July 2016. Frontiers in Artificial Intelligence and Applications, vol. 283, pp. 331–344. IOS Press (2016)
7. Millennium Ecosystem Assessment: Ecosystems and Human Well-Being, vol. 5. Island Press, United States of America (2005)
8. Balbi, S., et al.: The global environmental agenda urgently needs a semantic web of knowledge. Environ. Evid. **11**(1), 1–6 (2022)
9. Binder, C.R., Hinkel, J., Bots, P.W., Pahl-Wostl, C.: Comparison of frameworks for analyzing social-ecological systems. Ecol. Soc. **18**(4) (2013)
10. Blanco, G., et al.: Innovation, technology development and transfer. In: IPCC, 2022: Climate Change 2022: Mitigation of Climate Change. Contribution of Working Group III to the Sixth Assessment Report of the Intergovernmental Panel on Climate Change, pp. 2674–2814. Cambridge University Press (2022)
11. Borgo, S., Vieu, L.: Artefacts in formal ontology. In: Philosophy of Technology and Engineering Sciences, pp. 273–307. Elsevier (2009)
12. Bottazzi, E., Ferrario, R.: Preliminaries to a DOLCE ontology of organisations. Int. J. Bus. Process. Integr. Manag. **4**(4), 225–238 (2009)
13. Bunting, S.W.: Principles of Sustainable Aquaculture: Promoting Social, Economic and Environmental Resilience. Routledge, Milton Park (2013)
14. Caracciolo, C., et al.: The agrovoc linked dataset. Semant. Web **4**(3), 341–348 (2013)
15. Díaz, S.M., et al.: The global assessment report on biodiversity and ecosystem services: summary for policy makers. Technical report, IPBES (2019)
16. FAO: The state of world fisheries and aquaculture 2020. Sustainability in action. Technical report, FAO, Rome (2020)
17. Feenberg, A.: What is philosophy of technology? In: International Handbook of Research and Development in Technology Education, pp. 159–166. Brill (2009)

18. Feng, P., Feenberg, A.: Thinking about design: critical theory of technology and the design process. In: Kroes, P., Vermaas, P.E., Light, A., Moore, S.A. (eds.) Philosophy and Design, pp. 105–118. Springer, Dordrecht (2008). https://doi.org/10.1007/978-1-4020-6591-0_8
19. Franssen, M., Lokhorst, G.J., van de Poel, I.: Philosophy of Technology. In: Zalta, E.N., Nodelman, U. (eds.) The Stanford Encyclopedia of Philosophy. Metaphysics Research Lab, Stanford University, Winter 2022 (2022)
20. Gangemi, A.: Dolce-lite-plus. Technical report, W3C (2005)
21. le Gouvello, R., Brugere, C., Simard, F. (eds.): Aquaculture and Nature-based Solutions: Identifying synergies between sustainable development of coastal communities, aquaculture, and marine and coastal conservation. IUCN (2022)
22. Guizzardi, G., de Almeida Falbo, R., Guizzardi, R.S.: Grounding software domain ontologies in the unified foundational ontology (UFO): the case of the ode software process ontology. In: CIbSE, pp. 127–140. Citeseer (2008)
23. Haines-Young, R., Potschin, M.B.: Common international classification of ecosystem services (CICES) V5. 1 and guidance on the application of the revised structure (2018)
24. Hansson, S.O.: Technology and the notion of sustainability. Technol. Soc. **32**(4), 274–279 (2010)
25. Hinkel, J., Bots, P.W., Schlüter, M.: Enhancing the ostrom social-ecological system framework through formalization. Ecol. Soc. **19**(3) (2014)
26. Hinkel, J., Cox, M.E., Schlüter, M., Binder, C.R., Falk, T.: A diagnostic procedure for applying the social-ecological systems framework in diverse cases. Ecol. Soc. **20**(1) (2015)
27. Ihde, D.: The phenomenology of technics. In: Scharff, R.C., Dusek, V. (eds.) Philosophy of Technology: The Technological Condition: An Anthology, pp. 19–24. Wiley, Chichester (2013)
28. Ihde, D., Malafouris, L.: Homo faber revisited: postphenomenology and material engagement theory. Philos. Technol. **32**(2), 195–214 (2019)
29. IPCC: Climate Change 2022: Mitigation of Climate Change. Contribution of Working Group III to the Sixth Assessment Report of the Intergovernmental Panel on Climate Change. Cambridge University Press, Cambridge and New York (2022)
30. Johnson, T.R., et al.: A social-ecological system framework for marine aquaculture research. Sustainability **11**(9), 2522 (2019)
31. Kassel, G.: A formal ontology of artefacts. Appl. Ontol. **5**(3–4), 223–246 (2010)
32. Markard, J., Raven, R., Truffer, B.: Sustainability transitions: an emerging field of research and its prospects. Res. Policy **41**(6), 955–967 (2012)
33. Masolo, C., Borgo, S., Gangemi, A., Guarino, N., Oltramari, A.: WonderWeb deliverable D18 ontology library (final). Technical report, IST Project 2001-33052 WonderWeb: Ontology Infrastructure for the Semantic Web (2003)
34. McGinnis, M.D., Ostrom, E.: Social-ecological system framework: initial changes and continuing challenges. Ecol. Soc. **19**(2) (2014)
35. Outeiro, L., Villasante, S.: Linking salmon aquaculture synergies and trade-offs on ecosystem services to human wellbeing constituents. Ambio **42**, 1022–1036 (2013)
36. Paredis, E.: Sustainability transitions and the nature of technology. Found. Sci. **16**(2), 195–225 (2011)
37. Partelow, S.: A review of the social-ecological systems framework. Ecol. Soc. **23**(4) (2018)
38. Partelow, S., Senff, P., Buhari, N., Schlüter, A.: Operationalizing the social-ecological systems framework in pond aquaculture. Int. J. Commons **12**(1) (2018)

39. Partelow, S., Winkler, K.J.: Interlinking ecosystem services and ostrom's framework through orientation in sustainability research. Ecol. Soc. **21**(3) (2016)
40. Potschin, M., Haines-Young, R., et al.: Defining and measuring ecosystem services. In: Routledge Handbook of Ecosystem Services, pp. 25–44 (2016)
41. Preston, B.: Artifact. In: Zalta, E.N., Nodelman, U. (eds.) The Stanford Encyclopedia of Philosophy. Metaphysics Research Lab, Stanford University, Winter 2022 (2022)
42. Sanfilippo, E.M., et al.: Modeling manufacturing resources: an ontological approach. In: Chiabert, P., Bouras, A., Noël, F., Ríos, J. (eds.) PLM 2018. IAICT, vol. 540, pp. 304–313. Springer, Cham (2018). https://doi.org/10.1007/978-3-030-01614-2_28
43. Schmidt, W., et al.: Design and operation of a low-cost and compact autonomous buoy system for use in coastal aquaculture and water quality monitoring. Aquacult. Eng. **80**, 28–36 (2018)
44. Shukla, P., et al.: IPCC, 2019: climate change and land: an IPCC special report on climate change, desertification, land degradation, sustainable land management, food security, and greenhouse gas fluxes in terrestrial ecosystems. Technical report, Intergovernmental Panel on Climate Change (IPCC) (2019)
45. Soto, D., et al.: Applying an ecosystem-based approach to aquaculture: principles, scales and some management measures. In: Building an Ecosystem Approach to Aquaculture. FAO/Universitat de les Illes Balears Expert Workshop, vol. 7, p. e11 (2007)
46. Thomasson, A.L.: Public artifacts, intentions, and norms. In: Franssen, M., Kroes, P., Reydon, T.A.C., Vermaas, P.E. (eds.) Artefact Kinds. SL, vol. 365, pp. 45–62. Springer, Cham (2014). https://doi.org/10.1007/978-3-319-00801-1_4
47. UNFCCC: Climate technology centre and network programme of work 2023-2027. Technical report, UNCTCN (2022)
48. Verbeek, P.P.: Don ihde: the technological lifeworld. In: American Philosophy of Technology: The Empirical Turn, pp. 119–146 (2001)
49. Verbeek, P.P.: Toward a theory of technological mediation. In: Technoscience and Postphenomenology: The Manhattan Papers, p. 189 (2015)
50. Vermaas, P.E., Houkes, W.: Technical functions: a drawbridge between the intentional and structural natures of technical artefacts. Stud. Hist. Philos. Sci. Part A **37**(1), 5–18 (2006)
51. Zwier, J., Blok, V., Lemmens, P.: Phenomenology and the empirical turn: a phenomenological analysis of postphenomenology. Philos. Technol. **29**(4), 313–333 (2016)

Machine Learning and Analytics

Detection of Fishing Activities from Vessel Trajectories

Aida Ashrafi[1(✉)], Bjørnar Tessem[1], and Katja Enberg[2]

[1] Department of Information Science and Media Studies, University of Bergen, Bergen, Norway
aida.ashrafi@uib.no

[2] Department of Biological Sciences, University of Bergen, Bergen, Norway

Abstract. This work is part of a design science project where the aim is to develop Machine Learning (ML) tools for analyzing tracks of fishing vessels. The ML models can potentially be used to automatically analyse Automatic Identification System (AIS) data for ships to identify fishing activity. Creating such technology is dependent on having labeled data, but the vast amounts of AIS data produced every day do not include any labels about the activities. We propose a labeling method based on verified heuristics, where we use an auxiliary source of data to label training data. In an evaluation, a series of tests have been done on the labeled data using deep learning architectures such as Long Short-Term Memory (LSTM), Recurrent Neural Network (RNN), 1D Convolutional Neural Network (1D CNN), and Fully Connected Neural Network (FCNN). The data consists of AIS data and daily fishing activity reports from Norwegian waters with a focus on bottom trawlers. Accuracy is higher than or equal to 87% for all deep learning models. Example applications of the trained models show how they can be used in a practical setting to identify likely unreported fishing activities.

Keywords: Fishing Activity Detection · Deep Learning Models · Data Labeling

1 Introduction

One of the Sustainable Development Goals (SDGs) set by the United Nations focuses on life below water (SDG14)[1]. As oceans and marine resources play an important role in different aspects of our life such as economy, food, and ecosystem functioning, stability, and resilience, using them in a sustainable way is a must.

Our effort within this context is a design science project [4] in collaboration with the Norwegian Directorate of Fisheries (NDF). The main goal is to develop deep learning models to help the sustainable use of fish resources. These models can potentially support the surveillance of fishing vessels, by exploiting data about the fishing vessels' movements combined with their reports on fishing activities. The data are from Norwegian waters and are provided by NDF.

[1] https://www.globalgoals.org/goals/14-life-below-water/.

S. Nurcan et al. (Eds.): RCIS 2023, LNBIP 476, pp. 105–120, 2023.
https://doi.org/10.1007/978-3-031-33080-3_7

1.1 Problem Relevance

There are two approaches towards sustainable fisheries in the real world, and the best results are achieved when both of them are applied simultaneously. These two are: setting rules to regulate exploitation, and providing exploiters with incentives like ownership, for example in the form of quotas. This approach ensures that the long-term benefits of the resource become the priority of the exploiters [3].

In order to prevent Illegal, Unreported, and Unregulated (IUU) fishing, specific quotas are determined and surveillance is done regularly. However, as the high seas are not easily observable, rules are easily and sometimes violated. At the same time, the detection of IUU fishing by fishery inspectors is costly to implement.

Fishery inspectors have access to huge amounts of data about fishing vessels' movements obtained from Satellite-based Automatic Information Systems (S-AIS). However, no satisfactory automated approaches to support the detection of fishing activities are currently in use. While AIS data are mainly used to avoid collision between ships by tracking them, we envision the use of ML models as potential candidates to automate the detection of fishing activities from these data. This enables the inspectors to be aware of when and where fishing has taken or is taking place, and whether it is correctly reported. As a result, they would be able to focus their attention on vessels and locations where there is a higher risk of irregular fishing.

In Fig. 1 AIS data of 27 bottom trawlers on 38 fishing trips are shown. Red points show non-fishing activities and blue ones show fishing activities, based on labeling obtained from the fishing vessels' records.

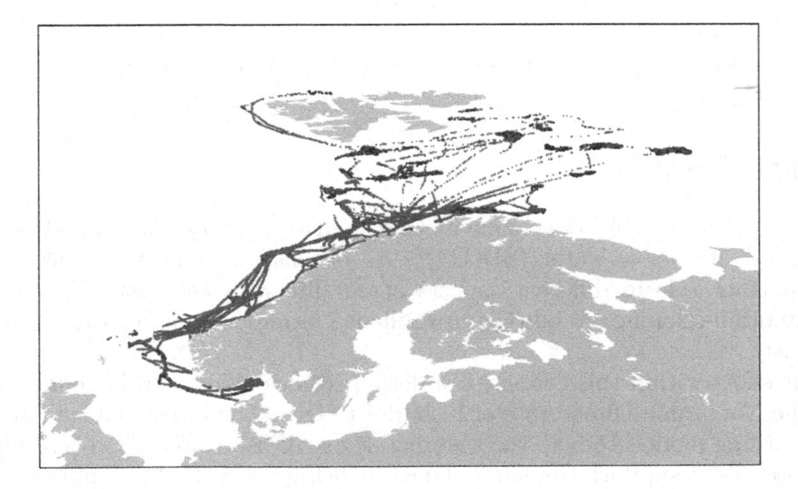

Fig. 1. AIS data for 27 bottom trawlers on 38 fishing trips, red points are non-fishing and blue points are fishing activities according to reports by fishermen used our labeling strategy (Color figure online)

1.2 The Contribution

AIS data include features such as latitude, longitude, speed, and course over ground in addition to the vessel's ID, the so-called Maritime Mobile Service Identity (MMSI). The normal frequency of reporting AIS data is every 5 min (with many exceptions). These discrete data points form the vessels' trajectories.

In our analysis, we segment these trajectories and label each segment as a fishing or non-fishing activity. Each segment is thus a sequence (time series) of AIS data and the task is to build a model that is able to classify these segments into fishing or non-fishing.

A main contribution of our work is a method to provide more labeled training data for the ML process. This is a significant problem, as it is too time-consuming for an expert to do the labeling. As an alternative, we label the data by using the records from fishing vessels' daily reports on fishing activities. Although this method is based on the reports by the fishermen, domain experts from NDF have verified that the obtained labels are accurate on examples, and that the accuracy is at a level that is sufficient to train a model. The ML models are the second contribution of this work. They should be seen as specific to the data from this region and for a particular fishing gear, as the experts believe that the movement patterns are highly related to these aspects. We focus here on bottom trawlers. However, the general approach should have validity also in other geographical areas and other types of fisheries.

In the next section, the research method is discussed. In Sect. 3 we describe the background for the problem and look into related work on fishery activities. We continue by describing the original data set used, and then the methods used to extract labels of the trajectory segments from the auxiliary data source, ending with a final selection of data (Sect. 4). In Sect. 5 we describe our evaluation approaches and show the models' results, as well as give an example of how the model performs on some interesting cases from the data. Some discussion and conclusions are presented in Sect. 6.

2 Research Method

The research method we have followed is standard design science as described by Hevner et al. [4]. We have created methods for labeling vessel tracking data, used for the further purpose of developing the ML models used to analyse vessel tracks. Two of the most important guidelines in design science are design as an artifact and design evaluation. The most significant design choices we have made for our tools and the sections in which they are explained are indicated in Table 1.

In the evaluations described in Sect. 5, different tests are done on unseen data using different ML architectures. Results are reported in the form of traditional performance metrics from the ML field. As a descriptive evaluation (in the terms of Hevner et al [4]), we provide visualization of speed and tracks of vessels combined with the output of the ML models.

Table 1. Overview of design choices

Design Choice	Related Section
Selection of data from bottom trawlers	Sect. 4.1
Obtaining labels for AIS data using daily catches	Sect. 4.2
Identifying and removing inaccurate data	Sect. 4.2
Selecting features for ML	Sects. 4.3 and 4.4
Representation of the data as track segments and labeling of segments	Sect. 4.5
ML architectures and performance metrics	Sect. 5.1

3 Background and Related Work

Commercial fishery is a complex activity dependent on many factors such as the target species, area of the sea, vessel size, and gear type. In Norwegian waters cod, haddock, pollock (also called saithe), mackerel, and herring are considered the most valuable species.

There are mainly two groups of fishing gears: passive gears such as long line and purse seine, and active gears such as trawl. The species most often targeted with trawl are cod, haddock, and pollock. Our focus in this paper is on recognizing fishing activities for vessels doing bottom trawling, which basically is performed by pulling a bag-formed net along the seafloor to catch the target fish.

There have been several attempts to classify fishing activities using ML. de Souza et al. [9] used ML models, i.e., the Hidden Markov Model (HMM) and data mining to identify fishing activities from AIS data. Three types of fishing gear were analyzed: trawl, longline and purse seine.

Jiang et al. [6] published the first work applying deep learning models for detecting the fishing activities of trawlers. A sliding window technique was used to divide the trajectory into shorter segments. They labeled each window with the same label as that of the middle point. To reduce the noise, undersampling was applied. Linear interpolation was utilized to recapture the trajectory which was then converted to an image matrix. Lastly, an autoencoder was used to detect fishing activities. According to Jiang et al. [5], AIS data is low-dimensional and heterogeneous making it hard to work with deep learning models on these data. They proposed Partition-wise Recurrent Neural Networks (pRNNs) to solve this issue. Their focused fishing gears were long-liners.

Global Fishing Watch (GFW)[2] has provided commercial fisheries datasets that are used in many studies. Kroodsma et al. [7] developed CNN models for the recognition of different vessels' features and also the detection of fishing

[2] https://globalfishingwatch.org/.

activities. In a more recent paper [1], they segmented the trajectories into smaller intervals and used the majority label as the label of each interval. Their proposed model (FishNET) is based on 1D CNN and new features extracted from main features are used to make the method independent of changes in gear type, vessel type, and location.

Shen et al. [8] used a multi-layer Bidirectional LSTM (BiLSTM) model to test the importance of different features in detecting fishing activities with both active and passive gear types used around Taiwan.

The most significant challenge in fishing activity detection tasks is the lack of labeled data [1]. However, the most common open and labeled data sets are limited, even though they are related to many gear types. Annotation of the data by experts is labor intensive and also difficult in reality since even experts are not aware of all fishing patterns.

A work close to ours is by Ferreira et al. [2]. They do not use true labels, but still classify sub-trajectories into fishing and sailing. k-means clustering based on speed and course changes is applied to find the labels in an unsupervised manner. They further use these labels with LSTM and Gated Recurrent Unit (GRU) units to do the classification task. The clustering method involves many tuning details and appears to be rather cumbersome.

4 Datasets and Preprosessing

4.1 Data

The data we use is an AIS dataset that provides fishing vessels' movement data, and DCA (Daily Catch) reports[3] which help to identify fishing activities. The AIS dataset includes information such as vessels' ID (MMSI), message time, latitude, longitude, speed over ground, and course over ground. These data are not labeled. However, DCA reports can be used to extract labels from the fishermen's reporting. DCA data consist of reported fishing intervals (start and stop time, duration, amount, and location of each catch), plus Call-sign for each vessel which itself could be matched to MMSI from the AIS dataset through another data table.

A main difference of our work from earlier studies is our ability to assess the data with the help of domain experts. According to them, the DCA reporting data regarding active gears (like bottom trawl) are cleaner and easier to work with since the patterns are more visible and the duration of fishing activities are recorded more accurately. Also, different types of active gears lead to different patterns. That is why, unlike most of the works in the literature, we have chosen to focus on only one gear type at a time. In this study, we only consider bottom trawlers. With a quick look at our dataset, we could conclude that there are enough data for this gear type to train and test the models. We have used more than 500 different fishing trips of almost 100 vessels between 2015 and 2020.

[3] A part of the electronic reporting by NDF: https://www.fiskeridir.no/Tall-og-analyse/AApne-data/elektronisk-rapportering-ers.

For each vessel, we chose a maximum of 5 fishing trips randomly. This gave us a subset of the data with complete trajectories from whole fishing trips. The vessels have provided departure and landing records (see footnote 3) which can be used to identify fishing trips. Using complete fishing trips allows us to get a representative data set, but visualizing the trips also helps to get a better understanding of the activities in a trip.

4.2 Obtaining Labels

We have used the DCA reports delivered by fishermen to label the AIS data. Some cleaning, however, has been needed. For example, it is immediately meaningless to include fishing activities reported to have zero duration. In addition, vessels are allowed to send correction messages, which are basically duplicates that are sent later. After eliminating these kinds of messages, there would be no overlaps among different records in DCA data.

In the next step, we have labeled the AIS data belonging to the chosen fishing trips. The labels are not certain to be true labels, as they are based on reports by fishermen, who are notoriously being inaccurate. But since the annotation process of AIS data by experts is very time-consuming and the DCA reports can be used to extract labels with a satisfactory level of trust, we chose to use the DCA data to obtain labels. This way we benefit from a larger training set and achieve good performance.

Still, some of the reporting indicated very long or short fishing activities. Such long or short fishing activities are considered to be most likely inaccurate. So, to remove noise from the DCA reports we eliminate the catch activities with a duration of less than 30 min or more than 400 min and define them as irregular messages. The histogram of duration for the most important species is shown in Fig. 2. Finally, we check each AIS data point from desired fishing trips to see if they are located inside any of the fishing intervals. If they are, we label them as fishing, and if not, they are labeled as non-fishing. The ones excluded due to them being outside the acceptable intervals will be used later for evaluation purposes.

4.3 Selecting Features for ML

The normal frequency of sending AIS messages is every 5 min, although this does not happen all the time. Therefore, we decided to add time difference (ΔT) which shows the difference between the time of the current message and the previous one's as a new feature. Messages with very long time-difference (more than 200 min) were also removed. Another feature that has been added is the month of the year that the message has been sent. This is a cyclic attribute and we believe since this feature is related to environmental factors, it can affect the patterns of vessels' movements. The rest of the features are mostly the same as the ones in [8] such as speed (SOG), average speed (S_{avg}), and change in the course (ΔCOG). We also consider changes in the speed (ΔSOG). In our work, instead of distance (distance between the current position and previous position,

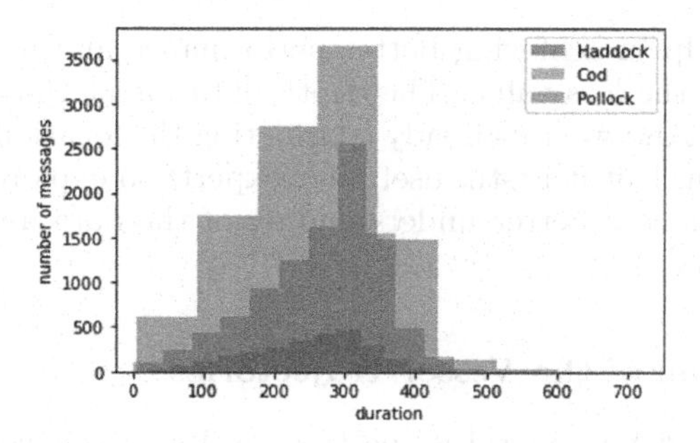

Fig. 2. Duration histogram of catch activities from 2018 for the most important species.

Table 2. definitions of different features used in our method, i is the index of ith data point in the sequence

feature	definition
ΔT	time difference: $T_{i+1} - T_i$
Month	month of the year coded as $Month_{sin}$, $Month_{cos}$
SOG	Speed over Ground
S_{avg}	Average Speed: $S_{i+1} + S_i/2$
ΔSOG	Speed over ground difference: $SOG_{i+1} - SOG_i$
ΔCOG	Course over ground difference: $COG_{i+1} - COG_i$
P	Position which includes latitude and longitude

ΔP), we have chosen to use latitude and longitude (position, P). Our focus is the Norwegian waters and we get a better result if we are more specific about the location. Furthermore, experts suggest that these patterns are very specific to regions, hence it is potentially better to build and train the models for each area separately. The features we used and their definitions are shown in Table 2.

4.4 Choosing Depth as a Feature

In our later sessions with experts, they suggest using depth as a feature to check for possible improvements plus providing explainability. Depth data is not part of AIS data and we needed to use publicly available bathymetry data sets provided by General Bathymetric Chart of the Oceans (GEBCO)[4]. Considering the position of each data point in AIS dataset as a center, we form a 0.01×0.01 latitude-longitude box around it, find the closest point to it using Haversine distance, and extract the depth value. If there are no points inside the box, we increase the box to $0.1 \times 0.1°$. Adding depth in combination with various other

[4] https://www.gebco.net/.

features did not help in achieving better performance and in some cases cause a slight increase in the loss value. The depth data might thus not be useful for developing models. And we are already considering the location, which obviously is a proxy for depth. But it is still useful for experts to explore the activities on depth maps to achieve a better understanding of the data and to evaluate the model's performance.

4.5 Segmentation of the Vessel Trajectories

The trajectory of a fishing vessel is the track of its movement in the form of a time series. Since it is long and includes different fishing patterns and intervals, it is preferable to divide it into shorter sequences (windows). This process is needed to prepare the input for LSTM, RNN, and 1D CNN. For FCNN, we represent the sequences as non-sequential data points.

It is very common in the literature to use the sliding window technique (Fig. 3) and the window size is mostly between 5 to 15 datapoints [1,6,8]. On the other hand, they have different approaches to label each sequence, either choosing the majority vote of the window or the label of the middle point of a window.

In our work, we tested different alternatives. We trained the model with sequences of length 8 and chose the label of the last point as the sequence's label. We also tried using the middle label and window size 10. No substantial difference in the performance among these choices has been observed, so we decided to keep the first setting.

5 Evaluation and Results

5.1 Using Different Architectures

To compare different deep learning architectures, we have used the same set of hyperparameters. For each of the architectures we trained 10 models, and the average scores of these 10 models are reported as the score. The hyperparameters used were

Number of epochs: 20
Neurons in hidden layers: 128
Optimizer: Adam
Activation functions: ReLU and Softmax
Batch size: 32
Loss function: binary cross entropy loss

The first model we tried was FCNN using the sequence data as a flattened input set. But since the data is in the form of a time series, we assumed that other architectures such as LSTM, RNN, and 1D CNN, which are designed for working with sequences, would be more suitable. Besides segmenting trajectories into sequences to feed into these models, we added the extra hidden specialized

Table 3. Performance on the test set from 2018 using different models

FCNN	Accuracy	Loss	F1-score	Precision	Recall
	90	0.27	89	87	92
LSTM	**92**	**0.22**	**91**	**90**	**94**
BiLSTM	92	0.23	91	90	92
RNN	91	0.23	91	89	93
1D CNN	91	0.25	91	88	93

recurrent or convolutional layer (LSTM, RNN or 1D CNN), as well as a dropout layer with 0.5 dropout rate for those models. We also tried bidirectional LSTM (BiLSTM).

We first tried all the candidate models on the 2018 dataset, with both training and testing data set from that year. All the models except FCNN consider temporal dependency. We, therefore, expected them to perform better than FCNN. The results were in line with that expectation. However, the difference was not substantial. Different metrics such as accuracy, loss, precision, recall, and F-score were calculated. The results are depicted in Table 3.

5.2 Some Observation Related to Overfitting

In our initial tests, the windows were not overlapped. To use the data in the best possible manner, we decided to extract overlapping windows by using the sliding window approach (Fig. 3). This way we can provide more sequences from the same dataset just by shifting the intervals halfway to the right. Adding these overlapped sequences to the original ones caused an overfit, though. We also observed that more complex models (more units per layer and more layers), and more epochs also result in an overfit.

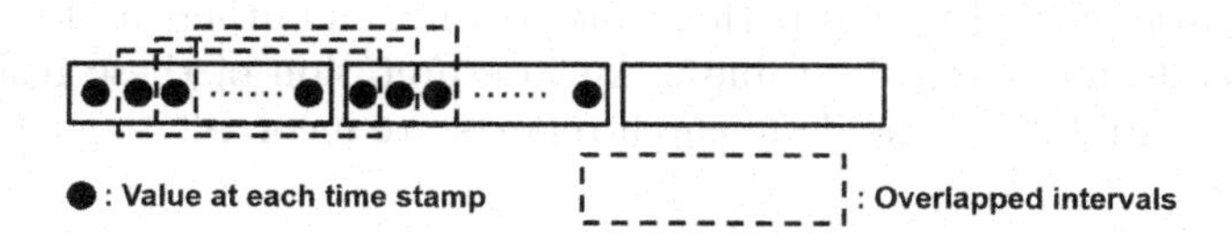

Fig. 3. Overlapped windows

5.3 Using Data from Different Years

To check if there are common fishing patterns among different years and if a model trained on the data from a specific year can perform reasonably on a test set from the following year(s), we did further tests using training and test sets belonging to different years.

For the rest of the tests, LSTM was chosen. First, the LSTM trained on 2018 data, was tested on data from 2019 and 2020. We also trained an LSTM on 2019 data and tested it on 2020 data. Finally, we trained a third LSTM on the data from 2015, 2016, 2017, 2018 and 2019 and tested it on 2020 data.

The accuracy of all the models was quite high on the test sets, higher than or equal to 87%. The scores are shown in Table 4. It seems that the fishing patterns do not change much over the years as the scores do not vary much. However, they are a bit lower when using a test set from a different year. The reason for this may be that the selection of vessels in the data sets is a bit different, and therefore will give a slightly lower performance.

Table 4. Performance on the test sets from different years using LSTM

train	test	Accuracy	Loss	F1-score	Precision	Recall
2018	2019	91	0.25	91	89	93
2018	2020	88	0.29	87	82	92
2019	2020	89	0.29	88	84	92
2015–2019	2020	87	0.32	86	81	92

5.4 Testing on Outliers

Outliers are, as we use the term, the DCAs which indicate very short or very long catch intervals. We have removed this data in the labeling step so that the model would be trained on accurately labeled fishing activities. However, at test time, we can check this part of the data set to see the difference between the output of the model and the reports by fishermen. Although fishermen labeled all these sequences as fishing, according to experts these reports are most likely incorrect. As expected, our model obtains less accuracy on this part of the data, as the prediction is probably more correct than the ground truth (labels by fishermen). The comparison of the LSTM's performance testing on outliers and non-outliers from 2018 is demonstrated in Table 5. An interpretation of these results could be that about 10–15% of the daily catch reports are incorrect.

Table 5. Comparison of performance on non-outliers vs outliers test set from 2018

	Accuracy	Loss	F1-score
Non-outliers	92	0.22	91
Outliers	85	0.45	91

5.5 Speed Analysis to Test Our Method

According to the experts, vessel speed is one of the most important contributing factors in detecting the fishing activities of bottom trawlers. The results of our tests also confirm this argument, the models can achieve comparable results (all the metrics are less than 1 or 2% below using LSTM) considering speed as the only feature. Therefore, we provide some visualizations of speed analysis which could help to see whether the model is learning and doing something reasonable.

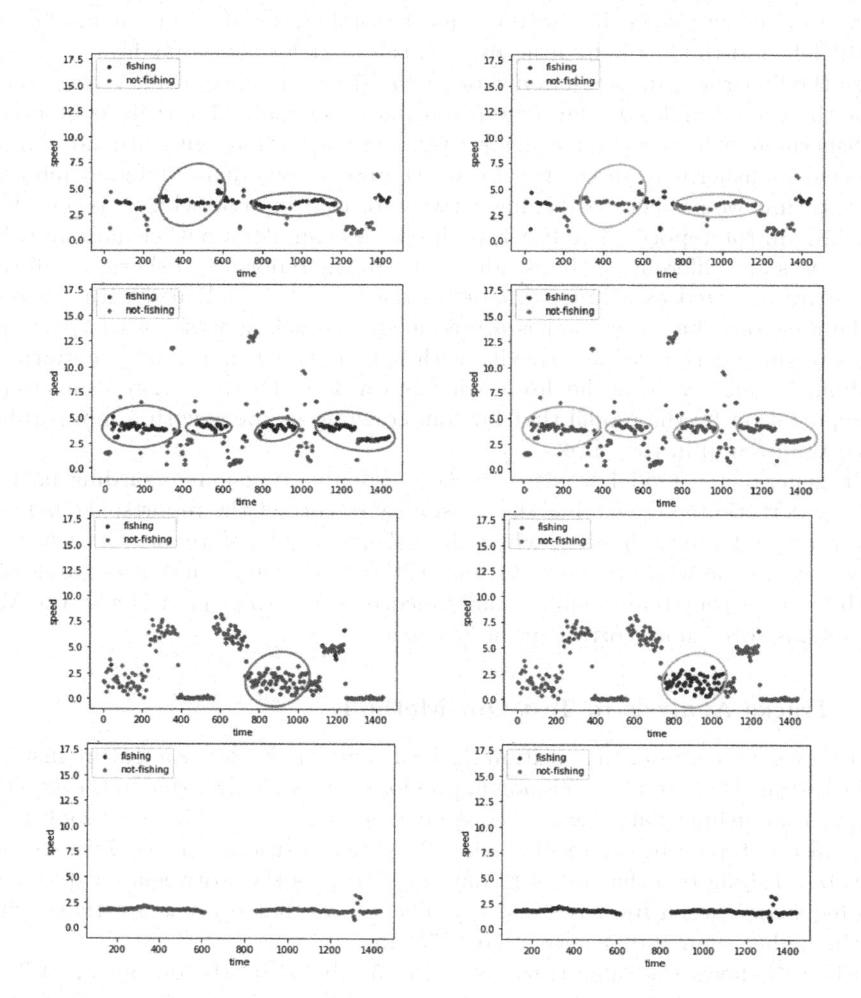

Fig. 4. speed analysis over time for a day: prediction (first column) vs truth or reports (second column)

The model was incorrect in about 10% of the sequences. The visualizations in Fig. 4 show a selection of those and where the discrepancy occurs. We hope that later by combining these graphs with the patterns in the tracks, domain experts could help us to get labels closer to the true labels. These visualizations are for the 2018 test set and predictions are the output of FCNN. The ground truth labels are the labels from using DCA data, not the true labels confirmed by experts.

Each row in Fig. 4 shows the speed analysis of different vessels on different days. Time is in minutes with a day being 1440 min. The blue dots are used to depict non-fishing operations and the red ones are used for fishing activities. The green and yellow circles illustrate the most visible differences in our prediction on the left and the labels by fishermen on the right for the same day.

In the first row, our prediction shows two fishing intervals inside green circles while the yellow circles on the right graph show non-fishing intervals as reported by fishermen. The second row shows the same situation but with no fishing reported by fishermen. In the third row, we observe one dense different interval but this time we predict non-fishing activity for that interval while it is reported as fishing in the reports. The last row shows an example of outlier data and the two figures are different at almost all points. In the reports by fishermen, all the points are reported as fishing which does not seem to be aligned with the rest of the data on which our models are trained. A quick analysis of the patterns seems to suggest that fishing activity with bottom trawls normally is performed at about 5 knots, cresting the discrepancy in row four. Further, there seems to be an expectation by the model that bottom trawling happens with a more stable speed than found in row three.

In general, the model seems to make reasonable decisions regarding fishing activity detection and even has the possibility to correct the reports. We believe that these differences happen when the fishermen do not report their fishing activities in time while they are fishing. Either they report activities carelessly much later, or they report intentionally incorrect intervals. That is why the ML predictions seem more correct in many cases.

5.6 Tracks Analysis to Test Our Method

In our tests, we consider the position (latitude and longitude) as a feature instead of the distance traveled (as common in previous works). This is due to the experts stating that fishing patterns are dependent on the region. This was confirmed as we also obtained better results using the exact positions. In Fig. 5 we see the track of a fishing trip that takes 12 days on a map of the Norwegian water. The trip has been chosen from the test set. The white-colored areas are those with depths higher than 100 or lower than −700 m.

Figure 6 shows the same track as in Fig. 5 labeled by the output of FCNN. On the bottom, we focused on two busy areas from the original map. Green data points are for True Positive (TP), meaning that the model's prediction for that data point was fishing and the label by fisherman's reports was also fishing. Blue points represent True Negative (TN), i.e., where the model output agrees on being non-fishing with the reports. As you can see the colors of the tracks are in line with the 90% accuracy of the model since most of the data points

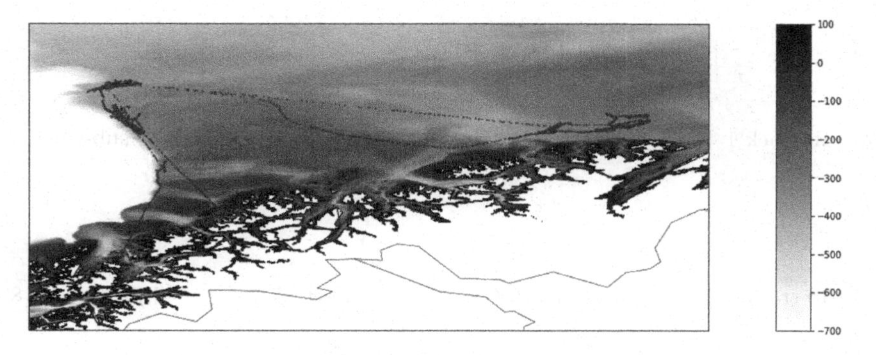

Fig. 5. Track of a fishing vessel from the test set on a map including depth data, white areas have depth lower than -700 or higher than 100 m.

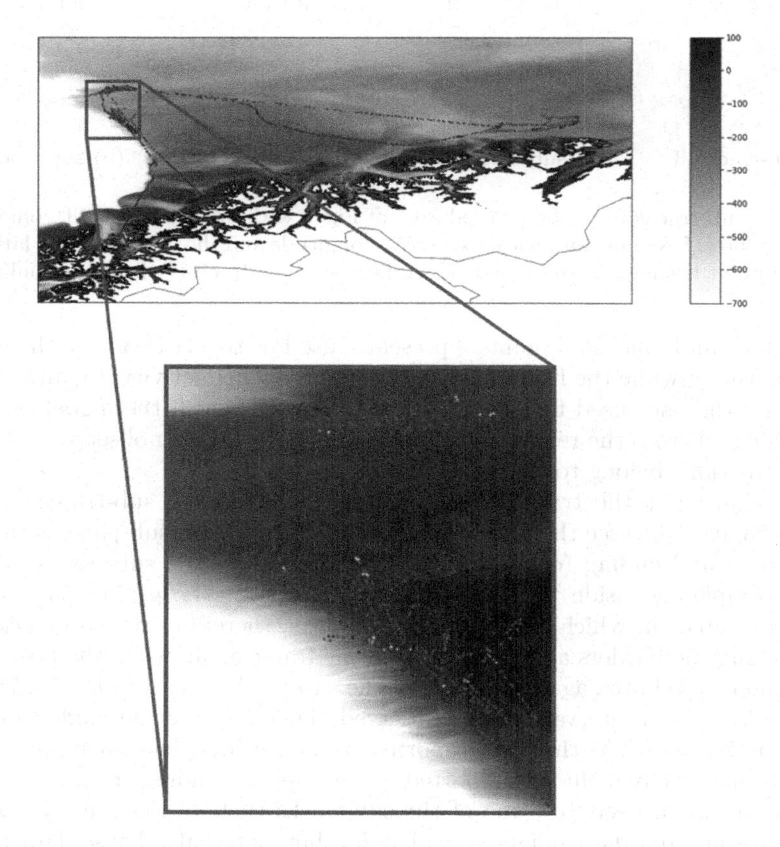

Fig. 6. Track of a fishing vessel with the labels by the model on the top, focus on the busy area on the bottom. Green for TP, blue for TN, red for FP, yellow for FN. (Color figure online)

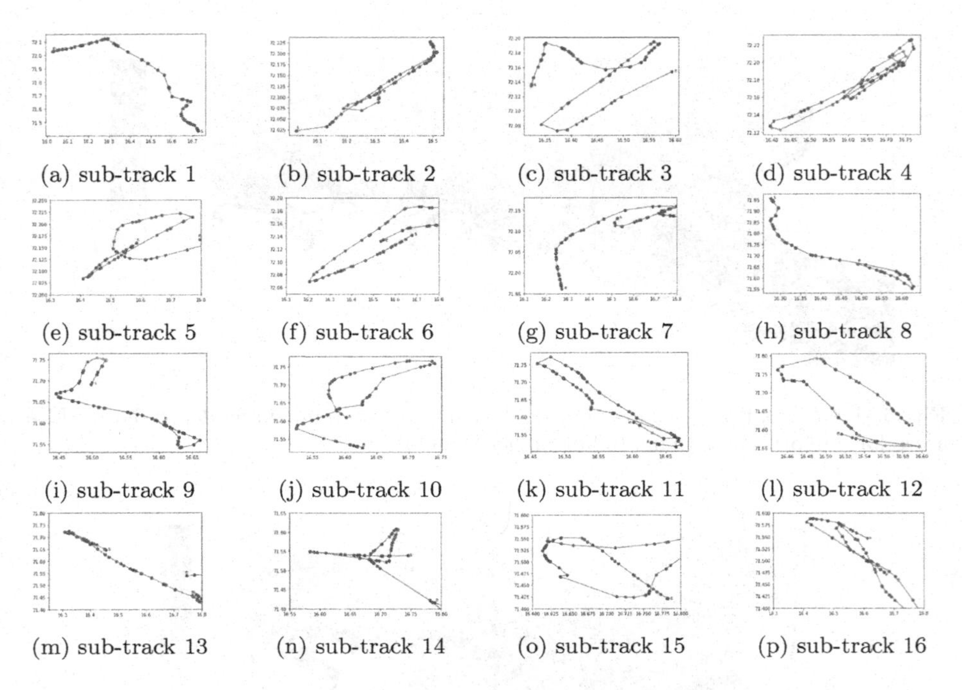

(a) sub-track 1 (b) sub-track 2 (c) sub-track 3 (d) sub-track 4

(e) sub-track 5 (f) sub-track 6 (g) sub-track 7 (h) sub-track 8

(i) sub-track 9 (j) sub-track 10 (k) sub-track 11 (l) sub-track 12

(m) sub-track 13 (n) sub-track 14 (o) sub-track 15 (p) sub-track 16

Fig. 7. 16 sub-tracks from the focused area of Fig. 6 with labels, green: TP, blue: TN, red: FP, yellow: FN. The horizontal axis shows longitude and the vertical one is latitude. Starting point is shown with 's' and the ending point with 'e'. (Color figure online)

are in green and blue. Red points represent False Positive (FP) where the model outputs fishing while the fisherman reported non-fishing activity. Finally, yellow points are the ones used for False Negative (FN) for which the model suggests non-fishing whereas the reports show fishing activity. We can observe that most false predictions belong to FP.

We again split the tracks in this focused area into 16 sub-trajectories of lengths 53 and 54 to see the patterns better. In Fig. 7, these sub-parts with their starting (s) and ending (e) points are shown. Some of the sub-tracks are not placed completely inside the focused area. Figures 7e, 7f, 7g, 7m, 7n, 7o, and 7p are examples in which this situation happens, therefore, the sub-tracks are cut reaching the borders and come back to the frame again when the position is inside the focused area again. In some of the sub-tracks such as Figs. 7e, 7f, and 7n some larger red intervals can be observed. This is where the model predicts fishing activities while they are reported as non-fishing by fishermen. These intervals most likely include unreported operations and finding them is the main purpose of our project. In some of the sub-tracks such as Figs. 7a, 7b, and 7c there are some red data points spread inside blue intervals. These data points are predicted as fishing by the model while the data points in their neighborhood are labeled as non-fishing by both the model and the fishermen. These points are most likely mistakes by the model but even in that case, they are very rare.

6 Conclusion and Discussion

Sustainable use of fish resources is highly important for the development of humanity's future, and the long-term goal is to avoid overfishing. While many fishing nations are active in the surveillance of fishing vessels' behavior on the ocean, there are still unreported activities going on in many areas. In this work, we tried to automate fishing activity detection.

The most common problem for tools like this is the limited number of labeled data and the expensive annotation process. We have proposed a method to extract labels using an auxiliary data source, namely the publicly available daily catch reports from NDF. Our alternative approach achieves accuracy at par or better than the works where they use expert labels. The labels are based on fishermens' reports and may be incorrect, but by removing outliers, they can be trusted to be at a sufficiently good level to be used in ML. Thus we can have significantly more data to train our model than in previous work.

Since the data is spatiotemporal we opted to use different models such as LSTM, RNN, and 1D CNN, and for that, we segmented data into smaller windows. We also explored the performance of FCNN. The difference was small. The 1 to 2% difference in accuracy might be due to the importance of temporal dependency. All the models are fairly simple and using more complex models or overlapped sequences ended in over-fitting. Testing the models on different years still gave a high performance. Those parts of the data which were left out during the training were also used to test the model. In this case, the model showed fairly good performance and seems to be able to identify irregular reporting.

We have developed models by using different features which were partly different from previous works. As experts suggested the impact of region on fishing patterns, we picked the exact position rather than the distance. Depth was omitted from the feature set, but can still be useful in order to help the expert get a better interpretation of the model output. Visualizing tracks and patterns, complemented with depth maps could help experts in the process of providing labels closer to true labels.

There are some limitations to our work. First, the models we used could only obtain around 90% accuracy by testing different sets of features, architectures, and hyperparameters. The various tests indicate no promise for substantial improvements in the performance. We believe this comes from the heuristics for removing inaccurate data being imperfect, which again created noise in the data labels. Second, our work is not totally comparable with the previous studies on the same topic since our focus is narrowed down to Norwegian waters and bottom trawlers. Although the method could be generalized to other regions and gears provided there are data, we cannot guarantee generalizability of the results.

In the future, we are going to extend our work to other active gear types and also to passive gear types. Further, we believe there are more major features such as seabed surface which can be included to increase the performance. We are also going to develop an application that presents results of the model, speed, and track analyses to domain experts. They will be invited to correct the labeling in an easy way and thus help us to obtain more true labels.

References

1. Arasteh, S., et al.: Fishing vessels activity detection from longitudinal ais data. In: Lu, C.T., Wang, F., Trajcevski, G., Huang, Y., Newsam, S., Xiong, L. (eds.) Proceedings of the 28th International Conference on Advances in Geographic Information Systems, pp. 347–356 (2020). https://doi.org/10.1145/3397536.3422267
2. Ferreira, M.D., Spadon, G., Soares, A., Matwin, S.: A semi-supervised methodology for fishing activity detection using the geometry behind the trajectory of multiple vessels. Sensors **22**(16), 6063 (2022). https://doi.org/10.3390/s22166063
3. Heino, M., Enberg, K.: Sustainable use of populations and overexploitation. In: Encyclopedia of Life Sciences (eLS). Wiley, Chichester (2008). https://doi.org/10.1002/9780470015902.a0020476
4. Hevner, A.R., March, S.T., Park, J., Ram, S.: Design science in information systems research. Manag. Inf. Syst. Q. **28**, 75–106 (2004)
5. Jiang, X., Liu, X., de Souza, E.N., Hu, B., Silver, D.L., Matwin, S.: Improving point-based AIS trajectory classification with partition-wise gated recurrent units. In: 2017 International Joint Conference on Neural Networks (IJCNN), pp. 4044–4051 (2017). https://doi.org/10.1109/IJCNN.2017.7966366
6. Jiang, X., Silver, D.L., Hu, B., de Souza, E.N., Matwin, S.: Fishing activity detection from AIS data using autoencoders. In: Khoury, R., Drummond, C. (eds.) AI 2016. LNCS (LNAI), vol. 9673, pp. 33–39. Springer, Cham (2016). https://doi.org/10.1007/978-3-319-34111-8_4
7. Kroodsma, D., et al.: Tracking the global footprint of fisheries. Science **359**, 904–908 (2018). https://doi.org/10.1126/science.aao5646
8. Shen, K., Chu, Y., Chang, S.J., Chang, S.: A study of correlation between fishing activity and AIS data by deep learning. TransNav Int. J. Mar. Navig. Saf. Sea Transp. **14**, 527–531 (2020). https://doi.org/10.12716/1001.14.03.01
9. de Souza, E.N., Boerder, K., Matwin, S., Worm, B.: Improving fishing pattern detection from satellite AIS using data mining and machine learning. PLoS ONE **11**(9), 1–2 (2016). https://doi.org/10.1371/journal.pone.0163760

A General Framework for Blockchain Data Analysis

Anh Luu([✉]), Tuan-Dat Trinh, and Van-Thanh Nguyen

Hanoi University of Science and Technology, Hanoi, Vietnam
luudamvietanh@gmail.com, datt@soict.hust.edu.vn

Abstract. Blockchain is a foundational technology that allows application paradigms to shift from trusting humans to trusting machines and from centralized to decentralized control. Along with its explosive growth, blockchain data analysis is getting increasingly important for both scientific research and commercial applications. The current blockchain analysis systems and frameworks have limitations and weaknesses; they have excessively focused on Bitcoin and a small set of features. This paper presents a framework for blockchain data analysis. The framework is general and can be applied to a wide range of data analyses. Our main contributions are as follows: (i) we formulate the requirements of the framework; (ii) we present the detailed design of the framework with multiple components to collect, extract, enrich, store, and do further processing with blockchain data; (iii) we implement the framework and evaluate its performance in a specific use case that analyzes token-transferring transactions. We also discuss the potential of the framework for a number of blockchain data analyses.

Keywords: Blockchain · Data Analysis · Cryptocurrency

1 Introduction

Since the launch of Bitcoin in 2009 [15], more than a decade has been marked by the rise of blockchain-based technologies. Blockchain is a distributed public ledger that stores transactions between parties without requiring a trusted central authority. On a blockchain, transactions created by parties are unmodifiable; the transactions are permanently recorded on the ledger to be seen by the public.

With the increasing development and use of blockchain, collecting, analyzing, and visualizing its data shows great potential. Blockchain data analysis has always been a fascinating problem for governments, organizations, companies, and individuals. For example, governments and industry bodies can work together to analyze the data in the crypto markets. Based on that, they can create regulations for the markets and enforce them. This can stop illicit transactions such as money laundering and fraud from being carried out. Data analysis can identify risks related to crypto assets and offers visibility and better reaction times for traders, businesses, and institutions.

S. Nurcan et al. (Eds.): RCIS 2023, LNBIP 476, pp. 121–135, 2023.
https://doi.org/10.1007/978-3-031-33080-3_8

Analyzing blockchain data is important, but it is also very challenging. We identify four challenges of analyzing blockchain data, including synchronization among nodes, data decoding, big data processing, and real-time data analysis.

Node Synchronization. A blockchain network consists of peer nodes. Several nodes can simultaneously mine new blocks; the branch with the most blocks is selected across the network. To this end, the nodes must exchange a large amount of information to synchronize data with each other. They thus often operate unstable, leading to the asynchronous situation among nodes. It causes the data collected from nodes to be nonidentical and not the latest one. This makes the data analysis process misleading and unreliable.

Blockchain Data Extraction and Processing. Blockchain data is stored in a raw format, which cannot be directly analyzed. It is either binary or encrypted data. We need to extract our desired data and decode it to obtain valuable information. It is non-trivial to do this because each application and chain have its own data encoding rules.

Big Data Processing. As of September 2022, an Ethereum *archive node* occupies 11 TB of hard drive data[1]. The *archive node* is a full node that stores the entire history of the Ethereum chain, from the genesis to the latest block [8]. This large amount of data is only the data of one among many public chains; the data processed in multiple chain analysis is much more massive.

Real-Time Data Analysis. Blocks are mined continuously. In chains like Fantom[2] and Solana[3], new blocks are created every second. It is hence challenging to collect, process, and visualize the data in real-time.

Existing studies and applications on blockchain data analysis often focus on dealing with a small set of problems. Although different systems may need the same data, they typically collect and store the data separately. Each time new blockchain developers have an analysis requirement, they might build a new system from scratch or reuse only a few components of other projects. This takes them a lot of resources and effort. It is hence urgent to have a general framework for blockchain data analysis, which can be applied to various analysis projects.

This paper will present such a framework. Our main contributions are as follows: (i) we formulate the requirements of the framework; (ii) we present the detailed design of the framework with multiple components to collect, extract, enrich, store, and do further processing with blockchain data; (iii) we implement the framework and evaluate its performance in a specific use case that analyzes token transferring transactions. We also discuss the potential of the framework for several data analysis use cases.

The rest of the paper is organized as follows. Section 2 reviews the existing studies and applications performing blockchain data analysis. The requirements of a general framework for blockchain data analysis are formulated in Sect. 3.

---

1 https://etherscan.io/chartsync/chainarchive.
2 https://ftmscan.com/.
3 https://solana.fm/.

The framework architecture is presented in Sect. 4. Section 5 discusses the theoretical evaluation of the framework. Section 6 presents the detailed implementation of the framework. Section 7 shows the benchmark results in a use case in which we analyze token-transferring transactions. Section 8 discusses further applications of the framework. Finally, Sect. 9 concludes our study.

2 Related Work

As blockchain data analysis can provide substantial benefits, there have been many studies and tools that focus on this topic. A few of them target a general framework that can be used for multiple data analysis projects at the same time.

Balakas and Franqueira [2]review existing blockchain data processing tools, and they found that all of their reviewed tools focus on finding illegal online activities and money laundering transactions. The tools are often implemented to run on Bitcoin and Ethereum but are rarely generalized to run on multiple chains.

Kalodner et al. [9] presents BlockSci, a framework that includes three components: a *data collection module* to collect blockchain data, a *structured data parser* to store in the database, and an *interface* to retrieve data for analysis. The system provides Python Jupiter Notebook and C++ *interfaces* as the two options for data analysts. The former *interface* is for intuitive exploration with small datasets. The latter is used for performance-critical tasks when we have to traverse all blocks or transactions of the data. The authors also list applications of BlockSci, including two applications related to security and two other applications related to the economics of cryptocurrency. The system can be used to analyze data of Bitcoin, Litecoin, and ZCash; it can not process data on Ethereum Virtual Machine (EVM) [7] chains like Ethereum or Binance Smart Chain (BSC).

XBlock-ETH [20] is a framework that analyzes data on EVM-based chains. It can extract raw data, including *block*, *trace*, and *receipt* data. (i) *Block data* is the basic type of blockchain data and can be collected easily. (ii) *Trace data* is the detailed run-time data that was generated in EVM. *Trace data* cannot be directly obtained or observed from the *block data* but can be recorded during the contract execution. (iii) After the transaction is executed, *data receipt* is the outcome of transaction execution on Ethereum. The outcome includes the hash of transactions, the gas used, the token ownership, etc. XBlock-ETH can be used to evaluate the price of cryptocurrencies, identify the fraud of contracts, or benchmark the performance of the Ethereum network. The framework lacks a mechanism to decode transactions; it hence cannot analyze the detailed activities of wallets in decentralized applications (DApps).

Galici et al. [5] follow a different approach that uses traditional data warehouse techniques on the Bitcoin network to build an ETL [1] workflow for data extraction, transformation, and loading. The data is stored on a relational database and provides custom queries to users. The authors collect wallet addresses and their transaction data to analyze the behavior of users on the

Bitcoin network; they perform clustering on user addresses based on the number of transactions the addresses involved in.

Chen et al. [4] develop DataEther framework to modify the internal mechanisms of Geth (which is an Ethereum full node) to extract and store data in an Elasticsearch[4] system. With the advantage of Elasticsearch, users can perform simple data queries with low latency to gather data for their analysis. The problem is that as soon as the system needs to regularly update data, Elasticsearch will expose the disadvantage of consuming extensive resources for data indexing. The framework works with the Ethereum chain only.

Along with those studies from scientists, many commercial data analysis tools have been developed and used widely. These tools can be divided into two groups: *market analysis tools* (e.g., Cryptoquant[5], Messari[6]) and *transactions analysis tools* (e.g., Dune[7]). The former tools are created for traders or financial institutions. They collect the data from both Centralized Exchanges (e.g., Binance, Coinbase) and Decentralized Exchanges (e.g., Uniswap, Pancakeswap). The interface of these tools typically includes a lot of charts. Users have to pay a fee to use more advanced features related to the analysis of token trading and the entire crypto market.

Dune is a *transaction analysis tool* for data analysts and individual users. It collects on-chain data from EVM-based chains. The data is divided into two types, i.e., *raw data* and *decoded data. Raw data* is encrypted data that is directly collected from blockchain networks. Some of the raw data is selected by the Dune development team to be decoded. The team usually selects the token-transferring transactions and transactions from DApps with a large number of users. The *decoded data* is stored in Dune's server, and users can access the data via SQL queries. User can write their own SQL queries on Dune websites to retrieve their desired data and display it in the form of customizable charts.

Dune is a framework for developers rather than for end users because they have to be an expert at SQL queries. However, developers can only interact with the decoded data selected by Dune teams. Developers cannot use this framework to build charts using data that has not been decoded. Dune users can query and get data in the form of tables displayed on the Dune website. They cannot download the data to their local computer.

In general, current blockchain data analysis tools and studies have limitations and drawbacks. They are typically designed and developed for certain tasks (e.g., smart contract fraud detection, crypto market analysis); they usually support one chain only and do not consider the issue of performance and reusability. There have been no analysis frameworks for blockchain data that developers can use to quickly initialize and develop applications for their own analysis topics.

[4] https://www.elastic.co/elasticsearch/.
[5] https://cryptoquant.com.
[6] https://messari.io.
[7] https://dune.com.

3 Requirements of a General Framework for Blockchain Data Analysis

Based on the challenges of analyzing blockchain data presented in Sect. 1 and the survey on related work presented in Sect. 2, we identify eight requirements for a general blockchain data analysis framework, including *accuracy*, *completeness*, *real-time processing*, *scalability*, *on-demand processing*, *generality*, *monitorability*, and *reusability*.

Accuracy. Blockchain data is accessible via node providers, but the synchronization process of these providers in many cases is inconsistent. Faulty or slow providers will affect the data collection process. There should be mechanisms to check the status of the providers to accurately collect the raw data and queries the smart contract states. Collecting, decoding, and processing data must be done precisely.

Completeness. A large amount of blockchain data and the complexity of the nodes make the data collection process unstable; the node providers sometimes do not return the complete set of events and transactions. The framework should check the completeness of the data and recollect all missing data.

Real-Time Processing. The flow of blockchain data is massive and continuous. Blocks and transactions are continuously created. The rate of processing a block should be higher than the rate of mining a block on the tracking chains. This allows us to collect and process data from the past to the present.

Scalability. The system must be scalable in terms of storing and processing data. (i) Once data grows to the limit of the system, the system needs to be able to increase its storage capacity without affecting the existing data; the rate of reading and writing data should not be affected. (ii) In terms of data processing capacity, the system should be used for multiple analysis tasks at the same time. New tasks can be added without affecting the running ones. The system must support horizontal scaling, meaning that they are capable of growing in device number instead of having to reside in one large device.

On-demand Processing. There is a large number of events, transactions, and blocks on blockchain networks. On the chains that support the concept of smart contracts, their DApps generate various types of transactions and events. These data records are in raw format and need to be decoded for further data analyses. It is unnecessary to decode all data because analysis projects typically concern certain types of transactions and events only. The data should be decoded and processed on-demand.

Generality. As the number of blockchain networks keeps increasing, the framework should be general so that it can collect and manipulate data on all chains, even for the chain that has not been created yet. The framework should also be able to serve multiple data analysis topics.

Monitorability. Big data processing always consumes a lot of resources. This leads to various kinds of unexpected problems: (i) the server runs out of disk space, memory, or internet bandwidth; (ii) the CPUs are overloaded; (iii) the database stops working; (iv) the blockchain nodes might stop their services because the system sends them too many requests at once. To address this problem and work stably, the framework should have a component that monitors the hardware and the operations of all components in the framework. The system administrators would get notifications as soon as there is any problem so that they can react promptly.

Reusability. Multiple analyses can be conducted in the same blockchain data analysis framework. To optimize the performance while using fewer resources, the framework should be designed in a reusable manner. For example, assuming that many analysis tasks need the same kind of blockchain data, the data should be collected, decoded, and processed single time and reused among all analysis tasks.

4 Framework Architecture

This section presents the detailed design of the blockchain data analysis framework that fulfills the requirements formulated above.

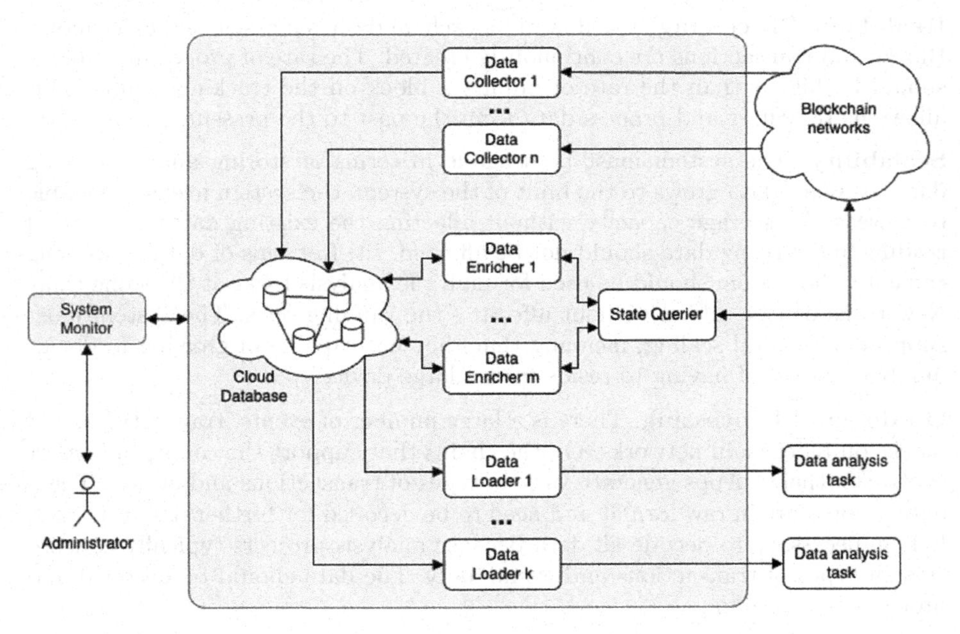

Fig. 1. Architecture of general framework for blockchain data analysis

The architecture of the framework is shown in Fig. 1. The *Data collector* first gathers raw data on concerning blockchain networks and stores it in a cloud database. The raw data, including blocks, transactions, and events, is the core of the networks and can be reused in multiple analysis projects. The cloud database makes the framework scalable in terms of data storage; it can store a large amount of data as required.

The *Enricher* takes certain records of data (which can be transactions, events, or anything needed in a specific analysis) and decodes them. The *Enricher* can enrich the records with further information. To this end, it queries the states of blockchain nodes to get the values of smart contract variables at specific blocks. The *Enricher* will not do this by itself, but it will ask the *State querier* for help.

For multiple analysis tasks, multiple *Enrichers* can run at the same time. The *State querier* collects all queries from the running *Enrichers* and sends queries **in batches** to the blockchain node providers. Because the providers set the limit number of requests in a batch, the *State queries* can run on multiple threads; each thread sends the providers a batch containing the maximum number of requests. These methods can speed up the process and facilitate real-time data processing. The methods also make it easy to develop and maintain the system.

The enriched data is then associated with the original records, and both are saved back into the database. The *Enricher* and the *Collector* create two data streams written to the database.

Depending on the analysis task, a single or multiple *Loaders* are created to gather the required data. For example, a *Loader* can query (i) all transactions related to a token or (ii) all transactions generated from a smart contract. The loaded data can be further processed and visualized in charts for the end users of the analysis task.

The whole system will be watched by the *Monitor*. It observes two aspects of the system, i.e., the hardware and the activity flows. (i) The *Monitor* regularly collects hardware parameters and visualizes the data in charts so that the system administrator can recognize any problem related to hardware. (ii) The *Monitor* needs to track the activity flow of the running *Collectors*, *Enrichers*, and *Loaders* to detect any fault in their operations. It should check the synchronization process of the service providers. It also logs various information such as the current block that is being processed, the rates of collecting and processing data, and other information related to the *State querier*.

To demonstrate the process of the framework, let us take a scenario that we need to analyze the transferring transactions of token $\mathcal{T}$. After the *Data Collectors* collect all transactions, the *Enricher* extract these tranffering transactions out of all collected transactions.

Structures of raw transactions are shown in Fig. 2. Let us consider a transaction in that Alice transfers 10 token $\mathcal{T}$ to Bob. In the raw transaction, **from_address** is Alice's address, and **to_address** is the address of the token $\mathcal{T}$. Because this transaction is generated as Alice calls the **transfer** function of the token $\mathcal{T}$'s smart contract, we can use the field *to_address* to as the filter to extract all transactions that transfer token $\mathcal{T}$.

The raw transaction data itself does not let us know to whom Alice transferred $\mathcal{T}$ and the amount of $\mathcal{T}$ that was transferred. To address this problem, we create an *Enricher* $\mathcal{A}$ that decodes the transaction based on Application Binary Interface (ABI) of the transferring method. Similar to API, ABI allows applications to understand the parameters of a smart contract method and decode

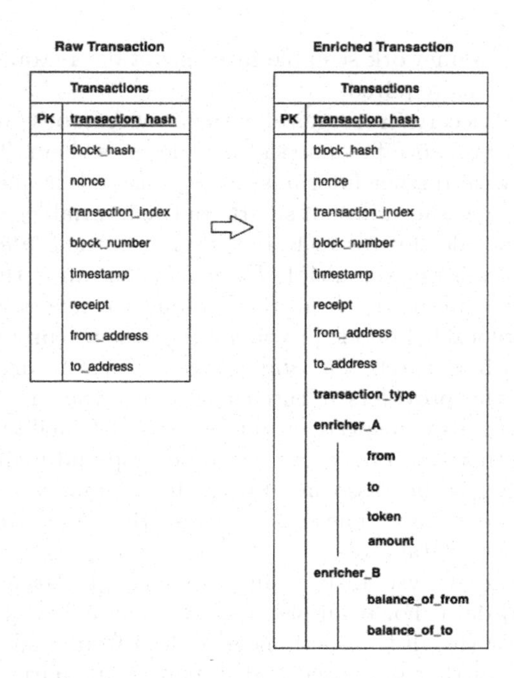

Fig. 2. Enrichment of Token Transfer Transaction

them. *Enricher* $\mathcal{A}$ decodes data and associates the original transaction with four new fields, including **from** – the address of Alice, **to** – the address of Bob, **token** – the address of token $\mathcal{T}$, and **amount** – the amount of $\mathcal{T}$ transferred (cf. Fig. 2).

Depending on their requirements, multiple *Enrichers* can add their enriched data to the same transaction. For example, to analyze all holders of token $\mathcal{T}$, developers would need to know the amount of $\mathcal{T}$ held by the receiver and sender after each transferring transaction. We can create *Enricher* $\mathcal{B}$ to obtain such information. This enricher sends requests to *State Querier*, which in turn gathers requests from other enrichers and sends queries in batches to the node providers. In the final step, *Enricher* $\mathcal{B}$ adds two pieces of information including **balance_of_from** and **balance_of_to** (which are respectively the total amount of $\mathcal{T}$ held by the sender and receiver) to the transaction.

Our detailed example shows that, depending on our needs, we can create an arbitrary number of enrichers to decode and add more data to the raw transaction. Data is thus increasingly enriched and becomes much more meaningful and useful. New analysis tasks can reuse the enriched information without wasting resources to "reinvent the wheel".

5 Theoretical Evaluation of the Framework

This section evaluates the design of the framework in a theoretical manner. The experimental evaluation of the framework is presented in Sect. 7.

The *Monitor* is responsible for the **accuracy**, **completeness**, and **monitorability** requirements. It checks the hardware status and tracks the activity flow of the running *Collectors*, *Enrichers*, and *Loaders*.

The tasks of collecting, decoding, enriching, and preparing data for analysis are separately assigned to the *Collector*, *Enricher*, *State querier*, and *Loader*. We can initialize many instances of these components, and those instances can run on multiple threads or devices. (i) This first speeds up all data handling processes and helps the framework to meet the **real-time processing** requirement. (ii) This is also proof that the system supports **horizontal scaling** in terms of data processing.

The chosen cloud database supports data replication and data distribution so that the system is horizontally **scalable** in terms of storing data. This is crucial because many components access the database at the same time.

Multiple instances of the *Enricher* can be initialized; each can select certain records (transactions, events, etc.) to decode or do further processing. This helps the framework to meet the requirement of **on-demand processing.**

Our approach to structuring raw transactions and their enriched data (see Fig. 2) allows the framework to fulfill the **reusability** requirements. The raw data and the enriched data can be reused in multiple analysis projects. These projects can create another enricher to add new data as they need. The new data, in turn, can be reused in other projects to facilitate reusability.

Finally, the design of the framework makes it **general**. *Collectors* can obtain records of data from arbitrary blockchain networks. *Enrichers* then freely decode and add more enriched data to the records. *Loaders* then take data as they need for any analysis task.

6 Implementation Discussion

We implement the framework to analyze data on four EVM-based chains, including Ethereum, Binance Smart Chain (BSC), Fantom, and Polygon. In the following, we discuss the technologies and techniques we used for the implementation.

We conducted a survey on node providers' services and decided to choose Chainstack because it is one of the rare services that have node providers for all EVM-based chains above. Chainstack node providers work more stably than others. They support querying states in batches; each batch can have a maximum of 2000 queries.

In the framework architecture, the database plays a crucial role. It must store a large amount of data; many components connect with it and establish read and write streams simultaneously. We identify three criteria that the database should fulfill. (i) It first should support flexible schema so that the enricher can add various types of data to the raw blockchain data. (ii) It should be distributed database with replication to facilitate horizontal scaling. (iii) Finally, the database should have a free version for the community.

We decide to use MongoDB as it meets the three criteria. MongoDB is one of the most prominent document-based databases on the market today. It is an ideal platform for building blockchain databases [16]. It is chosen by many blockchain enterprises such as BigchainDB [12], ProvenDB[8], and EthernityDB [6]. MongoDB has also developed an community version. This version is useful for small research teams and individual analysts who want to deploy the blockchain analysis framework.

We build a MongoDB cluster that consists of three shardings and three replications. Using shardings and replications increases fault tolerance and makes the system scalable. By increasing the number of shards and adding more hard drives, the system can still work efficiently even though blockchain data keeps increasing.

The *Collector* and *Enricher* are developed based on the ETH-ETL toolkit developed by Medvedev and the D5 team [13]. The toolkit is written in Python. It is an effective toolkit that has been used to crawl Google's Big Query Dataset [14].

Using Python, we develop the *State querier* from scratch. It can receive a large number of requests from multiple *Enrichers*. It is configured to work with a list of providers; all incoming requests will be packed in batches and sent simultaneously to those providers. The *State querier* then returns results to the *Enrichers*.

7 Experiments on Analysis of Transferring Transactions

The theoretical evaluation of the framework is presented in Sect. 5. The implementation of the framework allows us to empirically evaluate the framework.

As a use case for experiments, let us consider the analysis of token-transferring transactions on four blockchain networks, including Ethereum, BSC, Polygon, and Fantom. Because transferring tokens is a very basic operation on any chain, the number of transferring transactions is much larger than the number of other transaction types. Although this large amount of data makes analyzing transferring transactions challenging, the analysis provides us insights into the crypto market. For example, (i) the total number of transferring transactions per block indicates the activeness of the crypto market; (ii) the volume and the number of transferring transactions can be used to evaluate tokens; (iii) tracking transferring transactions of the whale wallets allows us to evaluate the safety of the market.

Each chain has a large number of tokens. According to Coingecko[9], as of September 1st, 2022, Ethereum, BSC, Polygon, and Fantom chains respectively contain 11162, 4308, 975, and 352 tokens. Among them, many tokens are rarely

[8] https://www.provendb.com/.
[9] https://www.coingecko.com/.

used. Using their statistics on the market cap, the total number of daily transactions, and the number of holders, we select 4000, 2000, 100, and 100 tokens on Ethereum, BSC, Polygon, and Fantom, respectively, and analyze their transferring transactions.

We do not evaluate the *Monitor* and *Loader* in this experiment. The *Monitor* ensures the accuracy of the system; it also notifies the administrator of hardware problems. The *Loader* reads the decoded and enriched data for visualization; it can also store the data in other temporary databases for further processing and analysis. These two components do not consume excessive resources; they should be evaluated for accuracy through test cases rather than be evaluated for performance.

The experiment focuses on the evaluation of the *Enricher* and *Collector*. These two essential components determine the applicability and the performance of the framework. They continuously collect, decode, and enrich a large amount of data on various blockchain networks. We deploy these two components on two Digital Ocean cloud servers[10]; one is for the *Enricher*, and the other is for the *Collector*. Each server has an 8-core CPU, 16 GB RAM, and 325 GB SSD.

We carry out the analysis of token-transferring for the latest one million blocks on each chain, as of September 1st, 2022. Table 1 shows the specification of the processed data. For example, the experiment needs to collect 180287332 Ethereum transactions from block 14449619 to block 15449618. Ethereum takes 3849.18 hours to generate this one million blocks; the average time to generate one block is 13.86 seconds.

Table 1. Specification of data processed on four chains

Chain	Start Block	End Block	Number of transactions	Time for the chain to generate one million blocks (hour)	Average time for the chain to generate one block (second)
Ethereum	14449619	15449618	180287332	3849.18	13.86
BSC	19936755	20936754	114111315	834.64	3.00
Fantom	45097923	46097922	10819625	347.81	1.25
Polygon	31534500	32534499	78230002	644.05	2.32

In the experiment, the *Collector* and *Enricher* continuously collect and process the data of one million blocks. As the server has an 8-core CPU, we let each component run on eight threads. The *Enricher* must add six pieces of information, including the sender address, the receiver address, the address of the transferred token, the transferring amount, and the balance of the receiver and sender after executing the transaction. The *Enricher* in this experiment is the combination of *Enricher* $\mathcal{A}$ and *Enricher* $\mathcal{B}$ in the example shown in Fig. 2.

[10] https://www.digitalocean.com/.

Table 2. Average time to collect and enrich blockchain data

Chain	Average time to collect one block (second)	Average time to enrich one block (second)
Ethereum	3.12	5.12
BSC	0.98	1.27
Fantom	0.82	0.91
Polygon	1.25	1.50

The result of the experiment is shown in Table 2. On Ethereum, for example, the average time to collect and enrich one block is respectively 3.12 and 5.12 seconds. Based on this result, we can evaluate the performance of the *Enricher* and *Collector*, as shown in Fig. 3.

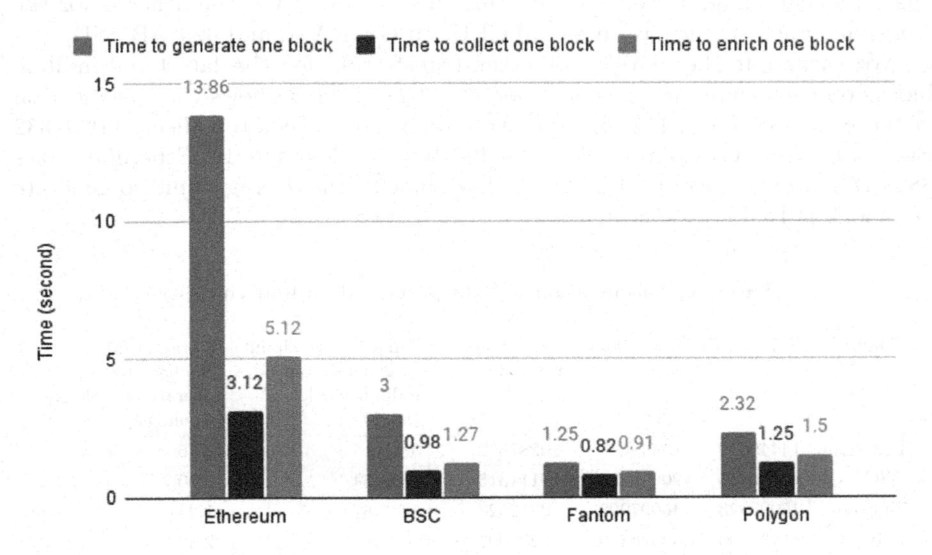

Fig. 3. Performance evaluation of the *Collector* and *Enricher*

Figure 3 shows that, on all four chains, the rate of collecting and enriching a block is always higher than the rate of generating a block. This is crucial for the framework to be able to handle both the latest blocks and the blocks that were generated in the past.

The collected and enriched data are useful for many analysis tasks. For example, we can analyze the percentage of the transferring transactions. On the EVM-based chains, there are two main types of crypto assets, including coins and crypto tokens [17]. Crypto coins are the native asset of the blockchain network like Ethereum or Fantom, whereas crypto tokens are created by platforms and applications that are built on top of the network. Based on the output of the experiment, we calculate the percentage of coin-transferring transactions and the percentage of token-transferring transactions (see Table 3). As shown in the table, it is interesting that the former is greater than the latter on Ethereum, Fantom, and Polygon. On Ethereum, 35.46% of transactions are coin-transferring transactions.

The experiment shows that the *Enricher* and the *Collector* can handle data in a real-time manner. Because of the horizontal scalability of the framework, we can add more computing resources to speed up the data handling process. We can use multiple node providers and servers to deploy many instances of *Enricher* and *Collector*. As the blocks can be collected and enriched independently, each instance is assigned to a certain number of blocks to process.

Table 3. Percentage of transferring transactions

Chain	Number of transactions	Number of coin transferring transactions	Number of token transferring transactions	Percentage of coin transferring transactions (%)	Percentage of token transferring transactions (%)
Ethereum	180287332	63927725	23294338	35.46	12.92
BSC	114111315	11023915	11325872	9.66	9.93
Fantom	10819625	90054	10254	0.83	0.09
Polygon	78230002	2890647	1827624	3.70	2.34

8 Applications of the Framework

The generality of the blockchain data analysis framework allows it to be used in various use cases. This section discusses some of them.

Analysis of Decentralized Applications. Decentralized Applications (DApps) [3] are applications that run on blockchain networks and thus can operate autonomously. With the launch of Ethereum, DApps have experienced widespread growth in recent years. DApps can be divided into many different categories, depending on the purpose of use. Wu et al. [18] divide DApps into 17 different categories. Among those categories, Decentralized Finance (DeFi) is the area where users invest the most. According to DefiLlama[11], as of August 2022, the total value of the DeFi market is $58 billion.

The framework is well-fit to analyze those DApps. The *Enricher* can extract the transactions of a selected DApp, decoding them and adding more meaningful

[11] https://defillama.com/.

data. The data allows us to identify the total number of users, analyze the user segmentation, or analyze the business performance of the DApp. For example, with a decentralized exchange (DEX) [11], we can identify the most exchanged token or figure out the safest and the most unsafe token pairs.

Token Analysis. Token analysis always attracts the attention of financial organizations and individuals. There are numerous tokens on blockchain networks. Finding a valuable token to invest in is a challenging task for investors.

The framework can collect all transactions related to the token. It then processes the data and offers statistics on the number of holders, the trading volume of tokens, the number of tokens distributed to whales, and the price volatility. These types of analyses can support investors in making decisions.

Wallet Evaluation. The framework is able to collect all activities of wallets on blockchain networks. Based on the transaction history of the wallets and their total asset, we can perform credit ratings for blockchain wallets. The ratings can facilitate lending protocols and other DeFi applications.

For example, lending protocols enable crypto asset holders to lend out their crypto and borrow against the crypto as collateral. Credit ratings are a useful index to evaluate the creditworthiness of borrowers; a user with a high credit score would have a chance to be approved for a loan with favorable interest rates.

9 Conclusion and Future Work

This paper presents a general framework that can collect, store, enrich, process, and analyze blockchain data. New blockchain data analyses can be quickly conducted by adding a few components to the framework rather than building a whole new system. The framework has been implemented, and our experiments show that it fulfills the requirements to be used in multiple blockchain analysis projects.

Our future works focus on experiments that collect and process data on blockchain networks other than EVM, such as Tron[12], Solana [19], and Cosmos [10]. This allows us to carry out in-depth analyses on multiple chains. For example, we can calculate the total asset of a wallet on multiple chains and monitor the token flow on these chains.

References

1. Ali, S.M.F., Wrembel, R.: From conceptual design to performance optimization of ETL workflows: current state of research and open problems. VLDB J. **26**(6), 777–801 (2017)

[12] https://tron.network/.

2. Balaskas, A., Franqueira, V.N.: Analytical tools for blockchain: review, taxonomy and open challenges. In: 2018 International Conference on Cyber Security and Protection of Digital Services (Cyber Security), pp. 1–8. IEEE (2018)
3. Cai, W., Wang, Z., Ernst, J.B., Hong, Z., Feng, C., Leung, V.C.: Decentralized applications: the blockchain-empowered software system. IEEE Access **6**, 53019–53033 (2018)
4. Chen, T., et al.: Dataether: data exploration framework for ethereum. In: 2019 IEEE 39th International Conference on Distributed Computing Systems (ICDCS), pp. 1369–1380. IEEE (2019)
5. Galici, R., Ordile, L., Marchesi, M., Pinna, A., Tonelli, R.: Applying the ETL process to blockchain data. prospect and findings. Information. **11**(4), 204 (2020)
6. Helmer, S., Roggia, M., Ioini, N.E., Pahl, C.: EthernityDB – integrating database functionality into a blockchain. In: Benczúr, A., Thalheim, B., Horváth, T., Chiusano, S., Cerquitelli, T., Sidló, C., Revesz, P.Z. (eds.) ADBIS 2018. CCIS, vol. 909, pp. 37–44. Springer, Cham (2018). https://doi.org/10.1007/978-3-030-00063-9_5
7. Hildenbrandt, E., et al.: KEVM: a complete semantics of the Ethereum virtual machine. University of Illinois Urbana-Champaign, United States, Technical report (2017)
8. Iyer, K., Dannen, C.: The Ethereum development environment. In: Building Games with Ethereum Smart Contracts, pp. 19–36. Apress, Berkeley (2018). https://doi.org/10.1007/978-1-4842-3492-1_2
9. Kalodner, H., et al.: {BlockSci}: Design and applications of a blockchain analysis platform. In: 29th USENIX Security Symposium (USENIX Security 20), pp. 2721–2738 (2020)
10. Kwon, J., Buchman, E.: Cosmos whitepaper. A Netw. Distrib, Ledgers (2019)
11. Lo, Y., Medda, F.: Uniswap and the rise of the decentralized exchange. University Library of Munich, Germany, Technical report (2020)
12. McConaghy, T., et al.: Bigchaindb: a scalable blockchain database. White paper, BigChainDB (2016)
13. Medvedev, E., the D5 team: Ethereum ETL (2018). https://github.com/blockchain-etl/ethereum-etl
14. Naidu, S., Tigani, J.: Google BigQuery Analytics. John Wiley & Sons, Hoboken (2014)
15. Nakamoto, S.: Bitcoin: A peer-to-peer electronic cash system. Decentralized Business Review p. 21260 (2008)
16. Raikwar, M., Gligoroski, D., Velinov, G.: Trends in development of databases and blockchain. In: 2020 Seventh International Conference on Software Defined Systems (SDS), pp. 177–182. IEEE (2020)
17. Victor, F., Lüders, B.K.: Measuring Ethereum-based ERC20 token networks. In: Goldberg, I., Moore, T. (eds.) FC 2019. LNCS, vol. 11598, pp. 113–129. Springer, Cham (2019). https://doi.org/10.1007/978-3-030-32101-7_8
18. Wu, K., Ma, Y., Huang, G., Liu, X.: A first look at blockchain-based decentralized applications. Softw. Pract. Expe. **51**(10), 2033–2050 (2021)
19. Yakovenko, A.: Solana: A new architecture for a high performance blockchain v0.8.13. Whitepaper (2018)
20. Zheng, P., Zheng, Z., Wu, J., Dai, H.N.: XBblock-eth: extracting and exploring blockchain data from Ethereum. IEEE Open J. Comput. Soc. **1**, 95–106 (2020)

Reinforcement Learning for Scriptless Testing: An Empirical Investigation of Reward Functions

Olivia Rodríguez-Valdés[1](✉), Tanja E. J. Vos[1,2], Beatriz Marín[2], and Pekka Aho[1]

[1] Open Universiteit, Heerlen, The Netherlands
{olivia.rodriguezvaldes,tanja.vos,pekka.aho}@ou.nl
[2] Universitat Politècnica de València, València, Spain
{tvos,bmarin}@dsic.upv.es

Abstract. Testing web applications through the GUI can be complex and time-consuming, as it involves checking the functionality of the system under test (SUT) from the user's perspective. Random testing can improve test efficiency by automating the process, but achieving good exploration is difficult because it requires uniform distribution over a large search space while also taking into account the dynamic content commonly found in web applications. Reinforcement learning can improve the efficiency of random testing by guiding the generation of test sequences. This is achieved by assigning rewards to specific actions and using them to determine which actions are most likely to lead to a desired outcome. While rewards based on the difference between consecutive states are commonly used in modern tools, they can lead to the Jumping Between States (JBS) problem, where large rewards are generated without significantly increasing exploration. We propose a solution to the JBS problem by combining rewards based on the change of state and a metric to estimate the level of exploration reached in the next state based on the frequency of actions executed. Our results show that this multi-faceted approach increases the exploration efficiency.

Keywords: GUI testing · Scriptless testing · Reinforcement learning

1 Introduction

Testing Web applications at the Graphical User Interface (GUI) is crucial since it allows checking quality from the user's point of view. However, manual testing via GUI is expensive and error-prone [9]. Automation of GUI testing is challenging due to the complexity of GUIs and difficulty in accessing them programmatically [21], as well as maintenance issues associated with script writing [3,7].

In addition to scripted automation, scriptless testing [21] has gained popularity [18]. Scriptless testing generates test sequences by automatically selecting and executing one action based on the current state of the System Under Test's (SUT) GUI. The simplest way to select actions is randomly, but it requires more

S. Nurcan et al. (Eds.): RCIS 2023, LNBIP 476, pp. 136–153, 2023.
https://doi.org/10.1007/978-3-031-33080-3_9

execution time to achieve good system coverage and may not reach all areas of the GUI. However, studies have shown random exploration to detect critical failures [1,21]. This paper explores using Reinforcement Learning (RL) [20] to implement more sophisticated ways to select actions. RL is a branch of machine learning that is directly applicable in scriptless testing since RL is trained using rewards after every interaction with an environment. RL uses *reward functions* to guide the selection of actions and explore the search space of the system being tested. These rewards are usually based on the difference between consecutive states, with high rewards given to actions that lead to very different states.

Using rewards based on the difference between consecutive states can lead to unwanted behavior such as *Jumping Between States* (JBS) instead of exploring other areas. Pan et al. [15] describe the problem: if a large reward is obtained as the result of moving between two very different states, the RL agent will frequently jump between these states instead of exploring other areas of the SUT. A solution to the JBS problem was proposed based on neural networks to extract main features from the states to detect if two states are different and saved them as a vector in a memory buffer, to avoid repeating states. However, neural networks require a lot of data and time to work well. Additionally, the inability to debug and explain the reasoning and evolution of a neural network over time is a disadvantage of this technique. Instead, our proposed solution uses state abstraction to compare states and includes memory, based on the frequency of executed actions, to the reward itself.

Testing Web applications have more diverse and complex states than mobile or desktop applications due to their frequent dynamic content updates and elaborated workflows. More effective exploration is required to test Web applications considering their huge search space and interactive nature. Therefore, besides an alternative solution to the JBS problem, we concentrate on testing Web applications in this paper. The contribution of this work is:

- an implementation of an RL framework that can be used to compare different rewards used during scriptless testing of Web applications;
- an empirical evaluation showing the reduction of the JBS problem for rewards containing memory based on the frequency of executed actions
- a comparison of the exploration effectiveness of the different rewards looking at URL coverage and state exploration.

The rest of the paper is organized as follows: Sect. 2 presents background about Q-Learning, Sect. 3 offers relevant related work, and Sect. 4 presents the main characteristics of the scriptless tool. Section 5 presents our approach for smart exploration, Sect. 6 presents the empirical evaluation, and Sect. 7 presents the results. Section 8 presents our conclusions and future works.

2 Q-Learning

Reinforcement Learning (*RL*) [20,24] is a machine learning technique that consists of an agent that learns to behave in an interactive environment. The agent executes actions within the environment and obtains information through trial-and-error interactions. RL algorithms contain four basic elements:

Algorithm 1. Q-Learning

```
Require: γ                                                    ▷ discount factor
Require: α                                                    ▷ learning rate
 1: initialiseQValues()
 2: s ← getStartingState()
 3: repeat
 4:     availableActions ← deriveActions(s)
 5:     a ← selectAction(availableActions)                    ▷ Select an action
 6:     executeAction(a)
 7:     s' ← getReachedState()
 8:     reward ← R(s, a, s')                                  ▷ Reward the action
 9:     learn(s, s', a, reward, γ, α)                         ▷ Learn from the experience
10:     s ← s'
11: until s' is the last state of the sequence
```

- The **state** describes the situation of the environment at every step.
- An **action** is a possible move to go from one specific state to another.
- The **policy** defines the strategy used to select an action in a given state and the learning approach of the algorithm.
- The **reward** defines the goal to achieve.

The environment can be formalised with a Markov Decision Process (MDP), defined as a 4-tuple $M = \langle S, A, T, R \rangle$, where S is the set of states, A is the set of actions. At each time step t, the agent executes an action $a_t \in A$ and will receive a reward $r_t \in R$. The transition probability function T describes the probability $P(s_{t+1}|s_t, a_t)$ of transitioning into state s_{t+1} from state s_t and executing action a_t. The goal of RL is to maximise the rewards obtained over time, known as *expected return* as in $R_t = \sum_t r_t \gamma^t$, where $\gamma \in [0, 1]$ is the discount factor used to define the balance between long-term and immediate rewards.

Q-learning [23] is a model-free RL approach to learn the value of an action in a particular state and, hence, find an optimal action-selection policy π. An action-value function estimates the expected return $Q^\pi(s_t, a_t) = E(R_t|s_t, a_t)$ after executing an action a_t in state s_t following policy π. We can define the action-value function in terms of its successor state-action pair as:

$$Q(s, a) = Q(s, a) + \alpha * [R(s, a, s') + \gamma * \max_{a' \in A} Q(s', a) - Q(s', a')] \quad (1)$$

The Q-function refers to the maximum Q expected for a given (state, action) pair, over all possible policies. However, Q can also be interpreted as the optimal strategy at each step maximising the sum of the immediate reward $R(s, a, s')$ of the current step and the Q-value $Q(s', a')$ of the next step. The parameter α (step size) defines the learning rate of the algorithm.

The idea behind using Q-learning for GUI testing is to reward each selection of possible actions over the SUT (see Algorithm 1). Choosing an action (line 5) and executing it (line 6) moves the agent from the current state s to a new state $s\prime$ (line 7). The agent is rewarded with a *reward* upon executing the action a (line 7). This reward is calculated by the reward function R. The main objective of Q-learning is to *learn* how to act in an optimal way that maximises the cumulative reward (in line 8). An approximation for the optimal Q-function that simplifies the problem and enables early convergence is shown in algorithm 1, where $learn(s, s', a, reward, \gamma, \alpha)$ is defined as the action-value in Eq. 1.

Table 1. Related work with state-based rewards

Publication	RL Algorithm	Reward	Policy	SUT types
Vos [21]	Q-Learning $\alpha = 1$ multiple γ	Frequency-based	Greedy	Windows desktop, Web
Adamo [2]	Q-Learning $\alpha = 1$ dynamic γ	Frequency-based	Greedy	Android
Mariani [13]	Q-Learning	Widget Tree difference between consecutive states	ϵ-Greedy, $\epsilon = 0.8$	Windows (java)
Vuong [22]	Q-Learning $\alpha = 1$ $\gamma = 0.9$	Combination of event distance between consecutive states and frequency-based rewards.	ϵ-Greedy ($\epsilon_{intial} = 1$) Gradually decreased up to 0.5	Android
Cao [10]	SARSA	Small reward (1) if crash or new activity is found. Low reward (-1) if activity is out of SUT. Very low reward (-10) otherwise	ϵ-Greedy	Android
Romdhana [19]	Q-Learning and Deep RL algorithms	Very large reward (1000) if crash or new activity is found. Large reward (100) if activity is out of SUT. Small reward (1) otherwise.	ϵ-Greedy $\epsilon = 0.8$ or $\epsilon = 0.5$	Android
Collins [6]	Deep RL algorithm	Large reward if code coverage increases or new activity is found. Small reward if state does not change		Android
Pan [15]	Q-Learning	Similarity between consecutive states (using Neural Network)	ϵ-Greedy, $\epsilon = 0.2$	Android
Degott [8]	multi-armed bandit [17]	1 if interface changes between consecutive states. 0 otherwise	ϵ-Greedy Thompson Sampling	Android

3 Related Work

RL has been used for GUI testing in different ways. Offline Q-learning, uses only previously collected offline data in [11,12]. These works learn from existing SUTs about actions that are useful to reach a particular objective and then apply it to new SUTs. Online Q-learning is applied independently for every SUT with state-based rewards and the approaches are summarised in Table 1. Since in this paper we concentrate on online learning, we discuss each of them next.

Vos et al. [21] present Testar: a GUI testing tool for web and Windows applications. Testar has implemented Q-learning with a frequency-based reward function for Java desktop applications. This reward aims to encourage the selection of the least executed actions, thus guiding the exploration towards unvisited areas of the application. Also applying Q-Learning, Adamo et al. [2] use the same frequency-based reward but in the scope of Android applications.

AutoBlackTest [13] also uses Q-Learning to test Java desktop applications. Their reward function favours actions that increase the difference between two consecutive states, measured by comparing their respective widget trees. Similarly, a combination of frequency and widget tree rewards can be found in [22].

AimDroid [10] uses RL to guide exploration of Android applications. A positive reward is obtained if a new activity or crash is observed. More recently, ARES [19] uses a similar reward to AimDroid (larger values) but with Deep Neural Network as a technique to learn the best exploration strategy. In the scope of Deep Reinforcement Learning, Collins et al. [6] use a reward based on the code coverage obtained during the execution.

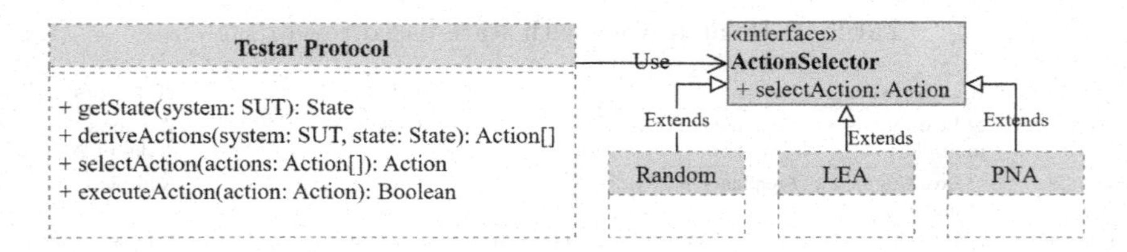

Fig. 1. Extending Testar with different ASMs.

Similar to AutoBlackTest, Pan et al. [15] give a large reward when a very different state is reached. They described the JBS problem that arises with this reward: if two states are too different, the agent will continuously receive high rewards, and consequently, the testing tool might frequently jump between them. As mentioned in the introduction, to address this problem, Pan et al. used a neural network to compare the reached state with the previously visited states.

Degott et al. [8] build a general model with the goal of sharing models between different apps. Their reward is either 1 or 0, according to visual changes.

We propose a different and simpler solution to the JBS problem by incorporating the memory based on the frequency of executed actions to the reward. For this, we keep track of the visited states and the executed actions. An additional novelty of our proposal resides in the use of RL for exploring web-based applications. Little work has been done on RL-based web exploratory testing without prior knowledge, with most works focusing on Android or Windows applications.

Our solution is built on top of Testar as it is a tool that can test Web applications. Moreover, Testar's action detection and state abstraction capabilities allow the generation of a state-model that can be used as the memory that keeps track of the visited states, needed to solve the JBS problem.

4 The TESTAR Tool

TESTAR[1] [21] is an open-source tool for scriptless testing. It does on-the-fly testing, meaning it generates test sequences consisting of (state, action)-pairs that start in the initial state of the SUT. After starting the SUT, Testar goes into a loop of selecting and executing actions to move the SUT from one state to another. A protocol (see Fig. 1) is responsible for executing the different parts of the test sequence generating the main thread. In each iteration of the Testar loop, the state of the GUI (i.e. its widgets and their properties) is obtained with (`getState`) and used to derive a set of possible actions with (`deriveActions`). One of these actions is selected by (`selectAction`) that is subsequently executed by (`executeAction`). The action selection determines the exploration effectiveness of the testing. After the selected action is executed and the SUT goes to a new state, oracles will check whether this state is valid. Testar continues the generation of test sequences until some predefined stopping criteria are met.

[1] https://testar.org/.

States and Widget Trees. In each state, GUI widgets are structured in a widget tree. The tree nodes represent the widgets and contain a record of their properties (e.g. type, position or title). The edges represent the parent-child relationship, as every child widget is displayed within the screen area corresponding to its parent widget. Given a widget tree from a specific state s, we define the set of nodes of the *widget tree* that integrate the GUI in that particular state s as $W(s) = \{w_1, w_2, \ldots, w_k\}$. The properties that can be associated with the widgets will be denoted by WPROP. For a widget $w \in W(s)$ and widget property p, we denote the value of that property p in state s by: $w.p(s)$. The values of all properties WPROP of all the widgets is the *concrete state.*

Action Derivation. Using the current state s, available actions are derived by `deriveActions`. For this, first, the *actionable widgets* are detected. To derive the actions that can be executed in a state, Testar loops through the widget tree and collects those actionable widgets.

Action Selection. Action Selection Mechanisms (ASM) can be added by implementing the `ActionSelector` interface (see Fig. 1). The most used ASM is *Random* (*RND*) [14,16]. It arbitrarily selects one action out of all possible actions in the current state. Others are *Least executed actions* (LEA) [5], selecting the least explored actions from the current state, or *Prioritize new actions* (PNA) [21], selecting an action in the current state that is not in the previous state.

State Abstraction. State abstraction is an important facet of scriptless GUI testing. Testar calculates state identifiers based on hashes over a selected subset of widget properties. This selected set, ABS_PROP, defines the abstraction level (ABS_PROP $\subseteq$ WPROP). The abstraction level determines the number of different states Testar can distinguish and the number of actions that can be derived. To illustrate this concept of abstraction, concrete actions can be "Press key 'q' " or "Press key 'w"', while both actions are represented abstractly as "Press key". Hence, certain actions may be considered equivalent and can be executed interchangeably. In the case of pressing a key, the specific key that is pressed may not be important at a high level of abstraction. Similarly, an abstract state depends on the attributes selected from each widget. ABS_PROP is configurable in the Testar's settings.

5 Approach

This section outlines our method for integrating Q-Learning into Testar. The SUT is the environment and states are determined by Testar along with all available actions that can be executed. Initially, Testar has no knowledge of the SUT, but as the tool learns to select the most optimal action at each step, it updates its knowledge to find the best policy. These actions generate test sequences. For the 4-tuple MDP $M = \langle S, A, T, R \rangle$:

- a state $s \in S$ is represented as the set $(w_1, ..., w_n)$ of widgets that together constitute the widget tree.
- an action $a \in A$ is represented as a 2-dimensional: (*action type*, *widget*).
- the execution of the SUT causes the transition T: Testar executes an action and observes the new state.

- each time Testar executes an action a in state s that results in state s', a reward $R(s, a, s')$ is calculated.

5.1 Rewarding Test Behaviours

To apply RL to automated GUI testing using Testar and define smart exploration strategies, rewards will need to be tailored towards the goal of improved testing. In the following, we will define 4 types of state-based rewards: frequency, state-change, state and combined.

Frequency-based Rewards. This reward is based on the previous work described by Vos et al. [21], where actions with low execution count are favoured (see 2). The reward function is inversely proportional to the number of times $ec(a, s)$ the action a has been executed in state s. The R_{max} parameter determines the initial reward assigned to unexplored actions. High values of R_{max} might bias the search towards executing new actions.

$$R_{frequency}(s, a, s') = \left\{ \begin{array}{ll} R_{max}, & \text{if } ec(a, s) = 0 \\ 1/ec(a, s) & \text{otherwise} \end{array} \right\} \tag{2}$$

Rewarding State Changes. By simply observing the interface of the state, it is possible to observe the changes, similar to how the user will experience the exploration of the SUT. Two screenshots corresponding to the previous and the current state are obtained and scanned pixel by pixel to compare their RGB value. The reward consists of calculating the ratio between the total number of different pixels $dp(s, s')$ and the total of pixels tp. If all pixels are equal, no observable state change is detected; hence the reward is 0. Otherwise, if all pixels are different, the maximum reward of 1 is returned.

$$R_{state-change}(s, a, s') = \frac{dp(s, s')}{tp}, \tag{3}$$

Rewarding Reached State. When tuning the RL parameters, we encountered a problem described by Pan et al. [15]. An agent is rewarded with large values after finding very different states, which might result in constantly jumping between them. The problem remains even if the *State Changes* rewards are combined with frequency-based ones since once most of the states/actions are visited/executed several times, we face the initial situation again. To solve this problem, in [15], a memory buffer was proposed, where the reached state is compared with a set of previously visited states instead of with only the immediately previous state. To solve this problem, our approach uses Testar's state model to keep a memory of which actions have been executed on every abstract state.

Initially, a single reward is calculated according to the level of exploration of the reached state, as shown in 4. The value is 1 (maximum possible) if none of the available actions has been executed, i.e. the state has not been explored yet. On the other hand, as the actions are executed, the reward decreases until it reaches the minimum value of 0. The main advantage of this reward is that it is independent of the previously visited states, acting as a pure measure of how *useful* the current state is. We hypothesise that this reward will have less incidence of the JBS problem.

$$R_{state}(s, a, s') = \frac{\sum_{a' \in A(s')} [ec(a') = 0]}{||A(s')||} \tag{4}$$

Algorithm 2. `selectAction` in class `QLearningLActionSelector`

Require: $RewardFunction$
Require: $QFunction$
Require: $Policy$
Require: s, a, s' ▷ Previous State, Executed Action, Current State
1: $reward \leftarrow RewardFunction.getReward(s, a, s')$
2: $q \leftarrow QFunction.getQValue(s, a, reward)$
3: $updateQValue(s, a, q)$
4: $a \leftarrow Policy.applyPolicy(s')$
5: return a

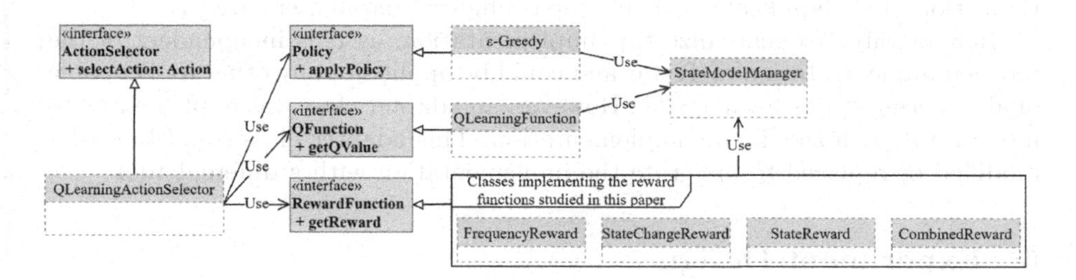

Fig. 2. `QLearningActionSelector` implements `selectAction` with Q-Learning.

Combining Rewards. We propose a final reward that combines all previous rewards in order to address the JBS problem and provide the agent with more information on how to effectively explore the state space of the SUT. However, the weights of this reward may need to be adjusted for each specific SUT. For the sake of simplicity, the reward function weights were assigned equal values.

$$R_{combined}(s, a, s') = w_1 * R_{frequency} + w_2 * R_{state-change} + w_3 * R_{state} \tag{5}$$

5.2 RL Framework

The interface-based architecture results in a modular and maintainable framework since the RL functionality is independent of the rest of the tool. The framework consists of an implementation of the `ActionSelector` interface (from Sect. 4) called `QLearningActionSelector` (see Fig. 2). This class implements `selectAction` and is in charge of selecting the following action using Q-Learning. The `QLearningActionSelector` class uses three main interfaces:

- `RewardFunction` is an abstraction for reward functions. In Fig. 2, the 4 implementations of the reward functions from Sect. 5.1 are shown.
- `Policy` is an abstraction for possible policies. For this study, we implemented ϵ-Greedy, which is a strategy that defines a probability ϵ of exploration.
- `QLearningFunction` is implemented to calculate Formula 1.

Algorithm 2 implements `selectAction` using the interfaces described earlier. Lines 1 and 2 in Algorithm 2 correspond to lines 8 and 9 in Algorithm 1, respectively. Once the Q value is computed, it is stored in the Testar State Model (see to Sect. 4) as a property of the action, since every action is represented as an outgoing transition of a state. Therefore, the Q values are associated with both the abstract representation of the state and the action. Finally, the `QLearningActionSelector` uses a policy to select the new action.

Configuration of a specific Q-Learning set-up is done through Testar's configuration file (`test.settings`). This is used to configure the reward function, policy and Q-function that the ASM will use and to define the parameters such as ϵ for the policy and δ and α for the Q-learning algorithm (1). Factories are used to select and initialise the applicable implementation class. During the initialization of the applicable classes, the configured parameters are set.

It is possible to generalize this implementation, as it is independent of the internal implementation of Testar and could be applied to any other tool capable of abstracting states from a SUT. However, an adapter class was implemented to integrate it with the Testar implementation. This adapter class could be easily modified or replaced to integrate the implementation with a different tool.

6 Experiment Design

We want to investigate the exploration effectiveness of the different rewards. Thus, we have formulated the following research questions:

RQ1: Which of the ASMs are most effective in exploring the SUTs?

RQ2: Do $R_{state-change}$ have a higher incidence of the JBS problem in comparison with $R_{frequency}$, R_{state} or $R_{combined}$?

Our experiment, based on guidelines presented in [4,25], aims to answer the research questions by analyzing the exploration effectiveness of four different rewards compared to random action selection. Additionally, we evaluate the effectiveness of frequency-based and state-based rewards as a memory-based solution for the JBS problem. We formulate null hypotheses to enable statistical analysis of the experiment.

$H0^1$: The evaluated ASMs do not show a statistical difference in exploration compared to each other.

$H0^2$: JBS problem has not higher incidence with $R_{state-change}$ than with R_{state}, $R_{combined}$ or $R_{frequency}$.

In this experiment, we will use the general design principle of blocking. This means we will block the testing capacity to detect errors, which could otherwise interrupt the test runs when discovering such faults. Consequently, oracles (that define the errors and exceptions that Testar will check) will not be used during this experiment because finding an error or exception will interrupt the test sequence, and the goal of this experiment is exploration. To deal with the randomness of Testar ASMs, we will repeat all test runs of our experiment 20 times using concurrent Virtual Machines (VMs) running tests. Each virtual Windows machine is configured with a 4.5 GHz CPU and 16 GB RAM.

6.1 Objects: Selection of SUTs

The SUTs selected for the experiment should comply with the following: 1) The SUTs have a GUI; 2) Testar is able to detect the widgets on the GUI of the SUTs; 3) The SUT contains a high difference between consecutive states to increase the probability of encounter loops. Since we focused on web exploration, the following three SUTs were selected.

Shopizer is an e-commerce sales management software that allows the creation of online stores, marketplaces or product listings. The home page consists of a search form, an item menu and a banner. Shopizer is an open-source website containing 126 Java packages for a total of 410 Java classes and 23330 lines of code. Shopizer was selected as a demo application for the initial experiments of this work. To increase the search space, 10 thousand fake products were added to 6 categories and 60 subcategories. To access the products, complex actions such as pagination and searching are required.

Craiglist is a classified advertisement website with more than 80 million new classifieds each month. Similar to Shopizer, Craigslist divides the products into multiple categories and subcategories. However, the product listing view is more complex: each category has specific search options for refining the displayed product list. During the execution of the experiments, a total of 105 search options were observed among all the categories.

Bol.com is a webshop with many different products. It was selected due to its similarity with both Shopizer and Craigslist. Nevertheless, Bol.com provides a more complicated interface: extensive sequences of complex actions are required to unblock certain areas of the application. Moreover, small images of the products are always displayed, making the comparison based on screenshots between states more sensitive. Furthermore, the home page consists of the more recent products visited by the user, adding extra dynamism to the website. Finally, as with Craigslist, specific search options are provided for every category.

6.2 Studying Rewards: Independent Variables and Factors

We want to compare ASMs based on the different rewards from Sect. 5.1 with random ASM. This means, the factors are the rewards, and all other independent variables are kept constant:

- State abstraction. The abstract state representation affects how Testar detects widgets. In this work $ABS_{PROP} = \{WidgetID, WidgetTextContext\}$. This abstraction was selected after several trials.

- Action derivation. The widget associated with the action is always part of the action's abstract representation. To differentiate actions originating from the same widget, we add the role (e.g. click or type) of the action to its representation.
- Filters. Different parts of the web applications were filtered out for every SUT, such as payment checkouts or registration, because they require specific sensitive information to work correctly.

Other independent variables for the experiment are constant values, such as 1 s for action execution and a minimum waiting time of 1 s between executed actions. A test-run consists of 300 sequences of 100 actions each. ϵ-Greedy was selected as the policy probability of exploration, with $\alpha = 1$, $\gamma = 0.7$ as the Q-learning parameters (based on the related work).

6.3 Dependent Variables

In order to answer RQ1, we should be able to measure exploration performance for every ASM. While the SUT is explored, new states and actions will be discovered and/or visited by the RL agent. To measure each RL algorithm's exploration performance in terms of state and action space size, we extract the available information from Testar's state model.

Even though code coverage is a good indicator of the exploration effectiveness of a testing tool, in the case of real web applications, this is not always available. Alternatively, we counted the number of unique URLs visited on each website.

The JBS problem occurs when two states are too different from each other. A state that appears often in short sequences is probably very different from the consecutive reachable states. It is possible to obtain a path $S_1, S_2, ..., S_n$ from every test sequence, such as S_i is the abstract state visited after executing action a_{i-1}. A loop in a path means that certain abstract state was revisited. We can define a *jumping state* as a state that appears in several loops, and, intuitively, this is an indicator of the presence of the JBS problem. When a test execution is finished, every loop found in every path is extracted. If a state s is the initial and final state of the loop, the length l_i of the loop is associated with that state. Thus, we obtain a tuple (s, l_i) for every loop in the test sequence.

The dependent variables measured to answer the RQs are: number of different abstract states visited, number of different abstract actions executed, number of different abstract actions discovered, average number of distinct URLs visited and number of pairs of (S_i, l_i) for every test sequence.

7 Results

7.1 RQ1: Exploration Effectiveness

To test the different ASMs' ability to explore a web application, we monitored the URLs visited during execution. Additionally, Testar's state model provides information about states and actions that have been discovered or visited.

We counted the number of distinct abstract states visited, different actions executed and unvisited actions, for each application. An action is considered unvisited when derived by Testar but has yet to be visited. Table 2 shows the average values per SUT and ASM, while Figs. 3, 4 and 5 show the results for URL performance and space-related variables. We conducted pairwise comparisons between every ASM according to established guidelines [4] and measured the effect size in each case. Table 3 presents a summary of our findings after applying the Mann-Whitney-U test to compare the exploration of abstract states.

In Fig. 3, we observe high variability of Random in every dependent variable for Shopizer, with executions showing great results but also the worst. Shopizer always lists the same products in the same order, making it challenging to browse different items. The pagination is based on a "load more button". Consequently, new actions appear only if that button is clicked. On the contrary, RL ASMs had less variability and obtained better results. Especially, R_{state} visited a larger amount of new abstract actions while also discovering more unvisited actions than any other ASM. The statistical results shown in table 3 confirm that R_{state} obtained the best performance. Also $R_{frequency}$ outperforms $R_{combined}$.

Table 2. Average values of dependent variables for every SUT

ASM	$R_{combined}$	Random	$R_{frequency}$	R_{state}	$R_{state-change}$
	Shopizer				
Abstract States (mean)	291.50	296.00	312.50	321.95	290.33
Abstract Actions (mean)	1097.85	1109.20	1146.17	1177.65	1065.67
Unvisited Actions (mean)	5135.65	4684.85	5551.56	5687.40	5180.39
URL (mean)	115.80	115.05	122.67	125.40	114.56
	Craigslist				
Abstract States (mean)	1010.30	1022.05	991.14	959.20	1003.26
Abstract Actions (mean)	2234.65	2464.25	2226	2229.70	2223.11
Unvisited Actions (mean)	35625.15	25781.40	37752	38824.10	36861.42
URL (mean)	1028.55	1031.70	963.10	946.50	1049.63
	Bol.com				
Abstract States (mean)	813.10	701.60	803.63	733.70	722.4
Abstract Actions (mean)	1544.47	1365.80	1546.42	1443.80	1422.70
Unvisited Actions (mean)	17445.26	14402.2	17455.95	15686.3	14748.5
URL (mean)	217.53	222.10	227.42	224.85	199.75

Table 3. p-values calculated pairwise for Abstract States. When $p < 0.05$, effect size is calculated with Cliff's delta.

	Shopizer				Craigslist				Bol.com			
ASM	*RC*	RND	*RF*	*RS*	*RC*	RND	*RF*	*RS*	*RC*	RND	*RF*	*RS*
RND	0.61	-	-	-	0.72	-	-	-	0.01 (l)	-	-	-
RF	*p*<0.01 (l)	0.07	-	-	0.33	0.02 (m)	-	-	0.20	0.18	-	-
RS	*p*<0.01 (l)	0.01 (l)	0.43	-	0.18	0.03 (l)	0.39	-	0.03 (m)	0.21	0.49	-
RSC	0.60	0.94	*p*<0.01 (l)	*p*<0.01 (l)	0.55	0.53	0.35	0.22	0.01 (l)	0.52	0.36	0.54

$RND = Random$, $RF = R_{frequency}$, $RC = R_{combined}$, $RS = R_{state}$, $RSC = R_{state-change}$
effect size $l = large$, $m = medium$

Figure 4 shows that Random ASM visits more unique abstract actions and executes more unique actions for Craigslist. Table 3 indicates that there is a significant statistical difference between Random and R_{state} or $R_{frequency}$ with regards to Abstract State exploration. This may be due to Craigslist's multiple search options available in most states, resulting in a vast set of possible actions to execute. Conversely, RL ASMs repeat many actions already executed to learn from the experience. Further research is needed to improve abstraction to handle multiple search options or increase the testing time to train the RL agent.

With the purpose of verifying if this is the case for Craigslist, we executed one run of 100 test sequences of 100 actions each, with Random and $R_{combined}$. Table 4 shows the results for state exploration. Both ASMs reached a similar number of abstract states, while $R_{combined}$ visited considerably fewer concrete states than Random, as a sign of better exploration. The model generated can be used in future test sessions or in different versions of the same SUT. Theoretically, the RL agents will need less execution time to reach the same exploration level.

Regarding Bol.com, we observe in Fig. 5 that the RL approaches performed better than Random in general. In particular, $R_{combined}$ outperformed R_{state} and $R_{state-change}$ in terms of search space exploration. Table 3 shows that there is a statistical difference at a significant level for the exploration of new abstract states. However, in terms of URL coverage, there are no significant differences. Observing the extracted URLs during the executions, we detected that most URLs contain multiple search parameters. Since this is an e-commerce site, new

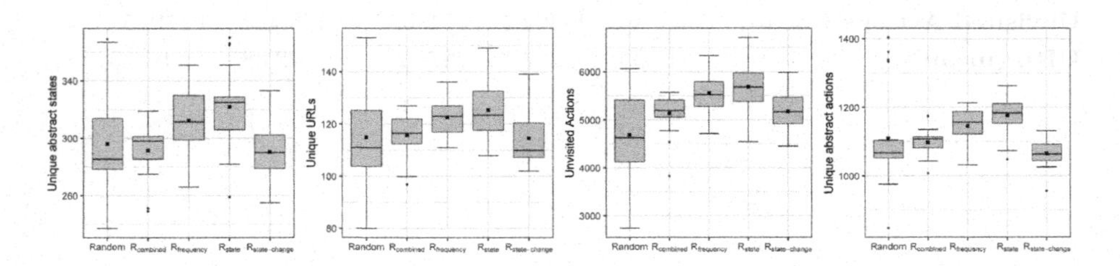

Fig. 3. Exploration performance of Shopizer

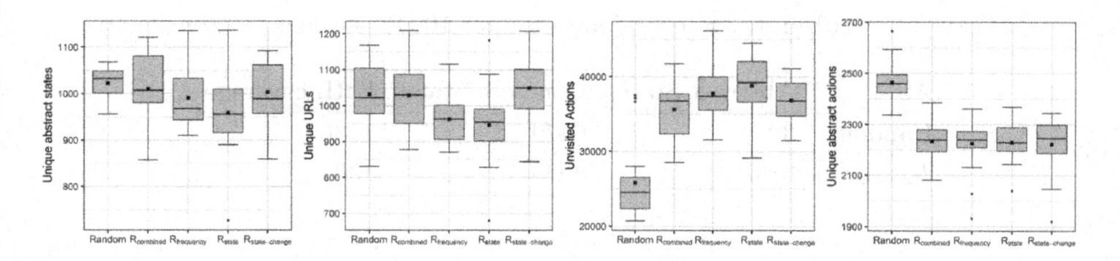

Fig. 4. Exploration performance of Craigslist

URLs are obtained after accessing a product view or by refining the search of products on the listing views. Normalization of URLs applied to this domain should be done to facilitate a better understanding of this metric.

7.2 RQ2: JBS Problem

Intuitively, jumping between states happens when states appear multiple times with short sequences of actions between each occurrence. However, the length and frequency required to classify a state as a jumping state are hard to predict for every SUTs. To analyze the distribution of jumping states, we extracted loops, focusing on the initial state and loop length. Figure 6 shows the relation between the number of times a state starts a loop and the average size of those loops. The pattern is consistent across ASMs. The interesting sections rely on the bottom right of every charts: states (represented as red dots) with many loops of small size.

Shopizer contains a more significant presence of jumping states, and in general, all the loops are of small size. There is no noticeable difference between the rewards. This can be explained by the nature of Shopizer, with smaller state space and simple navigability to access the main parts of the SUT.

For Craigslist and Bol.com, $R_{state-change}$ seems to have a higher accumulation of points in this area, while R_{state} contains the fewest states. The difference is more obvious for Bol.com. This result is expected: R_{state} is a reward based solely on the reached state, not depending on any characteristic of the consecutive states. Moreover,$R_{frequency}$ tends to have fewer loops per test sequence in Bol.com, with the states being visited fewer times. Additionally, the graphs

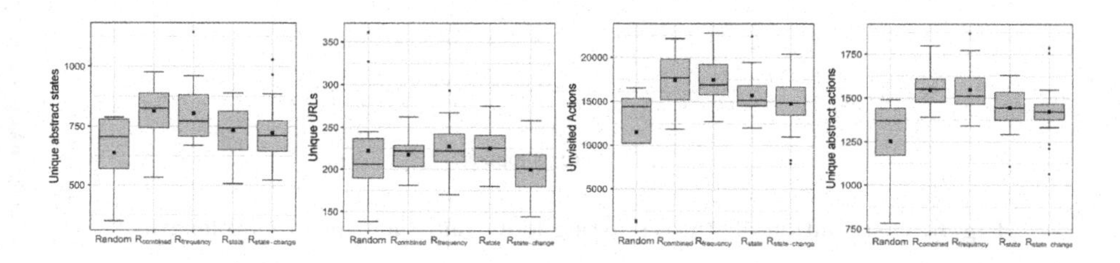

Fig. 5. Exploration performance of Bol.com

Table 4. State exploration after 10000 actions

ASM	Abstract States	Concrete States	URL coverage
Random	3274	6479	1136
$R_{combined}$	3166	4576	1301

show fewer loops for Bol.com. This could be an important factor in explaining the better performance of the rewards in this web application.

In large web applications such as Craigslist or Bol.com, the reward influences the frequency of jumping states. To further evaluate the incidence of the JBS problem on every ASM, we statistically analysed the number of jumping states using the Mann-Whitney U test. The goal is to measure if there is any statistical difference between the reward $R_{state-change}$ and the other 3 rewards, respectively. Table 5 summarises the p-values calculated for this comparison for Craigslist and Bol.com. $R_{state-change}$ is significantly different from $Rstate$ and $R_{frequency}$ ($p-value < 0.05$). $R_{state-change}$ tends to visit the same states more frequently with fewer actions in between. The hypothesis for *Rcombined* in Craigslist could not be rejected. We hypothesise that the weight parameter gives too much influence to the state-change reward within the combined reward.

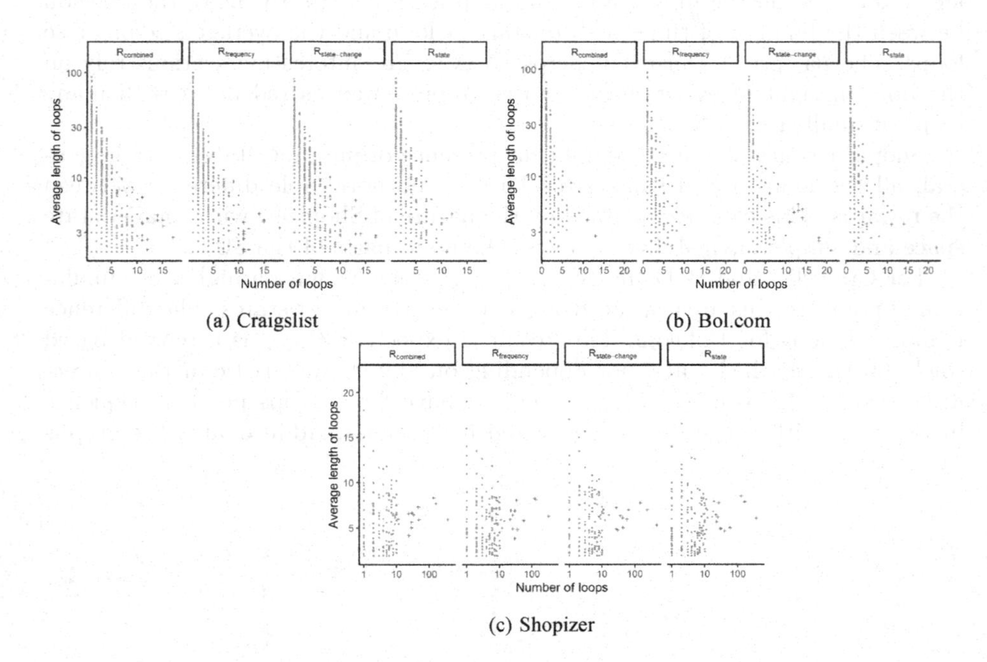

(a) Craigslist (b) Bol.com

(c) Shopizer

Fig. 6. Distribution of states starting loops and their mean loop length

Table 5. p-values calculated between every reward and $R_{state-change}$

	$R_{combined}$	$R_{frequency}$	R_{state}
Craigslist	0.48	$p < 0.001$	$p < 0.001$
Bol.com	0.03	$p < 0.001$	0.03

Several factors might affect the incidence of the JBS problem. Web applications with a high probability of accessing previously visited states can present this problem regardless of the selected reward. This can be observed, for example, when the state space is small, when there is a bottleneck to access other unvisited parts of the SUT or when there is high inter-connectivity between the main states. This is the case with Shopizer, where new pages are only accessible through a button that is not always visible, and the state space is small.

8 Conclusions and Future Work

We presented a Q-learning approach for automated GUI testing that utilizes rewards based on the state model of web applications. This is an understudied yet valuable area for understanding the behavior of an agent when executing actions to attain rewards. Our approach offers a unique advantage in this domain.

We conducted an experiment on 3 complex web applications to evaluate the performance of 4 rewards, using random as a baseline for comparison. The results showed that frequency, state, and combined rewards had the best performance for state exploration. The state-change reward, however, was outperformed by or did not improve upon the results of the other rewards. Additionally, we proposed a solution for the Jumping Between States (JBS) problem by incorporating memory information into the reward. The results demonstrated that rewards based solely on the reached state effectively circumvent the JBS problem. However, in domains with high connectivity between states, where most pages are accessible within a few clicks, the JBS problem has a higher incidence, regardless of the reward. Nevertheless, state or frequency rewards performed better in web applications with complex navigability.

Future work to improve the results includes increasing the number of SUTs and their diversity, as well as developing a suitable abstraction of states to tackle the limitation of multiple search options that makes exploration difficult. Additionally, a solution for the JBS problem will be studied by implementing a memory buffer similar to previous studies, with the appropriate size yet to be determined. The state models generated after executions can also be used as prior knowledge for other versions of the same SUT or combined with state models generated by other SUTs at a higher abstraction level.

References

1. The (google) monkey (2010). https://developer.android.com/studio/test/monkey.html. Accessed 27 Jan 2023
2. Adamo, D., Khan, M.K., Koppula, S., Bryce, R.: Reinforcement learning for android GUI testing. In: 9th ACM A-TEST Workshop (2018)
3. Alégroth, E., Feldt, R., Kolström, P.: Maintenance of automated test suites in industry: an empirical study on visual GUI testing. IST **73**, 66–80 (2016)
4. Arcuri, A., Briand, L.: A practical guide for using statistical tests to assess randomized algorithms in software engineering. In: 33rd ICSE. ACM (2011)
5. Borges, N., Hotzkow, J., Zeller, A.: Droidmate-2: a platform for android test generation. In: 33rd ACM/IEEE ASE, pp. 916–919 (2018)
6. Collins, E., Neto, A., Vincenzi, A., Maldonado, J.: Deep reinforcement learning based android application GUI testing. In: SBES, SBES 2021. ACM (2021)
7. Coppola, R., Ardito, L., Torchiano, M., Morisio, M.: Mobile testing: new challenges and perceived difficulties from developers of the italian industry. IT PROFESSIONAL (2019)
8. Degott, C., Borges, Jr., N.P., Zeller, A.: Learning user interface element interactions. In: 28th ACM SIGSOFT, ISSTA 2019, pp. 296–306. ACM (2019)
9. Grechanik, M., Xie, Q., Fu, C.: Creating GUI testing tools using accessibility technologies. In: ICST Workshops, pp. 243–250. IEEE (2009)
10. Gu, T., et al.: Aimdroid: activity-insulated multi-level automated testing for android applications. In: ICSME (2017)
11. Jabbarvand, R., Lin, J.-W., Malek, S.: Search-based energy testing of android. In: IEEE/ACM 41st International Conference on Software Engineering (ICSE)
12. Koroglu, Y., et al.: QBE: qlearning-based exploration of android applications. In: ICST (2018)
13. Mariani, L., Pezze, M., Riganelli, O., Santoro, M.: Autoblacktest: automatic black-box testing of interactive applications. In: 5th ICST (2012)
14. Nyman, N.: Using monkey test tools - how to find bugs cost-effectively through random testing. Software Testing & Quality Engineering, Jan/Feb:18–21 (2000)
15. Pan, M., Huang, A., Wang, G., Zhang, T., Li, X.: Reinforcement learning based curiosity-driven testing of android applications. In: 29th SIGSOFT. ASM (2020)
16. Patel, P., Srinivasan, G., Rahaman, S., Neamtiu, I.: On the effectiveness of random testing for android. In A-TEST (2018)
17. Robbins, H.: Some aspects of the sequential design of experiments. Bull. Am. Math. Soc. **58**(5), 527–535 (1952)
18. Rodríguez-Valdés, O., Vos, T.E.J., Aho, P., Marín, B.: 30 years of automated GUI testing: a bibliometric analysis. In: Paiva, A.C.R., Cavalli, A.R., Ventura Martins, P., Pérez-Castillo, R. (eds.) QUATIC 2021. CCIS, vol. 1439, pp. 473–488. Springer, Cham (2021). https://doi.org/10.1007/978-3-030-85347-1_34
19. Romdhana, A., Merlo, A., Ceccato, M., Tonella, P.: Deep reinforcement learning for black-box testing of android apps (2021). arXiv preprint arXiv:2101.02636
20. Sutton, R.S., Barto, A.G.: Introduction to Reinforcement Learning, 1st edn. MIT Press, Cambridge (1998)
21. Vos, T.E., Aho, P., Pastor Ricos, F., Rodriguez-Valdes, O., Mulders, A.: TESTAR - scriptless testing through graphical user interface. STVR **31**(3), e1771 (2021)
22. Vuong, T., Takada, S.: A reinforcement learning based approach to automated testing of android applications. In: 9th ACM A-TEST Workshop (2018)
23. Watkins, C.J., Dayan, P.: Q-learning. Mach. Learn. **8**(3), 279–292 (1992)

24. Wiering, M.A., Van Otterlo, M.: Reinforcement learning. Adapt. Learn. Optim. **12**(3), 729 (2012)
25. Wohlin, C., Runeson, P., Höst, M., Ohlsson, M.C., Regnell, B., Wesslén, A.: Experimentation in Software Engineering. Springer Science & Business Media, New York (2012). https://doi.org/10.1007/978-1-4615-4625-2

Conceptual Modeling and Semantic Networks

DBSpark: A System for Natural Language to SPARQL Translation

Laura-Maria Cornei(✉) and Diana Trandabat

Faculty of Computer Science, Alexandru Ioan Cuza University, Iaşi, Romania
cornei.laura10@gmail.com

Abstract. Knowledge bases offer clear advantages when compared to traditional databases, mainly due to semantic connections and automated reasoning over large datasets. However, limited knowledge of the specialized knowledge base query language (SPARQL) makes it difficult for most users to freely access these resources. To solve this issue, we propose a question-answering system able to translate natural language questions into SPARQL queries. The presented method is a rule-based approach that integrates information regarding dependency and constituency parsing, WordNet and named entity recognition to capture the structural and semantic representation of the question. The proposed solution is able to handle a wide variety of question types (list, count, yes/no, wh-questions, questions involving rankings, ordinals, and/or superlatives). Moreover, all involved steps except the phrase mapping phase (in which properties and entities from the ontological model are mapped to words from the natural language question) are knowledge base independent. Tests performed over the QALD-9 question-answering dataset using the DBpedia knowledge base have shown that our system obtains state-of-the-art results and a very good time-performance balance.

Keywords: Question-answering system · Rule-based system · Semantic Web

1 Introduction

The semantic web vision [1] has impacted the way we view data by introducing the concept of connecting it in order to convey meaning. Along with the linked data principles, the emergence of the open data initiative has enabled us to freely access, use and share numerous datasets structured in form of knowledge graphs. One can perform various interrogations over these knowledge bases by using the SPARQL query language. However, to make knowledge graphs available to a larger category of users, we need to provide a more intuitive way of interrogating them. One popular solution is to directly use natural language. Thus, the discovery of a method to translate natural language questions to SPARQL queries, that could be fairly easily adapted to any knowledge base, would unlock the full potential of (open) connected datasets.

Although researchers have developed throughout years many question answering systems that enable NL to SPARQL translation, there is still progress to be made in order to narrow the gap between the way the user formulates the questions and the way

S. Nurcan et al. (Eds.): RCIS 2023, LNBIP 476, pp. 157–170, 2023.
https://doi.org/10.1007/978-3-031-33080-3_10

the domain knowledge is structured. Furthermore, QA systems could be built or adapted to yield better results on knowledge bases that cover multiple domains, to handle more complex questions, to require minimal configurations when ported to new knowledge bases and to ensure a better user experience. Those improvements can significantly impact the development of applications in the semantic web field, nevertheless, none of them represents an easy task.

The problem of translating NL to SPARQL can be divided in smaller steps:

- *Processing and analyzing the given question.* The question processing step usually involves basic operations such as lemmatization, parts of speech tagging, building the constituency and/or the dependency parse tree associated to the question, performing named entity recognition. The question analysis step includes more complex subtasks (for e.g., question type and question focus identification), such that this information can be further used in the SPARQL query generation step. The analysis is often realized using either rule-based (utilized in Panto [2], Parot [3]) or machine learning approaches (utilized in QASparql [4], AutoSPARQL [5], COmQA [6]), both techniques presenting certain drawbacks. The rule-based method may be more prone to overlook certain cases, while the ML methods require relatively large and high-quality corpora in order to yield good results and might need a long training time.
- *Mapping between ontology concepts and question terms.* This mapping is usually done with the help of the dictionaries and phrase mapping systems. In some cases (onIQ [7], FREyA [8]), similarity distances (e.g., Jaccard, Levenshtein) are used. WordNet[1] is frequently utilized because it not only offers synonyms, but also definitions and the possibility to identify more complex relationships between words (meronymy, hyponymy, hypernymy, antonymy, etc.). Some QA systems (e.g., QASparql) utilize phrase mappers in order to associate concepts from well-known ontologies (DBpedia[2], Wikidata[3]) to question terms.
- *Generating the SPARQL query.* One of the most difficult steps when building the SPARQL query is to determine the triple patterns that must be included inside the WHERE clause. Most state-of-the-art techniques for identifying the triple patterns involve using heuristic algorithms that take into account information extracted from the question, the ontology structure and the mapping between ontology entities and question terms. Potential modifiers in the query structure are also usually identified using heuristic techniques. The results from the question focus identification subtask can be used to choose what variables to add after the SELECT clause. Because there are usually multiple queries that can be built using these algorithms, the resulted queries need to be ranked and the query with the highest score will be the final output.

In this paper, we propose a question-answering system named DBSpark that is able to perform translations from natural language to SPARQL. The presented system offers support for a wide variety of question types (list, count, yes/no, wh-questions, questions involving rankings, ordinals, and/or superlatives). It also ensures an increased portability

[1] https://wordnet.princeton.edu/documentation/.
[2] https://www.dbpedia.org/.
[3] https://www.wikidata.org/wiki/Wikidata:Main_Page.

as almost all of its components are knowledge base independent. Experimental results show that DBSpark obtains higher scores for precision, recall and F1 measure on the QALD-9[4] dataset when compared to two state-of-the-art systems.

The rest of the paper is organized as follows: Sect. 2 reviews related work in the context of systems performing translation from NL to SPARQL, Sect. 3 describes the proposed system's architecture and functionalities in detail, Sect. 4 analyses the obtained evaluation results and Sect. 5 discusses future improvements and states the conclusions.

2 Related Work

Below are detailed the used methods, the findings and the evaluation results for 9 systems that utilize natural language to SPARQL conversion.

Panto [2] is a natural language interface that accepts generic natural language questions and outputs the corresponding SPARQL queries. In order to build the SPARQL query, Panto generates the syntactic parse tree of the question and extracts pairs of nominal phrases (which will become part of future query triples) using an algorithm proposed by the authors. These nominal phrases are further mapped to ontology entities by performing queries over the knowledge base. Using the parse tree along with the knowledge base, the entities in the ontology are connected forming SPARQL triples. Lastly, modifiers are extracted from the parse tree and are combined with the SPARQL triples, forming the SPARQL query. On DBpedia's Geography dataset, the tool's precision was 88% and the recall was 85.9%.

QuestIO [9] is a natural language query interface converting NL questions into SPARQL queries. It contains a mapping component that associates the entities in the ontological model with the concepts identified in the question. WordNet and similarity metrics are used in order to ensure the mapping is performed correctly. The relations between concepts are extracted by performing queries over the knowledge base. In order to filter best relations, three different scores are used: a similarity score for assessing the similarity between the relation's name and a chunk of text from the question, a specificity score for determining how specific a relation is by measuring the depth from the ontology root concept to that relation and a distance score for determining the position of the domain and range classes of the property inside the ontology's hierarchy. SPARQL queries are formed and ranked by taking into account these scores. The highest-ranking query is outputted as a result.

FREyA [8] is a portable interactive Natural Language Interface for querying ontologies which uses disambiguation and mapping dialogues with the user, in order to enrich the domain lexicon from the user's vocabulary and solve ambiguities. It is developed as an improvement of QuestIO. In the disambiguation dialogue, the user is asked to map an ambiguous question term to one suggested ontology concept (the suggestion is made via a ranking algorithm). After the disambiguation step, a mapping dialogue follows, in which the user is asked to map question terms which could not have been previously identified as ontology concepts to a list of ontology concepts suggestions (chosen based on distance and similarity heuristics). The precision of testing on DBpedia dataset from

[4] https://qald.aksw.org/.

QALD-1 challenge (in which questions correctly answered after reformulation were not included/included) was 0.49/0.63, the recall was 0.42/0.54 and the f-measure score was 0.45/0.58.

AutoSPARQL [5] offers an interactive user interface for searching for resources and answering natural language questions. It uses Query Tree Learner, an active learning supervised machine learning algorithm, which uses positive and negative examples (RDF resources which should or should not be contained in the result set of the SPARQL query) in order to learn to generate the correct SPARQL query. The algorithm uses adapted operations from the Inductive Logic Programming field (least general generalization, negative based reduction) in order to operate on query trees, data structures which can be further converted into SPARQL queries. Authors provide time statistics for QALD-1 dataset, more exactly they highlight that a query can be learnt in approximately 7 s with a maximum amount of time of 77 s.

SPARKlis [10] is a SPARQL query builder system covering, in comparison to the previously mentioned systems, many SPARQL features such as union, negation, filtering, aggregations, ordering. Another important mention is the fact that the system builds the question through successive user choices rather than by receiving as input a natural language question. This limits the system's flexibility, however it enables it to handle complex SPARQL queries. The answer for the question is built at the same time with the question via a set of heuristic rules. The authors do not mention the performance metrics of the system, instead they highlight certain user metrics (for example, average number of steps for building a query).

ComQA [6] is a question answering system over knowledge bases. It includes a phrase mapping step in which entities from the ontology are mapped to concepts from the question using GloVe, a word vector representation algorithm. Multiple SPARQL candidate queries are built using information from the mappings and the dependency tree of the question. The final results are ranked taking into account the similarity between the dependency tree and the query structure. ComQA obtained on QALD-5 dataset a precision of 0.65, a recall of 0.39 and an F1-score of 0.4.

gAnswer [11] is a graph data-driven framework to answer natural language questions over RDF graphs. It obtains the SPARQL query in two stages: query understanding and query evaluation. In the query understanding phase a semantic query graph that captures the user's intention is built using the dependency tree structure of the question. During the query evaluation step, the semantic query graph is matched to RDF subgraphs from the ontological model and each RDF subgraph is assigned a score derived from the confidences of each edge and vertex mapping. The RDF subgraph with the highest score is then utilized in order to obtain the final SPARQL query. gAnswer obtained a precision of 0.293, a recall of 0.327 and a F1 score of 0.298 when evaluated on the QALD-9 dataset, being the winner [12] of the QALD-9 challenge.

Parot [3] is a tool for translating natural language to SPARQL, which supports features such as negation, opposing scalar adjectives and compound sentences. The system builds the dependency tree of the question and uses a set of heuristics based on this tree in order to map the question text to a SPARQL query. On QALD-9 dataset, authors stated that Parot obtained a precision of 0.43, a recall of 0.51 and the F-measure was approximately 0.47.

QASparql [4] is a system that translates natural language questions into SPARQL queries based on machine learning techniques. The first step that the system performs is building the dependency tree of the question. After that, the type of the question (list, yes/no, count) is identified using a Random Forest model. The mapping between ontology concepts and question terms is performed using predefined phrase mappers such as EARL [13], Falcon [14], TagMe [15] and DBpedia Spotlight [16]. This step is the only one that is knowledge base dependent because the above-mentioned phrase mappers work only for the DBpedia knowledge graph. The query generation is made by combining the resulting mapped entities and properties exhaustively. Multiple SPARQL queries are generated and the final query is chosen by ranking all queries based on the similarity between the query and the dependency tree structure using Tree-LSTM. On the QALD-7 dataset, QASparql obtained a precision of 0.81 for list questions, 0.83 for count questions and 0.8 for boolean questions. The recall was 0.59 for list questions, 0.71 for count questions and 0.13 for boolean questions.

Analyzing the presented applications, we concluded that there is still room for improvement in creating QA tools for the NL to SPARQL translation task. One major challenge in building such systems is simultaneously optimizing criteria such as: portability, correctness and generality of the approach, response time. Our system aims to come as a solution to this issue; it has the ability to handle a large variety of question types and achieve state of the art results, while ensuring an increased portability and a very good time-performance balance.

3 DBSpark's Architecture

Our proposed QA system has a modular design, being formed out of multiple components:

- The User Interface, which receives as input a question in natural language and displays the obtained formatted results for that question. Results are obtained by interrogating the SPARQL endpoint of the utilized knowledge base using the final selected SPARQL query.
- The question processing and analysis component, which extracts all useful information for generating the query from the question. The question type identification and the question focus identification subcomponents are part of it.
- The phrase mapping component, which is responsible for mapping question terms to entities and properties in the ontological model.
- The query construction and ranking component, in which multiple SPARQL queries are constructed based on the information collected from the previous two components, but only one final query is selected using a ranking mechanism.

Figure 1 depicts the architecture of the proposed application and its main components, which can be knowledge base dependent, knowledge base independent or external to the system.

The user submits a question in natural language via the interface and receives as output the formatted results (a sequence of one or multiple answers displayed in a table - for a list question, “yes” or”no” - in case of a yes/no question, a number - in case of a count question). For example, if the user introduces the question: "Which are the first four largest islands by population that belong to Japan?", he will receive as output four results: "Honshu, Kyushu, Hokkaido, Shikoku".

Next, we will discuss in detail the techniques used for performing each task and the functionalities offered by each component of DBSpark.

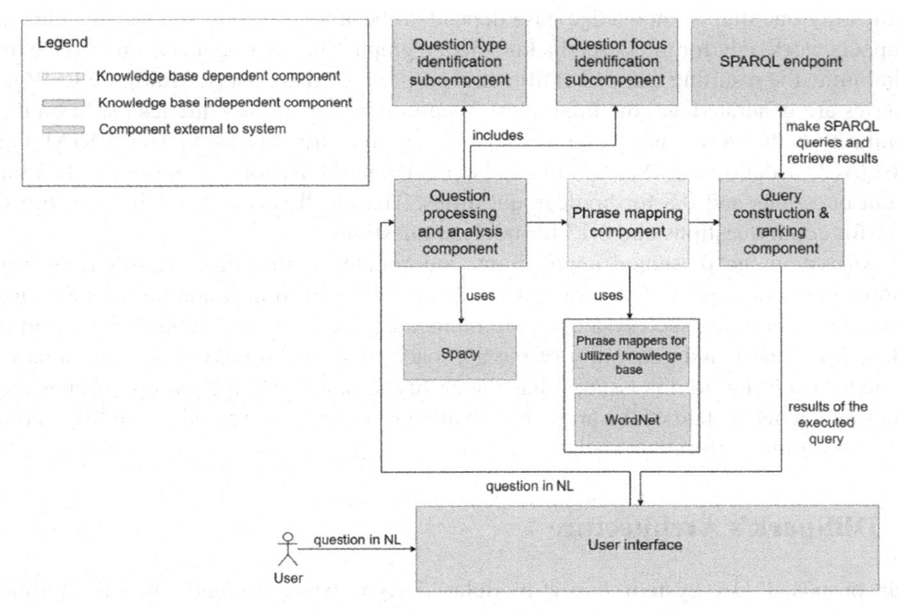

Fig. 1. DBSpark's architecture

3.1 Question Processing and Analysis Component

This component is responsible for processing and analyzing the given question, such that the information extracted from it can be further used to generate the corresponding SPARQL query.

The question processing step involved tokenization, lemmatization, POS tagging, building the constituency and the dependency trees and performing named entity recognition, all tasks being realized using the SpaCy[5] library.

The question analysis step consisted in identifying the question type and the question focus and extracting potential information concerning superlatives and rankings.

We categorized questions into three types: yes/no, list and count questions. Knowing the question's type, we could deduce the type of the SPARQL query (ASK query in case of a yes/no question, SELECT query in case of a list question, SELECT query using COUNT in case of a count question) and we could also infer the expected answer type (which was further used in the query ranking step).

[5] https://spacy.io/.

The question type was identified via a set of heuristic rules that took into account the words at the beginning of the question. If the question started with one form of the *be*, *do* or *have* verbs it was categorized as a yes/no question; if the question started with the adverb *how* followed by an adjective or an adverb indicating count or frequency (for e.g., many, much, often) it was classified as a count question; otherwise, it was typed as a list question.

The question focus, which captures what the question is about, was determined using rules over the dependency tree of the question. The question focus was chosen to be a noun who was in relation to the root.

Multiple nouns can be in relation to the root and there are also many types of relations in which the question focus can be with the root, for example *nsubj* ("What is the longest river in China?), *dobj* ("List top three cities by area in Japan"), *nsubjpass* ("Which movies were created by Walt Disney?), *attr* ("Who are the children of Michael Jackson?), etc.

In order to identify the question focus correctly, we took into account information regarding the question type (for example, in case of questions of type how many/much, the question focus was in an *amod* relation with the adjective many/much). We ensured that the found rules were suitable for all question types (count, yes/no, list) and also for questions starting with different wh-phrases (which, what, where, when, who).

The next step was extracting the potential superlative and ranking information from the question. This procedure was useful in selecting the parameters to include after the ORDER BY, LIMIT and OFFSET modifiers in the SPARQL query. We determined the superlative related to the question focus (e.g., "What is the largest city by population in Germany?"), by analyzing the terms in an *amod* relation with the question focus and picking up the term with a JJS part of speech tag.

To identify numerals indicating the number of results to be returned in a list question (e.g., "Retrieve top three mountains by height in France", "Give me the first 2 cities by area in Japan"), we searched for the term in a *nummod* relation with the question focus. If the term was a word, we converted it into a number with the use of the word2number[6] library in Python.

To determine ordinals that influenced both the LIMIT and OFFSET parameters, we searched for terms in an *advmod* relation with the selected superlative (e.g., "Which is the second largest country in the world?") or in an *amod* relation with the question focus (e.g., "Which is the 2nd country by area in the world?").

3.2 Phrase Mapping Component

This component performs the mapping between the words in the NL question and the entities and properties from the utilized ontology. It is the only component of the system that is not knowledge base independent. In our current implementation, we utilized four existent phrase mappers for the DBpedia knowledge base: DBpedia Spotlight, TagMe, Earl and Falcon. Each phrase mapper returned for each found entity or property the associated portion in text to which it was mapped and also a mapping score indicating how probable it is for the mapping to be correct. In case multiple mapping probabilities were

---

[6] https://pypi.org/project/word2number/.

returned for the same entity/property, the average between those probabilities represented the final mapping score.

We eliminated auxiliary verbs, verb phrases that weren't related to the question itself (e.g., "give me", "retrieve", etc.) and wh-phrases before passing the processed NL question to the phrase mappers in order to avoid redundant results. We converted the remained lemmatized verbs to nouns using WordNet's derivationally related forms property and passed again the resulted nouns to the phrase mappers in order to capture entities from the ontology present as verbs in the question (e.g., "birthdate" entity correlated with "born" verb). We collected all generated entities and properties in two separate lists. We further combined entities/properties referring to the same concept into groups, such that in the query construction step only one candidate would be used in a query at a time.

We also used the resulted entities of type GPE (countries, cities, states), LOC (non-GPE locations) and FAC (buildings, airports, highways, bridges) determined via NER from the question processing step, as well as the list of countries and cities provided by the geonamescache[7] Python library to enrich the current list of selected properties and entities.

Next, we generated synonyms of nouns and verbs converted to nouns from the question using WordNet and searched for close matches of these synonyms with properties from the Patty[8] dataset. We integrated the found matches in our existent set of properties.

During the testing phase, we discovered that the Earl and Falcon phrase mappers misclassified entities of form *dbo:X*, where *X* was a string starting with capital letter as properties. We introduced an additional verification in our solution in order to avoid this issue.

3.3 Query Construction and Ranking Component

This component performs the last steps for obtaining the SPARQL query out of the given NL question.

The query construction task involved choosing the correct select clause (SELECT/ASK/SELECT with COUNT) based on the answer type as explained earlier, generating the triple patterns and adding the modifiers if needed (ORDER BY, OFFSET, and LIMIT). Because there were multiple ways in which the triple patterns could be formed, multiple SPARQL queries were formed in the end.

Using the lists of entities and properties extracted by the phrase mapping component, we first generated patterns containing only one triple of form: *(e, p, ?uri)*, *(?uri, p, e)* or *(?uri, p, ?uri2)* for each entity *e* and each property *p*. Then, we generated patterns containing two triples of form:

[(e, p, ?uri), (?uri, p, ?uri2)], *[(e, p, ?uri), (?uri, p2, ?uri2)]* and.

[(?uri, p, e), (?uri, p2, e2)], where *p2* was a property different from *p*.

Lastly, in case a superlative was found in the question analysis task, we created new patterns by adding a third triple of form *(?uri, pSuperlative, ?x)* to existing patterns of form *[(?uri, p, e)]* and *[(?uri, p, e),(?uri, p2, e2)]*.

[7] https://pypi.org/project/geonamescache/.

[8] https://www.mpi-inf.mpg.de/departments/databases-and-information-systems/research/yago-naga/patty.

The property *pSuperlative* indicates the property based on which the sorting was performed and the *?x* variable was the variable added after the modifier ORDER BY. The property *pSuperlative* was chosen by combining two approaches.

First, the property was searched in a predefined list that mapped common superlatives with their corresponding properties (e.g., heaviest and lightest were mapped to *dbo:weigth* in case of a who-question and to *dbo:mass* otherwise); this first approach also defined a sorting order based on the superlative.

Secondly, if no match was found, the term from the question which was in an *amod* relation with the superlative was selected and *pSuperlative* became the property similar to that term. For example, in the question "Which company has the most employees?", the term "employees" was selected and compared to the current properties, resulting in choosing the property *dbo:numberOfEmployees* to become *pSuperlative*.

In order to further optimize the algorithm for generating the triple patterns, intermediary queries to the DBpedia endpoint were made after adding each triple to a pattern. In order to make the queries, we used the *sparqlwrapper*[9] Python library. This enabled us to discard intermediary patterns for which no answer was returned. A second optimization consisted in observing the fact that the only entity allowed after a *rdf:type* property should be of form < *KB_prefix* >:*X* (where < *KB_prefix* > was for e.g., *dbo*), where *X* was an entity similar to the question focus.

The ranking of the resulted candidate SPARQL queries was realized using a formula that combined 4 types of scores:

- a phrase mapping score (*PM*)
- a score for the number of triples in the pattern (*NT*)
- a score for the number of properties and entities in the pattern from unique groups (*U*)
- an answer types score (*AT*)

The used formula for ranking was:

$$\mathrm{PM} + \mathrm{NT} \ast 0.4 + \mathrm{U} \ast 0.9 + \mathrm{AT} \qquad (1)$$

The PM score for a given query was calculated by summing phrase mapping scores for each triple. The phrase mapping score for a triple was calculated as the product between the phrase mapping probabilities (returned by phrase mappers) for each entity and property part of the triple. URIs were also taken into account, receiving a mapping score of 0.1.

The NT score for a given query was equal to the number of triples in the pattern, with the slight modification that triples containing two URIs had a weight of only 0.5.

The U score was determined as the number of properties and entities in the pattern from unique groups for a given query (we previously mentioned that properties/entities referring to the same concept were grouped).

The AT score measured the degree to which the results from the query execution matched the expected answer type. We detail the calculation method for the AT score below.

[9] https://pypi.org/project/sparqlwrapper/.

First, we checked that the returned results matched the determined question type: a yes/no question should return an answer of type True/False, a count question should return a single resource with a numeric type, a list question should return at least one resource. AT was assigned -2 if there was a mismatch between the returned results' type and the question type.

In case there was a match, a second step was performed for questions of type list in order to check whether the type of the returned resources matched the question focus or a list of predefined datatypes. The list of the predefined datatypes was chosen based on the wh-phrase with which the question in natural language started (e.g., in case of where questions, the list contained datatypes referring to time, dates, different parts of a date). To see whether there was a match, we checked if the type of the first returned resource was in the predefined list of datatypes for the particular current wh-phrase or if it was related to the question focus. In case of a match, the associated AT score was set to 5.

These values for the AT score in different scenarios were empirically determined.

After calculating scores for each candidate SPARQL query, the query with the highest score was selected and used to interrogate the DBpedia endpoint. Retrieved results were formatted and returned to the user via the User Interface.

4 System Evaluation

We tested our system on the QALD-9 [17] cumulated training and testing datasets. Before performing the experiment, we removed duplicated questions. We also checked if the queries' results from the dataset matched the current real results by running the queries on Virutoso's DBpedia endpoint and updating them automatically in case of a mismatch. The testing results included for each question: the default query from the dataset and the query determined by our system, the default answers and the answers determined by our tool, the execution time and the query score assigned by our application.

Figure 2 contains performance metrics (accuracy, precision, recall and F-score) for each type of question: yes/no, count, list questions containing no wh-phrases, questions containing various wh-phrases (who, where, when, etc.), questions containing superlatives (and/or rankings). If the question contained a superlative but belonged also to other categories, then it was included only in the superlatives' category. The system obtained a total accuracy of 0.364, a total precision of 0.437 and a total recall of 0.685. The total F1-score was 0.5338.

Figure 3 contains the system's response time metrics (average, minimum and maximum) per one query, for each type of question. The system obtained a total average response time per query of 22.42 s, with a minimum of 1.48 s and a maximum of 274.46 s. In some cases, slow responses were due to the fact that after multiple subsequent calls made in a short period of time to a phrase mapper, the given responses were delayed. The response time metrics indicate a good time-performance balance.

We compared our proposed solution with two state-of-the-art systems: gAnswer (the winner of the QALD-9 challenge) and Parot (which obtained even better results than gAnswer when tested on the same dataset). Results from Table 1 indicate that our proposed application outperforms both systems.

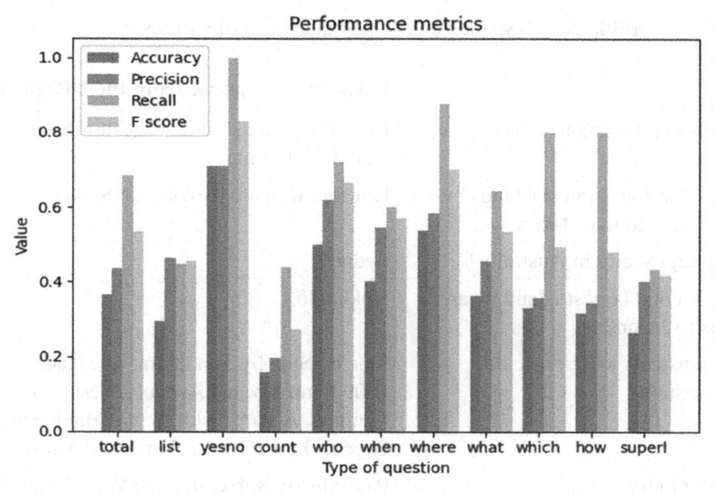

Fig. 2. Performance metrics for the proposed system on the QALD-9 dataset

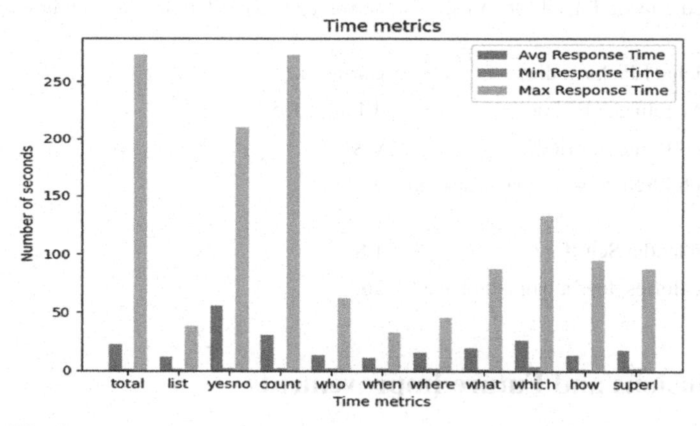

Fig. 3. Response time for the proposed system on the QALD-9 dataset

Table 1. Comparison of the proposed system DBSpark with gAnswer and Parot

System	Precision	Recall	F1 score
gAnswer	0.293	0.327	0.298
Parot	0.4321	0.512	0.4687
DBSpark	0.437	0.685	0.534

To complement the presented quantitative results, we highlighted in Table 2 the answers given by DBSpark to a variety of questions covering multiple types: list, count, yes/no, wh-questions, questions involving rankings, ordinals and superlatives.

Table 2. DBSpark's answers to questions of various types

Question	Formatted response from the DBSpark system
List 5 islands that belong to Japan	Denshima, Hokkaido, Mageshima, Geruma Island, Hateruma
Give me the first four largest islands by population that belong to Japan	Honshu, Kyushu, Hokkaido, Shikoku
What is the largest city in Australia?	Sydney
What is the second largest island by area that belongs to Japan?	Hokkaido
List all the musicals with music by Leonard Bernstein	Side by Side by Sondheim, The Race to Urga, 1600 Pennsylvania Avenue, West Side Story, Candide, A Party with Betty Comden and Adolph Green, On the Town, Wonderful Town
Who created Goofy?	Paul Murry, Bob Ogle, The Walt Disney Company
What is the official language of Suriname?	Dutch language
Which ingredients do I need for carrot cake?	carrot, egg as food, flour, backing powder, sugar
Why did Michael Jackson die?	cardiac arrest
When was Angelina Jolie born?	4 June 1975
Was Sigmund Freud married?	yes
How many children does Michael Jackson have?	3
How tall is Claudia Schiffer?	1.8
How many calories does a baguette have?	263

5 Conclusions and Future Improvements

After a comprehensive overview of the literature regarding translation of natural language to SPARQL, we have proposed in this paper a question answering system that enables users with no knowledge of SPARQL or of the underlying ontology model to access information from a knowledge base. In this way, the advantages that the knowledge bases possess in comparison to traditional database systems would be unlocked. The system is currently configured to function using the DBpedia knowledge base, however, it can be adapted to work with any knowledge base if the phrase mapping component is replaced accordingly.

In order to perform the translation from NL to SPARQL, we employed a rule-based approach that utilized the dependency parse tree of the question, the associated part of speech tags, WordNet and named entity recognition in order to analyze the given question and identify information useful for creating the SPARQL query. Our proposed solution is largely language-independent, supports a large variety of question types (list, count, yes/no, wh-questions, questions involving rankings, ordinals, and/or superlatives) and shows a good time-performance balance. Tests performed over the QALD-9 datasets showed that our proposed application clearly outperforms two state-of-the-art systems.

In the future, we consider generalizing our solution by allowing it to offer support for more complex SPARQL queries: queries containing more than three triples, queries supporting aggregation (GROUP BY and HAVING modifiers), queries with UNION, FILTER and REGEX modifiers. We also reflect on improving our system in order to handle more complex questions (such as compound questions, questions including comparatives and negation).

References

1. Berners-Lee, T., Hendler, J., Lassila, O.: The semantic web. Sci. Am. **284**(5), 34–43 (2001). https://doi.org/10.1038/scientificamerican0501-34
2. Wang, C., Xiong, M., Zhou, Q., Yu, Y.: PANTO: a portable natural language interface to ontologies. In: Franconi, E., Kifer, M., May, W. (eds.) The Semantic Web: Research and Applications. LNCS, vol. 4519, pp. 473–487. Springer, Heidelberg (2007). https://doi.org/10.1007/978-3-540-72667-8_34
3. Ochieng, P.: PAROT: translating natural language to SPARQL. Expert Syst. Appl. **176**, 114712 (2021). https://doi.org/10.1016/j.eswa.2021.114712
4. Liang, S., Stockinger, K., de Farias, T.M., Anisimova, M., Gil, M.: Querying knowledge graphs in natural language. J. Big Data **8**(1), 1–23 (2021). https://doi.org/10.1186/s40537-020-00383-w
5. Lehmann, J., Bühmann, L.: AutoSPARQL: let users query your knowledge base. In: Antoniou, G., et al. (eds.) The Semantic Web: Research and Applications. LNCS, vol. 6643, pp. 63–79. Springer, Heidelberg (2011). https://doi.org/10.1007/978-3-642-21034-1_5
6. Jin, H., Luo, Y., Gao, C., Tang, X., Yuan, P.: ComQA: question answering over knowledge base via semantic matching. IEEE Access **7**, 75235–75246 (2019). https://doi.org/10.1109/ACCESS.2019.2918675
7. Dorobăţ, I.C., Posea, V.: onIQ: an ontology-independent natural language interface for Building SPARQL queries. In: 2020 IEEE 16th International Conference on Intelligent Computer Communication and Processing (ICCP), September 2020, pp. 139–144 (2020). https://doi.org/10.1109/ICCP51029.2020.9266272
8. Damljanovic, D., Agatonovic, M., Cunningham, H.: FREyA: an interactive way of querying linked data using natural language. In: García-Castro, R., Fensel, D., Antoniou, G. (eds.) The Semantic Web: ESWC 2011 Workshops. LNCS, vol. 7117, pp. 125–138. Springer, Heidelberg (2012). https://doi.org/10.1007/978-3-642-25953-1_11
9. Damljanovic, D., Tablan, V., Bontcheva, K.: A text-based query interface to OWL Ontologies. In: Proceedings of the Sixth International Conference on Language Resources and Evaluation (LREC 2008), Marrakech, Morocco, May 2008. Accessed 18 Oct 2022. http://www.lrec-conf.org/proceedings/lrec2008/pdf/64_paper.pdf
10. Ferré, S.: Sparklis: an expressive query builder for SPARQL endpoints with guidance in natural language. Semant. Web **8**, 405–418 (2017). https://doi.org/10.3233/SW-150208
11. Hu, C., Ren, G., Liu, C., Li, M., Jie, W.: A Spark-based genetic algorithm for sensor placement in large scale drinking water distribution systems. Clust. Comput. **20**(2), 1089–1099 (2017). https://doi.org/10.1007/s10586-017-0838-z
12. Usbeck, R., Ngomo, A.-C., Haarmann, B., Krithara, A., Röder, M., Napolitano, G.: 7th Open challenge on question answering over linked data (QALD-7). In: Dragoni, M., Solanki, M., Blomqvist, E. (eds.) Semantic Web Challenges. CCIS, vol. 769, pp. 59–69. Springer, Cham (2017). https://doi.org/10.1007/978-3-319-69146-6_6

13. Dubey, M., Banerjee, D., Chaudhuri, D., Lehmann, J.: EARL: joint entity and relation linking for question answering over knowledge graphs. In: Vrandečić, D., et al. (eds.) The Semantic Web – ISWC 2018. LNCS, vol. 11136, pp. 108–126. Springer, Cham (2018). https://doi.org/10.1007/978-3-030-00671-6_7
14. Sakor, A., Singh, K., Patel, A., Vidal, M.-E.: Falcon 2.0: an entity and relation linking tool over Wikidata. In: Proceedings of 29th ACM International Conference on Information and Knowledge Management, pp. 3141–3148, October 2020. https://doi.org/10.1145/3340531.3412777
15. Ferragina, P., Scaiella, U.: TAGME: on-the-fly annotation of short text fragments (by wikipedia entities). In: Proceedings of the 19th ACM International Conference on Information and Knowledge Management - CIKM 2010, Toronto, ON, Canada, 2010, p. 1625. https://doi.org/10.1145/1871437.1871689
16. Mendes, M., Jakob, A., García-Silva, A., Bizer, C.: DBpedia spotlight: shedding light on the web of documents. In: Proceedings of the 7th International Conference on Semantic Systems - I-Semantics 2011, Graz, Austria, 2011, pp. 1–8. https://doi.org/10.1145/2063518.2063519
17. Unger, C., Usbeck, R.: QALD datasets. Semantic Computing Group@Bielefeld University, 02 October 2022. Accessed 18 Oct 2022. https://github.com/ag-sc/QALD

An Automated Patterns-Based Model-to-Model Mapping and Transformation System for Labeled Property Graphs

Pedro Guimarães[1,2(✉)], Ana León[2,3], and Maribel Yasmina Santos[2]

[1] Computer Graphics Centre, Campus de Azurém, Guimarães, Portugal
pedro.guimaraes@ccg.pt

[2] ALGORITMI Research Centre, University of Minho, Campus de Azurém, Guimarães, Portugal
maribel@dsi.uminho.pt

[3] Valencian Research Institute for Artificial Intelligence (VRAIN), Universitat Politècnica de València, Camí de Vera S/N, Valencia, Spain
aleon@vrain.upv.es

Abstract. Due to the increasing collection of highly interconnected and complex datasets, Labeled Property Graphs are gaining importance in extracting meaningful information for decision support. In addition, UML Class Diagrams are still a commonly used modeling technique for representing the main concepts of a domain. Although there are several model-to-model transformation approaches, these are mainly focused on moving from class diagrams to relational databases. Less work has been done on transforming class diagrams into labeled property graphs. This work constitutes a step forward in filling this gap by i) using a method that defines a set of patterns to improve the transformation process from class diagrams to labeled property graphs, considering the analytical requirements of a domain, and ii) proposing a technological system as an instantiation of the method, demonstrating its feasibility and enabling the assessment of its suitability. This system is grounded in a collection of templates for specifying the domain concepts and a library of transformation rules and patterns, and was evaluated using a widely known dataset exhibiting the proposed model-to-model transformation approach.

Keywords: Class Diagrams · Labeled Property Graphs · Model Transformation

1 Introduction

In recent years, Labeled Property Graphs (LPGs) have grown as a scalable and user-friendly method for representing complex real-world data. LPGs are a version of property graphs in which nodes and edges may be labeled with several categories, providing for more expressive and detailed data modeling [4,17]. For

S. Nurcan et al. (Eds.): RCIS 2023, LNBIP 476, pp. 171–186, 2023.
https://doi.org/10.1007/978-3-031-33080-3_11

representing domain concepts, UML Class Diagrams (CD) are a commonly used modeling technique in Software Engineering [24], and there has been a growing trend to transform CD into LPGs, based on Model-to-Model (M2M) transformations. Consequently, several works explore different techniques to automate M2M transformations, such as query-based [6], heuristic-based [14], or semantic and syntactic checking [23]. Nevertheless, the manual application of any transformation approach is a complex process that hinders the adoption of modeling techniques in any Information System design and development process.

This work proposes a technical solution that supports the mapping and transformation process, making available one or more LPGs for a given CD. We are aware that the aim of CDs goes beyond serving as mere data models. Nevertheless, LPGs are intended to model graph data schemas. This is why we focused only on this aspect of CD (data model representation) to make a CD-LPG equivalence based on the common aspects both representations address. The contributions of this work are i) the formalization of a meta-model for the patterns mapping proposed in [13], ii) the proposal of the logical and technological architecture that supports the mapping and transformation process, and iii) the implementation of a supporting system that shows the feasibility of the method and enables the assessment of its suitability to the intended purpose. This distinguishes our approach from previous studies and highlights the innovative nature of our contribution to the field. These contributions have been grounded in a research process guided by the Design Science Research Methodology for Information Systems [18].

This paper is organized as follows. Section 2 describes the related work. Section 3 presents the transformation method used in this work, highlighting the proposed mapping meta-model. Section 4 proposes the logical and technological architecture for the technical solution. Section 5 evaluates the implemented system using a widely known data model. Finally, Sect. 6 presents some remarks and guidelines for future work.

2 Related Work

According to [11], M2M tools transform one or more source(s) into one or more target(s) by adopting transformation languages that provide a set of constructs or mechanisms to apply transformations. Some works focus on meta-model definitions, discussing semantics using Domain Specific Languages [7,10,14], or using rule-based definitions [5,8,23] to drive the transformations between source model objects to target model objects. Focusing on transforming UML CD into NoSQL data structures, the most common method maps classes to tables and attributes to columns, and provides a set of transformation rules for relationships, compositions, and generalizations [5]. For example, [14] uses this method to transform a UML CD into an Apache HBase data model. The work of [16] proposes query execution methods from a conceptual schema and workload specification. Nevertheless, this approach is currently restricted to Apache Cassandra. In [1], the authors propose the UMLtoNoSQL approach to transform UML conceptual models into NoSQL physical models, taking a UML CD (conceptual PIM) and deriving a generic logical model (logical PIM). At the Platform Specific

Model (PSM) level, three different physical models corresponding to Cassandra, MongoDB, and Neo4j are considered. The authors use transformation rules to maintain the behavior and structure of the input model in the output NoSQL model.

The framework proposed on [8] suggests a meta-model for integrating graph databases by translating UML schemas into a graph representation model. The framework provides wide integration and is rule-driven. In [23], the authors present a methodology for automating M2M mapping and transformation, called Automated Model-to-Model (AMTM). The model transformation mappings step relies on the manual selection of potential mappings by domain experts. This selection process uses a semantic and syntactic checking technique that iteratively compares elements in the source and target meta-models.

The work of [15] presents ModelDrivenGuide, an MDA-based approach for generating logical models for NoSQL and relational databases. The authors rely on a meta-model to merge or split concepts, adapting them to all data models. The refinement relies on three transformation rules: merge rows between concepts, split rows to produce new columns in the same concept, and transform references into equivalent edges. Despite the authors do not explicitly mention using analytical requirements to drive their rules or transformations, they use a heuristic for generating data models that consider the use cases. In [21], the CM2KG platform is introduced, allowing the transformation of conceptual models into Knowledge Graphs. The architecture of CM2KG includes three components: Model Import, CM2KG Cloud, and Third-Party Tools. Similar to our approach, the input model is also an XML-formatted model, which serves as input for the CM2KG component. The core functionalities of the CM2KG platform include model transformation, knowledge base initialization, and knowledge graph visualization and analysis.

Although some of the related works share similarities with the approach proposed in this paper, they do not take into account the analytical requirements (queries of the domain) necessary for refining potential mappings, nor do they consider the analytical relevance of the input classes or a pattern-based approach during the transformation process. The systematic approach proposed in [13] defines a set of rules and patterns for the transformation of a UML CD into an LPG, considering the analytical requirements of a domain of interest. In this process, a single conceptual schema with knowledge about the main concepts of the domain can be transformed into different PSMs that conform to the analytical requirements of the domain. The use of patterns, along with a meta-model and rules, provides several advantages. Firstly, patterns offer a clearer and more user-friendly representation of the transformations. Secondly, incorporating patterns enables the reuse of common transformation solutions and improves process efficiency [12]. Finally, a patterns-based approach can handle complex transformation scenarios and accommodate dynamic changes to transformation requirements more effectively. Due to these advantages, this is the method selected as the basis of the work presented in this paper. The characteristics of the above-mentioned works are organized in Table 1, representing five dimensions to provide a comprehensive summary of the different approaches: (MM) Meta-Model, which refers to the use of a meta-model in the transformation process; (P) Patterns, denoting the presence of predefined patterns for guiding the transformations; (R) Rules, referring to a set of rules dictating the transformation process; (AR) Analytical Relevance, indicating the consideration of the analytical relevance of the input model classes in selecting appropriate rules or patterns; and (G) Graph Output, which shows compatibility with graph database systems.

Table 1. Summary of related work approaches and the proposed one.

Reference	Transf. approach	Transf. method	MM	P	R	AR	G
Billen et al. [5]	MDA-based	Automatic	yes	no	yes	no	yes
Daniel et al. [8]	MDA-based	Automatic	yes	no	yes	no	yes
Mior et al. [16]	Query/Cost-based	Semi-automatic	no	no	yes	no	no
Li et al. [14]	Heuristic-based	Manual	yes	no	yes	no	no
Wang et al. [23]	Semantic-checking	Automatic	yes	no	yes	no	no
Abdelhed et al. [1]	MDA-based	Automatic	yes	no	yes	no	yes
Mali et al. [15]	MDA/Heuristic-based	Automatic	yes	no	yes	no	yes
Smajevic et al. [21]	MDA-based	Automatic	yes	no	no	no	yes
Proposed Approach	Pattern-based	Automatic	yes	yes	yes	yes	yes

The approach presented in this paper encompasses all five dimensions, using not only a meta-model and rules but also patterns to guide the transformations, and taking into account the analytical requirements of the domain.

3 Transformation Patterns

A pattern is a collection of model elements arranged into a certain structure and specific constraints [20]. The systematic approach proposed in [13] includes the definition of the rules and patterns for the transformation of a CD into an LPG, and considers the domain's concepts and the analytical requirements to be met. A fundamental contribution of this work is the formalization of the patterns that can be applied when transforming classes, relationships, and attributes into nodes, edges, and properties of an LPG (Fig. 1). In [13], the patterns are formalized in a descriptive way, presenting and explaining their applicability and providing examples of their application. In the work here proposed, the meta-model links the main concepts of a CD with the main concepts of an LPG, supporting the M2M mapping and transformation approach that we explain in detail in further sections.

The main elements of a CD are classes, attributes, and associations. In [13], Classes can have analytical relevance, a fundamental distinction in the proposed

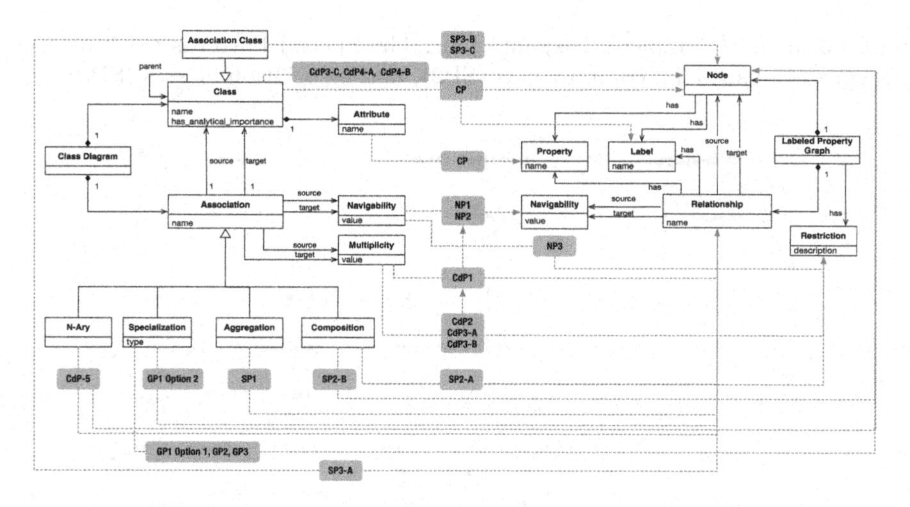

Fig. 1. Patterns mapping meta-model.

approach. The analytical relevance is evaluated by checking the classes that have i) high analytical value in the domain, representing the main concepts for analysis (C1 classes), ii) high cardinality, requiring efficiency in data processing (C2 classes), and iii) frequent access patterns, suggesting additional concerns with data organization (C3 classes).

LPGs include nodes and relationships that can also have labels and properties. In this work, LPGs can include restrictions, a concept not implemented in the graph database itself, but that must be considered by the Information System that uses the database to ensure data integrity. Restrictions are represented as notes in the graphical representation of the LPG.

The transformation patterns are grouped into five types: Class, Navigability, Cardinality, Generalization, and Special patterns. The Class Pattern (CP) expresses a direct transformation between a class and a node. The Navigability Patterns (NP) map the different possibilities for navigability of a relationship in the property graph. In a UML CD, navigability indicates that it is possible to navigate (uni-directionally or bi-directionally) across the association from objects of the source to the target class. NP1 applies in case of a uni-directional association between two classes, NP2 when the association is bidirectional or not specified, and NP3 when the association between the classes has a restriction. The Cardinality Patterns (CdP) consider the multiplicity of the associations. CdP1 applies when the target class in the association can have zero or more instances (0..*). CdP2 applies when the target class needs to verify at least one instance (1..*). CdP3 and CdP4 manage (0..1) and (1..1) cardinalities, also considering the analytical relevance of the classes. CdP5 applies for n-ary associations. The Generalization Patterns (GP) include four different constraints, usually represented as pairs. GP1 applies in the case of {Incomplete, Disjoint} or {Incomplete, Overlapping}, GP2 applies in the {Complete, Overlapping} pair, and GP3 in the {Complete, Disjoint} pair. The Special Patterns (SP) are used for aggregations (SP1), compositions (SP2), and association classes (SP3).

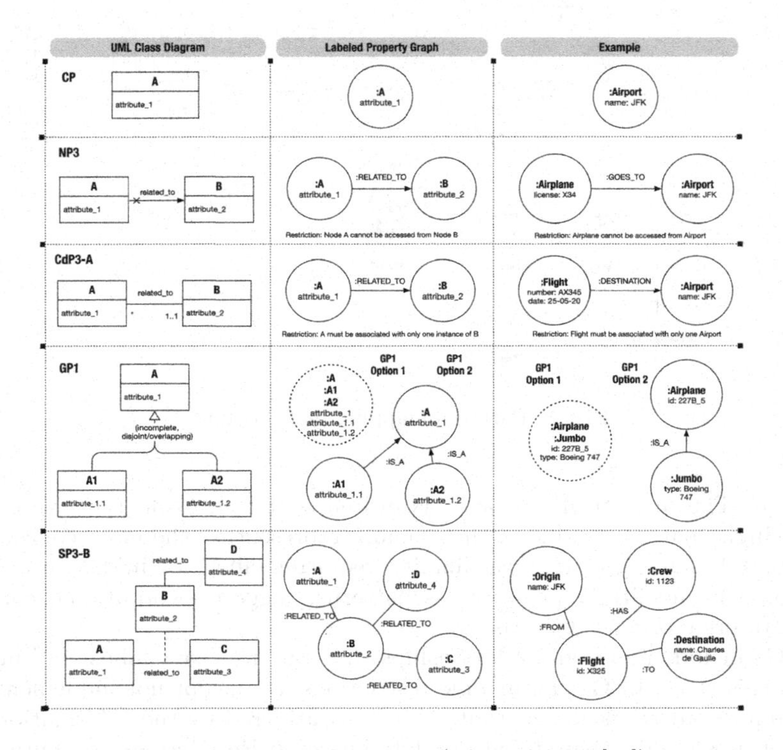

Fig. 2. Examples of patterns (adapted from [13]).

Figure 2 depicts an example for each pattern type, highlighting some of the patterns used afterwards in the evaluation case. For the identification of the most adequate patterns, as an iterative process, the work of [13] defines this set of systematic tasks: i) represent C1 classes as nodes and the relationships with other C1 classes; ii) represent C2 classes as nodes and the relationships with C1 and other C2 classes; iii) represent C3 classes as nodes and the relationships with C1, C2, and other C3 classes; iv) represent all the remaining classes and their associations with the other classes. This process is exemplified afterwards in the evaluation case, and it is embedded in the mapping and transformation approach next presented.

4 M2M Mapping and Transformation Approach

The logical and technological architectures proposed in this work (Fig. 3) include four components, two of them associated with the representation of the Domain

Knowledge and the Domain Expert as a valuable source of additional information, and the two technical modules that integrate the mapping and transformation process, the Transformation Module and the Validation Module.

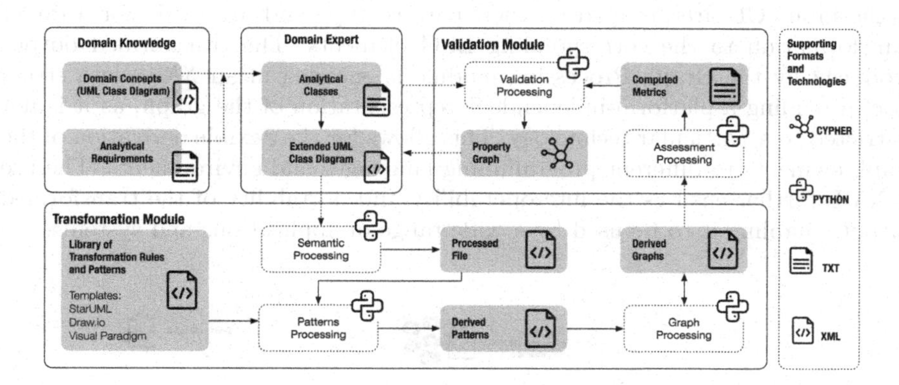

Fig. 3. Logical and technological architecture.

Regarding Domain Knowledge, it requires a UML CD representing the domain and a list of analytical requirements (queries) that the derived model must be able to solve. The Domain Expert provides the CD and the queries. The CD is represented using the main UML elements, and the analytical relevance of each class is represented using stereotypes (see example in Fig. 5). The stereotypes consider the Analytical Requirements and classify the High Analytical Value classes as C1, the High Cardinality classes as C2, and the Frequent Access classes as C3. The Transformation Module is the core of this approach and consists of multiple iterative read and write processes that take a source model and compares it to a target model. The Semantic Processing is supported by a Python library that provides a set of functions used to identify the source model and extract the UML elements according to the rules defined in [13].

The input of the Transformation Module is the CD represented in XML format since most modeling tools allow this representation. The input XML file structure depends on the modeling tool and commonly contains extra data referring to the graphical representation of the elements, such as colors, positions, notes added by the user, and other configuration data to manipulate the model in the tool. Therefore, the file must be cleaned to extract only the data related to the representation of the domain and ease the identification of the patterns. The elements of the CD are analyzed in an iterative way, identifying classes (and their analytical relevance), attributes, classifications, associations, and cardinalities, and omitting the rest of the unnecessary information. Once the cleaning is done, the newly created data structure is stored in the Processed File.

The Patterns Processing accepts the Processed File as input and initiates automatically the patterns identification process (Fig. 4). The Derived Patterns file contains all the UML elements that have been classified according to the type of patterns. The Graph Processing employs a straightforward semantic mapping function to transform UML elements into a graph data structure represented in the Derived Graphs file. This mapping function includes the nodes, edges, and restrictions of the graph. After the patterns have been identified, the elements of the input CD are grouped by their pattern type and are ready for a direct transformation to the corresponding LPG elements. The transformed output produced by the Graph Processing module is encoded in an XML data structure, providing a platform-independent representation of the graph, as it is not restricted to a particular technology. This allows for the flexible derivation of the graph elements to different programming languages and environments. The use of XML further ensures the interoperability and portability of the transformed output, enabling it to be used by a wide range of applications and systems.

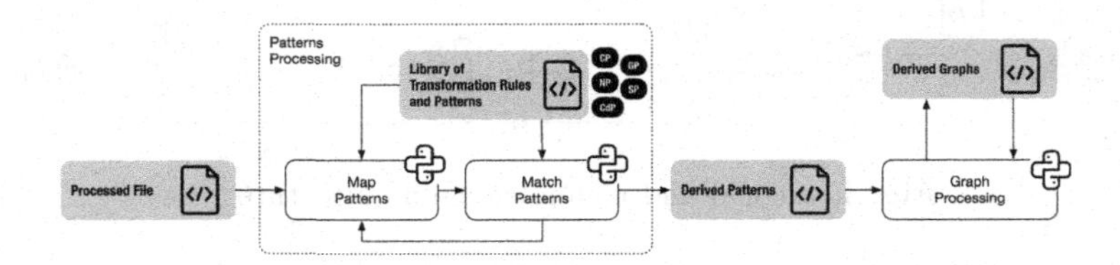

Fig. 4. Patterns and Graph Processing from the Transformation Module.

The technology underlying the Transformation Module employs an iterative approach to compare the elements of the CD to the rules and patterns library. This approach consists of a series of sequential systematic tasks that aim to identify similarities among the patterns. These tasks are as follows:

(T1) Identifying classes by reading the Processed File and (T2) Sorting the classes based on their analytical relevance. Algorithm 1 excerpts the main function of the Transformation Module. Using the BeautifulSoup library (Python), it first finds all the classes for the given input XML file (UML CD). The sort_classes function takes a list of classes as input and returns a new list of the same classes sorted based on the values of their 'C1', 'C2', and 'C3' attributes, following this priority order. Then, calls the sp3_b_patterns function to find the SP3-B pattern (detailed in Algorithm 2) and, if found, transforms it into an LPG, using the function to_labeled_property_graph (Algorithm 3).

(T3) Examining the number of associations between classes as source and target; (T4) Evaluating the cardinality between associations; and, (T5) Searching the Transformation Rules and Patterns Library for target models. The find_sp3_b_pattern function takes as input a parsed XML document representing a CD (Algorithm 2). The function aims to identify a specific pattern, in this case SP3-B, and returns a list of occurrences of this pattern. The pattern is defined by an association class (class_b) in a relationship with a multiplicity of '1..*' both for the source and target classes (class_a and class_c).

(T6) Grouping elements based on the matching pattern; (T7) Adding the pattern name to the group of elements that have been matched; and, (T8) Writing

```
1  main() begin
2      soup ← BeautifulSoup(class_diagram_xml,' lxml − xml')
3      classes ← soup.find_b_all('class')
4      sorted_b_classes ← sort_classes(classes)
5      sp3_b_patterns ← find_sp3_b_pattern(sorted_b_classes)
6      if sp3_b_patterns then
7          foreach (class_a, class_b, class_c) ∈ sp3_b_patterns do
8              sp3_b_pattern_classes ← [class_a, class_b, class_c]
9              sp3_b_graph ← to_labeled_property_graph(sp3_b_pattern_classes,' SP3 − B')
10         end
11     end
12 end
```

Algorithm 1: Excerpt of main function

```
   Input: Classes (from source file)
   Output: SP3-B Patterns
1  find_sp3_b_pattern(inputFile) begin
2      sp3_b_patterns ← ∅
3      forall the class_b ∈ inputFile do
4          b_id ← IdAttribute(class_b)
5          incoming_relations ← ∅
6          forall the relation ∈ inputFile ∧ TargetAttribute(relation) = b_id do
7              if ValueAttribute(relation) =' 1..*' then
8                  incoming_relations ← incoming_relations ∪ {relation}
9              end
10         end
11         if |incoming_relations| = 2 then
12             class_a_id ← SourceAttribute(incoming_relations[0])
13             class_c_id ← SourceAttribute(incoming_relations[1])
14             class_a ← FelementWattribute(inputFile, class_a_id)
15             class_c ← FelementWattribute(inputFile, class_c_id)
16             if class_a ≠ null ∧ class_c ≠ null then
17                 sp3_b_patterns ← sp3_b_patterns ∪ {(class_a, class_b, class_c)}
18             end
19         end
20     end
21     return sp3_b_patterns
22 end
```

Algorithm 2: Find SP3-B Patterns

the new data structure to the Derived Graphs file. The to_labeled_property_graph function takes a list of classes and a pattern name as input, and it returns a dictionary representing a graph with labeled nodes and edges (Algorithm 3). For each class in the `Derived Patterns` file, the function extracts information such as the class's label, ID, and attributes. It then creates a node with this information and appends it to the nodes list. Additionally, the function identifies relationships between classes by looking for relations where the source ID matches the target class ID. It creates edges between the nodes based on these relationships. Finally, the function builds a dictionary containing the pattern name, the list of nodes, and the list of edges, and writes it to the Derived Graphs file. This dictionary represents the LPG.

The Validation Module evaluates the generated Derived Graphs to guarantee the maintenance of the initial behavior and semantic equivalence between the input and output models. This evaluation considers several dimensions, such as the complexity of the output model in relation to the number of nodes and edges.

```
   Input: Derived Patterns File
   Output: Derived Graphs File
 1 to_labeled_property_graph(pattern_classes, pattern_name) begin
 2     nodes ← ∅
 3     edges ← ∅
 4     forall the class_element ∈ pattern_classes do
 5         label ← class_element['label']
 6         class_id ← class_element['id']
 7         nodes ← nodes ∪ {('id' : class_id, 'label' : label, 'properties' : properties)}
 8         relations ← class_element.find_all('relation', {'source' : class_id})
 9         forall the relation ∈ relations do
10             source ← relation['source']
11             target ← relation['target']
12             value ← relation['value']
13             edge_id ← relation['source'] + '_' + relation['target']
14             edges ← edges ∪ {('id' : edge_id, 'source' : source, 'target' : target, 'value' :
               value)}
15         end
16     end
17     graph ← ('pattern' : pattern_name, 'nodes' : nodes, 'edges' : edges)
18     return graph
19 end
```

Algorithm 3: Convert to Labeled Property Graph

The Domain Expert uses the metrics provided by the Assessment Processing module to evaluate the Derived Graphs. These metrics help in assessing different quality dimensions of the model:

- *Model Complexity*: Measures the size and complexity of the output model, taking into account the number of nodes, edges, and properties.
- *Behavior Preservation*: Measures how the original behavior of the UML model is preserved. This can be assessed by comparing the results of queries made on the original and transformed models.
- *Semantic Equivalence*: Measures how the semantic meaning of the UML model is preserved in the output model. This can be assessed by comparing the relationships and attributes in the original and transformed models.
- *Scalability*: Measures the ability of the output model to handle large amounts of data, and to maintain its performance as the size of the model grows.

The assessment results are documented in the Computed Metrics file. Since more than one LPG can be derived from a CD, this file provides a comprehensive overview of the output LPGs and enables the Domain Expert to make an informed decision about the most efficient or relevant model. The *Model Complexity* is already computed in the current implementation.

(T9) Generate the Cypher script. Once the Domain Expert has reviewed the Computed Metrics file and selected the LPG (Validation Processing), the model can be translated to the corresponding Cypher code. The generate_cypher_script function (Algorithm 4) takes the `Derived Graphs` file as input and generates a `Cypher script` for physically implementing the graph. First, the function iterates over the nodes in the graph, extracting each node's label and properties. It appends a Cypher command to `CREATE` a node with the corresponding label and properties to the cypher_script string.

```
Input: Derived Graphs File
Output: Cypher Script
1  generate_cypher_script(graph) begin
2      cypher_script ← "
3      forall the node ∈ graph['nodes'] do
4          label ← node['label']
5          properties ← node['properties']
6          cypher_script ← cypher_script + CREATE (: labelproperties);
7      end
8      forall the edge ∈ graph['edges'] do
9          source_id ← edge['source']
10         target_id ← edge['target']
11         relation_value ← edge['value']
12         cypher_script ← cypher_script + CREATE (a), (b)
           WHERE a.id = 'source_id' AND b.id = 'target_id'
           CREATE(a) − [: RELATED value : 'relation_value']− > (b);
13     end
14     cypher_script ← cypher_script + "
15     return cypher_script
16 end
```

Algorithm 4: Generate Cypher Script

It also appends a Cypher command to create a relationship between the nodes with matching IDs and the specified relationship value. Finally, the function returns the complete cypher_script string, which can be executed in a Neo4j database to create the nodes and relationships.

5 Evaluation Case: The TPC-H Dataset

The TPC-H benchmark (http://www.tpc.org/tpch) is a well-known decision support benchmark including a set of business-oriented ad-hoc queries usually used for benchmarking decision support systems. Its original data model includes eight tables modeling a business context with sales, customers, products and suppliers, and the associated entities [22]. In this section, we use the TPC-H dataset to show the transformation process. The reason for choosing this example is the availability of the queries established as analytical requirements by the benchmark. The analysis of the queries and the commonly used tables allowed the identification of the classes with high analytical value (`ORDER` and `LINEITEM`), the classes with high cardinality (`ORDER`, `LINEITEM`, and `PARTSUPP`), and the classes with frequent access patterns (`ORDER`, `LINEITEM`, `CUSTOMER`, `PART`, `SUPPLIER` and `NATION`) (Fig. 5).

The UML CD is the starting point for applying the proposed transformation approach. The analytical relevance of the classes is associated with the main indicators of the sales business process, which are integrated into `ORDER` and `LINEITEM`. The identification of the high cardinality required the evaluation of the dataset volume, looking into the number of rows in each table. Considering a dataset with 100 GB of data [19], the several tables have (number of lines): `ORDER` ($150x10^6$), `LINEITEM` ($600x10^6$), `CUSTOMER` ($15x10^6$), `PART` ($20x10^6$), `SUPPLIER` ($1x10^6$), `PARTSUPP` ($80x10^6$), `NATION` (25), and `REGION` (5). As we are analyzing a synthetic dataset whose data volume is generated attending to a specific scale

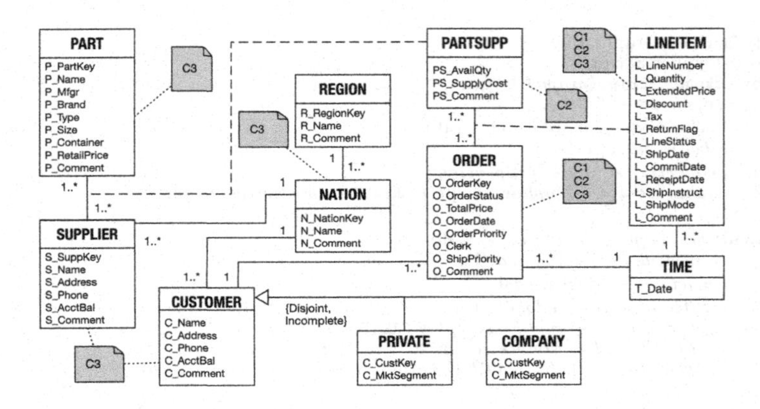

Fig. 5. TPC-H class diagram and analytical classes.

factor, the identification of the classes with high cardinality highlighted the ones that have more data volume as the benchmark was designed to test processing technologies under challenging data contexts. CUSTOMER, PART, and SUPPLIER could also have been identified as having high cardinality. In each application domain, the domain expert needs to do this classification.

The frequent access patterns are derived from the tables used in the 22 queries of the benchmark, looking into the entities usually queried. C1 classes, as the key concepts of the domain, are usually queried to know what happened or is happening, besides other classes, such as the C3, that provide the context to these queries (when or where it happened). After the analysis of the TPC-H data model and its analytical classes, the identification of the resulting property graph is performed. Figure 6 summarizes the internal process performed by the system and highlights how different property graphs can be identified, requiring the intervention of the domain expert.

The C1 classes identified in step 1 are ORDER and LINEITEM, and this last one is an association class between ORDER and PARTSUPP. As PARTSUPP is not considered in this step, pattern SP3-B is partially applied without considering the PARTSUPP class. Also, CdP3-A is applied considering the cardinality between LINEITEM and ORDER, additionally adding a restriction to the LineItem node. In the next step, the C2 PARTSUPP class is associated with the previously identified nodes, completing the SP3-B pattern and adding another restriction to the LineItem node. This dataset includes four C3 classes besides ORDER and LINEITEM, namely CUSTOMER, PART, SUPPLIER, and NATION. Starting with PART and SUPPLIER, as these are related to PARTSUPP, SP3-B completes the relationships to the Part and Supplier nodes. CdP3-A relates and adds navigability, and also adds two restrictions to the PartSupp node. Regarding NATION, this class is related to SUPPLIER and CUSTOMER. CP is used to create the Nation node and GP1 is used to apply the adequate generalization pattern to CUSTOMER. However, in this case, two possible transformations are allowed, identified in the figure with different line shapes. CdP3-A applies between SUPPLIER and NATION, and

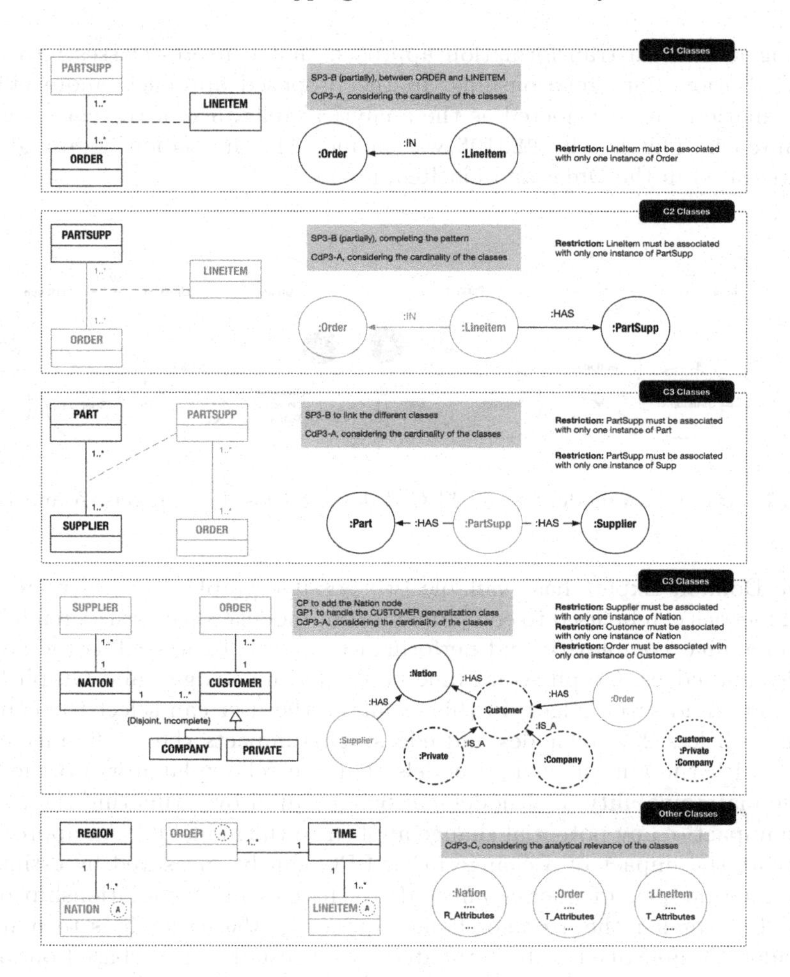

Fig. 6. Transformation process for the TPC-H dataset.

between CUSTOMER and NATION. The same CdP3-A pattern is used for ORDER and CUSTOMER. Three additional restrictions were identified in this process. Finally, two remaining classes need to be addressed, REGION and TIME. Starting with REGION, CdP3-C is applied between NATION and REGION. As NATION has analytical relevance, the Nation node includes REGION's attributes. In the case of TIME, the same reasoning is applied, and pattern CdP3-C allows the integration of TIME's attributes in the LineItem and Order nodes.

After applying the several steps, two possible graphs can be identified as a consequence of the transformation pattern for the CUSTOMER class (Fig. 7). One version, hereafter named Graph A, includes nine nodes for the 11 classes of the UML CD, while the more compact version, Graph B, includes seven nodes. Both possible graphs are different from the one that could be obtained following a classical transformation approach that converts classes into nodes and associations into relationships. In the proposed approach, more efficient transformations are considered as the analytical relevance of the classes guides the approach. This is why REGION was included in the Nation node and TIME was integrated in the Order and LineItem nodes.

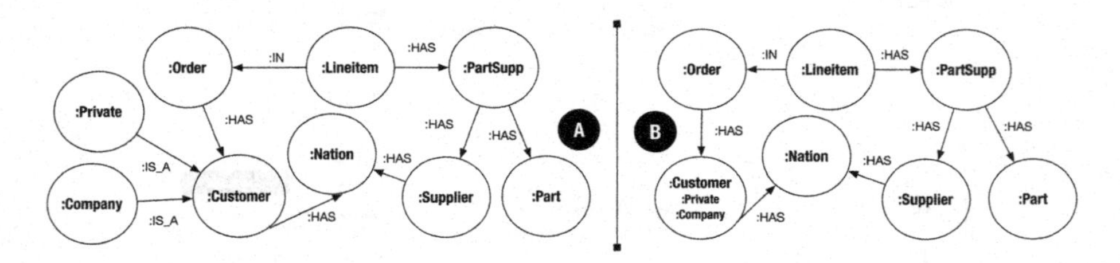

Fig. 7. Graph data model for the TPC-H dataset (possible versions, A and B).

The Domain Expert has available two possible graphs. As [13] points out, several metrics can be used to compare different models, such as size, readability, maintainability, complexity, and performance. In size, expressed as the number of nodes and edges, Graph A has nine nodes and nine edges, and Graph B has seven nodes and seven edges. In this example, the user can select the compact version with less 22% of nodes and edges. This impacts the readability of the models, with the aim of having models that are self-explanatory to the users [9]. The maintainability of a model can be measured by estimating the average change impact of any potential change applied to the model [2]. An approach to identifying the impact of a change in an LPG can be measured by estimating the average number of queries affected by changes in any relationship of the model. The smaller the average change impact is, the easier it is to maintain the model. Complexity can be estimated as suggested in [13] (based on [3]) by considering the path to solve a specific query, as nodes and edges define the route to obtain the result. Taking as an example query number three from the TPC-H benchmark (with 'FROM CUSTOMER, ORDER, LINEITEM' in the SQL version), Graph A requires five nodes and four relationships, with an average change impact of 0,80 in this portion of the model, while Graph B requires three nodes and two relationships, with an average change impact of 0,67 in this portion of the model. As the remaining model is the same, we can consider that Graph B is less complex than Graph A. Regarding performance, [13] performed a benchmark that measured two models with three different datasets of increasing sizes. In this, the compact graph performed best, with gains increasing as the size of the datasets also increased, which allows us to expect the same behaviour between Graphs A and B. These measures are devised to support the Domain Expert providing additional information about the obtained models, but the knowledge about the domain and the characteristics of the data are fundamental for making an informed decision.

6 Conclusions

This work proposes a technical solution based on the integration of several components and technologies that support the application of different transformation patterns to move from a UML Class Diagram to a Labeled Property Graph. The method used to perform the transformation considers the analytical requirements of the domain, which ensures that the obtained graph is optimized and is not a mere equivalence between classes and nodes. The system must provide a set of quality metrics (size, readability, maintainability, complexity, and performance), associated with the resulting graphs, so the user can evaluate the suitability of the result according to the analytical requirements. The proposed solution was evaluated using the TPC-H Dataset.

For future work, we highlight the need to consider additional quality metrics, the evaluation of the technical solution with more complex datasets and domains to prove the completeness and validity of the patterns, the need to develop User Interfaces for the domain expert, and the use of Artificial Intelligence approaches for optimizing complex transformation processes.

Acknowledgements. This work has been supported by *FCT - Fundação para a Ciência e Tecnologia* within the R&D Units Project Scope: UIDB/00319/2020, and by the Spanish Ministry of Universities and the Universitat Politècnica de València under the Margarita Salas Next Generation EU grant.

References

1. Abdelhedi, F., Brahim, A.A., Atigui, F., Zurfluh, G.: MDA-based approach for NOSQL databases. In: 19th International Conference on Big Data Analytics and Knowledge Discovery (DaWaK 2017), pp. 88–102 (2017)
2. Almasri, N., Korel, B., Tahat, L.: Toward automatically quantifying the impact of a change in systems. Softw. Qual. J. **25**(10), 3833–3861 (2017)
3. Almasri, N., Tahat, L., Korel, B.: Verification approach for refactoring transformation rules of state-based models. IEEE Trans. Softw. Eng. **48**(3), 601–640 (2022)
4. Angles, R., Gutierrez, C.: Survey of graph database models. ACM Comput. Surv. **40** (2008)
5. Billen, R.: Uml as a schema candidate for graph databases. University of Liège, Technical report (2014)
6. Chebotko, A., Kashlev, A., Lu, S.: A big data modeling methodology for apache cassandra. In: 2015 IEEE International Congress on Big Data, pp. 238–245. Institute of Electrical and Electronics Engineers Inc., August 2015
7. Cuadrado, J.S., Guerra, E., Lara, J.D.: A component model for model transformations. IEEE Trans. Softw. Eng. **40**, 1042–1060 (2014)
8. Daniel, G., Sunyé, G., Cabot, J.: UMLtoGraphDB: mapping conceptual schemas to graph databases. In: Comyn-Wattiau, I., Tanaka, K., Song, I.-Y., Yamamoto, S., Saeki, M. (eds.) ER 2016. LNCS, vol. 9974, pp. 430–444. Springer, Cham (2016). https://doi.org/10.1007/978-3-319-46397-1_33
9. Ehrlinger, L., Huszar, G., Wöß, W.: A schema readability metric for automated data quality measurement. In: Eleventh International Conference on Advances in Databases, Knowledge, and Data Applications (DBKDA), p. 12 (2019)

10. Favre, J.M., Nguyen, T.: Towards a megamodel to model software evolution through transformations. Electron. Notes Theoret. Comput. Sci. **127**, 59–74 (2005)
11. Kahani, N., Cordy, J.R.: Comparison and evaluation of model transformation tools. Technical report, School of Computing, Queen's University Kingston, Ontario, December 2015
12. Lano, K., Kolahdouz-Rahimi, S.: Model-transformation design patterns. IEEE Trans. Softw. Eng. **40**, 1224–1259 (2014)
13. León Palacio, A., Santos, M.Y., García, A., Casamayor, J.C., Pastor, O.: Model-to-Model Transformation: From UML Conceptual Schemas to Labeled Property Graphs. Accepted for publication, Business & Information Systems Engineering (2023)
14. Li, C.: Transforming relational database into Hbase: a case study. In: 2010 IEEE International Conference on Software Engineering and Service Sciences, pp. 683–687 (2010)
15. Mali, J., Atigui, F., Azough, A., Travers, N.: `ModelDrivenGuide`: an approach for implementing NoSQL schemas. In: Hartmann, S., Küng, J., Kotsis, G., Tjoa, A.M., Khalil, I. (eds.) DEXA 2020. LNCS, vol. 12391, pp. 141–151. Springer, Cham (2020). https://doi.org/10.1007/978-3-030-59003-1_9
16. Mior, M.J., Salem, K., Aboulnaga, A., Liu, R.: Nose: schema design for NOSQL applications. IEEE Trans. Knowl. Data Eng. **29**, 2275–2289 (2017)
17. OMG: Object management group model driven architecture (MDA) rev. 2.0. Technical report, The Object Management Group (2014). http://www.omg.org/mda/
18. Peffers, K., Tuunanen, T., Rothenberger, M.A., Chatterjee, S.: A design science research methodology for information systems research. J. Manag. Inf. Syst. **24**(3), 45–77 (2007)
19. Rodrigues, M., Santos, M.Y., Bernardino, J.: Big data processing tools: an experimental performance evaluation. WIREs Data Mining Knowl. Disc. **9**(2) (2019)
20. Di Ruscio, D., Eramo, R., Pierantonio, A.: Model transformations. In: Bernardo, M., Cortellessa, V., Pierantonio, A. (eds.) SFM 2012. LNCS, vol. 7320, pp. 91–136. Springer, Heidelberg (2012). https://doi.org/10.1007/978-3-642-30982-3_4
21. Smajevic, M., Bork, D.: From conceptual models to knowledge graphs: a generic model transformation platform. In: 2021 ACM/IEEE International Conference on Model Driven Engineering Languages and Systems Companion (MODELS-C), pp. 610–614 (2021)
22. TPPC: Transaction Processing Performance Council. (2017). TPC-H Specification (Decision Support), Standard Specification, Revision 2.17.2 (2017). http://www.tpc.org/tpc_documents_current_versions/pdf/tpc-h_v2.17.2.pdf
23. Wang, T., Truptil, S., Benaben, F.: An automatic model-to-model mapping and transformation methodology to serve model-based systems engineering. Inf. Syst. E-Bus Manage. **15** (2017)
24. Ziemann, P., Hölscher, K., Gogolla, M.: From UML models to graph transformation systems. Electron. Notes Theoret. Comput. Sci. **127**, 17–33 (2005)

Improving Conceptual Domain Characterization in Ontology Networks

Beatriz Franco Martins[1(✉)], José Fabián Reyes Román[1], Oscar Pastor[1(✉)], and Moshe Hadad[2]

[1] Valencian Research Institute for Artificial Intelligence (VRAIN), Universitat Politècnica de València, Camino de Vera s/n, 46022 Valencia, Spain
{bmartins,jreyes}@vrain.upv.es, opastor@dsic.upv.es

[2] Cyber R&D Lab, Accenture Israel, Ha-Menofim St 2, Tel Aviv-Yafo, 4672553 Tel Aviv, Israel
moshe.hadad@accenture.com

Abstract. The community of Conceptual Modeling (CM) perception that *Semantic Interoperability* cannot be achieved without the support of an ontology-driven approach has become increasingly consensual. Moreover, the more complex and extensive the domain of the application of conceptual models, the harder it is to achieve semantic consensus. Therefore, it has emerged the perception that ontologies built to describe complex domains should not be overly large or be used in isolation. Ontology Networks arose to cover this issue. The community had to deal with issues such as different ontologies of the network using the same concept with different meanings or the same term used to designate distinct concepts. We developed a framework for classifying ontologies that provides a stable and homogeneous environment to facilitate the ontological analysis process by dealing simultaneously with ontological and domain perspectives. This article presents our proposal where conceptualization is used to identify the relationships among the evaluated ontologies. Our goal is to facilitate semantic consensus, providing guidelines and best practices supported by a stable, homogeneous, and repeatable environment.

Keywords: Conceptual Modeling · Ontology Classification · Ontologies Network · Ontology

1 Introduction

The Ontology Engineering process became fundamental to dealing with knowledge domains in which data are heterogeneous, vast, and complex [25]. For instance, in Cybersecurity as presented in [34], Software Engineering [7,9], Value and Risks [10,39], Genomics and Biomedicine [4,12], etc. In this kind of scenario, engineers usually support the semantic consensus required by the ontologies in well-known standards used by their domain community of experts. For instance, standards like SEVOCAB[1] and ISO/IEC/IEEE 12207[2] (among others) support

[1] https://www.iso.org/standard/71952.html.
[2] https://www.iso.org/standard/63712.html.

S. Nurcan et al. (Eds.): RCIS 2023, LNBIP 476, pp. 187–202, 2023.
https://doi.org/10.1007/978-3-031-33080-3_12

ontologies for Software Engineering; likewise, ISO/IEC 27000[3] and ISO/IEC 27032[4] (among others) support ontologies for the Cybersecurity domain. Definitions used in standards, such as those, exist to clarify the interpretation of terms present in the knowledge domain they cover. However, the standards use natural (or technical) language, leaving room for more diverse interpretations. This results in disagreements among stakeholders about the conceptualization; even worse, they may think are having good communication and agreement without actually having. Indeed, the distinct viewpoints, approaches, and contexts of application of ontologies covering these domains move them away from the FAIR Principles [43]. This motivates us to provide ontology engineers and domain experts with a consensual, controlled, and traceable environment for performing ontological analysis to achieve interoperability and reuse among ontologies [34,35]. The final objective is to facilitate the ontological analysis process for integrating and reusing of ontologies, especially for intricate domains.

Meanwhile, it has emerged within the ontology engineering community the perception that ontologies built to describe vast or complex domains should not be overly large or be used in isolation. Therefore, the called Ontology Networks (ON) arose, in which ontologies covering subdomains of a larger domain are co-linked or interrelated [41]. In the academy, SEON [5] comprises a Software Engineering Ontology Network that provides a well-founded set of reference ontologies covering subdomains such as software process (SPO) [38], requirements (SwO, RSRO, and RRO) [8], among others. In the industrial environment, initiatives such as BRON [26,27] and OdTM [6] have been bringing a pragmatic view of the use of ON with the adoption of analytics from different and solid data sources covering subdomains such as vulnerabilities [30] and weaknesses [31].

In [35] we present the **Framework for Ontology Classification (F4OC)**, emphasizing the ontological perspective specifically working with the classification of cybersecurity ontologies. While in this work we generally present the domain perspective as well as extend the ontological perspective to contemplate the notions "I" and "R" of FAIR discussing the relationships between ontologies belonging to ON. Thus, the first contribution presented in this paper is to provide an ontology-driven framework that is in line with the"I" and "R" of FAIR principles, through good practices discussed within this context. The second contribution is to extend the F4OC to be used in ON. We explore the ability of the F4OC to unveil meta-characteristics of ontologies and to contextualize their cloud of concepts (in line with the ontological commitment notion) to provide interoperability and reuse of ontologies effectively [34,40]. To reach this goal, we are using a set of resources and information through the framework application and using the support of a prototype tool, and it's API [32,33]. This is an ontology-driven approach – in the light of Ontology for Ontological Analysis (O4OA), the meta-ontology that we developed. This way of working allows for clarifying points of semantic conflict between the views of domain specialists and ontology engineers. In this article, we are presenting a case study of

[3] https://www.iso.org/standard/73906.html.
[4] https://www.iso.org/standard/44375.html.

the application of our proposal where conceptualization is used to identify the relationships between the evaluated ontologies. Throughout the text, we explain how the framework allows for achieving this outcome.

We have organized the rest of this paper as follows: Sect. 2 presents a background regarding ON. Section 3 presents the Framework for Ontology Classification and details its domain perspective approach. Section 4 establish a set of relationships among ON. Section 5 presents an analysis of the BRON in light of the framework. Section 6 presents our conclusions and further research.

2 Conceptual Characterization in Ontology Networks

An *Ontology Network* "is a collection of ontologies related together through various relationships, such as alignment, modularization, and dependency" [41]. Besides, the metadata of the ontologies (i.e., their purpose, scope, granularity, etc.), together with the definitions that support the domain of knowledge applicable to these ontologies (and consequently its foundation), define the relationships present in this network. Aligned with this perception, in the work [11], the author explained and formalized a set of operators and postulates for ON based on its revision of beliefs. This approach aims to promote a better formal understanding of the relationships among ontologies.

Ontology networks are not just a set of isolated ontologies grouped together, merely because they act in a domain subdivided into smaller parts (subdomains). On the contrary, they are not limited to a single domain of knowledge, and neither the ontologies of networks are isolated islands of annotated knowledge. Ontologies can relate in different ways within a network. These relationships do not rely only on the domain, its application, the need to reuse these ontologies (or part of them), or if they accomplish some technical aspects (metadata) such as publicity, sharing, standardization, accessibility, and licensing, among others. However, all these aspects interfere with how ontologies are positioned and related within a network. Indeed, all these are requirements to reach FAIRability, and how to deal with these requirements is emerging as a vast subject of study, likewise with Requirements Engineering in the context of Software Engineering. We are talking about FAIRability Requirements Engineering as the body of knowledge that provides the set of requirements that an ontology (or an ON) relies on to be FAIR.

Several works aim to investigate which requirements in terms of metadata an ontology must meet to comply with the FAIR principles, most focusing on best practices. These works are centered in specific domains; for instance, the work [37] focuses on Core Ontologies covering the Security domain, [36] focuses on the Disaster Management domain, and [2] focuses on the Agronomy domain. Besides, the work [3] deals with this issue in quantitative terms of FAIRness, but it relies on prior established best practices notions over the FAIR Principles. However, we were not able to find any work that within the ontological perspective deals with the metadata of ontologies, and at the same time, deals with the cloud of concepts of ontologies at a relational and domain level. In this sense, the F4OA provides an environment where these two perspectives can be treated simultaneously and in a stable way. Moreover, even though our work focuses on the syntactic aspects of the cloud of concepts at the current moment of the investigation, this is not considered a limitation, since working on the semantic aspects is a natural part of the evolution of our proposal. Indeed, it is within the scope of future works.

3 The Framework for Ontology Classification

The **F4OC** [35] is strongly in line with the benefits of DWBP[5], more specifically regarding the "I" and "R". However, the framework also contributes to the other criteria for datasets on the web. Moreover, our approach uses ontological principles such as, that the notion of a foundational ontology [19] is essential in interoperating and reusing ontologies in ON, and is based on the ontological commitment notions present in [17,18,23]. The framework composes of five steps, as depicted in Fig. 1.

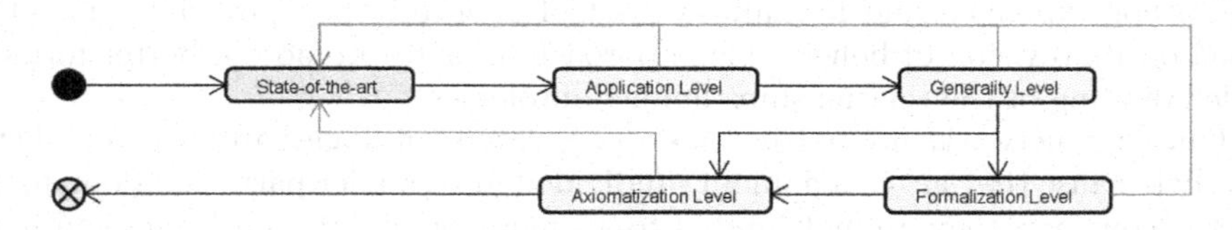

Fig. 1. Framework for Classifying Ontologies - Ontological perspective steps [35].

The classification process provided by the framework is depicted in Fig. 1. The *State-of-the-art* step comprises the search process for relevant information concerning the ontologies covering their domains. Then, the *Application Level* step provides the applicability level classification [23] and identifies if ontologies are well-defined or not. The *Generality Level* step uses the classification proposed in the works [17,42] and verifies whether the ontologies have any ontological grounding through some *Foundational Ontology*, identifying which ontologies are well-grounded or not. The *Formalization Level* step makes a possible classification based on the ontology formalization following the bi-dimensional approach of [13]. And, the *Axiomatization Level* step evaluates the ontologies based on their axiomatization level [14], using the previous bi-dimensional classification. Note in Fig. 1 the gray arrows indicate that during the framework execution, it is possible to return to step 1 to do additional state-of-the-art research for further details or just to fulfill any lack of information if needed. This process covers the ontological perspective and work [35] presents more details about each of these steps.

[5] https://www.w3.org/TR/dwbp/.

Regarding the domain perspective, we defined a process with five steps, as depicted in Fig. 2. The already consolidated umbrella of ontological analysis activities within the Ontology Engineering community supports these steps. Besides, they are in line with the *Support Process* of the SaBiO methodology [1]. The domain perspective of the framework strongly agrees with SaBiO and provides two additional contributions with bias in ontology networks. An important contribution focuses on identifying the cloud of concepts that each ontology of an ON encompasses. Another contribution refers to creating a strategy that links this cloud with the sources (standards, norms, etc.) established within the domain of knowledge of the ON ontologies. This embraces the *Terminological Verification and Validation* process described in [34,40].

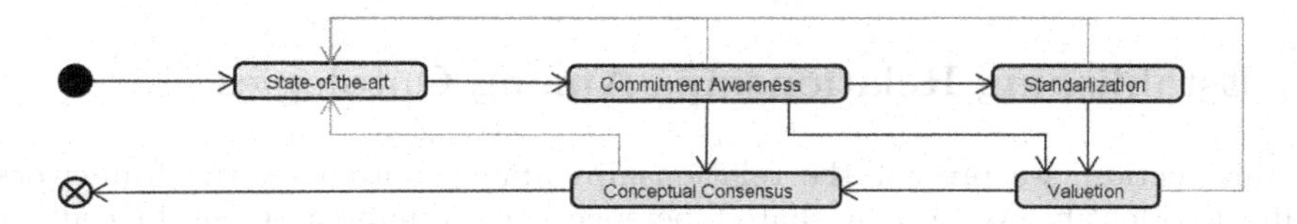

Fig. 2. Framework for Classifying Ontologies - Domain perspective steps.

This step of *State of the art* can be the same performed within the ontological perspective or a separate focusing mainly on the domain perspective aspects. We present more details about it in the work [35]. Then, in the *Commitment Awareness* step Ontology Engineers make contact with the domain and with the support of Domain Specialists identify ontologies' scope, granularity, purpose, competence questions, and any other information relevant to the domain. In the step of *Standardization*, stakeholders search for the best available and consolidated sources of information (classical books, international standards, glossaries, lexicons, classification schemes, and reference models as well-marked in SaBiO) for the various definitions of the terms present in the conceptualization. Each term and its definitions are confronted with the one used in the ontology; the objective is to identify other possible biases for the terms. Likewise, each definition present in the ontology can appear in sources to define other terms than those used in the ontology. The objective is to identify synonyms and terminology mismatching using the confronting the clouds of concepts of ontologies. Next, in the *Valuation* step we evaluate the compiled ontology information in search of modeling anti-patterns, misinterpretations, and flaws which are potential points of divergence or causes of errors. Ontology engineers and domain specialists must participate in comparing the expected representation with the one obtained for the ontology, both as a reference artifact and as an operational artifact. In this sense, as *Competence Questions* state the ontology requirements [1] and interfere in design decisions [15,16]. Finally, stakeholders come to the Conceptual Consensus step, when they reach a semantic consensus regarding the conceptualization and its functional requirements[6] in this step. This consensus must

[6] "Functional requirements refer to the knowledge to be represented" [1].

be well-documented to guarantee traceability and reproducibility. Note, all the domain perspective steps must preserve the principles of representation defended in [18,23].

All documentation gathered from the framework must depict the interoperability and reuse requirements, besides details of these conceptualizations. However, many other relational and inferred information can be reached after the proposed ontological analysis (for each ontology) and a cross-analysis among the ontologies. Indeed, the framework can provide details of interoperable aspects, relations, properties, and other needed information, including those supporting Analytics/AI. In Subsect. 5.2 we present the cloud concepts of BRON, and Subsect. 5.3 shows how these concepts emerge related within the ontologies of the network.

4 Establishing Relationships Among Ontologies

In this section, we present the relationships among ontologies the framework helps to identify. We use a shared perspective (domain and ontological) to describe the principles of those relationships and their implication in ON.

4.1 Grounding

The most accepted classification of ontologies according to its generalization level is Guarino's proposal [17]. Another important classification well accepted is Van Heijst's proposal [42]. The **F4OC** deals with both, providing five classification options. In this case, the framework deals with *Foundational Ontologies* (also called *Upper Ontologies*)[7] and *Non-Foundational Ontologies* (any other more specific regarding its granularity) differently. Thus, in this case Non-Foundational Ontologies can be *Domain Ontologies*[8], *Task Ontologies*[9], *Application Ontologies*[10], or *Core Ontologies*[11].

The notion of *well-grounded* ontologies is based on the statement the support of a Foundational Ontology avoids semantic interoperability problems in more specific ontologies [20]. In other words, Foundational Ontologies are fundamental for Ontology-Driven Modeling Languages (ODML) used to produce Domain, Task, and Application, and also for Core Ontologies, as well as providing ontological analysis for not grounded conceptualizations. Additionally, this step of classification provides stakeholders the ability to study the impact that a lack of ontological foundation can produce when it is necessary to interoperate concepts of this type of non-foundational ontologies and at the same time guarantee semantic clarification.

[7] *Foundational Ontologies* express very general concepts and their relations.

[8] *Domain Ontologies* conceptualize for specific domains.

[9] *Task Ontologies* conceptualize of domain tasks, processes, and activities.

[10] *Application Ontologies* encompass both contexts of Domain and Task Ontologies.

[11] *Core Ontologies* are more general than the other *Non-Foundational Ontologies*, but more specific than Foundational Ontologies.

The framework evidences the relationship between a Foundational Ontology, and any other ontologies grounded by it. Therefore, we define the `groundedOver` relation. It established that concepts and their relations defined in a non-foundational ontology specialize from more general conceptual (philosophical) notions from a Foundational Ontology, defining well-grounded ontologies and allowing stakeholders to make solid semantic considerations. Also, the `groundedOver` relation exists when Non-Foundational Ontologies inherit its foundational aspects through the language used to represent it, as in Subsect. 4.2.

4.2 Ontology-Driven Languages

The domain perspective of the framework helps the stakeholders identify the semantics adopted in different conceptualizations since domain descriptions use terms and their definitions which consequently describe ontological concepts, their relations, and their properties. Indeed, a domain description represents a conceptualization through some *Representation Language*. From this perspective, the framework provides an additional kind of relationship between ontologies and their representation language, we named this relation as `drives`.

As conceptual models, ontologies are described through languages, and these languages are specified through their metamodels. Therefore, as Ontology-Driven Metamodels specify Ontology-Driven Modeling Languages, ontologies described by these languages indirectly carry the ontological constraints imposed by the metamodel that defines them. When these ontological constraints come from domain ontologies, the specified languages are usually Ontology-Driven Modeling DSLs. When these ontological constraints come from foundational ontologies, they are Foundational Ontology-Driven Modeling Languages; for example, OntoUML. The relation `drives` derives from the relational dependency between an Ontology-Driven Metamodel and its concrete language, i.e. derives from the specification of that language. Furthermore, the specification of a Foundational Ontology-Driven Modeling Languages also yields the `groundedOver` relationship (already presented in Sect. 4.1).

4.3 Sub-Ontologies

Domains of knowledge in which the knowledge is vast tend to produce larger ontologies to represent them. Usually, sub-ontologies deal with parts of a larger domain to facilitate their description and comprehension. Thus, an ontology can have other minor ontologies to conceptualize smaller parts and/or contexts of the domain (subdomains), denoting a `part/whole` relationship that occurs between a composite ontology and its parts, i.e. its sub-ontologies. This relation appears in the *Weak Supplementation Pattern*, which states that every whole must be composed by at least two parts [22,24]. Besides, it is *Shareable* (an ontology can be part of many ontologies), *Irreflexive* (there is no ontology that is part of itself), *Asymmetric* (if an ontology B is part of an ontology A then A can not be part of B), *Acyclic* (an ontology can not be in its part-hood transitive closure, i.e. part of its parts, or parts of its parts, and so on), and *Transitive* (if ontology C is part of ontology B, and B is part of ontology A, then C is part of `A`). These are constraints to guarantee the correct interpretation and formalization for the relation that we name `subOntology`. Moreover, there are additional constraints imposed by the classification of the involved ontologies, they are: (i) ontologies composed of other ontologies must necessarily have the same application level; (ii) Foundational Ontologies can only have Foundational Sub-ontologies as parts; (iii) Core Ontologies can only have Core Sub-ontologies as parts; (iv) Non-Foundational Ontologies (Domain, Task, or Application) composed of other ontologies must necessarily have the same or wider granularity level [17], as its sub-ontologies.

4.4 Implementation

Ontologies as computational artifacts have different applications. According to [23], a *Reference Ontology* should be a conceptualization constructed to make the best possible description of the domain with respect to a certain level of generality and point of view. An *Operational Ontology* is the actionable version of a *Reference Ontology* that uses the most appropriate language in order to guarantee desirable computational properties without compromising the previously defined ontological commitment. Therefore, we identify a relationship among a *Reference Ontology* and its *Operational* version in the light of the O4OA that we named it `implementationFor`. This relation allows stakeholders to evaluate the relational characteristics[12], that a Reference Ontology can provide to its implementations (Operational Ontologies). Besides, this approach can also help ontology engineers deal with implementation language limitations by knowing which ontological aspects can (or can not) be implemented without losing ontological decidability. Note that the terms and their definitions (regarding the domain perspective – see the framework step of standardization) are the discretionary part of the structures that make up a domain description, regarding the possible adopted semantics. Therefore, stakeholders can compare the sources of information (from the state-of-the-art step) that support a conceptualization with the sources available about the domain. This helps ontology engineers identify the concepts, the definitions (aligned with domain specialists) that support these concepts, and the ontological commitment taken.

4.5 Reuse

The framework classification according to the ontology generality level provides another important feature. An ontology can reuse another ontology; in this case, we delimited the possible cases of reuse and named this relation as `reuses`. In general terms, this relation denotes an intersection among ontologies, and it requires that this intersection be semantically compatible (equivalent) with the involved ontologies. Therefore, the reuse of ontologies can occur in different ways concerning the concepts represented by an ontology.

[12] Given the ontological notion adopted in UFO [21].

Typically, the relation **reuses** requires at least the reuse of one concept (with its terminology, definitions, i.e. all its semantics) among two or more ontologies. One common way of reuse is when one or more ontologies specialize concepts in one or more concepts in another different ontology. Apart from the specializations of concepts defined in non-foundational ontologies from Foundational Ontologies notions (presented in Subsect. 4.1), ontologies can have specializations among those defined by concepts in non-foundational ontologies. Other kinds of relations, including part/whole relations, can appear to connect one ontology to another, denoting the reuse of its concepts (providing intersection).

Reusing ontologies can also occur by adding a more robust ontological foundation but maintaining the alignment of the domain definitions already adopted. This situation happens when the domain perspective about the definitions present in a conceptualization is aligned, but the ontological perspective must be reinforced. In other words, when the reused ontology lacks an ontological foundation and requires the grounding provided by a Foundational Ontology or the use of an ODML (which is provided by the ontology to be reused).

The notion of reuse adopted by the framework requires the control of some constraints, they are: (i) Foundational Ontologies can only reuse from Foundational Ontologies; (ii) Core Ontologies can reuse Core Ontologies; Domain Ontologies can reuse Domain Ontologies; Task Ontologies can reuse Task Ontologies; (iii) Application Ontologies can reuse Application, Domain, and/or Task Ontologies; and (iv) Application, Domain, Task Ontologies can reuse Core Ontologies. Last but not least, individually, no ontology can not reuse itself or any of its parents (see the notion of sub-ontologies in Subsect. 4.3).

4.6 Summary of the Relationships Among Ontologies

Table 1 summarizes the relationships among ontologies we describe thought the F4OC application, and the main formal aspect that describe them.

5 Conceptual Characterization of BRON

BRON is an open-source initiative[13] from ALFA group at CSAIL, MIT[14]. It is linking together offensive, defensive and vulnerability concepts to analyze potential attacks and their potential counter measurements [27]. BRON implements its ontological approach as a Knowledge Graph using OWL and RDF formats[15]. The implementation of BRON links together *ATT&CK*[16] and *CAPEC*[17] (for

[13] http://bron.alfa.csail.mit.edu/info.html.
[14] http://alfagroup.csail.mit.edu/.
[15] https://github.com/ALFA-group/BRON.
[16] https://attack.mitre.org/.
[17] https://capec.mitre.org/index.html.

Table 1. Summary of the Relationships among Ontologies.

Relationship	Description	Formalizm
`Drives`	When the language of the ontology is specified through an Ontology-Driven Metamodel	Relational Dependence
`GroundedOver`	When the ontology concepts are grounded over a Foundational ontology	Specialization
	When the language of the ontology is specified through a Foundational-Ontology-Driven Metamodel	Relational Dependence
`SubOntology`	When an ontology is composed by two or more ontologies	Part/Whole
`ImplementationFor`	When an Operational Ontology is defined by a Reference Ontology	Relational Dependence
`Reuses`	When ontologies have concept overlapping	Relational dependence

the offensive concepts), *D3FEND* [28] and *Engage*[18] (for the defensive concepts), and *CWE*[19] and *CVE*[20] (for the vulnerability concepts).

5.1 Classification of BRON

We classified each ontology of BRON based in the sources of information obtained during the state-of-the-art framework step. Later we repeated the framework steps, for BRON itself and already incorporating the information collected about each of its related ontologies. Table 2 summarizes the resulting characterization for BRON.

Table 2. Summary of BRON Ontologies Characterization.

	Applicability Level		Generality Level		Bisimensional Classification [13]	
Ontology	Ref. Ontology	Oper. Ontology	Found. Ontology	Classif. [18,42]	Formalization Level	Axiom. Level
ATT&CK	No	Yes	No	Task Ontology	Formal Ontology (XMLS)	Lightweight
D3FEND	No	Yes	No	Domain Ontology	Formal Ontology (XMLS)	Lightweight
CWE	No	Yes	No	Domain Ontology	Informal Ontology/Tesauri and Taxonomies	Lightweight
CVE	No	Yes	No	Domain Ontology	Informal Ontology/Tesauri and Taxonomies	Lightweight
CAPEC	No	Yes	No	Domain Ontology	Informal Ontology/Tesauri and Taxonomies	Lightweight

Finally, **BRON** is an *Application Ontology* [23] since it inbounds both a specific domain (Domain Ontology) and a set of tasks that scans new information from its network (Task Ontology). Besides, in the documentation studied

[18] https://github.com/mitre/engage.
[19] https://cwe.mitre.org/.
[20] https://www.cve.org/.

and also considering all its ontologies (previously classified), we are not able to identify any Reference Ontology supporting its operational conceptual model, therefore we consider that it is *not well-defined* (or *imprecise*). BRON is *not grounded* because there is no foundational ontology grounding its conceptual model (OWL), although their representation language is a W3C recommendation. Additionally, regarding the formalization level, it is considered a *Formal Ontology (Lightweight Ontology)* as well as regarding the axiomatization level it is a *Lightweight Ontology* according to the classification proposed in [13].

5.2 The Domain Perspective of BRON

BRON is an initiative within the CyberSecurity domain, and it is one of the cases of study of a research project in which we participate as members[21]. Therefore, we have the participation of a professional team of cybersecurity experts who actively work in providing their semantic agreement about the domain perspective application of the framework. We have executed the steps presented in the Sect. 3 and all information we reach in our state-of-the-art [34], standardization [33,40], and valuation [35] steps were documented thought an API over a NoSQL database we developed [32,33]. It is important to point that the framework is agnostic regarding the use of tools and automatic means to achieve good documentation and data administration, as long as the semantic consensus is reached and the information collected can be traceable. However, we believe that vast and complex domains require some automatization support.

5.3 Impact in Ontology Networks

The framework application outcomes provide a homogeneous base of comparison for the studied ontologies, which contemplates both the ontology engineer's and the domain specialist's perspectives. Besides giving an ontological classification outcome for the BRON ontology (and its related ontologies), it provides the stakeholders with the conceptualization definitions used in the ontology (domain perspective), and a broad view of how BRON is organized. Regarding the domain perspective, each documented concept present in the ontologies is being mapped through our team of domain specialists (cybersecurity experts that are members of the consortium). While regarding the ontological perspective, the broad view of how the ontology is organized provides the ontology engineering team with a better understanding of the domain and allows for more fluid communication with domain experts. This also helps us to have a better understanding of the conceptualization, their application context as well as the provided implementation. Moreover, this approach is going in line with the notions of *interpretation and implementation* provided in [25].

[21] Our research is part of a project to develop KGs (TKG and DTKGs) through a comprehensive solution within a project with Accenture LTD. The consortium also has research in partnership with other academic research centers.

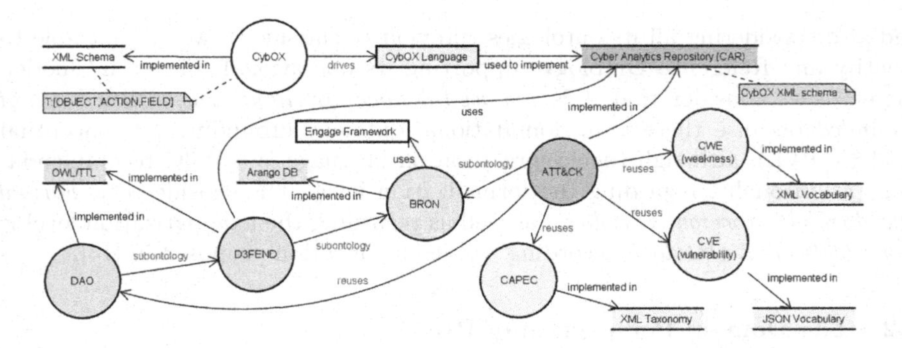

Fig. 3. Framework outcome showing relational aspects present BRON.

Fig. 3 shows a visual representation of the framework outcomes, showing a broad view of the organization of BRON. Note, the data instances present in this representation were obtained from queries made over the API we developed. We identify relations compounding BRON and (among its related ontologies) and its conceptualization.

In this case, BRON has two sub-ontologies, ATT&CK and D3FEND. Besides, we identify that DAO is a sub-ontology of D3FEND meanwhile reuses ATT&CK, mapping concepts from ATT&CK to D3FEND [28]. Moreover, ATT&CK reuse CAPEC, CVE, and CWE, such that its concept of *Technique* links to an existing CAPEC instance, likewise other standards such as CVE, CWE, and CAPEC itself [29]. Note, that the framework evidenced the absence of ontological patterns (semantic consensus) in those reuse relations adopted in BRON, i.e. there is a different approach for each case. For instance, comparing the definition used to conceptualize what is a *Vulnerability* in CVE and CWE, reused by ATT&CK:

CVE: A flaw in a software, firmware, hardware, or service component resulting from a weakness that can be exploited, causing a negative impact to the confidentiality, integrity, or availability of an impacted component or components.

CWE: An occurrence of a weakness (or multiple weaknesses) within a product, in which the weakness can be used by a party to cause the product to modify or access unintended data, interrupt proper execution, or perform incorrect actions that were not specifically granted to the party who uses the weakness.

Regarding a foundational approach (well-grounded), "**a flaw**" is not the same as "**an occurrence of weakness**". The first is a *Disposition* that depends on an object (*Endurant*) while the second is a *Perdurant* in the light of UFO [21], and as also described in OSDEF [9]. Indeed, we carried out the ontological analysis of both, CVE and CWE, during the cross-analysis process of the framework, and we used the OSDEF (grounded over UFO) to do it since CVE and CWE are not well-founded. However, it is not the purpose of this paper to detail these analyses, so we will not go into greater detail. In any case, the framework allows us to carry out a terminological analysis based on the cloud of concepts of these ontologies.

Thus, CVE and CWE are very good for identifying patterns of attack and possible elements of mitigation of these attacks, but they are not effective in tracking in-depth vulnerabilities and predicting failures in systems with requirements and coding traceable, such as those that OSDEF can treat. In this sense, BRON (ATT&CK) had to use programmatic resources to work around this issue. In this case, the necessary information is particularly present as properties[22] in both taxonomies. Thus, we conclude that integration of BRON (its ontologies) with an ontology such as OSDEF, aiming to track (or reason about) vulnerabilities to predict and avoid (or mitigate) weaknesses in software systems becomes more costly, given the analysis presented through the framework. This is a situation of lack of FAIRability.

This behavior makes the semantic resolution of those contained in integrating these ontologies concept-dependent, complex, and subject to exceptionalities. This should be done at the ontological level (of conceptual modeling), not through an implementation solution for data processing. Therefore, the relational structure identified has implications in their conceptualizations (and vice-versa) due to ontological perspective (and their ontological commitment) and considering its aim of making the semantics clear and accessible.

6 Conclusions

The F4OC demonstrates the ability to facilitate the identification and study of ontology metadata, going beyond the more explicit notions of ontologies. The holistic view approximates the specific metadata of the ontological perspective (accessibility, availability, sharing, aspects of modeling and implementation, etc.) with those of the domain perspective (cloud of concepts, domain structuring, granularity, etc.) present in ontologies and on ontology networks. The results have been promising, considering assessments by domain experts on the investigation team, including experts from Accenture LTD. The next step of our research is to validate the framework capabilities through an experiment run in partnership with Accenture, with subsequent presentation of more analytical results.

However, quantitative measurements are out of the scope of our research since we focus on a qualitative approach strongly related to FAIR Principles regarding its best practices. We are aware that this is an important step forward that can enhance the research since each mistake or misunderstanding that is not avoided in the reusability and interoperability process in complex domains can involve untold losses for organizations. An additional benefit of the framework is providing the reproducibility and traceability required for precise qualitative measurements. This opens opportunities for future research work in qualitative evaluations under the umbrella of the Ontology Engineering process.

[22] For example, see https://www.cvedetails.com/cve/CVE-1999-0067/. Note that CWE is not defined for this vulnerability in this example.

For instance, measuring the impact of misinterpretations caused by semantic interoperability issues in reusing or interoperating ontologies in an ON.

The broad view of relationships between ontologies that the framework provides, together with the domain perspective, allows us to deal with interoperability and reuse ontologies in ON since it clarifies the semantics of each ontology and describes its relationships. Moreover, this approach allows us to refine the outcomes presented in [35]. We believe the framework has good potential in facilitating ontological analysis in complex domains and helps to guarantee the notions advocated in [25]. As the next steps of our work, we will use the relationships established between the ontologies through the framework to stratify conceptualizations and their definitions, providing greater precision in the process of ontological analysis and cross-analysis of ontologies. Our goal is to facilitate reuse and semantic interoperability at the ontological level for knowledge domains where, due to their complex nature, this is increasingly important.

Acknowledgements. This work has been developed under the project Digital Knowledge Graph – Adaptable Analytics API with the financial support of Accenture LTD, the Generalitat Valenciana through the CoMoDiD project (CIPROM/2021/023), the Spanish State Research Agency through the DELFOS (PDC2021-121243-I00) and SREC (PID2021-123824OB-I00) projects, MICIN/AEI/10.13039/501 100011033 and co-financed with ERDF and the European Union Next Generation EU/PRTR.

References

1. de Almeida Falbo, R.: Sabio: Systematic approach for building ontologies. In: Onto. Com/odise@ Fois (2014)
2. Amdouni, E., Bouazzouni, S., Jonquet, C.: O'faire: ontology fairness evaluator in the agroportal semantic resource repository. In: Amdouni, E., Bouazzouni, S., Jonquet, C. (eds.) The Semantic Web: ESWC 2022 Satellite Events: Hersonissos, Crete, Greece, May 29-June 2, 2022, Proceedings, pp. 89–94. Springer, Cham (2022). https://doi.org/10.1007/978-3-031-11609-4_17
3. Amdouni, E., Jonquet, C.: FAIR or FAIRer? An integrated quantitative FAIRness assessment grid for semantic resources and ontologies. In: Garoufallou, E., Ovalle-Perandones, M.-A., Vlachidis, A. (eds.) MTSR 2021. CCIS, vol. 1537, pp. 67–80. Springer, Cham (2022). https://doi.org/10.1007/978-3-030-98876-0_6
4. Bernasconi, A., Guizzardi, G., Pastor, O., Storey, V.C.: Semantic interoperability: ontological unpacking of a viral conceptual model. BMC Bioinform. **23**(11), 1–23 (2022)
5. Borges Ruy, F., de Almeida Falbo, R., Perini Barcellos, M., Dornelas Costa, S., Guizzardi, G.: SEON: a software engineering ontology network. In: Blomqvist, E., Ciancarini, P., Poggi, F., Vitali, F. (eds.) EKAW 2016. LNCS (LNAI), vol. 10024, pp. 527–542. Springer, Cham (2016). https://doi.org/10.1007/978-3-319-49004-5_34
6. Brazhuk, A.: Towards automation of threat modeling based on a semantic model of attack patterns and weaknesses. arXiv preprint arXiv:2112.04231 (2021)
7. Duarte, B.B., de Almeida Falbo, R., Guizzardi, G., Guizzardi, R., Souza, V.E.S.: An ontological analysis of software system anomalies and their associated risks. Data Knowl. Eng. **134**, 101892 (2021)

8. Duarte, B.B., et al.: Ontological foundations for software requirements with a focus on requirements at runtime. Appl. Ontol. **13**(2), 73–105 (2018)
9. Duarte, B.B., Falbo, R.A., Guizzardi, G., Guizzardi, R.S.S., Souza, V.E.S.: Towards an ontology of software defects, errors and failures. In: Trujillo, J.C., Davis, K.C., Du, X., Li, Z., Ling, T.W., Li, G., Lee, M.L. (eds.) ER 2018. LNCS, vol. 11157, pp. 349–362. Springer, Cham (2018). https://doi.org/10.1007/978-3-030-00847-5_25
10. Engelberg, G., Fumagalli, M., Kuboszek, A., Klein, D., Soffer, P., Guizzardi, G.: Towards an ontology-driven approach for process-aware risk propagation. arXiv preprint arXiv:2212.11763 (2022)
11. Euzenat, J.: Revision in networks of ontologies. Artif. Intell. **228**, 195–216 (2015)
12. García S, A., Guizzardi, G., Pastor, O., Storey, V.C., Bernasconi, A.: An ontological characterization of a conceptual model of the human genome. In: Intelligent Information Systems: CAiSE Forum 2022, Leuven, Belgium, 6–10 June 2022, Proceedings, pp. 27–35. Springer, Cham (2022). https://doi.org/10.1007/978-3-031-07481-3_4
13. Giunchiglia, F., Zaihrayeu, I.: Lightweight ontologies. University of Trento, Technical report (2007)
14. Gómez-Pérez, A., Corcho, O.: Ontology languages for the semantic web. IEEE Intell. Syst. **17**(1), 54–60 (2002)
15. Gruninger, M.: Methodology for the design and evaluation of ontologies. In: International Joint Conference on Artificial Intelligence (1995)
16. Gruninger, M.: Designing and evaluating generic ontologies. In: 12th European Conference of Artificial Intelligence, vol. 1, pp. 53–64. Citeseer (1996)
17. Guarino, N.: Formal ontology in information systems. In: Proceedings of the 1st International Conference, pp. 6–8. IOS Press, Trento, Italy, June 1998
18. Guarino, N.: The ontological level. Philosophy and the Cognitive Sciences (1994)
19. Guarino, N., Poli, R.: The role of formal ontology in the information technnology. Int. J. Human Comput. Stud. **43**(5–6), 623–965 (1995)
20. Guizzardi, G.: The role of foundational ontology for conceptual modeling and domain ontology representation, keynote paper. In: 7th International Baltic Conference on Databases and Information Systems (DB&IS), Vilnius, IEEE Press (2006)
21. Guizzardi, G.: Ontological Foundations for Structural Conceptual Models. CTIT, Centre for Telematics and Information Technology (2005)
22. Guizzardi, G.: Modal aspects of object types and part-whole relations and the *de re/de dicto* distinction. In: Krogstie, J., Opdahl, A., Sindre, G. (eds.) CAiSE 2007. LNCS, vol. 4495, pp. 5–20. Springer, Heidelberg (2007). https://doi.org/10.1007/978-3-540-72988-4_2
23. Guizzardi, G.: On ontology, ontologies, conceptualizations, modeling languages, and (meta) models. Front. Artif. Intell. Appl. **155**, 18 (2007)
24. Guizzardi, G.: The problem of transitivity of part-whole relations in conceptual modeling revisited. In: van Eck, P., Gordijn, J., Wieringa, R. (eds.) CAiSE 2009. LNCS, vol. 5565, pp. 94–109. Springer, Heidelberg (2009). https://doi.org/10.1007/978-3-642-02144-2_12
25. Guizzardi, G.: Ontology, ontologies and the "I" of FAIR. Data Intell. **2**, 181–191 (2020)
26. Hemberg, E., et al.: Linking threat tactics, techniques, and patterns with defensive weaknesses, vulnerabilities and affected platform configurations for cyber hunting (2021)
27. Hemberg, E., O'Reilly, U.M.: Using a collated cybersecurity dataset for machine learning and artificial intelligence. arXiv preprint arXiv:2108.02618 (2021)

28. Kaloroumakis, P.E., Smith, M.J.: Toward a knowledge graph of cybersecurity countermeasures. Corporation (2021)
29. Kurniawan, K., Ekelhart, A., Kiesling, E.: An att&ck-kg for linking cybersecurity attacks to adversary tactics and techniques. In: In Proceedings of the International Semantic Web Conference (ISWC) (2021)
30. Mann, D.E., Christey, S.M.: Towards a common enumeration of vulnerabilities. In: 2nd Workshop on Research with Security Vulnerability Databases, Purdue University, West Lafayette, Indiana (1999)
31. Martin, R.A., Barnum, S.: Common weakness enumeration (CWE) status update. ACM SIGADA Ada Lett. **28**(1), 88–91 (2008)
32. Martins, B.F., Serrano, L., Reyes, J.F., Panach, J.I., Pastor, O.: Towards the Consolidation of Cybersecurity Standardized Definitions. Technical report. Version 2, Universidad Politecnica de Valencia (2021). http://hdl.handle.net/10251/163895
33. Martins, B.F., Serrano, L., Reyes, J.F., Panach, J.I., Pastor, O.: Towards the consolidation of cybersecurity standardized definitions: a tool for ontological analysis. In: Proceedings of the XXIV Iberoamerican Conference on Software Engineering, CIbSE 2021, San José, Costa Rica, 2021, pp. 290–303. Curran Associates (2021)
34. Martins, B.F., Serrano, L., Reyes, J.F., Panach, J.I., Pastor, O., Rochwerger, B.: Conceptual characterization of cybersecurity ontologies. In: Grabis, J., Bork, D. (eds.) PoEM 2020. LNBIP, vol. 400, pp. 323–338. Springer, Cham (2020). https://doi.org/10.1007/978-3-030-63479-7_22
35. Martins, B.F., et al.: A framework for conceptual characterization of ontologies and its application in the cybersecurity domain. Softw. Syst. Modeling **21**(4), 1437–1464 (2022)
36. Mazimwe, A., Hammouda, I., Gidudu, A.: Implementation of fair principles for ontologies in the disaster domain: A systematic literature review. ISPRS International Journal of Geo-Information **10**(5), 324 (2021)
37. Oliveira, Í., Fumagalli, M., Prince Sales, T., Guizzardi, G.: How FAIR are Security Core Ontologies? A Systematic Mapping Study. In: Cherfi, S., Perini, A., Nurcan, S. (eds.) RCIS 2021. LNBIP, vol. 415, pp. 107–123. Springer, Cham (2021). https://doi.org/10.1007/978-3-030-75018-3_7
38. Perini Barcellos, M., de Almeida Falbo, R.: Using a foundational ontology for reengineering a software enterprise ontology. In: Advances in Conceptual Modeling-Challenging Perspectives: ER 2009 Workshops CoMoL, ETheCoM, FP-UML, MOST-ONISW, QoIS, RIGiM, SeCoGIS, Gramado, Brazil, November 9–12, 2009. Proceedings 28. pp. 179–188. Springer (2009)
39. Sales, T.P., Baião, F., Guizzardi, G., Almeida, J.P.A., Guarino, N., Mylopoulos, J.: The Common Ontology of Value and Risk. In: Trujillo, J.C., Davis, K.C., Du, X., Li, Z., Ling, T.W., Li, G., Lee, M.L. (eds.) ER 2018. LNCS, vol. 11157, pp. 121–135. Springer, Cham (2018). https://doi.org/10.1007/978-3-030-00847-5_11
40. Serrano, L., Martins, B.F., Serrano, J.F., Panach, J.I., Pastor, O.: Una encuesta acerca de la Definición de Conceptos de Ciberseguridad. Tech. rep., Universidad Politecnica de Valencia (2021), https://riunet.upv.es/handle/10251/174756
41. Suárez-Figueroa, M.C., Gómez-Pérez, A., Motta, E., Gangemi, A.: Introduction: Ontology engineering in a networked world. Springer (2012)
42. Van Heijst, G., Schreiber, A.T., Wielinga, B.J.: Using explicit ontologies in kbs development. International journal of human-computer studies pp. 183–292 (1997)
43. Wilkinson, M.D., Dumontier, M., Aalbersberg, I.J., Appleton, G., Axton, M., Baak, A., Blomberg, N., Boiten, J.W., da Silva Santos, L.B., Bourne, P.E., et al.: The fair guiding principles for scientific data management and stewardship. Scientific data **3**(1), 1–9 (2016)

Business Process Design and Computing in the Continuum

Digital Technology-Driven Business Process Redesign: A Classification Framework

Kateryna Kubrak(✉), Fredrik Milani, and Juuli Nava

University of Tartu, Tartu, Estonia
{kateryna.kubrak,fredrik.milani}@ut.ee

Abstract. Organizations constantly seek ways to improve their business processes. This often involves using digital technologies to enable process improvements. However, simply substituting existing technology with newer technology has limited value as compared to using the capabilities of digital technologies to introduce changes to business processes. Therefore, process analysts need to understand how the capabilities of digital technologies can be used to redesign business processes. In this paper, we conducted a systematic literature review and examined 40 case studies where digital technologies were used to redesign business processes. We identified that, within the context of business process improvement, capabilities of digitalization, communication, analytics, digital representation, and connectivity can enable business process redesign. Furthermore, we note that these capabilities enable applying nine redesign heuristics. Based on our review, we map how each capability can facilitate the implementation of specific redesign heuristics to improve a business process. Thus, our mapping can aid analysts in identifying candidate redesigns that capitalize on the capabilities of digital technologies.

Keywords: Business process redesign · Digital technologies · Redesign heuristics

1 Introduction

Organizations constantly seek to improve their business processes to optimize the efficiency and quality of their products and services [14]. Business processes are commonly improved by enacting changes, i.e., redesigning the business processes [32]. Technology has always played an important role in how business processes are redesigned [9]. For instance, client-server architecture enabled the re-engineering of business processes [17]. Likewise, the internet made it possible to automate certain business processes by enabling online self-service [35].

In the past decades, digital technologies have ushered in new opportunities for business process improvements. This has caused a surge in initiatives to digitally transform organizations through redesigning existing business processes [48]. However, merely substituting existing technology with a newer one has limited value [9]. Rather, value is gained when capabilities of digital technologies are

S. Nurcan et al. (Eds.): RCIS 2023, LNBIP 476, pp. 205–221, 2023.
https://doi.org/10.1007/978-3-031-33080-3_13

used to innovate and redesign business processes [8]. Therefore, analysts working with improving business processes benefit from understanding how capabilities of digital technologies can enable the redesign of business processes [36].

Several studies analyzed digital technologies through the lens of business process management (BPM). For instance, Imgrund et al. [19] explain how to approach digitalization with business process management (BPM), Van Looy [50] explores synergies between BPM and digital innovation, and Löffler et al. [30] examine how to communicate business process change in the context of digital transformation. These works, however, consider digital transformation from an overall BPM perspective. They do not consider business process redesign, which is one of the main steps of BPM [14]. Furthermore, authors, such as [3,18,40,46] propose frameworks for digital technologies. These studies discuss the impact of specific digital technologies on business processes but not how to use the capabilities of digital technologies for process redesign. Thus, to the best of our knowledge, there is a lack of research on how to connect the capabilities of digital technologies with business process redesign.

We address this gap by eliciting a classification framework for digital technology-driven redesign of business processes. More specifically, we explore *what capabilities of digital technologies have been used to redesign business processes* and *how such capabilities have been used to redesign business processes.* Therefore, first, we consider the capabilities of digital technologies that can be used to redesign business processes. With this RQ, we seek to understand the distinctive capabilities of digital technologies in light of business processes. However, digital technologies can deliver value if the business process is redesigned to exploit these capabilities [9]. Therefore, we also consider how business processes can be redesigned through the capabilities of digital technologies.

Thus, the contribution of this paper is a classification framework for how capabilities of digital technologies can enable the redesign of business processes. To address the RQs and develop the framework, we conducted a systematic literature review (SLR) [22] to identify case studies where digital technologies are used to redesign business processes. Then, we identified the capabilities of digital technologies and mapped them against business process redesign heuristics proposed by Mansar & Reijers [32] and synthesized the results. The classification framework can aid analysts working with incorporating digital technologies in new or existing business processes. More specifically, it can help them consider redesigns that capitalize on the capabilities of employed digital technologies.

The rest of this paper is structured as follows. Background is outlined in Sect. 2, and related work in Sect. 3. In Sect. 4, we describe the systematic literature review protocol and present the results Sect. 5. In Sect. 6, we present and discuss the classification framework. Finally, Sect. 7 concludes the paper.

2 Background

As defined by Dumas et al. [14], "business process management (BPM) is a body of principles, methods, and tools to discover, analyze, redesign, implement and monitor business processes." Analysis provides input necessary for redesign. Analysts can use data-driven tools [25] to e.g., analyze batch behavior in a process [27], process performance [38], or other types of process analysis [37].

In the redesign phase, process redesigns are proposed based on the analysis conducted in the previous steps. There are different approaches to business process redesign. However, but the most widely adopted catalog of best practices in process redesigns are proposed in Mansar & Reijers [32] as business process redesign heuristics. There are 29 redesign heuristics that encompass different perspectives of a business process, such as customer, product, organizational, informational, and technological. Some of the heuristics can only be applied if a technology is used. For instance, *task automation* uses technology to automate repetitive manual tasks [32]. Other heuristics, such as *control addition*, can be applied manually. However, digital technologies enable applying such heuristics more efficiently. For instance, manual control addition can be replaced with a digital technology that enables data-driven and automated controls [11].

Currently, there is no consensus on the definition of digital technologies. Some researchers argue that "digital technology" are hype words that build on concepts that have been around for a long time [48] or that it is complex as it can be viewed from different perspectives (organizational, social, etc.) [20]. Nevertheless, Lipsmeier et al. [3] describe digital technologies as *"something that is made up of knowledge, skills and know-how for the creation, processing, transmission and use of digital data as well systems and procedures for practical implementation."* In this paper, we adopt this definition.

Similarly, different definitions have been proposed for a digital capability [18]. However, Korhonen et al. [24] define digital capability as the "capacity" of an enterprise *"to integrate and utilize digital data and information technologies in its products, services, business processes, and organizational systems and practices to create added value to its constituents and beneficiaries."* Here, we consider digital capabilities within the specific context of business process redesign. As such, we adopt this definition but define it as the capacity of a *digital technology* to enable the redesign of a business process to create added value.

Several attempts have been made to categorize digital technologies and their capabilities. Thus, Pousttchi et al. [40] categorize digital technologies into three categories: communication and other enabling technologies, those combining hardware and software in intelligent systems, and data technologies. Lipsmeier et al. [3], though, propose a more refined classification of digital technologies. They divide them into analytics, connectivity, fabrication, human-to-machine interface, interactivity, sensoring, storage, and visualization. Each category shares the commonality of having the same or similar digital capabilities. Henriette et al. [18] identified six capabilities of digital technologies: dematerialization, internet technologies, analytics, mobility, social network, knowledge, and skills.

3 Related Work

Studies that classify digital technologies according to different criteria are related to our work. For instance, Pousttchi et al. [40] provide a generic model to map the relationship between the use of digital technologies and their impact on companies. As a result, they propose a model that categorizes digital technologies. In Henriette et al. [18], the authors explore the nature of digital transformation. They report what digital capabilities are impacted by digital transformation and how digitalization transforms business models and user experience. Lipsmeier et al. [3], though, identify eight technology classes that are based on digital data operations. In da Silva et al. [46], the authors present 24 key factors that organizations should consider when implementing digital factoring in the context of Industry 4.0. These studies provide valuable insight into how digital technologies and capabilities can be classified. Therefore, they are complementary to our work. However, while these studies classify digital technologies, they do not connect them to BPM or provide insights into how processes can be redesigned.

A few studies have explored the intersection between digitalization and business processes. For instance, in Imgrund et al. [19], the authors propose a framework that can serve as the starting point for organizations seeking to embark on a digital transformation. The framework presents a set of organizational capabilities required for managing digital transformation and answers questions related to how BPM can be used to increase the feasibility of digital transformation.

Ahmad [1] addresses the question of how organizations can adopt technologies best fitted for specific business processes given their context. Mendling et al. [34] analyze previous works on BPM and digital innovation and conclude that they should be researched together rather than serve as complementary fields. Furthermore, Van Looy [50] identifies obstacles that companies face when using BPM for digital innovation. The study combines critical success factors and strategic dimensions of BPM with digital innovation and argues for combining them as they depend on the organization's context. These studies consider capabilities relevant at the organizational level to initiate digital transformation or incorporate digital technologies. Our contribution is complementary as we focus on the capabilities of digital technologies applicable to business process redesign.

4 Systematic Literature Review

In this paper, we aim to elicit a classification framework for categorizing how capabilities of digital technologies can be employed to redesign business processes. This objective can be achieved by addressing two research questions. The first aims at examining which capabilities of digital technologies have been used to redesign business processes. Then, we seek to understand how such capabilities have been used to redesign business processes. Thus, our research questions are as follows.

RQ$_1$. *What capabilities of digital technologies have been used to redesign business processes?*

RQ$_2$. *How have capabilities of digital technologies been used to redesign business processes?*

To address the research questions, we employ a systematic literature review method as it allows for summarizing existing relevant research [22] and, based on the review of existing research, developing a classification framework. In conducting the SLR, we followed the guidelines proposed by Kitchenham [22]. Therefore, we defined the search strings, selected electronic databases, defined exclusion and inclusion criteria, defined data extraction strategy, applied the search strings to identify a list of potentially relevant papers, examined the papers using exclusion and inclusion criteria to filter out irrelevant publications, and extracted data from the final list of relevant papers [22].

To identify studies that, in some form, used digital technologies for process redesign, we derived search string from the paper's objective and research questions. First, we applied a search string that contained terms such as digital, process, improvement, redesign, change, optimization, and case study. However, the derived search strings were too broad and resulted in highly irrelevant studies. Therefore, we narrowed the search string to include the key terms "business process" and "digital". We also added "case study" to capture studies that have implemented and, in some form, validated their findings. Thus, we applied the following search string: *"business process" AND "digital" AND "case study"* on Scopus and Web of Science. We selected these electronic databases as they index most publications within the domain of BPM.

The next step, defining selection criteria, serves to identify relevant studies that can provide sufficient information for addressing the research questions. The initial list of papers was filtered following exclusion and inclusion criteria. The first exclusion criterion served to clean the data for entries that were not papers (EC1). The second was to remove duplicates (EC2). The third exclusion criterion served to ensure that the study can be accessed and understood (EC3). Papers with open access, accessible via the University or in other ways accessible on internet, were considered accessible. As such, papers behind paywalls were considered inaccessible. Papers in any other language than English were also removed. The first inclusion criterion ensured that the papers explicitly discuss a particular digital technology in the context of business process redesign (IC1). Thus, papers that use digital technologies for other purposes than redesigning business processes, e.g., to enable a new business strategy or to increase employee satisfaction, were removed. The second inclusion criteria filtered out studies that do not explicitly discuss case studies and, therefore, have not been tested in practice (IC2). For instance, in [36], the authors examine blockchain technology for redesigning the timber-to-charcoal process and include a theoretical case study. However, the case study is not implemented in real-life and, therefore, is not included in our review. Likewise, we excluded papers that did not provide sufficient information to address the research questions. Thus, papers stating that business value was achieved through to digital transformation but did not discuss the implementation were discarded. Finally, as the technology landscape is shifting fast, we excluded papers dating before 2017.

Table 1. Paper selection process.

Selection Criteria	# Found	# Left
Search results	260	
(EC1) Data cleaning	23	237
(EC2) Filtering by duplicates	40	197
(EC3) Filtering by paper language	6	191
(IC1) Filtering by scope	108	83
(IC2) Filtering by case study criterion	70	13
Backward referencing	27	40
Total		**40**

Our primary search resulted in 260 papers (181 from Scopus and 79 from Web of Science). We used the search filter functions to remove studies published before 2017. Thus, the 260 were all published from 2017 and on-wards. We applied the exclusion criteria and, therefore, discarded 23 non-papers (e.g., proceedings volumes) (EC1), removed 40 duplicate studies (EC2), and 6 non-English papers (EC3). For inclusion criteria, we filtered out 108 papers that were clearly out of the scope of our study (IC1) and additional 70 papers that did not provide sufficient information or had not conducted a case study (IC2). We also conducted backward referencing, which resulted in adding 27 papers. In the end, the final list of papers consisted of 40 studies (Table 1). The final list of papers can be accessed at: https://doi.org/10.6084/m9.figshare.21947564.v2

We then proceeded with data extraction. In addition to extracting metadata of the paper (authors, title, etc.), we extracted data on the main technology used, the business process before and after the change, what changes were implemented, how the digital technology enabled the process change, the capability of the digital technology, and what redesign heuristics were applied. It should be noted that some publications did not explicitly state which redesign heuristic was used. In such cases, we examined the case and assigned the redesign heuristics they applied based on heuristics descriptions in [32].

5 Results

In this section, we present the results of our review. We begin with results regarding capabilities of digital technologies that enable business process redesign (**RQ**$_1$) and continue with results on how such capabilities have been used to redesign business processes (**RQ**$_2$).

5.1 Capabilities of Digital Technologies for Process Redesign

Our review showed that digital technologies are used for *digitalization* of paper-based processes. For instance, in Di Vaio and Varriale [13], a case is presented where paper-based documentation and tasks are digitalized in a supply chain process. In another example, an insurance company moved from paper to digital and then used Robotic Process Automation (RPA) technology to enable digital information processing [33]. Likewise, in Sobczak and Ziora [47], RPA was implemented in an electricity billing management process. Another way technology was used for digitalization was to introduce new digital devices in a gas turbine maintenance process [5]. In this case, they modeled and analyzed their processes. Then, having identified issues due to the limitations of paper-based processing, they introduced digital pens that enabled digital information processing. Similarly, in other cases, digital tools have been implemented. For instance, to introduce digital information processing, a library used RFID technology to redesign its processes to enable self-loan [20]. Similar solutions where capabilities of digital technologies are used for redesigning processes by introducing digital information processes are found in healthcare [16], fashion [29], and automotive industry [43]. Therefore, the first identified capability is **digitalization** (Table 2).

Table 2. Capabilities of digital technologies.

Capability	Description	Example
Digitalization	Transforming paper-based process into digital	Using RFID technology to enable self-loan at a library. [20]
Communication	Enabling effective flow of information between employees and stakeholders	Introducing a workflow management system that enables managing insurance-related documents across departments. [21]
Analytics	Collecting and analyzing data for the purpose of supporting control, transparency, and decision-making	Combining barcode applications, ERP systems, and mobile devices to improve automated inventory control. [11]
Digital Twin	Digitally representing a physical reality	Utilizing sensor-enabled wearable devices to move from a centralized information and equipment control to decentralized production monitoring. [44]
Connectivity	Enabling stakeholders or customers to gain access to systems	Implementing a blockchain-based digital platform to facilitate transparency for financial transactions in a supply chain. [41]

Another capability we noted is *communication*. Some technologies have the capability to enable effective communication and flow of information between employees and stakeholders. For instance, a Swiss insurance company implemented a workflow management system that enabled managing insurance-related documents across departments [21]. Likewise, a railway company introduced a process-based management system that enabled them to centralize a highly scattered accounting process for efficiency gains [31]. Similar solutions have also been applied in healthcare [39] and to enhance global communication between subsidiaries at an international production company [10]. Thus, we define the second capability as **communication**.

We also noted that digital technologies enable *analytics* in business processes. Analytics include collecting and analyzing data for the purpose of supporting control, transparency, and decision-making. The capability of technologies to automatically collect and analyze data has been used to enable automated inventory control [11], enable tractability in production processes [44], and for data-driven analysis of business processes [6,7,21,23,33,45]. In addition, this capability has been used to enable automated data collection and analysis for supporting decisions in manufacturing industries [5,15]. We refer to the third identified capability as **analytics**.

Our review also revealed that some technologies enable the capability to *digitally represent a physical reality.* For instance, digitally replicating business processes for supply chain management [45] and insurance products [33] has enabled efficiency gains. Likewise, the process for managing incidents has been digitally replicated to provide transparency [5]. Finally, in a paper production process, sensor-enabled devices were introduced to fully move from centralized information and equipment control to flexible, decentralized production monitoring [44]. Thus, digital technologies are used to digitally represent business processes, i.e., enable the capability of **digital twin**.

The last category of capability we identified is *connectivity*, i.e., enabling stakeholders or customers to gain access to systems. Digital technologies enable connectivity by providing possibilities for standardization and data management, such as sharing documents and enabling quick access to information via a web application [23,26,28,42]. For instance, in Becker et al. [6], a web-based modeling system was introduced to enable knowledge sharing in a manufacturing process. In Pufahl et al. [41], a blockchain-based digital platform was introduced to facilitate transparency for financial transactions in an agricultural supply chain. Thus, the fifth identified capability is **connectivity**.

5.2 Business Process Redesign

The capabilities of digital technologies can enable redesigning of business processes. We use business process redesigns heuristics [32] to structure the presentation of the results. We noted that nine of the listed 29 heuristics were applied in the reviewed papers. The applied heuristics are not exclusive as processes can be redesigned using different heuristics. Therefore, one capability can enable several redesign heuristics.

Digitalization. The capability of digital technologies to record, store, and process digital information enables redesign heuristics of *integral technology, task elimination*, and *task automation*. Integral technology redesign is when an information system is introduced that removes physical constraints in a business process [32]. To this end, RFID [20,29], digital platforms [13], and workflow systems [33,42] have been implemented to enable digitally managing a business process. Such technologies have enabled elimination of manual tasks, such as manual invoice processing [47] and manual confirmations [43]. Finally, digitalization of processes enables task automation, such as using RPA to automate invoice processing [47], automated scanning [29], and automated data entry [43].

Communication. The capability of communication enables redesigning business processes by applying the heuristics of *integration, task elimination, centralization*, and *specialist-generalist*. Implementing a new quality management system enabled applying integration as a heuristic, i.e., integrating a business process with that of a customer or supplier, to improve transparency [10]. Another company improved their logistics process by enabling resources to remove batch processing and adopt order-based workflow [39]. Likewise, an insurance company implemented an Adaptive Case Management solution enabling them to eliminate certain tasks and apply the specialist-generalist heuristic, i.e., making resources either more specialized or generalized for task execution [21] (Table 3).

Table 3. Business process redesign heuristics applied in identified case studies.

Redesign Heuristic	Description (adopted from [32])	Example
Centralization	Manage geographically spread resources as if they are centralized	Centralizing an accounting function into three Shared Service Centers to enable transparent reporting and communication. [31]
Control addition	Check the completeness and correctness of incoming materials and check the output before it is sent to customers	Enabling the stacker to check the address for pallet storage to reduce errors in shipping. [11]
Integral technology	Try to elevate physical constraints in a business process by applying new technology	Introducing RFID technology in a fashion retail company to automatically scan the passing items from the stockroom. [29]
Integration	Consider the integration with a business process of the customer or a supplier	Implementing a quality management system to align subsidiaries with a parent company. [10]
Interfacing	Consider a standardized interface with customers and partners	Creating a web application that can submit compliant data on compensation obligations of a train operator company to the railway authority. [42]
Re-sequencing	Move tasks to more appropriate places	Duplicating the process with integrated sensor-enabled devices, thus allowing the workers to control production. [44]
Specialist-generalist	Consider to make resources more specialized or more generalist	Modifying the insurance documents management application to enable the employees to create ad-hoc tasks that were not previously available. [21]
Task automation	Consider automating tasks	Using RPA to automate invoice processing in electricity billing management process. [47]
Task elimination	Eliminate unnecessary tasks from a business	Replacing manual health data management by enabling customers to do part of the work. [4]

Analytics. Analytics technologies can enable redesigning business processes by applying *integral technology, task automation, task elimination*, and *control addition* heuristics. A prerequisite for employing the analytic capability is to have the data in digital format. Therefore, other process redesigns heuristics are often enabled when the integral technology heuristic has been implemented. For instance, with integral technology heuristic in place it will be possible to apply task automation to, for instance, automate data collection [15] and eliminate certain tasks [5]. In addition, analytic capability enables control addition, i.e., checking completeness and correctness of inputs and outputs [32]. In the case of an inventory process, this redesign enabled increased control and traceability of an inventory process and, in addition, ensured better compliance [23].

Digital Twin. Technologies that can produce a digital representation of physical reality enable redesigning business processes by applying *integral technology* and *re-sequencing* heuristics. Integral technology is required to digitally represent physical reality. Similar to analytics, it is a prerequisite. By digitally duplicating the process with integrated sensor-enabled devices, the production process was made more efficient by enabling workers to apply re-sequencing, i.e., move the execution of an activity to a more appropriate place [44].

Connectivity. The capability of technologies that enable connectivity enables applying *integral technology, integration, task elimination, centralization, interfacing*, and *re-sequencing* redesign heuristics to business processes. As with the previous capability, integral technology is essential for connectivity. As such, companies have used web portals [28], document management systems [23], digital collaboration tools [49], quality management systems [10], and BPM systems [33] to redesign business processes by applying the integral technology heuristic. In another case, blockchain technology was used to redesign a business process by applying the integration heuristic [41]. Connectivity has enabled the elimination of tasks by replacing manual health data management by providing customers the necessary tools to do part of the work [4]. Likewise, technologies enabling connectivity have been used to redesign business processes by applying centralization [12,31], interfacing [42], and re-sequencing [4] heuristics.

6 Discussion

In this section, we discuss the RQs. Then, we contextualize the capabilities to business processes and, in particular, their redesign. Furthermore, we summarize the results as a classification framework that presents the redesign heuristics that each capability enables. Finally, we discuss the limitations of our study.

Our $\mathbf{RQ}_1$ concerns capabilities of digital technologies and, more specifically, capabilities used to improve business processes. In our review, we identified capabilities that enable (1) digitalization and digital information processing, (2) communication and flow of information, (3) analytics in business processes, (4) digital representation of physical reality, and (5) connectivity, i.e., enabling stakeholders or customers to gain access to systems. Our findings match those observed in earlier categorizations of capabilities [3,18,40]. The capabilities of "digitalization and digital information" and "communication and flow of information" correspond to data technologies and communication, respectively, in the framework of [40]. Likewise, "analytics in business processes" corresponds to analytics in [3] and [18]. Digital representation of a physical reality is also included in [40] and [18]. Finally, connectivity is found in both [3] and [18].

However, in previous studies [3,18], the authors listed capabilities of mobility, human-to-machine interface, sensoring, storage, and visualization. We did not identify these as separate capabilities in the context of business processes. Rather, capabilities of sensoring are included in the digital representation of physical reality. Sensors, such as IoT, provide the needed information to digitally represent a physical flow. Storage, in the same way, is a prerequisite for connectivity that enables stakeholders and customers to access data. Finally, human-to-machine interface and visualization are means that facilitate communication. These are, in the context of business processes, not separate capabilities but part of solutions to facilitate their usage.

Our $\mathbf{RQ}_2$ relates to how capabilities of digital technologies have been used to redesign business processes. The capabilities of digital technologies can enable nine redesign heuristics (see Fig. 1), the most common being integral technology, task elimination, and task automation. Mansar & Reijers [32] introduce a categorization for business process redesign. In this categorization, technology encompasses integral technology and task automation. However, in our study, we identify technology being used to enable eight additional redesign heuristics. At the same time, some redesign heuristics are not enabled by digital technologies. For instance, we did not find any case where parallelism heuristics, i.e., executing activities in parallel, was applied. However, it is reasonable to argue that analytics could indicate higher efficiency if certain cases are routed to specialists. One reason for not finding examples can be lack of academic research and case studies that enables technology enabling application of the parallelism heuristic. If this holds true, an implication for research is the need to explore more cases of technology-based redesign of business processes.

Furthermore, ML and AI based solutions have been proposed to optimize business processes [37]. In this regard, it might be valuable to study if optimizations solutions deliver higher value if they also consider dynamic process adaptations and changes based on time-tested redesign heuristics.

Based on the results of our review, we provide a classification framework for how capabilities of digital technologies can enable the redesign of business processes (see Fig. 1). The framework is read from left to right and begins with the capabilities of digital technologies in the context of business process redesign. Then, for each capability, we list which redesign heuristics they enable. As more than one capability can facilitate the same redesign heuristic, some heuristics occur under several capabilities. For instance, both analytics and digitalization enable task automation. Next, we present an example, and, finally, we provide additional references where the same capability has been used to enable the specific redesign heuristic in question (papers that have a number "F#" are not referenced in the paper but included in the final list of the publications).

Capability	Redesign	Example	Other references
Analytics	Control addition	Combining barcode applications, ERP systems, and mobile devices to improve automated inventory control to enable stackers to check addresses to reduce errors in shipping. [11]	7, 33, F2
	Integral technology	Implementing a system to monitor EDI processes, which enables identification and traceability of data processing errors in systems used by an automotive manufacturers. [7]	45, 23, 33, F4, F8
	Task automation	Implementing a monitoring and information system in automotive industry, thus eliminating the need for operators and industrial engineers to physically collect and analyze data. [15]	F11
	Task elimination	In an inventory management system, the generated barcode on the label identifies the stock's shelf address where the pallet will be stored, instead of the worker manually selecting a shelf to store the pallet. [11]	F12
Communication	Centralization	Centralizing an accounting function into three Shared Service Centers (SSCs) to enable transparent reporting and communication. [31]	
	Integration	Implementing a "Lightweight IT" system in an emergency unit, improving logistics within the organization and interaction between organizations, as well as in processes related to communication between clinical personnel, services (e.g., cleaners and personnel), and health units. [39]	10, F5
	Specialist-generalist	Instead of defining rigid process models that IT must implement with lengthy change-management cycles, the processes can be defined directly by the business administrators, which enables clerks to add ad hoc tasks at runtime under the control of the compliance rule system based on the current context. [21]	F9
	Task elimination	An IT system in an emergency unit has the functionality of a "whiteboard" where all decisions taken are displayed immediately, which eliminates the need for additional notification efforts between the different departments of the hospital. [39]	
Connectivity	Centralization	Sharing all the documents of the process in one digital worskpace, allowing the setting of workflows among the users. [12]	31
	Integral technology	Implementing a digital portal in a hospital which serves as a virtual collection point for all digital services supporting the medical practice of general practitioners, specialists and pharmacists. [28]	26, 23, 29, 33
	Integration	Implementing a digital platform in an agriculture process which supports farmers with an asset registry, secure payment, and provides buyers with access to financial products. [41]	
	Interfacing	Through implementing a web application for submitting data on compensation obligations to the Federal Railway Authority, internal staff responsible for compensation obligations have full access to the system, which was not possible before. [42]	6, 41, F10
	Re-sequencing	Introducing a queue management system in a hospital improved the efficiency and productivity of nursing staff by enabling easy access to treatment instructions from doctors. Access to the patient treatment information has helped the nursing staff improve their ward planning and scheduling activities. [4]	
	Task elimination	Interfaces reproducing manual procedures for reports is implemented in a phosphate extraction process. The interfaces allow reliable real-time data for all the stakeholders and traceability across the phosphate cycle within the mine. [26]	4, F15
Digitalization	Integral technology	The new digital platform replace paper and email enable exchange and display of data among port users, allowing public and private organizations to be connected. [13]	20, 16, 49, 5, 31, 33, F13
	Task automation	RPA is used to access the portals of the electricity supplier and the electricity distributor's electronic customer service centers which eliminates the need to manually rewrite the data from electricity invoices received in paper form. [47]	31, 39, F6
	Task elimination	Introducing an eletronic service where the patient can check the insurance status. Previously, as proof of paid contributions for healthcare insurance the Healthcare Fund was issuing a paper. [16]	39
Digital Twin	Integral technology	Implementing the Digital Twin approach in a supply chain process, thus enabling monitoring of the current KPI on three tiers and comparing results with simulated processes behavior. [49]	F1, F3, F14
	Re-sequencing	In a paper production process, a wearable production-information system was introduced that visualizes current process data and allows operators to control production. This allows the workers to control the production up-to-date and to impact production processes from anywhere in the plant. [44]	

Fig. 1. Classification framework for digital technology-driven business process redesign (papers that have a number "F#" are not referenced in the paper but included in supplementary material (sect. 4))

Analysts can use the framework in their redesign projects in two ways. Given a digital technology, analysts can identify what capabilities the particular digital technology has, and, thereby, identify which candidate redesign heuristics to apply to the business process. Thus, analysts might find additional improvement opportunities in their process analysis. In addition, analysts can also first consider redesigns to apply and, then, identify which technology needs to be introduced to the process to enable such a change. Analysts can also consider how existing digital technologies used in the business process can be used to facilitate the implementation of redesign heuristics. In this way, analysts can leverage existing technology for process redesign.

6.1 Limitations

There are limitations inherent to SLR studies. One is excluding relevant studies or not including relevant studies [2]. We mitigated these limitations by following the guidelines when constructing search strings [22] and applying them to the electronic databases that include scientific publications in BPM. Relevant studies could have been excluded due to being not in English or inaccessible. We searched for inaccessible papers using the internet and, when possible, requested full-length text from the authors. We addressed the limitation of not including relevant papers by applying backward referencing to include relevant studies not identified with the search string. As such, we did not eliminate these limitations but reduced threats to validity.

7 Conclusion

In this paper, we aimed to explore how the capabilities of digital technologies can be applied to business process redesign. To achieve this objective, we conducted a systematic literature review of case studies where specific digital technologies are used to change a business process. From these case studies, we identified five main capabilities of digital technologies, namely, analytics, communication, connectivity, digitalization, and digital twin. Having identified these capabilities, we mapped them against business process redesign heuristics and identified nine specific redesign heuristics that can be enabled by the capabilities of digital technologies. Then, we elicited a classification framework that can be used by analysts to identify ways by which digital technologies can be incorporated in new or existing business processes. Specifically, analysts benefit from the classification framework by considering examples of redesigns of business processes enabled by digital technologies capabilities. Although our classification can aid process analysts in their redesign efforts, it is not detailed enough to provide specific guidance for less experienced analysts. As such, we identify two possible directions for future work. The first is to provide more detailed examples and conceptualization of how each capability can enable redesign. The second is to identify, given a capability, redesign opportunities from an event log.

Acknowledgements. This research is supported by the Estonian Research Council (PRG1226) and the European Research Council (PIX Project).

References

1. Ahmad, T.: The impact of new IT adoption on business process management. In: 13th International Conference on Research Challenges in Information Science (RCIS), pp. 1–5. IEEE (2019)
2. Ampatzoglou, A., Bibi, S., Avgeriou, P., Verbeek, M., Chatzigeorgiou, A.: Identifying, categorizing and mitigating threats to validity in software engineering secondary studies. Inf. Softw. Technol. **106**, 201–230 (2019)
3. Andre, L., Michael, B., Daniel, R., Christian, K.: Framework for the identification and demand-orientated classification of digital technologies. In: 2018 IEEE International Conference on Technology Management, Operations and Decisions (ICTMOD), pp. 31–36. IEEE (2018)
4. Bandara, W., Syed, R., Ranathunga, B., Sampath Kulathilaka, K.B.: People-centric, ICT-enabled process innovations via community, public and private sector partnership, and e-leadership: the case of the Dompe eHospital in Sri Lanka. In: vom Brocke, J., Mendling, J. (eds.) Business Process Management Cases. MP, pp. 125–148. Springer, Cham (2018). https://doi.org/10.1007/978-3-319-58307-5_8
5. Barz, M., Poller, P., Schneider, M., Zillner, S., Sonntag, D.: Human-in-the-loop control processes in gas turbine maintenance. In: Mařík, V., Wahlster, W., Strasser, T., Kadera, P. (eds.) HoloMAS 2017. LNCS (LNAI), vol. 10444, pp. 255–268. Springer, Cham (2017). https://doi.org/10.1007/978-3-319-64635-0_19
6. Becker, J., Clever, N., Holler, J., Neumann, M.: Business process management in the manufacturing industry: ERP replacement and ISO 9001 recertification supported by the icebricks method. In: vom Brocke, J., Mendling, J. (eds.) Business Process Management Cases. MP, pp. 413–429. Springer, Cham (2018). https://doi.org/10.1007/978-3-319-58307-5_22
7. Blasini, J., Leist, S., Merkl, W.: Developing and implementing a process-performance management system: experiences from S-Y systems technologies Europe GmbH—A global automotive supplier. In: vom Brocke, J., Mendling, J. (eds.) Business Process Management Cases. MP, pp. 37–55. Springer, Cham (2018). https://doi.org/10.1007/978-3-319-58307-5_3
8. Brynjolfsson, E., Hitt, L.M.: Beyond computation: information technology, organizational transformation and business performance. J. Econ. Perspect. **14**(4), 23–48 (2000)
9. Brynjolfsson, E., McAfee, A.: The Second Machine Age: Work, Progress, and Prosperity in A Time of Brilliant Technologies. WW Norton, New York (2014)
10. Cee, K., Bruns, I., Schachermeier, A., Kaiser, L.F.: Adoption of globally unified process standards: the case of the production company Marabu. In: vom Brocke, J., Mendling, J., Rosemann, M. (eds.) Business Process Management Cases Vol. 2, pp. 249–259. Springer, Heidelberg (2021). https://doi.org/10.1007/978-3-662-63047-1_19
11. Dantas, R., Barbalho, S.C.M.: The effect of islands of improvement on the maturity models for industry 4.0: the implementation of an inventory management system in a beverage factory. Braz. J. Oper. Prod. Manage. **18**(3), 1–17 (2021)
12. Di Giuda, G.M., Marcandalli, G., Sanvito, L., Schievano, M., Paleari, F.: A workflow for building site digitalization. In: 11th International Structural Engineering and Construction Conference, ISEC-11 2021, pp. 1–6. ISEC Press (2021)
13. Di Vaio, A., Varriale, L.: Digitalization in the sea-land supply chain: experiences from Italy in rethinking the port operations within inter-organizational relationships. Prod. Plann. Control **31**(2–3), 220–232 (2020)

14. Dumas, M., La Rosa, M., Mendling, J., Reijers, H.A.: Fundamentals of Business Process Management. Springer, Heidelberg (2018). https://doi.org/10.1007/978-3-662-56509-4
15. Fernandes, J., Reis, J., Melão, N., Teixeira, L., Amorim, M.: The role of industry 4.0 and BPMN in the arise of condition-based and predictive maintenance: a case study in the automotive industry. Appl. Sci. **11**(8), 3438 (2021)
16. Gavrilov, G., Simov, O., Trajkovik, V.: Analysis of digitalization in healthcare: case study. In: Dimitrova, V., Dimitrovski, I. (eds.) ICT Innovations 2020. CCIS, vol. 1316, pp. 202–216. Springer, Cham (2020). https://doi.org/10.1007/978-3-030-62098-1_17
17. Hammer, M., Champy, J.: Reengineering the Corporation: Manifesto for Business Revolution, a. Zondervan (2009)
18. Henriette, E., Feki, M., Boughzala, I.: The shape of digital transformation: a systematic literature review. In: MCIS 2015 proceedings, vol. 10, pp. 431–443 (2015)
19. Imgrund, F., Fischer, M., Janiesch, C., Winkelmann, A.: Approaching digitalization with business process management. In: Proceedings of MKWI, pp. 1725–1736 (2018)
20. Karna, N., Pratama, D., Ramzani, M.: Self service system for library automation: case study at Telkom university open library. In: 2019 International Conference on Information and Communications Technology (ICOIACT), pp. 689–693. IEEE (2019)
21. Kim, T.T.T., Weiss, E., Ruhsam, C., Czepa, C., Tran, H., Zdun, U.: Enabling flexibility of business processes using compliance rules: the case of mobiliar. In: vom Brocke, J., Mendling, J. (eds.) Business Process Management Cases. MP, pp. 91–109. Springer, Cham (2018). https://doi.org/10.1007/978-3-319-58307-5_6
22. Kitchenham, B.A., Charters, S.: Guidelines for performing systematic literature reviews in software engineering. Technical report EBSE 2007–001 (2007)
23. Kloppenburg, M., Kettenbohrer, J., Beimborn, D., Bögle, M.: Leading 20,000+ employees with a process-oriented management system: insights into process management at Lufthansa Technik group. In: vom Brocke, J., Mendling, J. (eds.) Business Process Management Cases. MP, pp. 505–520. Springer, Cham (2018). https://doi.org/10.1007/978-3-319-58307-5_27
24. Korhonen, J.J., Gill, A.Q.: Digital capability dissected. In: ACIS 2018–29th Australasian Conference on Information Systems (2018)
25. Kubrak, K., Milani, F., Nolte, A.: Process mining for process improvement - an evaluation of analysis practices. In: Guizzardi, R., Ralyte, J., Franch, X. (eds.) RCIS. Lecture Notes in Business Information Processing, vol. 446, pp. 214–230. Springer, Cham (2022). https://doi.org/10.1007/978-3-031-05760-1_13
26. Lafquih, H., Elhaq, S.L., Krimi, I., Berquedich, M.: Modeling and analysis of a quality traceability framework for phosphate extraction process: evidence from morocco. Int. J. Qual. Reliab. Manage. **39**, 1412–1428 (2021)
27. Lashkevich, K., Milani, F., Chapela-Campa, D., Dumas, M.: Data-driven analysis of batch processing inefficiencies in business processes. In: Guizzardi, R., Ralyte, J., Franch, X. (eds.) RCIS. Lecture Notes in Business Information Processing, vol. 446, pp. 231–247. Springer, Cham (2022). https://doi.org/10.1007/978-3-031-05760-1_14
28. Laurenza, E., Quintano, M., Schiavone, F., Vrontis, D.: The effect of digital technologies adoption in healthcare industry: a case based analysis. Bus. Process Manage. J. **24**(5), 1124–1144 (2018)

29. Leitz, R., Solti, A., Weinhard, A., Mendling, J.: Adoption of RFID technology: the case of Adler—A European fashion retail company. In: vom Brocke, J., Mendling, J. (eds.) Business Process Management Cases. MP, pp. 449–461. Springer, Cham (2018). https://doi.org/10.1007/978-3-319-58307-5_24
30. Löffler, A., Prifti, L., Knigge, M., Kienegger, H., Krcmar, H.: Teaching business process change in the context of the digital transformation: a review on requirements for a simulation game. In: Proceedings of MKWI, pp. 759–770 (2018)
31. Ludacka, F., Duell, J., Waibel, P.: Digital transformation of global accounting at deutsche Bahn group: the case of the TIM BPM suite. In: vom Brocke, J., Mendling, J., Rosemann, M. (eds.) Business Process Management Cases Vol. 2, pp. 57–68. Springer, Heidelberg (2021). https://doi.org/10.1007/978-3-662-63047-1_5
32. Mansar, S.L., Reijers, H.A.: Best practices in business process redesign: use and impact. Bus. Process Manage. J. (2007)
33. Marek, J., Blümlein, K., Wehking, C.: Process automation at Generali CEE holding: a journey to digitalization. In: vom Brocke, J., Mendling, J., Rosemann, M. (eds.) Business Process Management Cases Vol. 2, pp. 19–28. Springer, Heidelberg (2021). https://doi.org/10.1007/978-3-662-63047-1_2
34. Mendling, J., Pentland, B.T., Recker, J.: Building a complementary agenda for business process management and digital innovation. Eur. J. Inf. Syst. **29**(3), 208–219 (2020)
35. Milani, F., García-Bañuelos, L., Dumas, M.: Blockchain and business process improvement. BPTrends newsletter (2016) (2016)
36. Milani, F., García-Bañuelos, L., Reijers, H.A., Stepanyan, L.: Business process redesign heuristics for blockchain solutions. In: EDOC, pp. 209–216 (2020)
37. Milani, F., Lashkevich, K., Maggi, F.M., Francescomarino, C.D.: Process mining: A guide for practitioners. In: Guizzardi, R., Ralyte, J., Franch, X. (eds.) RCIS. Lecture Notes in Business Information Processing, vol. 446, pp. 265–282. Springer, Cham (2022)
38. Milani, F., Maggi, F.M.: A comparative evaluation of log-based process performance analysis techniques. In: Abramowicz, W., Paschke, A. (eds.) BIS 2018. LNBIP, vol. 320, pp. 371–383. Springer, Cham (2018). https://doi.org/10.1007/978-3-319-93931-5_27
39. Øvrelid, E., Halvorsen, M.: Supporting process innovation with lightweight IT at an emergency unit. J. Integr. Des. Process. Sci. **22**(2), 27–44 (2018)
40. Pousttchi, K., Gleiss, A., Buzzi, B., Kohlhagen, M.: Technology impact types for digital transformation. In: 2019 IEEE 21st Conference on Business Informatics (CBI), vol. 1, pp. 487–494. IEEE (2019)
41. Pufahl, L., Ohlsson, B., Weber, I., Harper, G., Weston, E.: Enabling financing in agricultural supply chains through blockchain. In: vom Brocke, J., Mendling, J., Rosemann, M. (eds.) Business Process Management Cases Vol. 2, pp. 41–56. Springer, Heidelberg (2021). https://doi.org/10.1007/978-3-662-63047-1_4
42. Rau, I., Rabener, I., Neumann, J., Bloching, S.: Managing environmental protection processes via BPM at Deutsche Bahn. In: vom Brocke, J., Mendling, J. (eds.) Business Process Management Cases. MP, pp. 381–396. Springer, Cham (2018). https://doi.org/10.1007/978-3-319-58307-5_20
43. Schindlbeck, B., Kleinschmidt, P.: Integrate your partners into your business processes using interactive forms: the case of automotive industry company HEYCO. In: vom Brocke, J., Mendling, J. (eds.) Business Process Management Cases. MP, pp. 485–501. Springer, Cham (2018). https://doi.org/10.1007/978-3-319-58307-5_26

44. Schönig, S., Ermer, A., Jablonski, S.: Sensor-enabled wearable process support in the production industry. In: vom Brocke, J., Mendling, J., Rosemann, M. (eds.) Business Process Management Cases Vol. 2, pp. 29–40. Springer, Heidelberg (2021). https://doi.org/10.1007/978-3-662-63047-1_3
45. Shevtshenko, E., Mahmood, K., Karaulova, T., Raji, I.O.: Multitier digital twin approach for agile supply chain management. In: ASME International Mechanical Engineering Congress and Exposition, vol. 84492, p. V02BT02A012. American Society of Mechanical Engineers (2020)
46. da Silva, E.H.D.R., Angelis, J., de Lima, E.P.: In pursuit of digital manufacturing. Procedia Manuf. **28**, 63–69 (2019)
47. Sobczak, A., Ziora, L.: The use of robotic process automation (RPA) as an element of smart city implementation: a case study of electricity billing document management at Bydgoszcz city hall. Energies **14**(16), 5191 (2021)
48. Tomat, L., Trkman, P.: Digital transformation-the hype and conceptual changes. Econ. Bus. Rev. **21**(3), 351–370 (2019)
49. Vaia, G., DeLone, W., Arkhipova, D., Moretti, A.: Achieving trust, relational governance and innovation in information technology outsourcing through digital collaboration. In: Agrifoglio, R., Lamboglia, R., Mancini, D., Ricciardi, F. (eds.) Digital Business Transformation. LNISO, vol. 38, pp. 285–300. Springer, Cham (2020). https://doi.org/10.1007/978-3-030-47355-6_19
50. Van Looy, A.: On the synergies between business process management and digital innovation. In: Weske, M., Montali, M., Weber, I., vom Brocke, J. (eds.) BPM 2018. LNCS, vol. 11080, pp. 359–375. Springer, Cham (2018). https://doi.org/10.1007/978-3-319-98648-7_21

Supporting the Implementation of Digital Twins for IoT-Enhanced BPs

Pedro Valderas(✉)

PROS Research Centre, Universitat Politècnica de València, Camí de Vera S/N, 46022 Valencia, Spain

pvalderas@pros.upv.es

Abstract. IoT-Enhanced Business Processes make use of Internet of Things technology to integrate physical devices into the process as digital actors. Closely related to this topic arises the concept of Digital Twin, which is a virtual representation of real-world entities and processes that connect to the physical counterpart to represent, simulate, or predict changes in the physical system. There are many works that focus on supporting the high-fidelity implementation of Digital Twins for specific physical devices. However, few of them consider the process as a real-world entity to be integrated into the Digital Twin. In this work, we present a microservice architecture to support the implementation of Digital Twins for IoT-Enhanced Business Processes, considering not only the physical devices but also the process itself and the relationship among them. This architectural solution is supported by a model-driven development approach, which proposes (1) the construction of a BPMN model to represent an IoT-enhanced Business Process and (2) the application of model transformation to automatically generate both Digital Twin Definition Language (DTDL) models and microservice Java code templates. DTDL models are used in the implementation of the Digital Twins for the IoT-Enhanced Business Process. Java code templates are used to facilitate the implementation of the microservices required to deploy the IoT-enhanced Business Process and its Digital Twins into the proposed architecture and maintain the digital and physical parts synchronised.

Keywords: Internet of things · Business processes · Digital twins · Microservices

1 Introduction

A Digital Twin (DT) is a virtual representation of real-world entities (physical objects and systems, or immaterial things that exist over a significant time span) and processes (events or activities that occur in time), synchronised at a specified frequency and fidelity [1]. It connects with the physical part allowing data transfer from the physical to the digital part and vice versa, in such a way that it can represent, simulate, or predict changes

This work is part of the R&D&I project PID2020-114480RB-I00 funded by MCIN/AEI/1013039/501100011033

S. Nurcan et al. (Eds.): RCIS 2023, LNBIP 476, pp. 222–238, 2023.
https://doi.org/10.1007/978-3-031-33080-3_14

in the physical system. DTs are one of the key pillars of the fourth industrial revolution [2] and constitute a market that was valued at $3 billion in 2020 and is estimated to reach $73,5 billion by 2027 [3, 4].

IoT-enhanced Business Processes (hereafter, IoT-enhanced BPs) make use of IoT (Internet of Things) devices [5] to carry out the tasks required to achieve a process' goal [6]. Although many efforts have been done in recent years to create high-fidelity digital twins of individual IoT devices, little attention has been focused on DTs of IoT-enhanced BPs [7]. A DT for an IoT-enhanced BP should maintain, as typical DTs, a virtual representation of each IoT device that participates in the process. However, it must also maintain a representation of the process itself (i.e., the state of activities to be done and the logic flow in which they must be done), and the relationships between the IoT Devices and the process (i.e., which activities are the responsibility of each IoT device). Note that all these virtual representations should be connected with their real-world counterparts to represent their state in real time.

DTs are implemented as a system of systems, each of them supporting one digital twin feature [8]. This work considers two main features of a digital twin system: (1) the representation and storage of the characteristics of real-world entities in a structured form, and (2) the runtime synchronisation between digital and real-world entities.

The contributions of this work are twofold. On the one hand, we present a distributed microservice architecture to support the implementation of DTs for IoT-Enhanced BPs by considering not only the physical devices but also the process itself and the relationship among them. This architectural solution is based on decoupled communication mechanisms, which provide a technology-independent solution to face the runtime synchronisation between the IoT devices and their digital counterparts.

On the other hand, we propose a Model-Driven approach to support the creation of the software artefacts required to represent, store and synchronise at runtime the characteristics of real-world entities in IoT-enhanced BPs DTs. Note that considering the complexity of developing a DT, the use of models is a recommended option [9, 10]. The current solution is based on previous work presented in [11] and [12], which proposed an interdisciplinary approach based on the Business Process Model and Notation (BPMN) [13] to create IoT-enhanced BPs. In this work, we evolve this approach by integrating models defined with the Digital Twin Definition Language (DTDL) [14] proposed by the Digital Twin Consortium [15]. The current approach proposes (1) the construction of a BPMN model to represent an IoT-enhanced Business Process and (2) the application of model transformations to automatically generate both DTDL models and microservice Java code templates. DTDL models are used in the implementation of the DTs for the IoT-Enhanced BPs. Java code templates are used to facilitate the implementation of the microservices required to deploy the IoT-enhanced Business Process and its DTs into the proposed architecture and maintain the digital and physical parts synchronised.

The rest of the paper is organised as follows: Sect. 2 introduces the microservice architecture proposed to support the synchronisation issue at runtime as well as some rationale about the decision taken in its definition. Section 3 presents a Model-Driven approach to create a DT implementation of an IoT-Enhanced BP based on BPMN, DTDL and microservices. Section 4 analyses the related work. Conclusions and further work are commented on in Sect. 5.

2 A Microservice Architecture for IoT-Enhanced BP Digital Twins

A DT must provide users with a faithful digital representation of real-world entities. To do so, two of the most important features to be faced in a DT system are (1) the use of a proper structure to register and manage the particular characteristics of these real-world entities, and (2) the runtime synchronisation of this structure with the data of the real world. In this section, we present a microservice architecture to support these two issues. It is graphically presented in Fig. 1. Note that the use of adequate visual representations that allow users to operate with DTs is also an important aspect to be considered. However, as we explain below, this is out of the scope of this paper.

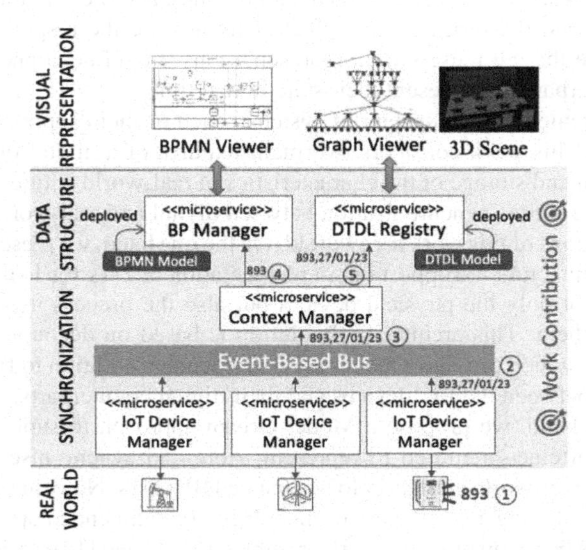

Fig. 1. Microservice architecture to support IoT-Enhanced BP DTs

2.1 Data Structure for Managing Real-World Entity Characteristics

In the case of IoT-Enhanced BPs, the real-world entities to be considered are the process and each IoT device that participates in the process. To manage and store the characteristics of a process (i.e., the tasks and events or activities that occur in time [1]), we can rely on any of the multiple process engines that support the management of process instances (i.e., each of the different executions of a process). For each instance, these engines store the activities of the process, those that are already performed and those that are still waiting to be done, the actors that are in charge of each activity, the received and produced data, the starting and ending (if ended) date and time, the produced errors (if any), and so on. We can take advantage of BP engines to manage all this data, which can be used to provide a virtual representation of a running process instance.

Regarding the characteristics of the IoT devices, we need to provide a solution that allows us to manage both the static data of IoT devices (e.g., geometrical dimensions, bill of materials, components, etc.) and dynamic data (the one that changes with time due to the participation of the IoT device in the process). We also need to consider the relationships of these IoT devices with the process (i.e., which tasks are the responsibility of each IoT Device). To create data structures with the aim of representing DT characteristics, the Digital Twin Consortium [15] has proposed the Digital Twin Definition Language (DTDL) [14]. DTDL is an open-source modelling language based on JSON-LD that allows describing the data structure that a DT needs to manage and store the characteristics of its physical counterpart. We can find DTDL registries that create instances of DT whose state can be managed at runtime using specific APIs.

2.2 Runtime Synchronization

The BP engine and the DTDL repository that are proposed to maintain the data of DTs need to be synchronised with the state of their real-world counterparts in real time. Considering the distributed nature of IoT environments and the heterogeneity of technologies that support IoT devices, we propose a microservice architecture to face this synchronisation issue. Microservices [16] propose an architectural style where systems are decomposed into small independent building blocks (the microservices), each of them focused on a single business capability. They communicate with each other with lightweight mechanisms and can be deployed and evolved independently, which leads to more agile developments and technological independence among them.

Based on [11], we propose a microservice architecture to manage the runtime synchronisation between DTs and their real-world counterparts. The previous architecture is extended by (1) integrating a microservice that oversees the management of DTDL descriptions to provide runtime data to specific viewers, and (2) adapting the rest of the architectural elements to interact with this registry. Thus, the architecture (see Fig. 1) proposes an *IoT Device Manager* microservice per each IoT Device to oversee the interaction with them. There is no restriction on how these microservices must be implemented. Any operating system or implementation technology can be used. Any type of IoT device can be managed by these microservices, independently of its supporting technology or manufacturer. There are only two requirements that an *IoT Device Manager* must satisfy to participate in the proposed architecture: (1) It must provide a REST API that allows the *BP Manager* microservice (presented below) to request the execution of its operations, and (2) it must publish any change in the IoT device state to an event-based bus, which will be used to update their digital counterpart. A change in an IoT device's state is produced either by the execution of an operation requested by the BP Manager or by an alteration produced in the physical environment that is sensed by the device. Note that *IoT Device Manager* microservices hide the technology heterogeneity of IoT devices from the rest of the architectural elements. This facilitates the replacement of a device by another that, providing the same characteristics (e.g., light sensing, door control, presence detection, etc.), relies on a different technology.

The *BP Manager* is a microservice that is endowed with an existing BPMN engine that is in charge of managing the activity flow of the process. This microservice oversees the execution of the process as it is modelled (further details about the modelling issue are introduced in the next section). To do so, the BP engine uses the REST API published by each *IoT Device Manager* to execute the operations of IoT Devices. The execution of

IoT-Enhanced BPs is out of the scope of this paper, so this interaction is not graphically represented in Fig. 1 to not overload it. Details about this can be found in [11]. For the current work, the most interesting aspect to be considered is that the BP engine maintains the state of the execution of each process.

The *DTDL Registry* is a microservice that oversees maintaining the state of the IoT device's DTs. This microservice receives any changes produced in the real world from the *Context Manager* microservice and updates it in its DTDL-based structure.

The *Context Manager* is a microservices that is subscribed to the event bus to monitor any change published by the IoT Device Managers. Context is any relevant data from the physical world [17]. Thus, when an IoT device state changes (e.g., the temperature of a sensor has increased, a door has been opened, etc.), the IoT Device Manager that manages the corresponding IoT Device must publish the change into the event bus. This change is received by the *Context Manager* microservice, which injects it into the BPMN engine to be considered in the process execution, and also sends it to the *DTDL Registry* in order to be updated in the DTDL structure.

As a representative example, consider the red labels shown in Fig. 1. When a CO2 Sensor produces a CO2 level reading (e.g., 893ppm, see number 1), the microservice that manages this IoT device oversees getting this reading. Once the microservice has the CO2 reading, it is published with a time stamp on the event bus (see number 2). Then, the Context Manager microservice, which is subscribed to the event bus, receives this data (number 3). It sends the CO2 reading to the BP engine (see number 4) to be considered in the logic flow of the process (which may update its state). Also, the Context Manager registers the CO2 reading and the time stamp into de DTDL registry (see number 5) to update the state to the CO2 Sensor digital twin. As a result, the corresponding viewers (as shown below in Fig. 2) can provide an updated visual representation of the IoT-Enhanced BP DT in real-time.

2.3 Visual Representation

Once we have the characteristics of the real-world entities stored in a structured repository and synchronized with the real world, a visual representation that faithfully reflects their real-world counterparts is needed. To do so, there are different options depending on the realism we want to obtain in the representations. We can use visual 2D models that provide a graphical representation of the structured data that represent a DT; 3D scenes that help better represent physical characteristics; or virtual or augmented reality environments that provide a more realistic interaction with the DT. This work does not focus on defining new visual representations for DTs. Thus, we use the visual 2D models that are natively supported by the solutions taken to store DT characteristics (see Fig. 2).

On the one hand, we have justified the use of BP engines to store the characteristics of a process. Currently, most of the commercial engines (Camunda[1], Activi[2], Bonita[3], etc.) are based on BPMN, which is a widely accepted visual notation for representing BPs. This notation is not only used by process designers who are experts in the notation

[1] https://camunda.com/.
[2] https://www.activiti.org/.
[3] https://es.bonitasoft.com/.

to define processes, but also by other process stakeholders such as customers, marketing professionals, or finance employees that just need to analyse them [18–20]. In addition, BPMN is the most used and preferred modelling language to face the integration of BPs and IoT [6]. Thus, using this modelling language seems to be an adequate selection to provide a visual representation of a process. On the other hand, DTDL is a language based on JSON-LD, which is a format that, typically, is graphically represented as a graph. Thus, we use visual graphs for the purpose of this work.

2.4 Realisation

In this subsection, we introduce a realisation of the architectural solution presented above as a prototype involving mapping technology choices onto the solution concepts. The microservices included in the architecture were implemented by using Java/Spring Boot[4] technology. The BP Manager microservice was endowed with a Camunda[1] engine. The DTDL Registry used the Azure Java API[5] in order to manage an Azure Digital Twin instance, which supports DTDL. The Event Bus was supported by a RabbitMQ[6] message broker, so the Context Manager and the IoT Device Manager microservices used the corresponding client Java libraries to be subscribed (in the case of the former) and to publish messages (in the case of the latter) to this broker.

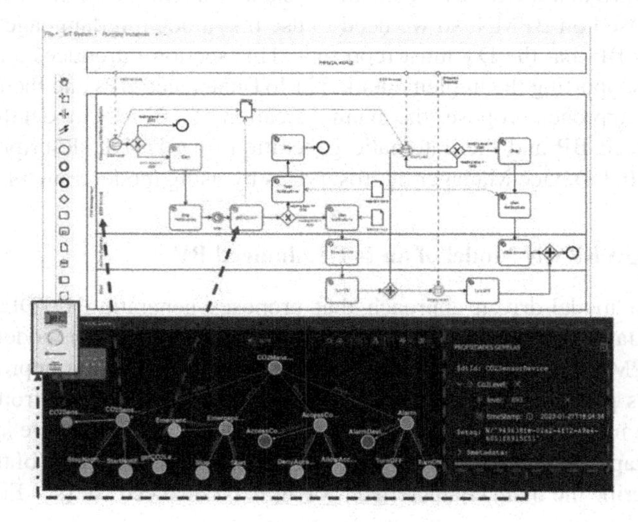

Fig. 2. Visual representation of the DT for CO2 Management IoT-Enhanced BP

In order to provide a visual representation of a DT, we used: (1) a BPMN-based tool[7] we developed to create IoT-Enhanced BPs, which was extended in order to visually show

[4] https://spring.io/projects/spring-boot.

[5] https://github.com/Azure/azure-sdk-for-java.

[6] https://www.rabbitmq.com/.

[7] Tool URL: http://pedvalar.webs.upv.es/iot-enhanced-bp-modeller/. Source code available at the following Github repository: https://github.com/pvalderas/iot-enhanced-bp-modeller.

the state of each process instance; and (2) the graph-based viewer provided by Azure Digital Twin Explorer[8]. As a proof-of-concept, we implemented a case study to support the management of CO2 levels in a library. The domain-dependent artefacts required by this case study were developed by following the Model-Driven approach presented in the next section. As a representative example, Fig. 2 presents two snapshots of the visual representation our current approach supports. Figure 2A shows the BPMN viewer that provides the state of a running process instance. Elements in red are those that are already executed and represent the state of a process instance. Figure 2B shows the visual graph of the DT managed by Azure. The root node of the graph represents the process, and its children are the actors that participate in the process. Each actor is associated with the tasks it must do and the IoT Device that supports these tasks. We can see the current state of each IoT device at any time. For instance, the right side of Fig. 2B shows the properties that define the state of a CO2 sensor: the last reading of the CO2 level was on 27–2-2023 at 18:04:34, and the level was 893 ppm.

3 Model-Driven Development Approach

As explained above, we have decided to use a BP engine as a mechanism to maintain the data structure that describes the characteristics of a real-world process. Mostly BP engines are based on BPMN, so we need to use this modelling language to create the IoT-Enhanced BP that the DT must represent. This section introduces a model-driven approach for supporting the implementation of IoT-enhanced BPs and their corresponding DTs. This approach proposes the manual creation of a BPMN model that represents an IoT-Enhanced BP and the automatic generation of a DTDL description and Java templates for IoT Device Manager microservices by using model transformations[9].

3.1 Creating a BPMN Model of an IoT-Enhanced BP

We propose a model-driven approach that proposes generating DTDL models and microservice Java templates from BPMN models by applying two model transformations. The BPMN models are created by following the modelling approach for IoT-Enhanced BPs presented in [11]. Modelling approaches are defined from an abstract syntax, which indicate the concepts that must be defined and a concrete syntax, which define the metaphors used to create these concepts. The abstract syntax of the modelling approach describe the main characteristics of an IoT-Enhanced BP (see Fig. 3):

- As traditional BPs, the tasks that is required to achieve the goal of the BP.
- The IoT devices that participate in the BP. In particular, IoT devices are classified into: (1) actuators, which can control environmental conditions (e.g., air conditioners, heating, watering systems, security systems, etc.) [21]; and sensors, which are used to collect and transfer data about the process execution context (e.g., temperature sensor, camera, heart rate sensor, etc.) [22].

[8] https://learn.microsoft.com/es-es/azure/digital-twins/concepts-azure-digital-twins-explorer.
[9] Implementation available in the app/bpmn folder of the tool's Github repository.

- The interaction between the BP and the physical world (the IoT devices), which can be of two types:
- On-demand: The BP decides when and how to interact with IoT devices. For instance, this interaction is done when the BP, according to its business logic, decides to activate an air conditioner or request the temperature of a room at a specific time.
- Autonomous: IoT devices autonomously inject some data into the BP without an explicit request. For instance, this interaction is done when a presence detector automatically informs a BP about the detection of a person.

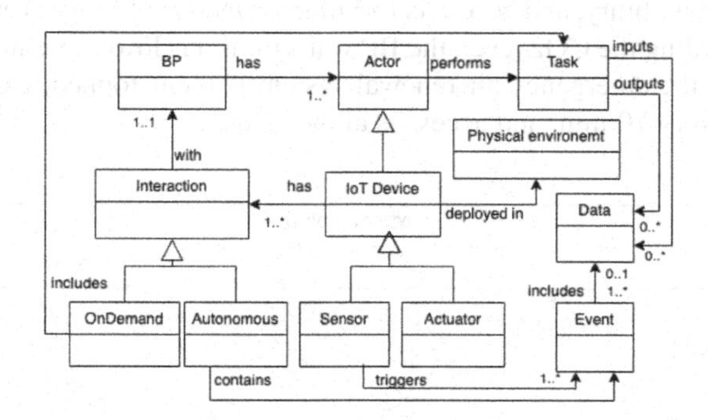

Fig. 3. The abstract syntax of the modelling approach

To create the concepts of this abstract syntax, we propose a concrete syntax based on BPMN, the guidelines proposed in [11] are used:

1. A pool is used to represent the whole IoT-enhance BP within an organisation.
2. Each *IoT Device* that participates in the process is represented by a lane within this pool. The BPMN internals of the model (which are not graphically shown) also store if an IoT Device is a Sensor or an Actuator. Graphically, the names of sensors are depicted in square brackets.
3. Each *IoT Device's Task* required by the BP is defined as a Service Task in the corresponding lane. This supports *On-demand Interactions* between the BP and the IoT devices (i.e., when the BP executes one of these tasks, it will be demanding an IoT device to perform an action).
4. The *Physical Environment* is represented by a collapsed pool.
5. *Autonomous Interactions* between IoT Devices and the BP are represented by means of message flows drawn between the physical environment pool (Guideline 4) and the pool that represents the IoT-enhance BP (Guideline 1). Each message flow is labelled with the name of the IoT device that injects the data and is connected to a message *Event* that represents this data.
6. The data injected by autonomous interactions and the inputs and outputs of on-demand interactions are described with BPMN data objects. Although it is not graphically shown in a BPMN model, they are defined as a set of tuples key-value.

To support this modelling approach, we developed the web tool introduced in the previous section. As a representative model, Fig. 4 shows an example that describes the process of managing the CO2 level in a smart library. Let's suppose we have a CO2 sensor informing the BP about the CO2 level every ten minutes. When the CO2 level is lower or equal to 2000 ppm, the BP does nothing. However, if the CO2 level is greater than this value, an emergency protocol is executed. First, an emergency air renewal system is started, and the notification of CO2 levels is stopped in the sensor. After waiting for five minutes, the BP directly requests the current level of CO2 to the CO2 Sensor. If it is lower or equal to 1500ppm, the emergency air renewal system is stopped, and the sensor notifications are configured every ten minutes again. Otherwise, the BP asks the CO2 Sensor to notify the CO2 level every minute, an alarm is activated to warn people that they must leave the library, and access is forbidden. Once the library is empty, the alarm is stopped. Regarding the CO2 level, the BP waits until it is lower or equal to 1000 ppm. At this moment, the emergency air renewal system is then stopped, CO2 notifications are configured every 10 min, and access is allowed again.

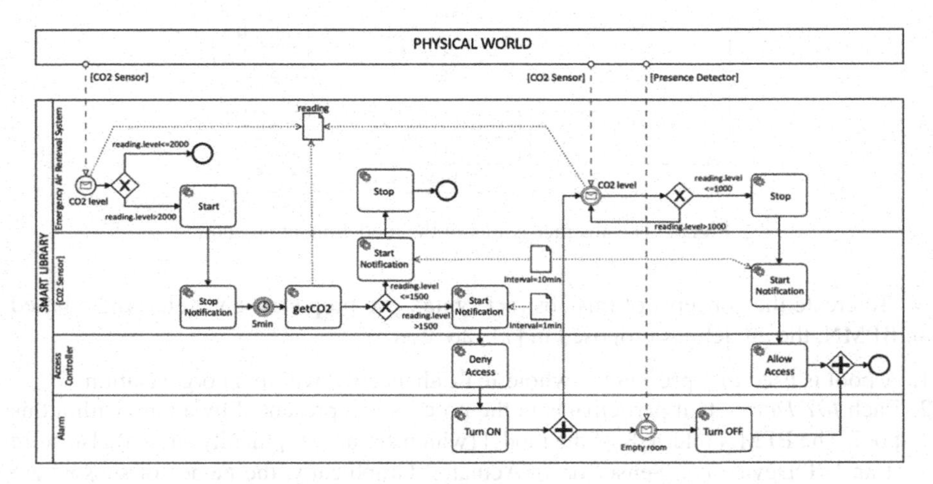

Fig. 4. An IoT-enhanced BP for controlling CO2 level

To support the tasks of this process, five IoT devices are used: Emergency Air Renewal System, CO2 Sensor, Access Controller, Alarm and Presence Detector. Note how the first four devices are represented by BPMN lanes since an on-demand interaction between the BP and them is needed. The on-demand requests for executing operations are represented by the Service Tasks that are included in each lane. Regarding the autonomous interaction of IoT devices, note how three flow messages are defined between the Physical World pool and the process pool in order to represent that the CO2 Sensor must inject the CO2 level and that the Presence Detector must inform when the room is empty. Regarding data, note how the *CO2 level* injected into the BP as well as the *getCO2* and the *Start Notification* tasks, are associated with a BPMN Data Object, which defines the data injected into the BP or the inputs or outputs of the operations.

3.2 Generating DTDL Models from BPMN

DTDL [14] is made up of a set of metamodel classes that are used to define the structure and behaviour of digital twins. It is a textual language based on JSON-LD. The main concepts of its metamodel are represented through a UML class diagram in Fig. 5. These concepts allow describing the structure and behaviours of general DTs. The main concept is Interface which describes the contents (Property, Telemetry, Command, Relationship, or Component) of any digital twin. *Components* enable interfaces to be composed of other interfaces. A *Relationship* describes a link to another digital twin and enables graphs of digital twins to be created. Property, Telemetry and Command are the concepts that allow us to describe the static and dynamic structure of an IoT Device. A *Property* describes the read-only or read/write state of any digital twin. For example, a device serial number may be a read-only property, and the desired temperature on a thermostat may be a read-write property. Telemetry describes the data emitted by any digital twin as a stream of data. It is a set of data messages that have a short duration and need an event listener to be processed. They cannot be read time after they are emitted. Finally, A *Command* describes a function or operation that can be performed on any digital twin. Each command can have a *CommandPayload* that describes the request data (i.e., the input arguments that the operation needs) and another that describes the response data.

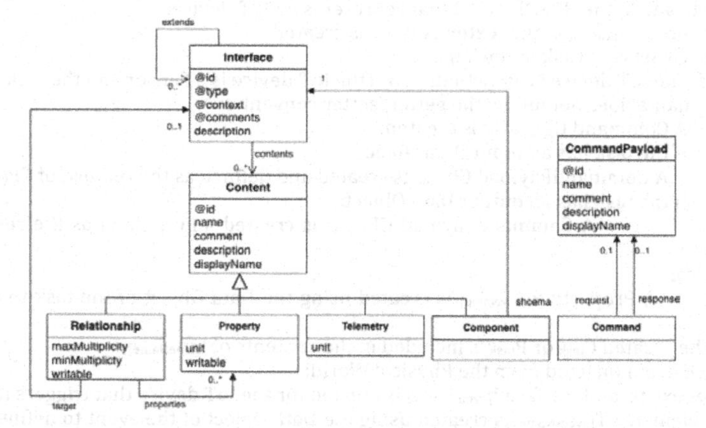

Fig. 5. The DTDL metamodel represented in a UML class diagram

We have defined a model transformation to generate DTDL descriptions from BPMN. First, the DTDL Interfaces $I_{Process}$, I_{Actor}, $I_{Activity}$ and I_{Device} are generated to represent the concepts of Process, Actor, Activity and IoT Device. They are associated with each other by means of DTDL Relationships. Then, Algorithm 1 is applied:

- For each BPMN Lane associated with an IoT Device, we generate a DTDL Interface $I_{\{laneName\}}$ that represents the device by extending I_{Device} (line 1–2). The content of each of these interfaces is defined as follows:
  - If the BPMN lane defines an actuator (lines 4–9), each of its Service Tasks defines a Command in the corresponding Interface. If the task is associated with some

input or output Data Object, it is used to define the request or the response. The Command is included in the content of $I_{\{laneName\}}$ (line 12).

- If the BPMN lane represents a sensor, we use a name convention to generate DTDL Properties or Commands from the defined Service Tasks. If the name of a task is defined as a getter or setter (lines 10–11), we understand the BP is reading or setting some property of the device, and then a DTDL Property is defined. The property structure is obtained from the associated Data Object. If the task name does not follow the commented convention, it is transformed into a Command as explained for actuators (lines 4–9). The Command or Property created is included in the content of $I_{\{laneName\}}$ (line 12).

- Each event injected from the Physical World (defined by BPMN Message Flows) is transformed into a Telemetry of the IoT device that triggers it (lines 13–16).

Algorithm 1. From BPMN models to DTDL descriptions

INPUT: a BPMN model that represents an IoT-Enhanced BP

OUTPUT: a DTDL description

1. **For each lane in the BPMN model that represents an IoT device:**
2. **An Interface $I_{\{laneName\}}$ that extends I_{Device} is created;**
3. **For each Service Task in the lane:**
4. **If (the IoT device is an Actuator) or (the IoT device is a Sensor and the task name does not follow the getter/setter convention):**
5. **A Command $C_{\{taskName\}}$ is created;**
6. **If the task has an input Data Object:**
7. **A CommandPayload CP_{Input} is created and defined as the request of $CP_{\{taskName\}}$;**
8. **If the task has an output Data Object:**
9. **A CommandPayload CP_{output} is created and defined as the response of $CP_{\{taskName\}}$;**
10. **Else:**
11. **A Property $P_{\{taskName\}}$ is created using the Data Object of the task to define its schema;**
12. **The created C_{task} or P_{task} is included in the contents of $I_{\{laneName\}}$;**
13. **For each event injected from the Physical World:**
14. **If not exists: an Interface $I_{\{eventDevice\}}$ is created for the IoT device that triggers it;**
15. **A Telemetry $T_{\{eventName]}$ is created using the Data Object of the event to define its schema;**
16. **$T_{\{eventName]}$ is included in the contents of $I_{\{eventDevice\}}$;**

As a representative example, Fig. 6 shows the DTDL specification generated for the CO2 Sensor IoT Device. As we can see, it includes: (1) one property, generated from the getCO2Level task (getter/setter convention name); (2) two commands, generated from the other two Service Tasks that do not follow the getter/setter name convention; and (3) one telemetry generated from the CO2 Level event injected from the Physical World. This specification was loaded into the Azure Digital Twin Explorer[10], which validated it to create the visual graph presented in Fig. 2. This shows that the model transformation generates DTDL models that are syntactically correct.

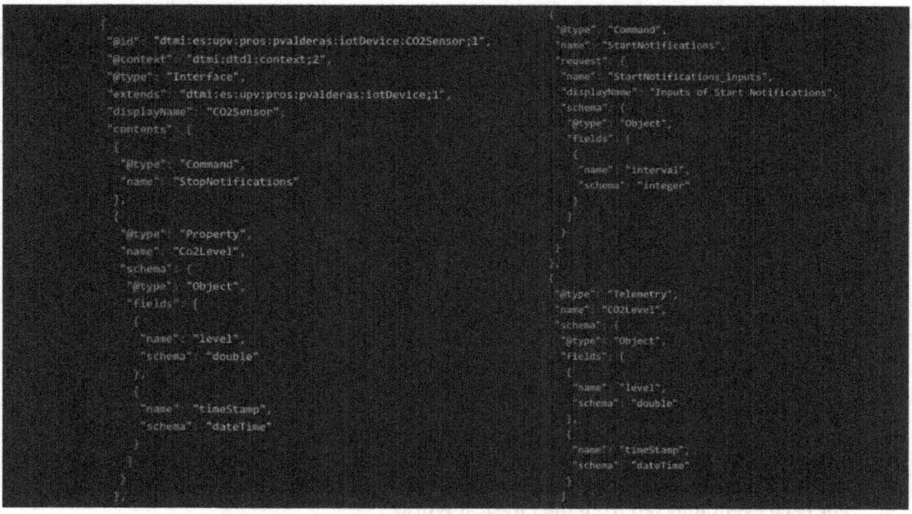

```
{
"@id": "dtmi:es:upv:pros:pvalderas:iotDevice:CO2Sensor;1",
"@context": "dtmi:dtdl:context;2",
"@type": "Interface",
"extends": "dtmi:es:upv:pros:pvalderas:iotDevice;1",
"displayName": "CO2Sensor",
"contents": [
 {
  "@type": "Command",
  "name": "StopNotifications"
 },
 {
  "@type": "Property",
  "name": "Co2Level",
  "schema": {
   "@type": "Object",
   "fields": [
    {
     "name": "level",
     "schema": "double"
    },
    {
     "name": "timeStamp",
     "schema": "dateTime"
    }
   ]
  }
 },
 {
  "@type": "Command",
  "name": "StartNotifications",
  "request": {
   "name": "StartNotifications_inputs",
   "displayName": "Inputs of Start Notifications",
   "schema": {
    "@type": "Object",
    "fields": [
     {
      "name": "interval",
      "schema": "integer"
     }
    ]
   }
  }
 },
 {
  "@type": "Telemetry",
  "name": "CO2Level",
  "schema": {
   "@type": "Object",
   "fields": [
    {
     "name": "level",
     "schema": "double"
    },
    {
     "name": "timeStamp",
     "schema": "dateTime"
    }
   ]
```

Fig. 6. DTDL specification for the CO2 Sensor

3.3 Generating Java Code from BPMN

IoT Device Manager microservices oversee managing the interaction with real-world IoT devices. They play the role of gateways between physical IoT devices and the rest of the microservice architecture. They must provide a REST API that allows the BP Manager to request the execution of IoT device operations. They must also publish every change in the state of the IoT device into the event bus. To facilitate the implementation of these aspects, we have implemented a model transformation that generates a Java/Spring template of IoT Device Manager microservices from the BPMN Model.

Algorithm 2, presented below, is applied to generate the following Java code:

- A Java class that creates DTDL instances of the digital twins for a specific repository. Currently, Azure Digital Twin is supported (lines 1–3).
- A Java class that implements the REST interface required by each IoT Device Manager microservice. This class implements an endpoint per each Service Task defined in the BPMN model considering the input and output Data Objects (lines 4–8).

- For each event that an IoT Device must inject into the BPMN model (any telemetry defined for the DT), a piece of code that supports the publication of this message into the Event bus is generated (lines 9–10). Currently, we support RabbitMQ.

Algorithm 2. From BPMN models to Java code

INPUT: a BPMN model that represents an IoT-Enhanced BP

OUTPUT: a set of Java classes

1. A Digital Twin Creator Java class is generated;
2. **For each** lane in the BPMN model that represents an IoT device:
3. An Azure Digital twin instantiation is defined in the Digital Twin Creator;
4. An IoT Device Controller Java class is created;
5. **For each** Service Task in the lane:
6. A REST endpoint is included in the IoT Device Controller.
7. The Input Data Object of the Service Task defines the arguments of the endpoint;
8. The Output Data Object of the Service Task defines the response of the endpoint;
9. **For each** event injected from the Physical World:
10. An event publication handler is generated;

4 Related Work

Many works face the development of digital twins from different perspectives. In [23–25], three literature reviews are presented to precisely characterise, among other aspects, the concept of a digital twin, its related notions, and the underlying technologies. [26] presents an analysis of the requirements to be considered to propose a modelling notation for a digital twin and discusses the existing modelling solutions that can be used to support each requirement.

Most of the existing works focus on the implementation of digital twins for manufacturing systems. For instance, [27] proposes the use of a UML class diagram to model the physical setup of these types of systems, paying little attention to the processes they must be involved in. In addition, once the UML model is created, it is not clear how it can be transformed into executable software artefacts. [28] presents a model-based framework with five dimensions to support reconfigurable digital twins. This work uses models to manage the reconfiguration issue in manufacturing systems. However, models are not used to drive the development process. In [29] and [30], the authors define a model-driven architecture to realise digital twins of cyber-physical systems and a model-driven engineering method to systematically generate the supporting information systems and interactive digital twin cockpits that facilitate interaction with them. This work does not consider the process as a real-world entity and only focuses on digital twins that represent physical machinery and its environment. In addition, the proposed architecture seems to follow a monolithic design missing the benefits of scalability and maintenance that provide distributed architectures such as microservices. [31] proposes a model-driven approach and the modelling language AutomationML to model physical devices. This work also focuses on modelling the physical structure of devices. It pays little attention to how integrating devices and processes into a DT and how to manage the synchronisation of data at runtime.

Focusing on DTs for IoT-Enhanced BPs, [7] proposes an interesting solution based on BPMN, as ours, but complemented with 3D scenes developed with the Gazebo Simulator. However, beyond the proposal of using these two technologies in a complementary way, the authors provide few details about the software infrastructure required to maintain the runtime synchronisation with the real world and the development process that must be followed to implement the DT.

Finally, many research efforts have already been made to integrate Business Process Management (BPM) with IoT, as is analysed in [6]. None of them explicitly considers the challenge of creating digital twins for IoT-Enhanced BPs. However, it is worth remarking on some of them due to their close relation to our proposal. For instance, [22] presents a manifesto that analyses the benefits and challenges of integrating both paradigms of BPM with IoT. However, any specific solution to support the integration of BPM and IoT is proposed. Other works provide solutions based on BPMN as ours. Some extend the original BPMN notation with new concepts to model requirements imposed by IoT systems. For instance, [32] introduced the concept of Event Stream Processing Units to integrate real-time data generated in Cyber-physical systems into BPs. [33] introduced new event annotations to specify the binding points between external events and the BP model. [34] extended the BPMN metamodel with three new classes (PhysicalEntity, SensingTask and ActuatingTask) to explicitly represent IoT devices and the interaction with them. The main difference between these works and ours is that they all extend the BPMN's metamodel, which requires adapting BP engines to execute such extensions, preventing current existing engines from being used as we do in our solution. Other works, such as [35, 36] or [37] propose to use the original BPMN constructs to model IoT-enhanced BPs. In this case, however, the BPMN notation needs to be transformed into another language/technology to be executed. Our solution to model IoT-Enhanced BPs allows the direct execution of BPMN models.

To conclude this section, let us summarise the main contributions of this work in relation to the state of the art. As explained above, most of the approaches that investigate the creation of digital twins focus only on the representation of physical machinery. Our work improves state-of-the-art by considering digital twins for IoT-enhanced BPs as a combination of physical IoT objects and a real-world process. In addition, we also demonstrate how BPMN models and DTDL descriptions can be integrated to support a model-driven development of the proposed digital twins. The developed tool and the implemented model transformations are available online to be used by the research community. Finally, in contrast to the analysed solutions that are mostly based on monolithic architectures, we improve state-of-the-art by proposing a distributed architecture based on microservices. This solution provides several benefits, such as an easy adaptation when technological requirements change, support for integrating technically heterogeneous IoT devices, or separation of development responsibilities.

5 Conclusions and Further Work

A digital twin is implemented as a system of systems, each of them supporting one digital twin feature. Two of these features are (1) the definition and storage of the characteristics of a digital twin and (2) its runtime synchronization with the state of its physical counterpart. In this work, we have proposed a solution to support them in the implementation of DTs for IoT-enhanced BPs.

On the one hand, we have presented a distributed microservice architecture based on decoupled communication mechanisms to support the runtime synchronization between the digital and physical parts. This solution provides a technology independency between the IoT devices and their digital counterparts, allowing that the replacement of IoT devices has little impact on the rest of the digital twin system. On the other hand, we have proposed a solution based on BPMN engines and DTDL registries to create a structured storage for the characteristics of an IoT-Enhanced BP DT. Finally, these two contributions are complemented with an MDD approach to support the development of the domain-dependent artefacts required to implement an IoT-Enhanced BP DT (i.e., the BPMN model, the DTDL model, and the IoT Device Manager microservices).

As further work, we plan to extend the developed BPMN web tool to simulate the execution of an IoT-Enhanced BP and be able to analyse the interaction of an IoT-Enhanced BP with the real world previously to its deployment. In addition, we also want to investigate the visualisation of DTs for IoT-Enhanced BP in order to integrate high-fidelity 3D scenes in our approach.

References

1. Digital Twin Consortium, "Digital Twin Definition." https://www.digitaltwinconsortium.org/glossary/glossary/#digital-twin. Accessed 22 Mar 2023
2. Singh, M., Fuenmayor, E., Hinchy, E.P., Qiao, Y., Murray, N., Devine, D.: Digital twin: origin to future. Appl. Syst. Innov. **4**(2), 36 (2021)
3. Velosa,A., Middleton, P.: Emerging Technologies: Revenue Opportunity Projection of Digital Twins, Gartner, Stamford, Conn (2022). https://www.gartner.com/en/documents/4011590. Accessed 22 Mar 2023
4. Markets and Markets, "Digital Twin Market by Enterprise, Application (Predictive Maintenance, Business optimization), Industry (Aerospace, Automotive & Transportation, Healthcare, Infrastructure, Energy & Utilities) and Geography - Global Forecast to 2027" (2022). https://www.marketsandmarkets.com/Market-Reports/digital-twin-market-225269522.html. Accessed 22 Mar 2023
5. I. Society, "The Internet of Things (IoT): An Overview." https://www.internetsociety.org/resources/doc/2015/iot-overview/
6. Torres, V., Serral, E., Valderas, P., Pelechano, V., Grefen, P.: Modeling of IoT devices in business processes: a systematic mapping study. In: Proceedings - 2020 IEEE 22nd Conference on Business Informatics, CBI 2020, vol. 1, pp. 221–230 (2020)
7. Corradini, F., Pettinari, S., Re, B., Rossi, L., Tiezzi, F.: An approach to support digital process twin. In: 1st International Workshop on Digital Process Twins. In conjuntion with the DASC/PiCom/CBDCom/CyberSciTech conference, pp. 1–4 (2022)
8. Digital Twin Consortium, "Digital Twin System." https://www.digitaltwinconsortium.org/glossary/glossary/#digital-twin-system. Accessed 22 Mar 2023

9. Rasheed, A., San, O., Kvamsdal, T.: Digital twin: values, challenges and enablers from a modeling perspective. IEEE Access **8**, 21980–22012 (2020)
10. Bordeleau, F., Combemale, B., Eramo, R., van den Brand, M., Wimmer, M.: Towards model-driven digital twin engineering: current opportunities and future challenges. In: Babur, Ö., Denil, J., Vogel-Heuser, B. (eds.) ICSMM 2020. CCIS, vol. 1262, pp. 43–54. Springer, Cham (2020). https://doi.org/10.1007/978-3-030-58167-1_4
11. Valderas, P., Torres, V., Serral, E.: Modelling and executing IoT-enhanced business processes through BPMN and microservices. J. Syst. Softw. **184**, 111139 (2022)
12. Valderas, P., Torres, V., Serral, E.: Towards an interdisciplinary development of IoT-enhanced business processes. Bus. Inf. Syst. Eng. **65**(1), 25–48 (2023)
13. BPMN, Business Process Model and Notation Concepts (2010). http://www.omg.org/spec/BPMN/20100501. Accessed 22 Mar 2023
14. Digital Twins Definition Language. https://github.com/Azure/opendigitaltwins-dtdl/blob/master/DTDL/Docs/en-US/DTDL.v2.md. Accessed 22 Mar 2023
15. Digital Twin Consortium, Digital Twin Consortium. https://www.digitaltwinconsortium.org/. Accessed 22 Mar 2023
16. Lewis, J., Fowler, M.: Microservices (2014). https://martinfowler.com/articles/microservices.html. Accessed 22 Mar 2023
17. Abowd, G.D., Dey, A.K., Brown, P.J., Davies, N., Smith, M., Steggles, P.: Towards a better understanding of context and context-awareness. In: Gellersen, H.-W. (ed.) HUC 1999. LNCS, vol. 1707, pp. 304–307. Springer, Heidelberg (1999). https://doi.org/10.1007/3-540-48157-5_29
18. Nysetvold, A.G., Krogstie, J.: Assessing business processing modeling languages using a generic quality framework. Adv. Top. Database Res. **5**, 79–93 (2005)
19. Harmon, P., Wolf, C.: Business process modeling survey. Business process trends (2011). http://www.bptrends.com/surveys/Process_Modeling_Survey-Dec_11_FINAL.pdf
20. Leopold, H., Mendling, J., Günther, O.: Learning from quality issues of BPMN models from industry. IEEE Softw. **33**, 26–33 (2015)
21. Beverungen, D., et al.: Seven paradoxes of business process management in a hyper-connected world. Bus. Inf. Syst. Eng. **63**(2), 145–156 (2020). https://doi.org/10.1007/s12599-020-00646-z
22. Janiesch, C., et al.: The internet of things meets business process management: a manifesto. IEEE Syst. Man Cybern. Mag. **6**(4), 34–44 (2020)
23. Semeraro, C., Lezoche, M., Panetto, H., Dassisti, M.: Digital twin paradigm: a systematic literature review. Comput. Ind. **130**, 103469 (Sep.2021)
24. Jones, D., Snider, C., Nassehi, A., Yon, J., Hicks, B.: Characterising the digital twin: a systematic literature review. CIRP J. Manuf. Sci. Technol. **29**, 36–52 (May2020)
25. Lim, K.Y.H., Zheng, P., Chen, C.H.: A state-of-the-art survey of digital twin: techniques, engineering product lifecycle management and business innovation perspectives. J. Intell. Manuf. **31**(6), 1313–1337 (2020)
26. Corradini, F., Fedeli, A., Polini, A., Re, B.: Towards a digital twin modelling notation. In: 1st International Workshop on Digital Process Twins. In Conjuntion with the DASC/PiCom/CBDCom/CyberSciTech Conference, pp. 1–6 (2022)
27. Azangoo, M., Taherkordi, A., Olaf Blech, J.: Digital twins for manufacturing using UML and behavioral specifications. In; IEEE International Conference on Emerging Technologies and Factory Automation, ETFA, vol. 2020, pp. 1035–1038 (2020)
28. Zhang, C., Xu, W., Liu, J., Liu, Z., Zhou, Z., Pham, D.T.: A reconfigurable modeling approach for digital twin-based manufacturing system. Procedia CIRP **83**, 118–125 (2019)

29. Dalibor, M., Michael, J., Rumpe, B., Varga, S., Wortmann, A.: Towards a model-driven architecture for interactive digital twin cockpits. In: Dobbie, G., Frank, U., Kappel, G., Liddle, S.W., Mayr, H.C. (eds.) ER 2020. LNCS, vol. 12400, pp. 377–387. Springer, Cham (2020). https://doi.org/10.1007/978-3-030-62522-1_28
30. Kirchhof, J.C., Michael, J., Rumpe, B., Varga, S., Wortmann, A.: Model-driven digital twin construction: synthesizing the integration of cyber-physical systems with their information systems. In: MODELS 2020, vol. 12, no. 20, pp. 90–101 (2020)
31. Schroeder, G.N., Steinmetz, C., Rodrigues, R.N., Henriques, R.V.B., Rettberg, A., Pereira, C.E.: A methodology for digital twin modeling and deployment for industry 4.0. Proc. IEEE **109**(4), 556–567 (2021)
32. Appel, S., Kleber, P., Frischbier, S., Freudenreich, T., Buchmann, A.: Modeling and execution of event stream processing in business processes. Inf. Syst. **46**, 140–156 (2014)
33. Mandal, S., Hewelt, M., Weske, M.: A framework for integrating real-world events and business processes in an IoT environment. In: OTM 2017 Conferences: Confederated International Conferences: CoopIS, C&TC, and ODBASE 2017, pp. 194–212 (2017)
34. Meyer, S., Ruppen, A., Hilty, L.: The things of the internet of things in BPMN. In: CAiSE 2015: Advanced Information Systems Engineering Workshops, 2015, vol. 215, pp. 285–297 (2015)
35. Baresi, L., Meroni, G., Plebani, P.: A GSM-based approach for monitoring cross-organization business processes using smart objects. In: Reichert, M., Reijers, H.A. (eds.) BPM 2015. LNBIP, vol. 256, pp. 389–400. Springer, Cham (2016). https://doi.org/10.1007/978-3-319-42887-1_32
36. Domingos, D., Martins, F.: Using BPMN to model internet of things behavior within business process. Int. J. Inf. Syst. Proj. Manag. **5**(4), 39–51 (2017)
37. Dar, K., Taherkordi, A., Baraki, H., Eliassen, F., Geihs, K.: A resource oriented integration architecture for the internet of things: a business process perspective. Pervasive Mob. Comput. **20**, 145–159 (2015)

Context-Aware Digital Twins to Support Software Management at the Edge

Rustem Dautov(✉) and Hui Song

SINTEF Digital, Oslo, Norway
{rustem.dautov,hui.song}@sintef.no

Abstract. With millions of connected edge gateways, there is a pressing challenge of remote maintenance of containerised software components after the initial release. To support remote update operations, edge software providers have been increasingly adopting digital twin-based device management platforms for run-time monitoring and interaction. A common limitation of these solutions is the lack of support for modelling the multi-dimensional context of edge devices deployed in the field, which hinders the software management in a tailored and context-aware manner. This paper aims to address this lack of context-awareness in digital twins required for edge software assignment by introducing two modelling principles, which allow focusing on the device fleet as a whole and capturing the diverse cyber-physical-social context of individual devices. As part of proof of concept, these principles were incorporated in an existing digital twin platform. This prototype implementation demonstrates the viability of the proposed modelling principles via a running example in the context of a telemedicine application system.

Keywords: Digital Twin · Context Awareness · Edge Computing · IoT · Device Fleet · Eclipse Ditto · Remote Patient Monitoring

1 Introduction

With the increasing computing and networking capabilities, IoT devices and edge gateways have become part of a larger IoT-edge-cloud computing continuum, where processing and storage tasks are distributed across the whole network hierarchy, not concentrated only in the cloud. At the same time, this also introduced continuous delivery practices to the development of software components for network-connected edge gateways. These devices are placed on end users' premises and are characterised by changing multi-dimensional contexts, forcing developers to maintain multiple software versions and frequently re-deploy them on a distributed fleet of devices with respect to their updated contexts. Unlike the traditional cloud model, where computing resources are homogeneous and concentrated in a single physical location, an edge fleet may be distributed across thousands of heterogeneous devices, each with a unique context in terms of hardware capacity, surrounding physical environment, network connection,

S. Nurcan et al. (Eds.): RCIS 2023, LNBIP 476, pp. 239–255, 2023.
https://doi.org/10.1007/978-3-031-33080-3_15

user preferences, *etc.* To address such heterogeneous contexts, software components are also becoming increasingly diverse, and edge software providers often simultaneously maintain multiple active versions of the same application. Taken together, the increasing diversity on both sides, *i.e.* edge devices and software components, raises a challenge of *assignment* implemented in a precise, reliable and scalable manner.

Doing this correctly and efficiently goes beyond the manual capabilities of software vendors and requires an intelligent and reliable automated solution. This has given rise to device management cloud platforms, which allow collecting information from distributed devices and provide a near real-time view on the overall fleet. This functionality is often underpinned by the prominent concept of *Digital Twins* (DTs). However, existing platforms often require increased coding and modelling effort from edge software providers in order to implement software assignment for the two main reasons. First, the existing solutions typically focus on individual devices in isolation from each other, thus neglecting their interrelations. Second, the default DT modelling support mainly considers the traditional context metrics, such as hardware and software resources, to assign software updates, neglecting a much more diverse and dynamic multi-dimensional context of each edge device. As a result, this lack of modelled context information forces edge application providers to manually map multiple versions of their software, which eventually becomes a bottleneck in their agile development processes.

Accordingly, this paper aims to address the following high-level research question – **how to model the dynamic multi-dimensional device context in order to support correct and targeted assignment of software management updates on edge infrastructures?** The contribution of the paper is threefold. First, by exploiting the DT concept we elevate the level of modelling abstraction to the *fleet level*, as opposed to the state of practice primarily focusing on individual devices. Second, we enhance digital representations of devices within a fleet with a notion of *multi-dimensional context*, as opposed to the existing IoT device management platforms primarily focusing only on hardware and, occasionally, software aspects. Third, we demonstrate the applicability of the proposed methodology with a proof of concept from the telemedicine domain.

The rest of the paper is organised as follows. Section 2 provides motivation behind this research by explaining the challenges existing in the edge software management. Section 3 briefs the reader on the state of practice and related research works in the domains of DTs and software assignment on edge infrastructures, highlighting the existing limitations. Section 4 explains how the prominent DT paradigm can be applied to address these limitations. Section 5 puts theory into practice and describes how the proposed research ideas have been implemented on top of the existing DT platform Eclipse Ditto. Section 6 summarises the results and concludes the paper with an outlook for future work.

2 Research Context and Motivation

2.1 Software Maintenance at the Edge

The ubiquitous connectivity has given rise to a new sector within the ICT market comprised by companies specialising in software development for cloud-connected edge infrastructures. Typically, such software providers are not directly involved in hardware manufacturing, but rather rely on some third-party gateways shipped directly to end users, along with various IoT sensors and actuators. While downstream IoT devices are not always Internet-connected and are not equipped with sufficient computing resources, edge gateways have traditionally served to collect sensor data, transfer it to the central cloud back-end, and relay actuation commands back to actuators, as illustrated by Fig. 1.

More recently, edge gateways have also become fully-functional processing units in their own right, going beyond the passive transferring to the cloud. Indeed, edge gateways are equipped with relatively powerful computing (*i.e.* CPU and RAM) and networking capabilities to run some business applications on top of data collected from downstream IoT sensors. Thanks to these increased capabilities, edge devices are not only able to support intensive data exchange between the IoT and the Cloud layers, but also run complex software on top of Linux OS, including a user interface for direct interaction with end users. With the recent advancements in virtualisation and containerisation technologies, edge software is often packaged and released as Docker containers – the *de facto* standard for enterprise-level software containerisation. The hardware and software counterparts of the described edge-based installations comprise the next generation of connected cyber-physical systems across a wide range of intelligent scenarios, such as various smart spaces, transportation, Industry 4.0, telemedicine [10], *etc.*[1]

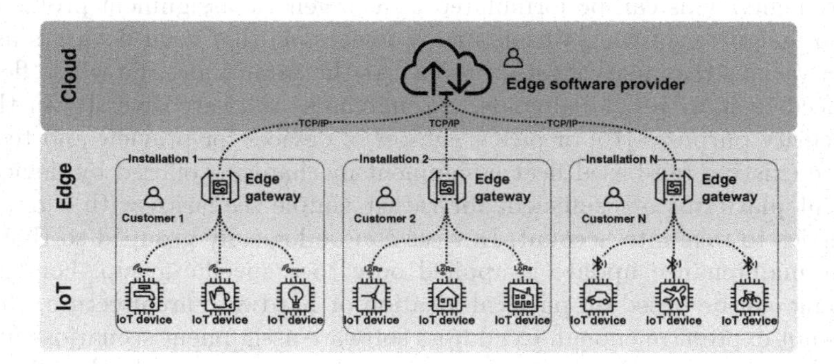

Fig. 1. Application system spanning across the Cloud-Edge-IoT infrastructures.

[1] Throughout this paper, and in Sect. 5 in particular, we will use telemedicine as the main reference example. Nevertheless, the described concepts apply to other scenarios dealing with edge software management.

Like all software, edge applications may contain bugs or security flaws, which will only emerge after the initial deployment. At the same time, software providers also frequently release functional updates in incremental manner following agile development practices. Unlike in traditional data centres, maintenance of edge software needs to take place remotely, since edge devices are installed and connected on end users' premises. For a relatively small fleet of devices, this might not be a challenge. Larger fleets, comprised of thousands of units distributed around the globe, need an automated device management platform to remotely provision, monitor, reset or lock every device, as well as to push updates whenever a new software version is released. Automating functions through a device management platform saves time and resources, and minimises human errors. Examples of such platforms include mainstream commercial solutions such as Azure IoT Hub[2] and AWS IoT Greengrass,[3] niche market offerings such as Balena Cloud,[4] and open-source community-driven toolkits such as Eclipse hawkBit.[5] All these platfroms focus on container-based software management and rely on some device-side monitoring agents for collecting metrics at run-time.

2.2 Challenge: Context-Awareness for Software Assignment

Assignment is an important part of all software management activities, be it the initial release or the follow-up maintenance. Unlike the centralised cloud model, where computing resources are relatively homogeneous, a fleet of edge gateways consists of distributed and heterogeneous devices, which have diverse multi-dimensional contexts in terms of hardware capacity, network connection, physical deployment, user preferences, *etc.* [8]. Developers often need to maintain several software variants to fit such different device contexts. For example, some devices may require a variant tailored to a specific hardware capacity, while some others with limited connectivity may need a variant with lower bandwidth requirements. This can be formulated as as a generic assignment problem – *i.e.* how to assign m software variants to n devices, so that each device is assigned with a variant that matches its context. At the same time, the whole fleet may also need to meet its global goals, *e.g.* maximise software diversity in the fleet for security purposes [16] or pick a sub-set of devices for preview and testing.

The existing tag-based fleet assignment mechanisms offered by device management platforms are sufficient for rather simple scenarios with a few device properties to take into account. Devices can be logically grouped so that a particular maintenance update is applied only to a specific group. For example, grouping can be based on physical location or hardware architecture. However, this is not expressive enough to address software assignment scenarios with hundreds and thousands devices, each having its own continuously changing multi-dimensional context – on the one hand, and multiple software versions – on

[2] https://azure.microsoft.com/products/iot-hub/.
[3] https://aws.amazon.com/greengrass/.
[4] https://www.balena.io/.
[5] https://www.eclipse.org/hawkbit/.

the other. In these circumstances, correctly assigning software components in a precise and scalable manner heavily depends on up-to-date contextual information about managed devices, which seems to be not immediately available in existing solutions, which rely on general-purpose monitoring agents collecting a limited number of hard-coded performance metrics. In the absence of richer context-aware information about the managed fleet, using the default available tag-based assignment mechanisms is still possible, but would result in a complex collection of chained fine-grained $if - then$ conditions – an error-prone and difficult-to-maintain solution.

3 State of the Art and Practice

3.1 Digital Twins for Edge Fleet Management and Context Modelling

As already explained, by using a device management platform, edge application providers are able to remotely provision, monitor, and maintain their device, as well as push software updates whenever a new version is released. In recent years, all these tasks have been underpinned by the prominent concept of DTs. DTs are virtual models designed to accurately reflect physical objects equipped with various sensors related to certain aspects of their functionality. They enable to create rich digital models of anything physical or logical, from simple assets or products to complex cyber-physical environments. The collected sensor data is relayed to a processing system and applied to the DT. Once updated with such data, the DT can be used to run simulations, explore performance issues, and identify possible improvements – which can then be applied back to the original physical twin using bi-directional IoT connections.

Among many research efforts on DTs applied to the IoT [20,21] an important part belongs to research on DT interoperability and standards [14,22]. An industry-driven effort in this context was done by Microsoft's DT Definition Language (DTDL)[6] to provide more semantics to modelled physical environments. Albeit open-source, it can primarily be used only within the Azure ecosystem. A similar attempt was made by the now-retired Eclipse Vorto.[7] W3C Web of Things (WoT) has come up with another effort to standardise the modelling domain of IoT systems its Things Description model [6] – a formal information model and a common representation to describe the metadata and interfaces of IoT devices. In parallel, there is also an on-going effort to integrate the Semantic Web ontologies to provide more formal semantics and reasoning to the WoT.

Another relevant research field is *context modelling*, which has already been active for a few decades [5,15]. Context modelling produces a formal or semi-formal description of the context information to create a context-aware system. Some recent attempts [7,11] have started considering the cyber-physical dimension of the IoT and edge environments. In particular, there is a demand for

6 https://learn.microsoft.com/azure/digital-twins/concepts-models.

7 https://www.eclipse.org/vorto/.

context-awareness to assess trustworthiness of devices [18], as well as to complement application-specific analytics with enriched context information, *e.g.* disease diagnostics based on extra information from IoT sensors [12].

In summary, engineering of context-aware DTs is currently *ad-hoc* to a great extent, which is a challenge for quality-controlled development, deployment, and operation. Most works focus on a limited set of conventional metrics, not sufficient and suitable for edge software maintenance scenarios. A common limitation of the existing approaches is the hard-coded focus only on modelling the cyber-aspects of the managed physical entities, while more extensive modelling and coding is left to system administrators to be done manually.

3.2 Software Assignment

The existing works on software assignment at the edge primarily focus on resource usage optimisation [4,13,17,19], in which the objective function is typically based on generic infrastructure-level metrics, such as network latency [25], hardware resources utilisation, energy consumption, *etc.* and less frequent metrics, such as CO_2 footprint, monetary costs, number of tenants, *etc.* Optimisation metrics and related challenges are extensively surveyed in [2]. In all these works, edge software is typically treated as a black box, owned or managed by some external third party. As opposed to this, this proposed work tackles the software assignment problem from the perspective of an application provider, who owns and manages the whole vertical application system, including edge and cloud components. This means that software placement is mainly driven by application- and business-level requirements and evaluated against continuously monitored device-specific contexts, not strictly limited to the infrastructure level.

In this light, implementing software assignment requires modelling the application domain (*i.e.* software requirements and device context) and corresponding multi-dimensional constraints. Research approaches to solving complex constraints have resulted in several theories and tools, such as Satisfiability Modulo Theories (SMT) [1], which offered several fully automated fleet assignment solutions based on various constraints and conditions. The SMT-based approaches are specifically popular and efficient due to their expressive and rich modelling language [3,23,24]. Nevertheless, they still require manually modelling the target system with some contextual information and testing the constraints.

Managing software updates across a large fleet of devices is an everyday task for the leading mobile OS providers, such as Apple and Google. At present, their adopted over-the-air update mechanism implements a publish-subscribe model, where mobile devices first get notified of and then fetch available updates through the centralised marketplace. Since smartphone fleets are rather homogeneous in terms of hardware and OSs, there is no challenge of multi-criteria software assignment as such, and the compatibility check is performed only once, upon the initial installation. Furthermore, in a fleet of smartphones there are typically no global system goals, such as even distribution of components or A/B testing.

4 Modelling Principles for Context-Awareness

As we have argued, existing approaches lack the support for context-aware software assignment at the edge, which we aim to tackle by this paper. We now proceed with a description of the two main underpinning design principles.

I. Device Fleet as the Main Level of Abstraction. This design principle represents one step up from the conventional approach to maintain DTs of individual devices. By lifting the abstraction level, we are thus able to maintain a live view not only on the managed individual devices, but also on the dynamic inter-relations between them – something that can potentially play an important role when deciding what software management commands need to be applied to specific devices. This is especially important when considering some global software management policies within a fleet, where assigning a software version to one device will depend on what version has already been assigned to another. For example, in DevOps, A/B testing is often implemented to test newly-added features on a limited sub-set of staging devices before pushing them to production. Another example is the artificial *software diversity* – *i.e.* a common technique to improve system resilience, robustness, and security by deploying functionally-equivalent, yet different software versions on multiple systems [16]. In all such cases, updated software versions are required to be deployed only on a sub-set of the device fleet. Another common scenario is to limit the simultaneous use of some licensed software to a specific number of devices. Assignment in such circumstances will depend on whether other devices belonging to the same user are already running this software or not.

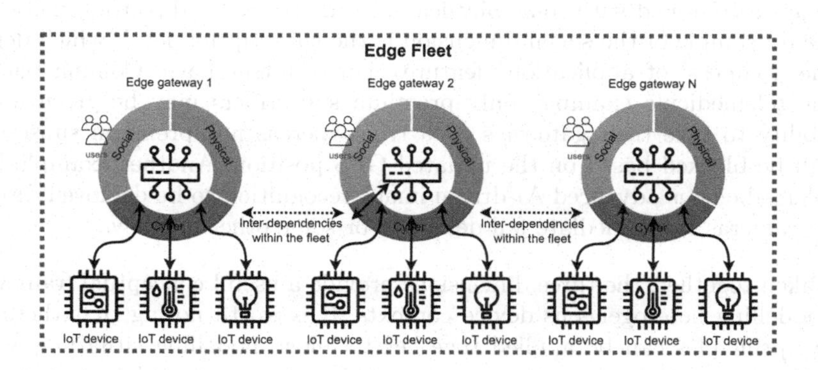

Fig. 2. Multi-dimensional context of edge gateways within a fleet.

II. Multi-dimensional Context Awareness. In the context of rather static and homogeneous computing nodes (*e.g.* a computing cluster or a data centre), software management typically takes into consideration only the cyber-dimension of the target nodes, such as available hardware, networking configuration, OS version, already installed software, *etc.* As opposed to this, correctly assigning software components at the edge often needs to take into account a much richer and more diverse context of edge infrastructures [9] (Fig. 2). To this end, we propose adopting a three-fold context model, consisting of the **cyber**, **physical**, and **social** dimensions, as explained below:

1. **Cyber** dimension covers the hardware and software aspects of edge gateways. For example, software assignment on a specific gateway may depend on connected sensors/actuators, OS version, security patches, networking interfaces (*e.g.* wireless/wired), power source (*e.g.* battery/constant power supply).
2. **Physical** dimension primarily refers to where exactly an edge gateway is deployed physically. As explained in Sect. 3, in this paper we consider connected edge infrastructures deployed on end users' premises. For many scenarios (*e.g.* precision agriculture or environmental monitoring), this means that edge gateways may be exposed to changing and possibly harsh weather conditions, such as high/low temperatures or extreme humidity. Accordingly, vendors may opt for not assigning compute intensive AI-driven applications to those nodes already exposed to high temperatures not to cause overheating. Another example comes from telemedicine, where monitored patients are advised to take gateways with them even when travelling (*e.g.* in a camping trailer or in a cruise). In such cases, the updated physical placement or even the GPS location may be taken into account to assign a matching software update. The last but not the least, the physical placement is crucial for software assignment as far as security-related software management is concerned due to the direct affect on the current trustworthiness of the target device.
3. **Social** dimension captures the human-related aspects of edge gateways, which, as previously explained, are deployed on end users' premises and may be even equipped with some physical user interfaces for direct interaction. A good example of the social dimension is the subscription level, which defines the 'richness' of applications features a user might have. Coming back to the telemedicine example, only premium subscribers may be given a possibility to take their gateways on a trip, whereas non-premium subscribers will be blocked based on the updated GPS position. Another example from telemedicine is advanced AI-driven image recognition to be deployed directly on cameras for immediate fall detection or some other urgency.

Taken together, the three dimensions provide a useful conceptual framework for modelling heterogeneous device contexts. It is worth noting that these two design principles can be applied together to allow modelling situations where one device is part of a context of another device within a fleet. This way, we are able to capture the contextual inter-dependencies existing between the devices at the fleet level. For example, assigning a new software version to a device may depend on what other devices are connected and what software they are already running. We will discuss such relevant use cases in more detail in Sect. 5.

5 Proof of Concept

We now proceed with explaining how the proposed context modelling principles can be put in practice. For convenience, we will base our demonstration on a relevant example of remote patient monitoring (RPM), which we will explore through several use case scenarios related to day-to-day software maintenance.

5.1 Running Example: Edge Gateways for Remote Patient Monitoring

The use case scenario refers to a telemedicine provider, who offers RPM services. For each customer (typically – elderly people living at their own residences), they provide a healthcare gateway, *i.e.* a small single-board computer similar to Raspberry Pi, along with a set of medical sensors, cameras, and wearable emergency beepers. Each gateway collects measurements from Bluetooth-connected downstream sensors (*e.g.* blood pressure, glucose and oxygen levels) or cameras (*e.g.* video stream for fall detection) and, before sending them to the back-end cloud service, may also perform some local (pre-)processing and aggregation. The patients and their caretakers have access to the data via a Web interface and a mobile app. Some patients also opt for battery-powered portable gateways to have a possibility to carry them with the essential sensors whenever they are away, *e.g.* relocating to a summer house, travelling in a camper trailer or on a cruise ship. During these periods, the gateway connectivity switches from the residential WiFi network to a mobile one (3G or 4G). The gateways also differ in terms of their hardware capacity – *i.e.* the older versions are less powerful than the modern ones in terms of CPU, RAM, and available disk storage. At present, the device fleet comprises more than 500 gateways installed on the customers' premises, and there also exist several versions of the edge RPM software. More specifically, there is a version dependent on additional disk storage to accumulate sensor measurements and occasionally send them in batches, whenever a device is not connected to a WiFi network (*e.g.* in a trailer camper). Another version requires increased CPU and RAM resources, since it not just transfers data, but also implements local data pre-processing. There is also a version enhanced with support for image processing for fall detection, and therefore requires several cameras to be installed and connected. Depending on the subscription level, some versions provide only basic limited features, while some other are richer in terms of offered functionality. All these examples demonstrate how the heterogeneous context determines what RPM software version has to be assigned and installed on each individual gateway in the fleet.

The development team continuously updates the edge software by adding new features and patching bugs, following agile practices. After each DevOps cycle, typically executed on a weekly basis, there is a mass re-deployment involving many or sometimes all the devices in the fleet. To perform the assignment, it is required to parameterise each deployment and match them to devices and their cyber-physical-social context properties.

5.2 Technological Baseline: Eclipse Ditto

As the technological baseline on top of which we have developed our proof of concept implementation, we have used Eclipse Ditto[8] – an open-source framework for building DTs of Internet-connected devices with extensible modelling and built-in querying languages. Ditto acts as middleware, providing an abstraction layer for IoT solutions interacting with physical devices via the DT pattern. It can be seen as a toolkit, providing some core functionality (*e.g.* meta-model, database, different messaging protocols and connectors, REST APIs, *etc.*), while some other features have to be written by users on top of them (*e.g.* domain-specific DT models, graphical user interfaces, device-side monitoring agents, *etc.*). Being part of a larger open-source ecosystem, it can be relatively easy integrated with some other technologies from the Eclipse stack, including various communication protocols, pub-sub messaging and load balancing.

Eclipse Ditto Meta-Model. Ditto offers developers a meta-model, which in its simplest form enables modelling physical entities (*i.e.* things) using a JSON schema with the following key concepts (as depicted in Fig. 3):

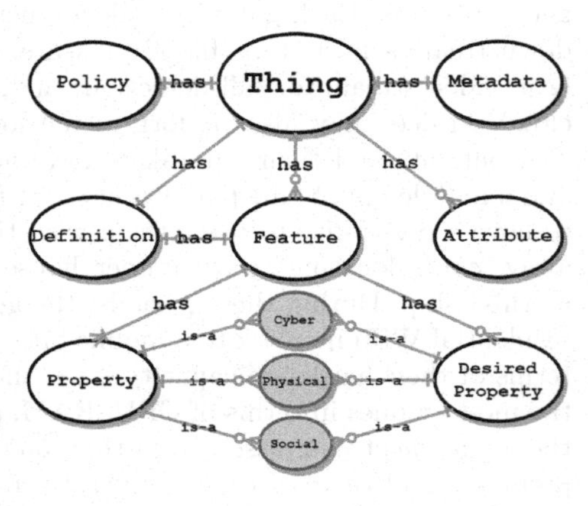

Fig. 3. Eclipse Ditto meta-model.

- `Thing` is the top-level modelling concept for describing physical assets.
- `Definition` is included in every `Thing` (and optionally in `Features`) and essentially represent a URI linking to an external WoT model. It describes how a `Thing` is structured and which behaviour/capabilities can be expected in a interoperable and standard manner.[9]
- `Policy` enables to configure fine-grained access control. A specific policy defines who and how can access a specific resource. Although very important for the overall security trustworthiness of the DT system, `Policies` are beyond the scope of this work.
- `Metadata` is Ditto's internal field to store technical information, *e.g.* version or creation/modification timestamps.
- `Attributes` are used to model rather static properties of a `Thing`, *i.e.* values that do not change as frequently as `Features`. They can be of any type and can be used to search for `Things`.

[8] https://www.eclipse.org/ditto/.

[9] We further discuss this functionality in Sect. 6 as part of future work.

- **Features** are the central modelling concept to capture all run-time data and functionality of a **Thing** in a given application system. Users are allowed to define their own **Features** or extend existing WoT definitions. This is a key enabler for modelling the multi-dimensional context through more fine-grained **Properties**.
- **Properties** are used within **Features** to model individual run-time indicators of a **Thing**, *e.g.* to manage the status, the configuration or any fault information. Each **Property** can be either a simple scalar value or a complex JSON object. By using **Properties**, it is possible to implement the prominent *desired-reported* pattern widely adopted in DTs, wherein sensor measurements are reported upstream, while desired configuration updates are pushed downstream, until eventually **Properties** and **DesiredProperties** are in sync.

Extending the Default Ditto Functionality. The default Ditto functionality was extended in the following ways:[10]

- Graphical user interface (based on the NodeJS/React stack) to offer the RPM provider the required application-specific software management features.
- Device-side monitoring agents for collecting contextual information and receiving updates.
- Back-end service for exchanging and synchronising desired (*i.e.* pushed by the RPM provider) and reported (*i.e.* collected by monitoring agents) properties.
- Extended DT definitions of RPM gateways including the cyber-physical-social context information (*i.e.* **Properties**) received from monitoring agents, as part of the application-specific fleet model. The gray circles in Fig. 3 represent these extensions to the **Property** concept.
- Software assignment logic based on Resource Query Language (RQL)[11] making use of the enriched DT definitions.

The latter two extensions are particularly relevant in the context of this paper. The simplified JSON snippet in Listing 1.1 demonstrates the overall structure of the DT. Please note the three blocks within **Features**, corresponding to cyber, physical and social context properties of the managed gateways. For clarity purposes, we omit extensive definitions and will focus on some practical use cases in the next subsection. In practice, the definition of a DT is expected to be much longer in order to accommodate all possible contextual information required for software assignment.

Listing 1.1. A sample of an RPM gateway digital twin.

```
1  "thingId": "no.sintef.sct.giot:rpm_gateway_01",
2  "policyId": "no.sintef.sct.giot:rpm_policy",
3  "definition": "no.sintef.sct.giot:rpm_gateway:1.0.0",
4  "attributes": {
5      "manufacturer": "RPM Inc.",
6      "cpu_model": "Broadcom BCM2711"
7      "cpu_arch": "arm_v8",
```

[10] https://github.com/SINTEF-9012/ditto-fleet.
[11] https://github.com/persvr/rql.

```
8           "os": {
9               "name": "Raspberry Pi OS 11 (Bullseye)",
10              "kernel_version": "5.15.84"
11          }
12      },
13      "features": {
14          "cyber": {
15              "properties": {
16                  "monitoring_agent": {
17                      "enabled": true,
18                      "version": "1.0.0"
19                  },
20                  "docker_engine": {
21                      "enabled": true,
22                      "version": "containerd.io_1.2.0-1_arm64.deb"
23                  },
24                  "proxy": {
25                      "enabled": false,
26                      "host": "",
27                      "port": ""
28                  },
29                  "ssh": {
30                      "enabled": false,
31                      "port": 443
32                  },
33                  "dev_env": "testing",
34                  "fingerprint_sensor": true,
35                  "2fa": false
36              }
37          },
38          "physical": {
39              "properties": {
40                  "deployment_site": "hospital",
41                  "gps_location": {
42                      "latitude": "59.945283167835846",
43                      "longitude": "10.713121330182533"
44                  }
45              }
46          },
47          "social": {
48              "properties": {
49                  "user_subscription": {
50                      "type": "premium",
51                      "expires":"2023-12-31T23:59:59Z"
52                  }
53              }
54          }
55      }
```

5.3 Use Case Scenarios

Ditto comes with an RQL-based query language to search for devices within the managed fleet using their attributes and features and a collection of logical and arithmetical operators. We now present a series of common use cases scenarios serving to demonstrate the use of context-aware DTs in day-to-day software maintenance activities of the RPM provider. The examples use RQL to search for target devices within the fleet and use conventional logical operators. Please note that the code snippets with queries are somewhat simplified for demonstration and clarity purposes.

Use Case 1: Targeted Maintenance of a Single Device. A very simple, yet common task is to perform remote maintenance on a single device, *e.g.* during live interaction with a customer over the phone. In such cases, the device ID is known, and some test commands can be executed using this query.

```
eq(thingId,"no.sintef.sct.giot:rpm_gateway_01")
```

Use Case 2: Maintenance of Devices Based on a Single Context Property. Another common scenario is to target a specific subset of devices based on some single attribute or feature. It can be, for example, a new version release for all ARM-based gateways:

```
eq(attributes/cpu_arch,"arm_v8")
```

Similarly, it is also a common practice to release some experimental features to a testing environment, before pushing it to production:

```
eq(features/cyber/properties/dev_env,"testing")
```

To a great extent, the first two use case scenarios represent the current state of practice as far as DT-based software assignment is concerned.

Use Case 3: Maintenance of Devices Based on a Multi-dimensional Context. This is a more advanced scenario, which needs to take into account several features to release agile updates only to those devices falling into the target scope of the update command. Depending on the strictness of the target conditions, it can be none, a few or all devices from the fleet. The following query refers to a situation when a new software version with mandatory fingerprint-based authentication needs to be pushed to a device equipped with such a sensor (**cyber**) and where the two-factor authentication is not yet enabled (**social**).

```
and(eq(features/cyber/properties/fingerprint_sensor,"true"),
  eq(features/social/properties/2fa, "false"))
```

Another example is releasing new AI-driven wellness recommendation features only to GPU-enabled devices (**cyber**), do not get overheated (**physical**), and have a premium subscription (**social**). For simplicity, we define 'overheating' as an exceeding of the mean value of recent CPU temperature observations (40°).

```
and(eq(features/cyber/properties/gpu,"true"),lt(mean(features/physical/
    properties/cpu_temp),"40"),eq(features/social/properties/user,"premium"))
```

Use Case 4: Releasing an Experimental Feature to a Limited Number of Devices This scenario refers to a situation when new software features should be safely tested only on a very small number of devices in a production environment. For example, the assignment logic can target a single device within a larger installation of 10+ gateways in the same medical institution. Accordingly, this combined query will first check if there are more than 10 gateways installed, and, if yes, will return a single device among the available. This is a demonstration of an assignment logic at the fleet level, wherein the decision is based taking into account other neighbour devices.

```
and(first(eq(features/physical/properties/deployment_site,"hospital")),
  gt(count(eq(features/physical/properties/deployment_site,"hospital")),10))
```

Use Case 5: Evenly Distributing 2 Software Versions Within the Fleet The next related example refers to the software diversification strategy, often applied to increase the overall security of the fleet. Briefly, the idea is to run multiple software versions (functionally identical, yet different at the code level). This can be achieved by first querying the total number of target gateways (*e.g.* running a specific Raspbian OS), and then equally splitting them into two.

```
num = count(eq(attributes/os/kernel_version, "5.15.*"))                    (1)
limit(count(eq(attributes/os/kernel_version, "5.15.*"), 0, num/2)          (2)
```

5.4 Assumptions and Limitations

The described query snippets serve to demonstrate the feasibility of the proposed approach. Admittedly, in practice they will be somewhat more complex and expressive, as well as the definition of the DT in Listing 1.1. Despite the fact that current proof of concept is based on Eclipse Ditto and its internal modelling and querying languages, the high-level design concepts can be applied to and implemented in other DT platforms allowing extending the definition of DTs with cyber-physical-social context properties.

An important assumption to make, however, is the need to design and implement own monitoring agents. These agents will be deployed on devices to collect run-time multi-dimensional metrics about the managed devices, to be collected by the centralised DT platform. In many occasions, these monitoring agents would also need to have some elevated access rights to be able to probe low-level information about hardware or host OS. Admittedly, as with many modelling approaches, there is usually more than one way of representing the surrounding world with digital models. While the proposed design principles can provide a high-level modelling framework, some extensive modelling is still needed to make the DTs usable in practice. To this end, some deep knowledge of the target application system is required.

6 Conclusion and Future Work

In this paper, we answer the research question posed in Sect. 1 and aimed to address the challenge of limited context-awareness in the existing DT platforms – an important pre-requisite to perform assignment for software maintenance in edge application systems. We brought forward the two design principles required for modelling DTs. First, it is important to focus on the overall fleet, rather than on individual devices. This way, it is possible to capture and evaluate possible inter-dependencies between the devices, as well as global goals of the overall edge system. Second, it is required to go beyond the traditional hardware and software context properties, but also take into consideration the physical and social dimensions of edge devices, which are often installed on end users' premises and provide an interface for physical interaction. By enabling such multi-dimensional cyber-physical-social context awareness, we provide a much richer foundation for performing software assignment across a wide range of business scenarios. The latter is demonstrated in the context of an RPM edge application, using the existing DT framework Eclipse Ditto. In this proof of concept implementation, we we were able to model the device fleet following the two design principles, which supported a series of software maintenance use case. Although the overall results are positive, the proposed solution to a certain degree still requires manual modelling effort, since proposed design principle are too high-level, while the described prototype implementation is specific to the RPM scenario and underlying Ditto implementation. Another related limitation is the need for device-specific monitoring agents, which will collect multi-dimensional context information required for DTs. Nevertheless, this is a work-in-progress research effort, which we ar eplanning to further improve in the following directions:

1. **Integration with WoT Thing Description and semantic modelling**: WoT Thing Description is used to model the metadata and interfaces of physical things to enable integration of heterogeneous devices and interoperability across diverse applications. They are encoded in a JSON format that also allows JSON-LD processing. The latter provides a promising foundation to represent knowledge managed devices in a machine-readable way. Even further possibilities for automated reasoning can be unleashed with the adoption of Semantic Web ontologies, which are based on Description Logics and come with multiple automated reasoners and modelling editors. Ditto can support both technologies, meaning that assignment may be implemented using more expressive tools, going beyond the the simple RQL reported in this paper.
2. **Using a graph-based database**: As we advanced with the experiments on Eclipse Ditto, it became apparent that entities and relationships describing the multi-dimensional context of devices within a fleet do not always fit into the fixed nested structure of JSON documents, adopted by Eclipse Ditto and its underlying document-oriented MongoDB database. A promising research direction is to adopt a *graph-based* representation of a managed fleet. Graph-based DTs is a prominent paradigm, mainly due to a more intuitive representation of data, which will then be easier to query for a human by traversing the graph elements. Arguably, a graph is a more natural abstraction for a device fleet, with all the inter-dependencies and heterogeneous contexts of individual devices.
3. **Empirical evaluation**: Even though we have used the running RPM example to demonstrate the viability of the proposed approach, it still needs to undergo a proper empirical validation by DevOps engineers to prove its applicability in enterprise-level production environment. This will be implemented as part of the ongoing work in the R&D projects acknowledged below.

Acknowledgements. This work is co-funded by the European Commission's HEU and H2020 Programmes under grant agreements 101070455 (DYNABIC), 101095634 (ENTRUST) and 101020416 (ERATOSTHENES), and the Research Council of Norway's BIA-IPN programme under grant agreement 309700 (FLEET).

References

1. Barrett, C., Tinelli, C.: Satisfiability modulo theories. In: Handbook of Model Checking, pp. 305–343. Springer, Cham (2018). https://doi.org/10.1007/978-3-319-10575-8_11
2. Bellendorf, J., Mann, Z.Á.: Classification of optimization problems in fog computing. Future Gener. Comput. Syst. **107**, 158–176 (2020)
3. Bonacina, M.P., Graham-Lengrand, S., Shankar, N.: Satisfiability modulo theories and assignments. In: de Moura, L. (ed.) CADE 2017. LNCS (LNAI), vol. 10395, pp. 42–59. Springer, Cham (2017). https://doi.org/10.1007/978-3-319-63046-5_4
4. Brogi, A., Forti, S., Guerrero, C., Lera, I.: How to place your apps in the fog: state of the art and open challenges. Softw. Pract. Experience **50**(5), 719–740 (2020)
5. Brown, P., et al.: Context-awareness: some compelling applications. In: Proceedings the CH12000 Workshop on the What, Who, Where, When, Why and How of Context-Awareness (2000)
6. Charpenay, V., Käbisch, S.: On modeling the physical world as a collection of things: the W3C thing description ontology. In: Harth, A., et al. (eds.) ESWC 2020. LNCS, vol. 12123, pp. 599–615. Springer, Cham (2020). https://doi.org/10.1007/978-3-030-49461-2_35
7. Da Silva, D.M.A., Sofia, R.C.: A discussion on context-awareness to better support the IoT cloud/edge continuum. IEEE Access **8**, 193686–193694 (2020)
8. Dautov, R., Distefano, S.: Stream processing on clustered edge devices. IEEE Trans. Cloud Comput. **10**(2), 885–898 (2020)
9. Dautov, R., Distefano, S., Bruneo, D., Longo, F., Merlino, G., Puliafito, A.: Data processing in cyber-physical-social systems through edge computing. IEEE Access **6**, 29822–29835 (2018)
10. Dautov, R., Distefano, S., Buyya, R.: Hierarchical data fusion for smart healthcare. J. Big Data **6**(1), 1–23 (2019)
11. Gu, T., Wang, X.H., Pung, H.K., Zhang, D.Q.: An ontology-based context model in intelligent environments. arXiv preprint arXiv:2003.05055 (2020)
12. Gubert, L.C., da Costa, C.A., Righi, R.D.R.: Context awareness in healthcare: a systematic literature review. Univers. Access Inf. Soc. 19, 245–259 (2020)
13. Guerrero, C., Lera, I., Juiz, C.: Evaluation and efficiency comparison of evolutionary algorithms for service placement optimization in fog architectures. Future Gener. Comput. Syst. **97**, 131–144 (2019)
14. Jacoby, M., Usländer, T.: Digital twin and internet of things-current standards landscape. Appl. Sci. **10**(18), 6519 (2020)
15. Kaenampornpan, M., O'Neill, E., Ay, B.B.: An integrated context model: bringing activity to context. In: Proceedings of the Workshop on Advanced Context Modelling, Reasoning and Management (2004)
16. Larsen, P., Brunthaler, S., Franz, M.: Security through diversity: are we there yet? IEEE Secur. Priv. **12**(2), 28–35 (2013)
17. Leivadeas, A., Kesidis, G., Ibnkahla, M., Lambadaris, I.: VNF placement optimization at the edge and cloud. Future Internet **11**(3), 69 (2019)
18. Magdich, R., Jemal, H., Ben Ayed, M.: Context-awareness trust management model for trustworthy communications in the social internet of things. Neural Comput. Appl. **34**, 1–26 (2022)
19. Merlino, G., Dautov, R., Distefano, S., Bruneo, D.: Enabling workload engineering in edge, fog, and cloud computing through OpenStack-based middleware. ACM Trans. Internet Technol. (TOIT) **19**(2), 1–22 (2019)

20. Minerva, R., Lee, G.M., Crespi, N.: Digital twin in the IoT context: a survey on technical features, scenarios, and architectural models. Proc. IEEE **108**(10), 1785–1824 (2020)
21. Qian, C., Liu, X., Ripley, C., Qian, M., Liang, F., Yu, W.: Digital twin-cyber replica of physical things: architecture, applications and future research directions. Future Internet **14**(2), 64 (2022)
22. Semeraro, C., Lezoche, M., Panetto, H., Dassisti, M.: Digital twin paradigm: a systematic literature review. Comput. Ind. **130**, 103469 (2021)
23. Song, H., Dautov, R., Ferry, N., Solberg, A., Fleurey, F.: Model-based fleet deployment of edge computing applications. In: Proceedings of the 23rd ACM/IEEE International Conference on Model Driven Engineering Languages and Systems, pp. 132–142 (2020)
24. Song, H., Dautov, R., Ferry, N., Solberg, A., Fleurey, F.: Model-based fleet deployment in the IoT-edge-cloud continuum. Softw. Syst. Model. **21**(5), 1931–1956 (2022)
25. Xu, D., et al.: A survey of opportunistic offloading. IEEE Commun. Surv. Tutorials **20**(3), 2198–2236 (2018)

Adoption of Virtual Agents in Healthcare E-Commerce: A Perceived Value Perspective

Claire Deventer[1,2](✉) and Pietro Zidda[1]

[1] NaDI-CeRCLe, University of Namur, Namur, Belgium
{claire.deventer,pietro.zidda}@unamur.be
[2] SkalUP, Namur, Belgium

Abstract. Virtual agents help their users find what they need thanks to an interactive dialog. In the healthcare e-commerce market, virtual agents allow "virtual consultations" available on the web, that lead to a recommendation for personalized treatment. While the adoption intention of these virtual agents by their users is critical for many organizations, the traditional explanatory models such as UTAUT-2 miss key elements specific to the virtual agents in healthcare e-commerce. Filling this gap can help organizations better understand the factors leading to the adoption of such solutions and take them into account in the design and launch of their virtual agents. This paper adopts a perceived value perspective, and proposes an extended model explaining the adoption of virtual agents and of their recommendation in a healthcare context. We test this model with 903 observations collected via an online survey in collaboration with a major European actor in the food supplement market. Our model provides highly actionable recommendations for practitioners and offers a complementary view on the adoption mechanisms of virtual agents, leading to further research recommendations.

Keywords: Virtual Agent · E-health · E-commerce · Technology Adoption

1 Introduction

Healthcare e-commerce is on the rise. The market size is expected to grow from $309.62 billion in 2022 to $732.3 billion in 2027 at an annual growth rate of nearly 20% [1]. Parapharmaceutical products such as food supplements are a good example of this growing popularity [2]. New players entering this market both in USA and Europe rely on new technologies and especially on personalized advice as a differentiating factor [3]. These new actors offer on their website virtual agents that, via an interactive dialog about their users' needs, suggest personalized sets of food supplements. Such virtual agents greatly help customers make their purchase decision [4].

Consistent with the stream of research on market-driven requirement engineering, understanding the users' motivations for technology adoption can help design and develop better applications, that are actually used. Until now, when considering user adoption of information technology in healthcare, past research focused mainly on TAM or UTAUT-2 [5]. The latter model is considered as one of the most relevant in the domain [6, 7]. However, originally created as a generalist model that explains the adoption of

S. Nurcan et al. (Eds.): RCIS 2023, LNBIP 476, pp. 256–271, 2023.
https://doi.org/10.1007/978-3-031-33080-3_16

many kinds of technologies, UTAUT-2 could in the case of virtual agents for healthcare e-commerce miss important constructs and therefore miss important insights for effective virtual agents' design and development.

At first, in a healthcare context, risks are more salient. There are strong consequences to a wrong choice: an overdose or an incompatibility with a physical state can be really harmful and generate a perceived health risk [8]. In addition, in order to provide the personalized advice, the virtual agent tends to ask very personal questions. This could trigger perceived risks for the privacy of the information [9]. Moreover, the UTAUT-2 model [7] puts forward the utilitarian and hedonic values but misses other aspects such as learning (epistemic benefits), self-expression and uniqueness, which could be relevant for the specific case of virtual agents in healthcare e-commerce. Indeed, virtual agents give a lot of information about the product category, helping customers define their own preferences [10, 11]. In addition, the provision of personalized recommendations helps people feel listened to and unique. This could be reflected by perceived self-expressiveness and uniqueness benefits [12, 13].

Perceived value theory, including various benefits, costs and risks as drivers of behavioral attitudes [14] could help complete UTAUT-2 model in explaining the adoption of virtual agents in a healthcare e-commerce context. We therefore aim to integrate perceived value theory elements that are specific both to virtual agents and to healthcare e-commerce context to the initial UTAUT-2 model. More specifically, we propose a model incorporating to UTAUT-2 perceived health and privacy risks, perceived epistemic, self-expressiveness and uniqueness benefits to predict the intention to adopt a virtual agent and its recommendation. We test this model with 903 observations via an online survey thanks to a collaboration with a major player in the food supplement market. The results highlight the role of epistemic and self-expressiveness benefits to explain the adoption of virtual agents and their recommendations. We also show that the virtual agent's adoption is not equivalent to the adoption of its recommendation and that different drivers underlie these two behaviors.

Our contribution is multifaceted. First, we provide a new perspective on the adoption of virtual agents and their recommendations with a model integrating constructs relevant for virtual agents in the healthcare e-commerce context. Second, this study highlights the importance of epistemic and self-expressiveness benefits, often overlooked in the technology adoption literature. Third, we enhance the understanding on the interaction between the adoption of virtual agents and the adoption of their recommendations. Our model can serve as a basis for further research on the design of virtual agents in a healthcare e-commerce context by highlighting the most important user perceptions to drive adoption. At last, we provide several recommendations for practitioners aiming to develop and launch a virtual agent on a healthcare e-commerce website.

The remainder of this paper is organized as follows. First, Sect. 2 reviews related work on information systems adoption in the contexts of healthcare and virtual agents, and identifies their limitations. Then, Sect. 3 represents the new model that we propose to address these limitations and explain the adoption of virtual agents in healthcare e-commerce. After that, Sect. 4 details the methodology used to validate the proposed model. Section 5 then presents the results and the adapted model by considering the statistical results on the initial model proposition. Then, a discussion on the results, their limitations, its theoretical contribution and managerial implications are provided in Sect. 6. General conclusions are provided in Sect. 7.

2 Related Work and Research Gap

2.1 Technology Adoption in a Healthcare Context

Traditionally, research on the adoption of Health Information Technology (HIT) by customers focus on well-established, general technology adoption models. As mentioned in the integrative review of Gücin and Berk [15], the most used theories to explain HIT adoption are Technology Acceptance Model (TAM), the Diffusion of Innovations Theory, and the Unified Theory of Acceptance and Use of Technology (UTAUT) and UTAUT-2, an adaptation of UTAUT.

In general, the explanatory power of UTAUT and UTAUT-2 is better than the explanatory power of the other technology adoption models [16]. Research about UTAUT and UTAUT-2 has thrived and these models are considered by many as the most relevant for predicting technology adoption [7, 17].

However, in a healthcare context, important aspects have to be incorporated to the original UTAUT-2 model [18, 19]. In particular, studies examine the HIT adoption through the lenses of health behavior models [8, 20, 21]. A common point to those health behavior models is the threat appraisal: when confronted to the decision to use or not use a health technology, customers tend to evaluate the threat to their health of using the technology vs. not using the technology.

In addition to this threat appraisal, privacy is another very important aspect of HIT adoption [22, 23]. Health information is indeed particularly sensitive for individuals [9]. A well-established theory about privacy concerns is privacy-calculus. When a user is confronted to the choice of giving away personal information, he experiences a perceived risk that malicious individuals steal this information to use it against the user's will. They therefore operate a trade-off between this risk and the possible benefits they could derive from disclosing their information.

Recent studies argue for integrating traditional technology adoption theories with privacy theories [24] or health behavior theories [21]. Gao, Li and Luo [24] indicate that merging these elements leads to a more effective technology adoption model in the context of healthcare. It corroborates the position of Miltgen, Popovic and Oliveira [25] that HIT adoption is too complex to be explained from a single perspective.

These elements confirm the interest of tailoring the original UTAUT-2 framework to the specific object of the study: virtual agents in a healthcare e-commerce context. When efforts have been made for wearables [24] or mobile health technology [19], to the best of our knowledge no studies have tried to build on the traditional UTAUT-2 adoption model to extend it to the specific case of virtual agents.

2.2 Perceived Value Perspective on Virtual Agents' Adoption

Considering the adoption of virtual agents, we might take a closer look at what we consider exactly by "adoption". Indeed, in the case of a "virtual consultation" by a virtual agent, the user actually faces two adoption stages: the process adoption and the recommendation adoption. We define "the process adoption" as the choice to use the virtual agent to obtain a personal recommendation. On the other hand, "recommendation adoption" refers to the choice to follow the given recommendation. If process adoption is necessary to have recommendation adoption, it is not sufficient as the user might follow all the steps of the dialog and decide at the end not to follow the recommendation. For these reasons, we consider process adoption distinct from recommendation adoption.

To account for the specificities of virtual agents in our proposed adoption explanatory model, we rely on perceived value theory. Central concept in marketing [26], perceived value is defined as the "the consumer's overall assessment of the utility of a product based on perceptions of what is received and what he is given" [27, p. 14]. Put in simple words and applied to technology adoption, perceived value theory says that when a user chooses to adopt or not a technology, he does an unconscious trade-off between what is received (the perceived benefits of the technology usage) and what is given (the perceived costs and risks associated with the technology usage).

While extended research about perceived value attempted to holistically define all types of benefits, costs and risks a customer might perceive [28–31], recent research advise to use only perceived benefits, costs and risks that are relevant for our particular object of study [14].

Comparing constructs from perceived value theories, applied to the adoption of virtual agents, to constructs from technology adoption models, we observe some strong commonalities.

At first, we can consider utilitarian benefits as amongst the perceived benefits of virtual agents. Utilitarian benefits of a product/service refer to the fact that this product/service allows the user to reach another goal [31]. In the case of virtual agents for healthcare e-commerce, the other goal is to find the perfect treatment for our health concerns. Utilitarian benefits concept, when applied to virtual agents, is therefore really close to the "Performance expectancy" of UTAUT and UTAUT-2, defined as "the degree to which an individual believes that using the system will help him or her to attain gains in job performance" [32, p. 447].

In addition, hedonic benefits, recognized as the entertainment customers experience with a particular product/service [31] could also be relevant in the case of virtual agents. Indeed, as mentioned in Hess et al. [33], the use of embodied virtual agents induces fun and enjoyment for users. Applied to virtual agents, the hedonic benefits concept is also very close, if not equivalent, to the "Hedonic motivation" of UTAUT-2, defined as "the fun or pleasure derived from using a technology" [17, p. 161].

On the other hand, some benefits from the perceived value literature might be relevant to the case of virtual agents and differ from the constructs of UTAUT.

At first, as virtual agents often give a lot of information about the product they recommend, they might help their users better know what they want [10, 11]. By remembering the information given by the virtual agent, their users experiment epistemic (or learning) benefits. They can be defined as the feeling of a better understanding of the world [31]. Novelty or curiosity arousal is also considered as epistemic benefits [29].

Then, with their interactive dialog, virtual agents set the space for personal discussions that could lead their users to feel listened to. This feeling is translated in perceived value theory in self-expressiveness benefits. These benefits are the perception that the service, i.e. the discussion in the case of virtual agents, reflects who we are [12, 13].

In addition, as each recommendation and dialog are unique, depending on the personal combination of answers of the user, the usage of virtual agents could lead to uniqueness benefits: the feeling of having a special treatment, of being unique compared to the other users [12].

On the side of the perceived costs, as the virtual agent usage requires an interactive dialog, it requires some time (to answer the questions) and effort (to understand the questions and choose the appropriate answer) to their users. When the UTAUT and UTAUT-2 models only consider effort expectancy, i.e. "the degree of ease associated with the use of the technology", we argue that in the case of virtual agents it is also interesting to take the perceived time cost into account.

At last, perceived value theory is also aware of the perceived privacy and health risks associated with a particular product or service. It makes perceived value theory highly consistent with the privacy-calculus and health behavior models described in the previous section.

3 Proposed Model

Building on this theoretical background, we propose an explanatory model adapting the original UTAUT-2 framework with elements from healthcare context and perceived value of virtual agent and describe the hypotheses that are tested.

As exposed in Sect. 2.2., our model incorporates two adoption stages: process adoption and recommendation adoption. If a user has high process recommendation intention, s/he presumably thinks that the process will lead to a useful recommendation. We therefore expect that process adoption intention precedes and drives recommendation adoption intention.

H1. Process adoption intention increases recommendation adoption intention

Starting from UTAUT-2, the original model proposes five variables (performance expectancy, effort expectancy, social influence, hedonic motivations and price value) that affect behavioral intention to use a technology, two variables (facilitating conditions and habits) that affects actual usage, and three moderators (age, gender and experience).

As we previously explained, performance expectancy and hedonic motivations are really close to utilitarian and hedonic benefits. We therefore choose to incorporate them in the model, along with epistemic, self-expressiveness and uniqueness benefits, the other benefits specific to virtual agents presented in Sect. 2.2. We hypothesize that the benefits positively influence the intention to adopt the virtual agent process as well as the intention to adopt the recommendation.

H2. The perceived (a) utilitarian, (b) hedonic, (c) self-expressiveness and (d) uniqueness benefits increase the process adoption intention

H3. The perceived (a) utilitarian, (b) hedonic, (c) self-expressiveness and (d) uniqueness benefits increase the recommendation adoption intention

Likewise, effort expectancy is close to the concept of perceived effort costs so we incorporate this along with perceived time costs, identified as relevant for virtual agents in Sect. 2.2. However, as virtual agents used for healthcare e-commerce are most of the time free of use, we set aside the price value. Consistently with Sect. 2.1 about the specificities of healthcare context, we incorporate perceived health and privacy risks. We expect that perceived time and effort costs, along with perceived health risks, negatively impact the intention to adopt the virtual agent process and the recommendation.

H4. Perceived (a) time and (b) effort costs decrease the process adoption intention

H5. Perceived (a) time and (b) effort costs decrease the recommendation adoption intention

H6. Perceived health risks decrease the process adoption intention

H7. Perceived health risks decrease the recommendation adoption intention

As the data collection only concerns the process of the virtual agent usage, we hypothesize that perceived privacy risks negatively impact the intention to adopt the virtual agent process only.

H8. Perceived privacy risks decrease the process adoption intention

Defined as "the degree to which an individual perceives that important others believe he or she should use the new system" [17, p. 159], we expect that, as in the original UTAUT-2 model, social influence increases the process adoption and the recommendation adoption intention.

H9. Social influence increases the process adoption intention

H10. Social influence increases the recommendation adoption intention

We decided not to incorporate habits in our model. Indeed, as "virtual consultations" to obtain automatically a personalized treatment are very new, it is difficult for the population of our study to develop habits. Moreover, facilitating conditions originally refer to the resources, knowledge and support available to facilitate the use of a system [32]. While it can be of tremendous importance when the system requires rare hardware or specific skills, it is less the case with virtual agents for healthcare e-commerce. Indeed, these agents are directly accessible on the e-commerce website. Most of people in our studied population (Western Europe) have internet and have already used chats applications. We therefore propose not to integrate facilitating conditions in our model neither. Since the focus on the study is not on the moderating variables, we did not include any moderation effect. Figure 1 illustrate our proposed model.

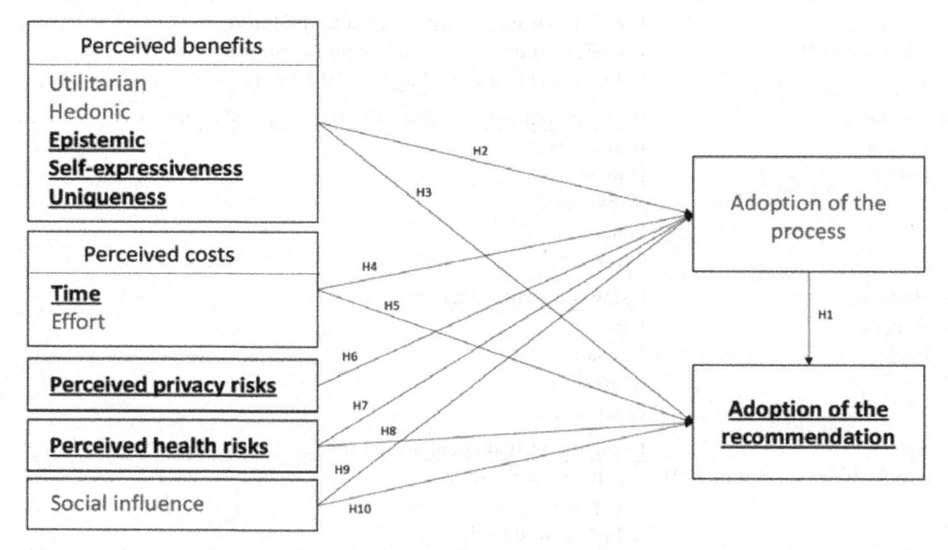

Fig. 1. Adoption of virtual agents in healthcare e-commerce conceptual model. Constructs added to the original UTAUT-2 framework are in bold and underlined.

4 Methodology

To test our model, we developed a virtual agent with the collaboration of a major actor in the food supplement market. We then launched an online survey during which the participants were told to evaluate a new virtual agent: a real chatbot. The respondents answered the questions of the chatbot in order to obtain a personalized recommendation. According to their personal profile, the participants had to answer from 7 questions (simple profile with only one health concern) to 13 questions (more complex profile with several health concerns). Then, they received a personal recommendation.

After that, we evaluated the intention to buy the recommendation, that will be our recommendation adoption construct in our model. We measured it with a 7-points Likert scale from "I would definitely not buy it" to "I would definitely buy it". Then, we measured the other constructs with items adapted from scales existing in the literature. All items were measured with a 7-points Likert agreement scale, in French. As virtual agents for healthcare e-commerce are particularly new, it is more likely that the user's friends and family advice to be cautious and not use the virtual agent. For this reason, we reversed all items for social influence to measure negative social influence. We measured also a set of control variables such as trust towards the virtual agent and the company, familiarity with the virtual agent, loyalty to the company as well as expertise and involvement in the product category with traditional scales used in the literature. All scales were pre-tested with target population before survey launch to ensure the items were well understood. Table 1 illustrates all scale items used in the survey, obtained via a double-translation process.

Table 1. Survey items

Construct and source	Scale
Process adoption [17]	I intend to use this virtual agent in the future I predict I'll use this virtual agent in the future I plan to use this virtual agent in the future
Utilitarian benefits [34]	In my search for a product, I find this virtual agent: Is effective Is useful Works well Is necessary Is practical
Hedonic Benefits [34]	Using the virtual agent is…: Fun Pleasant Interesting Entertaining
Self-expressiveness benefits [12]	Using the virtual agent allows me…: to reflect my lifestyle to express my way of life to be true to myself

(*continued*)

Table 1. *(continued)*

Construct and source	Scale
Uniqueness benefits [13]	Thanks to the virtual agent: I know that I don't have the same advice as other people I get advice that other people will not have I have my little difference compared to other people
Epistemic benefits [35]	Using the virtual agent: Makes me try new things Makes me discover new things Satisfies my curiosity Is teaching me things I didn't know
Time costs [36]	Using this virtual agent: Takes a lot of my time Takes me longer than expected
Effort costs [37]	Using this virtual agent: Takes a lot of effort Seems easy to me Seems complicated to me
Privacy risks [38]	By using this virtual agent I fear: I'll lose control of the confidentiality of my personal information My personal information will be used without my consent People will take advantage of my personal data
Health risks [39]	I am nervous about following the agent's recommendation due to health concerns Following the agent's recommendation is a risky decision for my health I feel uncomfortable following the agent's recommendation for health reasons I think there is a good chance that following the agent's recommendation will lead to health problems
Social Influence [17]	People who are important to me would find it a bad idea to use a virtual agent for food supplements People who matter to me would advise me to avoid virtual agents for food supplements People whose opinion I value would prefer that I use something other than virtual agent for food supplements

The final sample includes 903 respondents from Belgium and France. The sample is mainly women (85%) and middle-aged (45 – 50 years old), which is consistent with the profile of food supplement customers [40].

5 Results

We performed validity and reliability checks before testing our conceptual model. We discarded the following items ("Using the virtual agent is interesting.") and effort cost ("I find the virtual agent easy to use") from the scales according to the minimum communality, minimum loading and cross-loading rules [41]. The final confirmatory factor analysis (CFA) show that all factor loadings are above 0.7 and statistically significant ($p < .01$). The composite reliabilities (CR) and the alpha coefficients are all over the 0.7 recommended threshold [42]. The average variance extracted (AVE) being higher than 0.5 for every construct, the convergent validity is verified. As the square root of AVE is greater than the correlation between any pair of constructs, the discriminant validity is also supported [42]. The scales are therefore reliable and valid. All our constructs follow a standard normal distribution. Table 2 summarizes the results.

Table 2. Confirmatory, reliability and validity analysis (n = 903)

	Constructs	AVE	Composite Reliability	Cronbach's Alphas	Mean	Std. Dev
Adoption	Process	0.932	0.976	0.964	5.01	1.56
	Recommendation	/	/	/	4.54	1.36
Benefits	Utilitarian	0.833	0.961	0.949	5.25	1.31
	Hedonic	0.797	0.940	0.916	4.41	1.51
	Epistemic	0.843	0.956	0.938	5.25	1.36
	Self-expressiveness	0.909	0.968	0.950	4.66	1.52
	Uniqueness	0.873	0.954	0.927	4.41	1.60
Costs	Time	0.912	0.954	0.905	2.24	1.41
	Effort	0.877	0.934	0.860	2.27	1.31
Risks	Privacy	0.910	0.968	0.951	3.27	1.85
	Health	0.733	0.916	0.880	2.47	1.47
	Negative social inf	0.870	0.953	0.926	3.10	1.74

As shown in Table 2, utilitarian (mean of 5.25) and epistemic (also mean of 5.25) benefits are the highest, followed by self-expressiveness (mean of 4.66), hedonic and uniqueness (means at 4.41) benefits. The perceived costs are quite low and equivalent (2.24 on average for time cost, 2.27 for effort cost). On the perceived risk side, privacy risks are higher (mean of 3.27) than health risks (mean of 2.47). Process adoption intention is also high (mean of 5.02), immediately followed by recommendation adoption intention (mean of 4.54). Finally, we note that the negative social influence is quite low (3.10), indicating that most of the respondents have not experienced situations where their loved ones advised them against using the virtual agent.

We then used SmartPLS to test our hypotheses. The choice of SmartPLS was made due to the complexity of the model. We obtained a SRMR of 0.072, indicating that our model correctly reflects our data. The adjusted R-square is 0.542 for the process adoption intention and 0.361 for the recommendation adoption intention.

As indicated in Table 3, regarding the direct effects on the adoption of the process, contrary to what was expected, the perceived risks, whether for privacy or health, do not influence the adoption of the process. Similarly, the costs in time and effort do not have a significant influence on the adoption of IRS as a process guiding the choice. As shown in Table 3, hedonic, uniqueness, and self-expression benefits also do not have significant impacts on process adoption. The results also indicate that negative social influence does not have significant impacts on the adoption of the process.

On the other hand, we note that utilitarian benefits are the most impactful predictors of the process adoption intention ($\beta = 0.567$), followed by epistemic ($\beta = 0.116$).

Table 3. Analysis of the direct effects on the adoption of the process

Hyp	Path	Coeff	p-value
H2.a	Utilitarian benefits → Process adoption	**$\beta = 0.567$**	**< 0.001**
H2.b	Hedonic benefits → Process adoption	$\beta = 0.024$	= 0.535
H2.c	Epistemic benefits → Process adoption	**$\beta = 0.116$**	**= 0.023**
H2.d	Self-expressiveness benefits → Process adoption	$\beta = 0.037$	= 0.428
H2.e	Uniqueness benefits → Process adoption	$\beta = 0.030$	= 0.427
H4.a	Time costs → Process adoption	$\beta = -0.010$	= 0.792
H4.b	Effort costs → Process adoption	$\beta = -0.019$	= 0.657
H6	Health risks → Process adoption	$\beta = 0.010$	= 0.704
H7	Privacy risks → Process adoption	$\beta = -0.026$	= 0.333
H9	Negative social inf. → Process adoption	$\beta = -0.017$	= 0.572

Regarding the direct effects of predictors of the recommendation adoption intention, Table 4 indicates that the process adoption intention has a big significant impact on the adoption of the recommendation ($\beta = 0.408$, p-value < 0.001). On the side of perceived benefits and costs, we realize that only self-expressiveness benefits have a significant impact ($\beta = 0.166$). This confirms the importance of personalization in this type of software. Moreover, unsurprisingly, perceived health risks have a significant negative impact on the intention to adopt the recommendation. More interestingly, we notice that the negative social influence seems to have a significant positive impact on the intention to adopt the recommendation. An explanation could be a "rebellion" effect, or even a desire to test for yourself instead of listening to opinions.

Finally, in order to have the whole picture, Table 5 establishes the mediation analysis. As only utilitarian and epistemic benefits have significant effect on the process adoption intention, we only analyzed the "Utilitarian benefits → Process adoption → Recommendation adoption" and the "Epistemic benefits → Process adoption → Recommendation adoption" paths. We note that neither the utilitarian benefits nor the epistemic benefits

Table 4. Analysis of the direct effects on the adoption of the recommendation

Hyp	Path	Coeff	p-value
H1	Process adoption → Recommendation adoption	**β = 0.408**	**< 0.001**
H3.a	Utilitarian benefits→ Recommendation adoption	β = 0.079	= 0.215
H3.b	Hedonic benefits → Recommendation adoption	β = −0.053	= 0.214
H3.c	Epistemic benefits → Recommendation adoption	β = 0.044	= 0.429
H3.d	Self-exp. Benefits → Recommendation adoption	**β = 0.166**	**= 0.002**
H3.e	Uniqueness benefits → Recommendation adoption	β = 0.015	= 0.708
H5.a	Time costs → Recommendation adoption	β = 0.028	= 0.550
H5.a	Effort costs → Recommendation adoption	β = −0.038	= 0.437
H8	Health risks → Recommendation adoption	**β = −0.142**	**< 0.001**
H10	Negative social inf. → Recommendation adoption	**β = 0.072**	**= 0.055**

have a direct effect on the adoption of the recommendation. This leads us to conclude that there is full mediation through the adoption of the process.

Table 5. Mediation analysis through adoption of the process

Relationship	Indirect effect	Direct effect	Total effect
Utilitarian benefits → Process adoption → Recommendation adoption	β= 0.231 (p < 0.001)	*No effect* (p = 0.215)	β= 0.231
Epistemic benefits → Process adoption → Recommendation adoption	β= 0.047 (p = 0.025)	*No effect* (p = 0.429)	β= 0.047

Looking at the total impacts of predictors on the recommendation adoption intention, we observe that process adoption intention is the most impactful (β = 0.408) followed by utilitarian benefits (β = 0.231, fully mediated by process adoption intention), self-expressiveness benefits (β = 0.166, direct effect) and epistemic benefits (β = 0.047, fully mediated by process adoption intention). Figure 2 summarizes the significant relationships in our model.

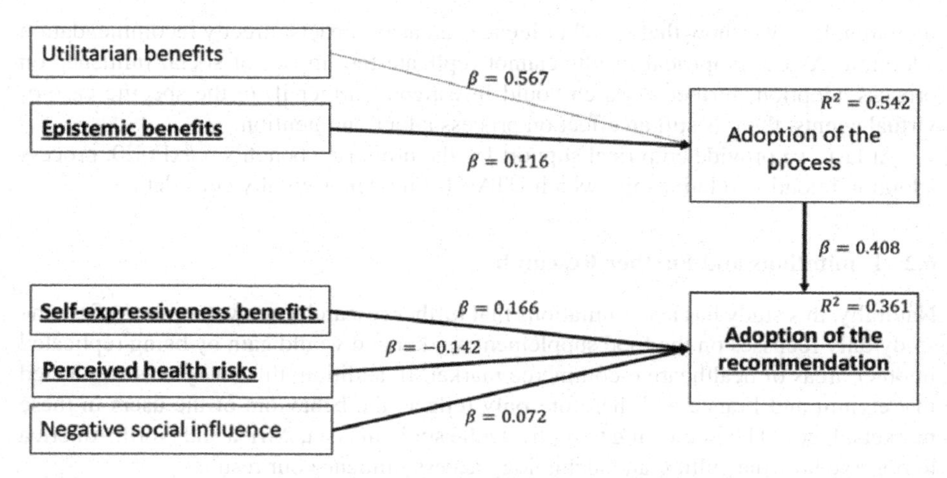

Fig. 2. Summary of the significant effects in our proposed adoption model, the new constructs and relationships, compared to UTAUT-2, are in bold and underlined

6 Contributions to Theory and Practice

This study contributes to information systems research community by providing a new adoption framework, specific to the case of virtual agents for healthcare e-commerce. While COVID-19 has fostered the growth of e-commerce in many new areas, including healthcare, our framework gives important insight in this area.

6.1 Theoretical Contributions

From a theoretical point of view, we show that process adoption intention is different from recommendation adoption intention and that their drivers can be different. It also stresses the fact that process adoption is not sufficient for recommendation adoption and research on virtual agent adoption should consider these two aspects and search for their specific drivers.

In addition, we highlight the importance of two new constructs closely related to the specific type of technology: epistemic and self-expressiveness benefits.

As in the case of product search many individuals need clear information to define their preferences and increase their knowledge, we incorporated epistemic benefits to the original UTAUT-2 model. Epistemic benefits have indeed a significant impact on the process and recommendation adoption intention. When it impacts directly process adoption intention, it also impacts recommendation adoption intention through a full mediation via process adoption intention.

On the other hand, self-expressiveness benefits only impact recommendation adoption intention, meaning that as users feel they are listened to, they are more convinced in following the recommendation. Further research may build on our results to investigate how the virtual agent can generate this self-expressiveness benefits perception.

Our results also highlight the relationship between social influence and recommendation adoption intention. When the original UTAUT-2 only considers process adoption intention, here we show that social influence can also impacts directly recommendation adoption. As our empirical results cannot replicate the impact of social influence on process adoption, further research could investigate further if, in the specific case of virtual agents, there is still an effect on process adoption intention.

At last, we provide empirical support for the utilitarian benefits  process adoption intention relationship, which UTAUT-2 model originally considers.

6.2 Limitations and Further Research

Naturally, this study has also limitations that further research can overcome. At first, the study only focusses on the food supplement market and would gain by being replicated in other areas of healthcare e-commerce market. In addition, this study was conducted in Belgium and France and therefore only reflects the behaviors of the users in these markets. It could be interesting to replicate the study in Asia, Africa and North America to observe how the culture and technology access influence our results.

Despite being representative of the food supplement market, women mainly compose our sample (85%), further research could also replicate the study with a sample more representative of the population in general. In addition, the scope of our model was limited as we only considered the intention of adoption (not the actual behavior) and did not incorporate any moderation effect.

As our study is one of the first attempt to incorporate in the model the process and recommendation adoptions, further research could investigate this relationship deeper by providing more qualitative insights on the two constructs. Moreover, the absence of effects for several well-known UTAUT-2 relationships could also be investigated.

6.3 Managerial Implications

This study shows that the process and the recommendation adoptions are two distinct constructs, driven by different, though linked, elements. In order to design successful virtual agents for healthcare e-commerce, practitioners should then consider these two aspects.

At first, to maximize recommendation adoption, self-expressiveness benefits are important to consider. The personal recommendation should be tailored to the needs and lifestyle of the user. In particular, effort should be put to make sure that every element important for the users is considered in the virtual agent dialog. As an anecdote, we received an email from a respondent that told us she was vegan and that it was extremely important for her to obtain a vegan recommendation. As we did not take it into account in the virtual agent dialog, she told us she was disappointed and did not feel listened to. The self-expressiveness benefit was absent for her. Previous research also showed that other aspects such as allowing to modify previous answers could also be a good way to enhance self-expressiveness perception [13].

Then, the process adoption itself is extremely important for increasing the recommendation adoption. To design a successful virtual agent process, it is important to make it as efficient as possible in order to increase the utilitarian benefits perception. Indeed, utilitarian benefits are highlighted in this study as the most important predictor of process adoption intention and second most important predictor of recommendation adoption intention. While it is important to consider every aspect that is key to the user decision making, it is also important to make the dialog to the point and to avoid unnecessary steps as much as possible.

Finally, epistemic benefits are also interesting to consider while designing the virtual agent process. Allowing the users to learn about their preference is key. A way to do it is for example to adopt Wang and Benbasat [43] insights about questions explanations. The authors indicate that to enhance the trust in a recommendation agent, it is interesting to give three types of explanations: *how*, *why* and *trade-off*. The *how* refers to how the virtual agent will reason to come up with his conclusions, regarding the answers given by the customer. The *why* deals with the importance of every questions to come up with the final recommendation. At last, the *trade-off* explanations give indications on the consequences of the user's choices (i.e. if he chooses one option, he will lose the opportunity to choose another one). Beyond increasing trust, this could also help users learn about the product category and their preferences, increasing their perceived epistemic benefits and therefore their process and recommendation adoption intention.

While those recommendations are suggestions from our findings, an important avenue for further research is to uncover new ways to maximize utilitarian, epistemic and self-expressiveness. Empirical validation of our suggestions might also be of great interest.

7 Conclusion

This study investigates the adoption of virtual agents by customers in a healthcare e-commerce context. On basis of the technology adoption and perceived value literatures, we propose an explanatory model specific to the adoption of the process and of the recommendation given by virtual agents. Contrary to the other generalist technology adoption models in the literature, our model provides insights on particularly relevant constructs for virtual agents in healthcare e-commerce. We test this model with 903 observations via an online survey thanks to a collaboration with a major actor in the food supplement market. Our findings indicate that process adoption intention is distinct from recommendation adoption intention and is driven by different perceived benefits. While utilitarian benefits are the most important benefits to predict process adoption and recommendation adoption, the results also highlight the important roles of epistemic and self-expressiveness benefits, often overlooked in technology adoption literature. We contribute to research on technology adoption by building on perceived value theory to propose and empirically test a model specific to the particular context of healthcare e-commerce. Our model allows for actionable managerial recommendations and can be used as a basis for further research on virtual agent adoption. We give managerial recommendations and indicate avenues for further research about the design and adoption of virtual agents.

End note: The authors would like to thank gratefully Eva Nalmpantidis, a graduate student at the University of Namur for her help in collecting the data and for her comments.

Funding. The authors thank the Win4Doc funding program from Wallonia Belgium.

References

1. Business Research Company, "Healthcare E Commerce Global Market Report" (2023). https://www.thebusinessresearchcompany.com/report/healthcare-ecommerce-global-market-report
2. Allied Market Research, "Personalized Nutrition Market" (2022). https://www.alliedmarketresearch.com/personalized-nutrition-market-A16650
3. Journal du net, "L'essor sans précédent de la nutrition personnalisée" (2020). https://www.journaldunet.com/economie/sante/1489265-l-essor-sans-precedent-de-la-nutrition-personnalisee/
4. Häubl, G., Trifts, V.: Consumer decision making in online shopping environments: the effects of interactive decision aids. Mark. Sci. **19**(1), 4–21 (2000)
5. Tao, D., Wang, T., Wang, T., Zhang, T., Zhang, X., Qu, X.: A systematic review and meta-analysis of user acceptance of consumer-oriented health information technologies. Comput. Hum. Behav. **104**, 106147 (2020)
6. Blut, M., Chong, A., Tsiga, Z., Venkatesh, V.: Meta-analysis of the unified theory of acceptance and use of technology (UTAUT): challenging its validity and charting a research agenda in the red ocean. J. the Assoc. Inf. Syst. (2021)
7. Venkatesh, V., Thong, J.Y., Xin, X.: Unified theory of acceptance and use of technology: a synthesis and the road ahead. J. Assoc. Inf. Syst. **17**(5), 328–376 (2016)
8. Prentice-Dunn, S., Rogers, R.W.: Protection motivation theory and preventive health: beyond the health belief model. Health Educ. Res. **1**(3), 153–161 (1986)
9. Bansal, G., Gefen, D.: The impact of personal dispositions on information sensitivity, privacy concern and trust in disclosing health information online. Decis. Support Syst. **49**(2), 138–150 (2010)
10. Adomavicius, G., Bockstedt, J.C., Curley, S.P., Zhang, J.: Do recommender systems manipulate consumer preferences? A study of anchoring effects. Inf. Syst. Res. **24**(4), 956–975 (2013)
11. Häubl, G., Murray, K.B.: Preference construction and persistence in digital marketplaces: the role of electronic recommendation agents. J. Consum. Psychol. **13**(1–2), 75–91 (2003)
12. Merle, A., Chandon, J., Roux, E., Alizon, F.: Perceived value of the mass-customized product and mass customization experience for individual consumers. Prod. Oper. Manag. **19**(5), 503–514 (2010)
13. Sandrin, E., Trentin, A., Grosso, C., Forza, C.: Enhancing the consumer-perceived benefits of a mass-customized product through its online sales configurator: an empirical examination. Ind. Manage. Data Syst. **117**(6), 1295–1315 (2017)
14. Leroi-Werelds, S.: An update on customer value: state of the art, revised typology, and research agenda. J. Serv. Manage. **30**(5), 650–680 (2019)
15. Gücin, N.Ö., Berk, Ö.S.: Technology acceptance in health care: an integrative review of predictive factors and intervention programs. Procedia Soc. Behav. Sci. **195**, 1698–1704 (2015)
16. Lee, J., Rho, M.J.: Perception of influencing factors on acceptance of mobile health monitoring service: a comparison between users and non-users. Healthc. Inf. Res. **19**(3), 167–176 (2013)
17. Venkatesh, V., Thong, J.Y., Xu, X.: Consumer acceptance and use of information technology: extending the unified theory of acceptance and use of technology. MIS Q. 157–178 (2012)
18. Holden, R.J., Karsh, B.-T.: The technology acceptance model: its past and its future in health care. J. Biomed. Inform. **43**(1), 159–172 (2010)
19. Duarte, P., Pinho, J.C.: A mixed methods UTAUT2-based approach to assess mobile health adoption. J. Bus. Res. **102**, 140–150 (2019)

20. Weinstein, N.D.: Testing four competing theories of health-protective behavior. Health Psychol. **12**(4), 324 (1993)
21. Sun, Y., Wang, N., Guo, X., Peng, Z.: Understanding the acceptance of mobile health services: a comparison and integration of alternative models. J. Electr. Commer. Res. **14**(2), 183 (2013)
22. Angst, C.M., Agarwal, R.: Adoption of electronic health records in the presence of privacy concerns: the elaboration likelihood model and individual persuasion. MIS Q. 339–370 (2009)
23. Li, H., Gupta, A., Zhang, J., Sarathy, R.: Examining the decision to use standalone personal health record systems as a trust-enabled fair social contract. Decis. Support Syst. **57**, 376–386 (2014)
24. Gao, Y., Li, H., Luo, Y.: An empirical study of wearable technology acceptance in healthcare. Ind. Manage. Data Syst. (2015)
25. Miltgen, C.L., Popovič, A., Oliveira, T.: Determinants of end-user acceptance of biometrics: Integrating the "Big 3" of technology acceptance with privacy context. Decis. Support Syst. **56**, 103–114 (2013)
26. American Marketing Association (AMA), "Definitions of Marketing." https://www.ama.org/the-definition-of-marketing-what-is-marketing/
27. Zeithaml, V.A.: Consumer perceptions of price, quality, and value: a means-end model and synthesis of evidence. J. Mark. **52**(3), 2–22 (1988)
28. Holbrook, M.B., Hirschman, E.C.: The experiential aspects of consumption: consumer fantasies, feelings, and fun. J. Consum. Res. **9**(2), 132–140 (1982)
29. Sheth, J.N., Newman, B.I., Gross, B.L.: Why we buy what we buy: a theory of consumption values. J. Bus. Res. **22**(2), 159–170 (1991)
30. Sweeney, J.C., Soutar, G.N.: Consumer perceived value: the development of a multiple item scale. J. Retail. **77**(2), 203–220 (2001)
31. Aurier, P., Evrard, Y., N'goala, G.: Comprendre et mesurer la valeur du point de vue du consommateur. Rech. Appl. Mark. (French Edition) **19**(3), 1–20 (2004)
32. Venkatesh, V., Morris, M.G., Davis, G.B., Davis, F.D.: User acceptance of information technology: toward a unified view. MIS Q. 425–478 (2003)
33. Hess, T.J., Fuller, M., Campbell, D.E.: Designing interfaces with social presence: using vividness and extraversion to create social recommendation agents. J. Assoc. Inf. Syst. **10**(12), 1 (2009)
34. Voss, K.E., Spangenberg, E.R., Grohmann, B.: Measuring the hedonic and utilitarian dimensions of consumer attitude. J. Mark. Res. **40**(3), 310–320 (2003)
35. Pihlström, M., Brush, G.J.: Comparing the perceived value of information and entertainment mobile services. Psychol. Mark. **25**(8), 732–755 (2008)
36. Tam, J.L.: Customer satisfaction, service quality and perceived value: an integrative model. J. Mark. Manag. **20**(7–8), 897–917 (2004)
37. Van Ittersum, K., Pennings, J.M., Wansink, B.: Trying harder and doing worse: how grocery shoppers track in-store spending. J. Mark. **74**(2), 90–104 (2010)
38. Featherman, M.S., Pavlou, P.: Predicting e-services adoption: a perceived risk facets perspective. Int. J. Hum. Comput. Stud. **59**(4), 451–474 (2003)
39. Shin, H., Kang, J.: Reducing perceived health risk to attract hotel customers in the COVID-19 pandemic era: Focused on technology innovation for social distancing and cleanliness. Int. J. Hospitality Manage. **91**, 102664 (2020)
40. CRN, "Consumer Survey on Dietary Supplements" (2017). https://www.crnusa.org/resources/2017-crn-consumer-survey-dietary-supplements
41. Hair, J.: Multivariate Data Analysis, vol. 7. Prentice Hall, Upper Saddle River (2009)
42. Fornell, C., Larcker, D.F.: Evaluating structural equation models with unobservable variables and measurement error. J. Mark. Res. **18**(1), 39–50 (1981)
43. Wang, W., Benbasat, I.: Recommendation agents for electronic commerce: effects of explanation facilities on trusting beliefs. J. Manag. Inf. Syst. **23**(4), 217–246 (2007)

Requirements and Evaluation

Addressing Trust Issues in Supply-Chain Management Systems Through Blockchain Software Patterns

Eddy Kiomba Kambilo, Irina Rychkova, Nicolas Herbaut(✉), and Carine Souveyet

Centre de Recherche en Informatique, Université Paris 1 Panthéon-Sorbonne, 75013 Paris, France
{Eddy.Kambilo,Irina.Rychkova,Nicolas.Herbaut, Carine.Souveyet}@univ-paris1.fr

Abstract. Blockchain technology is a decentralized and distributed ledger that allows for secure, transparent, and immutable tracking of transactions. However, it is not a one-size-fits-all solution for addressing trust issues in the supply chain. In software engineering, design patterns provide a blueprint that developers can follow to solve a specific problem in a structured and efficient manner. In this paper, we identify and discuss the reusable blockchain software patterns that can be applied to design trustworthy solutions in supply chain management (SCM). Based on the literature analysis, we define a comprehensive taxonomy of SCM-specific trust issues. Then we apply requirement engineering technique to translate these issues into trust requirements and demonstrate how these requirements can be met by the specific blockchain software patterns.

Keywords: Trust · Blockchain · Software patterns · Supply Chain Management

1 Introduction

The supply chain is one of the most important economic systems [13] because it enables the production and delivery of goods and services to customers. In [15], the authors define supply chain as *"the network of organizations involved, through upstream and downstream linkages, in the different processes and activities that produce value in the form of products and services delivered to the ultimate consumer"*. Trust is a vital aspect with far-reaching consequences across various domains. In supply chains, trust plays a key role in shaping relationships between stakeholders: timely identification and elimination of trust issues is crucial for successful collaborations. Gambetta [14] defines trust as *"the expectation that another person (or institution) will perform actions that are beneficial or at least not detrimental, to us regardless of our capacity to monitor those actions"*. Following this definition, a *trust issue* can be defined as a lack of trustor's belief that another party (trustee), for one reason or another, will actually meet these expectations.

S. Nurcan et al. (Eds.): RCIS 2023, LNBIP 476, pp. 275–290, 2023.
https://doi.org/10.1007/978-3-031-33080-3_17

In modern society, where interpersonal or inter-organizational relations are often mediated by technology, trust becomes multidimensional: Mayer [27] defines trust between social entities (individuals or organizations), McKnight [10] specifies trust between humans and technology (Artificial Intelligence, Business intelligence), Andrew [31] discusses trust between humans depending on technology, Pietrzak et. al. [32] addresses digital trust as a determinant of interpersonal and inter-organizational relationships in the digital world. As a result, trust issues related to digital security, privacy of data, process transparency and performance gain a lot of attention.

Today, blockchain is considered a de facto technology to address trust issues in the supply chain. Blockchain is a distributed ledger system supported by a network of peers, each of whom maintains a copy of the ledger [28]. Blockchain is particularly attractive in supply chains due to its hacker-proof architecture and cryptographic algorithms(aspects), such as consensus algorithms that allow to verify and validate transactions on the network. In the context of SCM, this means that all parties involved in the supply chain can trust that the data recorded on the blockchain is accurate and has been agreed upon by the network. Additionally, blockchain can control access to information through smart contracts by defining specific conditions that must be met in order for the information to be accessed. These self-executing contracts, which are tamper resistant and traceable, can monitor the activity of each participant based on hash and signatures of each transaction [34].

Implementing blockchain-based solutions in SCM, organizations aim to address trust issues. However the success of this endeavor is contingent on the architectural model and implementation of blockchain technology.

Software patterns provide a blueprint that software engineers can follow to solve a specific problem in a structured and efficient manner. Bushman [4] defines software patterns as *"a function-form relationship that occurs in a context where the function is described in terms of unresolved trade-offs or forces in the problem domain. The form is a structure described in the solution domain that achieves a good and acceptable equilibrium among those forces."* Software patterns focus on capturing and systematizing successful experiences and techniques used in software development.

Blockchain software patterns are discussed in the literature [35] [39]. To the best of our knowledge, there is a lack of research exploring blockchain software patterns focusing on trust. In this paper, we investigate how trust issues expressed in the supply chain domain (problem domain) can be efficiently addressed by blockchain technology (solution domain) using specific blockchain software patterns.

In this work, we develop the following contributions: 1) We construct a taxonomy of trust issues in Supply Chain Management(SCM) based on the analysis of 18 research publications in the domain. 2) Following the guidelines of requirements engineering, we propose a technique for translating trust issues into trust requirements. 3) We define the mapping between the formulated trust requirements and the blockchain software patterns. This mapping can guide decision-making in the design of trustworthy solutions in SCM.

The paper is structured as follows: In Sect. 2, we provide the background on blockchain technology, software patterns, supply chain, and discusses the concept of trust. In Sect. 3, we define the taxonomy of trust issues in SCM based on related literature, than we translate these issues into trust requirements. In Sect. 4, we discuss how the trust requirements from the previous section can be met by the specific blockchain software patterns. We illustrate our findings on the example of Sect. 5. This example also serves as a preliminary validation of our findings. In Sect. 6 we present our conclusions.

2 Background

2.1 Trust

In social sciences, trust is defined as *"the willingness of one party (trustor) to be vulnerable to the actions of another party (trustee), based on the expectation that the other party will perform the expected action"* [27]. *Social trust* reflects (subjective) trustor's beliefs that the trustee has suitable attributes for performing as expected in a specific situation. These attributes include ability, benevolence and integrity [14]. Zheng [40] stated that social trust is a product of experiences and perceived trustworthiness.

Advances in technology, such as Artificial Intelligence (AI) and robotics have led to the need for organizations to establish processes to regulate trust in technology [9]. *Trust in technology* can be defined as trustor's confidence in technology (trustee) to accomplish the task at hand. In [10], the authors present three essential elements that can help build trust in technology: reliability, functionality, and helpfulness. *Digital trust* emerges in interpersonal or inter-organizational relations where technology plays a role of mediator. Jeffrey(2020) defines digital trust *"as the confidence users have in the ability of people, technology, and processes to create a secure digital world"*[1]. Trust in technology is a precursor to digital trust, as people must trust technology before using it.

2.2 Blockchain in Supply Chain Management

Supply chain management (SCM) aims to ensure that goods and services are delivered to consumers promptly, cost effectively, and efficiently [15]. In SCM, trust plays an important role. Tradelens [19] is one example of blockchain in SCM. It provides transparency, efficiency, and accountability in global trade by digitizing and streamlining the flow of information and documents among supply chain participants. Another example of practical applications is traceability of drugs that can be provided through a blockchain solution [21].

Blockchain is a decentralized, distributed ledger technology widely recognized as a critical enabler for the secure, transparent, and immutable tracking of

[1] https://www.techtarget.com/whatis/definition/digital-trust.

transactions [36]. Blockchain has several intrinsic features that make it relevant for supply chain management. It can create a *decentralized* and *tamper-proof* ledger of all transactions that occur throughout the supply chain. The ledger could track the movement of goods from their origin to their final destination, providing complete *transparency* and *traceability.* It can also help increase supply chain data's *integrity* by enabling *monitoring* and *auditing* of all transactions. Smart contracts can be used to *automate* specific processes within the supply chain. This can improve efficiency and reduce the risk of human errors. Finally, blockchain can be used for *the real-time identification* of goods, particularly for perishable goods with a limited shelf life.

2.3 Software Patterns

Pattern-based design is widely adopted by the software engineering community since the mid-1990s. The resulting software patterns describe recurring designs used in software development [4]. A software pattern is considered as *"a function-form relation that occurs in a context, where the function is described in problem domain terms as a group of unresolved trade-offs or forces, and the form is a structure described in solution domain terms that achieve a good and acceptable equilibrium among those forces."* [4]

According to [39], software patterns play a vital role in addressing trust issues. Blockchain software pattern is a repeatable design solution to a recurring problem in blockchain development [35]. In SCM, blockchain software patterns can provide a systematic way of tackling trust-related concerns, such as ensuring data authenticity and integrity, promoting transparency and accountability, and maintaining the privacy and security of the system.

In [35], authors identify a set of 120 unique patterns. 104 of them have been classified as design patterns, 3 of them as architectural patterns, and 14 as idioms. These blockchain software patterns come from a range of fields, including agriculture and industries and address a number of generic issues. In this work, we review the software patterns in [35] and identify twelve patterns that can be used to address the specific trust issues in SCM.

3 From Trust Issues to Trust Requirements in SCM

A trust issue refers to a challenge, problem, or disagreement that affects the level of trust between individuals or parties [22]. They can be grounded on explicit evidence (frauds, contract violations, bad user experience) or on implicit beliefs. They are subjective and hard to grasp. In order to be explicitly analyzed and addressed by the software solutions, trust issues need to be translated into requirements. A requirement is a statement which translates or expresses a need and its associated constraints and conditions [1].

In this section, we define a taxonomy of trust issues based on our analysis of related literature and translate these issues into trust requirements.

3.1 Taxonomy of Trust Issues in SCM

The work in [30] presents a methodology for building a taxonomy and discusses problems associated with taxonomy development. To establish our taxonomy, we define the following research protocol:

1 Identification: We conducted the search for primary research publications in the two major databases: Scopus and Google Scholar. We used the following key words: supply chain, trust, issues, requirements(38 papers for Scopus and 52 google scholar).
2 Selection: We selected the articles on the literature that met the following criteria:
C1: Evoke the issues related to social, digital trust or trust in technology
C2: Propose a solution that addresses trust issues or aims to improve trust in SCM. We identified (non-systematically) 18 research studies published between 2018 and 2022 (we filter by date to ensure the research is consistent and avoid irrelevant or outdated papers.) by screening abstracts and full texts.
3 Extraction: We extracted two types of text evidences
(a) evidence evoking (explicitly and implicitly) trust issues and
(b) evidence on the proposed technological solutions, indicating technological, architectural, design choices.
4 Synthesis: The extracted data was revised and discussed by several researchers (authors of this paper) to reduce the interpretation bias. Eventually the 21 extracted trust issues were grouped into 7 categories to form a taxonomy (Fig. 1).

We applied the protocol to identify and formulate trust issues in all selected sources. Here is an example: In [37], the authors highlight the difficulties faced in Supply Chain Management in verifying the authenticity of goods and conducting investigations into illegal activities. Three main issues are identified as "insufficient auditing", "opacity - lack of transparency" and "lack of oversight".

Figure 1 groups trust issues extracted in the literature into seven categories, *"Traceability"* focuses on the challenge of tracking and monitoring products in real time along the supply chain. *"Cost Control"* addresses the stakeholders' concern about reducing costs associated with blockchain transactions(cost limit and cost reduction). It is essential to minimize the cost of these transactions to ensure their practicality. *"Lack of Auditability"* highlights the difficulties in auditing blockchain transactions, and stakeholders need to be able to audit them at any time. *"Security"* emphasizes the importance of keeping data and transactions confidential, secure, and tamper-proof. *"Data Governance"* concerns the users' control over data shared with other institutions and the need to anticipate scalability to avoid additional fees. *"Lack of Accountability"* highlights the responsibility of stakeholders to be accountable for their actions, as it is essential for everyone to take responsibility for their decisions and actions. The final category, *"Acceptance"* that concerns the user acceptance.

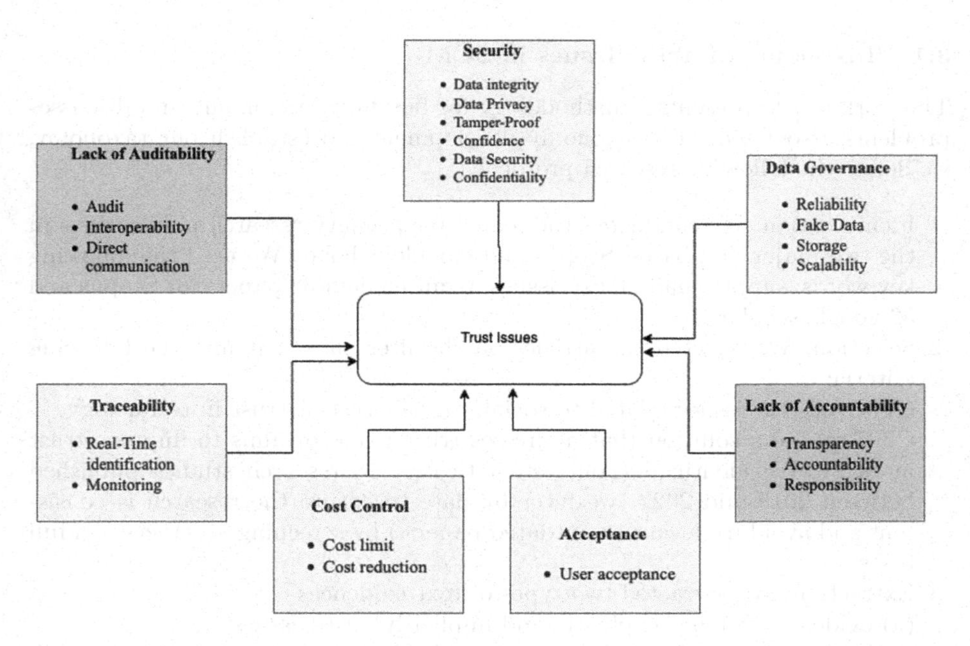

Fig. 1. Taxonomy of Trust Issues

3.2 Translating Trust Issues into Trust Requirements.

Trust issues are subjective and sometimes stem from stakeholders' intentions. Defining requirements starts with understanding the stakeholders' intentions, needs, goals, or objectives, as outlined in ISO [1]. In requirements engineering, a requirement is defined as a statement which identifies an operational, functional or design features or constraint of the product or process, which is unambiguous, testable or measurable, and necessary for the product or process to be accepted by consumers or internal quality assurance guidelines [1]. A set of explicit, clearly stated requirements facilitates communication between stakeholders: it justifies technological and design decisions and provides a basis for solution validation. When expressed in natural language, the statement of requirements should include a subject (e.g., system, software, etc.), an active verb and other elements necessary to specify the information content of the requirement.

The guidelines for writing requirements are specified by ISO/IEC standard [1]. Transforming trust issues into trust requirements involves thoroughly analyzing the needs and expectations of all stakeholders and examining current systems and practices. The subjective needs of stakeholders are then transformed into objective needs or objectives.

Requirements Engineering(RE) as a mediator between the acquirer and supplier domains, establishing and maintaining the requirements for the desired system, software, or service [33]. RE covers the discovery, elicitation, development, analysis, determination of verification methods, validation, communication, documentation, and management of requirements [1]. It is a crucial part of the software development process and involves stakeholder collaboration to guarantee that the end product fulfills stakeholders' needs and adheres to project constraints.

In this paper, we adapt the RE process and follow the ISO/IEC standard [1] for transforming trust issues into trust requirements. Our process consists of the following steps:

(1) Elicitation of user trust issues: In this step, the evidences of trust issues has to be gathered. Various techniques defined in the fields of requirements engineering and knowledge management can be used to collect the empirical data, including interviews, case studies, workshops, action research, etc. The outcome is a collection of trust issues expressed by end users or stakeholders. In this work, the data about trust issues has been collected through the literature review.

For instance, in [17], an evidence of lack of accountability (I1) is expressed as follows: "it is important to listen to the interactions and responsibility between the suppliers and the OEM to maintain the transparency between the different vendors". In [29], the same issue is expressed as follows: "The use of accountability and incentive structures to punish and encourage dishonest or trustworthy individuals was a strategy to increase trust and confidence in the data".

(2) Analysis: The purpose of this step is to analyze each expressed trust issue in order to identify a subject of trust (actor, system, process, technological component), an object of trust (e.g., data, activity, function, etc.) and an expected relationship that must be established between the former and the latter in order to mitigate the issue. For example: Lack of accountability (I1) issue addresses a business partner in the supply chain (the subject) and the transaction data (the object). The issue expresses a trustor's belief that, in case of dispute, the partner can avoid responsibility for his actions unless the formal proof of such actions is provided. To mitigate the issue, the transaction has to be non-reputable (the relationship). In the field, such analysis has to be conducted iteratively, confirming and validating the results with users.

(3) Specification of requirements: In this step, the requirements are documented based on the analysis from the previous step and following the recommendations from [1]. The outcome of this step is a formalized requirement specification. Example: For the Lack of accountability (I1) issue, we formalize the corresponding trust requirement as follows: "System must guarantee non-repudiation of data".

The taxonomy of 21 trust issues and their corresponding trust requirements defined following the process above is presented Table 1. This taxonomy provides a decision-making support for requirements engineers and designers and guides the design of the prospective trustworthy SCM solution. The proposed process for translating issues into requirements can potentially support designers in identifying new issues and requirements.

Table 1. Taxonomy & Mapping trust issues into trust Requirements

Taxonomy of Trust Issues		Trust issues to Trust requirements	
Issue	**References**	*Requirement Specification*	*Requirement Name*
Lack of accountability (I1)	[5,7,8,16,22,24,25,29,37]	System must guarantee non repudiation of data.	accountability
Lack of auditing(I2)	[8,16,18,24]	System must have log files to facilitate auditing	Audit
Lack of security (I3)	[3,7,17,18,37,38]	System must guarantee a no intrusion of external stakeholders	Security
Lack of interoperability(I4)	[2,6,7,12,18,26,37,38]	System must be flexible to insert other modules or SI.	Interoperability
Lack of transparency(I5)	[2,11,16,17,22,24,25]	System must save and share data in readable and transparent manner for everyone.	Transparency
High cost of transaction(I6)	[2,3,11,12,12,37,38]	System doesn't cost much by utilization	Cost control
Lack of reliability(I7)	[16,29]	System must be reliable	Reliability
Lack of confidence(I8)	[29]	System must guarantee data confidence	Confidence
Lack of confidentiality(I9)	[11,22]	System must guarantee process confidentiality	Confidentiality
Lack of privacy(I10)	[2,3,6,8,11,17,22]	System must guarantee data privacy	Privacy
Lack of responsibility(I11)	[25]	System must save a trace of each transaction	Responsibility
Lack of traceability(I12)	[2,3,26]	System must guarantee tracking of data	Traceability
Lack of direct communication(I13)	[16,29]	System must be decentralised with Peer-to-Peer communication	Decentralized
Lack of RT identification(I14)	[5,7,8,17,26]	System must track data on real time	R-T identification
Lack of data integrity(I15)	[5,7,8,22,26,37]	System must be guaranteed data integrity	Integrity
Lack of User acceptance(I16)	[3,5]	System must be guaranteed a User satisfaction.	User Acceptance
Lack of monitoring(I17)	[2,6,12,24]	System must give the possibility to monitor all processes	Monitoring
Fake data(I18)	[3,6]	System must guarantee data origin	Fake Data
Lack of scalability(I19)	[8,22,24,37]	System must guarantee a High availability of data	Scalability
Lack of tamper-proof(I20)	[12,22,24,37]	System must guarantee data tamper-proof	Tamper-proof
Lack of good storage(I21)	[5,6,17,24]	System needs Good support for storage of data	Storage

4 Use of Blockchain Software Patterns for Meeting Trust Requirements in SCM

Blockchain is often considered a de facto trust enabler. We argue that, while offering a number of key features, stock blockchain may not provide a complete solution for the specific trust requirements in a given context. In this section, we discuss the current limitations of stock blockchain solutions and propose the use of blockchain software patterns to efficiently address the specific trust requirements in SCM.

According to the literature, intrinsic features of blockchain(public) technology address a number of requirements in SCM including trustworthiness. However, this technology also has limitations that can overshadow the benefits and have a negative impact on trust. These limitations have to be taken into account when making *design decisions.* Challenges related to privacy, scalability, trust and interoperability are some examples relevant to the SCM domain. These challenges can be efficiently addressed using specific blockchain software patterns. Here are some examples.

Blockchain is not suitable for storing and managing large amounts of data. Keeping images and other large data sets in blockchain can be expensive due to the high cost of transaction fees. This undermines the performance and credibility of the solution and negatively impacts user's trust in this technology. To overcome this, off-chain data storage (patterns) such as the Interplanetary File System (IPFS) can be implemented in conjunction with blockchain. Using blockchain to store hashes of the data and off-chain storage to store extensive data can reduce transaction costs and add scalability to the system.

Along the same lines, if a blockchain solution is not designed with security in mind, it can be vulnerable to cyberattacks or other forms of malicious activity. This can result in the loss of funds or the compromise of sensitive information. Additionally, if the consensus mechanism used in the blockchain is not well adapted, it could lead to issues with trust in the network. For privacy, if a blockchain is not designed with privacy in mind, it can lead to the exposure of sensitive information because ledger is public for all stakeholders. The use of encryption-on chain(patterns) data is recommended to address these challenges.

Table 2 presents the mapping between the trust requirements in SCM, the key blockchain features discussed in Sect. 2.2 that are recurrently used to meet these requirements, and the blockchain software patterns from [35], which we identify to complement the features. We consider three cases:

CASE 1: blockchain features provide a complete solution for specific requirements. For example: transparency, as the transparent ledger offered by blockchain inherently meets this requirement without the need for additional patterns (indicated with a ✔ symbol in BC features column in the table).

CASE 2: blockchain features are insufficient to meet a requirement and blockchain software patterns are proposed. For example, a public blockchain's (transparency) cannot ensure data privacy. In this case, a private blockchain or encryption on-chain data patterns can encrypt data during transit to maintain privacy on the blockchain (indicated with a — symbol in BC software patterns column in the table).

Table 2. Mapping of Trust requirements on BC features & Patterns ✔:: Satisfied Req., !:: partially satisfied Req., —:: Unsatisfied Req.

Requirements	Blockchain Features	Blockchain Software Patterns [35]
Accountability	✔Accountability	✔ Identifier Registry
Audit	✔ Audit	—
Interoperability	—	✔ Contract Observer
Transparency	✔ Transparency	—
Cost control	!Automating	! Minimize On-Chain data, Flyweight, Off-chain data storage
Reliability	✔ Immutability	—
Privacy	—	✔ Encryption on-chain
Responsibility	—	✔ Identifier Registry
Traceability	✔ Traceability	—
Decentralized	✔ Decentralized	—
RT identification	✔ RT identification	—
Integrity	! Hash, Integrity	! Hash secret
Monitoring	✔ Monitoring	! Event Log, Publisher-Subscriber
Scalability	—	✔ State channel[50%], off-chain data storage[50%]
Tamper-proof	✔ Tamper-proof	! Embedded permission
Storage	—	✔ Off-chain data storage, Limit-Storage

CASE 3: blockchain features offer only a partial solution and can be complemented by using software patterns to meet a requirement. For example, the automation of information systems in blockchain can lead to increased costs if the data transit is extensive in terms of storage. To address this issue, minimize On-chain data or using Flyweight patterns can limit the data size in each transaction and reduce costs (indicated with a ! symbol in the table).

Implementing blockchain solutions by using software patterns can help to improve scalability, good storage, privacy and trust in the supply chain management environment. In Table 2, we present the mapping of 16 out of 21 trust requirements in SCM . According to our analysis, Security, Confidence, Confidentiality, Fake data, and User acceptance requirements are not fully met by the current blockchain solutions (features and/or patterns). These complex problems require more research and development and need to be addressed by the blockchain community in the future.

5 Illustrative Example and Discussion

To illustrate our proposed mapping of trust issues/requirements into specific blockchain software patterns and to provide the initial validation of this mapping, we consider an example from [7]. This paper discusses the use of blockchain for supporting traceability in a food supply chain for food safety risk management and compliance. It describes in details the design and implementation and validates the approach.

5.1 Running the Process on the Illustrative Example

Following the process defined in Sect. 3.2, we identified the following trust issues from this case [7] and mapped them on our taxonomy in Table 1:

Lack of Interoperability (I4): According to the case, "The aim is to develop an interoperable, autonomous systems", where heterogeneous stakeholders can collaborate. *High cost of transactions(I6):* Stakeholders are preoccupied by the transaction costs: "Ipfs help to store large amount of data to reduce cost transaction". *Lack of Scalability(I19):*"We must use decentralized storage such as IPFS to guarantee the integrity of the information", stakeholders need systems with scalability of data. *Lack of Traceability(I12):* "Traceability-related information is not shared between participants, since they have their own traceability mechanisms and inevitably store their unique traceability records" *Lack of Data integrity and privacy (I15, I10):* "It is essential to guarantee the privacy of the transactions and the involved actors by using SC".

We map the identified trust issues on trust requirements. For exemple: "Stakeholders need to be reassured of the source of provenance and the authenticity of the products." correspond to Traceability requirement in Table 1. We analyzed the architecture and patterns proposed by the case authors and compared them to our recommendation based on the mapping in Table 2. Table 3 shows the comparison results.

5.2 Results

In this example, we were able to identify the trust issues from the case text and to map them on our taxonomy. Though the patterns indicated by the case authors are not expressed explicitly, we were able to match them with the patterns

Table 3. Blockchain features and Patterns resolving trust issues !:: Satisfied Req., ✔:: partially satisfied Req., ━:: Unsatisfied Req.

Requirements (extracted from the case)	BC intrinsic features	Patterns (extracted from the case)	BC software patterns (our proposal)
Interoperability	━	━	Contract Observer
Cost control	!	Off-chain data storage	Limit Storage
Privacy	━	Encryption On-chain	Encryption On-chain
Traceability	✔	━	━
Integrity	!	Hash Secret	Hash Secret
Storage	━	Off-chain data storage	Off-chain data storage, Limit storage

provided by [35]. The authors in the case use IPFS to store vast amounts of data, which generates a hash secret on a smart contract address to reduce transaction costs and enable `Off-chain data storage`. The authors are also concerned with ensuring the privacy and traceability of data through encryption and the use of hash (`Encryption On-chain` & `Hash secret`).

The patterns used in the case correspond to our recommended blockchain software patterns for the following four requirements: Privacy, Integrity, Storage and Cost control. For the storage and Cost c requirement, we propose the `Limit storage` pattern to set a gas limit for transactions in addition to the `Off-chain data Storage`, already identified by the case authors.

The Traceability requirement does not require a specific pattern as it is directly provided by blockchain. This corresponds to our mapping in Table 2.

The Interoperability requirement is not addressed in the paper. We propose the following blockchain software patterns to meet these requirements: the `Contract observer` pattern to guarantee interoperability and confirm that data written to the blockchain comes from a trustworthy source, and the `Identifier registry` pattern to track transactions and hold stakeholders accountable in the event of any problems.

In this example, we used a mapping proposed in the previous section to extend the solution proposed by the authors with two specific blockchain software patterns. This proposal completes the solution by addressing more trust requirements.

5.3 Discussion

Software patterns provide developers with technical best practices. However the trust implications of the patterns are not explicit. Our literature analysis shows that trust issues are not explicitly addressed in the design of the SCM solutions.

Trustworthiness is often taken for granted by the mere use of a blockchain and cannot by validated.

The blockchain often meets trust requirements through a goal-oriented approach, which many researchers have explored. For example, authors in [23] focused on studying and identifying trust requirements in blockchain systems and created a trust engineering taxonomy to meet blockchain systems' trust requirements and goals. However, they did not provide evidence of how their taxonomy can be used or how to meet trust requirements using goals. In [20], a goal-oriented approach for business process reengineering is discussed. Here trustworthiness concerns are explicitly represented as (soft) goals and mapped to the relevant trust-enhancing features of blockchain, supporting business process reengineering. These works use a goal-oriented approach to focus on requirements and how specific blockchain features meet trust requirements.

The uniqueness of our approach resides in combining specific blockchain patterns from [35] and intrinsic blockchain features to address trust issues and to create trustworthy solutions in the field of SCM.

We make a first attempt to create a taxonomy of *trust issues* and *trust requirements* and define their design implications for blockchain solutions. This taxonomy will guide organizations to create trustworthy solutions in SCM.

Despite the promising results obtained from our study, it is important to note that there are several limitations and opportunities for improvement that should be addressed. These include:

- We identified trust issues using a sample of 18 research articles on SCM. We plan to conduct Systematic Literature Review to validate our findings and to extend our taxonomy.
- We limited our study to SCM, and more work is required to generalize this approach to other domains.
- Both trust issues and pattern were often not explicit in the literature, we had to rely on our expertise and interpretation to extract them. Our main effort is to promote standardization and the use of a common language (taxonomy of trust issues and requirements) to alleviate the interpretation bias in the future.

For the practical and effective use of the proposed taxonomy and mapping, we plan to establish a recommendation system application where developers can select the issues encountered as input and have access to the blockchain software patterns they can leverage to ensure a design that inspires trust for collaboration.

6 Conclusion

While blockchain technology is often considered as the de facto trust enabler, some limitations persist. These limitations has to be systematically addressed by improved design practices. In this article we consider blockchain software design patterns to address the trust issues in SCM domain. First, we provided an overview and developed a taxonomy of trust issues based on literature in the supply chain. Following the recommendations from requirement engineering, we translated the trust issues into explicit trust requirements. We examined the existing solutions that address these trust requirements in the literature and identified their limitations: We argue that the use of intrinsic features of blockchain often provides only a partial solution for trust issues in SCM. To complete this solution, we propose the use of blockchain software patterns.

We identified 12 patterns from the blockchain software pattern literature that can support the trust requirements in SCM and evaluated our proposal on one example from the literature. This preliminary evaluation shows the relevance of the trust issues taxonomy defined in this work. The proposed blockchain software patterns extend the solution from the case, demonstrating the potential interest and added value of our mapping.

This work aims at helping enterprises to better understand their trust-related requirements and to improve the design of their SCM systems.

References

1. AWARE, T.A., DOCUMENTATION, T.P.S.: ISO/IEC/IEEE international standard - systems and software engineering - life cycle processes - requirements engineering. ISO/IEC/IEEE 29148:2018(E), pp. 1–104 (2018). https://doi.org/10.1109/IEEESTD.2018.8559686
2. Baralla, G., Pinna, A., Tonelli, R., Marchesi, M., Ibba, S.: Ensuring transparency and traceability of food local products: a blockchain application to a smart tourism region. Concurrency Comput.: Pract. Exp. **33**(1), e5857 (2021)
3. Biswas, D., Jalali, H., Ansaripoor, A.H., De Giovanni, P.: Traceability vs sustainability in supply chains: the implications of blockchain. Eur. J. Oper. Res. **305**(1), 128–147 (2023)
4. Buschmann, F., Meunier, R., Rohnert, H., Sommerlad, P., Stal, M.: Software patterns (1996)
5. Caro, M.P., Ali, M.S., Vecchio, M., Giaffreda, R.: Blockchain-based traceability in agri-food supply chain management: a practical implementation. In: 2018 IoT Vertical and Topical Summit on Agriculture - Tuscany (IOT Tuscany), pp. 1–4 (2018). https://doi.org/10.1109/IOT-TUSCANY.2018.8373021
6. Casino, F., Kanakaris, V., Dasaklis, T.K., Moschuris, S., Rachaniotis, N.P.: Modeling food supply chain traceability based on blockchain technology. IFAC-PapersOnLine **52**(13), 2728–2733 (2019). https://doi.org/10.1016/j.ifacol.2019.11.620
7. Casino, F., et al.: Blockchain-based food supply chain traceability: a case study in the dairy sector. Int. J. Prod. Res. **59**(19), 5758–5770 (2021). https://doi.org/10.1080/00207543.2020.1789238
8. Chang, S.E., Chen, Y.: When blockchain meets supply chain: a systematic literature review on current development and potential applications. IEEE Access **8**, 62478–62494 (2020). https://doi.org/10.1109/ACCESS.2020.2983601
9. Chopra, K., Wallace, W.A.: Trust in electronic environments. In: Proceedings of the 36th Annual Hawaii International Conference on System Sciences 2003, pp. 10-pp. IEEE (2003)
10. Mcknight, D.H., Carter, M., Thatcher, J.B., Clay, P.F.: Trust in a specific technology: an investigation of its components and measures. ACM Trans. Manage. Inf. Syst. **2**(12), 1–25 (2011)

11. De Giovanni, P.: Blockchain and smart contracts in supply chain management: a game theoretic model. Int. J. Prod. Econ. **228**, 107855 (2020)
12. Figorilli, S., et al.: A blockchain implementation prototype for the electronic open source traceability of wood along the whole supply chain. Sensors **18**(9) (2018)
13. Flores-González, L., Vargas Florez, J., Monteza-Valdivia, L., Cáceres-Cansaya, A., García-Salinas, J., Silva-Alarco, L.: Urban road network resilience assessment on freight logistics by simulating disruptive events. In: Production and Operations Management, pp. 427–450. Springer International Publishing, Cham (2022)
14. Gambetta, D., et al.: Can we trust trust. Trust: Making Breaking Coop. Relat. **13**(2000), 213–237 (2000)
15. Giannakis, M., Croom, S., Slack, N.: Supply chain paradigms. Understanding supply chains, pp. 1–22 (2004)
16. Hameed, H., Zafar, N.A., Alkhammash, E.H., Hadjouni, M.: Blockchain-based formal model for food supply chain management system using VDM-SL. Sustainability **14**(21) (2022). https://doi.org/10.3390/su142114202
17. Hasan, H.R., Salah, K., Jayaraman, R., Ahmad, R.W., Yaqoob, I., Omar, M.: Blockchain-based solution for the traceability of spare parts in manufacturing. IEEE Access **8**, 100308–100322 (2020). https://doi.org/10.1109/ACCESS.2020.2998159
18. Imeri, A., Agoulmine, N., Feltus, C., Khadraoui, D.: Blockchain: analysis of the new technological components as opportunity to solve the trust issues in supply chain management. In: Arai, K., Bhatia, R., Kapoor, S. (eds.) Intell. Comput., pp. 474–493. Springer International Publishing, Cham (2019)
19. Jensen, T., Hedman, J., Henningsson, S.: How Tradelens delivers business value with blockchain technology. MIS Q. Executive **18**(4) (2019)
20. Johng, H., Kim, D., Park, G., Hong, J.E., Hill, T., Chung, L.: Enhancing business processes with trustworthiness using blockchain: a goal-oriented approach. In: Proceedings of the 35th Annual ACM Symposium on Applied Computing, pp. 61–68 (2020)
21. Kambilo, E.K., Zghal, H.B., Guegan, C.G., Stankovski, V., Kochovski, P., Vodislav, D.: A blockchain-based framework for drug traceability: Chaindrugtrac. In: Proceedings of the 37th ACM/SIGAPP Symposium on Applied Computing, pp. 1900–1907 (2022)
22. Kamble, S.S., Gunasekaran, A., Sharma, R.: Modeling the blockchain enabled traceability in agriculture supply chain. Int. J. Inf. Manage. **52**, 101967 (2020). https://doi.org/10.1016/j.ijinfomgt.2019.05.023
23. Khalifa, D., Madjid, N.A., Svetinovic, D.: Trust requirements in blockchain systems: a preliminary study. In: 2019 Sixth International Conference on Software Defined Systems (SDS), pp. 310–313. IEEE (2019)
24. Kuhn, M., Funk, F., Zhang, G., Franke, J.: Blockchain-based application for the traceability of complex assembly structures. J. Manuf. Syst. **59**, 617–630 (2021). https://doi.org/10.1016/j.jmsy.2021.04.013
25. Manning, L., et al.: Artificial intelligence and ethics within the food sector: developing a common language for technology adoption across the supply chain. Trends Food Sci. Technol. **125**, 33–42 (2022). https://doi.org/10.1016/j.tifs.2022.04.025
26. Masudin, I., Rahmatullah, B.B., Agung, M.A., Dewanti, I.A., Restuputri, D.P.: Traceability system in halal procurement: a bibliometric review. Logistics **6**(4), 67 (2022)
27. Mayer, R.C., Davis, J.H., Schoorman, F.D.: An integrative model of organizational trust. Acad. Manage. Rev. **20**(3), 709–734 (1995)

28. Mohanta, B.K., Panda, S.S., Jena, D.: An overview of smart contract and use cases in blockchain technology. In: 2018 9th International Conference on Computing, Communication, and Networking Technologies (ICCCNT), pp. 1–4 (2018)
29. Jayashri, N., Rampur, V., Gangodkar, D., Abirami, M., Balarengadurai, C., Kumar, A.: Improved block chain system for high secured IoT integrated supply chain. Meas.: Sensors **25**, 100633 (2023). https://doi.org/10.1016/j.measen.2022.100633
30. Nickerson, R.C., Varshney, U., Muntermann, J.: A method for taxonomy development and its application in information systems. Eur. J. Inf. Syst. **22**, 336–359 (2013)
31. Patrick, A.S., Briggs, P., Marsh, S.: Designing systems that people will trust. Secur. Usability **1**(1), 75–99 (2005)
32. Pietrzak, P., Takala, J.: Digital trust-asystematic literature review. Forum Sci. Oeconomia **9**(3), 59–71 (2021)
33. Pohl, K.: Requirements Engineering: Fundamentals, Principles, and Techniques. Springer Publishing Company, Cham (2010)
34. Sahai, A., Pandey, R.: Smart contract definition for land registry in blockchain. In: 2020 IEEE 9th International Conference on Communication Systems and Network Technologies (CSNT), pp. 230–235. IEEE (2020)
35. Six, N., Herbaut, N., Salinesi, C.: Blockchain software patterns for the design of decentralized applications: a systematic literature review. Blockchain: Res. Appl. 100061 (2022)
36. Viriyasitavat, W., Hoonsopon, D.: Blockchain characteristics and consensus in modern business processes. J. Ind. Inf. Integr. **13**, 32–39 (2019)
37. Wei, Y.: Blockchain-based data traceability platform architecture for supply chain management. In: 2020 IEEE 6th Intl Conference on Big Data Security on Cloud (BigDataSecurity), IEEE Intl Conference on High Performance and Smart Computing, pp. 77–85 (2020). https://doi.org/10.1109/BigDataSecurity-HPSC-IDS49724.2020.00025
38. Xu, X., Choi, T.M.: Supply chain operations with online platforms under the cap-and-trade regulation: impacts of using blockchain technology. Transp. Res. Part E: Logistics Transp. Rev. **155**, 102491 (2021)
39. Xu, X., Pautasso, C., Zhu, L., Lu, Q., Weber, I.: A pattern collection for blockchain-based applications. In: Proceedings of the 23rd European Conference on Pattern Languages of Programs, pp. 1–20 (2018)
40. Yan, Z., Holtmanns, S.: Trust modeling and management: from social trust to digital trust. In: Computer Security, Privacy and Politics: Current Issues, Challenges and Solutions, pp. 290–323. IGI Global (2008)

Evaluating Process Efficiency with Data Envelopment Analysis: A Case in the Automotive Industry

Rutger Kerkhof(✉), Luís Ferreira Pires, and Renata Guizzardi

University of Twente, PO Box 217, 7500 AE Enschede, The Netherlands
r.g.kerkhof@alumnus.utwente.nl

Abstract. In some industries, small improvements to processes and profit margins may lead to a significant change in profit, and this holds especially in the automotive industry. A popular approach to achieving process improvement is benchmarking, in which the execution of a process is measured and compared between different work units so that improvement opportunities can be identified. This paper reports on our efforts to improve car dealership benchmarking by designing a benchmarking tool for the automotive workshop department, such that it calculates the efficiency of its main process, which we call the Standard Service Process (SSP) in this paper. We achieved this by designing a Data Envelopment Analysis (DEA) Network Slacks-Based Measure (NSBM) model and used this model to measure the SSP efficiency for each workshop by considering its sub-processes. This model was programmed using R, after which it was verified and extended based on the literature, and the results of the verified model were validated using a real case. In this paper, we show that this has enabled a more insightful assessment of the workshops so that suggestions for improvement can be automated. In this way, we demonstrate that our approach is appropriate to rank the efficiency of work units that perform a certain process.

Keywords: Process Efficiency · Data Envelopment Analysis · Car Dealerships

1 Introduction

The automotive industry is one of the most important industries in terms of contribution to global turnover and regarding the number of jobs [1]. Since this industry is characterised by a low-profit margin, small improvements to their processes can result in a significant change in profit.

Benchmarking is a popular approach for improving one or more processes by measuring and comparing them to other (similar) processes. The benchmarking approach in this paper is based on the traditional but effective perspective in Camp [2], who argued that one should not exclusively focus on performance measures but, instead, take the underlying process(es) into account. According to them, measures are overemphasised and processes are overlooked; their motto is *"What without how is an empty statement."* Some examples of approaches found in the literature that do not follow this are [3–6],

S. Nurcan et al. (Eds.): RCIS 2023, LNBIP 476, pp. 291–307, 2023.
https://doi.org/10.1007/978-3-031-33080-3_18

and [7], which all view a car dealership as a 'black box'. Camp continues that the process must be considered, as it is the basis for delivering the output that gave the result (the benchmark/performance measure) in the first place. Since benchmarking is not a new concept by any means, a wide variety of different benchmarking techniques can be applied. Due to the improvements in technology and the rapid growth of available data, it is expected that the benchmarking techniques currently in use can be improved to yield even better results.

Data Envelopment Analysis (DEA) is a known and appropriate approach that can be used for *benchmarking a standardised process*. DEA has a variety of models that calculate the relative efficiency of the assessed process with a slightly different set of constraints. Based on a systematic literature review, we identified that the Network Slacks-Based Measure (NSBM) model was closely aligned with the requirements of the benchmarking tool we aimed to design in our research. Our goal was not only to rank workshop departments by evaluating their process but also to provide tailored suggestions to improve their rank in the future. The NSBM model was first proposed by Tone and Tsutsui [8], but the benchmarking tool that we implemented in this paper uses a modified version of this model, proposed and applied by Kao [9, 10].

When determining the scope of this paper, we considered all departments that a car dealership branch typically consists of, namely, new car sales, used car sales, workshop, parts, and other/overhead. These departments have different processes, each with different relevant performance measures. This paper focuses on measuring, comparing, and improving the performance and efficiency of the workshop (also known as the service department) at dealership branches, as this is the only dealership department where we identified a standardised process.

In this paper, we explain the main design choices that we made in our research, and we also present a validation using a real case. This case shows that the proposed method is promising and that it may be further improved to be adopted by the car importer investigated in this work. To accomplish our goals, this paper is further structured as follows: Sect. 2 gives the background on DEA and its application in dealership assessment; Sect. 3 forms the foundation for the required domain knowledge of the automotive industry; Sect. 4 elaborates on the R implementation of the DEA NSBM model and how it was verified; Sect. 5 describes the real-world case, process model, data, and factor selection used; Sect. 6 evaluates the proposed tool by analysing and validating the results from the implementation using expert opinion; and Sect. 7 poses some final considerations and discusses future work directions.

2 Background

This section introduces Data Envelopment Analysis (DEA) and discusses its application in the automotive industry.

2.1 DEA Models

DEA is a non-parametric approach to efficiency measurement, which means that it does not require any information on the relationship between inputs and outputs [11]. DEA is used to find the relative efficiency of Decision-Making Units (DMUs) that have the same (multiple) inputs and outputs, based on a linear programming problem. This allows one to determine whether a single DMU is performing relatively efficiently or inefficiently (i.e., compared to the full set of DMUs), given the set of input and output factors. Usually, DMUs are separate firms, branches, or departments, however, in some cases, even units or entities like employees can be viewed as DMUs in DEA.

If a DMU is deemed efficient, it is part of the *efficiency frontier*, which is a subset of DMUs that are identified as the (relative) best practice. When a DMU is deemed inefficient, the degree of its inefficiency can then be determined based on its position relative to the efficiency frontier. An example of this efficiency frontier can be seen in Fig. 1. Here, the efficiency frontier (the curved line going from axis to axis) consists of DMU1, DMU2, and DMU 3, while DMU4 and DMU5 are deemed inefficient. The inefficient DMUs get a composite unit marked on the frontier, marked as DMU'4 and DMU'5, respectively. From this, it can be concluded that DMU5 is more efficient than DMU4, as it is closer to (its projection on) the efficiency frontier. The position of DMU'4 is determined based on DMU1 and DMU2, while the distance between DMU'5 and DMU1 represents the *slack* that must be overcome to be considered efficient.

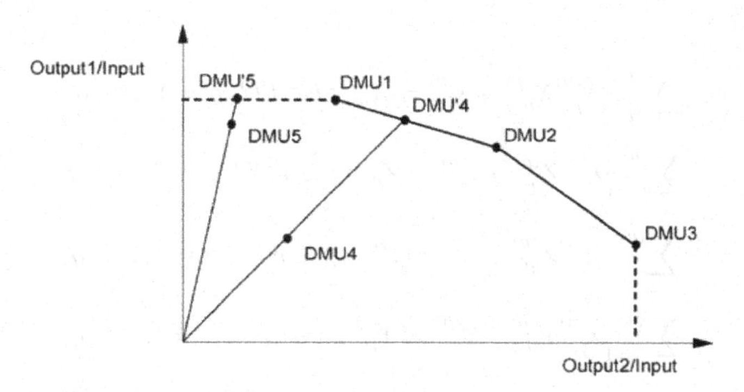

Fig. 1. DEA efficiency frontier (one input, two outputs), taken from Melão [11].

In the CCR model [12] (1978), the *technical efficiency* is calculated, assuming constant returns to scale (CRS). The original primal CCR model is a nonlinear fractional programming problem, used to determine the relative efficiency for all *n* DMUs. For the BCC model [13] (1984), the CCR model was extended to estimate the *pure technical efficiency*, excluding the *scale efficiency*, assuming variable returns to scale (VRS).

The CCR and BCC model both use the radial measure of efficiency, causing two major problems [9, 14]: (1) the inputs and outputs need to undergo proportional changes and (2) the efficiency of weakly efficient DMUs cannot be measured appropriately.

These problems were initially solved using a (non-radial) slacks-based measure (SBM) model, by directly dealing with input excess and output shortfall. However, there still was a problem with the CCR, BCC, and SBM models: they all treat the DMU as a 'black box', disregarding the internal structure and assuming a positive correlation between input and output [9]. When taking the operations of system components into account, the resulting efficiency measurement is more meaningful and informative when compared to its black box counterpart [10]. *Network DEA* is used to measure the efficiency of such a network of components or sub-processes, which led to the introduction of the Network Slacks-Based Measure (NSBM) model.

While conventional models only support *exogenous* (i.e., external) inputs and outputs (X_i, Y_r), network models also support *endogenous* inputs and outputs (Z_f, Z_g). In essence, these endogenous factors are intermediate products that are an output of one sub-process and an input for another. This allows the NSBM model to explain why some DMUs with a perfect CCR or BCC efficiency score are in reality inefficient, by considering their internal processes. The NSBM objective function, which is the foundation for all other equations, can be seen in Eq. 1, followed by the constraints it is subjected to. This equation defines how E_k^S is calculated, which corresponds to the System Efficiency (weighted average of all sub-process efficiencies) of DMU k [9].

$$E_k^S = \min \frac{\sum_{p=1}^{q}\left[1-\left(\sum_{i\in I^{(p)}}\frac{s_i^{(p)-}}{X_{ik}^{(p)}}+\sum_{f\in M^{(p)}}\frac{t_f^{(p)-}}{Z_{fk}^{(p)}}\right)/\left(\hat{i}^{(p)}+\hat{m}^{(p)}\right)\right]}{\sum_{p=1}^{q}\left[1+\left(\sum_{r\in O^{(p)}}\frac{s_r^{(p)+}}{Y_{rk}^{(p)}}+\sum_{g\in N^{(p)}}\frac{t_g^{(p)+}}{Z_{gk}^{(p)}}\right)/\left(\hat{o}^{(p)}+\hat{n}^{(p)}\right)\right]} \tag{1}$$

Subject to:

$$\sum_{j=1}^{n}\lambda_j^{(p)}X_{ij}^{(p)}+s_i^{(p)}=X_{ik}^{(p)};\ i\in I^{(p)};\ p=1,\ldots,q$$

$$\sum_{j=1}^{n}\lambda_j^{(p)}Z_{fj}^{(p)}+t_f^{(p)}=Z_{fk}^{(p)};\ f\in M^{(p)};\ p=1,\ldots,q$$

$$\sum_{j=1}^{n}\lambda_j^{(p)}Z_{gj}^{(p)}-t_g^{(p)}=Z_{gk}^{(p)};\ g\in N^{(p)};\ p=1,\ldots,q$$

$$\sum_{j=1}^{n}\lambda_j^{(p)}Y_{rj}^{(p)}-s_r^{(p)}=Y_{rk}^{(p)};\ r\in O^{(p)};\ p=1,\ldots,q$$

$$\lambda_j^{(p)}\geq 0;\ j=1,\ldots,n;\ p=1,\ldots,q$$

The proposed network model is non-radial and can support both VRS and CRS by either including or excluding the constraint in Eq. 2 for all q internal processes.

$$\sum_{j=1}^{n}\lambda_j^{(p)}=1;\ j=1,2,\ldots,k,\ldots,n;\ p=1,\ldots,q \tag{2}$$

2.2 Related Work: DEA in Dealerships

To start our research, we performed a Systematic Literature Review on the application of DEA in car dealerships. In this review, only six studies fulfilled the (quality) criteria we defined, like, e.g., explicitly listing used factors. Of these six studies, five applied the CCR model, while the BCC and NSBM models were applied in only one study we found, indicating the popularity in this industry. Table 1 summarises our findings.

Table 1. Overview of mentioned and applied DEA models.

Study	Mentioned DEA models				Applied DEA models			
	CCR	BCC	SBM	NSBM	CCR	BCC	SBM	NSBM
Biondi et al. [3]	X	X			X	X		
Gencer et al. [4]	X	X			X			
Sadat Rezaee et al. [5]	X				X			
Tan et al. [6]	X	X			X			
Toloo and Ertay [7]	X	X			X			
Yang [14]	X	X	X	X			X	X

The two studies that stood out from this list are Biondi et al. [3] and Yang [14] since the former compared the two (more popular) radial DEA models by applying them and the latter was the only study to mention and apply an NSBM model.

Biondi et al. explain and demonstrate the difference between the CCR model (with CRS) and the BCC model (with VRS). They argue that *technical efficiency* (TE, using CCR) is a global measurement of efficiency that looks at both managerial efficiency and efficiency of scale. *Pure technical efficiency* (PTE, using BCC), on the other hand, determines how efficiently a dealership utilises the available resources, without considering the efficiency of scale. Therefore, they argue that the PTE exclusively quantifies managerial efficiency, while the *scale efficiency* (TE $\div$ PTE) allows for the evaluation of the decision-makers responsible for sizing. However, despite this elaborate explanation of the differences, they did not explain why they compared the traditional measure of market share performance to the CCR model, rather than the BCC model.

Yang argued that there was a problem with the commonly used models (CCR, BBC, and SBM), namely that they treat a DMU as a black box. The Network Slacks-Based Measure (NSBM) model takes the internal structure of a DMU into account and was, therefore, the model of choice for Yang. In the results, the SBM efficiency is compared to the NSBM efficiency, showing that including the internal sub-processes results in a stronger discriminating power than a (mathematically comparable) 'black box'-model.

Yang [14] is the most similar work compared to ours, however, they focused on measuring the efficiency of the parts department, while we analysed the workshop department. Moreover, they compared only 27 DMUs, and they did not discuss any details regarding their implementation (using LINGO) or the challenges they encountered.

3 Domain Analysis

This section analyses the automotive domain by identifying its stakeholders, processes, and performance measures that are relevant for benchmarking purposes.

3.1 Stakeholders

We defined an extensive list of stakeholders, using the stakeholder taxonomy [15] and onion model [16], both proposed by Ian Alexander. The identified stakeholders can be divided into three distinct groups: the car importer, the dealerships, and the implementation partner.

- The *car importer* imports and distributes cars to affiliated car dealerships. They are also the *normal operator* (a generic term used for end user [15]) of the benchmarking tool since they actively recommend improvements to dealerships, primarily by monitoring their output and can be considered both financial and functional beneficiaries.
- The *dealerships* include the *workshop*, which is the department subject to the benchmarking tool being designed. In this group, the stakeholders likely to experience immediate impact are the *workshop employees* that perform the process being assessed. In the longer term, effects may be experienced by the workshop *department head* and *branch management*, the *dealership management* is only affected indirectly. Except for the workshop employees, all stakeholders in this group may be considered both functional and financial beneficiaries.
- The *implementation partner* implements and maintains the benchmarking tool, while also supporting and advising the normal operator in the operation of the tool. This group consists of the more common stakeholders, like the *developer*, *architect*, *operation staff*, and *maintenance staff*. Even though they are not the normal operator of the tool, their requirements are still relevant to the tool.

The main goal of the car importer stakeholder group is to increase the dealerships' profitability by using data for monitoring and advising them on their operations. This is primarily done by fulfilling the workshop's goals of optimizing processes by minimizing costs (input) and maximizing results (output). These goals were translated into requirements, which had to be fulfilled in the proposed benchmarking tool. For example, the assessed process should be standardised; the benchmarking is closely related to reality; and the tool provides improvements for units that were deemed inefficient.

3.2 Standard Service Process

To align benchmarking with reality, we based it on the main standardised process that is carried out at the workshop department. The guidelines for this process, the *Standard Service Process* (SSP), are prescribed by the car manufacturer and are consequently followed up by the car importer and the affiliated car dealerships.

The SSP is a process that consists of seven sequential steps (i.e., sub-processes) that all types of services (e.g., repair, maintenance, and problem diagnosis) follow. A single completion of the SSP is called a *pass-through* from this point onward. The seven steps in the SSP are (1) making an appointment; (2) preparing the appointment; (3) receiving/accepting the car; (4) performing the repair; (5) quality control/preparing the car return; (6) returning the car/invoicing; and (7) aftercare.

3.3 Performance Measures

In our work, we analysed which performance measures are used in the automotive industry for measuring process efficiency. We noticed that measures related to employees (e.g., count, salary, labour utilization) are widely used as process input. Another recurring type of factor was measures regarding the physical aspects of the assessed unit (i.e., the DMU). Supposedly, this was done to reflect these aspects in the placement on (or behind) the efficiency frontier, since these factors are typically uncontrollable. Some examples are the building size and industrialization of the surrounding area.

Output measures predominantly consisted of financial measures, with the majority being focused on profit (margin) and revenue; each with a slightly different aggregation (e.g., per employee, per pass-through, and per billed hour). Some examples of non-financial measures are the time per pass-through and the number of pass-throughs (total, per day, or a percentage of the capacity). Lastly, customer surveys for satisfaction, complaints, and service quality were also found to be common for output measures.

Table 2 shows the identified measures for the workshop department at car dealerships from the literature, and was further extended by using expert opinion (indicated in the table with '*'). An example of the latter is the inclusion of measures aimed at the electrification of cars. These experts are the same four experts that gave their opinion as part of the result validation discussed in Sect. 7. Each measure was associated with the relevant sub-process of the aforementioned SSP (P1–P7) and then marked with an 'I' or 'O', indicating whether it was considered to be an input or output measure to the sub-process, respectively.

4 DEA Model Implementation

This section explains the implementation, verification, and extension of the DEA NSBM model that we used in our work.

4.1 Model Verification

While a wide variety of DEA models have publicly available implementations, no readily available implementation of the NSBM model in Kao [9, 10] or of any other network DEA model was found. Therefore, we decided to program the NSBM model using R, as this language offers customizability and has well-documented DEA libraries. For this, the R implementation by Soteriades [17] of the DEA SBM model by Tone [18] was used as a basis, since the SBM model and NSBM model have a similar underlying logic. Soteriades built their model using two linear programming packages, lpSolve [19] and lpSolveAPI [20], therefore, we also used these in our R implementation.

The correctness of the R implementation of the NSBM model was verified by comparing it to the model created in Kao [9]. This was achieved by using the same raw input data as presented in [9] and comparing the resulting efficiencies and weights of our R implementation with theirs. Kao used the same verification approach, to compare their model to the one in Tone and Tsutsui [8], to support their proposed modifications.

When comparing the output of the R implementation to the output shown in [9], one of the 30 process efficiencies (process 1 of DMU 2) showed a discrepancy of 0.0006 or 0.28%, while all 30 process weights were identical. The discrepancy is likely caused by two slightly different local optima or by a rounding error in either of the two NSBM models. Since the difference found in this verification is considered to be negligible, and all other values were identical between the two NSBM implementations, we concluded that our implementation was correct, when compared to the model in Kao [9].

Table 2. Potential performance measures for the Standard Service Process.

Measure	Unit	P1	P2	P3	P4	P5	P6	P7
Service advisors	FTE	I	I	I		I	I	
Marketing/advertisement costs	Euro	I						
Local businesses (industrialization)	Count	I						
Appointments/pass-throughs	Count	O	I	I	I	I	I	
Rescheduled appointments*	Count	O	I					
Planning accuracy*	Minutes			I				
Usage of replacement vehicle fleet*	Percentage			O				
Additional services sold	Euro			O				
Profit margin replacement vehicles*	Percentage			O				
Mechanics	FTE				I			
Electric vehicle (EV) pass-throughs*	Percentage				I			
Combustion vehicle pass-throughs*	Percentage				I			
Unique car brands (actual) *	Count				I			
Unique car brands (contractual) *	Count				I			
Unique car models*	Count				I			
Car bridges*	Count				I			
Average car weight*	Kilogram				I			
Average car age*	Years				I			
Hour efficiency (allowed ÷ taken)	Percentage				O		I	
Billed hours total	Hours				O			
Average time per pass-through	Hours				O			
Average time per EV pass-through*	Hours				O			
Warranty costs per pass-through*	Euro				O			
Net wage cost price per billed hour	Euro				O			

(continued)

Table 2. (*continued*)

Measure	Unit	P1	P2	P3	P4	P5	P6	P7
Cost price parts per pass-through	Euro				O			
Warranty jobs/services*	Count				O			
Additional work performed	Euro				O			
Budgeted minus realised margin*	Percentage				O			
Time spent on quality control	Hours					I		
Work to do after quality control	Hours					O		
Net parts revenue per pass-through	Euro						O	
Net wage revenue per pass-through	Euro						O	
Net wage revenue per billed hour	Euro						O	
Earnings before tax (EBT)	Percentage						O	
Total revenue	Euro						O	
Profit margin (w.r.t. cost price)	Percentage						O	
Gross profit margin (w.r.t. revenue)	Percentage						O	
Service absorption rate	Percentage						O	
Customer satisfaction	Percentage							O
Response rate*	Percentage							O

4.2 Additional Features

To allow for a more flexible model, and to make it easier to use real-life data when running the model, additional features had to be added. The first was the ability to *handle negative data*, this is not natively supported in DEA models, but it is reasonable to assume that this will occur when the model is used in practice (e.g., negative profit). To extend the verified NSBM model, the method proposed by Tone et al. [21] was used. If the lowest value of an input or output factor is zero or negative, this method defines a new *base point* for this factor by translating each value by the lowest value, so that all other values are positive, plus a small margin (σ) to prevent division by zero.

The second feature enables *undesirable output*, which was added since the default assumption in DEA is that all output factors are desired and are, therefore, increased in case of an inefficient DMU. This type of data is also assumed to commonly occur in real-life implementation, for example, when dealing with the resulting costs of a process. Tone [22] proposed an approach to deal with undesirable outputs in SBM by adjusting the respective 'factor constraint'. In their approach, they replaced the minus sign in front of the output slack decision variable with a plus sign. Applying the same technique to input slack (i.e., desired input) is possible, but results in negative efficiency scores when the slack value is too large compared to its current value (X_{pk}). This problem is also addressed by Tone [23].

Lastly, *non-controllable factors* can be included to provide 'context' to a DMU. This means that these factors should not receive any suggested improvement in the form of slack either, in case of an inefficient DMU. This is achieved by excluding the slack decision variable from the respective 'factor constraint' and objective function by setting it to zero. However, altering the objective function also affects the resulting efficiency. The goal of adding non-controllable context factors is to alter the reference set (i.e., lambda values ($\lambda_{DMU}^{(p)}$)) of a DMU, for a more realistic placement or projection onto the efficiency frontier.

5 Real Case

This section discusses the real case we explored, in which we used data supplied by a car importer to rank the workshop departments in their dealership network, based on the Standardised Service Process (SSP). This case was selected since it matched a question the car importer had, regarding a (more) automated benchmarking approach.

5.1 Process

Table 2 showed that there is limited coverage of identified *potential* (i.e., not all data was available) performance measures for some of the seven SSP sub-processes listed in Sect. 3.2. Therefore, we chose to only focus on three steps, which still cover the essence of a pass-through in the SSP when combined and can be regarded as a more compact representation of the original 7-step SSP. These sub-processes are: (1) making an appointment; (2) performing the repair; and (3) returning the car/invoicing.

These sub-processes ($p1, p2$, and $p3$, , respectively) were represented as the network structure shown in Fig. 2. Here, X_p and Y_p are the exogenous input and output factors of each sub-process, respectively. Z_p^{in} and Z_p^{out} are the endogenous (i.e., intermediate) input and output factors for each sub-process, respectively.

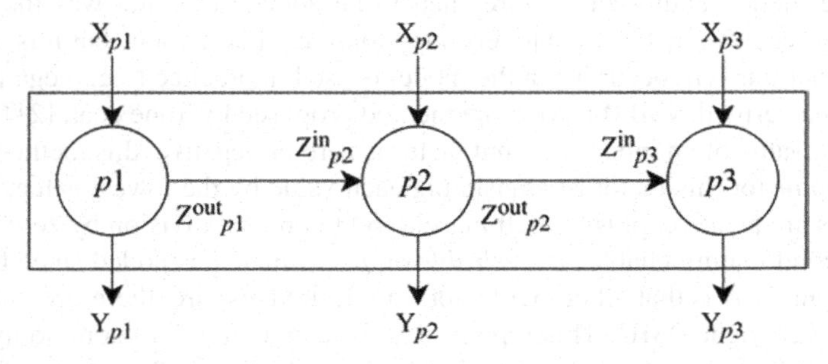

Fig. 2. Network structure to represent the simplified Standard Service Process model.

5.2 Data

The data used only includes dealership branches that are affiliated with the car importer, which implies that a dealership holding may also have other branches that are not included in this dataset. Altogether, the full dealership network of the importer consists of more than 200 branches (subdivided between different holdings). Since, in our case, the Decision-Making Units (DMUs) correspond to the car dealership's workshop department, and not all branches have (or are) a workshop, we made a selection in our dataset of *155 DMUs* to be analysed.

The data used in this case study originated from the same dataset that the car importer currently uses for benchmarking car dealerships. This means that little time had to be spent on pre-processing and cleaning the dataset, as all data was already in a usable format (i.e., numeric). We only had to remove some extreme outliers and blank entries. The data used for this case study and to validate our results is from 2021, as this was the most recent full calendar year at the time this research was performed. Another reason for selecting the same dataset was to enable comparison in the result validation.

5.3 Model Implementation

For each of the three sub-processes we considered, relevant measures were selected from Table 2 based on their availability in the dataset. Although this is often recommended, we did not perform a (sensitivity) analysis, since DEA does not have any formal guidelines for doing this [11]. The final factor selection is shown in Fig. 3. Initially, marketing costs were also included as an exogenous input of $p1$, , based on the literature, but was later removed in accordance with the feedback we received from some of the experts during the validation. They argued that it could lead to a skewed assessment (and easy manipulation) for the efficiency of $p1$, since these expenses generally are for a dealership holding instead of their underlying branches.

To deal with negative data, our implementation uses $\sigma = 10^{-2}$ for the feature explained in Sect. 4.2, as this is also the value proposed in Tone et al. [21]. Since service advisors are included in both $p1$ and $p2$, , according to the SSP guidelines, but the data does not reflect how much time is spent on one or the other, an equal split of hours spent was assumed ((50%) in Fig. 3). Furthermore, six input factors were considered as *non-controllable* context, namely, Electric vehicle (EV) pass-throughs, Combustion vehicle pass-throughs, Unique car brands (actual), Appointments/pass-throughs (for both $p2$ and $p3$), and Hour efficiency. Three output factors were considered as *undesirable output*, namely, Net wage cost price per billed hour, Cost price parts per pass-through, and Warranty costs per pass-through.

Altogether, the full linear programming problem (LPP) solved for the 155 DMUs consists of three sub-processes with 25 factors in total: six exogenous input factors, three endogenous input factors, fourteen exogenous output factors, and two endogenous output factors. For this LPP, variable returns to scale was assumed. Following the generic NSBM objective function in Eq. 1 leads to Eq. 3 for this LPP.

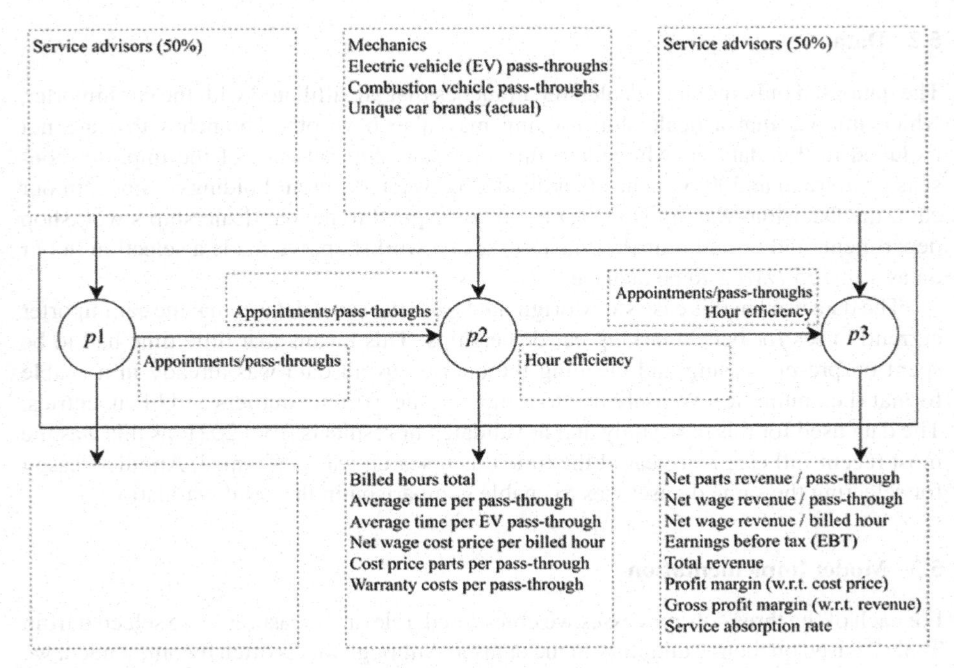

Fig. 3. Process model (Fig. 2) combined with the final factor selection.

$$E_k^S = min\left(\frac{\left[1-\left(\frac{s_1^-}{X_{1k}}\right)\right]+\left[1-\left(\frac{s_2^-}{X_{2k}}+\cdots+\frac{s_5^-}{X_{5k}}+\frac{t_1^-}{Z_{1k}^{in}}\right)/5\right]+\left[1-\left(\frac{s_6^-}{X_{6k}}+\frac{t_2^-}{Z_{2k}^{in}}+\frac{t_3^-}{Z_{3k}^{in}}\right)/3\right]}{\left[1+\left(\frac{t_1^+}{Z_{1k}^{out}}\right)\right]+\left[1+\left(\frac{s_1^+}{Y_1}+\cdots+\frac{s_6^+}{Y_6}+\frac{t_2^+}{Z_{2k}^{out}}\right)/7\right]+\left[1+\left(\frac{s_7^+}{Y_7}+\cdots+\frac{s_{14}^+}{Y_{14}}\right)/8\right]}\right) \tag{3}$$

Similar to Tone [18] and Yang [14], this objective function was linearised by letting:

$$\left[1+\left(\frac{t_1^+}{Z_{1k}^{out}}\right)\right]+\left[1+\left(\frac{s_1^+}{Y_1}+\cdots+\frac{s_6^+}{Y_6}+\frac{t_2^+}{Z_{2k}^{out}}\right)/7\right]+\left[1+\left(\frac{s_7^+}{Y_7}+\cdots+\frac{s_{14}^+}{Y_{14}}\right)/8\right]=\frac{1}{Q} \tag{4}$$

6 Results

This section presents and explains the main results we obtained with our benchmarking tool by showing an example of a DMU reference set ($p1$ for DMU 90) and the resulting process efficiency of one DMU ($p3$ for DMU 31). Lastly, the overall System Efficiency is discussed, which is the weighted average of all three sub-processes.

6.1 Reference Set

The DMU reference set was elaborately explained to the experts in the result validation session to further clarify how the model works. More specifically, we explained how the model determines which DMU(s) is/are relatively comparable to the DMU being assessed and where the improvement suggestions originate from.

The reference set of a DMU is determined based on n lambda values for each sub-process ($\lambda_j^{(p)}$), as defined in the constraints of Eq. 1. The variable returns to scale (VRS) assumption dictates that the sum of all lambda values must be equal to one (Eq. 3), which is the case for each of the sub-processes separately. Hence, each lambda value can be interpreted as the percentual degree/intensity of reference.

An example is shown in Table 3, in which the reference set of DMU 93 for $p1$ consists of DMU 40 (for 37.71%) and DMU 125 (for 62.29%). Based on these two reference units, the process efficiency of $p1$ for DMU 90 ($E_{90}^{(p1)}$) is 69.45%. The model suggests decreasing the input factor X_1 (FTE service advisors) by 30.55%, for DMU 90 to become 100% efficient for $p1$; Z_1^{out} does not need to be optimised. It can also be seen why the model selected these two DMUs to be in the reference set since the value for both X_1 and Z_1^{out} are in the middle of these DMUs, while they are deemed 100% efficient. Hence, the model concludes that the optimum of DMU 90 lies in the middle.

Table 3. Reference set of DMU 93 for $p1$, , including lambda values and suggested change (%).

DMU	$E_{DMU}^{(p1)}$	$\lambda_{DMU}^{(p1)}$	$X_{1;DMU}$	$Z_{1;DMU}^{out}$
40	100%	37.71%	9.30 FTE	22,770 pass-throughs
			(0.00%)	*(0.00%)*
90	69.45%	0.00%	9.01 FTE	19,650 pass-throughs
			(−30.55%)	*(0.00%)*
125	100%	62.29%	4.42 FTE	17,761 pass-throughs
			(0.00%)	*(0.00%)*

The example given here and during the result validation was deliberately one where the reference set consisted of only two other units. Theoretically, a reference set could consist of an arbitrary number of different units.

6.2 Process Efficiency and Improvement

Figure 4 illustrates the kind of results we obtained by showing all factors (input and output factors) related to a selected process ($p3$ for DMU 31). For each factor, Fig. 4 shows the current value, suggested value (current ± slack), and the percentual difference (i.e., suggested percentual change) between these two. This overview allows for a comprehensible comparison of where a DMU (or a selection of several DMUs) should focus to improve their process efficiency.

As can be seen on the top left in Fig. 4, the process efficiency of DMU 31 for $p3$ is 32%; in order to increase this score, the suggestions on the right must be implemented. Here, DMU 31 is suggested to reduce the number of service advisors FTE by 57% and maintain the other input factors (hour efficiency and the pass-throughs) as-is. Simultaneously, their reference set, which consists of three DMUs, suggests that they should (almost) double their net revenue on parts (per pass-through). While (exactly) executing all suggestions simultaneously is not realistic, this does give a detailed insight into where a DMU should focus, based on comparable DMUs in the dealership network.

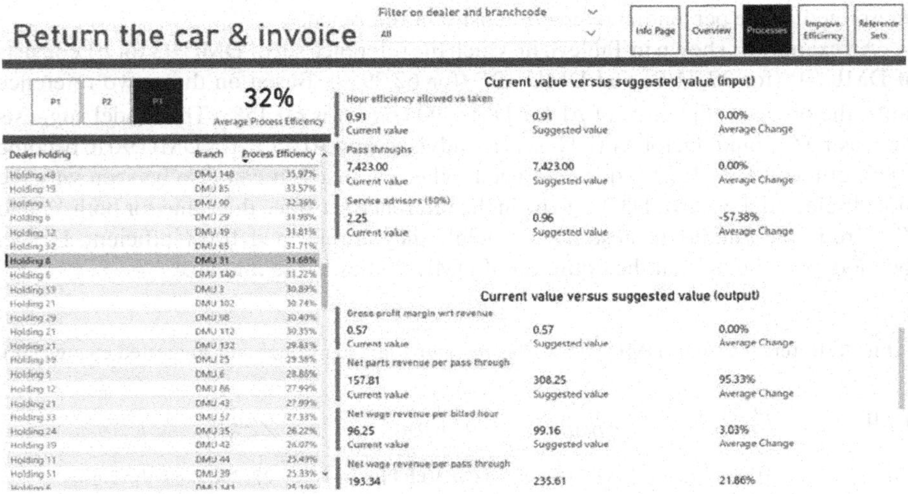

Fig. 4. Overview of the current and suggested value of all factors of DMU 31 for $p3$.

6.3 System Efficiency

Lastly, Fig. 5 shows an overview page where all three sub-processes can be compared at once, in contrast to the pages discussed above. This allows for the assessment of the full process ($p1$, $p2$, and $p3$) performed by the workshop department of all branches. The three process efficiencies shown above the bar chart are the average process efficiency of the selection, which currently is all 155 DMUs.

Figure 5 shows the *System Efficiency* in the third column of the table on the left. This is the weighted average of all process efficiencies, as each process weight is separately calculated for each process of each DMU, which can also be seen in the stacked bar chart on the right. The formulas used in [9] for calculating the process efficiency and weight can be found in Eq. 5 and Eq. 6, respectively.

$$E_k^{(p)} = \frac{1 - \left(\sum_{i\in I^{(p)}} \frac{s_i^{(p)-*}}{X_{ik}^{(p)}} + \sum_{f\in M^{(p)}} \frac{t_f^{(p)-*}}{Z_{fk}^{(p)}}\right)/\left(\hat{i}^{(p)} + \hat{m}^{(p)}\right)}{1 + \left(\sum_{r\in O} \frac{s_r^{(p)+*}}{Y_{rk}^{(p)}} + \sum_{g\in N^{(p)}} \frac{t_g^{(p)+*}}{Z_{gk}^{(p)}}\right)/\left(\hat{o}^{(p)} + \hat{n}^{(p)}\right)}; p = 1, \ldots, q \quad (5)$$

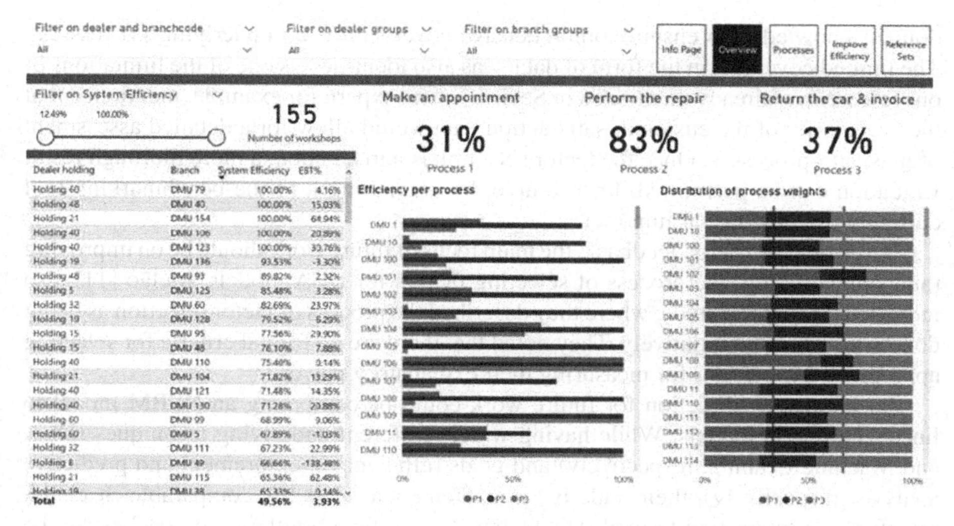

Fig. 5. Overview page showing the process efficiency of all SSP sub-processes per DMU.

$$\omega^{(p)} = \frac{1 + \left(\sum_{r\in O} \frac{s_r^{(p)+*}}{Y_{rk}^{(p)}} + \sum_{g\in N^{(p)}} \frac{t_g^{(p)+*}}{Z_{gk}^{(p)}}\right) / \left(\widehat{o}^{(p)} + \widehat{n}^{(p)}\right)}{\sum_{h=1}^{q}\left[1 + \left(\sum_{r\in O} \frac{s_r^{(h)+*}}{Y_{rk}^{(h)}} + \sum_{g\in N^{(p)}} \frac{t_g^{(h)+*}}{Z_{gk}^{(h)}}\right) / \left(\widehat{o}^{(h)} + \widehat{n}^{(h)}\right)\right]}; p = 1, \ldots, q \tag{6}$$

7 Final Considerations

This paper reported on our efforts to develop a technique to perform process benchmarking and apply it to rank and improve the workshop department of car dealerships. The tool was designed based on requirements like calculating the efficiency of the main standardised process and giving tailored improvement suggestions to inefficient units.

The validation using expert opinion was performed by first giving a demo of the results and dashboards discussed above, followed by a Q&A round, and finally a discussion on the concluding remarks and (future) applicability. The four experts are all highly qualified employees, each with more than ten years of experience in or with the automotive industry. To summarise, the experts concluded that the proposed approach was interesting and promising. Especially when they took this approach into account, in contrast to their current method to compare and analyse car dealerships, they mentioned that this approach is more insightful, since it is closely aligned with the real-life process (SSP). Furthermore, they recognized that the comparisons are more consistent because all factors are considered at the same time, instead of one or two in isolation. Their main critique was that the factor selection, as presented in Sect. 5.3, should be improved. Having an optimal factor selection was, however, not in the scope of our research, since optimising the factor selection would not only require more design iterations but also domain knowledge to ensure comprehensive coverage of the underlying process(es). The limited coverage, in the form of data, was also identified as one of the limitations of our research, as already mentioned in Sect. 5.1. One expert, for example, mentioned that the granularity of the customer satisfaction data would allow for a detailed assessment of most sub-processes. Once the factor selection is agreed upon, a more thorough result validation can be performed, for example, by comparing it to the benchmarking tool currently used by the car importer.

Based on the experts' feedback, the main focus of future work should be on improving (and standardizing) the process of selecting factors for DEA models. Melão [11] also indicated this in their study, where they described the process of factor selection as being done arbitrarily and intuitively. They noted that there are no formal criteria for selecting appropriate factors, nor for measuring their explanatory power.

An interesting direction for future work could be comparing an NSBM model to linear regression models. While having widely different underlying techniques (DEA and Machine Leaning, respectively) and goals (efficiency measurement and predictive analysis, respectively), their underlying mathematical model is comparable. It could, therefore, be interesting to explore whether the gap between these approaches can be bridged to improve the current DEA NSBM model based on established Machine Learning concepts. Additionally, this may allow the factor selection to be viewed from an alternative perspective, similar to the selection of regression coefficients.

References

1. Jerenz, A.: Revenue Management and Survival Analysis in the Automobile Industry, 1st edn. Springer Gabler, Wiesbaden (2008)
2. Camp, R.C.: Business Process Benchmarking: Finding and Implementing Best Practices, 1st edn. ASQC Quality Press, Milwaukee (1995)
3. Biondi, S., Calabrese, A., Capece, G., Costa, R., Di Pillo, F.: A new approach for assessing dealership performance: an application for the automotive industry. Int. J. Eng. Bus. Manag. **5**(1), 1–8 (2013)
4. Gencer, Y.G., Akkucuk, U.: Measuring aftersales productivity by multi attribute decision making methods: an application in the automotive sector. Int. J. Adv. Appl. Sci. **5**(9), 88–95 (2018)
5. Sadat Rezaee, M., Haeri, A., Noori, S.: Automotive vendor's performance evaluation and improvement plan presentation by using a data envelopment analysis. Int. J. Eng. **31**(2), 374–381 (2018)
6. Tan, Y., Zhang, Y., Khodaverdi, R.: Service performance evaluation using data envelopment analysis and balance scorecard approach: an application to automotive industry. Ann. Oper. Res. **248**(1–2), 449–470 (2016). https://doi.org/10.1007/s10479-016-2196-2
7. Toloo, M., Ertay, T.: The most cost efficient automotive vendor with price uncertainty: a new DEA approach. Measurement **52**, 135–144 (2014)
8. Tone, K., Tsutsui, M.: Network DEA: a slacks-based measure approach. Eur. J. Oper. Res. **197**, 243–252 (2009)
9. Kao, C.: Efficiency decomposition in network data envelopment analysis with slacks-based measures. Omega **45**, 1–6 (2014)
10. Kao, C.: Network data envelopment analysis: a review. Eur. J. Oper. Res. **239**(1), 1–16 (2014)
11. Melão, N.: Data envelopment analysis revisited: a neophyte's perspective. Int. J. Manag. Decis. Mak. **6**(2), 158–179 (2005)

12. Charnes, A., Cooper, W.W., Rhodes, E.: Measuring the efficiency of decision making units. Eur. J. Oper. Res. **2**(6), 429–444 (1978)
13. Banker, R.D., Charnes, A., Cooper, W.W.: Some models for estimating technical and scale inefficiencies in data envelopment analysis. Manage. Sci. **30**(9), 1078–1092 (1984)
14. Yang, F.-C.: Efficiency decomposition in dealers from the perspectives of demand forecasting, sales force, and inventory control: a case study. Prod. Plann. Control **27**(16), 1334–1343 (2016)
15. Alexander, I.F.: A taxonomy of stakeholders: human roles in system development. Int. J. Technol. Human Interact. **1**(1), 23–59 (2005)
16. Alexander, I.F.: A better fit - characterising the stakeholders. in: CAiSE 2004 Workshops in Connection with The 16th Conference on Advanced Information Systems Engineering, pp. 215–223. Faculty of Computer Science and Information Technology, Riga Technical University, Riga, Latvia (2004)
17. Soteriades, A. D.: Package 'additiveDEA': dea.sbm function (2019). https://CRAN.R-project.org/package=additiveDEA
18. Tone, K.: A slacks-based measure of efficiency in data envelopment analysis. Eur. J. Oper. Res. **130**(3), 498–509 (2001)
19. Berkelaar, M., et al.: lpSolve: Interface to 'Lp_solve' v. 5.5 to Solve Linear/Integer Programs (2020). https://CRAN.R-project.org/package=lpSolve
20. lp_solve, Konis, K., Schwendinger, F., Hornik, K.: lpSolveAPI: R Interface to 'lp_solve' Version 5.5.2.0 (2022). https://CRAN.R-project.org/package=lpSolveAPI
21. Tone, K., Chang, T.S., Wu, C.H.: Handling negative data in slacks-based measure data envelopment analysis models. Eur. J. Oper. Res. **282**(3), 926–935 (2020)
22. Tone, K.: Dealing with undesirable outputs in DEA: a Slacks-Based Measure (SBM) approach. Nippon Opereshonzu, Risachi Gakkai Shunki Kenkyu Happyokai Abusutorakutoshu **2004**, 1–18 (2003)
23. Tone, K.: Dealing with desirable inputs in data envelopment analysis: a slacks-based measure approach. Am. J. Oper. Manag. Inf. Syst. **6**(4), 67–74 (2021)

A Model of Qualitative Factors in Forensic-Ready Software Systems

Lukas Daubner[1], Raimundas Matulevičius[2](✉), and Barbora Buhnova[1]

[1] Masaryk University, Brno, Czechia
{daubner,buhnova}@mail.muni.cz
[2] University of Tartu, Tartu, Estonia
raimundas.matulevicius@ut.ee

Abstract. Forensic-ready software systems enhance the security posture by designing the systems prepared for potential investigation of incidents. Yet, the principal obstacle is defining their exact requirements, i.e., what they should implement. Such a requirement needs to be on-point and verifiable. However, what exactly comprises a forensic readiness requirement is not fully understood due to distinct fields of expertise in software engineering and digital forensics. This paper describes a forensic readiness qualitative factor reference model that enables the formulation of specific requirements for forensic-ready software systems. It organises the qualitative properties of forensic readiness into a taxonomy, which can then be used to formulate a verifiable requirement targeted at a specific quality. The model is then utilised in an automated valet parking service to define requirements addressing found inadequacies regarding a potential incident investigation.

Keywords: Forensic Readiness · Forensic-by-Design · Forensic-Ready Software Systems · Requirements Engineering · Risk Management

1 Introduction

With the ever-looming threats to software systems, security, safety, survivability, and robustness factors are becoming more and more critical. These are focused on defence against the threats of making it operable even in exceptional conditions [22]. However, their implementation does not guarantee complete protection, as attacks might still succeed, and disasters might disable the systems [19]. For this reason, there is a growing interest in applying measures allowing for post-incident investigation [25], for which digital forensic methods are utilised.

Often, results from digital forensic investigation are relied upon in a court of law [8]. Those typically involve investigations of criminal cases but also internal investigations as part of incident response [38] or post-mortem analysis of disasters [19]. The results, called digital evidence, must conform to a high standard of soundness to be admissible. As such, forensic investigation is a laborious, costly, and delicate process to avoid the spoliation of digital evidence. Still, its success is not assured as the evidence might be inconclusive or completely missing.

S. Nurcan et al. (Eds.): RCIS 2023, LNBIP 476, pp. 308–324, 2023.
https://doi.org/10.1007/978-3-031-33080-3_19

Forensic readiness was formulated to mitigate such issues, aiming to increase the value of evidentiary data while lowering the cost of the investigation [42]. Traditionally, it is approached as a series of organisational processes [26,38]. Recently, however, an idea of approaching forensic readiness from a software engineering perspective was formulated, coining the term forensic-ready software systems [35]. Such a system can, in anticipation of an incident, proactively collect potential digital evidence[1] and soundly conduct forensic processes.

Currently, there is a rough idea about which requirements the forensic-ready software systems should address [35], how they should be elicitated [14], and represented [16]. However, the understanding of factors and their relationships that form forensic-ready software systems is insufficient. As such, it is challenging to define proper implementable and verifiable requirements for such systems.

1.1 Research Questions

The goal of this paper is to define a reference model of forensic readiness qualitative factors, to enable the definition of on-point and verifiable requirements for forensic-ready software systems. Based on the initially defined requirements, literature survey, and interviews conducted with forensic experts [11], a taxonomy of factors is defined. Then, the model is demonstrated by formulating verifiable requirements for an automated valet parking service that addresses inadequacies regarding a potential incident investigation. To reach this goal, we formulate the following research questions:

RQ1: *What are the factors that form the forensic readiness requirements?* The aim is to create a reference model of its components starting from the preliminary requirements. Additionally, the model should provide more insight into the relationship between the requirements, including their classification.

RQ2: *How can the factors be manifested as a concrete requirement?* The aim is to describe the procedure of defining a concrete forensic readiness requirements, using the reference model concerning a particular system.

1.2 Research Method

To answer the research questions, we advised a research method depicted in Fig. 1. As a starting point, we gathered three principal inputs. The first is a list of preliminary requirements by Pasquale et al. [35], listed in Table 1, which are explicit for forensic-ready systems. The second is requirements and obligations surveyed from related literature, including general forensic readiness. The third is the findings from the interviews with digital forensic experts (e.g., investigators, lawyers) on their problems, ideas of a forensic-ready software system, and evidence quality factors [11], conducted as part of previous work.

With the inputs in place, the candidate factors are elicited by inspecting the requirements, properties, and needs of forensic readiness. Then the reference model is iteratively created, grouping the factors and creating a hierarchy.

[1] Note the difference: potential digital evidence – potentially useable for future investigation, and digital evidence – used to satisfy or refute the investigation hypothesis.

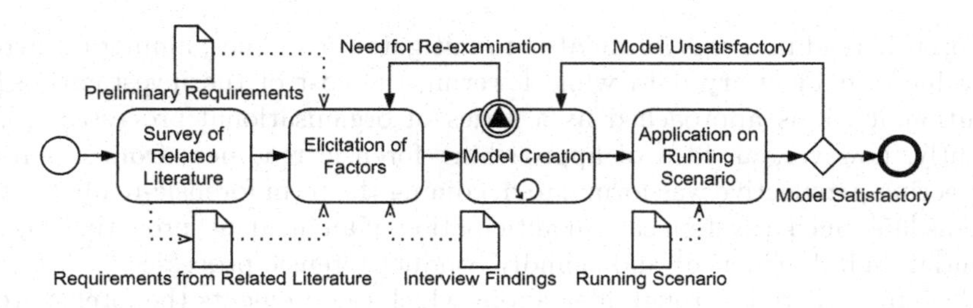

Fig. 1. Research Method

In a case of ambiguity, a particular factor is re-examined. The resulting model corresponds to **RQ1**. Then, the model is demonstrated on a running scenario, formulating concrete forensic readiness requirements for the system corresponding to **RQ2**. If the model is found unsatisfactory, it is redesigned.

The paper is structured as follows: After introducing the topic and aim, Sect. 2 summarises the related work. Then Sect. 3 presents the forensic readiness qualitative factor reference model. It is demonstrated in Sect. 4 on a running example, where it is used to manifest concrete requirements. Finally, Sect. 5 answers the research questions and concludes the paper.

2 Related Work

Design of forensic-ready software systems, or sometimes systems forensic-by-design, was approached by several domain-specific frameworks. For example, cyber-physical cloud systems [1], medical cyber-physical systems [24], and smart buildings [5]. A common component of the frameworks is the utilisation of a risk-based approach and emphasis on validation and verification. On the other hand, the frameworks typically do not go into detail regarding the implementation.

Focusing on the software engineering point of view, a survey was conducted to explore the practice of considering forensic readiness in the development of software systems [23]. It culminated in publishing the idea of a forensic-ready software system [35], including the preliminary requirements and open challenges. Some of the requirements were already addressed. For example, a preservation of minimal and relevant evidence [2] and automated logging instrumentation [37]. Furthermore, the representation of forensic-ready software systems and incidents in cyber-physical environments has been explored [3,4].

2.1 Forensic Readiness Software Requirements

Forensic-ready software systems take the concept of forensic readiness and apply it as a high-level qualitative aspect, or non-functional requirement, of a software system. However, to implement it, a solid foundation is needed to guide the formulation, validation, and verification of the specific requirements [13]. A preliminary set of requirements was formulated with the idea of a forensic-ready software system [35]. They are summarised in Table 1.

In a sense, forensic readiness is similar to other non-functional requirements like dependability (especially security [22] and safety [21]), which share some attributes. The security requirement is defined as a condition over the phenomena of the environment that we wish to make true by installing the system in order to mitigate risks [17,31]. They can be classified based on the area they specify (e.g., intrusion detection, integrity, authentication, immunity) [22]. Mostly, however, they are discussed based on their qualitative factor [20] (attribute, characteristics, property, aspect), a subject of its condition. They can be the security criteria violated by the mitigated risk (e.g., confidentiality, integrity, availability) [34]. In this regard, the violated properties of STRIDE [40] are also relevant. Other works use requirement categories mapped on the incident-handling process, creating a security checklist [6]. Notably, the factors can be organised in a quality model [20], which creates a taxonomy-like structure of the qualitative (i.e., non-functional) aspects to address the measurable requirements.

Table 1. Preliminary Requirements for Forensic-Ready Software Systems [35]

Availability	All useful potential evidence is preserved, prepared and retrievable if needed.
Relevance	Preserved potential evidence is relevant to considered incidents and scenarios.
Minimality	Potential evidence unnecessary for the expected investigation is not preserved.
Linkability	Preserved pieces of potential evidence can be linked with other pieces.
Completeness	Preserved potential evidence is sufficient to satisfy or refute the considered investigation hypothesis.
Non-Repudiation (Admissibility)	Preserved potential evidence should conform as admissible evidence. Its integrity and authenticity are ensured.
Data Provenance	The handling process of potential evidence records all operations made.
Legal Compliance	The potential evidence handling process is compliant with laws and regulations.

Risk management is considered a reasonable approach for the identification of forensic-ready requirements. This is corroborated by its utilisation in secure design [29] but also in guidelines for implementing organisational forensic readiness [9,38]. In the case of forensic-ready software systems, there is an extension of Information Systems Security Risk Management (ISSRM) [17,31] which adds forensic readiness concepts [14,15] and accompanying metrics for evaluation. Additionally, the work utilises a custom notation for modelling [16].

As noted with the risk management case, there is an overlap with security and forensic readiness scopes [25], which includes requirements. While there was little done in terms of requirements analysis and representation on the forensic readiness side, the security side contains numerous approaches. A good example of security requirements is a list by Firesmith [22]. For a model-based example, UMLsec [28] captures the security concerns into the UML model and analyses them. On the risk management side, Secure Tropos [30] and Misuse Cases [41] are the well-known approaches.

2.2 Summary

Various works have capabilities of forensic-ready systems, but their requirements are rarely addressed or their implementation verified. While there are preliminary requirements defined, they have several limitations. The relationship between them is unclear as there can overlap in some contexts (e.g., a provenance record of potential evidence contributes to its non-repudiation). Furthermore, the requirements can be satisfied in multiple ways (e.g., non-repudiation might combine corroborating evidence integrity assurance). Lastly, there is a gap in connecting risk-based techniques to formulating concrete requirements that would verifiably address the identified risks.

3 Factors of Forensic Readiness Requirements

This section presents the identified qualitative factors (i.e., attributes, characteristics, properties, aspects) of forensic-ready software systems, which make up the requirements. Defining the factors aims to facilitate the elicitation of precise, implementable, and verifiable forensic readiness requirements from forensic readiness scenarios [14]. In this regard, a reference quality model is proposed, describing components of a qualitative requirement [20], depicted in Fig. 2.

The reference model is created iteratively, following the set method. During each step, the elicited factors from the source material are examined for similarity in purpose or target. The resulting hierarchy is checked for consistency so that one factor is not duplicated and factors with the same overarching purpose are aggregates of a higher one.

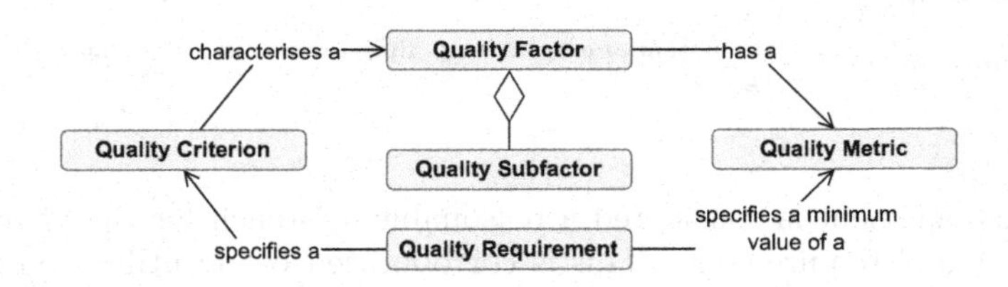

Fig. 2. Excerpt from Firesmith's Quality Model Definition [20]

Per the definition, a quality requirement specifies a minimum amount of a quality factor [20], which can be decomposed into subfactors. It refers to a quality criterion, which is evidence for or against the existence of a factor or subfactor. A quality metric then quantifies the factor. Thus, the implementation of such a requirement (a control [29]) is verifiable. The factors discussed below are all special cases of forensic readiness that may be composed of further subfactors.

3.1 Evidence Factors

The evidence factors are a group of qualities of a single piece of potential evidence, essentially any information that can support legal proceedings, demonstrate due diligence, manage the impact of risk, and support a claim or dispute [38]. Therefore, its qualities should be aligned with such actions, especially in being convincing to a 3rd party. Figure 3 depicts the relationships between the evidence factors. The rest of the section describes them in more detail.

Digital evidence is commonly associated with legal context [8,32] and is obliged to follow legal requirements. While they differ based on the legal framework [39], they do follow common principles [11]. However, proactively collected "potential" evidence can be, in many cases, more lenient, as it is not (yet) in the scope of a formal investigation. Still, high-quality potential evidence greatly improves the evidentiary value of the proper evidence when it is required.

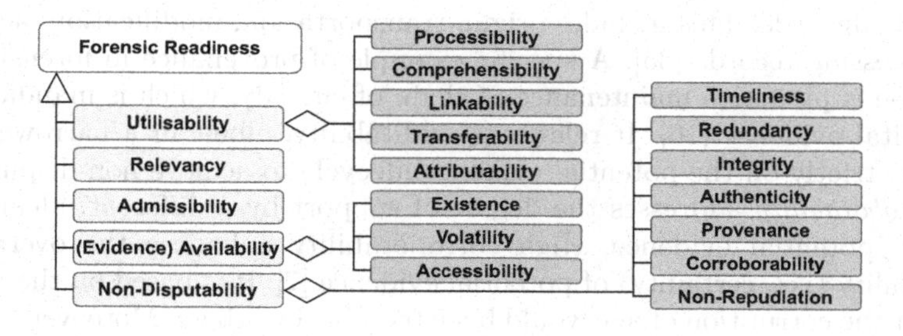

Fig. 3. Forensic Readiness Qualitative Factors: Evidence Factors

Non-Disputability addresses the prevention of disputes regarding the potential evidence. In this sense, the fundamental disputes aim at admissibility, i.e., acceptance in a court of law. In other words, it gives assurances on the genuineness of potential evidence. Generally, the purpose of non-disputability in forensic readiness is to safeguard the evidentiary value of the data and increase its confidence [7]. Consequently, non-disputability should allow for confident disputation of non-genuine potential evidence. It concerns the possible dangers of tampering or corruption of (potential) evidence [8,39].

As a result, the non-disputability influences the evidentiary value of the potential evidence. It is further decomposed into subfactors, complementing and potentially substituting one another. For example, even if the potential evidence is not protected against tampering (Integrity), it could have high value due to multiple independent pieces of evidence supporting it (Corroborability) [7]. It is usually referred to as "Non-Repudiation" [35,36], but we instead consider Non-Repudiation as a factor closer to security and capture it here only as a subfactor.

Timeliness addresses the accuracy of time of origin, or generally, any time information associated with the potential evidence. It motivates a timely creation of potential evidence relative to an action or event and assurance of the correctness of the time information. The reliability of the time is related to the Integrity [43] or can be corroborated by other time information.

Redundancy addresses the extent and manner of storing duplicities of potential evidence. It is related to Integrity as a way to provide integrity assurance and Corrborability as the copy corroborates the original and vice versa [38].

Integrity addresses the assurance that the potential evidence was not tampered with [42]. In other words, it addresses the non-disputability of potential evidence corruption based on its unauthorised creation, modification, or deletion. It directly relates to integrity in a security sense [22].

Authenticity addresses the non-disputability of potential evidence origin. It relates to Authentication from a security perspective, which often implies Integrity [33]. Additionally, authenticity might relate to Attributability, concerning binding the person or device of origin but must give stronger guarantees making the attribution non-disputable.

Provenance addresses a record-keeping of the actions on potential evidence during its lifecycle. This includes origin, transportation, modification, accessing, and processing records [35]. A specific example of provenance in forensic readiness sense is proactive maintenance of chain of custody, which is mandatory for true digital evidence [10]. It relates to auditability, albeit in a narrower sense, focusing strictly on the potential evidence lifecycle to achieve non-disputability.

Corroborability addresses the degree of support by a different, ideally independent, potential evidence. High corroborability enhances the overall non-disputability (i.e., certainty) of potential evidence [7]. It is based on the assumption that the corruption of one would be detectable by others. Moreover, an undetectable corruption of one should require the corruption of another. It relates to Linkability, albeit in a narrower sense, focusing on the non-disputability.

Non-Repudiation addresses the extent to which any aspect of potential evidence is prevented from repudiation. Thus, there is a significant overlap with the non-repudiation from a security perspective [22]. As noted, non-repudiation for digital evidence is often a synonym for the whole non-disputability [35,36]. However, here it is considered as its part, as the ability of one party to repudiate potential digital evidence arguably does have an impact on disputes.

Evidence Availability addresses the extent of potential evidence availability during an investigation. It encompasses the existence, preservation, and ability to retrieve the potential evidence [35]. Typically, the requirements specify which, where, and when the potential evidence is created, what it should contain, and its retention. Its subfactors are *Existence* addressing the actual presence of potential evidence; *Volatility* addressing the window of when it can be preserved or accessed and; *Accessibility*, addressing the ease of access for an investigator.

Admissibility the concern of acceptability in a court of law. Generally, admissibility is a set of legal tests a judge performs to assess the formal evidence [7]. However, their exact nature is dependent on the legal framework. It could be considered the most important quality of digital evidence, typically involving the proper handling and Non-Disputability qualities. While it is not a primary concern of potential evidence, it might be considered based on its intended use.

Relevancy addresses the need for the potential evidence and its fitness in a forensic readiness scenario [14]. From an investigation point of view, it specifies the ability of the potential evidence to support or refute investigation hypotheses [35]. Thanks to the mapping of scenarios, the relevancy addresses the specific reason why it is preserved and its value. As such, it acts as a counterweight to privacy regulations demanding that the evidence collection is not excessive by explicitly stating the reason and extent.

Utilisability addresses the assistance in the forensic investigation itself to ease the work of the investigator or cybersecurity response team. In other words, it is a factor that addresses the helpfulness and contribution of potential evidence to the effectiveness of the investigation. It refers to one of the principal aims of forensic readiness. Different aspects of the support are reflected by its subfactors.

Processability addresses the degree of ease of automated processing of the potential evidence. It concerns aspects like data format, which influences its effectiveness and reliability [12], but also the availability of tools to process it.

Comprehensibility addresses the knowledge the potential evidence can provide to the investigation. In other words, the knowledge about the semantics of potential evidence. Arguably, the apriori knowledge of the information the potential evidence can provide (e.g., from documentation [38]), including its limits, can help accelerate the investigation. Additionally, it concerns possible errors [32] and allows validation of the results [12].

Linkablility addresses the degree to which the potential evidence can be linked or correlated with others. Presence and awareness of the existing links are essential in reconstructing events during the investigation. Moreover, strong linkability should allow for easier explorative analysis [35].

Transferability addresses the degree to which the potential evidence can be transferred to another custody. Typically, the evidence is transferred to local law enforcement based order for evidence release [38]. However, the release might also demand a cross-border or cross-organisational transfer. It concerns the easiness of the transfer, including the need for supplementary material and privacy.

Attributability is a factor describing a degree of ability to bind potential evidence with an entity, meaning a person, device, place, or application. Specifically, establishing attributability to a person is considered highly important [11].

3.2 Scenario Factors

The scenario factors are a group of qualities of a set of potential evidence referring to a particular forensic readiness scenario [14]. In contrast to the qualities of single potential evidence, these factors address how well they work together towards forensic readiness. Figure 4 depicts the relationships between the scenario factors. The rest of the section describes them in more detail.

Completeness addresses the degree to which the scenario includes sufficient potential evidence for its investigation. From an investigation point of view, it

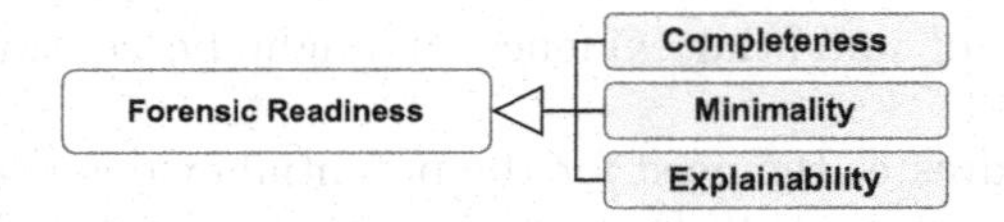

Fig. 4. Forensic Readiness Qualitative Factors: Scenario Factors

specifies whether the potential evidence is sufficient to support or refute investigation hypotheses [35].

Minimality addresses the degree to which the scenario includes only the potential evidence important for the investigation. Its purpose is to limit the amount of potential evidence an investigator must go through [35]. Essentially, it refers to a proactive forensic triage [27].

Explainability addresses the degree of the ability to explain conclusions based on the scenario clearly [38]. It deals with the number of phenomena that could be effectively described and the manner of doing so.

3.3 Process Factors

The process factors are a group of qualities of the digital forensic investigation process, scoped on those that can be reasonably affected by proactive measures. As such, it refers to the qualitative factors of forensic-ready software systems conducting forensic processes [35]. Figure 5 depicts the relationships between the process factors. The rest of the section describes them in more detail.

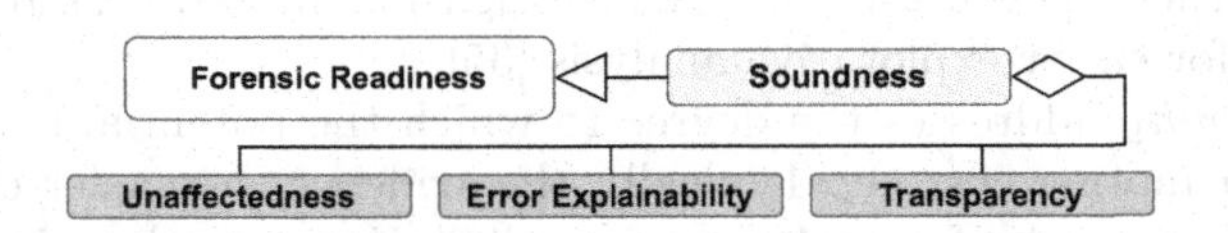

Fig. 5. Forensic Readiness Qualitative Factors: Process Factors

Process Soundness addresses the degree of soundness of the conducted forensic processes. In other words, it deals with the assurance that the process does not diminish the value of potential evidence [32]. It is further specified by subclasses focusing on a different part of soundness.

Unaffectedness addresses the degree of assurance that the conducted processes do not affect the meaning or interpretation of the potential evidence.

Error Explainability addresses the degree to which the errors in the forensic process can be detected and explained.

Transparency addresses the degree to which the process can be re-examined and verified by an independent party.

3.4 Cross-Cutting Factors

The cross-cutting factors are a group of qualities influencing forensic readiness in general, or they address a principle applicable to all other factor groups. For example, Legal Compliance must be accounted for in virtually all aspects of a forensic-ready software system. Figure 6 depicts the relationships between the cross-cutting factors. The rest of the section describes them in more detail.

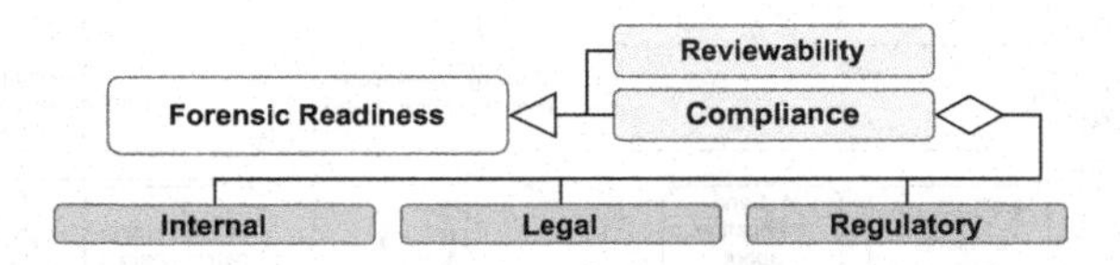

Fig. 6. Forensic Readiness Qualitative Factors: Cross-Cutting Factors

Reviewability addresses the degree to which the system's actions, components, or data involving potential evidence and forensic processes support audit and legal review. The goal is twofold. The first is to enable legal advice on an incident [38]. The second is allowing the inspection of forensic-ready controls to analyse nominal and abnormal behaviour in a similar sense to security auditing [22]. That could result in meta-potential evidence, similar to Provenance.

Compliance addresses the degree to which is the forensic-ready software system compliant with the law, regulations, or organisational policies [35]. It is further specified by subfactors focusing on the specific part, namely *Legal*, *Regulatory*, and *Local*. The factor places constraints on others. For example, demand a level of Transferability to law enforcement, or on the other hand, forbid the existence of potential evidence.

4 Running Example

This section presents a running scenario used to demonstrate the reference model of qualitative factors by manifesting the forensic readiness requirements based on it. The scenario describes issuing a parking permit within Automated Valet Parking (AVP), which is a service allowing the user to leave an autonomous vehicle in a drop-off area to park it automatically. It is represented by a BPMN process model, depicted in Fig. 7. The process model is enhanced by BPMN4FRSS [16] notation to describe evidence sources (magnifying glass symbol) and potential evidence (green BPMN data objects). Previously, the process model was utilised to demonstrate a risk-oriented approach for eliciting forensic readiness requirements [15] and the design of privacy-preserving services [18].

There are three entities present in the execution of the AVP scenario. Each entity has different assumptions regarding its cooperability during a possible investigation [15]. The description of the entities is as follows:

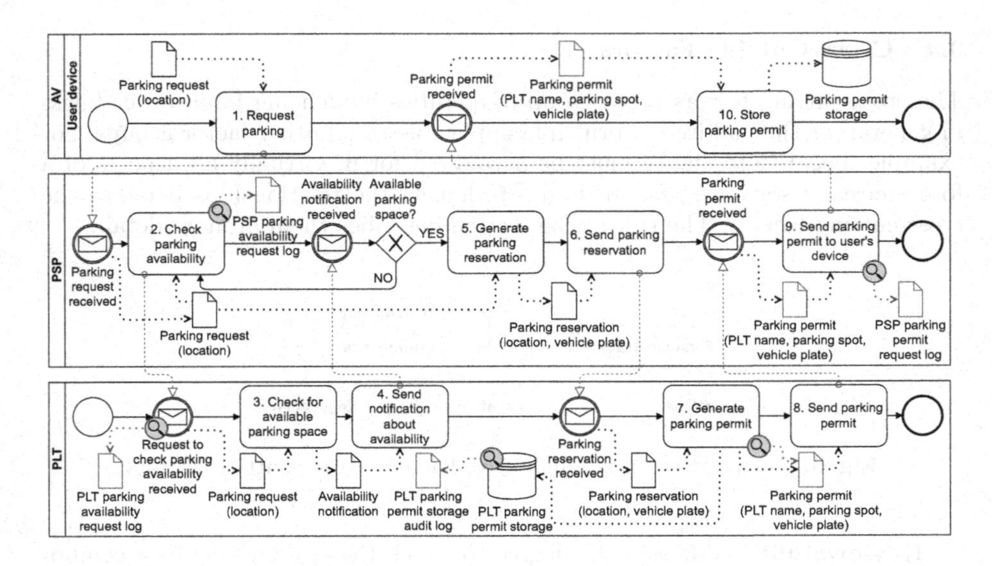

Fig. 7. Automated Valet Parking (AVP) Scenario: Issuing a Permit

- **Autonomous Vehicle (AV)**: Representing the vehicle. It contains a User Device that handles communication with the Parking Service Provider and stores the issued parking permit. From the forensic readiness point of view, the entity is semi-cooperative, meaning that evidence is fully under the user's custody but could be obtained if it benefits the user.
- **Parking Service Provider (PSP)**: Representing a contact point of the parking service deployed in the cloud. It acts as an intermediary between AV and Parking Lot Terminals. It orchestrates the search, reservation, and parking permit delivery. As it is entirely under the stakeholder's control, the entity is cooperative, and its data can be used during an investigation.
- **Parking Lot Terminal (PLT)**: Representing an edge IoT device physically located at the parking lot, which provides relatively easy physical access. It controls access to the particular parking lot by issuing and checking the parking permits and is the source of authoritative data regarding the parking lot. The entity is cooperative and under stakeholder control.

The BPMN process model in Fig. 7 describes a nominal execution of the scenario. It also highlights several sources of potential evidence generated during the execution as part of the business process and logging. Still, there are several deficiencies in effectively investigating incidents regarding this system. While the risk of these incidents could be mitigated by deploying security controls, they do not entirely prevent them. In this sense, forensic readiness can complement security [15], where it handles situations where security controls fail.

4.1 Forensic Readiness Scenarios

The scenario-based planning is utilised in general forensic readiness [9,38]. Also, the forensic-ready risk management approach FR-ISSRM [14] defines *Forensic Readiness Scenario* (FR Scenario) as a focal point for the assessment, aiming to elicit *Forensic Readiness Requirements* (FR Requirements). The FR Scenario is a combination of a *Risk*, *Forensic Readiness Goal* (FR Goal) capturing the reason why it is considered, and *Evidence* relevant to investigate the occurrence of Risk given the FR Goal. The FR Requirements enhance the FR Scenarios to aid the investigation (e.g., adding missing evidence, enhancing their quality). Subsequently, the FR Requirements are implemented by concrete *Forensic Readiness Controls* (FR Controls). The BPMN4FRSS [16] notation captures the FR-ISSRM concepts as a process model to assist the FR Scenario assessment.

Table 2. Forensic Readiness Scenarios

	S1: Permit injection	S2: Covert disruption	S3: Permit repudiation
Risk	A malicious insider access PLT injects a Parking permit into the PLT parking permit storage due to their ability to access it. Leading to a loss of parking permit integrity.	An attacker intercepts the parking availability requests to PLT and forges a negative reply at random times due to insecure protocol usage, leading to a loss of reservation process availability.	A dishonest customer repudiates a Parking permit received from PSP, demanding a reimbursement, leading to a financial loss.
FR Goal	Prove loss of parking permit integrity	Prove loss of reservation process availability	Defend against repudiation of parking permit
Evidence	– PLT parking permit storage audit log – Parking permit	– PSP parking availability request log – PLT parking availability request log	– Parking permit – PSP parking permit request log

For this paper, three FR Scenarios are considered, described in Table 2, capturing incidents to investigate. Figure 7 depicts their composite BPMN4FRSS model. However, through their assessment, several inadequacies can be found:

S1: How the *Parking permit* was stored cannot be determined, as the *PLT parking permit storage audit log* is created both in nominal execution and during the incident. Moreover, it could be tampered with by the attacker.
S2: The link between the *PSP parking availability request log* and the *PLT parking availability request log* uses only circumstantial data (time, IP address).
S3: Correct delivery of the *Parking permit* cannot be proven only by the *PSP parking permit request log.*

The identified inadequacies can be addressed by appropriate FR Requirements. That means manifesting a particular forensic readiness quality factor.

4.2 Manifestation of the Factors

Table 3 contains the description of the FR Requirements, addressing the found inadequacies. The correctness of their implementation by FR Control can be verified due to the establishment of the minimal acceptable value of a metric.

Table 3. Forensic Readiness Requirements

	S1.FRR1	S1.FRR2	S2.FRR1	S3.FRR1
Factor	Completeness	Non-Disputability	Utilisability	Availability
Criterion	Known scenarios of parking permit generation tampering shall contain Evidence.	PLT parking permit storage audit log shall be non-disputable.	PLT parking availability request log shall be linkable with PSP's potential evidence.	Evidence of Parking permit reception shall be available.
Metric	Scenario Coverage	Relative Evidentiary Value	Number of links	Existence
Requirement	Known scenarios of parking permit generation tampering shall contain Evidence, with at least 90% Scenario Coverage.	PLT parking permit storage audit log shall be non-disputable, with at least 2 Relative Evidentiary Value.	PLT parking availability request log shall be linkable with PSP's potential evidence with at least 1 link.	Evidence of Parking permit reception shall be available for at least 99.9% of delivered Parking permits.

The FR Requirements are established by the following approach. First, a factor (or subfactor) referring to the inadequacy or enhancement of the FR Scenario is selected from the reference model. Then a criterion is defined, which maps the selected factor to the concrete inadequacy of the FR Scenario. Essentially, an FR Control that meets such criteria shall resolve the inadequacy. However, that is practically impossible to reach and verify (e.g., complete satisfaction of "System shall be available" criterion is not feasible).

Then, a metric, which quantifies the factor, is selected to define the minimum acceptable value for the criterion. The value needs to be realistic [29]. For the AVP, the following metrics are used: (1) Scenario Coverage [14], a model-based metric quantifying the coverage of Evidence, (2) Relative Evidentiary Value [14], quantifying the confidence in the Evidence, (3) Number of links with the non-circumstantially linkable Evidence, and (4) Existence which is a factor of expected and actual occurrences. Together, the criterion and its minimum acceptable value make up the requirement.

As factors can aggregate multiple subfactors, the metric and control can address only their subset. It is acceptable, as the satisfaction of a requirement is tied to the metric, which could be influenced differently by different subfactors. For example, to satisfy S1.FRR2 is enough to increase the Redundancy by an independent copy. Otherwise, the requirement might target a subfactor directly.

5 Conclusion

This paper presented a forensic readiness qualitative factor reference model as a foundation for formulating verifiable requirements. The model was shown to supplement the FR-ISSRM, an existing forensic-ready risk management approach, by filling the gap in defining the concrete form of the Forensic Readiness Requirements. Consequently, the set research questions are answered as follows:

RQ1: *What are the factors that form the forensic readiness requirements?* Based on the inputs, a reference model of forensic readiness qualitative factors was established. The factors are organised in a taxonomy, describing desired qualities of a forensic-ready software system. Notably, several factors (e.g., Non-Disputability) share a common aim, which can be addressed by their combination, creating sub-factors. Such hierarchical organisation is important for enabling alternative implementations and applying metrics. While the aim was to make the model consistent, its completeness is not guaranteed. However, adding new factors, sub-factors, or their further decomposition is possible.

RQ2: *How can the factors be manifested as a concrete requirement?* The reference model was used to define forensic readiness requirements based on the results from forensic-ready risk management. Concretely, an automated valet parking service with several inadequacies in forensic readiness was presented. These were transformed into specific criteria with associated minimum acceptable values, forming verifiable requirements. Moreover, the model's hierarchical nature does not limit the implementing controls other than meeting the minimum acceptable value of a given metric.

Future Work branches into three directions, which tackle a challenge in the assurance of forensic-ready software systems. First and foremost, the reference model needs to be further validated in a more in-depth study involving empirical analysis of a system. Secondly, richer metrics shall be explored to ascertain how to measure the factors and identify possible gaps. Thirdly, checklist-like guidelines of the systems based on the reference model shall be investigated.

Acknowledgement. This research was supported by ERDF "CyberSecurity, CyberCrime and Critical Information Infrastructures Center of Excellence" (No. CZ.02.1.01/0.0/0.0/16_019/0000822). It was also co-founded by the European Union under Grant Agreement No. 101087529. Views and opinions expressed are however those of the author(s) only and do not necessarily reflect those of the European Union or European Research Executive Agency. Neither the European Union nor the granting authority can be held responsible for them.

References

1. Ab Rahman, N.H., Glisson, W.B., Yang, Y., Choo, K.K.R.: Forensic-by-design framework for cyber-physical cloud systems. IEEE Cloud Comput. **3**(1), 50–59 (2016)
2. Alrajeh, D., Pasquale, L., Nuseibeh, B.: On evidence preservation requirements for forensic-ready systems. In: Proceedings of the 2017 11th Joint Meeting on Foundations of Software Engineering, ESEC/FSE 2017, pp. 559–569. ACM (2017)
3. Alrimawi, F.: Software engineering for forensic-ready cyber-physical systems. Theses, University of Limerick (2020). https://hdl.handle.net/10344/9294
4. Alrimawi, F., Pasquale, L., Nuseibeh, B.: Software engineering challenges for investigating cyber-physical incidents. In: 2017 IEEE/ACM 3rd International Workshop on Software Engineering for Smart Cyber-Physical Systems, pp. 34–40 (2017)

5. Bajramovic, E., Waedt, K., Ciriello, A., Gupta, D.: Forensic readiness of smart buildings: preconditions for subsequent cybersecurity tests. In: 2016 IEEE International Smart Cities Conference, pp. 1–6 (2016)
6. Bierska, A., Buhnova, B., Bangui, H.: Integrated checklist for architecture design of critical software systems. Ann. Comput. Sci. Inf. Syst. **31**, 133–140 (2022)
7. Casey, E.: Error, uncertainty and loss in digital evidence. Int. J. Digit. EVid. **1** (2002)
8. Casey, E.: Digital Evidence and Computer Crime, 3rd edn. Academic Press, Waltham (2011)
9. CESG: Good Practice Guide No. 18: Forensic Readiness. Guideline, National Technical Authority for Information Assurance, United Kingdom (2015)
10. Cosic, J., Baca, M.: Do we have full control over integrity in digital evidence life cycle? In: Proceedings of the ITI 2010, 32nd International Conference on Information Technology Interfaces, pp. 429–434. IEEE (2010)
11. Daubner, L., Buhnova, B., Pitner, T.: Forensic experts' view of forensic-ready software systems: a qualitative study. J. Softw. Evol. Process (2023, under review)
12. Daubner, L., Macak, M., Buhnova, B., Pitner, T.: Towards verifiable evidence generation in forensic-ready systems, pp. 2264–2269. IEEE (2020)
13. Daubner, L., Macak, M., Buhnova, B., Pitner, T.: Verification of Forensic Readiness in Software Development: A Roadmap, pp. 1658–1661. ACM (2020)
14. Daubner, L., Macak, M., Matulevičius, R., Buhnova, B., Maksović, S., Pitner, T.: Addressing insider attacks via forensic-ready risk management. J. Inf. Secur. Appl. **73**, 103433 (2023)
15. Daubner, L., Matulevičius, R.: Risk-oriented design approach for forensic-ready software systems. In: The 16th International Conference on Availability, Reliability and Security. ACM (2021)
16. Daubner, L., Matulevičius, R., Buhnova, B., Pitner, T.: Business process model and notation for forensic-ready software systems. In: Proceedings of the 17th International Conference on Evaluation of Novel Approaches to Software Engineering, pp. 95–106. SciTePress (2022)
17. Dubois, É., Heymans, P., Mayer, N., Matulevičius, R.: A systematic approach to define the domain of information system security risk management. In: Nurcan, S., Salinesi, C., Souveyet, C., Ralyté, J. (eds.) Intentional Perspectives on Information Systems Engineering, pp. 289–306. Springer, Heidelberg (2010). https://doi.org/10.1007/978-3-642-12544-7_16
18. Dzurenda, P., et al.: Privacy-preserving solution for vehicle parking services complying with EU legislation. PeerJ Comput. Sci. **8**, e1165 (2022)
19. Erol-Kantarci, M., Mouftah, H.T.: Smart grid forensic science: applications, challenges, and open issues. IEEE Commun. Mag. **51**(1), 68–74 (2013)
20. Firesmith, D.: Common concepts underlying safety, security, and survivability engineering. Technical report, CMU/SEI-2003-TN-033, Software Engineering Institute, Carnegie Mellon University, Pittsburgh, PA (2003)
21. Firesmith, D.: Engineering safety requirements, safety constraints, and safety-critical requirements. J. Object Technol. **3**(3), 27–42 (2004)
22. Firesmith, D., et al.: Engineering security requirements. J. Object Technol. **2**(1), 53–68 (2003)
23. Grispos, G., García-Galán, J., Pasquale, L., Nuseibeh, B.: Are you ready? Towards the engineering of forensic-ready systems. In: 2017 11th International Conference on Research Challenges in Information Science, pp. 328–333 (2017)

24. Grispos, G., Glisson, W.B., Choo, K.K.R.: Medical cyber-physical systems development: a forensics-driven approach. In: 2017 IEEE/ACM International Conference on Connected Health: Applications, Systems and Engineering Technologies, pp. 108–113 (2017)
25. Grobler, C.P., Louwrens, C.P.: Digital forensic readiness as a component of information security best practice. In: Venter, H., Eloff, M., Labuschagne, L., Eloff, J., von Solms, R. (eds.) SEC 2007. IIFIP, vol. 232, pp. 13–24. Springer, Boston, MA (2007). https://doi.org/10.1007/978-0-387-72367-9_2
26. Grobler, C., Louwrens, C., von Solms, S.: A framework to guide the implementation of proactive digital forensics in organisations. In: 2010 International Conference on Availability, Reliability and Security, pp. 677–682 (2010)
27. Hitchcock, B., Le-Khac, N.A., Scanlon, M.: Tiered forensic methodology model for digital field triage by non-digital evidence specialists. Digit. Invest. **16**, S75–S85 (2016). dFRWS 2016 Europe
28. Jürjens, J.: UMLsec: extending UML for secure systems development. In: Jézéquel, J.-M., Hussmann, H., Cook, S. (eds.) UML 2002. LNCS, vol. 2460, pp. 412–425. Springer, Heidelberg (2002). https://doi.org/10.1007/3-540-45800-X_32
29. Matulevičius, R.: Fundamentals of Secure System Modelling. Springer, Cham (2017). https://doi.org/10.1007/978-3-319-61717-6
30. Matulevičius, R., Mayer, N., Mouratidis, H., Dubois, E., Heymans, P., Genon, N.: Adapting secure tropos for security risk management in the early phases of information systems development. In: Bellahsène, Z., Léonard, M. (eds.) CAiSE 2008. LNCS, vol. 5074, pp. 541–555. Springer, Heidelberg (2008). https://doi.org/10.1007/978-3-540-69534-9_40
31. Mayer, N.: Model-based Management of Information System Security Risk. Theses, University of Namur (2009). https://tel.archives-ouvertes.fr/tel-00402996
32. McKemmish, R.: When is digital evidence forensically sound? In: Ray, I., Shenoi, S. (eds.) DigitalForensics 2008. ITIFIP, vol. 285, pp. 3–15. Springer, Boston, MA (2008). https://doi.org/10.1007/978-0-387-84927-0_1
33. Menezes, A.J., Van Oorschot, P.C., Vanstone, S.A.: Handbook of Applied Cryptography. CRC Press, Boca Raton (2018)
34. Mohammadi, N.G., et al.: An analysis of software quality attributes and their contribution to trustworthiness. In: Proceedings of the 3rd International Conference on Cloud Computing and Services Science, pp. 542–552. SciTePress (2013)
35. Pasquale, L., Alrajeh, D., Peersman, C., Tun, T., Nuseibeh, B., Rashid, A.: Towards forensic-ready software systems. In: Proceedings of the 40th International Conference on Software Engineering: New Ideas and Emerging Results, pp. 9–12. ACM (2018)
36. Richter, J., Kuntze, N., Rudolph, C.: Security digital evidence. In: 2010 Fifth IEEE International Workshop on Systematic Approaches to Digital Forensic Engineering, pp. 119–130 (2010)
37. Rivera-Ortiz, F., Pasquale, L.: Automated modelling of security incidents to represent logging requirements in software systems. In: Proceedings of the 15th International Conference on Availability, Reliability and Security. ACM (2020)
38. Rowlingson, R.: A ten step process for forensic readiness. Int. J. Digit. Evid. **2**, 1–28 (2004)
39. Sethia, A.: Rethinking admissibility of electronic evidence. Int. J. Law Inf. Technol. **24**(3), 229–250 (2016)
40. Shostack, A.: Threat Modeling: Designing for Security. Wiley, Hoboken (2014)

41. Soomro, I., Ahmed, N.: Towards security risk-oriented misuse cases. In: La Rosa, M., Soffer, P. (eds.) BPM 2012. LNBIP, vol. 132, pp. 689–700. Springer, Heidelberg (2013). https://doi.org/10.1007/978-3-642-36285-9_68
42. Tan, J.: Forensic readiness. Technical report, @stake, Inc. (2001)
43. Ćosić, J., Bača, M.: (Im)proving chain of custody and digital evidence integrity with time stamp. In: The 33rd International Convention MIPRO, pp. 1226–1230 (2010)

Monitoring and Recommending

Monitoring Object-Centric Business Processes: An Empirical Study

Lisa Arnold(✉), Marius Breitmayer, and Manfred Reichert

Institute of Databases and Information Systems, Ulm University, Ulm, Germany
{lisa.arnold,marius.breitmayer,manfred.reichert}@uni-ulm.de

Abstract. Monitoring dashboards provide an appropriate way of presenting a multitude of information on running business processes to involved actors. Essential components of a monitoring dashboard are charts that visualise this information in an aggregated, intuitive and useful way. A popular representative is the Sunburst Chart, which constitutes a pie chart with several colour-coded circles that can visualise a hierarchical structure. This visualisation technique seems to be particularly suited for monitoring object-centric processes. In this paper a procedure for automatically deriving a sunburst chart from the patterns of a relational process structure, describing an object-centric process, is presented. To investigate the readability, comprehension, and general acceptance of sunburst chart in the context of a monitoring object-centric process, an empirical study with 157 participants was conducted. As key observation of this study, the majority of the participants can read and comprehend the sunburst chart very well, e.g., on average more than 90% of the multiple-choice questions were answered correctly. Overall, sunburst charts offer promising perspective for the monitoring of large object-centric process.

Keywords: business process monitoring · dashboard · object-centric process · sunburst chart · empirical study

1 Introduction

In order to remain competitive, companies need to continuously improve their dynamically evolving processes and eliminate process-related problems. Monitoring these dynamic processes is a suitable way of recognising emerging problems and errors at an early stage [1]. In this context, visual dashboards seem to be an appropriate means for process monitoring [2]. However, a significant challenge from the perspective of an end-user of a process monitoring component concerns the readability and comprehensibility of dynamic dashboards. The latter may comply a multitude of information on one screen. This information, in turn, should be understandable, legible and properly structured for all user groups. Otherwise, the dashboard is only of low value for end-users.

In principle, a sunburst chart [3] allows presenting the progress, status, and current risks of an object-centric process, which may comprise interacting objects of different types. First, the relational process structure and its patterns need to

S. Nurcan et al. (Eds.): RCIS 2023, LNBIP 476, pp. 327–342, 2023.
https://doi.org/10.1007/978-3-031-33080-3_20

be identified. Second, rules to transfer this process structure into the sunburst chart need to be developed. Third, the created sunburst chart needs to be evaluated regarding its comprehensibility, readability, and acceptance by end-users through an empirical study.

The remainder of this paper is structured as follows. Section 2 gives backgrounds on the process paradigm presumed by this paper, i.e., object-centric business processes. Section 3 defines two fundamental research questions addressed by this work. Section 4 identifies the pattern of the relational process structure that describes the object lifecycle processes involved in an object-centric process and defines a procedure to translate them into a sunburst chart. In Sect. 5, the research design of the empirical study is described. Section 6 analyses, evaluates, and discusses the answers of the 157 participants of the study. Related work is discussed in Sect. 7 and Sect. 8 concludes the paper.

2 Backgrounds

We illustrate basic concepts and artefacts of object-centric processes along a running example. This example describes a recruitment business process from the perspective of a human resource management department. The example is also used in the context of the empirical study to visualise the evaluated process as well as the sunburst chart.

Example 1: (Recruitment Business Process) To fill a vacancy, management advertises and publishes it (e.g. on the company's homepage or on a job portal). Afterwards, applicants may prepare their job application and then submit it to the company. All received applications are then assessed by at least 3 reviewers. For this purpose, the application documents are sent to suitable employees (i.e., domain experts). The latter need to review the application until a given date. As a result, they provide an assessment stating whether or not the applicant is suited for the job. If the majority of the reviewers consider the respective applicant being suitable for the job, he or she is invited for an interview. In case the applicant does not have the majority of votes, two additional reviews may follow-up, who evaluate the application as well. If the majority of the reviewers do not considers the applicant being suitable for the job, he or she is rejected. Otherwise, the applicant is invited for an interview. In this case, an interview is prepared, i.e. an appointment date is agreed upon and questions are formulated. The interview may then be conducted and a decision be made on whether or not to hire the applicant. Once the position is filled, the job offer is closed. Subsequent applicants must then be rejected.

Object-centric business process are based on interacting objects. The interactions between these objects, including the object hierarchy and the semantic relations (including cardinalities) between the objects is defined by a *Relational*

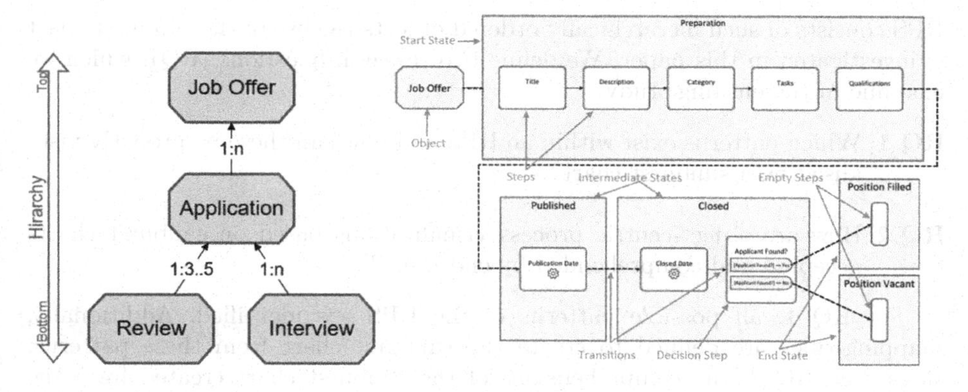

Fig. 1. RPS from Example 1. **Fig. 2.** Lifecycle of *Job Offer* from Example 1.

Process Structure (RPS) [4]. A simple view of the RPS without semantic relations is shown in Fig. 1. This RPS represents the recruitment business process from Example 1 with the top-level object *Job Offer* and three lower-level objects.

During run time, object instances are created from the object type defined in the RPS [5]. The behaviour of each object instance is defined by a *lifecycle*. Consider the object *Job Offer* as shown in Fig. 2. Furthermore, a lifecycle is defined in terms of states. Thereby, each lifecycle has exactly one start state (*Preparation*) and at least one end state (*Position Filled* and *Position Vacant*), as well as an arbitrary number of intermediate states (*Published* and *Closed*). The states of a lifecycle may be refined by steps. In turn, these steps correspond to object attributes (e.g., *Title*) defined by the RPS. During run time, form sheets are auto-generated from the states. Hence, the steps represent the data input fields of the respective form [6].

3 Research Questions

Monitoring object-centric processes during run time is a challenging task:

- A varying number of object instances and, thus, of lifecycle processes interactions may exist.
- The overall relation process structure is not known at build time, but dynamically evolves during execution (i.e., run time).
- Due to dynamic changes of object-centric processes (as supported by, for example, PHILharmonicFlows [7]) their behaviour may vary significantly.

Overall the challenge is to enable the monitoring of dynamically evolving and varying process structures that may comprise hundreds or thousands of interacting object lifecycle processes. Note that this is not trivial as an ambiguous or non-readable visualisation will be not accepted by end-users. Sunburst charts seem to be well suited for presenting hierarchically structured objects. As the

RPS consists of such hierarchically ordered objects the use of the sunburst chart is investigated in this paper. We define two research questions (RQ), which we examine in the ensuing study:

RQ 1 Which patterns exist within an RPS and how can these be properly visualised by a sunburst chart?

RQ 2 How are object-centric process visualisations based on sunburst charts accepted and comprehended by end-users?

For RQ 1, all possible patterns of the RPS are identified. Additionally, mapping-rules are defined to create the sunburst chart from these patterns. Regarding RQ 2, the comprehensions of the sunburst chart created from the defined mapping-rules (cf. RQ 1) is investigated. This includes the evaluation of whether or not sunburst charts are well accepted by end-user in general.

4 Creating Charts for Object-Centric Processes

We first address RQ 1 and discus how a sunburst chart can be automatically derived from a relational process structure (RPS), which describes a potentially large object-centric process. Basically, a sunburst chart can be derived from an RPS by applying the following three mapping-rules. First, the *innermost ring* is created as the core of the sunburst chart. Second, the *additional rings* are created in the hierarchical order set out by the RPS from the top to bottom. Third, the sunburst chart created at design time is transformed into one sunburst chart reflecting all object lifecycle process instances at run time.

Innermost Ring. First of all, the base of the sunburst chart respectively the innermost ring needs to be set.

Mapping-Rule 1: Innermost Ring

The top-level object(s) of the RPS define(s) the innermost ring of the sunburst chart. Two cases need to be distinguished:

RPS with exactly one top-level object: The top-level object is represented by the innermost ring of the sunburst chart.
RPS with more than one top-level objects: The innermost ring comprises all top-level objects in equal parts.

In the following, two examples for creating of the innermost ring of a sunburst chart based on Mapping-Rule 1 are given: the innermost ring of a sunburst chart based on an RPS with one top-level object (cf. Fig. 3) and the innermost ring of a sunburst chart based on an RPS with three top-level objects (cf. Fig. 4). For example, *Job Offer* (cf. Fig. 1) is the top-level object and therefore represents *Object A* in Fig. 3.

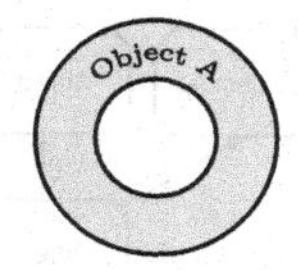

Fig. 3. Innermost ring of the sunburst chart with one top-level *Object A*.

Fig. 4. Innermost ring of the sunburst chart with three top-level objects.

Additional Rings. All other objects of the RPS are added as additional rings to the initialised sunburst chart in hierarchical order beginning with the second-level object(s). For each additional level, the pattern (combination and relation of a group of objects) of the RPS has to be identified to apply the corresponding rule. A pattern consists of two consecutive hierarchy levels, the higher level containing the parent object(s) and the lower level with the child object(s). The relevant patterns are depicted in Fig. 5. Pattern 1 corresponds to a *chain* with one parent object for one child object (cf. Fig. 5a), Pattern 2 to one *common parent* object for multiple child objects (cf. Fig. 5b), and Pattern 3 to multiple parent objects for one *common child* object (cf. Fig. 5c).

Mapping-Rule 2: Additional Ring(s)

For each hierarchy level, from high to low, an additional outer (fraction of a) ring is added to the sunburst chart initially created with Mapping-Rule 1. For this purpose, Patterns 1-3 (consisting of one or more parents) are applied as follows:

Pattern 1 (Chain): The child object is added on top of the parent object as the next outer ring.
Pattern 2 (Common parent): The child objects are added on top of the common parent object as the next outer ring sharing the parent object's fraction ring equally.
Pattern 3 (Common child): The common child object is added on top of all parent objects as the next outer ring stretching over all parent objects' fraction rings.
No child: If no further child object exists, no additional ring will be added on top of the parent.

In the following, three examples of simple RPSs, each covering one of the patterns, are shown to illustrate the mapping-rules. The inner ring is created according to Mapping-Rule 1, whereas the outer ring is based on Mapping-Rule 2. Note that these patterns may be combined to form larger and more complex RPS.

Mapping RPS Instances to a Sunburst Chart at Run Time. At run time, multiple object instances can be created for each object of an RPS. The

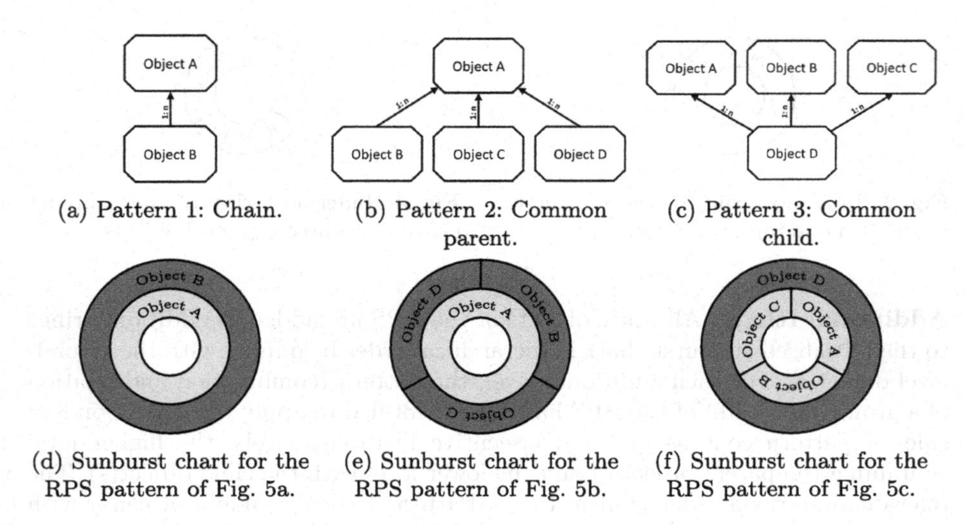

(a) Pattern 1: Chain. (b) Pattern 2: Common parent. (c) Pattern 3: Common child.

(d) Sunburst chart for the RPS pattern of Fig. 5a. (e) Sunburst chart for the RPS pattern of Fig. 5b. (f) Sunburst chart for the RPS pattern of Fig. 5c.

Fig. 5. Possible patterns within an RPS and their corresponding sunburst charts.

following mapping-rule sets out how the sunburst chart previously created for an RPS can be used to represent all object instances at run time (mind the Limitations).

Mapping-Rule 3: Run Time

The sunburst chart created at design time (based on Mapping-Rule 1 and 2) can be used to represent an RPS instance at run time. Thus, the sunburst chart is transformed hierarchically beginning with the innermost ring. Thereby, each (fraction of a) ring representing an object is split evenly into as many parts as there are object instances. If no instance of an object exists, the (fraction of a) ring remains empty and is not shown.

Figure 6a shows the sunburst chart created for the RPS of the running example (cf. Fig. 1) at design time. Figure 6b, in turn, shows the sunburst chart of a possible instance of this RPS at run time. In this example, four *Job Offer* instances *JO1* - *JO4* exist. Further, for *JO1* there are two *Application* instances *A1* - *A2*. Finally, for *A2* three *Review* instances *R3* - *R5* and one *Interview* instance *I1* exist. Note, that at run time different colours are used by the sunburst chart to cluster instances belonging to the same *Job Offer* instance, whereas at design time colours are used to visually distinguish the different rings from each other.

Limitations. Concerning the described mapping approach two limitations exist. However, the first is (currently) not implemented in PHILharmonicFlows anyway, and experience shows that the second rarely occurs and can usually be avoided by suitable restructuring.

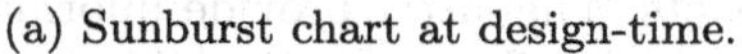
(a) Sunburst chart at design-time.

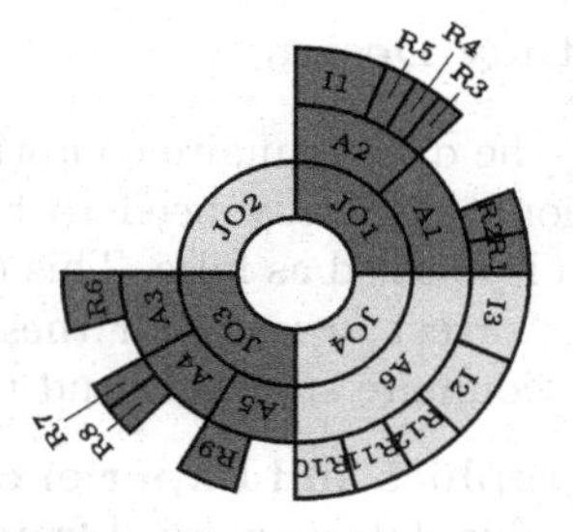

(b) Example sunburst chart at run time.

Fig. 6. Sunburst charts of the example RPS *Job Offer* (cf. Fig. 1).

m:n relations (multiple common parents with multiple common children) between two objects of an RPS instance cannot be realised based on Mapping-Rule 3. For example, more than two instances of a child object would need to overlap to match to more than two instances of a parent object and, thus, cannot be displayed in a 2-dimensional chart. However, this limitation can be neglected as the only existing object-centric business process management system (i.e., PHILharmonicFlows [7]) does not implement such relations. Furthermore, the combination of the *common parent* and the *common child* patterns, (i.e., multiple common child objects with multiple common parents objects) forms an m:n relation as well. They, therefore, cannot be represented with a sunburst chart at design time based on Mapping-Rule 2.

An object that can be associated with several hierarchical levels can not be included in a sunburst chart as it would need to be placed in multiple rings simultaneously. As example consider an RPS with objects A, B and C with A being the parent of B and C, and B being the parent of C. In this case, C can be associated with both hierarchical levels 2 and 3.

5 Research Methodology

5.1 Data Collection Method

The readability and comprehensibility of sunburst charts in the context of monitoring object-centric process is evaluated in an empirical study. The latter is performed based on an anonymous online questionnaire[1] that leverages the web-based tool `Unipark` for data collection. The study is available in both German and English (with the English part being provided first and the German translation being given in brackets). Therefore, language options do not differ with respect to content or structure. The questionnaire was available over a period of one month.

[1] The questionnaire and the responses from the 157 participants are available on Researchgate: https://www.researchgate.net/publication/368296351_Comprehension_of_sunburst_charts.

5.2 Study Design

In total, the questionnaire comprises 31 different questions, whereby 5 questions are optional and only need to be answered when the respective dichotomous question is marked as false. This results in a total of 28 questions, if participants give the correct answers for these dichotomous true-false questions. Moreover, the questionnaire structure and the type of questions are based on [8].

Demographics and Experience: In the first part of the online questionnaire, demographic data is queried from the participants. The latter provide information about the participants and their background. In detail, the demographic questions query information on gender, age, current profession (student, academic, and others), and their professional field, e.g., MINT (Mathematics - Computer Science - Natural Science - Technology), services, healthcare, economics, and others. Furthermore, the experiences of the participants with respect to process modelling, entity-relationship (ER) models, and (business process) monitoring tools are queried. Based on these data we can divide the participants into an expert group (experience in process modelling and ER) and a non-expert group.

Comprehending RPS and Sunburst Charts: The next section of the questionnaire comprises three blocks, of which each comprises three statements about a given figure. With the dichotomous questions, participants are asked to answer whether the statement is correct (true) or incorrect (false). The three blocks are described in the following.

Block 1: RPS (for this block, Fig. 1 is depicted in the questionnaire)

a) At most five Reviews are written for each *Application.*
b) For every *Job Offer* there is at least one *Application.*
c) At run time, the number of *Reviews* for the total process is arbitrary.

Based on Fig. 1 each participant has to answer each of the three statements with either true or false. Statement a) investigates whether the participants recognise the cardinality in the RPS. More precisely, this question refers to the cardinality between *Application* and *Review* (i.e., 1 : 3..5) – each *Application* requires three to five *Reviews*. Statement b) investigates whether the participants recognise that no *Application* exists when initially creating a job offer application. Note that this is indicated through the use of cardinality 1 : n. Statement c) investigates whether the participants are able to recognise from the RPS that the number of *Reviews* may be arbitrary at run time as zero to infinite *Applications* with zero to five *Reviews* per application exist.

Block 2: Sunburst chart at design time (for this block, Fig. 6a is depicted in the questionnaire)

a) The associated process of the given sunburst chart has a total of 5 different objects.
b) The *Job Offer* has a direct top-down relationship to the *Interview.*
c) *Application* has two child objects.

The second block considers the sunburst chart at design time. Statement a) investigates whether the participants are able to recognise the number of objects directly from the sunburst chart. Statement b) investigates whether the participants recognise the hierarchies within the sunburst chart at design time. Statement c) investigates whether the participants recognise the hierarchies within the sunburst chart at design time in connection with parent and child relations.

Block 3: Sunburst chart at run time (for this block, Fig. 6b is depicted in the questionnaire)

a) There are 15 *Reviews* throughout the process.
b) A total of 4 *Reviews* were carried out for *Job Offer* 3.
c) Overall, there are more *Reviews* than *Interviews.*

The third block investigates the legibility of a sunburst chart that comprises multiple instantiated objects of different levels at run time. Concerning Statement a) the total number of *Reviews* needs to be identified. In Statement b) the creation of corresponding object instances is vetted. In particular, the number of *Reviews* for object instance *Job Offer 3* needs to be determined. Finally, concerning Statement c), the participants need to either take position on whether more *Interviews* than *Reviews* may exist at any time during run time or simply count the *Reviews* and *Interviews* in the given figure. Note that the former cannot be the case as at least three *Reviews* are needed to trigger an*Interview.*

Comprehending Sunburst Charts During Run Time: A sunburst chart may be used to visualise the current state of the entire object-centric process at run time by individually colouring object instances. For this purpose, the sunburst chart may be used in different ways. One use case is to visualise the current progress of each single object instance. In Fig. 7a, an example of a sunburst chart visualising the progress of the various objects of a process instance is shown. Another use case is to visualise the current status of each business object instance. In this context, the term status may be interpreted in different ways. The status may be considered as the state of execution, e.g., with options *waiting*, *running*, *completed*, and *skipped.* Or it may be considered as the state of punctuality with a range from *on time* to *on risk* to *overdue* (cf. Fig. 7b).

This section of the questionnaire consists of two parts of which one addresses the sunburst chart for visualising the progress of each object instance (cf. Fig. 7a) and the other addresses the sunburst chart visualising the punctuality of each object instance (cf. Fig. 7b). In both parts, the respective figure is shown and two questions are asked without any text-based explanation. First, participants are asked what kind of information the depicted sunburst chart displays. For this purpose, a closed multiple choice question with the following options is provided:

- Overview of the execution status (waiting, running, completed, ..) of each process instances.
- Progress of each process instance
- Overview of the schedule status (on time, risk or overdue) of each process instances.

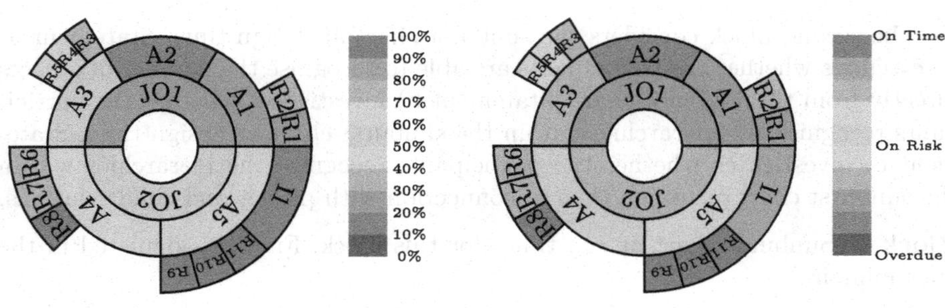

(a) Sunburst chart visualising the current progress (from 0% to 100%) for each object instance.

(b) Sunburst chart visualising the current status (on time, on risk or overdue) for each object instance.

Fig. 7. Two use cases of sunburst charts during run time based on the example RPS *Job Offer* (cf. Fig. 1).

The second question addresses the intuitive comprehensibility and the suitability of the shown sunburst chart. Here, a 5 point Likert-scale ranging from 1 (*completely false*) to 3 (*neutral*) to 5 (*completely true*) is used.

Comprehensibility of the Transformation of RPSs to Sunburst Charts: In the last section of the questionnaire, the comprehensibility concerning the transformation of an RPS to a sunburst chart is investigated. The section is divided into five similar parts. Each part includes two figures side by side. Thereby, the first figure shows an RPS and the second one a sunburst chart. An example can be seen in Fig. 8. Through the use of dichotomous questions, participants are asked to answer whether the sunburst chart was correctly created from the given RPS (true) or not (false). If the participant states that the RPS and the sunburst chart did not correspond to each other, an explanation may be provided by the participants in an optional text field.

Note that the participants neither have the mapping-rules at their disposal nor any previous training in general. This design decision was made as the study aims to investigate the intuitive comprehensibility and any previous training could falsify the study results.

5.3 Data Analysis Method

The data analysis and validation of the study is based on the checklist defined in the empirical cycle from [9]. All collected data are analysed and evaluated in a structured way to answer RQ 2. For data analysis, we apply the methodology presented in [8,9]. In detail, an exploratory analysis using *Cross Tabulations* to compare the quantitative results from different participant groups is applied. Expand with open-ended questions are added in the last part of the questionnaire to get the "why" of the data responses. Most of the participants are students from Ulm University. About half are students from the first bachelor semester in economics with no experience in process modelling and ER diagrams in general.

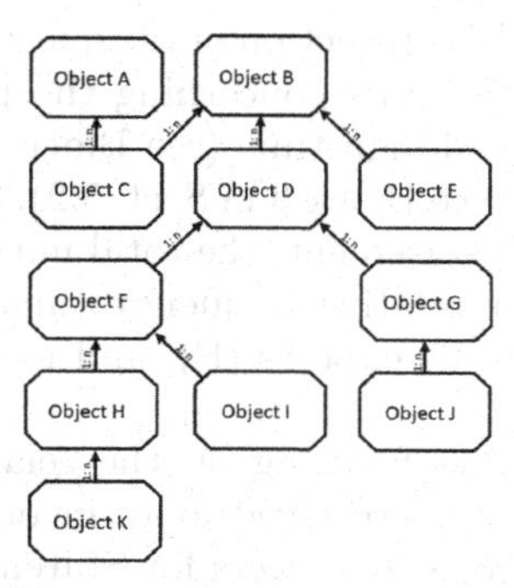

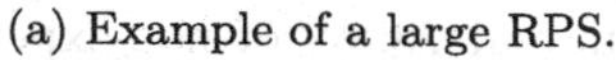
(a) Example of a large RPS.

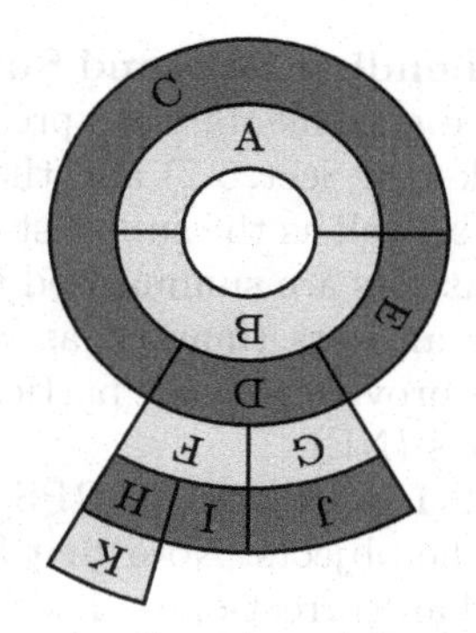

(b) Associated sunburst chart of the RPS shown in Fig. 8b.

Fig. 8. Example of two diagrams used to investigate the comprehensibility of the transformation from an RPS to a sunburst chart.

The other half are master students with a focus on business process modelling and PhD students with relevant prior knowledge. This selection of participants should avoid confounding variables such as general school education or major age differences as far as possible and at the same time allow for a comparison between experts with domain knowledge (e.g., experience with modelling processes) and non-experts without such prior knowledge. First of all, the answers of all participants are analysed. Then, among the participants the expert group is defined based on the answers given in the demographic section of the questionnaire to allow for a comparison between the results of experts, non-experts, and all participants.

6 Evaluation

Demographics and Experience: Overall,157 participants completed the questionnaire. Thereby, more male (103 | 65.6%) than female (54 | 34.3%) participants took part in the study. The majority of the participants are between 18 and 26 (151 | 96.2%) years old, with their average age being 22.4 years. Most participants study economics (122 | 77.7%). The remaining participants study on MINT (Mathematics - Computer Science - Natural Science - Technology) programs (35 | 22.3%). More than two third (105 | 66.9%) of the participants have made experiences with process models, and more than half of them (91 | 58.0%) have experiences with entity relationship diagrams. Only few participants (22 | 14.0%) are experienced with monitoring tools in general and business process monitoring tools in particular (21 | 13.4%).

For the evaluation, we define an expert group among the participants being familiar with both process models and entity relationship charts (77 | 49.0%). The non-experts are composed of all other participants. In the following, the results from the experts (E) are compared with the ones of the non-experts (NE) to enable a profound analysis.

Comprehending RPS and Sunburst Charts: In this section of the questionnaire three statements were presented to the participants concerning the RPS (see Block 1 in Sect. 5.2) and the sunburst chart at design time (see Block 2 in Sect. 5.2) as well as the sunburst chart at run time (see Block 3 in Sect. 5.2). The correct answers are summarised in Tab. 1. For each statement, the total number of correct answers (marked as #) as well as their relative frequency (marked as %) are provided for all participants (Total), for the experts (E), and for the non-experts (NE).

Block 1 addresses an RPS with the statements focusing on the relation between the objects. Note that Statement 1b) and 1c) received more incorrect answers than correct ones. The high number of incorrect answers for Statement 1b) (e.g., correctly answered by 15.6% of experts and 23.8% of non-experts) indicates that the RPS notion used in our modelling tool (cf. Fig. 1) is not intuitively comprehensible. The familiar notation of relations and cardinalities in ER diagrams might have confused the experts in trying to understand the RPS, whose notation is similar but not the same. The same applies to Statement 1c). Based on these results, the notation of the cardinalities in an RPS should be revised.

Block 2 deals with sunburst charts at design time. Statement 2a) is answered correctly by most participants (~88%) of both groups. The other two statements are correctly answered only by 58.0% and 54.8% of the participants respectively. As a possible explanation of this result, in both statements unknown terminology like *direct top-down relationship* and *child objects* are used, i.e., terminology that might not be intuitively comprehensible. This explanation is supported by the significant difference of 15% correctly answered questions between the experts and the non-experts regarding Statement 2c).

Block 3 deals with sunburst charts at run time, which may be used in a similar way in a real-time monitoring dashboard later. Note that this part of the questionnaire is of particular interest. The high percentages of correct answers for all participant (between 91.1% and 96.2%) indicates that real-time sunburst charts are very well comprehended by participants. This presumed "intuitive" comprehensibility is further supported by the relatively small gap of at maximum 7.4% between expert and non-expert answers.

Table 1. Correct answers of *Comprehension of RPS and Sunburst Charts.*

		1. Block			2. Block			3. Block		
		a)	b)	c)	a)	b)	c)	a)	b)	c)
Total	#	133	31	54	139	91	86	147	143	151
(157)	in %	72.0	19.8	34.4	88.5	58.0	54.8	93.6	91.1	96.2
E	#	58	12	28	68	46	50	75	69	76
(77)	in %	75.3	15.6	36.4	88.3	59.7	65.0	97.4	89.6	98.7
NE	#	55	19	26	71	45	36	72	74	75
(80)	in %	68.8	23.8	32.5	88.8	56.3	45.0	90.0	92.5	93.8

Comprehending Sunburst Charts During Run Time: In this section of the questionnaire, the participants were confronted with two real-world scenario sunburst charts in succession: one visualising *Progress* (cf. Fig. 7a) and one *Status* (cf. Fig. 7b). For each chart, the participants had to identify its content by answering multiple choice questions, of which each offered three options. The results for *Progress* and *Status* are depicted in Table 2 broken down into the number of correct answers in absolute number (#) and in % for all participants, for the group of experts, and for the group of non-experts. Moreover, the participants evaluate the sunburst charts on whether they are intuitively comprehensible using a 5 point Lickert-scale (from 1 (*completely false*) to 3 (*neutral*) to 5 (*completely true*)). The results of these questions with respect to all participants (L_a) as well as to all participants who had correctly answered the corresponding multiple choice question ($L_\#$) can be seen in Table 2 as well.

The results from both sunburst charts differ significantly. Almost 20% more correct answers were given for the *Status* sunburst chart (82.2%) compared to the *Progress* sunburst chart (62.4%). There are several possible explanations of this result. First, as the *Status* sunburst chart was the second chart of this type the participants were facing, they were more experienced with reading *Status* sunburst charts. Second, the provided caption might be more intuitively comprehensible for the second chart (cf. Fig. 7b). Third, the participants are most likely familiar with a progress bar in the context of visualising the progress of a process. Thus, they might not have recognised the advantages of the new visualisation but have noticed its increased complexity. As opposed to the presentation of the *Status* no generally known standard exists. This explanation is supported by the results of the second question, for which the *Progress* sunburst chart is assessed as neutral (~3 in the 5 point Lickert-scale) and the *Status* sunburst chart is rated as being more intuitively comprehensible (~4 in the 5 point Lickert-scale). Finally, experts rated both sunburst charts being more comprehensible than non-experts. Consequently, RQ 2 can be answered positively for the *Status* sunburst chart.

Table 2. Results about *Comprehending Sunburst Charts during Run Time* (#: Number of correct answers; L: Score on a 5 point Likert-scale; a: Number of all answers)

	Progress (cf. Fig. 7a)				Status (cf. Fig. 7b)			
	#	# in %	$L_\#$	L_a	#	# in %	$L_\#$	L_a
Total (157)	98	62.4	2.9	2.9	129	82.2	4.1	3.9
E (77)	47	61.0	3.1	3.1	68	88.3	4.2	4.1
NE (80)	51	63.8	2.7	2.7	61	76.3	4.0	3.7

Comprehensibility of the Transformation of RPSs to Sunburst Charts: In this section of the questionnaire, participants should decide whether or not the five transformations of an RPS to a sunburst chart were correct (true or false) and, if not, where the error is located (text field). The results are summarised in Table 3 by showing the correct answers in absolute and relative (in %) numbers for all participants, for the expert group, and for the non-experts group.

The number of correct answers are high for all questions. Example 4 shows the worst results with 84.7% of correct answers. Example 1 shows the best results with 98.1% of correct answers for all participants. In a direct comparison, the experts achieved a slightly higher number of correct answers than the non-experts (~1.2% to 5.8%), except for Example 5 with 0.3% fewer correct answers. The latter indicates almost no difference between the experts and the non-experts regarding the comprehension of the transformation of an RPS into a sunburst chart (and the other way around). Examples 3 and 4 were based on small RPSs with only two hierarchy levels and only five respectively six objects. However, the number of relations between the objects was high. This might have caused confusion on the participants' side, especially concerning non-experts. In turn, Example 5 (cf. Fig. 8a) has five hierarchy levels and comprises eleven different objects. This example is significantly larger than all others. However, Example 5 was answered correctly by 93.6% of all participants. This justified the assumption that the complexity of RPSs is not determined by the number of objects and hierarchy levels, but rather by the number of relations between the objects. Furthermore, the error explanation of the participants who correctly identified the sunburst chart as mistransformed was mostly correct (94.5% and 98.2% cf. Table 3 Expl. of 3 and 4.). Note that not all participants (109 of 157) answered these optional questions. Consequently, the intuitive comprehensibility investigated in RQ 2 can be answered positively for sunburst charts in general.

Table 3. Correct answers of *Comprehension and Readability of RPSs and Sunburst Charts.*

		Question						
		1	2	3	Expl. of 3	4	Expl. of 4	5
Total	#	154	140	134	103/109	134	107/109	147
(157)	in %	98.1	89.2	85.4	94.5	85.4	98.2	93.6
E	#	76	70	68	54/58	66	54/55	72
(77)	in %	98.7	90.9	88.3	93.1	85.7	98.2	93.5
NE	#	78	70	66	49/51	67	53/54	75
(80)	in %	97.5	87.5	82.5	96.1	83.8	98.2	93.8

7 Related Work

Sunburst charts can be versatility used for dashboard visualisations. Particularly, they allow demonstrating the patterns, relations, and dependencies of business process data. Thus, big data can be read quickly and easily. In [10], various charts for visualising big data are presented, including sunburst charts.

Sunburst charts are used in other research disciplines as well, e.g., distributed networks for geographic information systems (e.g., weather online risk assessment) [11] and intrusion detection systems. Regarding the latter, the network is represented in terms of a sunburst chart. Colour-coded alerts may be categorised to quickly identify the location in the network and the type of the intrusion [12].

Related to our work are empirical studies dealing with readability and comprehensibility of charts. In [13], a case study investigates the application of layered diagrams in software engineering. The study has shown that the latter is both easy to comprehend and easily applicable by modellers. In [14], the aesthetics of automatically generated graphs (e.g., node-arc graphs, especially the unified modelling language class and collaboration chart) are investigated in a usability study. Corresponding results have shown that the usability of these charts and graphs are significantly affected when only using some of the investigated aesthetics.

8 Summary and Outlook

Two research questions on the use of sunburst charts in monitoring object-centric process were addressed in this paper. Regarding RQ 1, all possible RPS patterns were identified and a procedure to derive a sunburst chart from the given RPS was developed based on three mapping-rules. To address RQ 2, an empirical study was conducted that investigates the comprehension, readability, suitability, and acceptance of the created sunburst charts. The obtained study results allowed for the following interpretations. First, there is a potential to improve the presentation of the cardinality notation in the RPS. Second, the comprehension and readability of sunburst charts at run time achieves a correctness of over 90% for all participants (i.e., experts and non-experts). Experts even performed better with a correctness of up to 98.7%. Based on these results, it may be concluded that experts need not be provided with additional information (e.g., caption or additional introductory training) to read and understand the sunburst chart on a monitoring dashboard. Third, in almost all cases, the results of experts and non-experts differ by less than 5 percent making the developed sunburst charts suitable for non-experts as well. Fourth, the use of the sunburst chart for visualising the progress of single object instances was rated about neutrally suitable, whereas the visualisation of the status of each object instance was rated generally suitable. This indicates that the suitability of sunburst charts is significantly affected by the visualised property.

In future work, the mentioned limitations of RPS should be further investigated to find suitable solutions. In addition, it should be investigated, which run time properties (e.g., progress) of object instances are suitable for being presented by a sunburst chart in order to enable the creation of dashboards that are as suitable as possible.

References

1. Andrews, K., Steinau, S., Reichert, M.: Enabling ad-hoc changes to object-aware processes. In: IEEE 22nd International Enterprise Distributed Object Computing Conference (EDOC 2018), pp. 85–94. IEEE (2018)
2. Breitmayer, M., Arnold, L., Reichert, M.: A dashboard-based approach for monitoring object-aware processes. In: Central European Workshop on Services and their Composition (ZEUS 2021), pp. 29–33 (2021)
3. Le Guen, F.: Sunburst chart (2022). https://www.excel-exercise.com/sunburst-chart/
4. Steinau, S., Andrews, K., Reichert, M.: The relational process structure. In: Krogstie, J., Reijers, H.A. (eds.) CAiSE 2018. LNCS, vol. 10816, pp. 53–67. Springer, Cham (2018). https://doi.org/10.1007/978-3-319-91563-0_4
5. Steinau, S., Andrews, K., Reichert, M.: Coordinating large distributed relational process structures. Softw. Syst. Model. **20**(5), 1403–1435 (2021)
6. Steinau, S., Andrews, K., Reichert, M.: Executing lifecycle processes in object-aware process management. In: Ceravolo, P., van Keulen, M., Stoffel, K. (eds.) SIMPDA 2017. LNBIP, vol. 340, pp. 25–44. Springer, Cham (2019). https://doi.org/10.1007/978-3-030-11638-5_2
7. Andrews, K., Steinau, S., Reichert, M.: Enabling runtime flexibility in data-centric and data-driven process execution engines. Inf. Syst. **101**, 101447 (2021)
8. Wieringa, R.: Design Science Methodology for Information Systems and Software Engineering. Springer, Heidelberg (2014). https://doi.org/10.1007/978-3-662-43839-8
9. Brace, I.: Questionnaire Design: How to Plan, Structure and Write Survey Material for Effective Market Research. Kogan Page Publishers, London (2018)
10. Sun, G., Li, F., Jiang, W.: Brief talk about big data graph analysis and visualization. J. Big Data **1**(1), 25 (2019)
11. Leite, J., Mantovani, J., Kezunovic, M.: Use of distribution network topological fractality and sunburst charts in the online risk assessment. In: IEEE PES Innovative Smart Grid Technologies Conference-Latin America (ISGT - LA 2019), pp. 1–6. IEEE (2019)
12. Patton, R., Beaver, J., Steed, C., Potok, T., Treadwell, J.: Hierarchical clustering and visualization of aggregate cyber data. In: 7th International Wireless Communications and Mobile Computing Conference (IWCMC 2011), pp. 1287–1291. IEEE (2011)
13. Störrle, H.: Improving model usability and utility by layered diagrams. In: Proceedings of the 10th International Workshop on Modelling in Software Engineering (ICSE 2018), pp. 59–66 (2018)
14. Purchase, H., Carrington, D., Allder, J.-A.: Empirical evaluation of AES-thetics-based graph layout. Empir. Softw. Eng. **7**(3), 233–255 (2002)

A Peek into the Working Day: Comparing Techniques for Recording Employee Behaviour

Tea Šinik, Iris Beerepoot(✉), and Hajo A. Reijers

Department of Information and Computing Sciences, Utrecht University, Princetonplein 5, Utrecht, The Netherlands
t.sinik@students.uu.nl, {i.m.beerepoot,h.a.reijers}@uu.nl

Abstract. Detailed recordings of employee behaviour can give organisations valuable insights into their work processes. However, recording techniques each have their advantages and disadvantages in terms of their obtrusiveness for participants, the richness of information they capture, and the risks that are involved. In an effort to systematically compare recording techniques, we conducted a multiple-case study at a multinational professional services organisation. We followed six participants for a working day, comparing the outcomes from non-participant observation, screen recording, and timesheet techniques. We generated 136:04 h of data and 849 records of activities. We identified 58 differences between the techniques. The results show that the use of only one technique will not produce a complete and accurate record of the activities that occur on the screen (online), in the hallway (offline), and in the extra hours (overtime). Therefore, it is vital to choose a technique wisely, taking into account the type of information it does not capture. Furthermore, this study identifies some open challenges with respect to accurately recording employee behaviour.

Keywords: Employee Behaviour · Work Patterns · Data Collection Techniques · Observation · Screen Recording · Timesheet

1 Introduction

"What did you do today?". This is a simple question that may be presented to an employee by co-workers, management, or even the employee themselves. The behaviour of employees in the workplace is directly related to the success and operations of an organisation [9]. There is an assumption that there might be a discrepancy between what employees said they have done and what is actually observed throughout the working day [22]. Mills et al. [17] argue that employees tend to omit records that reflect negatively on their behaviour or only record the records that they deem to be important. Therefore, recording employee behaviour whilst carrying out business-related activities has become indispensable [1]. Nowadays, business processes are increasingly supported by information systems [14]

S. Nurcan et al. (Eds.): RCIS 2023, LNBIP 476, pp. 343–359, 2023.
https://doi.org/10.1007/978-3-031-33080-3_21

which record detailed trails of the execution of tasks in databases, system logs, or records [2]. Within Business Process Management (BPM), process mining has gained a lot of interest, both in research and in practice [5]. This family of techniques provides organisations with an opportunity to reveal exactly how processes are executed, in addition to how they should be executed [7].

The main input for the process mining techniques is event data [15]. The assumption is that these records are a truthful representation of the actual employee behaviour to discover work patterns [16]. However, Baier et al. [3] suggest that tasks are not recorded (properly) by the employee or occur outside of the information systems altogether. Thus, it is not always evident to what extent event logs or employee recordings reflect the actual behaviour of the employees. There is a sizeable gap in the literature on recording employee behaviour within an organisational context by using multiple data collection techniques [16]. The issue related to this knowledge gap is that previous studies have almost exclusively focused on one stand-alone data collection technique. This is of particular concern because each data collection technique yields unique results but also has its shortcomings which may impact the accuracy or completeness of the results [13]. Therefore, we aim to answer the following research question with this study: *"How do different data collection techniques compare in discovering work patterns of employees within an organisation?"*. We contribute to the existing body of knowledge by comparing three data collection techniques with varying characteristics to record employee behaviour and discover work patterns within an organisational context: non-participant observation, screen recording, and the timesheet. We present the (dis)advantages, commonalities and differences between the data collection techniques to examine the level of confidence that should be placed in the analysis of this type of data. In addition, we sketch new lines of research that are required to arrive at better recording techniques.

The structure of this paper is as follows. In Sect. 2, we define work patterns and examine the characteristics of the techniques used to record the work patterns. In Sect. 3, we describe the set-up of our multiple case study. The results of the case study are reported in Sect. 4. In Sect. 5 we discuss the limitations of our study as well as the research opportunities it has revealed. Finally, we conclude this paper with Sect. 6.

2 Related Work

2.1 Work Patterns

The study of recording employee behaviour to discover work patterns has gained a lot of interest, both in research and in practice [16]. There is a great amount of variation in how work patterns are recorded depending on the sector (e.g., health or education) or the scope of the research. In general, work patterns are defined as *"the (characteristics) of work activities performed by the organisational members to execute specific activities, and accomplish practices of interest related to a task"* [19], against a set of predetermined classifications. This refers to the *"everyday nature of the work activities exhibited by the organisational members*" [6, p. 24]. The characteristics are the *where* and *when* the work activities are performed (location and time), *how* (the mediums and documents), but also by *whom* (the involved members) [10].

The classifications introduced by Mintzberg [18] are among the most commonly used for recording employee behaviour and discovering work patterns [10]. Mintzberg used a predetermined classification scheme consisting of (1) desk work, (2) scheduled meetings, (3) unscheduled meetings, (4) telephoning, and (5) tours in the organisation.

2.2 Data Collection Techniques

Lethbridge et al. [13] provide an extensive overview of data collection techniques for studying employee behaviour. They categorise the data collection techniques based on the required degree of contact between the researcher and participant:

1. Direct Technique: The researcher must have direct access to the participant.
2. Indirect Technique: The researcher must have direct access to the working environment of the participants (e.g., (home) office). In comparison to the direct technique, the indirect technique does not require the researcher to interact with the participant.
3. Independent Technique: Involves a retrospective study of work artefacts such as event logs or archival sources. The records are not created by or for (the purpose of) the study.

We examined eight data collection techniques suitable to record employee behaviour in an organisational context[1]. For this study, we selected one from each category: observation, screen recording and timesheet. Not only do they differ in terms of the categorisation, they also differ in terms of obtrusiveness, richness of information, and associated risks. We illustrate their characteristics in Table 1 and further explain them below. Additionally, the data collection took place during a peak period ("busy season") which limited the availability of the participants. We favoured the aforementioned techniques because they do not require any active participation from the participants.

Table 1. Characterisation of the Data Collection Techniques.

Characteristic	Observation	Screen Recording	Timesheet
Categorisation	Direct technique	Indirect technique	Independent technique
Obtrusiveness	Obtrusive and invasive	Less obtrusive, highly invasive	Unobtrusive, possibly invasive
Richness of information	Detailed information, but fast-paced	Highly detailed information	Less detailed information
Risks	Change in behaviour	Incomplete recording	Omitted behaviour

[1] See folder "Literature Review Results": https://doi.org/10.5281/zenodo.7535574.

Observation. This is a direct technique [13] that allows a real-time representation of the studied phenomena [21]. The data consists of subjective information which is influenced by the observer's perspective and what they deem to be important [20]. This technique requires direct contact between the researcher and the participant. The advantage is that the records contain detailed information about the online setting (on the screen) and the offline setting (physical environment). The disadvantage is its obtrusive and invasive nature which is the cause of the Hawthorne effect. This suggests that participants may change their behaviour due to their awareness of being observed [16].

Screen Recording. This is an indirect technique [13] that allows for a retrospective observation of the studied phenomena [8]. This technique is less obtrusive because it only requires access to the working environment of the participant. The advantage is that the records contain highly detailed information about the online setting because the recordings are permanent records of interactions that can be viewed repeatedly for later analysis [20]. The disadvantage is its dependency on a recording application to provide a complete recording (of the screen and sound) [12].

Timesheet. This independent technique [13] consists of existing material not created by the researcher or for the purpose of the research. The advantage of this self-reporting technique is that the participants can record their own activities (e.g., overtime). The disadvantage is that participants can omit activities that reflect negatively on their behaviour or only record the activities that they deem to be important [17]. Moreover, the records do not provide detailed information because the researcher has no control over the details of the data (e.g., the timesheet must be filled in according to the guidelines provided by the organisation) [20].

3 Research Method

We conducted a multiple-case study and compared the non-participant observation, screen recording, and timesheet technique to record employee behaviour. According to Yin [23], the multiple-case study is a suitable method since it investigates and provides a deeper understanding of a contemporary phenomenon in its own real context (e.g., the organisational context), by using multiple sources of evidence. We performed the case study from May 2022 to February 2023.

We followed the case study method by Yin [23]. Our research method consists of five phases: case study selection, technical pilot, participant selection, data collection, and data analysis.

3.1 Phase 1: Case Study Selection

The organisation investigated in this study is a multinational professional services firm located in the Netherlands. The context of this study was set within a team of data consultants, which is part of the Assurance service line. The team is divided into four self-managing "squads". Each squad is composed of members and one squad lead who is assigned to the planning activities.

3.2 Phase 2: Technical Pilot

We were required to use software that was approved by the organisation. Therefore, we used *Snagit*, which is a screen capture and recording application created by TechSmit[2]. We conducted a pilot test to confirm that the application can record multiple screens. In addition, we created a set of guidelines because the collected data consisted of sensitive information such as client-specific data.

3.3 Phase 3: Participant Selection

To create a diversified group of research participants with varying profiles, we selected an employee from each squad. The profiles of our six participants are shown in Table 2.

Table 2. Overview of Research Participants.

ID	Squad	Squad Lead	Rank	Work experience	Employment
P1	C	No	Staff	1,5 years	Part-time
P2	B	No	Staff	2,5 years	Full-time
P3	C	No	Staff	2,5 years	Full-time
P4	B	Yes	Senior	2,5 years	Full-time
P5	D	Yes	Senior	4 years	Full-time
P6	A	No	Staff	2 years	Full-time

3.4 Phase 4: Data Collection

We used a standardised form [18] for all data collection techniques per research participant to preserve uniformity, and allow the comparison and analysis of the findings across the techniques [10]. The standardised form used to record employee behaviour consisted of the following classifications:

- **Time:** The time the activity was recorded.
- **Category:** The overall category of activity, e.g., Desk Work (i.e., all work-related activities), or Personal (i.e., use of the personal mobile phone).
- **Activity:** The specific activity that the employee spent time on, e.g., Organisational Work (i.e., extraction, transformation, and validation of a dataset).
- **Sub-Activity:** The sub-activity within the main activity (e.g., incoming and outgoing calls, messages, and emails).
- **Medium:** The medium used to perform the activities (e.g., Microsoft Teams or Alteryx).
- **Participants:** The individual with whom the employees interacted throughout the working day (e.g., Audit Team).

[2] Website TechSmith: https://www.techsmith.com/snagit-features.html.

Time	Category	Activity	Sub-Activity	Medium	Participants	Initiated	Duration	Field Notes	Client	Participant
09:39:00	Desk Work	Scheduling & Administration	Planning	Microsoft Outlook	Independent	Employee	00:01:00	The employee opened their calender and planned hours for client 6A. It seems that they use their calender for the planning of hours (sort of a diary).	6A	P6
09:40:00	Desk Work	Scheduling & Administration	Planning	In-House Medium 5	Independent	Employee	00:01:00	The employee opened In-House Medium 5. They changed the ticket status of client 6A.	6A	P6
09:41:00	Desk Work	Scheduling & Administration	Planning	In-House Medium 5	Independent	Employee	00:03:00	The employee opened In-House Medium 5. They search on client 6D, they are assigning hours to themselves. They changed the ticket status of client 6A.	6A	P6
09:42:00										
09:43:00										
09:44:00	Desk Work	Scheduling & Administration	Mail Incoming	Microsoft Outlook	Data Team - One	Other party	00:04:00	Received an email from a Data Team participant regarding the planning of client 6D.	6D	P6
09:45:00										
09:46:00										
09:47:00										
09:48:00	Desk Work	Technical Problem Solving	Technical Issues	In-House Medium 7	Independent	Employee	00:07:00	The employee opened In-House Medium 7, with the intention to process the data from a client, however, they decided to first update the Datawasher version.	Non-Client	P6

Fig. 1. Snippet of an Observation Record.

- **Initiated:** The individual who initiated the activity (e.g., the employee or the other party).
- **Field Notes:** The purpose of notes is to aid in recalling the (context of the) activities during the transcription and coding.
- **Client:** Each client was given a unique identifier to be able to record the time spent on (non-)client-related work.
- **Participant:** The research participants were given a unique identifier (e.g., P1 or P2).

Each participant was subjected to three data collection techniques: observation, screen recording and timesheet on one working day of their choosing. Only one researcher conducted the data collection and data analysis. The case study time frame was set between 09.00 and 17.00. During the observation, we used two screens. On the screen turned towards the participant, we opened an application unrelated to our study e.g., Microsoft Outlook. We opened the standardised form (see Fig. 1 for an example snippet) on the screen turned away from the participant. The aim was to minimise the Hawthorne effect by pretending that we are not actively observing the participant. While the observation took place, the participant recorded their screens including (system) audio. Moreover, each participant recorded their behaviour in a timesheet using an in-house timesheet system. The timesheet consists of an engagement ID, activity ID, and a short description. The result of the data collection was the following: six full-day screen records, timesheets and observation records.

3.5 Phase 5: Data Analysis

To minimise the risk of being influenced by using the data gathered from one technique to another technique, we rearranged the data analysis through two steps.

1. Predefined the data analysis order of the data collection techniques. First, we analysed the data collected from the observation. We recorded our observations in the standardised form, meaning that we had already transcribed (a part of the data). Second, we analysed the data collected from the timesheets that provided a broader picture of the studied phenomena [4] from the perspective of the research participant [16]. Lastly, we analysed the data from the screen recording that presented rich empirical data regarding the behaviour and work patterns of the employees [11]. This technique allowed us to (re-)watch the recordings until the entire working day was transcribed and coded.

2. Randomised the data analysis order of the research participants per data collection technique. To minimise the risk of creating connections between the data of a research participant across the data collection techniques, we randomised the analysis order of the collected data of the research participants per data collection technique (archival analysis and screen recording). For the observation, we used the order taken during the data collection. We used a random number generator tool to generate a unique and randomised order for the archival analysis and screen recording. Table 3 illustrates the predefined analysis order of the data collection techniques (see column "Order") and the randomised order of the research participants per data collection technique (see column "Participant order").

Table 3. Order of Analysis.

Order	Data Collection Technique	Participant order
1	Observation	1 2 3 4 5 6
2	Archival Analysis (timesheet)	2 5 1 3 6 4
3	Screen Recording	4 1 2 6 3 5

To reduce errors and improve the reliability of the data analysis, we used two tools. First, we used Alteryx[3] to decrease the number of manual actions (e.g., merging files by copy-pasting). Second, to summarise the large amounts of data, we generated pivot tables in Microsoft Excel.

4 Results

After following six participants over the course of six working days, we collected 136:04 h of data during the non-participant observation technique (43:49 h), screen recording technique (43:25 h), and the timesheet technique (48:50 h). In total, the techniques recorded 849 activities[4].

We compared the records of all classifications in the standardised form. For the observation, screen recording, and timesheet technique we recorded the duration and the count (i.e., the total number of times the classification is observed).

[3] Website Alteryx: https://www.alteryx.com/.

[4] See folder "Results Multiple Case Study": https://doi.org/10.5281/zenodo.7535574.

Table 4 contains the durations and counts of the categories and Table 5 those of the activities. By analysing the records, we found 58 differences related to *timestamps, online versus offline activities, brief activities, overtime activities, and uncategorised activities.* We will discuss each of these differences in the following sections.

Table 4. Results per Category (Obs = Observation, SR = Screen recording, TS = Timesheet, NOB = not observed)

	Duration (in Hours)			Category (Count)		
Category (Total)	Obs	SR	TS	Obs	SR	TS
Desk Work	23:17	22:34	39:55	322	324	37
Personal	01:24	01:44	NOB	18	15	NOB
Telephone	09:50	09:47	08:25	32	33	13
Tours	08:23	07:42	NOB	26	21	NOB
Meeting	00:55	NOB	00:30	3	NOB	1
Unable to categorise	NOB	01:38	NOB	NOB	4	NOB
Grand Total	**43:49**	**43:25**	**48:50**	**401**	**397**	**51**

Table 5. Results per Activity (Obs = Observation, SR = Screen recording, TS = Timesheet)

	Duration (in Hours)			Activity (Count)		
Activity (Total)	Obs	SR	TS	Obs	SR	TS
Breaks	07:17	07:25	NOB	19	19	NOB
Giving Information	01:44	00:49	00:30	24	21	1
Organisational Work	14:59	14:31	26:00	145	144	16
Personal	01:11	00:37	NOB	12	6	NOB
Receiving Information	00:57	57:00	01:15	14	14	2
Requests & Solicitations	00:48	00:42	NOB	22	22	NOB
Scheduling & Administration	12:27	12:10	16:50	124	124	30
Set-up time	00:24	NOB	NOB	1	NOB	NOB
Socialising	01:22	01:56	NOB	13	14	NOB
Technical Problem Solving	02:40	02:40	04:15	27	29	2
Unable to categorise	NOB	01:38	NOB	NOB	4	NOB
Grand Total	**43:49**	**43:25**	**48:50**	**401**	**397**	**51**

4.1 Timestamp

The timesheet aggregates the duration of the activities, therefore the records do not contain a timestamp. This restricts the ability to use these records for additional time analyses. The records of the observation and screen recording technique both contain timestamps. However, the observation technique is less accurate than the screen recording technique. We observed 15 differences in the timestamps between the observation records and the screen recording (Table 6). The observation technique relies on the researcher capturing the exact time the activity is executed. As opposed to the screen capturing technique, the observation technique does not allow going back in time to check the exact starting time, and as such, the timestamp may be inaccurate. In our study, we observed *time lags* in 13 instances, i.e., where the timestamp of the activity was set at a time later than in the screen recording. We consider the screen recording the ground truth, as we can check exactly when the activity started or ended. *Overlapping (parallel) activities* are especially difficult to deal with when using the observation technique. For example, participant 1 joined a weekly squad call via Microsoft Teams from 16.05 until 16.30, but at the same time started working on a second task on a different screen. The screen recording technique showed that the second task started at 16.07. Using the observation technique, we only observed the switch in tasks at 16.10.

Table 6. Difference in Timestamps between Observation and Screen Recording.

Description	Count
Observation recorded the activity 1 min before screen recording	2
Observation recorded the activity 1 min later than screen recording	9
Observation recorded the activity 2 min later than screen recording	2
Observation recorded the activity 3 min later than screen recording	2

4.2 Online Versus Offline Activities

Offline activities, i.e., activities that did not occur on the screen, were only recorded with the observation technique. Here, we distinguish three categories: Meeting, Personal, and Tours.

Meeting: The category Meeting was adopted from Mintzberg's categorisation and refers to *offline* meetings, as opposed to the category of Telephoning for *online* meetings. Table 7 shows the results of recorded offline meetings for each of the techniques. The observation technique allows for complete recording of offline meetings. It recorded three meetings with a total duration of 55 min. The opposite is true for the screen capturing technique: no offline meetings could be recorded, as by definition, the technique does not capture activities taking place outside of the screen. The timesheet technique partially includes offline meetings, i.e., only when employees choose to include them in their time registration. It recorded one meeting with a total duration of 30 min. This means that this technique failed to record two meetings with a total duration of 25 min (55%).

Table 7. Offline Meetings Recorded per Technique.

	Observation	Screen Recording	Timesheet
Duration recorded (in hours)	00:55	00:00	00:30
Number of meetings recorded	3	0	1
Percentage recorded	100%	0%	55%

Note: We used the 55 min recorded by observation as the ground truth, i.e., the time that should have been recorded by the techniques.

Over the course of six working days, the six participants spent 2,09% of their time (55 min of 43:49 h) on meetings. Although this percentage (2,09%) may seem insignificant, offline meetings used to be a substantial part of the daily activities. The case study was conducted while COVID-19 restrictions were still in effect. As a result, the employees switched from *offline meetings* to online *telephoning* via Microsoft Teams. Once the employees return to their five-day workweeks in the office (or client sites), the meetings are expected to again become a larger component of the daily activities (24,53%), as shown in Table 8. Therefore, the inability of the screen recording and timesheet techniques to (completely) record the meetings could have a significant impact.

Table 8. Potential Significance of the Meetings.

	Meeting	Telephoning	Combined total
Duration recorded (in hours)	00:55	09:50	10:45
Activities recorded	3	32	35
Total (%)	2,09%	22,44%	24,53%

Personal: The activities that fall under the Personal category are *the use of a personal mobile phone, socialising, and the use of the internet browser (not work-related).* As shown in Table 9, we observed differences in the ability of the techniques to record these activities. With the observation technique, we were able to record all (online and offline) activities. The screen recording technique, however, failed to record three activities (15/18). Although screen recording provides an accurate recording of *online* activities, *offline* activities such as the use of a personal mobile phone or socialising cannot be recorded. The timesheet technique did not record any activities related to the category Personal. As it is a self-reporting technique, employees can omit records that reflect negatively on their behaviour, or can choose to only record the activities they deem important. Taking the observation records as the ground truth, we note that participant 3 spent 58 min on activities related to using the internet browser, their personal mobile phone, and socialising. However, based on the timesheet records, we know that the employee *chose* not to include this in their timesheet.

Table 9. Personal Activities Recorded per Technique.

	Observation	Screen Recording	Timesheet
Duration recorded (in hours)	01:24	01:44	00:00
Activities recorded	18	15	0

Tours: The activities that fall under the category of Tours are *set-up time and breaks (e.g., restroom, coffee, and lunch breaks).* As they are offline activities, the observation technique is the only one that can record all activities within this category. For example, the observation technique recorded that participant 5 started their working day with a 24-minute fire drill at 10.25. The screen recording and timesheet technique failed to record this activity. Due to the sound recording, the screen recording technique was able to *partially* record activities related to Tours. For example, the technique was able to record that participant 1 asked their colleague to go on a (smoke) break. As shown in Table 10, for all six participants over the course of their working days, the observation technique recorded 08:23 h of activities related to Tours. The timesheet technique does record any information about Tours and thus omits a significant part of employee behaviour.

Table 10. Tours Recorded per Technique.

	Observation	Screen Recording	Timesheet
Duration recorded (in hours)	08:23	07:42	00:00
Activities recorded	26	21	0

4.3 Brief Activities

One type of activity that was only recorded by the screen recording technique was short activities with a duration of only 1 or 2 min. Based on the records of the screen recording technique, we observed that there were 133 activities with a duration of 1 min and 55 activities with a duration of 2 min. In total, 47,36% of the activities had a duration of 1 or 2 min. This means that it is vital for a data collection technique to be able to record these brief activities. The screen recording technique produces permanent records of the activities (the screen recording and sound) that can be viewed repeatedly [20]. This aids in the recording of brief activities such as the use of the internet browser to resolve technical issues or incoming and outgoing emails, (chat) messages, and calls. Using the observation technique, the researcher might miss the activity because they are busy categorising and/or creating field notes of the previous activity.

The timesheet technique was particularly incomplete in terms of recording brief activities. The shortest activity recorded in the timesheet had a duration of 10 min. Employees might feel that recording each brief activity takes too much time. However, choosing not to record them might affect the number of hours that can be billed to the client. Table 11 shows the clients that were recorded for each of the techniques.

Table 11. Clients Recorded per Technique.

	Observation	Screen Recording	Timesheet
Total recorded (count)	43	44	24
Total unrecorded (count)	2	1	21
Total unrecorded (%)	4.44%	2.22%	46.67%

Note: The calculation is based on 45 being the total number of recorded Clients, calculated based on the 44 recorded and 1 unrecorded Client (screen recording).

It shows that the timesheet technique failed to capture a large number of these activities where employees worked for clients. According to the records of the screen recording technique, there were 60 activities related to the 21 clients that the timesheet technique failed to record. The far majority of these missing activities are brief activities of 1 or 2 min, as shown in Table 12.

Table 12. Clients-Related Activities Unrecorded by the Timesheet Technique.

	More than 10 min	5–10 min	3–4 min	1–2 min
Activities (count)	2	7	9	42
Activities (%)	3.33%	11.67%	15%	70%

Note: The calculation is based on 60 being the total number of client-related activities.

4.4 Overtime Activities

Comparing the total amount of time spent on different categories of activities, we observed a difference of more than 5 h between the observation and screen recording technique on the one hand, and the timesheet technique on the other. We classify this difference as *overtime activities* because only the timesheet technique was able to record these activities. This means that the employee performed the activities outside the study's time frame (before 09.00 or after 17.00). The observation technique is a direct technique and as such, requires direct contact between the researcher and the participant. A disadvantage of the observation technique is that it cannot record any activities performed outside of the time frame of the study because the physical presence of the researcher is required. The screen recording technique is an indirect technique. This means that it only requires direct access to the working environment (laptop). Compared to the observation technique, it is more flexible in terms of the recording of online behaviour [8]. The participants can record all online activities, from the office or their homes, as long as the screen recording application works. However, the disadvantage is that the offline activities that occur during overtime (e.g., meetings or use of a personal mobile phone) cannot be recorded by the screen recording technique. The participants could have performed the overtime activities at the office, on the train, or at home. However, the timesheet technique can record overtime activities. As the timesheet is an independent self-reporting technique, participants can record their own activities, whether performed during or outside of work hours. As a result, this technique produces a record that includes overtime activities.

4.5 Uncategorised Activities

A final difference between the techniques that can be observed relates to the number of activities that could not be categorised, i.e., the fields indicated by *NOB* in Table 4 and Table 5. The observation technique can categorise both online and offline activities. As such, 100% of activities could be categorised. The screen recording technique was able to categorise the activities for almost all recorded hours. However, one of the participants (participant 3) experienced technical issues during the recording of their working day which caused the tool to produce a partial screen and sound recording. As a result, the technique was unable to categorise 1,01% of the activities. The percentage of the activities that the timesheet technique failed to categorise is higher. Due to a lack of information that can be extracted from the timesheet caused by the design of the timesheet, and the limited information provided in the description, many activities could not be categorised.

5 Discussion

The previous sections have illustrated how each of the techniques has strengths and weaknesses in terms of the completeness of recording. In Fig. 2, we provide an overview of our comparison of recording techniques. The screen recording technique provides accurate timestamps and is an excellent choice when brief activities are an important aspect of daily work. Observations provide valuable information when important employee behaviour takes place offline. The timesheet technique provides the least information of the three, but is readily available for analysis in many organisations that perform billable work.

From the overview, we conclude that choosing the most valuable technique heavily depends on the purpose of the analysis. Evidently, there is a trade-off involved in terms of the obtrusiveness of the technique and the richness of information that the technique can offer. Generally, techniques that provide rich information seem to be more obtrusive, typically, and a combination of techniques would result in an even more complete picture of employee behaviour. However, it would also be highly invasive to the employee to collect this information and would ask for a significant effort from the collector. Therefore, choosing a technique or combination of techniques requires careful consideration of the recording impacts on the employee and the extent to which these are balanced by the positive outcomes for the employee. Regardless of the selection of techniques, it is crucial to take into account the shortcomings of the chosen techniques when drawing conclusions about employee behaviour.

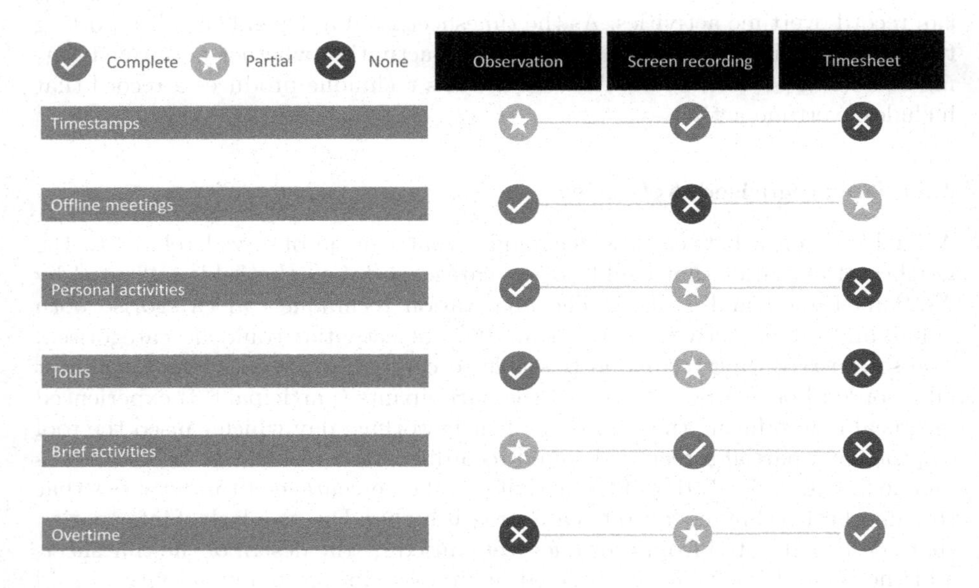

Fig. 2. Overview of Recording Techniques.

5.1 Limitations

Participant 3 faced technical issues during the use of the recording tool. We became aware of this once the working day had ended. We analysed the data and concluded that we missed 01:38 h of the screen recording. We did have sound (microphone and system audio), and thus, considered the following options. First, to accept the partial recording if we could demonstrate that the partial data was still reliable and valid for our data analysis based on the audio recording or our observation notes. Second, if the reliability of the data could not be ensured, then we would plan a new observation (either a whole or partial working day). The reason why we accepted the partial recording is that we had the sound recording, and we did not want to risk the participant exhibiting different work patterns from those noted during the initial observation.

5.2 Future Work

We see several avenues for future research in this area. First, future work might expand on the data collection techniques that were used in this study. Techniques such as the analysis of event logs or the use of work diaries would be valuable additions to the comparison. Probably, by considering additional techniques, we will become aware of how to better combine existing techniques to achieve the objectives for recording user behaviour. Second, other studies might validate the found patterns in different types of organisations and study employee behaviour for a longer period of time. It is fair to expect that by doing so, we may encounter types of activities unseen so far with their own 'fit' with the data collection techniques.

Finally, future studies might focus on the implications of choosing a particular data collection technique and the consequences of missing activities for insights that can be drawn from the analysis. This insight is arguably the most important line for future research since it may help us better understand the impact of relying on one or the other data collection technique.

In conclusion, there are major opportunities for studying employee behaviour. We trust that the present study provides a starting point for better understanding these techniques and making better decisions in selecting these techniques for actual application.

6 Conclusion

In this study, we followed six employees during their working days, recording their behaviour using three data collection techniques: non-participant observation, screen recording, and timesheet. By systematically analysing the differences in the 136:04 h of data that the techniques recorded, we show that each technique yields different results in terms of the activities that were recorded and the level of detail at which employee behaviour can be analysed. The use of one of the techniques will not produce a complete and accurate record of the activities that occur on the screen (online), in the hallway (offline), and during the extra hours (overtime). Depending on the purpose of the analysis, researchers or practitioners may select the best-fitting technique. However, it remains vital to reflect on the behaviour the chosen technique cannot capture. There are also opportunities to improve and extend existing techniques to better capture employee behaviour.

References

1. Anderson-Gough, F., Grey, C., Robson, K.: Tests of time: organizational time-reckoning and the making of accountants in two multi-national accounting firms. Acc. Organ. Soc. **26**(2), 99–122 (2001). https://doi.org/10.1016/S0361-3682(00)00019-2
2. Augusto, A., et al.: Automated discovery of process models from event logs: review and benchmark. IEEE Trans. Knowl. Data Eng. **31**(4), 686–705 (2019). https://doi.org/10.1109/tkde.2018.2841877

3. Baier, T., Di Ciccio, C., Mendling, J., Weske, M.: Matching events and activities by integrating behavioral aspects and label analysis. Softw. Syst. Model. **17**(2), 573–598 (2017). https://doi.org/10.1007/s10270-017-0603-z
4. Espedal, G., Jelstad Løvaas, B., Sirris, S., Wæraas, A.: Researching Values: Methodological Approaches for Understanding Values Work in Organisations and Leadership. Springer, Cham (2022). https://doi.org/10.1007/978-3-030-90769-3
5. Ghasemi, M., Amyot, D.: Process mining in healthcare: a systematised literature review. Int. J. Electron. Healthc. **9**(1), 60–88 (2016). https://doi.org/10.1504/IJEH.2016.078745
6. Gratton, L.: The Shift: The Future of Work is Already Here. William Collins, Glasgow (2014)
7. Grisold, T., Mendling, J., Otto, M., vom Brocke, J.: Adoption, use and management of process mining in practice. Bus. Process. Manag. J. **27**(2), 369–387 (2021)
8. Ho, W.Y.J.: 'I knew that you were there, so I was talking to you': the use of screen-recording videos in online language learning research. Qual. Res. **21**(1), 120–139 (2021). https://doi.org/10.1177/1468794119885044
9. Ind, N.: Inside out: how employees build value. J. Brand Manag. **10**(6), 393–402 (2003). https://doi.org/10.1057/palgrave.bm.2540136
10. Johnson, B.M.: Understanding the work of top managers: a shadowing study of Canadian healthcare CEOs. Ph.D. thesis, University of Warwick (2013). https://ethos.bl.uk/OrderDetails.do?uin=uk.bl.ethos.589919
11. Krieter, P.: Can I record your screen? Mobile screen recordings as a long-term data source for user studies. In: Proceedings of the 18th International Conference on Mobile and Ubiquitous Multimedia, pp. 1–10 (2019)
12. Lammers, W.J., Badia, P.: Fundamentals of Behavioral Research Textbook. Wadsworth Publishing, Belmont (2004). https://uca.edu/psychology/fundamentals-of-behavioral-research-textbook/
13. Lethbridge, T.C., Sim, S.E., Singer, J.: Studying software engineers: data collection techniques for software field studies. Empir. Softw. Eng. **10**(3), 311–341 (2005). https://doi.org/10.1007/s10664-005-1290-x
14. Mannhardt, F., de Leoni, M., Reijers, H.A., van der Aalst, W.M.P., Toussaint, P.J.: From low-level events to activities - a pattern-based approach. In: La Rosa, M., Loos, P., Pastor, O. (eds.) BPM 2016. LNCS, vol. 9850, pp. 125–141. Springer, Cham (2016). https://doi.org/10.1007/978-3-319-45348-4_8
15. Marin-Castro, H.M., Tello-Leal, E.: Event log preprocessing for process mining: a review. Appl. Sci. **11**(22), 10556 (2021). https://doi.org/10.3390/app112210556
16. McDonald, S.: Studying actions in context: a qualitative shadowing method for organizational research. Qual. Res. **5**(4), 455–473 (2005). https://doi.org/10.1177/1468794105056923
17. Mills, A.J., Durepos, G., Wiebe, E.: Encyclopedia of Case Study Research. Sage Publications, Thousand Oaks (2009). https://doi.org/10.4135/9781412957397
18. Mintzberg, H.: Structured observation as a method to study managerial work. J. Manage. Stud. **7**(1), 87–104 (1970). https://doi.org/10.1111/j.1467-6486.1970.tb00484.x
19. Nicolini, D.: Practice Theory, Work, and Organization. Oxford University Press, Oxford (2012)
20. Robson, C., McCartan, K.: Real World Research: A Resource for Social Scientists and Practitioner-Researchers. Blackwell Publishing, Hoboken (2002)
21. Shull, F., Singer, J., Sjøberg, D.I.: Guide to Advanced Empirical Software Engineering. Springer, London (2007). https://doi.org/10.1007/978-1-84800-044-5

22. Stickdorn, M., Hormess, M.E., Lawrence, A., Schneider, J.: This is Service Design Doing: Applying Service Design and Design Thinking in the Real World. O'Reilly, Sebastopol (2018)
23. Yin, R.K.: Case Study Research and Applications: Design and methods. Sage Publications, Thousand Oaks (2018)

Context-Aware Recommender Systems: Aggregation-Based Dimensionality Reduction

Elsa Negre[1(✉)], Franck Ravat[2], and Olivier Teste[3]

[1] Paris-Dauphine University, PSL Research University, CNRS UMR 7243, LAMSADE, Paris, France
elsa.negre@lamsade.dauphine.fr

[2] IRIT (UMR5505) - Université Toulouse Capitole, Toulouse, France
ravat@irit.fr

[3] IRIT (UMR5505), Université Toulouse 2, Toulouse, France
teste@irit.fr

Abstract. Context-aware recommender systems (CARS) rest on a multidimensional rating function: Users × Items × Context → Ratings. This multidimensional modelling should improve the quality of the recommendation process, but unfortunately, it is rare or even impossible to have ratings for all possible cases of context. Our objective is therefore twofold: (i) to reduce the dimensionality of the contextual information (in order to reduce the sparsity), which leads to (ii) propose a technique for aggregating the ratings associated with the aggregated dimensions. To do this, we organize, in the CARS utility matrix, the contextual information according to hierarchical dimensions as is done in OLAP (OnLine Analytical Processing) and we use a regression-based approach for the rating aggregation according to previously defined hierarchies. Our approach supports multiple dimensions and hierarchical aggregation of ratings. It was validated on two real world datasets.

Keywords: Context-aware recommender system · Dimensionality reduction

1 Introduction

Recommender systems (RS) filter information sources for helping users in their searches of relevant information. Typically, the recommendation process predicts the rating the user would give to each item, and recommends the items with the highest ratings. Traditionally, the problem of recommendation can be summarized as the problem of estimating/predicting ratings for items that have not been seen by a user, it is a rating function: $R_{RS} : Users \times Items \rightarrow Ratings$ [5]. This is represented by a matrix, called the utility matrix.

Nowadays, new recommender systems increasingly take into account the context of the user in order to offer ever more relevant recommendations [14] and

S. Nurcan et al. (Eds.): RCIS 2023, LNBIP 476, pp. 360–377, 2023.
https://doi.org/10.1007/978-3-031-33080-3_22

become context-aware recommender systems (CARS). The preferences for items within one context may be different from those in another context [3]; e.g., during a lockdown, the movies searched by users may be different than in another context. Thus, the more contextual information we have, the more precise the recommendations will be [5]. Nevertheless, CARS may face different limitations. The main one is that contextual information is not organized, ordered so as to be able to provide a sound recommendation even when contextual information is missing. This paper addresses this drawback. To address this problem of organisation, a promising solution consists in multidimensional modelling related to contextual information.

In the multidimensional modelling approaches, data associated to analysis axes (also called dimensions) are organized through hierarchies for efficiently supporting OnLine Analytical Processing (OLAP). In CARS, we can also organise contextual information through multiple dimensions. If we have a temporal dimension for the specific CARS and if we aggregate the daily ratings into monthly ratings (the rating of the month corresponds to the aggregate rating of the days of this month), the number of different contexts is reduced, and we could resolve the problem of the lack of ratings for all possible contexts in order to improve the relevance of recommendations. Many studies agree that addressing this problem of lack of ratings increases the recommendation accuracy [9,11].

Contrary to the usual multidimensional systems, this multidimensional modelling context induces a problem as for the ratings aggregation. Indeed, the context involves many associated dimensions (cf. Fig. 3) leading to a large dimensionality of the information to be taken into account in the CARS. This generates a sparsity problem (in CARS data processing). It should be noted that not all the contextual information is necessarily known (as detailed in the experiments of Sect. 5, the sparsity linked to the context exceeds 90%). Our objective is therefore twofold: (i) to reduce the dimensionality of the contextual information in order to reduce the sparsity, which leads to (ii) propose a technique for aggregating the ratings associated with the aggregated dimensions. To do this, we organize, in the CARS utility matrix, the contextual information according to hierarchical dimensions and we use a regression-based approach for the rating aggregation according to previously defined hierarchies.

The remainder of the paper is organized as follows. In Sect. 2, we present related works on CARS. In Sect. 3, we describe our approach to model contextual information for RS. In Sect. 4, we explain ratings aggregation in multidimensional CARS context. Finally, we detail some experimental assessments in Sect. 5.

2 Related Works

In this section we present some related works about CARS and multidimensional modelling for CARS.

The probably most widely accepted definition of context is the one of [10] where: "*Context is any information that can be used to characterize the situation of an entity*". A key accessor to the context in any context-aware system is

Table 1. 7-dimensional (utility) matrix, M_u

	Users	Movies	Cities	Countries	Dates	Months	Years	Ratings
line 1	U_1	F_1	Paris	France	Jan. 10, 2019	Jan. 2019	2019	3
line 2	U_1	F_1	Marseille	France	Mar. 3, 2019	Mar. 2019	2019	4
line 3	U_1	F_2	Paris	France	Jan. 17, 2019	Jan. 2019	2019	7
line 4	U_1	F_3	Paris	France	Feb. 18, 2019	Feb. 2019	2019	6
line 5	U_2	F_1	.	.	Feb. 15, 2019	Feb. 2019	2019	7
line 6	U_3	F_1	.	.	.	.	.	2
line 7	U_4	F_1	Lille	France	.	.	.	2
line 8	U_5	F_1	Marseille	.	Feb. 18, 2019	Feb. 2019	2019	8
line 9	U_8	F_5	Lille	France	Jan. 10, 2019	Jan. 2019	2019	5
line 10	U_8	F_5	Lille	France	Jan. 17, 2019	Jan. 2019	2019	2
line 11	U_8	F_5	Paris	France	Feb. 18, 2019	Feb. 2019	2019	4
line 12	U_8	F_6	Paris	France	Jan. 15, 2019	Jan. 2019	2019	8
line 13	U_8	F_6	Lille	France	Feb. 18, 2019	Feb. 2019	2019	3

a well-designed model. To reach a complete and appropriate context model for CARS, [21] proposed a context factors categorization. Some context modelling approaches exist [8]. CARS incorporate context to generate more relevant recommendations [5,13]. Considering the context, the rating function for CARS is $R_{CARS} : Users \times Items \times Context \rightarrow Ratings$ [5]. Among the existing methods, there is no single winner [18], and the experiments showed effectiveness depends on the dataset and the application domain. However, despite good performance, all approaches suffer from data sparsity [14,18].

[4] were the first to indicate that in some cases the 2D recommendation ($Users \times Items$) is not enough and that some side information must be taken into account. While [20] focuses on the integration of personalization rules in a multidimensional context, conversely [3] proposed to use an OLAP-based multidimensional model for CARS. The authors in [3] indicate that the limitations are (i) a lack of an appropriate aggregation function (when drilling within the hierarchies, the ratings are not additive) and (ii) a need to reduce the number of dimensions. Although an OLAP-based approach was advocated by [3,5], researches interested in such approaches have often been limited to having analysis dimensions with only one hierarchical level. [16] have proposed analysis dimensions with several hierarchical levels allowing the OLAP operations named drilling (rollup/drilldown). Unfortunately, the approach of [16] is travel-specific. Finally, let us note that [15] proposed a CARS specific to OLAP analyzes by integrating contextual information by post-filtering in order to recommend exploitable OLAP queries.

Regarding the rating aggregation function, the most used is the average but without considering the hierarchy/functional dependency between certain contextual information [2,14]. However, let's mention that [1] propose to use a weighted mean aggregation and [23] a hierarchical average. Finally, to the best of our knowledge, none of the works indicate a condition to stop aggregating; i.e., if we aggregate "too much", we will lose the contextual characteristics. Let us also mention [22] which proposes to predict the context-aware multi-criteria ratings and learn the aggregation function by applying Deep Neural Network. The authors deal with the aggregation function but do not address the problem of dimensionality.

Regarding the dimensionality reduction, the utility matrix of 2D recommender system is often sparse and this is even more true when there are more than two dimensions, as it is the case in CARS. [9,11], all supported the fact that reducing sparsity enhances recommendation accuracy.

Therefore, to overcome the limitations specified in [3] knowing that many challenges remain [14] and in order to improve the quality of the recommendations returned by CARS, in this paper we define an adapted aggregation function (with a stopping condition) as well as a domain-agnostic approach, to reduce the dimensionality of the utility matrix; i.e., the dimensionality of contextual information based on an OLAP model (to drill).

3 Modelling Context Cube

3.1 Use Case

Take the example of recommending movies to users, knowing that the contextual information is the city and the country where the user saw the movie and the date, month and year on which the movie was seen. So, in this example, we have (i) *Users*, (ii) *Items* that are Movies and (iii) *Context* which is composed of Cities, Countries, Dates, Months and Years. The CARS rating estimating function $R : Users \times Items \times Context \rightarrow Ratings$ [5] can be formulated as follows $R : Users \times Movies \times Cities \times Countries \times Dates \times Months \times Years \rightarrow Ratings$. Table 1 shows the utility matrix of R and illustrates three types of sparsity:

1. There is a lack of contextual information; e.g. lines 5, 7 and 8.
2. Even a lack of all contextual information; e.g. line 6.
3. In addition, it should be noted that given the raw data, in Table 1, there are 6 users, 5 movies, 3 cities and 6 dates, which means that the matrix should contain 540 ratings; i.e., 6 Users $\times$ 5 Movies $\times$ 3 Cities $\times$ 1 Country $\times$ 6 Dates $\times$ 3 Months $\times$ 1 year to be complete. However, we only have 13 ratings, which represents a density of 2.4% and therefore a sparsity of 97.6%. Therefore we have another type of sparsity linked to the lack of ratings.

Suppose now that we want to recommend a movie to the user U_6, who is in the city of Toulouse (France) on April 6, 2019. We need to estimate U_6 user's rating for each movie. Take the example of F_1 movie, we have no other user who saw the F_1 movie in the same city on the same date, so it is difficult to get a relevant rating. One possibility is to predict for U_6 a rating of 2 for movie F_1, considering line 6 of Table 1 in which U_3 gives 2 to F_1 whatever the date and the location. Note that missing data are not "penalized" because it is very difficult to obtain contextual information [14]. It can cause bias[1]. An interesting avenue will be to propose an aggregation strategy to "penalize" the lack of contextual data; e.g., a weighted average according to the contextual data.

3.2 Conceptual Modelling

Multidimensional modelling represents data as points in a multidimensional space. More precisely, a multidimensional schema is based on a *subject of analysis* (composed of indicators) related to different *analysis dimensions*. An analysis dimension models an analysis axis; it represents information according to which subjects of analysis are to be dealt with. Analysis dimension attributes are organized according to one or more hierarchies. Each hierarchy is a path from the lowest level to the higher level of granularity and expresses a point of view of the analysis dimension. Thus, the attributes are grouped by points of view - which correspond to the axes of analysis. All the attributes of the same axis are organized in a hierarchy. Each hierarchy corresponds to a view of the considered axis which organizes the attributes according to their functional dependencies.

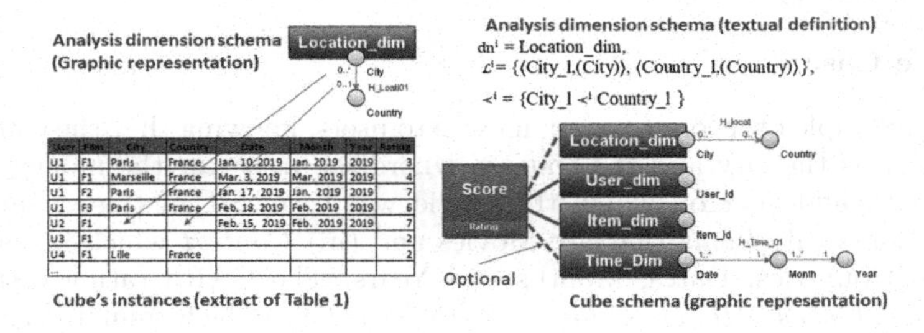

Fig. 1. Analysis dimension schema (graphic representation and textual definition), cube's instances and cube schema.

Our objective is to apply this multidimensional modelling to the context of RS. In the field of CARS, we define a *cube* composed of *analysis dimensions* and one *measure* corresponding to the rating. Each component (cube and analysis dimension) is modeled through a schema and instances [19].

An *analysis dimension* represents an analysis axis composed of attributes. There are at least two analysis dimensions: *Users* and *Items*. These two analysis dimensions are composed of only one attribute: its identifier. The other analysis dimensions are related to the context and may contain several hierarchies. We define an *analysis dimension schema* as a triple $D_i = \langle dn^i, \mathcal{L}^i, \prec^i \rangle$ where dn^i is the analysis dimension's name; $\mathcal{L}^i$ is a finite set of pairs $\langle I_X, \mathcal{A}_X \rangle$ such that I_X is a level and $\mathcal{A}_X$ is a set of attributes associating to a level; $\prec^i$ is a set of asymmetric binary relations which reveals the aggregation path between a

[1] We do some interpretation on the data. Through this, we introduce a bias.

pair of levels (hierachy definition). In the upper part of Fig. 1, we represent the *analysis dimension* named Location_dim. The upper right part of the figure gives the textual definition of this *analysis dimension*. This *analysis dimension* is composed of two levels (City_l, Country_l); the level City_l contains the attribute City and the level Country_l contains the attribute Country. This dimension is composed of one hierarchy where the lowest level is City_l and the higher level is Country_l. Thus, we can analyse the rating by cities or by countries.

We complete these textual definitions by graphic notations. Each *analysis dimension schema* is represented by a tree structure whose root contains the name of the dimension. Each node in this graph represents an aggregation level ($l_m \in \mathcal{L}^i$) and is annotated with a list of attributes. To represent the fact that contextual information may be missing, we complete this graphical representation with cardinalities. These cardinalities may translate the fact that a son may possibly have a father or that a father may have some sons or none at all. We define 4 cardinality values (0..*, 1..*, 0..1 or 1). An instance d_j of an analysis dimension $D_i = \langle dn^I, \mathcal{L}^I, \prec^I \rangle$, consists of a finite set of instances of attributes. The upper left part of Fig. 1 contains the graphical representation of the *analysis dimension* Location_dim. The cardinalities express the fact that a city may be associated with a country and that a country may possibly be subdivided into cities. In the lower part of Fig. 1, we represent some instances of the cube. For example, the fifth instance contains no information for the dimension location_dim.

Table 2. Utility matrix at the most aggregated level, M_u^{aggr}

Users	Movies	Countries	Years	Ratings
U_1	F_1	France	2019	3.5
U_1	F_2	France	2019	7
U_1	F_3	France	2019	6
U_2	F_1	.	2019	7
U_3	F_1	.	.	2
U_4	F_1	France	.	2
U_5	F_1	France	2019	8
U_8	F_5	France	2019	3.66
U_8	F_6	France	2019	5.5

A *cube schema* is a tuple $\langle nameCS, D, M \rangle$ where $nameCS$ is the name of the cube, D is a finite set of analysis dimensions, with $|D| = d$, and M is the measure related to the rating. D is at least composed of two mandatory analysis dimensions ($User\_dim$, $Item\_dim$) and possibly some dimensions related to context characteristics. In order to identify these contextual dimensions, we use the definition proposed in [21] where the physical context represents all aspects that can be influenced by the geographic position of the user; the personal context represents personal information about the user; and the technical context gathers characteristics of the devices used by the user to access the application.

A *cube schema* is represented by a star schema. In the center, we have M (the rating) and around, we have the analysis dimensions. A cube may contain two types of dimensions: mandatory dimensions (users and items) and optional dimensions dedicated to contextual characteristics. In the graphic representation, the mandatory dimensions are related to M with solid lines whereas the contextual/optional dimensions are related to M through dotted lines. For instance, the lower part of Fig. 1 illustrates a schema and instances of a cube. The schema is composed of two mandatory dimensions (*User_dim* and *Item_dim*) and two optional dimensions (*Location_dim* and *Time_dim*). From the instance point of view, the sixth instance of the cube does not contain any contextual characteristics and the fifth instance contains no information for the dimension location_dim. In fact, cube instances represent a utility matrix. This multidimensional cube serves as a support for the elaboration of other cubes containing more aggregated information from the utility matrix that will be used to support the recommendations.

4 Ratings Aggregation

4.1 Use Case

Now we complete the previous use case. We may organise the contextual information with two multidimensional dimensions: 1) a dimension named *Location_dim* with a hierarchy containing city and country where a country corresponds to the aggregation of all the cities of this country, and 2) a dimension named *Time_dim* with a hierarchy containing day, month and year.

Table 3. Utility matrix from which ratings will be aggregated (extracted from Table 1)

Users	Movies	Cities	Countries	Dates	Months	Years	Ratings
U_8	F_5	Lille	France	Jan. 10, 2019	Jan. 2019	2019	5
U_8	F_5	Lille	France	Jan. 17, 2019	Jan. 2019	2019	2
U_8	F_5	Paris	France	Feb. 18, 2019	Feb. 2019	2019	4
U_8	F_6	Paris	France	Jan. 15, 2019	Jan. 2019	2019	8
U_8	F_6	Lille	France	Feb. 18, 2019	Feb. 2019	2019	3

At the most aggregated level, the matrix (corresponding to Table 1) becomes the one of Table 2 (only rows containing ratings are displayed and aggregation function is average - used for users U_1 and U_8 who respectively have seen the movies F_1, F_5 and F_6 several times). Note that we took the average as an example, other aggregation functions are possible [12] and the aggregation function has no impact on the process (only on the recommendations returned). In Table 2, there is not a "loss" of information but rather a "loss" of detail. We make the following choice: rather than having nothing on some users/items, we prefer to aggregate and have something to the detriment of the detail of those which contained contextual information.

In addition, it should be noted that given these data, there are 6 users, 5 movies, 1 country and 1 year, which means that the matrix should contain 30 ratings; i.e., 6 Users × 5 Movies × 1 Country × 1 year to be dense and we have 9 ratings, which represents a density of 30% (i.e. a sparsity of 70%). If we assume that the aggregation function of the ratings is the average then we can now estimate (with more relevance) the rating of U_6 for the F_1 movie by aggregating the ratings of users who have seen the F_1 movie in *France* in 2019 and we get: 4.5 ($\frac{3{,}5+7+2+2+8}{5}$). Finally, through this example, our proposal allows the improvement of the matrix sparsity (from 99.198% to 70%) to better predict the scores (from considering only one other user with missing values, to considering many other users with similarities to the user for which we want to estimate a score) and thus improve the recommendations.

In order to improve the quality of the recommendations, we overcome the problem of recommendation when it requires to estimate ratings for items that have not yet been seen by a user. We will show in this section how to take advantage of the multidimensionality of attributes defining the recommendation context. We propose an approach to reduce the sparsity of the utility matrix.

4.2 Aggregating Ratings

In this section we overcome the problem of recommendation when it requires to estimate ratings for items that have not yet been seen by a user. In order to reduce the sparsity of the utility matrix, the idea consists in taking advantage of the multidimensionality of attributes defining the recommendation context.

The recommendation process is modelled as a function:

$$R : Users \times Items \times Context^p \rightarrow Ratings \tag{1}$$

$Context^p$ is the contextual information, and p represents the number of analysis dimensions that are related to the context. The context is considered as multidimensional because it incorporates several analysis dimensions, $Context^p = c^1 \times ... \times c^p$ where $\forall i \in [1, p], c^i$ is a fixed attribute from one analysis dimension. Then $Context^p$ defines a multidimensional context formed from p analysis dimensions. Thus the Eq. (1) is as follows.

$$R : Users \times Items \times c^1 \times ... \times c^p \rightarrow Ratings \tag{2}$$

As mentioned in previous sections, the higher the number of dimensions, the more it may form a sparse space; i.e., the more missing information may occur. For instance, if we consider the following function as introduced in Sect. 1, $R : Users \times Movies \times Dates \times Months \times Years \times Cities \times Countries \rightarrow Ratings$ where the context is $Dates \times Months \times Years \times Cities \times Countries$ as illustrated by Fig. 1, the sparsity of the multidimensional space,due to missing information of some combinations, is important.

To limit the sparsity during the recommendations we use the hierarchical organization of the analysis dimension attributes. In the previous example, the context consists of 5 attributes that can be organised in 2 analysis dimensions ($p = 2$): $Dates \times Months \times Years$ and $Cities \times Countries$.

At conceptual level, this context corresponds to 2 dimensions respectively defined as $Time\_dim$ and $Location\_dim$ with $Time\_dim = \langle Time, \{\langle I_{Dates}, Dates\rangle, \langle I_{Months}, Months\rangle, \langle I_{Years}, Years\rangle\}, I_{Dates} \prec^{Time} I_{Months} \prec^{Time} I_{Years}\rangle$ and $Location\_dim = \langle Location, \{\langle I_{Cities}, Cities\rangle, \langle I_{Countries}, Countries\rangle\}, I_{Cities} \prec^{Location} I_{Countries}\rangle$.

Then at the most detailed level, each analysis dimension is defined by $Dates \times Cities$, which corresponds to the root levels respectively annotated I_{Dates} and I_{Cities} of each analysis dimension into this context. This possibly sparse multidimensional context could be densified by replacing an attribute with a hierarchically higher attribute; e.g. along the first analysis dimension the attribute named $Cities$ could be replaced by $Countries$ while along the second analysis dimension the attribute named $Dates$ could be replaced by $Months$, and this latter could be replaced again by $Years$. This process follows the OLAP querying principle known as "roll-up".

Based on this idea of replacing attributes with higher hierarchical attributes, the challenge is to determine the attributes of the p analysis dimensions to minimize the sparsity of $Context^p$. The approach we introduce consists in starting from the context using root levels. At each iteration the "most informative attribute" of the context is replaced by its hierarchically higher attribute. In order to determine the more informative attribute we use a linear regression model, which consists in considering the following linear regression equation : $f(x) = \theta_1 x_1 + \theta_2 x_2 + \theta_3$ where θ_1 and θ_2 are the coefficient respectively associated to $Dates$ and $Cities$, whereas θ_3 is a bias coefficient. This equation is transformed in the following matrix equation

$$Y = X\theta \tag{3}$$

where Y is composed of the values we want to predict ($Ratings$) while $X = Context^p \times 1$ is a matrix of p context variables from which we want to approximate Y. This simple model is powerful because it supports different types of regression, which can be simple (when $p = 1$) or multiple linear (when $p > 1$) regressions as well as polynomial regressions such as $f(x) = \theta_1 x_1 + \theta_4 x_1^2 + \theta_2 x_2 + \theta_5 x_2^2 + \theta_3$ (these regressions make it possible to refine the model so that the values can be separated by a curve instead of being limited to a straight line). In order to extend to different regression Eq. 3, it is sufficient to both describe in additional columns in the X matrix the variables whose coefficients are to be found and add corresponding θ_i coefficients.

For instance, the conversion of $f(x) = \theta_1 x_1 + \theta_2 x_2 + \theta_3$, according to Table 3, gives equation/formula 4. For the sake of readability we consider only non-empty rows and categorical variables are transformed into numerical values. Then Y is composed of *Ratings* values, and $X = Context^2 \times 1 = Dates \times Cities \times 1$. This model based on a matrix equation consists in determining the θ_i coefficients that best approximate the values of Y (the *Ratings* in this example) from X.

$$\begin{bmatrix} 5 \\ 2 \\ 4 \\ 8 \\ 3 \end{bmatrix} = \begin{bmatrix} 1\,1\,1 \\ 2\,1\,1 \\ 3\,2\,1 \\ 4\,2\,1 \\ 3\,1\,1 \end{bmatrix} \begin{bmatrix} \theta_1 \\ \theta_2 \\ \theta_3 \end{bmatrix} \tag{4}$$

To determine the θ_i coefficients we use the loss function $J(\theta) = \frac{1}{2m} \times \sum(X\theta - Y)^2$ where m is the number of samples used to learn the predictive model, and we apply the process of gradient descent based on $\frac{\partial J(\theta)}{\partial \theta} = \frac{1}{m} X^T(X\theta - Y)$, from which we determine iteratively θ using the gradient descent $\theta = \theta - \alpha \frac{\partial J}{\partial \theta}$ where α is an hyper-parameter known as the learning rate. Considering Table 3, we can learn the model using the process described above. We obtain $\theta_1 = 0.295, \theta_2 = 1.972, \theta_3 = 0.864$ coefficients using $\alpha = 0.01$ and 1000 iterations of the gradient descent. Thus, the linear model solution of our problem is then $f(x) = 0.295x_1 + 1.972x_2 + 0.864$. When the learning model is obtained, we calculate its determination coefficient according to the formula $R^2 = 1 - \frac{\sum_{i=1}^{p-1}(y_i - \widehat{y_i})^2}{\sum_{i=1}^{p-1}(y_i - \overline{y_i})^2}$. In our running example, the determination coefficient we obtained is $R^2 = 0.389$ that is a down value; an R^2 of 1 indicates that the regression predictions perfectly fit the data while an R^2 near to 0 indicates that the predictions do not fit good the data.

The consequence of sparse matrices is that not many samples are available to build an efficient prediction model. We argue that recommendations can be improved by replacing detailed attributes with more aggregated attributes to reduce missing values in the matrix. We calculate several regression models where one context attribute is replaced by its directly higher attribute. For each new regression model, we calculate its determination coefficient, and we retain the more significant having the best coefficient. Considering the example above, we can create two regression models. At each iteration the best context is kept.

- The first model is learnt using $X = Months \times Cities \times 1$. Using the process of the gradient descent described above with the same hyper-parameters ($\alpha = 0.01$ and 1000 iterations), we obtain $\theta_1 = -1.238, \theta_2 = 3.087, \theta_3 = 1.742$ coefficients. The corresponding determination coefficient is $R^2 = 0.587$.
- The second one is learnt using $X = Dates \times Countries \times 1$. In the same manner, we obtain $\theta_1 = 0.966, \theta_2 = 0.937, \theta_3 = 0.937$ and a determination coefficient is $R^2 = 0.208$.

Thus we determine that the first regression model has the best R^2. The context attribute is then replaced by the best regression model's context, *Months* $\times$ *Cities* in our case. This new multidimensional context should be less sparse because generally we have less dispersed information at higher aggregation levels; i.e., the resulting matrix usually has fewer missing values, allowing more samples to be taken into account in the context.

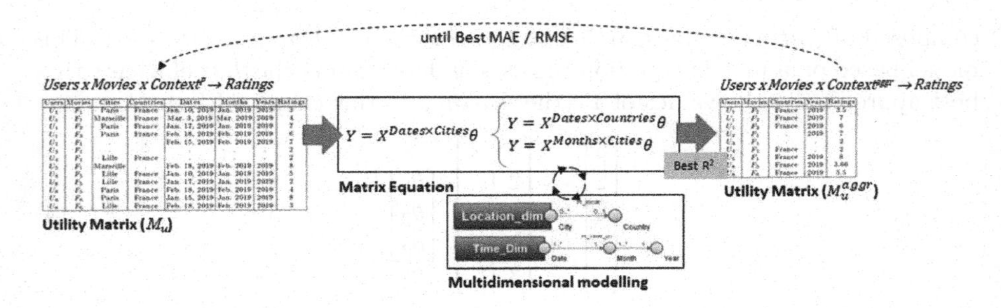

Fig. 2. Steps of our process.

This process, as illustrated in Fig. 2, is repeated until finding the best rating estimation; e.g. in the following experiments we use MAE (Mean Absolute Error[2]) and RMSE (Root Mean Square Error[3]) metrics to determine the best rating estimation.

5 Experimental Assessments

5.1 Datasets and CARS

We evaluated the approach on two real world datasets, which are well-known among the CARS community: *LDOS-CoMoDa* [17], a contextual dataset for movie recommendation and *Music* [6] that contains ratings for contextual music recommendation.

- *LDOS-CoMoDa* contains *UserIDs*, *ItemIDs*, *ratings* (between 1 and 5), and 27 context characteristics including *User's city*, *User's country*, *Time* (*Morning*, *Afternoon*, *Evening*, *Night*), *Daytype* (*Working day*, *Weekend*, *Holiday*), *Season* and *Location* (*Home*, *Public place*, *Friend's house*).
- *Music* contains *UserIDs*, *ItemIDs*, *ratings* (between 1 and 5), and 8 context characteristics including *Landscape* (*coast line*, *country side*, *mountains/hills*, *urban*), *Mood* (*active*, *happy*, *lazy*, *sad*), *Natural phenomena* (*afternoon*, *day time*, *morning*, *night*), *Traffic Conditions* (*free road*, *lots of cars*, *traffic jam*), and *Weather* (*cloudy*, *rainy*, *snowing*, *sunny*).

Table 4 illustrates some descriptive statistics about these datasets. Note that we calculated the sparsity and UI ($User \times Item$) sparsity as:

sparsity: $1 - \frac{\#ratings}{\#users \times \#items \times \#context\ characteristics}$

UI sparsity: $1 - \frac{\#ratings}{\#users \times \#items}$

[2] The Mean Absolute Error measures the average deviation (error) in the predicted rating versus the true/actual rating.

[3] The Root Mean Square Error is the standard deviation of the residuals (prediction errors). Residuals are a measure of how far from the regression line data points are. RMSE is similar to MAE, but places more emphasis on larger deviation.

Table 4. Dataset's descriptive statistics.

	LDOS-CoMoDa	Music
#ratings	2295	4012
#users	121	42
#items	1232	139
rating scale	1-5	0-5
rating's mean	3,83	2,37
#context characteristics	27	8
sparsity	99,9429%	91,4097%
UI sparsity	98,4604%	31,2778%

LDOS-CoMoDa. There exist functional dependencies between *Location*, *City* and *Country* and also between *Time*, *Daytype* and *Season*. This is why we consider them as attributes of analysis dimensions that can be ordered according to hierarchies such as:

- *Location* $\prec$ *City* $\prec$ *Country* meaning that each location is associated to a specific city and each city belongs to one country (a country may have several cities and a city may have several locations).
- *Time* $\prec$ *Daytype* $\prec$ *Season* meaning that several times of the day (Time) belong to the same type of day (Daytype) and that several types of day belong to the same season (reciprocally, a season contains different types of day and a type of day contains different times of the day).

Using our approach on the LDOS-CoMoDa dataset, we will be able to reduce the dimensionality of the context via these 6 attributes[4]. Therefore, the utility matrix with our approach will change from 27 context characteristics to 23 analysis dimensions. Note that if we represent the LDOS-CoMoDa data through a cube (as presented in the previous section), we obtain the schema of Fig. 3.

Music. With an OLAP point of view, there does not exist functional dependencies between context characteristics. On the other hand, by observing the possible values that the context characteristics can have, we see that we can group/aggregate some of them. So, for *Landscape*, *Mood*, *Natural phenomena*, *Traffic conditions* and *Weather*[5], we can create categories where *Landscape category* can be urban or non-urban, *Mood category* can be positive or negative, *Natural Phenomena category* can be during day or during night, *Traffic Conditions category* can be free or busy and *Weather category* can be good or bad.

Using our approach on the Music dataset, we will be able to reduce the dimensionality of the context via these 5 new characteristics. Therefore, the

[4] Note that for these 6 attributes, we have 3 different values for *Location*, 20 for *City*, 5 for *Country*, 4 for *Time*, 3 for *Daytype* and 4 for *Season*.

[5] Note that for these 5 attributes, we have 4 different values for *Landscape*, 4 for *Mood*, 5 for *Natural Phenomena*, 3 for *Traffic Conditions* and 4 for *Weather*.

utility matrix with our approach will change from 10 context characteristics to 10 analysis dimensions (with different levels of aggregation). The representation of the Music data through a cube is the schema of Fig. 4.

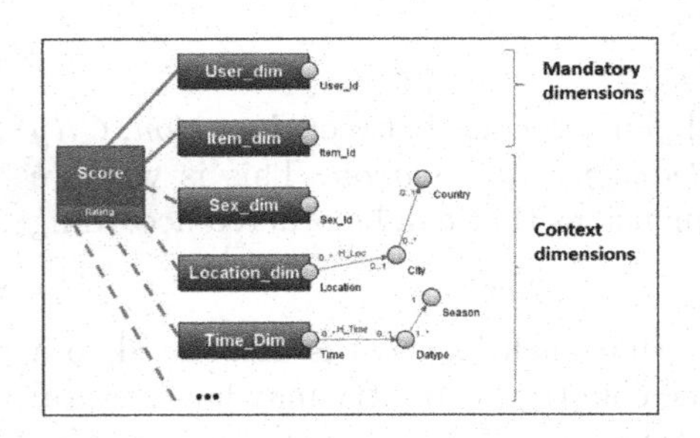

Fig. 3. Cube schema for the LDOS-CoMoDa data.

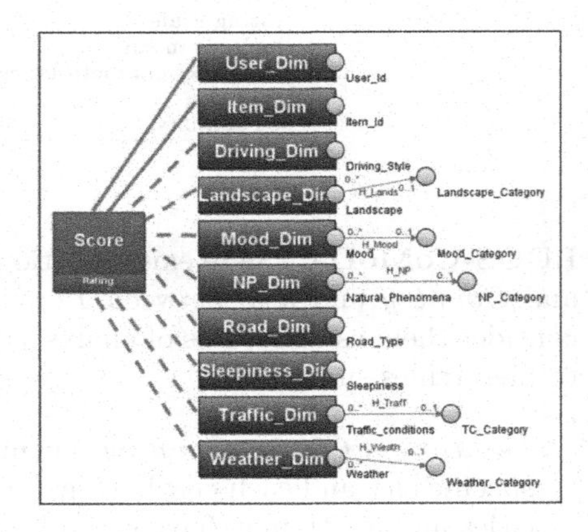

Fig. 4. Cube schema for the Music data.

Table 5. MAE/RMSE (and standard deviations - std) for LDOS-CoMoDa and Music according to four approaches.

Matrix	LDOS-CoMoDa MAE	LDOS-CoMoDa RMSE	Music MAE	Music RMSE
(1) Original	0,751 (0,0086)	0,979 (0,0064)	0,666 (0,0009)	0,942 (0,0023)
(2) $Users \times Items$	0,849 (0,0040)	1,069 (0,0042)	1,202 (0,0226)	1,497 (0,0254)
(3) Average	0,749 (0,0100)	0,984 (0,0110)	0,165 (0,0091)	0,319 (0,0078)
(4) Our approach	**0,730** (0,0071)	**0,963** (0,0066)	**0,663** (0,0018)	**0,934** (0,0052)

Our goal here is not to evaluate the quality of a given CARS but to show that our proposal achieves better results by resizing the utility matrix. For this purpose, we will use CARSKit[6], an open-source Java-based context-aware recommendation engine, specifically designed for context-aware recommendations. Many recommender algorithms are implemented. According to [24], among them, CAMF-C [7] is one of the methods that get the best results (particularly on the LDOS-CoMoDa dataset). So we will use this algorithm[7].

[6] https://github.com/irecsys/CARSKit.

[7] Some other approaches exist, such as SVD (Singular Value Decomposition), Matrix Factorization (MF), Factorization Machines (FM), ... All are derived from MF (e.g. FM is a generalization of MF (specific to (2D) collaborative filtering recommender systems), and CAMF-C is an extension of MF, specific to CARS approaches).

5.2 Experiments

Our first experiment consists in comparing our approach to existing approaches. Thus, we test the two real world datasets presented previously, namely LDOS-CoMoDa and Music, according to four approaches and so, we compare the results obtained with the CAMF-C algorithm on the four following matrices:

1. The original utility matrix (unmodified): $Users \times Items \times Context$
2. A 2D utility matrix, without taking into account contextual information: $Users \times Items$
3. A utility matrix averaging ratings for close/most similar $(User, Item)$ pairs, because the most used techniques is aggregate ratings using the average function according to the context [2].
4. The utility matrix resized by our approach, the one obtains the best MAE/RMSE.

Table 6. Best MAE/RMSE (and std) for LDOS-CoMoDa according to our approach.

	Matrix	MAE	RMSE
LDOS-CoMoDa	$Users \times Items \times 21\ dimensions \times$		
	$Location \times Time$	**0,734** (0,0057)	**0,959** (0,0070)
	$Location \times DayType$	0,749 (0,0056)	0,972 (0,0052)
	$Location \times Season$	**0,730** (0,0071)	**0,963** (0,0066)
	$City \times Time$	0,748 (0,0044)	0,982 (0,0081)
	$City \times DayType$	0,745 (0,0061)	0,979 (0,0093)
	$City \times Season$	0,749 (0,0070)	0,981 (0,0032)
	$Country \times Time$	0,749 (0,0097)	0,977 (0,0110)

Table 7. Best MAE/RMSE (and std) for MUSIC according to our approach. (L(c): Landscape (Category), M(c): Mood (Category), NP(c): Natural Phenomena (Category), T(c): Traffic conditions (Category), W(c): Weather (Category))

	Matrix	MAE	RMSE
Music	$Users \times Items \times DrivingStyle \times RoadType \times Sleepiness \times$		
	$L \times Mc \times NP \times T \times Wc$	0,663 (0,0007)	0,939 (0,001)
	$Lc \times Mc \times NP \times T \times Wc$	0,664 (0,0013)	0,939 (0,0025)
	$Lc \times Mc \times NPc \times Tc \times Wc$	**0,663** (0,0018)	**0,934** (0,0052)

As many researches in the domain, we used MAE (see footnote 1) and RMSE (see footnote 2) metrics to evaluate the rating estimation. Table 5 shows the results obtained for MAE and RMSE (and their standard deviations) of the four used matrices.

Here, because CAMF-C is not deterministic, we run CAMF-C five times on each tested matrix and we present in Tables 5, 6 and 7 the average values (and standard deviations). As a reminder, for MAE and RMSE, the smaller the values, the better the results.

Tables 6 and 7 show the detail of the matrices resulting from our approach which obtained MAE/RMSE values higher (except for one[8]) than the best obtained by the approaches (1-Original), (2-Users×Items) and (3-Average) of Table 5. Indeed, Table 5 shows that our approach promises us to surpass competing approaches but we get there with several tested matrices. Thus, Table 6 presents the 7 matrices among the 33 554 432 possible of our approach whose MAE/RMSE surpasses other approaches on the LDOS-CoMoDa dataset. Table 7 presents the 3 matrices among the 1944 possible of our approach whose MAE/RMSE surpasses other approaches on the Music dataset.

Concerning LDOS-CoMoDa, the tested matrices correspond to $Users \times Items \times\ Context$ (23 *analysis dimensions*) where the context characteristics *Location*, *City* and *Country* are considered as attributes of a single analysis dimension with 3 hierarchical levels and where context characteristics *Time*, *DayType* and *Season* are considered attributes of a single analysis dimension with 3 hierarchical levels. Thus, there are 1(Users) * 1(Items) * 21 (flat dimensions) with 2 levels (1 attribute + All) * 4 (Location, City, Country, All) * 4 (Time, DayType, Season, All) = 33 554 432 possible combinations.

In the same way, for Music, the tested matrices correspond to $Users \times Items \times$ *Context* (8 *analysis dimensions*) where the analysis dimensions *DrivingStyle*, *RoadType* and *Sleepiness* are flat dimensions with 2 levels (1 attribute + All) and each of the other 5 analysis dimensions have 3 levels (including All). Thus, there are 1(Users) * 1(Items) * 3 (flat dimensions) with 2 levels * 5 dimensions with 3 levels = 1944 possible combinations.

As stated in the literature, recommendations (rating estimations) are of better quality when contextual information is taken into account. Indeed, according to Table 5, the values of MAE and RMSE[9] are much better on the initial matrix (Matrices: Original) than on a 2D matrix (Matrices: $Users \times Items$).

Similarly, as indicated by previous work, aggregating the ratings with the average makes it possible to improve the recommendations (rating estimations). Indeed, according to Table 5, the values of MAE and RMSE by aggregating with the average (Matrix: Average) are better than on the initial matrix (Matrix: Original). Unfortunately, with regard to matrix sparsity, for which we obtain an improvement varying between +0.9 and +4.95 $^0/_{00000000}$ for LDOS-CoMoDa and between +0.034 and +1.613 $^0/_{00}$ for Music, the more contextual dimensions, the greater the sparsity. It should be noted that these weak improvements in the matrix sparsity are linked to the difficulty of generating/creating deep hierarchies

[8] The good results obtained with Average on the Music dataset can be explained: Music is relatively dense (UI sparsity $\approx$ 31%) in the sense that it contains many $User \times Item$ duplicates and because it has a particular structure. In fact, each user evaluates repeatedly the same item where a single context characteristic is filled.

[9] The lower the MAE/RMSE, the better the recommendations.

with real contextual data which does not lend themselves to it. Indeed, LDOS-CoMoDa contains 23 dimensions where 2 dimensions have at least 4 hierarchical levels, i.e. 8.7% of "deep" dimensions, while Music contains 8 dimensions but none has at least 4 hierarchical levels, i.e. 0% of "deep" dimensions. Our approach, as expected, is more effective when the dataset contains deep dimensions.

Concerning LDOS-CoMoDa, our approach allows us to act on 6 context characteristics that are dependent (Location/City/Country and Time/DayType/Season) resizing the initial matrix of 27 context characteristics to 23 analysis dimensions. We have chosen to show in Table 6, the "best" results obtained for 7 possible crossings between the hierarchical levels (i.e. attributes). Note that, all the crossings tested make it possible to densify (slightly) the matrix (compared to Average/Original matrices). The best MAE is obtained (with a very slight advance) with *Location* × *Season*.

Concerning Music, our approach allows us to act on 5 context characteristics by aggregating data along hierarchies (Landscape, Mood, NaturalPhenomena, TrafficConditions, Weather).

We have chosen to show in Table 7, the results obtained for the 3 "best" crossings between the hierarchical levels (i.e. attributes). It should be noted that the last line of Table 7, i.e. $Users \times Items \times DrivingStyle \times Roadtype \times Sleepiness \times Lc \times Mc \times NPc \times Tc \times Wc$, corresponds to the most aggregated matrix that we can obtain with our OLAP schema and it is this crossing which gives the best results in terms of MAE/RMSE and sparsity.

Finally, as expected, our approach enables to improve the recommendation quality (rating estimations) and the matrix sparsity. The gain may seem small but it is a gain. Indeed, in the field of CARS, with very sparse real world datasets, our results are noteworthy. Finally, let us note that in general, the initial data are at a low level of aggregation, which makes it difficult to find similar contexts; by aggregating, our approach makes it possible to obtain contexts with similarities and improves predictions.

6 Conclusion and Perspectives

In this paper, we propose a contextual modelling specific to CARS based on a data modelling used in decision-making and which is proven, namely the analysis of ratings according to different dimensions with attributes modelled as hierarchies. In order to limit the data sparsity, our proposal is to reduce the dimensionality of the context by aggregating the most detailed contextual data. Our approach supports multiple dimensions and hierarchical aggregation of ratings. It should be noted that our approach will also allow the pre-aggregation of the ratings thanks to the hierarchization of the data by dimension, this should save time in calculating recommendations. Through the first experiments carried out, our approach improves the quality of the recommendations and (slightly) the utility matrix sparsity. A next step will be to test our approach on different datasets with different density/sparsity with more complex dimensions. We plan to find real world datasets as sparse as those tested here, but which will allow us to build deeper hierarchies (of contextual data) in order to show the true added value of our approach. We will also compare our approach to those such as [22].

In addition, it should be noted that our proposal which makes it possible to hierarchize the contextual information according to certain dimensions seems exploitable for group recommendations. Indeed, if a hierarchical level corresponding to a given group of users exists/is created in an analysis dimension, then it would be possible to estimate, according to our approach, a rating for this hierarchical level/attribute. Finally, our goal will be to propose a generic and therefore reproducible approach regardless of the type of CARS (the lack of genericity is a real challenge for RS).

We explored the issue of reducing the sparsity of the utility matrix by incorporating a linear regression model to determine the most informative attributes based on multidimensional modeling. Our current work involves extending the approach with polynomial regressions to divide the data space by curves instead of straight lines. We also plan to take advantage of the multidimensional modeling by pre-aggregating some of the calculations to speed up recommendations.

References

1. Abbar, S., Bouzeghoub, M., Lopez, S.: Context-aware recommender systems: a service-oriented approach. In: VLDB 2009 (2009)
2. Adomavicius, G., Mobasher, B., Ricci, F., Tuzhilin, A.: Context-aware recommender systems. AI Mag. **32**(3), 67–80 (2011)
3. Adomavicius, G., Sankaranarayanan, R., Sen, S., Tuzhilin, A.: Incorporating contextual information in recommender systems using a multidimensional approach. ACM Trans. Inf. Syst. **23**(1), 103–145 (2005)
4. Adomavicius, G., Tuzhilin, A.: Extending recommender systems: a multidimensional approach. In: Proceedings of the International Joint Conference on Artificial Intelligence (IJCAI 2001), pp. 4–6 (2001)
5. Adomavicius, G., Tuzhilin, A.: Context-aware recommender systems. In: Ricci, F., Rokach, L., Shapira, B., Kantor, P.B. (eds.) Recommender Systems Handbook, pp. 217–253. Springer, Boston (2011). https://doi.org/10.1007/978-0-387-85820-3_7
6. Baltrunas, L., et al.: InCarMusic: context-aware music recommendations in a car. In: Huemer, C., Setzer, T. (eds.) EC-Web 2011. LNBIP, vol. 85, pp. 89–100. Springer, Heidelberg (2011). https://doi.org/10.1007/978-3-642-23014-1_8
7. Baltrunas, L., Ludwig, B., Ricci, F.: Matrix factorization techniques for context aware recommendation. In: ACM Conference on Recommender Systems, RecSys, pp. 301–304 (2011)
8. Bolchini, C., Curino, C.A., Quintarelli, E., Schreiber, F.A., Tanca, L.: A data-oriented survey of context models. SIGMOD Rec. **36**(4), 19–26 (2007)
9. Chen, Y., Wu, C., Xie, M., Guo, X.: Solving the sparsity problem in recommender systems using association retrieval. JCP **6**, 1896–1902 (2011)
10. Dey, A.K.: Understanding and using context. Pers. Ubiquitous Comput. **5**(1), 4–7 (2001)
11. Guo, G.: Resolving data sparsity and cold start in recommender systems. In: User Modeling, Adaptation, and Personalization, pp. 361–364 (2012)

12. Hassan, A., Ravat, F., Teste, O., Tournier, R., Zurfluh, G.: Differentiated multiple aggregations in multidimensional databases. In: Cuzzocrea, A., Dayal, U. (eds.) DaWaK 2012. LNCS, vol. 7448, pp. 93–104. Springer, Heidelberg (2012). https://doi.org/10.1007/978-3-642-32584-7_8
13. Kulkarni, S., Rodd, S.F.: Context aware recommendation systems: a review of the state of the art techniques. Comput. Sci. Rev. **37**, 100255 (2020). https://doi.org/10.1016/j.cosrev.2020.100255
14. Le, Q.H., Vu, S.L., Nguyen, T.K.P., Le, T.X.: A state-of-the-art survey on context-aware recommender systems and applications. Int. J. Knowl. Syst. Sci. **12**, 1–20 (2021)
15. Negre, E., Ravat, F., Teste, O.: OLAP queries context-aware recommender system. In: Hartmann, S., Ma, H., Hameurlain, A., Pernul, G., Wagner, R.R. (eds.) DEXA 2018. LNCS, vol. 11030, pp. 127–137. Springer, Cham (2018). https://doi.org/10.1007/978-3-319-98812-2_9
16. Nhat Vinh, M., Nhat Duy, N., Thi Hoang Vy, H., Nguyen Hoai Nam, L.: an approach for integrating multidimensional database into context-aware recommender system. In: CISIM, pp. 231–242 (2014)
17. Odić, A., Tkalčič, M., Kunaver, M., Požrl, T., Tasič, J.F., Košir, A.: LDOS-CoMoDa dataset (2012). https://www.lucami.org/en/research/ldos-comoda-dataset
18. Panniello, U., Tuzhilin, A., Gorgoglione, M., Palmisano, C., Pedone, A.: Experimental comparison of pre- vs. post-filtering approaches in context-aware recommender systems. In: Proceedings of RecSys 2009, pp. 265–268. ACM (2009)
19. Ravat, F., Song, J., Teste, O.: Designing multidimensional cubes from warehoused data and linked open data. In: IEEE Tenth International Conference on Research Challenges in Information Science (RCIS), pp. 1–12 (2016)
20. Ravat, F., Teste, O.: Personalization and OLAP databases. In: Kozielski, S., Wrembel, R. (eds.) New Trends in Data Warehousing and Data Analysis, vol. 3, pp. 1–22. Springer, Boston (2009). https://doi.org/10.1007/978-0-387-87431-9_4
21. Vahidi Ferdousi, Z., Negre, E., Colazzo, D.: Context factors in context-aware recommender systems. In: Recommender Systems, Paris, France (2017)
22. Vu, S.L., Le, Q.H.: A deep learning based approach for context-aware multi-criteria recommender systems. Comput. Syst. Sci. Eng. **44**(1), 471–483 (2023)
23. Weng, S.S., Lin, B., Chen, W.T.: Using contextual information and multidimensional approach for recommendation. Expert Syst. Appl. **36**, 1268–1279 (2009)
24. Zheng, Y., Burke, R., Mobasher, B.: Splitting approaches for context-aware recommendation: an empirical study. In: 29th Annual ACM Symposium on Applied Computing, SAC 2014, pp. 274–279. ACM (2014)

Business Processes Analysis and Improvement

Discovery of Improvement Opportunities in Knock-Out Checks of Business Processes

Katsiaryna Lashkevich(✉), Lino Moises Mediavilla Ponce, Manuel Camargo, Fredrik Milani, and Marlon Dumas

University of Tartu, Narva mnt 18, 51009 Tartu, Estonia
{katsiaryna.lashkevich,linomoises.mediavillaponce,manuel.camargo, fredrik.milani,marlon.dumas}@ut.ee

Abstract. Overprocessing is a source of waste that occurs when unnecessary work is performed in a process. Overprocessing is often found in application-to-approval processes since a rejected application does not add value, and thus, work that leads to the rejection constitutes overprocessing. Analyzing how the knock-out checks are executed can help analysts to identify opportunities to reduce overprocessing waste and time. This paper proposes an interpretable process mining approach for discovering improvement opportunities in the knock-out checks and recommending redesigns to address them. Experiments on synthetic and real-life event logs show that the approach successfully identifies improvement opportunities while attaining a performance comparable to black-box approaches. Moreover, by leveraging interpretable machine learning techniques, our approach provides further insights on knock-out check executions, explaining to analysts the logic behind the suggested redesigns. The approach is implemented as a software tool and its applicability is demonstrated on a real-life process.

Keywords: Business process improvement · Process mining · Overprocessing waste

1 Introduction

Minimizing waste is a common process improvement goal [19]. Overprocessing is a type of waste that occurs when effort is spent performing some activities in the process, but no value is provided to the customer or the business [24]. Overprocessing waste is typical for processes that contain *knock-out checks* [2, 24], i.e., activities that classify cases as either "accepted" or "rejected" and that may then lead to premature termination of a case. When a case is rejected, the work performed on it up to its rejection is the overprocessing waste. Knock-out checks are commonly found in application-to-approval processes in banks (e.g., loan origination processes), insurance companies (e.g., claim-to-settlement processes), and government agencies [2,24].

S. Nurcan et al. (Eds.): RCIS 2023, LNBIP 476, pp. 381–397, 2023.
https://doi.org/10.1007/978-3-031-33080-3_23

Process mining enables data-driven analysis of business processes by analyzing process execution data captured in the form of event logs [1]. By using process mining techniques, analysts can discover the structure of their processes and analyze their performance [13]. In particular, process mining supports the identification and analysis of wastes in business processes [18]. However, few existing process mining-based techniques consider overprocessing waste. In [23], overprocessing is considered, but the approach does not provide transparency as it is based on a black-box model. Thus, existing data-driven techniques for optimizing business processes with knock-out checks either do not consider overprocessing waste or do not provide interpretable results for analysts to consider when seeking to improve the business process.

In this paper, we address this gap by tackling the following research questions (RQ): (1) *How can improvement opportunities related to knock-out checks be identified from event logs?*, and (2) *How can knock-out checks be optimized in a process to reduce the overprocessing waste?* To answer these questions, we propose an interpretable process mining-based approach to (1) discover knock-out checks, their decision rules and dependencies from an event log, (2) identify overprocessing waste associated with knock-out checks, and (3) identify improvement opportunities in knock-out checks and suggest redesigns to reduce overprocessing wastes. The approach has been implemented as a software tool that allows process analysts to upload an event log and obtain overprocessing waste analysis results and redesign suggestions w.r.t. discovered knock-out checks. Our approach is particularly useful for analysts aiming to reduce overprocessing in processes with multiple knock-out checks.

The rest of the paper is structured as follows. Section 2 presents the background and related work. Section 3 describes the proposed approach. Section 4 presents the implementation of the approach. Section 5 covers the evaluation of the approach, and Sect. 6 concludes the paper.

2 Background and Related Work

In this section, we present the concepts of knock-out checks and overprocessing waste. Then, we position our contribution w.r.t. existing approaches to discover, analyze, and redesign knock-out checks.

Knock-Out Checks and Overprocessing. Knock-out checks are activities in business processes that classify cases into two groups: "accepted" and "rejected". When the case is "accepted", it proceeds forward. When "rejected," the case is directed to a designated point of the process known as an anchor. An anchor can be any point in the process, i.e., when the case is rejected, it can be returned to the earlier stage of the process, sent to a later stage, or to process completion [24]. In this paper, we consider knock-out checks that directly (the anchor is the end event of the process) [24] or eventually (the anchor is the activity at a later stage of the process followed by the end event) conclude the case with a negative outcome. A typical example is application-to-approval processes [7], such as a loan application process, where a customer applies for a loan, and the application goes through several eligibility checks. If a check fails, the application is rejected, and the process execution ends [24].

When a knock-out check results in the rejection of a case, the work performed on the case up to the rejection is considered unnecessary since it delivers value neither to the customer nor to the business [7]. Such unnecessary performed work is a manifestation of an overprocessing waste [22] and, thereby, an inefficiency that results in increased process time and costs [4].

Related Work. A number of studies focus on methodologies for improving business processes [12], in particular, for process redesign. For example, Netjes et al. [14] presented an approach for "Process Improvement by Creating and Evaluating process alternatives" (PriCE), as a tool to support the BPM lifecycle, including the redesign phase. This approach allows analysts to identify applicable redesigns for the selected process parts and to generate alternative process models, alongside their performance evaluation. Likewise, Niedermann and Schwarz [16] introduced a "Deep Business Optimization Platform" that, given a process model and optimization goals, computes recommended changes using a redesign pattern catalog. Souza et al. [20] proposed heuristics to automate the redesign pattern selection. Following the idea of supporting analysts in process redesign, Fehrer et al. [8] proposed a conceptualization of assisted Business Process Redesign (aBPR) and a classification of redesign recommendations by the automation level. These approaches consider collections of redesigns targeting multiple issues, whereas we focus on optimizing knock-out checks to reduce overprocessing. Furthermore, these approaches require an as-is process model as input, whereas we use event logs.

In order to minimize overprocessing waste with knock-out checks, several redesign options have been proposed [11,15,17]. For instance, van der Aalst suggests reordering knock-out checks based on their rejection rate and processing time, combining checks into composite tasks, and placing subsequent checks in parallel [2]. Another heuristic, called the "knock-out principle" [11], proposes ordering knock-out checks according to the "least effort to reject" ratio, i.e., in decreasing order of effort and in increasing order of the rejection rate [17]. Redesign heuristic "early knockout" suggests moving knock-out checks to the earliest possible point of a process [15]. In our approach, we analyze the process to identify when these redesign heuristics can be applied.

Several studies specifically focus on knock-out check optimization. In [2], a set of heuristics for the knock-out process redesign was described, and an implementation of the redesign approach was presented that requires a process model in Petri-Nets notation as input. Verenich et al. [24] proposed a run-time knock-out check reordering technique with predictive models, as opposed to the design-time [2]. Verenich et al. [24] take an event log and information about knock-out checks (the knock-out activity names and the disallowed permutations) as input and suggest how to re-order the knock-out checks on a case-by-case basis. However, this approach is based on black-box predictive models, i.e., does not provide reasoning for the recommended changes.

Process mining is increasingly focusing on the interpretability of the obtained results [10]. Interpretability appears essential for analysts to understand the logic behind suggested redesigns and predicted performance [6] to gain confidence in making decisions based on recommendations [10]. Thus, Lee [10] proposed an approach for interpretable prediction of process outcomes at run-time from event logs that explains the predicted outcomes. Lashkevich et al. [9] presented an interpretable approach for discovering and analyzing batch processing inefficiencies with insights on batch processing behavior and associated waiting times. In this paper, we also develop an interpretable approach but for the discovery, analysis, and redesign of business processes w.r.t. knock-out checks.

3 Knock-Out Check Discovery and Analysis

In this section, we describe our approach to discover and analyze knock-out check improvement opportunities in a business process. Our approach takes an event log as input to produce a report comprising candidate improvement opportunities relating to knock-out checks and recommended redesigns. Optionally, an analyst can provide insights regarding the process structure as additional input.

Figure 1 depicts an overview of the main steps of the approach. In the first step, we discover knock-out checks, their knock-out rules, and data dependencies from the event log. In the second step, we identify how much overprocessing waste each knock-out check produces and compute their effort-per-rejection rates. In the third and final step, we determine which knock-out checks can be redesigned and, if so, suggestions on how they can be redesigned.

3.1 Knock-Out Check Discovery

The first step of our approach is to discover knock-out checks, their decision rules, and dependencies. As input, we require an event log containing at least case ID, activity label, and one timestamp. With these minimum required data, the approach allows for discovering knock-out checks, identifying improvement opportunities, and presenting redesign options. If the event logs include both start and end timestamps, the approach can also calculate overprocessing waste and, if case attributes are available, knock-out decision rules and dependencies.

Knock-Out Check Discovery. We propose a semi-automated approach for discovering knock-out checks from event logs. First, given an event log, the approach requests analysts to specify either *post-knock-out activities* (one or more

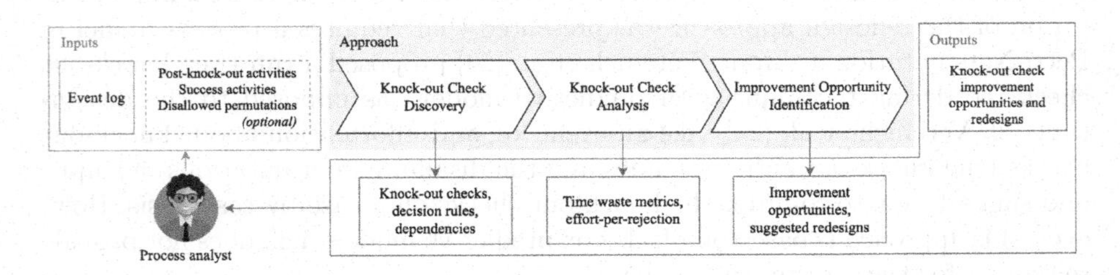

Fig. 1. Overview of the proposed approach.

activities performed immediately after cases are knocked out) or *success activities* (one or more activities performed only for cases that are not knocked out), or both. In addition, analysts can specify *disallowed permutations* (prohibited activity orders). Disallowed permutations, such as activity C cannot be executed before activity B, are useful when dependencies cannot be automatically discovered from the event log (when no relevant case attributes are available). However, manually entered parameters are optional, and analysts can skip this step.

Second, the knock-out check discovery is performed.

(1) If the post-knock-out and/or success activities are specified, the technique identifies knock-out checks as activities preceding post-knock-out and success activities with a directly-follow relation with them.
(2) If the post-knock-out and/or success activities are not specified, the technique extracts the process variants based on the recorded unique pathways and sorts them according to their length. Subsequently, it extracts all transitions between activities in each variant. The transitions are filtered to extract those differentiating for each variant, i.e., those that do not occur in other variants. Among these transitions, the most frequent for each variant is selected and marked as a potential transition that leads to a knockout. These possible transitions are filtered to extract only those activities that led to an early end of the process, i.e., knock-out checks.
(3) If the knock-out checks are known a priori, the analyst can manually mark them. In such cases, tagging post-knock-out and success activities for knock-out discovery is not needed.

As such, the approach supports three modes of operation: (1) semi-automatic discovery with known post-knock-out or success activities, (2) automatic discovery (no input from an analyst is provided, and the discovery is performed exclusively from the event log data), and (3) known knock-out checks (no discovery is performed, and analysts specify the knock-out checks). These operation modes aim to provide flexibility to analysts, allowing them to integrate their domain knowledge of the process.

Decision Rule Discovery. When knock-out checks are identified, we discover the decision rules used to determine which cases are "rejected". For each knock-out check, we obtain a decision rule model and answer the question *"given this case, will the knock-out check reject it?"*, i.e., for every knock-out check, we solve a binary classification problem. We use RIPPER [5], a rule discovery algorithm, to solve the binary classification problem. Other commonly used classification algorithms, such as C4.5 and IREP, were also considered for this task. However, RIPPER has been shown to achieve high accuracy rates in many real-world applications and is generally more efficient than C4.5 and IREP when working with large datasets [5]. This is because RIPPER generates a smaller set of more accurate rules, reducing the computation time needed to classify new instances. Moreover, RIPPER produces a set of rules in natural language with lower complexity (i.e., easier for humans to understand). Since we aim for interpretability in data analysis and decision-making, this algorithm is suitable.

Table 1. Example event log showcasing data dependency.

Case ID	Activity	Start Timestamp	End Timestamp	Amount	Risk Score
1	Check Documents	07/09/2022 16:36	07/09/2022 16:46	35000	—
1	Assess Application	07/09/2022 16:50	07/09/2022 17:30	35000	—
1	Assign Risk Score	07/09/2022 18:10	07/09/2022 18:45	35000	0.56
1	Check Risk Score	07/09/2022 18:45	07/09/2022 18:55	35000	0.56
1	Notify Rejection	07/09/2022 19:00	07/09/2022 19:01	35000	0.56

We create feature vectors by sorting events by end timestamp in ascending order and aggregating event data at the case level, taking the last available value of the attributes of each case. Then, we train the RIPPER decision rule model for every knock-out check and obtain a set of rules in disjunctive normal form, such that if it evaluates to *True* on a given case encoded as a feature vector, the case is labeled as "rejected". For example, consider a knock-out check "Check Liability" and the discovered set of rules `(Monthly Income < 800) V (Owns Vehicle = False)`. If a case has attributes `{Monthly Income: 1200, Owns Vehicle: False, ...}`, the decision rule model of "Check Liability" indicates that this case will be "rejected" in this knock-out check.

Then, the discovered decision rules are filtered based on their confidence; if it is lower than a given threshold, the rules are not taken into consideration for the rest of the analysis. Analysts can also choose to keep them in the analysis with relevant warnings instead. As a result, for each identified knock-out check, we obtain a collection of rules that capture (up to some level of confidence and support) the conditions under which a case is rejected by a knock-out check.

Dependency Discovery. Activities might depend on each other for data and objects. These dependencies must be considered when reordering and relocating activities [2,17,24]. We identify dependencies by using case attributes that appear in the discovered decision rules to perform a search in the log and determine which activity *produces* the value of each case attribute. If the decision rule of a knock-out check K involves a case attribute that is available (or stops changing) after activity A, we consider that the knock-out check K depends on A. For example, consider a loan application process with a knock-out check "Check Risk Score" and knock-out rule `Risk Score > 0.5`. Given a log as in Table 1, we can identify that `Risk Score` is produced by "Assign Risk Score" and "Check Risk Score" depends on "Assign Risk Score" for the `Risk Score` case attribute. Additionally, when the data dependencies cannot be detected from the event log, we allow the user to specify disallowed knock-out check permutations, as in [24].

Thus, the first step results in discovered knock-out checks, their decision rules, and dependencies on other activities based on the case attributes.

3.2 Knock-Out Check Analysis

Once the knock-out checks, their decision rules, and dependencies are discovered, we conduct the knock-out check analysis to answer the questions: (1) *how much*

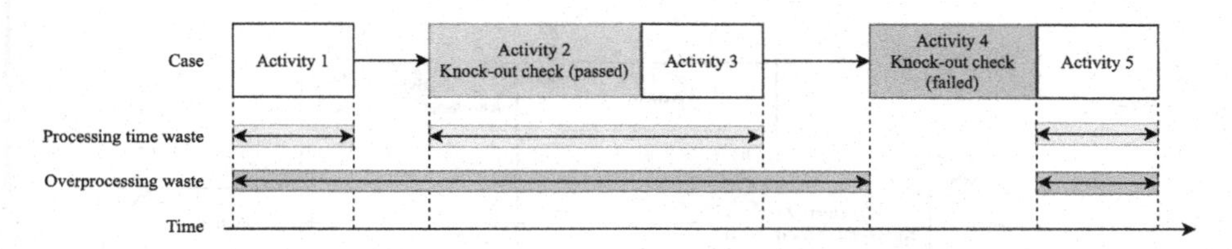

Fig. 2. Processing time waste and overprocessing waste.

waste is associated with cases rejected by each knock-out check? and (2) *what is the mean effort associated with each knock-out check?* This allows us to assess how much overprocessing waste is attributed to each knock-out check and how efficiently they are ordered w.r.t. the "knock-out principle" [11].

When the case is rejected, the effort spent on the case becomes a waste [24]. Therefore, to address question (1), we propose measuring the following metrics: *overprocessing waste*, *processing time waste*, and *waiting time waste.*

Processing time waste is the total effort (processing time) spent on a rejected case, except for the processing time of the check by which it was knocked out. The knock-out check that rejected the case is value-adding since it allows for the termination of an unnecessary case and, thus, is not considered waste [24]. Therefore, the ideal knock-out situation is when an unnecessary case is "rejected" in the first activity (i.e., no overprocessing waste).

Given an application-to-approval process P, a set of resources $R \in P$, a set of activities $A \in P$, a set of knock-out checks $K \subset A$, a set of cases C, a particular case C_i, and a particular knock-out check K_i that rejects C_i,

Definition 1. *The* ***processing time waste*** *for a case C_i due to knock-out check K_i is the sum of the processing times of all activities performed on case C_i, excluding waiting times and the processing time of K_i itself (Fig. 2).*

Overprocessing waste is the total time spent on a rejected case (processing and waiting time), except for the processing time of the check by which it was knocked out. This metric indicates how long the case was in processing before being knocked out.

Definition 2. *The* ***overprocessing waste*** *for a case C_i due to knock-out check K_i as the time elapsed since the case started until it finished, including processing and waiting times but excluding the processing time of K_i itself (Fig. 2).*

Apart from spending effort (processing time) on the rejected cases, the waiting times of "accepted" cases can increase due to the resource being busy processing the cases that are eventually "rejected". Hence, we use the **waiting time waste** metric that defines how much waiting time is induced to the "accepted" cases while the resource processes the "rejected" cases.

Definition 3. *Given the set of cases C_R rejected by K_i and the set of cases C_{NR} not rejected by K_i, we define the* ***waiting time waste*** *associated to K_i*

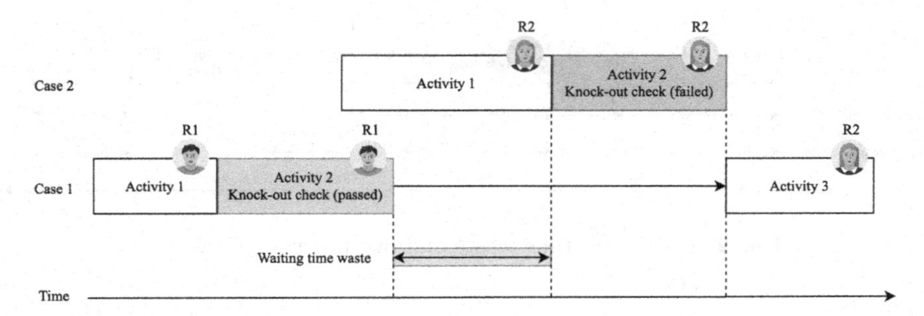

Fig. 3. Waiting time waste. *Case 1* (a non-knocked out case) needs *Resource R2* for advancing, but it has to wait because *Resource R2* is busy on *Case 2* (a case to be knocked out).

as the sum of the duration of the intervals (excluding the processing time of K_i*) during which* C_{NR} *cases are held on standby because the resources responsible for performing activities required to advance them are busy performing work on* C_R *cases (Fig. 3).*

To address question (2), we use the **effort-per-rejection** of each knock-out check that indicates the ratio between the effort spent on the check execution and its rejection probability [11,17]. It describes how much time is spent on the check and how frequently the cases are terminated.

Definition 4. *We define the* ***effort-per-rejection*** *of a knock-out check as the ratio between its average processing time and its rejection rate.*

When event logs include only one timestamp, the waste metrics calculation is omitted. In addition, the constant value of processing time is assumed. That is, the effort-per-rejection of knock-out checks in these situations becomes simply the inverse of its rejection rate, similar to what is proposed in [24].

After computing the time waste metrics and effort-per-rejection ratio, we calculate the descriptive statistics for the knock-out checks: (1) *total frequency:* the total number of cases in which the knock-out check is performed, (2) *case frequency:* the proportion of cases in which the knock-out check is performed, (3) *mean duration:* the average duration (including processing and waiting time) of the given knock-out check across all the cases where it has been performed. These statistics provide insights into how frequently the knock-out checks are executed, for how many cases, and how long they take.

3.3 Improvement Opportunity Identification

The final step of our approach is to identify improvement opportunities w.r.t. knock-out checks and suggest possible redesigns. We identify improvement opportunities by examining if there are knock-out check orders, positions of knock-out checks, or different decision rules that can reduce overprocessing. Therefore, we consider the following redesigns:

- **Knock-out reordering**: ordering the knock-out checks by least effort to reject, as in [24].
- **Knock-out relocation**: moving the knock-out checks as early in the process as the case attributes required by their knock-out rules are available (based on the "early knockout" pattern [15]).
- **Knock-out rule adjustment**: changing the value (or range) of numerical attributes of knock-out rules based on the actual distribution of the values observed in the event log.

The knock-out check reordering options are obtained by computing the optimal and dependency-aware ordering of the checks. We do so by applying the *knock-out principle* [11,17], (i.e., in ascending order by their effort-per-rejection value), taking into account any dependencies detected between activities and disallowed permutations if specified by analysts. We, then, use the data dependencies to relocate the knock-out checks in the process, i.e., as early as the case attributes required by their knock-out rules are available. Thus, the technique automatically provides a suggestion on how the knock-out checks should be ordered and relocated to obtain overprocessing reduction.

Finally, for every numerical case attribute appearing in the decision rules of the knock-out checks, we display the distribution of the attribute's values captured in the log and highlight the values of the cases knocked out by a specific check. Although we do not provide suggestions on how the rule should be changed, this data could help analysts in adjusting the knock-out rule values. The result of this step is a report specifying alternative knock-out check orders and positions, and the data on the decision rules that would help to achieve higher temporal efficiency and, in particular, reduce overprocessing waste.

4 Implementation

The proposed approach has been implemented as a software tool and is available on GitHub[1]. Analysts can use this tool to upload event logs with knock-outs and obtain recommendations on how the processes can be redesigned to reduce overprocessing waste. In this section, we illustrate how the approach can be applied on a real-life event log.

For that, we use the environmental permit application process [3]. The log has 1230 cases, 18 activity types, including 3 knock-out checks: "T02-check confirmation of receipt," "T06-determine necessity of stop advice," and "T10-determine necessity to stop indication". These checks have dependencies: "T10" can only be done after either "T02" or "T06" has been performed [24]. Cases in this log contain the following data attributes: the channel by which the case has been lodged, the department that is responsible for the case, the responsible resource, and its group. The log only contains end timestamps.

We uploaded the event log and specified the disallowed permutations (dependencies). Figure 4 depicts the discovered knock-out checks, their total and case

[1] https://github.com/AutomatedProcessImprovement/knockouts-redesign.

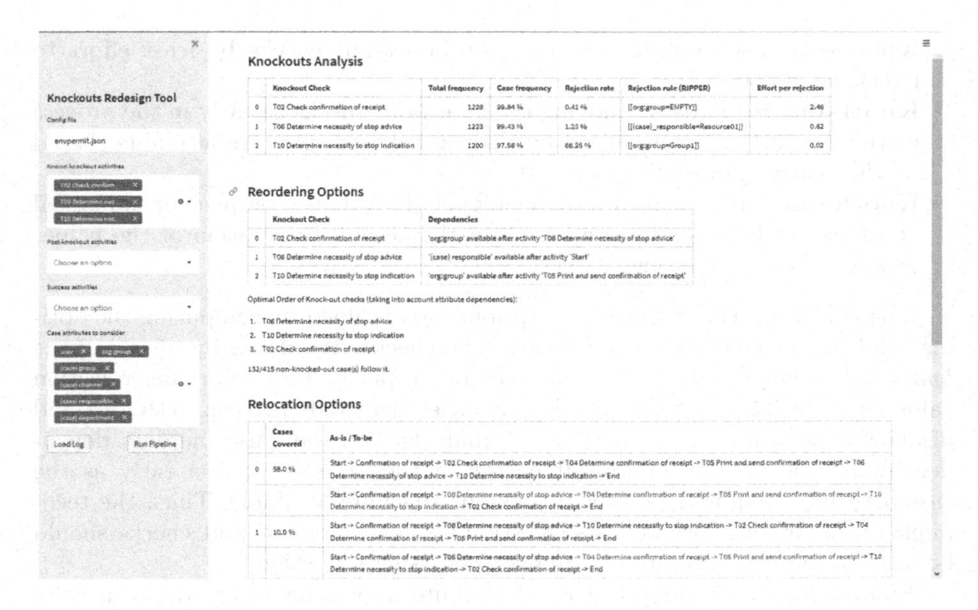

Fig. 4. Screenshot of the tool interface depicting the knock-out check analysis results for the environmental permit application process.

frequencies, rejection rate, decision rule, and effort-per-rejection. The time waste metrics are not calculated since the log has only end timestamps.

Further, the tool provides possible redesign options. The tool computes the optimal ordering of the knock-out checks considering constraints, specifically "T06" -> "T10" -> "T02" (Fig. 4). If "T10" did not require either "T06" or "T02" to be performed before, it would be suggested as the first knock-out check to perform, given its very high rejection rate and consequently low effort-per-rejection value compared to the other checks.

As for the relocation redesign, we observe that the knock-out checks have been placed as early as possible in the process, considering the discovered data dependencies. Figure 4 depicts the relocation option for the most frequent process variants. Theoretically, the knock-out checks could be placed right after the process start since such an order does not violate any discovered dependencies. However, we observed that "Confirmation of receipt" was always performed after the start. Therefore, we marked "Confirmation of receipt" as the start activity of the process, and thus, the relocation did not affect this activity. This showed that in our approach, the dependency detection between activities is limited by the availability and granularity of data.

In this event log, the decision rules of the knock-out checks are based on categorical case attributes (e.g., `org:group=Group1` in Fig. 4). The knock-out rule adjustment redesign focuses on amending numerical case attributes based on which decision rules are formulated (e.g., `Loan Amount > 10000`). Therefore, for this log, no data for the knock-out rule adjustment is presented.

5 Evaluation

In this section, we present the evaluation of our proposed approach, based on the implementation described in the previous section. We evaluate the approach by answering the following evaluation question: (EQ1) *To what extent is the technique able to correctly discover knock-out checks, associated overprocessing waste, and relevant redesigns?* We used synthetic data to address this question and, thereby, validate the ability of the technique to accurately rediscover knock-out checks, overprocessing waste, and improvement opportunities known to be present in the event log (Subsect. 5.1). Further, we compare the accuracy of our approach with the results of the approach proposed in [24], which we consider as baseline (Subsect. 5.2), to answer the following evaluation question: (EQ2) *To what extent the proposed approach sacrifices accuracy (in favor of interpretability) w.r.t. the baseline approach?*

The datasets used in the evaluation and the detailed results are available at https://github.com/AutomatedProcessImprovement/knockouts-redesign.

5.1 Evaluation of Rediscovery Accuracy

To answer EQ1, we created a synthetic event log of a credit application process. The log includes 3000 cases, 6 activity types with resources, start and end timestamps, including 4 knock-out checks: "Check liability", "Check risk", "Check monthly income" and "Assess application" (see Fig. 5). Cases that successfully pass all checks move to "Make credit offer", those that fail any of the checks move to "Notify rejection" and the process is terminated. For all knock-out checks, we assigned rejection rates (see R.R. in Fig. 5) and decision rules. For instance, "Check risk" was assigned a rejection rate of 30% and a decision rule of `(Loan Amount > 10000)`. For that, we post-processed the log to add case attributes with values that reflected the rules and rejection rates.

To test the technique's ability to identify data dependencies, we replicated the situation described in Table 1, namely, "Assess application" can be executed only after "Check risk" is performed. We did so by selectively removing the value

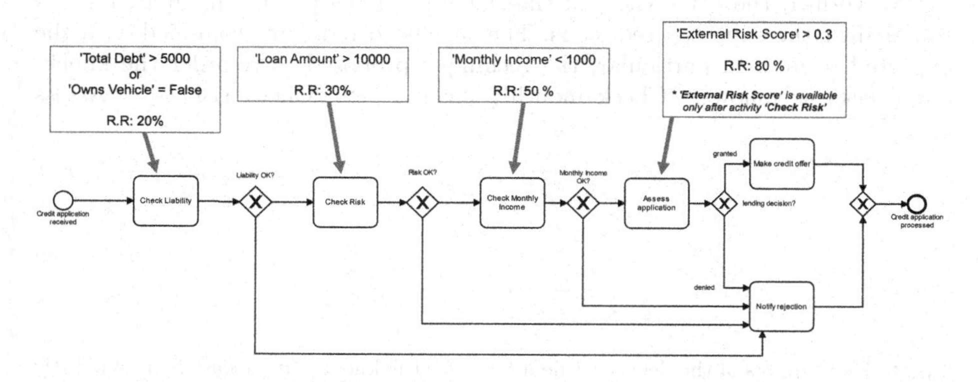

Fig. 5. Synthetic credit application process.

Table 2. Discovered knock-out checks and their decision rules.

Knock-out check	Decision rule	Confidence	Support
Assess application	[[External_Risk_Score=0.36-0.64] V [External_Risk_Score=>0.64]]	1.000	0.732
Check liability	[[Total_Debt=>5219.85] V [Owns_Vehicle=FALSE]]	1.000	0.193
Check monthly income	[[Monthly_Income=<564.21] V [Monthly_Income=564.21-830.79] V [Monthly_Income=830.79-1020.15]]	0.933	0.536
Check risk	[[Loan_Amount=11647.32-16709.71] V [Loan_Amount=>16709.71]]	1.000	0.252

of the "External risk score" attribute such that it becomes available only after "Check risk" is executed. We assess the approach's performance through time-series cross-validation using a temporal split of 80% of the cases for training and 20% for testing on each partition as recommended for time-series data [21].

The technique correctly discovered all four knock-out checks without any input provided by the analyst (the discovery was performed solely from the event log data) and their decision rules (see Table 2). To quantify the performance of the decision rule models obtained for the knock-out checks, we use standard metrics for binary classification performance: the ROC curve and the area under it (AUC). We report the resulting ROC curves and AUC values averaged over five cross-validation folds in Fig. 6. We can observe ROC curves and AUC values near 1.00 (perfect classifier). This observation, together with the high values of confidence (Table 2) and coherence between discovered rules and injected patterns in the log, confirms that the classification models perform as expected.

The technique then calculated the effort-per-rejection rate and the time waste metrics – total overprocessing waste, total processing time waste, and total and mean waiting time waste (see Fig. 7). The correctness of the metrics calculation was verified with a set of unit tests included in the code repository of the approach's implementation.

We verified the correctness of the suggested redesigns by manual comparison against the expected redesigns. The suggested redesigns complied with the expected results. In particular, the technique proposed to re-order the knock-out checks as follows: "Check monthly income" -> "Check risk" -> "Assess

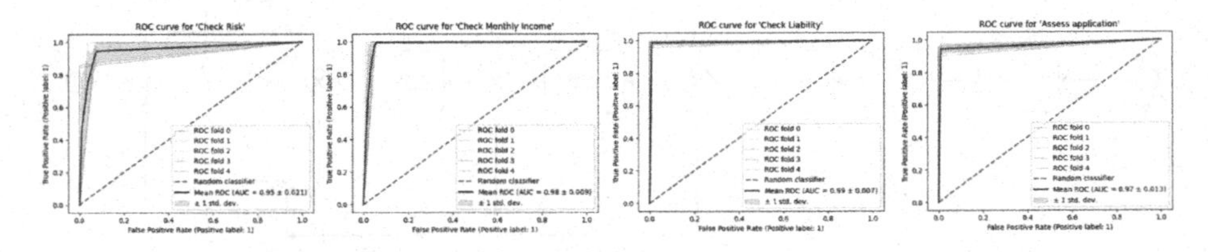

Fig. 6. ROC curves of the decision rule models for the knock-out checks of the synthetic event log.

	Knockout Check	Total frequency	Case frequency	Rejection rate	Rejection rule (IREP)	Effort per rejection	Mean Duration	Total Overprocessing Waste	Total PT Waste	Total Waiting Time Waste	Mean Waiting Time Waste
0	Assess application	836	27.87 %	80.02 %	[[External_Risk_Score=0.35-0.64] V	50.2	1:06:57	178 days, 0:35:20	139 days, 6:45:45	0:00:00	0:00:00
					[External_Risk_Score=>0.64]]						
1	Check Liability	3000	100.0 %	20.17 %	[[Owns_Vehicle=FALSE] V	202.83	1:08:10	55 days, 18:56:48	33 days, 15:40:30	0:00:00	0:00:00
					[Total_Debt=>5200.1]]						
2	Check Monthly Income	1674	55.8 %	50.06 %	[[Monthly_Income=555.77-830.79] V	43.08	0:35:56	139 days, 4:24:45	114 days, 18:13:13	0:00:00	0:00:00
					[Monthly_Income=<555.77] V						
					[Monthly_Income=830.79-1019.68]]						
3	Check Risk	2395	79.83 %	30.1 %	[[Loan_Ammount=11693.71-16840.45] V	136.54	1:08:30	143 days, 1:12:24	93 days, 4:40:17	0:00:00	0:00:00
					[Loan_Ammount=>16840.45]]						

Fig. 7. Knock-out analysis report for the synthetic event log.

application" -> "Check liability". The proposed order indicated that the technique was able to capture the inserted data dependency ("Assess application" depends on the case attribute produced by "Check risk") and consider it for computing the redesigns. If there were no dependencies, "Assess application"' would be ordered before "Check risk" and "Check liability" due to a smaller effort-per-rejection.

All decision rules of the knock-out checks include numerical case attributes (that are expressed as ranges of numerical values). Therefore, for each numerical case attribute of decision rules, the technique provided a graph with the distribution of numerical attribute values and highlighted values of the knocked-out cases. The graphs are coherent with the assigned case attributes and decision rules. For instance, Fig. 8 depicts the graph for the "Total dept" case attribute based on which the knock out in "Check liability" is performed. Using these data, analysts might consider adjusting the knock-out rule by, e.g., rejecting cases with a lower debt amount and, thus, reducing the number of cases processed further in the process and, subsequently, the overprocessing waste. In this way, we have verified that the approach identifies improvement opportunities in the form of redesign recommendations considering data dependencies.

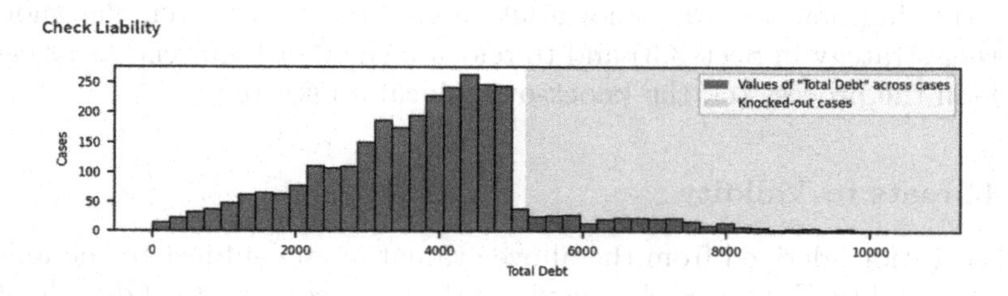

Fig. 8. Data for the knock-out rule adjustment for the synthetics event log.

5.2 Evaluation of Classification Rule Quality

To answer EQ2, we compare our approach with the one proposed by Verenich et al. [24] that we consider the baseline. More specifically, we compare the classification accuracy of the decision rules for knock-out checks discovered by the approach, relative to the classification accuracy of the predictive models used by the baseline [24]. Note that the baseline focuses on run-time optimization using *black-box* models, with an emphasis on classification accuracy. In this research, we focus on design-time optimization and place emphasis on *interpretability*. Accordingly, we expect a priori that the proposed approach would achieve a lower classification accuracy than the baseline, as it sacrifices accuracy in favor of interpretability. What we seek to determine here is the magnitude of the accuracy loss relative to the baseline.

To address this question, we use the same event log as [24], which is an environmental permit application process [3] introduced in Sect. 4. To replicate the experimental conditions of the baseline, we use the same dataset splitting proportions, namely 80% of the cases for training and 20% for testing. Verenich et al. [24] observed that the environmental permit application event log is highly imbalanced regarding the proportions of accepted and rejected cases. Therefore, they performed class balancing by undersampling the accepted (non-knocked-out) cases. We followed this same strategy.

We use the ROC curve and AUC metric to compare the performance of our models with the baseline. To do so, same as Verenich et al. [24], we performed the 5-fold cross-validation procedure using stratified random-sampling-based splitting. We observe AUC values ranging from 0.571 ("T10") to 0.744 ("T02"). These values seem encouraging to a certain extent, especially considering that the AUC values in the baseline ranged from 0.527 ("T06") to 0.645 ("T10") [24]. However, we also observe relatively high standard deviations, e.g., the classifier for "T02" obtained the highest mean AUC value of 0.744 but a standard deviation of 0.37. This can be due to a few positive examples available for training the rule model of this knock-out check: "T02" rejected only 0.4% of all cases, as opposed to "T10", which rejected 64.6%. Nonetheless, the performance of our classification model is comparable to that of the baseline, and no significant performance loss is observed by using an interpretable decision rule-based model instead of black-box models.

This observation suggests that the discovered rules can be used as a proxy for determining how to order knock-out checks in a process (cf. the knock-out reordering strategy in Sect. 3.3) and to relocate the knock-out checks as early as possible in the process (cf. the knock-out relocation strategy).

5.3 Threats to Validity

The observations derived from the above evaluation are subject to the following threats to validity. First, we acknowledge a threat to construct validity due to the choice of classification accuracy measures (AUC) to address the evaluation questions. While the proposed approach discovers decision rules with relatively high accuracy, comparable to the baseline, this may or may not translate into comparable overprocessing reductions in practical scenarios. Second, we acknowledge a threat to external validity (generalizability) due to the fact that the evaluation relies on one synthetic and one real-life dataset. As such, the conclusions should be seen as preliminary and subject to additional validation.

6 Conclusion

This paper outlines an interpretable process mining approach to identify improvement opportunities in processes containing knock-out checks and recommend redesigns to reduce overprocessing waste. Given an event log, the approach discovers knock-out checks, their decision rules, and dependencies, and calculates time wastes associated with knock-out checks. Next, the approach identifies improvement opportunities and suggests redesigns for reducing overprocessing waste. We present the approach's implementation as a software tool that allows analysts to upload an event log and obtain suggested redesigns. The evaluation shows that our approach can accurately discover and quantify improvement opportunities and suggest relevant redesigns considering data dependencies. Compared to the baseline, we did not lose performance, but we gained explainability. However, the effectiveness of our approach on real-life logs is limited by the availability and granularity of event log data.

In future work, we aim to extend the proposed approach with a simulation phase to uncover further potential improvement opportunities to reduce overprocessing waste. We foresee that a simulation phase would enable us to capture the impact of reordering and relocating knock-out checks for different subsets of cases. Another avenue for future work is to extend the proposed technique to address other types of waste, including waiting waste, transportation waste, and defect waste.

Acknowledgments. This research is funded by the European Research Council (PIX Project).

References

1. van der Aalst, W.M.P.: Process Mining: Data Science in Action, 2nd edn. Springer, Heidelberg (2016). https://doi.org/10.1007/978-3-662-49851-4
2. van der Aalst, W.M.: Re-engineering knock-out processes. Decis. Support Syst. **30**(4), 451–468 (2001)
3. Buijs, J.: 3TU. DC dataset: receipt phase of an environmental permit application process (WABO) (2015)
4. Chahal, V., Narwal, M.: Impact of lean strategies on different industrial lean wastes. Int. J. Theor. Appli. Mech. **12**(2), 275–286 (2017)

5. Cohen, W.W.: Fast effective rule induction. In: Machine Learning Proceedings 1995, pp. 115–123. Elsevier (1995)
6. Du, M., Liu, N., Hu, X.: Techniques for interpretable machine learning. Commun. ACM **63**(1), 68–77 (2019)
7. Dumas, M., La Rosa, M., Mendling, J., Reijers, H.A., et al.: Fundamentals of Business Process Management, vol. 1. Springer, Heidelberg (2013). https://doi.org/10.1007/978-3-642-33143-5
8. Fehrer, T., Fischer, D.A., Leemans, S.J., Röglinger, M., Wynn, M.T.: An assisted approach to business process redesign. Decis. Support Syst. **156**, 113749 (2022)
9. Lashkevich, K., Milani, F., Chapela-Campa, D., Dumas, M.: Data-driven analysis of batch processing inefficiencies in business processes. In: Guizzardi, R., Ralyté, J., Franch, X. (eds.) RCIS 2022, pp. 231–247. Springer, Cham (2022). https://doi.org/10.1007/978-3-031-05760-1_14
10. Lee, S.: A rule-based framework for interpretable predictions of business process outcomes using event logs. Master's thesis, Ulsan National Institute of Science and Technology, Ulsan (2021)
11. Lohrmann, M., Reichert, M.: Effective application of process improvement patterns to business processes. Softw. Syst. Model. **15**(2), 353–375 (2016)
12. Malinova, M., Gross, S., Mendling, J.: A study into the contingencies of process improvement methods. Inf. Syst. **104**, 101880 (2022)
13. Milani, F., Lashkevich, K., Maggi, F.M., Di Francescomarino, C.: Process mining: a guide for practitioners. In: Guizzardi, R., Ralyté, J., Franch, X. (eds.) RCIS 2022, pp. 265–282. Springer, Cham (2022). https://doi.org/10.1007/978-3-031-05760-1_16
14. Netjes, M., Reijers, H., Aalst, van der, W.: The price tool kit: tool support for process improvement. In: La Rosa, M. (ed.) Proceedings of the BPM 2010 Demonstration Track, pp. 58–63. CEUR Workshop Proceedings, Springer (2010)
15. Niedermann, F., Radeschutz, S., Mitschang, B.: Design-time process optimization through optimization patterns and process model matching. In: 2010 IEEE 12th Conference on Commerce and Enterprise Computing, pp. 48–55. IEEE (2010)
16. Niedermann, F., Schwarz, H.: Deep business optimization: making business process optimization theory work in practice. In: Halpin, T., et al. (eds.) BPMDS/EMMSAD -2011. LNBIP, vol. 81, pp. 88–102. Springer, Heidelberg (2011). https://doi.org/10.1007/978-3-642-21759-3_7
17. Reijers, H.A., Mansar, S.L.: Best practices in business process redesign: an overview and qualitative evaluation of successful redesign heuristics. Omega **33**(4), 283–306 (2005)
18. Reinkemeyer, L.: Process mining in action. Process Mining in Action Principles, Use Cases and Outlook (2020)
19. Rohleder, T.R., Silver, E.A.: A tutorial on business process improvement. J. Oper. Manag. **15**(2), 139–154 (1997)
20. Souza, A., Azevedo, L.G., Santoro, F.M.: Automating the identification of opportunities for business process improvement patterns application. Int. J. Bus. Process. Integr. Manag. **8**(4), 252–272 (2017)
21. Teinemaa, I.: Predictive and prescriptive monitoring of business process outcomes. Doctoral thesis, University of Tartu, Tartu, Estonia (2019)
22. Thürer, M., Tomašević, I., Stevenson, M.: On the meaning of 'waste': review and definition. Prod. Plan. Control **28**(3), 244–255 (2017)

23. Verenich, I.: Explainable Predictive Monitoring of Temporal Measures of Business Processes. Ph.D., Queensland University of Technology (2018). https://doi.org/10.5204/thesis.eprints.124037
24. Verenich, I., Dumas, M., La Rosa, M., Maggi, F.M., Di Francescomarino, C.: Minimizing overprocessing waste in business processes via predictive activity ordering. In: Nurcan, S., Soffer, P., Bajec, M., Eder, J. (eds.) CAiSE 2016. LNCS, vol. 9694, pp. 186–202. Springer, Cham (2016). https://doi.org/10.1007/978-3-319-39696-5_12

Persuasive Visual Presentation of Prescriptive Business Processes

Janna-Liina Leemets[1], Kateryna Kubrak[1], Fredrik Milani[1], and Alexander Nolte[1,2(✉)]

[1] University of Tartu, Tartu, Estonia
{kateryna.kubrak,fredrik.milani,alexander.nolte}@ut.ee
[2] Carnegie Mellon University, Pittsburgh, PA, USA

Abstract. Prescriptive process monitoring methods recommend interventions during the execution of a case that, if followed, can improve performance. Research on prescriptive process monitoring so far has focused mainly on improving the underlying algorithms and providing suitable explanations for recommendations. Empirical works indicate, though, that process workers often do not follow recommendations even if they understand them. Drawing inspiration from the field of persuasive technology, we developed and evaluated a visualization that nudges process workers towards accepting a recommendation, following a design science approach. Our evaluation points towards the feasibility of the visualization and provides insights into how users perceive different persuasive elements, thus providing a basis for the design of future systems.

Keywords: Persuasive visualization · Prescriptive process monitoring · User interface

1 Introduction

Companies continuously seek ways to improve their efficiency. For this, they use business process management (BPM) to discover, analyze, redesign, and implement process changes [8]. In the past decade, process mining, i.e., data-driven methods supporting BPM [1], have gained popularity [28]. Process mining methods use event logs, i.e., recorded data on process execution, to discover, analyze, check, and monitor business processes [20]. Recently, process mining has increasingly been used to predict process outcomes and prescribe interventions [17].

Predictive process monitoring methods use event logs to predict outcomes of ongoing process instances [17]. Prescriptive process monitoring methods build upon these predictions to recommend actions that, if followed, reduce the probability of negative outcomes of ongoing cases [17].

Existing work on prescriptive process monitoring mainly focused on improving underlying algorithms or explaining recommendations [17]. At the same time, empirical works indicate that process workers often do not follow recommendations, even when they understand them [7]. The potential gains of prescriptive process monitoring approaches can, therefore, remain unused in practice.

S. Nurcan et al. (Eds.): RCIS 2023, LNBIP 476, pp. 398–414, 2023.
https://doi.org/10.1007/978-3-031-33080-3_24

To address this gap, we designed and evaluated PERSEVERE – a PERsuasive System for prEscriptiVe businEss pRocEss monitoring. We state the following research goals.

RG$_1$. *Create a persuasive visualization for prescriptive process monitoring.*
RG$_2$. *Evaluate the visualization to understand how different persuasive components are perceived.*

To fulfill these goals, we followed a design science approach [27]. Drawing inspiration from the field of persuasive technology [11] – which focuses on the design of systems that nudges users towards specific behaviors – we developed PERSEVERE. Our evaluation points towards the feasibility of the approach. It also indicates the importance of fostering trust in the displayed information, supporting the discovery of additional information, and the adaption of incentives for a corporate context. The contribution of this paper is thus twofold. First, we created the first persuasive visualization for prescriptive process monitoring. Second, our evaluation provides insights into how persuasive elements are perceived and, thereby, provides a basis for the design of future systems.

2 Background

2.1 Prescriptive Process Monitoring

Prescriptive process monitoring methods prescribe recommendations that, if followed, optimize process performance such as cycle time [4], processing time [26], or success rate [10,30]. Prescriptive process monitoring methods predict the outcomes of an ongoing case and prescribe recommendations to mitigate or avoid negative outcomes [17]. For example, a method may predict that a case is likely to violate a deadline and recommend adjusting resources to avoid this negative outcome [13]. Existing prescriptive process monitoring methods differ w.r.t. several aspects such as their objective, recommendations they prescribe, or algorithms employed [17]. For example, a method may recommend a next task to execute or a resource to assign a task to.

Existing work mainly focuses on developing algorithms to generate recommendations rather than communicating recommendations to process workers [17]. There are, however, examples in, e.g., the medical domain that introduce graphical user interfaces displaying recommendations for a process trace (representing a treatment) based on patient data [33]. Such interfaces are expected to be used by medical workers. In another example [15], a user interface to review case goals and recommendations is proposed, which targets a process analyst. Commercial process mining tools also recently started developing interfaces for prescriptive process monitoring. For instance, Celonis' Action Engine[1] generates recommendations for continuous improvement based on data that can be reviewed and implemented by process analysts.

[1] https://www.celonis.com/.

2.2 Persuasive Technology

Persuasive technology is an umbrella term for systems that aim to nudge users towards a certain behavior [11]. Grounded in research on human-computer interaction and social psychology, such systems have been developed for and successfully been deployed in a variety of contexts, including healthcare [16,25], sustainability [19], and education [21].

To aid creators of persuasive technology, researchers have developed design frameworks such as the Patient-Clinician-Designer (PCD) [18], and the health recommender framework [31], which were developed in the context of healthcare, and the Persuasive System Design (PSD) framework [32] which was developed in the context of information systems. The frameworks include guidelines (e.g., *"Reward target behaviors"* [32] and *"One should minimize the barriers of following the recommendation"* [31]) to follow, and questions (e.g., *"Will the system fit into the daily lives of users?"* [18]) to ask when creating a persuasive system. We utilized them as the basis to create PERSEVERE, thus expanding their use to a novel context.

3 Methodology

The aim of this work was to design ($\mathbf{RG}_1$) and evaluate ($\mathbf{RG}_2$) a persuasive system for prescriptive process monitoring. We followed the design science approach proposed by Peffers et al. [27]. We first identified an existing problem – process workers not following recommendations from prescriptive process monitoring systems – that has been observed in prior work (step 1, Sect. 1). We then selected a common process from the finance domain as a test case and described it as a scenario (step 2, Sect. 3.1). As a basis for the following requirements elicitation, we utilized the scenario together with existing literature from the field of persuasive technology (step 3, Sect. 3.2). Based on the scenario and requirements, we created a prototype (step 4, Sect. 3.3) and evaluated it with eight practitioners from a financial institution (step 5, Sect. 3.4). This paper represents part of our findings report (step 6). In the following, we will describe steps 2 to 5 in detail (Fig. 1 provides an overview).

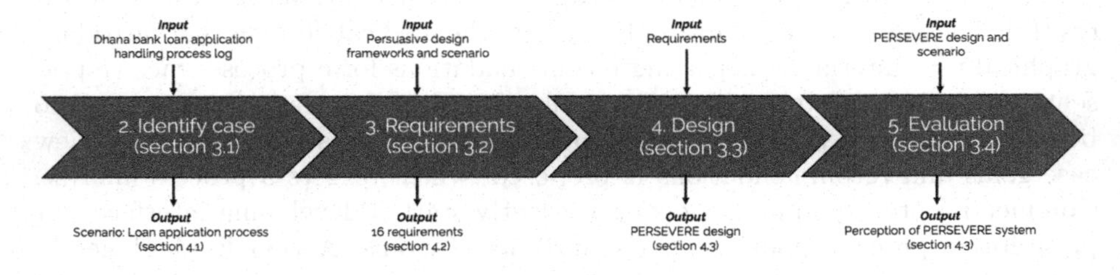

Fig. 1. Methodological approach (steps 2 to 5)

3.1 Setting

As our study setting, we chose a loan application process in a financial institution. This process is suitable because the credit score of applicants can be automatically calculated and thereby prescribed, but the decision to approve or reject an application is manually taken by a credit administrator. We discovered the details of this process using the open Dhana bank loan application handling process log as a basis[2]. It contains information about loan applications, including the requested amount, the application type, the reason for an application, the offered amount, the first withdrawal amount, monthly repayment costs, the number of payback terms, the customer's credit score, and information about if the offer was accepted. The case description can be found in Sect. 4.1.

For this work, we focus on the execution of one particular process instance and one recommendation within that instance. We chose this focus because our aim was to understand how different persuasive components are perceived by process workers and which role they play in their decision process ($\mathbf{RG}_2$).

3.2 Requirements

We based the initial requirements for PERSEVERE on existing literature by searching for visualization frameworks in the context of persuasive technology. The search was conducted in IEEE Explore, the ACM digital library, and Google Scholar utilizing the terms "persuasion", "persuasive technology", and "visualization" and yielded the three frameworks, namely introduced in Sect. 2.2. We proceeded to collaboratively extract guidelines from these three frameworks, reflected them on the aforementioned case (Sect. 3.1), and elicited requirements for the design of PERSEVERE (Sect. 4.2).

3.3 Design

Using Figma[3], we created a design based on the elicited requirements (Sect. 3.2) and the setting (Sect. 3.1), thus addressing our first research goal ($\mathbf{RG}_1$). We based our design on the design of the system that the test bed company use (Sect. 3.4). The design can be found in Sect. 4.3.

3.4 Evaluation

To evaluate PERSEVERE, assess how it is perceived, and identify means for improving it ($\mathbf{RG}_2$), we conducted semi-structured observational online interviews with eight individuals (5 female and 3 male) from the company serving as our test bed (Table 1 provides an overview). This sample size can be considered common [23] when conducting qualitative studies of a specific system – in this case, a visualization for prescriptive process monitoring.

---

[2] The log is available here.

[3] https://www.figma.com/.

We conducted the evaluation within one financial institution. This ensures that all participants are familiar with the same system and process. It also allows us to assume a similar background for all participants, thus ruling out intervening aspects that could compromise our findings, such as different conventions, social norms, and policies. At the same time, we deliberately chose individuals that were familiar with the loan application process but had different levels of experience and fulfilled different roles within that process (Table 1).

Table 1. Overview of study participants, their experience in years, and their position within the financial institution we utilized as a test bed.

Interviewee	Experience	Position
I1	1 year	loan specialist
I2	1.5 months	loan specialist
I3	1 year	loan specialist
I4	10 years	solicitor
I5	8 years	deputy head of retail banking
I6	21 years	credit manager
I7	20 years	team lead
I8	20 years	credit risk expert

The interviews were conducted by one of the authors and were divided into two phases. During the first phase, we focused on observing participants while they utilized PERSEVERE. For this phase, the interviewer introduced PERSEVERE before explaining the study scenario. S/he then asked participants to decide (approve or reject a car loan) based on the available information. S/he also asked participants to explain their thoughts while interacting with PERSEVERE ("thinking aloud", c.f. [9]). During this phase, the interviewer remained passive, only answering questions from participants as they arose.

The second phase focused on understanding the participants' perceptions of their interaction with PERSEVERE. For this phase, the interviewer first asked each participant questions related to their decision process covering the decision process in general (e.g., *"Please explain your decision process to me."*, *"Which information did you base this decision on and why?"*) before focusing on the different components of PERSEVERE (*"How did you use [component] for your decision?"*). The interviewer then continued to ask questions about the general perception of the participant about the different components of PERSEVERE (*"What did you think about [component]?"*, *"What would you change?"*) and the system as a whole (*"What is your perception about the design?"*), including questions related to missing information (*"Is there any information that you missed?"*, *"Is there any additional information that would have helped you?"*). At the end of this phase, the interviewer asked the participant about their confidence in their decision (*"How confident are you in your decision?"*).

All sessions were video recorded. They lasted between 26 and 45 min (m = 40 min). To transcribe the interviews, we used the AI transcription tool otter.ai before correcting potential mistakes manually.

Afterward, we also asked each participant to fill out a short post-questionnaire. This questionnaire focused on their perception of the usefulness and ease of use of PERSEVERE. These aspects have been found in prior work to be important for user acceptance of a system in an organizational context [6]. Both aspects were assessed using established 5-point Likert scales [6]. In addition, we also asked the participants about their overall satisfaction with PERSEVERE utilizing an established 5-point Likert scale [3].

Our analysis mainly focused on the video recordings, which included an observation of the participants while they interacted with PERSEVERE and the follow-up interviews. The collected questionnaire data served as an additional qualitative data point.

For the analysis, we combined deductive and inductive coding. We started with a set of codes related to our two main research goals, which included the creation of a persuasive visualization for prescriptive process monitoring ($\mathbf{RG}_1$) and the perception of process workers about their relevance ($\mathbf{RG}_2$). We included codes that related to the different components of PERSEVERE, such as *"risks"* and *"additional information"*. The coding was done in two steps. First, one of the authors conducted an initial coding utilizing the aforementioned codes. Afterward, we conducted a second round of collaborative inductive coding, during which we added codes such as *"missing information"* and *"analytics"* and modified the coding until everyone agreed.

4 PERSEVERE - PERsuasive System for PrEscriptiVe BusinEss pRocEss Monitoring

In this section, we present PERSEVERE. The section is organized along the methodological steps introduced in Sect. 3. The outcomes of step 1 have already been presented. Therefore, we discuss the outcomes of steps 2 to 5 in the following Sects. 4.1 to 4.4.

4.1 Setting

As a basis for the development of PERSEVERE (Sect. 4.3) and the subsequent evaluation (Sect. 3.4), we utilized a scenario that describes the process of handling a car loan application. This is a common process in financial institutions, such as the one that served as our test bed. During this process, the process worker has to decide whether or not to grant a loan for a car that a customer asks for. This is a lengthy process with many intricate details that sometimes are difficult for a process worker to oversee. PERSEVERE can aid their decision by recommending to accept or reject an application.

Scenario: Anne is working as a loan officer at Dhana bank. She currently works on a car leasing application. She has worked with car leasing cases many times before. The customer has submitted an application via the bank's website, but the application is missing the required amount. The customer has now emailed Anne to let them know the sum they need, which is 5700 euros. Anne updates the application's details and now has to decide whether or not to make an offer. Normally, Anne would analyze the application, the customer's income and cost structure, repayment ability, and credit habits to assess if an offer is feasible. This is a lengthy process, and sometimes she would miss an important detail or under- or overestimate a customer's solvency. Luckily, PERSEVERE has a recommendation for her to accept the loan application. Looking at the information provided by PERSEVERE, she now has to decide whether to abide by this recommendation and accept the application or reject it.

4.2 Requirements

We based the requirements for PERSEVERE on the previously identified frameworks (Sect. 2.2). They include principles and questions that can be grouped into three general categories related to (1) when to show a message, (2) the structure of the message that contains a recommendation, and (3) the message content. We identified a total of 16 requirements.

Timing. The timing when to show a message is critical for it to have the desired persuasive effect [24]. It should be consistent with the usual process that the user follows [24] (R1) and aid the user in making a decision [29] (R2).

Structure. Similar to timing, it is important that the overall structure of the recommendation is consistent with what the user is used to seeing to be persuasive [24] (R3). Moreover, the design should emphasize the most important details to make a decision [5] (R4). It should also be possible to access additional details about the recommendation [12,24] (R5). Finally, accepting or dismissing the recommendation should require minimal additional input [24] (R6).

Content. For a recommendation to be persuasive, it is important that its content is understandable. This means that the content should be in line with what the user is used to seeing when following their normal routine [24] (R7). It also means that the system should provide information so the user can understand the reasoning behind a recommendation [18] (R8). The understandability of a recommendation can be aided by combining text and visualizations [2] (R9).

In addition, a system needs to outline the consequences for following or not following a recommendation to be persuasive [18] (R10). The system should emphasize the advantages of following a recommendation [24] (R11) but also provide information about potential risks [31] (R12).

Moreover, it is important for a recommendation to be trustworthy in order for it to be persuasive. Existing work suggests multiple approaches to aid trust. These commonly include providing evidence for a recommendation [18] (R13) and providing information about its source [31] (R14).

Existing work on persuasive design also emphasizes the social component of a recommendation for it to have a persuasive effect. Related work suggests providing indications for how others have successfully used a recommendation [32] (R15). Moreover, prior work also provides evidence that rewarding a user for following a recommendation can aid future use [22] (R16).

4.3 Design

Based on the previously elicited requirements (Sect. 4.2), we designed the PERSEVERE prototype, thus addressing our first research goal (**RG**$_1$). The overall design of PERSEVERE is based on the design that is used by process workers at the financial institution that served as our test bed (R3, Fig. 2). A recommendation is shown (Fig. 3) after a process worker updates information of a loan application (R1, c.f. scenario in Sect. 4.1). The process worker then subsequently needs to decide whether to approve or reject the application (R2). To accept the recommendation and approve the loan application, the process worker needs to press the *"Accept and approve"* button (bottom right in Fig. 3). To reject the recommendation, the process worker needs to press the *"Reject"* button next to it. This design ensures minimal additional input on the part of the process worker (R6).

Fig. 2. Process worker entering a new loan amount

The design clearly indicates the recommendation through its headline and through a green area (R4, Fig. 3) which shows information about the positive effects of following the recommendation (e.g., an increase in sales and a reduction of the overall case time, R10). Utilizing a green background emphasizes that it would be advantageous to follow the recommendation (R11). The following text (under the headline *"Why you should follow this recommendation"*) further emphasizes the advantages of abiding by the displayed recommendation. It mentions that a *"positive outcome is likely"* and that the *"last recommendation you followed decreased the average case length by 1 day"* (R11).

The red area (next to the green one) provides information about potential negative consequences of not following this recommendation (e.g., an increase in churn rate, R10). The potential risks of following a recommendation are also accessible by hovering over the red warning sign (R12, Fig. 4 left).

The visualization also contains information that is common for loan applications, such as the credit score and overdues (R7, Fig. 3 bottom) and uses a combination of text and visuals to aid understandably (R9). That information, together with the explanation that the *"credit score is the most important feature"* (under the headline *"Why you should follow this recommendation"*) serves

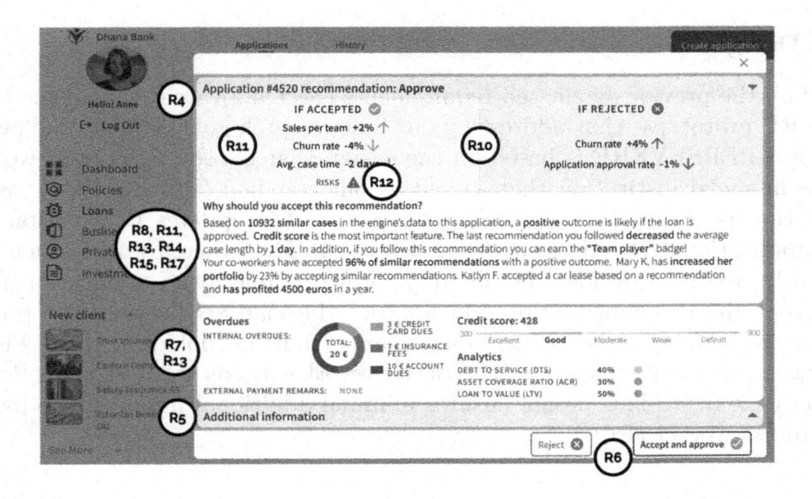

Fig. 3. Visualization of the recommendation to approve the loan application

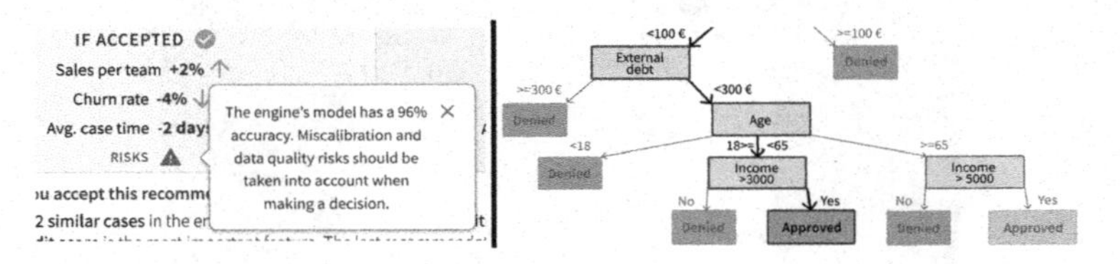

Fig. 4. Risk popup (left) and decision tree (right)

as evidence for the recommendation (R13). The text also includes information about the basis of the recommendation (*"based on 10932 similar cases"*, R14).

The user can access additional information through the *"Additional information"* tab near the bottom of the interface (R5). Clicking on this tab opens a visualization of a decision tree (Fig. 4 right). This decision tree visualizes the decision process of the algorithm, thus providing information about how it reached the presented recommendations (R8).

The text in the center also provides information about how following a recommendation had a *"positive outcome"* for most co-workers, including specific information for two of them (text about *"Mary"* and *"Katlyn"* in the center of Fig. 3, R15). Moreover, it states that by following the recommendation, the user can *"earn the Team Player badge"* (R17).

4.4 Evaluation

In this section, we elaborate on the findings of the evaluation, thus addressing our second research goal (**RG**$_2$). The participants generally found PERSEVERE useful (m = 4.19, SD = 0.47) and easy to use (m = 4.38, SD = 0.28), as evident by their survey responses (Fig. 5). Their satisfaction (m = 3.94, SD = 0.25) was, in comparison, slightly lower, indicating room for improvement. Here, we discuss how participants approached their decision-making process, followed by outlining their perception of the different components of PERSEVERE and the information that they missed.

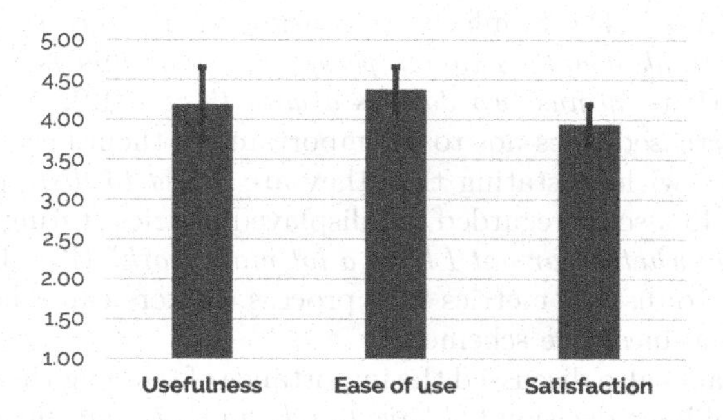

Fig. 5. Questionnaire responses by participants. All responses were given on a 5-point scale which was anchored between strongly disagree (1) and strongly agree (5). The bars indicate the mean (m) and standard deviation (SD).

Decision-Making. Seven out of eight participants accepted the recommendation. Only I2 rejected it, pointing out that s/he would prefer to *"look through the application manually"* (I2) before making a decision. The participants took between one and seven minutes each to make a decision. This difference can potentially be attributed to some participants taking their time to understand all parts of the interface while others focused on components that were familiar to them, such as the credit score (*"concentrated on this credit score"*, I6).

Their decision was mainly based on metrics that are common for loan applications (Fig. 3 bottom), such as *"the credit score"* (I1), which all participants referred to when making their decision with some perceiving it to be the *"most important thing"* (I3). I7 even went as far as touting it to be *"the only relevant number"* (I7). All other participants, however, utilized additional metrics when making a decision. Some utilized *"loan to value"* (I5), while others used *"debt to service"* (I6), *"asset coverage rate"* (I1) or overdues (*"overdues are small"*, I3).

Apart from the metrics, participants also included other aspects of the visualization in their decision-making. I1, along with I3, I4, I6, and I8, pointed towards the accuracy of the model (Fig. 4, left) to support their decision (*"I think it helps me"*, I1, *"it's very good to know that how accurate the model is"*, I8). They also considered it reassuring that the displayed recommendation is based on *"a lot of similar cases"* (I3) or, as I8 put it, that *"it's using sufficient information"* (I8) (*"based on 10932 similar cases"* in the center of Fig. 3). However, while the risk part was reassuring for the aforementioned participants, I7 stated it as one of their reasons to reject the recommendation (*"this part that says that there is a risk makes me want to reject [the recommendation]"*, I7). Thus, these findings point towards the necessity to provide details about the quality of the displayed information.

In addition, some participants also mentioned the potential benefits of accepting the recommendation, as displayed in the green area at the top of Fig. 3 during their decision-making process. For example, I8 mentioned that it is important for them *"to know how much this case is impacting the sales volumes"* (I8), and I1 stated that they *"looked at increases sales per term"* (I1) and at the *"application approval rate"* (I1). Similarly, I7 considered the *"decreased average case time, because I think about customer perspective that this is very important"*, with I2 stating that *"minus two days is a good thing"* (I2). At the same time, I4 mentioned increased sales not to be important for them because *"I am not a salesperson"* (I4), with I6 stating that they are *"used to disregarding the sales objectives"* (I6). I3 also disregarded the displayed metrics stating that *"I have to analyze correctly, whether or not I have a lot more work"* (I3). Thus, it appears to be important to display metrics that process workers can relate to, and that fit their individual incentive schemes.

The participants also discussed the importance of trusting the displayed information when making a decision (*"I have to rely on that, that [the system] works"*, I7). I4 mentioned they trusted the information *"because it has high accuracy"* (I4), while for I6 and I8, it was more important that the displayed information matched their personal experience (*"the text here [referring to the middle part of Fig. 3] is confirming my historical experience that credit score is the most important single feature"*, I8). Others were *"not so sure about analytics"* (I3), with I2 stating that according to their *"previous work experience, the automatic engines usually miss a lot of important things"* (I2). To build trust in the presented information, participants made different suggestions. I2 discussed observing *"the recommended and my decisions, how they match or mismatch"* (I2) over a longer period of time, while I5 stated that a *"third party needs to [...] accept this scoring model"* (I5) to increase their trust. More concretely, some participants referred to a lack of information about how the statistics were calculated, stating that they knew *"too little about this credit score"* (I7), that *"debt to services [...] I don't know how this is calculated"* (I6) or that it remains unclear *"what is considered a similar case"* (I6).

PERSEVERE Components. In this section, we discuss the perception of the participants about the different components of PERSEVERE, starting from the green and red areas at the top of Fig. 3.

Participants generally appreciated the *"good color code"* (I3) of this area, which they perceived to *"draw attention first"* (I2). They also thought that it was *"good that it has positives and negatives"* (I3), showing *"the result if I accepted or rejected"* (I3). While some participants utilized the information presented in this area for their decision-making, as discussed in the previous section, there were also critical voices stating that it *"doesn't give me any information for my final decision"* (I5). As a potential reason for discarding this information, I3 mentioned that *"it's not specific to this case; it's like an average"* (I3). This, again, points towards the necessity to identify indicators that are suitable for each process worker. Our participants also showed different usage patterns related to the displayed risks. The risks were generally perceived to be *"important"* (I5) but sometimes overlooked, as evident by I4 who *"would suggest that the [visualization of the] risks part is kind of bigger"* (I4) because they *"didn't see the risk part"* (I4).

The text headlined *"Why should you accept this recommendation?"* in Fig. 3) was perceived to be lengthy (*"it's a lot of text"*, I7) which led to some participants reading the text *"a little bit"* (I1) while others discarded it completely (*"I don't like to read text"*, I5). Participants that included the text in their decision-making mainly focused on the reference to similar cases and the stated role of the credit score, as discussed previously. The remainder of the text, which focused on how colleagues have benefited from accepting recommendations, was met with skepticism. For I3, it did not become clear *"how this can profit me"* (I3), and I6 thought that *"most arguments written here are personal benefits for me which is not in case of making great decisions"*, even going as far as stating that they perceive them to *"decrease the trustworthiness"* (I6). I2 voiced a similar sentiment, stating that they *"did not trust the text"* (I2) and felt pressured (*"The part that says Mary Kay has increased her portfolio by 23% that kind of puts pressure to accept the automatic decision"*, I2). The same participant, however, also stated that their perception might be different if *"this text is based on my previous work"* (I2). These findings thus point towards the need for a shorter description that fosters trust in the presented information and outlines – if at all – suitable benefits.

The promise of a *"Team player" badge* in the same section was met with a similarly negative sentiment. Only I1 mentioned that they *"liked the badge"* but stated at the same time that they *"wouldn't make decisions based on [the badge]"* (I1). This perception was echoed by I8, who stated that *"it's nice to have a badge, but it shouldn't impact the decision"* (I8). I3 did not see the benefit of the badge (*"unclear what this team player batch will give me"*, I3), I4 found it *"distracting"* (I4), and I2 thought that *"the badge kind of like makes it seem like work is a game"* (I2). This indicates that in order for a badge or a similar reward to have a persuasive impact, it is necessary to identify a reward that is perceived to be advantageous by process workers.

Finally, it should be noted that participants only opened the *"Additional information"* (Fig. 3 bottom) part after being prompted. At the same time, all of them found the information that it contained (Fig. 4, right) *"definitely useful"* (I3). As the main reason, they mentioned that this part *"gives me additional confidence in my decision"* (I8). Some participants consequently mentioned that they *"would prefer for this to be part of the starting visualization"* (I2) because they perceived it to contain *"more relevant information"* (I6). This sentiment was echoed by I4, who suggested changing the visualization and first show this *"additional information and then the result if accepted if rejected"* (I4). Regarding the way that the additional information was visualized – in the form of a graph – there were mixed reactions. I7 found them *"quite complicated to read"* (I7). At the same time, I3 stated that s/he preferred *"that the additional information is graphical"* (I3). These findings point toward the usefulness of the presented additional information. They, however, also indicate that the design should emphasize their existence and that the graphical visualization should potentially be amended to aid understandability.

Missing Information. Our participants also mentioned information that they are typically *"used to having"* (I1) and that were not part of PERSEVERE. This information includes details about the customer, such as their *"age"* (I8), *"income"* (I5) and potential *"other obligations"* (I3), such as *"other loans"* (I2) and *"other applications"* (I1). In addition, I6 also mentioned that they would like to *"check current accounts of the customer with the bank"* (I6) and I2 missed information about the timeline of the loan that the customer applied for (*"time for that loan"*, I2). These findings point towards differences related to how individual process workers deal with loans and indicate the potential necessity to connect PERSEVERE to additional information sources.

5 Discussion

In this section, we first present our findings, discuss the limitations of our study (Sect. 5.1), and elaborate on implications and future work (Sect. 5.2).

We developed a persuasive visualization for prescriptive process monitoring ($\mathbf{RG}_1$) and evaluated it with eight process workers from one financial institution ($\mathbf{RG}_2$). Our findings indicate the feasibility of this approach. Only one of eight participants rejected the recommendation. This rejection appeared to be driven more by personal preferences rather than the displayed visualization, though.

We observed that process workers – in this case, loan processors – mainly focused on information displayed in the interface they were familiar with when making a decision. Despite displaying information that was familiar to them, some participants asked for additional information to increase their confidence. The unfamiliar interface that included both familiar and unfamiliar information might thus have led to reduced confidence in the displayed information. This seems to indicate that simply visualizing commonly used metrics alongside a recommendation might not be sufficient for process workers to follow it, thus pointing towards the feasibility of persuasive interface components. This finding is in line with prior work in the context of prescriptive process monitoring [7].

We also found that process workers asked for more information, but they did not necessarily access the additional information that was available in PERSEVERE. Moreover, when presented with additional information, some process workers had trouble understanding it or asked for different information than the information provided. Thus, it appears important to display a minimum amount of information clearly – as reported by Oinas-Kukkonen and Harjumaa [24] and Marcu et al. [18] – while at the same time providing access to additional information based on the preferences or practices of individual process workers.

Our findings also highlight trust as an important aspect when accepting a recommendation. Trust, in this context, relates to both trust in the displayed information and trust in the recommendation as such. To foster trust, participants referred to transparency in that they found it important to understand what information a recommendation was based on. This finding is in line with

Gedikli et al. [12]. Moreover, participants also referred to developing trust over time – a concept that is well established in work related to continued use [14].

Finally, prior work suggests presenting individual and social benefits of following a recommendation to foster persuasiveness [22]. Our findings extend this work by pointing towards the necessity of selecting suitable benefits that fit within the work context of a process worker. Presenting benefits that relate to other departments (e.g., sales in our case) or to the process worker themselves might hinder rather than foster persuasiveness. Similarly, social benefits, such as badges, might not have the desired effect, as they can be perceived as an unacceptable gamification element in a corporate context.

5.1 Limitations

Our goal was to develop and evaluate a persuasive visualization for prescriptive process monitoring. Thus, it made sense to follow a design science methodology [27]. There are, however, several limitations associated with our approach. First, we based the initial requirements for PERSEVERE on literature rather than an empirical study of existing work practices. We mitigated this limitation by eliciting the requirements based on a real-life scenario and conducting a follow-up evaluation. Second, we focused on one specific process within one institution. Despite making a deliberate selection, utilizing a different process in a different institution might yield different results. Third, we focused on one specific recommendation that affected the process it was designed for. In a real setting, recommendations might depend on or might have implications for other processes. Fourth, we chose a recommendation that the process worker can fulfill themselves. Decisions might be different if others are involved. We accepted these limitations because our focus was on assessing the perception of process workers related to the visualization. Fifth, we evaluated PERSEVERE in an artificial setting. Evaluating it in a real-life process might yield different results. Sixth, the participants only used PERSEVERE once. The observed behavior might change over time when participants utilize the same system multiple times and become more familiar with it. Finally, qualitative data is subject to interpretation bias during analysis. We attempted to mitigate this bias by collaboratively coding the data. We also abstain from making causal claims, instead providing a rich description of the reported perceptions of our participants.

5.2 Implications and Future Work

Our study has implications for the design of prescriptive process monitoring systems and for the use of persuasive technology in organizations. It provides indications of how process workers perceive different persuasive elements, how they use them for decision-making, and how the organizational context in which process workers operate can affect their perception and, thus, ultimately, the usefulness of persuasive elements for nudging workers toward a desired behavior.

In the future, we aim to adapt PERSEVERE to be applicable across domains and processes. In addition, we aim to further study the influence of trust and of repeated use on process workers' abidance to recommendations. Finally, for this study, we adapted the stance that the provided recommendations are always the best way to go for a process worker. We aim to challenge this assumption by, e.g., studying user behavior and decision-making in situations where persuasive principles are applied to false recommendations.

Acknowledgements. This research is supported by the Estonian Research Council (PRG1226) and the European Research Council (PIX Project).

References

1. van der Aalst, W.M.P.: Process Mining - Data Science in Action, 2nd edn. Springer, Heidelberg (2016). https://doi.org/10.1007/978-3-662-49851-4
2. Anagnostopoulou, E., Magoutas, B., Bothos, E., Mentzas, G.: Persuasive technologies for sustainable smart cities: The case of urban mobility. In: Companion Proceedings of the 2019 World Wide Web Conference, pp. 73–82 (2019)
3. Bhattacherjee, A.: Understanding information systems continuance: an expectation-confirmation model. MIS Q. 351–370 (2001)
4. Bozorgi, Z.D., Teinemaa, I., Dumas, M., Rosa, M.L., Polyvyanyy, A.: Prescriptive process monitoring for cost-aware cycle time reduction. In: ICPM, pp. 96–103. IEEE (2021)
5. Chih, C.H., Parker, D.S.: The persuasive phase of visualization. In: Proceedings of the 14th ACM SIGKDD International Conference on Knowledge Discovery and Data Mining, pp. 884–892 (2008)
6. Davis, F.D.: Perceived usefulness, perceived ease of use, and user acceptance of information technology. MIS Q. 319–340 (1989)
7. Dees, M., de Leoni, M., van der Aalst, W.M.P., Reijers, H.A.: Accurate predictions, invalid recommendations: lessons learned at the Dutch social security institute UWV. In: vom Brocke, J., Mendling, J., Rosemann, M. (eds.) Business Process Management Cases Vol. 2, pp. 165–178. Springer, Heidelberg (2021). https://doi.org/10.1007/978-3-662-63047-1_13
8. Dumas, M., Rosa, M.L., Mendling, J., Reijers, H.A.: Fundamentals of Business Process Management, 2nd edn. Springer, Heidelberg (2018). https://doi.org/10.1007/978-3-662-56509-4
9. Ericsson, K.A., Simon, H.A.: Protocol Analysis: Verbal Reports as Data. The MIT Press, Cambridge (1984)
10. Fahrenkrog-Petersen, S.A., et al.: Fire now, fire later: alarm-based systems for prescriptive process monitoring. Knowl. Inf. Syst. **64**(2), 559–587 (2022)
11. Fogg, B.J.: Persuasive technology: using computers to change what we think and do. Ubiquity **2002**(Dec), 2 (2002)
12. Gedikli, F., Jannach, D., Ge, M.: How should i explain? A comparison of different explanation types for recommender systems. Int. J. Hum. Comput. Stud. **72**(4), 367–382 (2014)
13. Gröger, C., Schwarz, H., Mitschang, B.: Prescriptive analytics for recommendation-based business process optimization. In: Abramowicz, W., Kokkinaki, A. (eds.) BIS 2014. LNBIP, vol. 176, pp. 25–37. Springer, Cham (2014). https://doi.org/10.1007/978-3-319-06695-0_3
14. Hoehle, H., Huff, S., Goode, S.: The role of continuous trust in information systems continuance. J. Comput. Inf. Syst. **52**(4), 1–9 (2012)

15. Huber, S., Fietta, M., Hof, S.: Next step recommendation and prediction based on process mining in adaptive case management. In: S-BPM ONE, pp. 3:1–3:9. ACM (2015)
16. IJsselsteijn, W., de Kort, Y., Midden, C., Eggen, B., van den Hoven, E.: Persuasive technology for human well-being: setting the scene. In: IJsselsteijn, W.A., de Kort, Y.A.W., Midden, C., Eggen, B., van den Hoven, E. (eds.) PERSUASIVE 2006. LNCS, vol. 3962, pp. 1–5. Springer, Heidelberg (2006). https://doi.org/10.1007/11755494_1
17. Kubrak, K., Milani, F., Nolte, A., Dumas, M.: Prescriptive process monitoring: quo vadis? PeerJ Comput. Sci. **8**, e1097 (2022)
18. Marcu, G., Bardram, J.E., Gabrielli, S.: A framework for overcoming challenges in designing persuasive monitoring and feedback systems for mental illness. In: PervasiveHealth, pp. 1–8. IEEE (2011)
19. Midden, C., McCalley, T., Ham, J., Zaalberg, R.: Using persuasive technology to encourage sustainable behavior. Sustain. WS Pervasive **113**, 83–86 (2008)
20. Milani, F., Lashkevich, K., Maggi, F.M., Francescomarino, C.D.: Process mining: a guide for practitioners. In: Guizzardi, R., Ralyté, J., Franch, X. (eds.) RCIS 2022. LNBIP, vol. 446, pp. 265–282. Springer, Cham (2022). https://doi.org/10.1007/978-3-031-05760-1_16
21. Mintz, J., Aagaard, M.: The application of persuasive technology to educational settings. Edu. Tech. Res. Dev. **60**(3), 483–499 (2012)
22. Mumm, J., Mutlu, B.: Designing motivational agents: the role of praise, social comparison, and embodiment in computer feedback. Comput. Hum. Behav. **27**(5), 1643–1650 (2011)
23. Nielsen, J., Landauer, T.K.: A mathematical model of the finding of usability problems. In: INTERCHI, pp. 206–213. ACM (1993)
24. Oinas-Kukkonen, H., Harjumaa, M.: Persuasive systems design: key issues, process model, and system features. Commun. Assoc. Inf. Syst. **24**(1), 28 (2009)
25. Orji, R., Moffatt, K.: Persuasive technology for health and wellness: state-of-the-art and emerging trends. Health Informatics J. **24**(1), 66–91 (2018)
26. Park, G., Song, M.: Prediction-based resource allocation using LSTM and minimum cost and maximum flow algorithm. In: ICPM, pp. 121–128. IEEE (2019)
27. Peffers, K., Tuunanen, T., Rothenberger, M.A., Chatterjee, S.: A design science research methodology for information systems research. J. Manag. Inf. Syst. **24**(3), 45–77 (2007)
28. Reinkemeyer, L.: Status and future of process mining: from process discovery to process execution. In: van der Aalst, W.M.P., Carmona, J. (eds.) Process Mining Handbook. LNBIP, vol. 448, pp. 405–415. Springer, Cham (2022). https://doi.org/10.1007/978-3-031-08848-3_13
29. Ricci, F., Rokach, L., Shapira, B.: Recommender systems: introduction and challenges. In: Ricci, F., Rokach, L., Shapira, B. (eds.) Recommender Systems Handbook, pp. 1–34. Springer, Boston, MA (2015). https://doi.org/10.1007/978-1-4899-7637-6_1
30. Shoush, M., Dumas, M.: Prescriptive process monitoring under resource constraints: a causal inference approach. In: Munoz-Gama, J., Lu, X. (eds.) ICPM 2021. LNBIP, vol. 433, pp. 180–193. Springer, Cham (2022). https://doi.org/10.1007/978-3-030-98581-3_14
31. Torkamaan, H., Ziegler, J.: Integrating behavior change and persuasive design theories into an example mobile health recommender system. In: UbiComp/ISWC Adjunct, pp. 218–225. ACM (2021)

32. Torning, K., Oinas-Kukkonen, H.: Persuasive system design: state of the art and future directions. In: Proceedings of the 4th International Conference on Persuasive Technology, pp. 1–8 (2009)
33. Yang, S., et al.: A data-driven process recommender framework. In: KDD, pp. 2111–2120. ACM (2017)

TraVaG: Differentially Private Trace Variant Generation Using GANs

Majid Rafiei(✉), Frederik Wangelik, Mahsa Pourbafrani, and Wil M. P. van der Aalst

Chair of Process and Data Science, RWTH Aachen University, Aachen, Germany
majid.rafiei@pads.rwth-aachen.de

Abstract. Process mining is rapidly growing in the industry. Consequently, privacy concerns regarding sensitive and private information included in event data, used by process mining algorithms, are becoming increasingly relevant. State-of-the-art research mainly focuses on providing privacy guarantees, e.g., differential privacy, for trace variants that are used by the main process mining techniques, e.g., process discovery. However, privacy preservation techniques for releasing trace variants still do not fulfill all the requirements of industry-scale usage. Moreover, providing privacy guarantees when there exists a high rate of infrequent trace variants is still a challenge. In this paper, we introduce TraVaG as a new approach for releasing differentially private trace variants based on Generative Adversarial Networks (GANs) that provides industry-scale benefits and enhances the level of privacy guarantees when there exists a high ratio of infrequent variants. Moreover, TraVaG overcomes shortcomings of conventional privacy preservation techniques such as bounding the length of variants and introducing fake variants. Experimental results on real-life event data show that our approach outperforms state-of-the-art techniques in terms of privacy guarantees, plain data utility preservation, and result utility preservation.

Keywords: Process Mining · Event Data · Differential Privacy · GANs · Machine Learning · Autoencoder

1 Introduction

Process mining is a family of data-driven techniques for business process discovery, analysis, and improvement. Process mining techniques require event data, which are widely available in most information systems, including ERP, SCM, and CRM systems. During the last decade, process mining has been successfully deployed in many industries, and it has become a crucial success factor for any type of business. Similar to any data-driven technique in the larger area of data science, concerns about the privacy of people whose data are processed by process mining algorithms are developing as the amount of event data and their usage rise. Thus, privacy regulations, e.g., GDPR [10], restrict data storage and process, which motivates the development of privacy preservation techniques.

Modern privacy preservation methods are mostly based on Differential Privacy (DP), which provides a privacy definition by introducing noise into data.

S. Nurcan et al. (Eds.): RCIS 2023, LNBIP 476, pp. 415–431, 2023.
https://doi.org/10.1007/978-3-031-33080-3_25

Table 1. A simple event log from the healthcare context, including trace variants and their frequencies.

Trace Variant	Frequency
$\langle register, visit, blood\text{-}test, visit, release\rangle$	15
$\langle register, blood\text{-}test, visit, release\rangle$	12
$\langle register, visit, hospitalization, surgery, release\rangle$	5
$\langle register, visit, blood\text{-}test, blood\text{-}test, release\rangle$	2

This is because of its significant properties, including its ability to ensure mathematically proven privacy and protect against PSO (predicate-singling-out) attacks [5]. The purpose of DP-based approaches is to inject noise into the released output in order to conceal the involvement of an individual. State-of-the-art research in process mining leveraging privacy preservation techniques based on DP focuses on releasing distributions of trace variants, which serve as the foundation for core process mining techniques such as process discovery and conformance checking [1]. A trace variant refers to a complete sequence of activities performed for an individual that is considered to be sensitive and private information. In the healthcare context, for instance, a trace variant shows a complete sequence of treatment-related activities performed for a patient that is private information itself and can also be exploited to conclude other sensitive information, e.g., the disease of the patient. Table 1 shows a small sample of a trace variant distribution in the healthcare context. Note that in a trace variant distribution, each trace variant is associated with an individual, a so-called case. Moreover, each case has precisely one trace variant.

To achieve DP for trace variants, conventional so-called *prefix-based* approaches inject noise drawn from a *Laplacian distribution* into the variant distribution obtained from an event log [11,21]. These approaches need to generate all possible unique variants based on a set of activities to provide differential privacy for the original distribution of variants. Since the set of possible variants that can be generated given a set of activities is infinite, prefix-based techniques need to limit the length of generated sequences. Also, to limit the search space, these approaches typically include a pruning parameter to exclude less frequent prefixes. Such a process to obtain DP has a high computational complexity and results in the following drawbacks: (1) *introducing fake variants*, (2) *removing frequent true variants*, and (3) *having limited length for generated variants.*

Several approaches have been proposed to partially or entirely address the aforementioned drawbacks. A method, called SaCoFa [11], aims to mitigate drawbacks (1) and (2) by gaining knowledge regarding the underlying process semantics from the original event data. However, the privacy quantification of all extra queries to gain knowledge regarding the underlying semantics is not discussed. Moreover, the third drawback still remains since this work itself is a prefix-based approach. In [9] and a technique called Libra [8], which is based on [9], trace variants are converted to a DAFSA (Deterministic Acyclic Finite State Automata) representation to avoid such drawbacks. However, Libra introduces a clipping parameter for removing infrequent variants. This clipping parameter grows based on the number of unique trace variants and the strength of privacy guarantees. Thus, depending on the number of unique trace variants and privacy parameters, Libra may even remove all the variants and return empty outputs. A recent work called TraVaS [25] proposes an approach based on *differentially private partition selection strategies* to overcome the above-mentioned drawbacks. Similar to Libra, TraVaS also removes infrequent trace variants. However, in TraVaS, the threshold for removing infrequent variants is only dependent on the input privacy parameters and does not grow with the number of unique variants or the size of event data. Yet, for small event data with a high rate of unique trace variants, TraVaS may not be able to provide strong privacy guarantees.

In this paper, we introduce TraVaG to generate differentially private trace variants from an original variant distribution by means of GANs (Generative Adversarial Networks) [13]. The main idea of TraVaG is to privately learn important event data characteristics. The trained GAN enables the generation of new synthetic anonymized variants that are statistically similar to the original data. Trained generative models work without data access. Thus, as long as the statistical characteristics of the original data do not significantly change, one does not need to apply DP directly to the original event data. For industry-scale big event data, this property can considerably improve the computational complexities [22]. Moreover, TraVaG is based on DP-SGD (Differentially Private - Stochastic Gradient Descent) [2] optimization techniques that avoid thresholding on training data or released network outputs. Hence, TraVaG can generate infinite and arbitrarily large anonymized synthetic trace variants even if the original variant frequencies are comparably small. Moreover, our experiments on real-life event logs demonstrate a better performance of TraVaG compared to state-of-the-art techniques in terms of data utility preservation for the same privacy guarantees.

The remainder of this paper is structured as follows. In Sect. 2, we provide a summary of related work. Preliminaries are provided in Sect. 3. In Sect. 4, we present the details of TraVaG. Section 5 discusses the experimental results based on real-life event logs, and Sect. 6 concludes the paper.

2 Related Work

Privacy-preserving process mining is recently growing in importance. Several techniques have been proposed to address privacy issues in process mining. In the following, we provide a summary of the work focusing on *releasing differentially private event data* and *generating differentially private event data*.

2.1 Releasing Differentially Private Event Data

In [21], the authors apply an (ϵ, δ)-DP mechanism to event logs to privatize *directly-follows relations* and trace variants. The underlying principle uses a combination of an (ϵ, δ)-DP noise generator and an iterative query engine that allows an anonymized publication of trace variants with an upper bound on their length. In [11], SaCoFa has been introduced as an extension of [21], where the goal is to optimize the query structures with the help of underlying semantics. All the aforementioned techniques follow the so-called prefix-based approach that suffers from the drawbacks explained in Sect. 1. To deal with such drawbacks, in [9], the authors introduced an approach that transforms a trace variant distribution into a DAFSA representation. This approach aims to keep all the original trace variants that may result in high noise injection during the anonymization process. Libra [8] is a recent work that employs the approach proposed in [9] and aims to increase utility using subsampling and composing privatized subsamples to release differentially private event data. TraVaS [25] introduces a novel approach based on differentially private partition selection to address the mentioned drawbacks in Sect. 1.

2.2 Generating Differentially Private Synthetic Data

Although DP-based generative Artificial Neural Networks (ANNs) have been quite extensively researched in the major field of data science and machine learning, they have not been used in the context of process mining. Thus, we mainly focus on some of the work outside the domain of process mining. In [4], the authors adopted a so-called *variational autoencoder*, DP-VAE, which assumes that the mapping from real data to the Gaussian distribution can be efficiently learned. A different direction was then chosen by [12], where the authors used a *Wasserstein* GAN (WGAN) to generate differentially private mixed-type synthetic outputs employing a Wasserstein-distance-based loss function. Finally, in [27], the concepts of WGAN and DP-VAE were combined to first learn a private data encoding and then generate respective encoded data. We adapted this principle for our work to cope with the large dimensionality of event data.

Research in non-private generative models for process mining, primarily focuses on exploiting ANNs and GANs to predict the next state of processes such as [17], and [19]. Note that the approach in [17] only provides synthetic event data without any privacy guarantees.

3 Preliminaries

We start the preliminaries by introducing basic notations and mathematical concepts. Let A be a set. $\mathcal{B}(A)$ is the set of all multisets over A. Given B_1 and B_2 as two multisets, $B_1 \uplus B_2$ is the sum over multisets, e.g., $[a^2, b^3] \uplus [b^2, c^2] = [a^2, b^5, c^2]$. We define a finite sequence over A of length n as $\sigma = \langle a_1, a_2, \ldots, a_n \rangle$ where $\sigma(i) = a_i \in A$ for all $i \in \{1, 2, \ldots, n\}$. The set of all finite sequences over A is denoted with A^*.

3.1 Event Log

Process mining techniques employ event data that are typically collections of unique events recorded per activity execution and characterized by their attributes, e.g., *activity* and *timestamp*. Events in an event log have to be unique. A *trace* is a single process execution represented as a sequence of events belonging to a case (individual) and having a fixed ordering based on timestamps. An event cannot appear in more than one trace or multiple times in one trace. Our work focuses on the control-flow aspect of an event log that only considers the activity attribute of events in a trace, so-called a *trace variant*. Thus, we define a simple event log based on activity sequences, so-called *trace variants*.

Definition 1 (Simple Event Log). *Let $\mathcal{A}$ be the universe of activities. A simple event log L is defined as a multiset of trace variants $\mathcal{A}^*$, i.e., $L \in B(\mathcal{A}^*)$. $\mathcal{L}$ denotes the universe of simple event logs.*

In a simple event log representing a distribution of trace variants, one case, which refers to an individual, cannot contribute to more than one trace variant.

3.2 Differential Privacy (DP)

The main idea of DP is to inject noise into the original data in such a way that an observer who sees the randomized output cannot with certainty tell if the information of a specific individual is included in the data [7]. Considering simple event logs, as our sensitive event data, we define differential privacy in Definition 2.

Definition 2 ((ϵ,δ)-DP for Event Logs). *Let L_1 and L_2 be two neighboring event logs that differ only in a single entry, i.e., $L_2 = L_1 \uplus [\sigma]$ for any $\sigma \in \mathcal{A}^*$. Also, let $\epsilon \in \mathbb{R}_{>0}$ and $\delta \in \mathbb{R}_{>0}$ be two privacy parameters. A randomized mechanism $\mathcal{M}_{\epsilon,\delta}: \mathcal{L} \rightarrow \mathcal{L}$ provides (ϵ, δ)-DP if for all $S \subseteq B(\mathcal{A}^*)$: $Pr[\mathcal{M}_{\epsilon,\delta}(L_1) \in S] \leq e^{\epsilon} \times Pr[\mathcal{M}_{\epsilon,\delta}(L_2) \in S] + \delta$.*

In Definition 2, ϵ specifies the probability ratio, and δ allows for a linear violation. In the strict case of $\delta = 0$, $\mathcal{M}$ offers ϵ-DP. The randomness of respective mechanisms is typically ensured by the noise drawn from a probability distribution that perturbs original variant-frequency tuples and results in non-deterministic outputs. The smaller the privacy parameters are set, the more noise is injected into the mechanism outputs, entailing a decreasing likelihood of tracing back the instance existence based on outputs.

3.3 Generative Adversarial Networks (GANs)

A generative adversarial network (GAN) represents a special type of ANN compound to synthesize similar data to its original input. It comprises two separate ANNs, a *generator* and a *discriminator* [13]. The training principle follows a two-player game: a generator tries to fool the discriminator by generating authentic fake data while a discriminator tries to distinguish between real and fake results. A generator $gen : \mathbb{Z}^m \rightarrow \mathbb{R}^n$ and a discriminator $dis : \mathbb{R}^n \rightarrow \{0, 1\}$ can be described as highly parametrizable functions. Here, a generator gen is seeded with random multivariate Gaussian noise $z \in Z^m$ of user-defined dimension m that is translated into a synthetic desired output. A discriminator dis aims to determine whether its input originates from the generator's output. In a simple form, it outputs a binary decision variable, where 0 means the input is fake and 1 means the input is original data. In our work, we apply a GAN architecture to synthesize event data.

3.4 Autoencoder

An *autoencoder* is a certain type of ANN structure used to learn efficient encodings of unlabeled data [15]. The respective encoding is validated and optimized by attempting to regenerate the input from the encoding by decoding. The autoencoder learns the encoding for a set of data to typically provide dimensionality reduction. As a result, an autoencoder always consists of two separate ANNs, an encoder $enc : \mathbb{R}^n \rightarrow \mathbb{R}^d$ and a decoder $dec : \mathbb{R}^d \rightarrow \mathbb{R}^n$. These components allow for transforming high-dimensional data $x \in \mathbb{R}^n$ to a compact representation within the so-called latent space $\mathbb{R}^d$ and vice versa (typically $d \ll n$). The specific mappings of *enc* and *dec* are characterized by the network's weights and learned from the data during the training phase. For our work, we employ an autoencoder structure to achieve a compressed encoding of input event data.

4 TraVaG

As presented in Sect. 2, DP-based generative networks have been extensively researched outside of the process mining context. Typical approaches either adopt variational autoencoder architectures that leverage both encoder and decoder components or GAN architectures employing a discriminator and a generator part. When transferring these ideas to event data, one crucial aspect is the high-dimensional structure that turns out to be challenging during training, particularly if strong DP is added. Thus, we follow the approach of the novel work [27] and [14] that combines the compression functionality of autoencoders with the flexibility of GANs and demonstrated superior performance for general high-dimensional mix-type input data [27]. Instead of directly generating new event logs, we first learn a compressed encoding and then train a GAN to reproduce data within the encoded latent space. Final datasets are obtained by decoding back the dimension-reduced intermediate format. This principle mitigates the complication of GANs when extracting statistical properties from feature-rich data that is limited in size. Particularly, sparse features can be compressed without significant loss of information, while generator networks improve their learning performance due to the lower dimension. Moreover, no Gaussian Mixture distribution is enforced on the latent space, as is the case for typical generative stand-alone autoencoder methods [4].

4.1 The TraVaG Framework

Different components and the workflow of our framework are shown in Fig. 1. We start with preprocessing a simple event log that contains variant distributions in the form of variant-frequency pairs. There are two common possibilities. The first option considers the activities within variants and extracts all subsequences of direct neighbors, i.e., Directly-Follows Relations (DFRs). These DFRs are then mapped to a binary or number space and either fed into a GAN as a single feature or as two features along with their frequencies. A downside of this method is that the generator serves as a sequence constructor which allows the

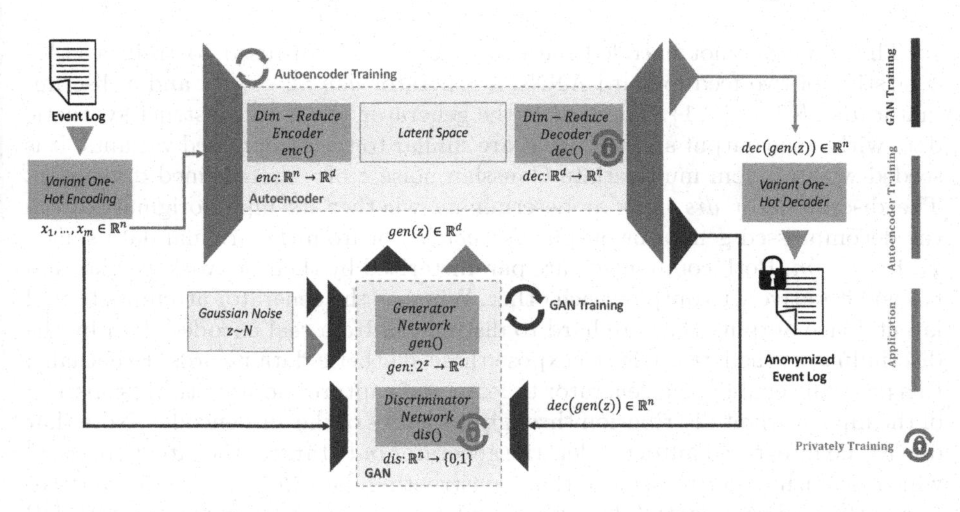

Fig. 1. A simplified workflow diagram of the TraVaG training and application processes.

creation of artificial variants in the postprocessing phase where all generated activity pairs are linked back together. To avoid creating fake trace variants, we choose the second option, where only complete variants are considered as inputs. Therefore, a simple event log L with n variants and m cases is binary-encoded as follows. Within a $m \times n$ matrix, each variant represents a binary feature column and each case denotes a row instance that contains 1 at the respective variant column and 0 elsewhere (sparse matrix). Analogously, this transformation can be inverted back to the original data space. Thus, TraVaG never produces fake trace variants. Also, one-hot encoding does not influence the data statistics and hence does not incur any privacy costs. We refer to this preprocessing procedure as *one-hot encoding* and *one-hot decoding*.

We perform two main training phases including autoencoder training (blue parts) and GANs training (purple parts). Since the focus of this work is on the privacy aspect, we describe the privately trained components in more detail. A detailed algorithmic explanation of the training components including the structure of the networks, parameter tuning, activation functions, loss measures, and optimizations is provided in our supplementary document.[1] After the preprocessing, the sparse binary variant vectors $x_1 \dots x_m \in \mathbb{R}^n$ are forwarded to the autoencoder training phase, including an encoder and a decoder component. These components allow for transforming high-dimensional data $x_i \in \mathbb{R}^n$ to a compact representation within the so-called latent space $\mathbb{R}^d$ and vice versa, s.t., $d \ll n$. The dimension d is a hyperparameter of the autoencoder and needs to be selected w.r.t. the GANs configuration. Since the encoder does not participate in the process of training the GAN or synthesizing new event data, it does not need to be optimized privately [3,4]. The decoder is strongly involved in the anonymization process and is released to the public. Thus, the training of the decoder is performed privately by means of DP-SGD (see Sect. 4.2).

[1] https://github.com/wangelik/TraVaG/blob/main/supplementary/TraVaG.pdf.

The same one-hot encoded data $x_1 \dots x_m \in \mathbb{R}^n$ are used to train a GAN consisting of two feed-forward ANNs; a generator $gen : 2^{\mathbb{Z}} \to \mathbb{R}^d$ and a discriminator $dis : \mathbb{R}^n \to \{0, 1\}$. The goal of the generator *gen* is to construct synthetic data within the output space $\mathbb{R}^d$ that are similar to the compressed variants. It is seeded with random multivariate Gaussian noise z of a user-defined dimension. The discriminator *dis* aims at determining whether its input originates from the decompressed generator output $dec(gen(.))$ or from the original data source $x_i, 1 \leq i \leq m$. Both components are parameterized by their network weights and trained iteratively to outplay each other. Whereas the generator attempts to find latent space outputs that are hard to distinguish from real encoded data by the discriminator, the latter tries to expose these synthetic data records. Eventually, this principle enables the generator to learn and capture the statistical properties of the input variant distribution through the lenses of the autoencoder. Note that due to the integrated autoencoder, the generator only targets the latent space $\mathbb{R}^d$ which is much easier to achieve than constructing data in $\mathbb{R}^n$. Also, it averts to access the real confidential data space and does not need to be trained with DP as opposed to the discriminator that is again privately optimized with DP-SGD algorithms [27].

Once both the autoencoder and GAN are trained, one can generate new synthetic anonymized event data (orange parts). The underlying mechanism equals the training step of the generator. Starting with a random Gaussian noise sample z, this noise becomes digested by the generator, yielding $gen(z)$. From the latent space, the decoder then maps $gen(z)$ to $dec(gen(z))$. Finally, the synthetic one-hot encoded result is transformed back to the variant universe. One compelling advantage of TraVaG lies in the underlying data format. Since the feature space represents the different variants of the original data, TraVaG considers them as given and only has to learn their distribution during training. When applied, the framework reconstructs an anonymized version of this distribution over multiple runs without introducing new variants. The more synthetic data are created, the better the consolidated TraVaG output, i.e., new anonymized variants better approximate the original variant distribution. Note that this process does not converge to the true variant frequencies, but to the TraVaG-internal learned anonymous version. Thus, it is recommended to run TraVaG at least as often as the number of cases in the original event log. In case smaller privatized datasets are needed, the output can be down-sampled during postprocessing rounds.

4.2 Differentially Private - Stochastic Gradient Descent (DP-SGD)

To render SGD differentially private, Abadi et al. [2] proposed the following two steps. Given a dataset $X = \{x_i \in \mathbb{R}^n \mid 1 \leq i \leq m\}$, f as a loss function, and θ as the model parameter. First, the gradient $g_i = \nabla_\theta f_\theta(x_i)$ of each data sample x_i is clipped at some real value $C \in \mathbb{R}_{>0}$ to ensure its L^2-norm of the gradient does not exceed the clipping value. For our work, we refer to the following clipping function[2]: $\text{clip}(g_i, C) = g_i \cdot \min(1, C/||g_i||_2)$.

[2] Note that also other clipping strategies exist, as highlighted in [22].

Then, as Eq. 1 shows, multivariate Gaussian noise parametrized by a noise multiplier $\Phi \in \mathbb{R}$ is added to the clipped gradient vectors before averaging over the batch $B \subseteq X$. We further denote the identity matrix as I and the Gaussian distribution of unspecified dimension as $\mathcal{N}$.

$$g_B \leftarrow \frac{1}{|B|} \left(\sum_{i \in B} \text{clip}(\nabla_\theta f_\theta(x_i), C) + \mathcal{N}(0, C^2 \Phi^2 I) \right) \tag{1}$$

The noisy-clipped-averaged gradient g_B is now differentially private and can be used for conventional descent steps: $\theta \leftarrow \theta - \eta \cdot g_B$, where η is the so-called *learning rate.* Note that clipping the individual gradients as in Eq. 1 can also be replaced by instead clipping gradients of groups of more data points, so-called *microbatches* [22]. Instead of the common DP parameters ϵ and δ, DP-SGD uses the related noise multiplier Φ. When translating between these two types of settings, novel research has demonstrated a tighter privacy bound if the batch sampling process for B is conducted according to a specific procedure [2]. This procedure independently selects each data point of X with a fixed probability q, the so-called *sampling rate*, in each step.

4.3 Privacy Accounting

To evaluate the exact privacy guarantee provided by DP-SGD algorithms, we employ the so-called *Renyi Differential Privacy* (RDP) [23], a different notion of DP typically used for private optimization. RDP is defined based on the concept of *Renyi divergence.* Given two probability distributions P and Q, the Renyi divergence of order α is defined as follows: $D_\alpha(P||Q) := \frac{1}{\alpha-1} \log \mathbb{E}_{x \sim Q} \left(\frac{P(x)}{Q(x)} \right)^\alpha$.

Definition 3 ((α, ϵ)-RDP for Event Logs).
Let L_1 and L_2 be two neighboring event logs that differ only in a single entry, e.g., $L_2 = L_1 \uplus [\sigma]$ for any $\sigma \in \mathcal{A}^*$. Given $\alpha > 1$ and $\epsilon \in \mathbb{R}_{>0}$, a randomized mechanism $\mathcal{M}_{\alpha,\epsilon}: \mathcal{L} \rightarrow \mathcal{L}$ provides (α, ϵ)-RDP if $D_\alpha(\mathcal{M}(L_1)||\mathcal{M}(L_2)) \leq \epsilon$.

To obtain the final (ϵ, δ)-DP parameters, we employ the following two propositions on the composition of (α, ϵ)-RDP mechanisms and the conversion of (α, ϵ)-RDP parameters to (ϵ, δ)-DP parameters.

Proposition 1 (Composition of RDP [23]**).** *If $\mathcal{M}_1$ and $\mathcal{M}_2$ are two (α, ϵ_1)-RDP and (α, ϵ_2)-RDP mechanisms for $\alpha > 1$, respectively. Then, the composition of $\mathcal{M}_1$ and $\mathcal{M}_2$ satisfies $(\alpha, \epsilon_1 + \epsilon_2)$-RDP.*

Proposition 2 (RDP Parameter Conversion [23]**).** *If a mechanism $\mathcal{M}$ satisfies (α, ϵ)-RDP with $\alpha > 1$, then for all $\delta > 0$, $\mathcal{M}$ satisfies $(\epsilon + (\log 1/\delta)/(\alpha - 1), \delta)$-DP.*

During an iterative application of Gaussian mechanisms, as is the case in DP-SGD, the Renyi divergence allows more tightly capturing of the corresponding privacy loss than standard (ϵ, δ)-DP. To compute the final (ϵ, δ)-DP parameters from multiple runs of DP-SGD, the following three steps are followed.

1. **Subsampled RDP.** Given a sampling rate q and noise multiplier Φ, the RDP privacy parameters for one iteration of DP-SGD can be derived as a non-explicit integral function of $\alpha \geq 1$ [23]. This function is standardized in many privacy-related optimization packages and will be referred to as $\text{RDP}_1(q, \Phi)$ [2].
2. **RDP Composition.** Since DP-SGD is most likely to run iteratively, we need to compose Step 1 over all executions according to Proposition 1. Hence, the resulting RDP parameters of T iterations are obtained by computing $\text{RDP}_T(q, \Phi, T) := \text{RDP}_1(q, \Phi) \cdot T$.
3. **Conversion to (ϵ, δ)-DP.** After retrieving an expression for the overall RDP privacy parameters with RDP_T, we need to convert the respective (α, ϵ) tuple to a (ϵ, δ) guarantee according to Proposition 2. Since the ϵ parameter of RDP is also a function of α, Step 3 involves optimizing for α to achieve a minimal ϵ and δ.

We apply this procedure to obtain the respective privacy guarantees (ϵ, δ)-DP on both the autoencoder and the GAN-based discriminator of TraVaG. The resulting values are then combined into a final privacy cost by the *composition theorem* of DP [7]. According to the composition theorem, different (ϵ, δ)-DP mechanisms can be easily combined into more complex algorithms at the cost of a directly measurable cumulative privacy loss, and the result still promises (ϵ, δ)-DP independent of the exact form of composition or query structure.

5 Experiments

We evaluate the performance of TraVaG on real-life event logs. We select two event logs of varying sizes and trace uniqueness. As we discussed in Sect. 1 and stated in other research such as [11,21], and [8] infrequent variants are challenging to privatize. Thus, trace uniqueness is an important analysis criterion. The Sepsis log describes hospital processes for Sepsis patients and contains many rare traces [20]. In contrast, BPIC13 has significantly more cases at a four times smaller trace uniqueness [6]. BPIC13 describes an incident and problem management system called VINST. Both logs are realistic examples of confidential human-centered information where the case identifiers refer to individuals. Table 2 shows detailed log statistics.

We perform our evaluation for a wide range of the main privacy parameters $\epsilon \in \{0.01, 0.1, 1, 2\}$ and $\delta \in \{10^{-6}, 10^{-5}, 10^{-4}, 10^{-3}, 0.01\}$. These ranges are selected in accordance with typical values employed at industrial applications

Table 2. General statistics of the event logs used in our experiments.

Event Log	#Events	#Cases	#Activities	#Variants	Trace Uniqueness
Sepsis	15214	1050	16	846	80%
BPIC13	65533	7554	4	1511	20%

as well as state-of-the-art DP research [8,11,21,26]. We particularly note that extreme settings such as $\epsilon = 2, \delta = 0.5$ are not chosen due to practical relevance, but to demonstrate how the anonymization methods behave when starting from a weak- or non-private environment. Due to the probabilistic nature of (ϵ, δ)-DP, we run the TraVaG generator 100 times on all input event logs and all privacy parameters and report the average values. We compare our results with TraVaS [25] as a state-of-the-art technique and the original prefix-based framework called benchmark [21].[3] The sequence cutoff for the benchmark method is set to the length that covers 80% of variants in each log, and the remaining pruning parameter is adjusted such that on average anonymized logs contain a comparable number of variants with the given original log. The ANNs of TraVaG are configured by a semi-automated tuning approach w.r.t the different input logs. Whereas most design decisions and hyperparameters are tweaked according to results of manual tests as well as research experience, the settings: *batch size* (B), *number of iterations* (I) and *noise multiplier* (Φ) are automatically optimized via a grid-search [18] for fixed privacy levels. A detailed list of the derived settings for each event log, the concrete network designs, and configuration values are available on GitHub.[4]

5.1 Evaluation Measures

Suitable evaluation measures are required to assess the performance of an (ϵ, δ)-DP mechanism in terms of data (result) utility preservation. The *data utility* perspective measures the similarity between two logs independent of future applications. For evaluating data utility we employ the following measures: *relative log similarity* [24,25] and *absolute log difference* [25]. *Relative log similarity* measures the *earth mover's distance* between two trace variant distributions, where the normalized *Levenshtein* string edit distance is used as a similarity function between trace variants. This measure quantifies the degree to which the variant distribution of an anonymized log matches the original variant distribution on a scale from 0 to 1. *Absolute log difference* accounts for the situations where distribution-based measures provide misleading expressiveness [25]. Exemplary cases are event logs possessing similar variant distributions, but significantly different sizes. To calculate an absolute log difference value, we use the approach introduced in [25], where input logs are converted to a *bipartite graph* of variants as vertices. Then, a *cost network flow* problem is solved by setting demands and supplies to the absolute variant frequencies and utilizing a *Levenshtein* distance

[3] Note that in [25], TraVaS was already compared with SaCoFa [11] and benchmark [21] and showed better performance. Here, the benchmark method is included for easier comparison. Moreover, Libra [8] does not take ϵ as an input parameter but computes it based on α as an RDP parameter and its sampling strategy. This makes the comparison based on exact ϵ and δ parameters very difficult. Nevertheless, an important observation in contrast to TraVaG is that Libra returns an empty log for event logs with many infrequent variants, such as Sepsis when $\delta \leq 10^{-3}$.

[4] https://github.com/wangelik/TraVaG/blob/main/supplementary/TraVaG.pdf.

between variants as an edge cost. Thus, the result of this measure shows the minimal number of *Levenshtein* operations to transform variants of an anonymized log into variants of the original log. Details of the exact algorithms are available.[5]

We additionally evaluate the performance of TraVaG in terms of *result utility preservation* for *process discovery* as a specific application of trace variant distribution. In this respect, we use the *inductive miner infrequent* [16] with a default noise threshold of 20% to discover process models from the privatized event logs for all (ϵ, δ) settings under investigation. Then, we compare the models with the original event log to obtain token-based replay *fitness* and *precision* scores [1].

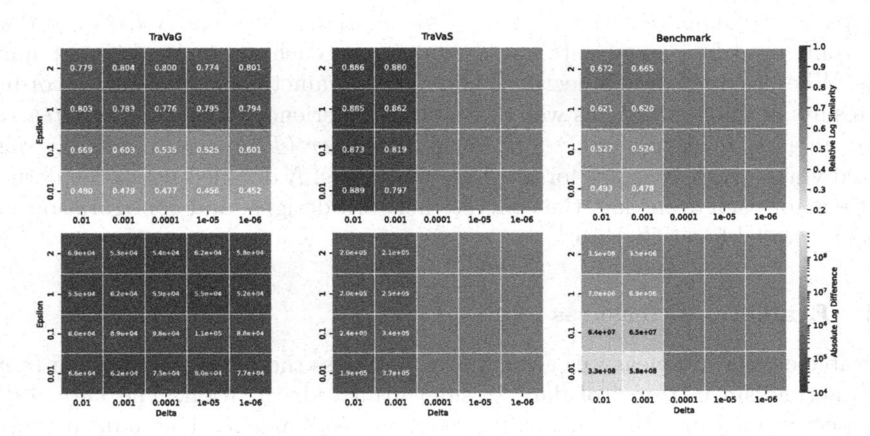

Fig. 2. The *relative log similarity* and *absolute log difference* results of anonymized BPIC13 logs generated by TraVaG, TraVaS, and the benchmark method. Each value represents the mean of 100 generations for TraVaG and 10 algorithm runs for TraVaS and the benchmark method.

5.2 Data Utility Analysis

In this subsection, the results of the two aforementioned data utility metrics are presented for both real-life event logs. Figure 2 shows the average results on BPIC13 in a six-fold heatmap. The gray fields at the TraVaS and benchmark methods denote an unsuccessful algorithm execution. For $\delta < 10^{-3}$, the thresholding of TraVaS becomes too strict and removes many variants in the anonymized outputs. On the contrary, the benchmark method introduces artificial variants and noise to an extent that is unfeasible to average within reasonable time and accuracy. In opposition, TraVaG successfully manages to generate anonymized outputs for $\delta < 10^{-3}$. More importantly, both results of *relative log similarity* and *absolute log difference* do not illustrate clear decreasing trends on lower δ within the investigated parameter range. We explain this expected

[5] https://github.com/wangelik/TraVaG/blob/main/supplementary/metrics.pdf.

observation by the fact that TraVaG avoids any pruning mechanism on its output and implements less δ-dependent Gaussian noise via RDP into the gradients (see Sect. 4.3 and [23]).

Whereas the absolute log difference results maintain a rather stable output for the different (ϵ, δ) values, the TraVaG relative log similarity presents a strong positive ϵ-dependency. As a result, the absolute statistics (absolute Levenshtein distances and absolute frequencies) of the anonymized event data seem to be more similar to the original logs as the variant distributions. A rationale for this discrepancy lies in the still comparably small dataset with 7554 instances over 1511 variants (features). By construction, TraVaG accomplishes reproducing equally sized event logs containing many original variants but fails to pick up some characteristics of the underlying distribution once the input data or the training iterations are limited. Hence, we expect this diverging trend to diminish with increasing training data.

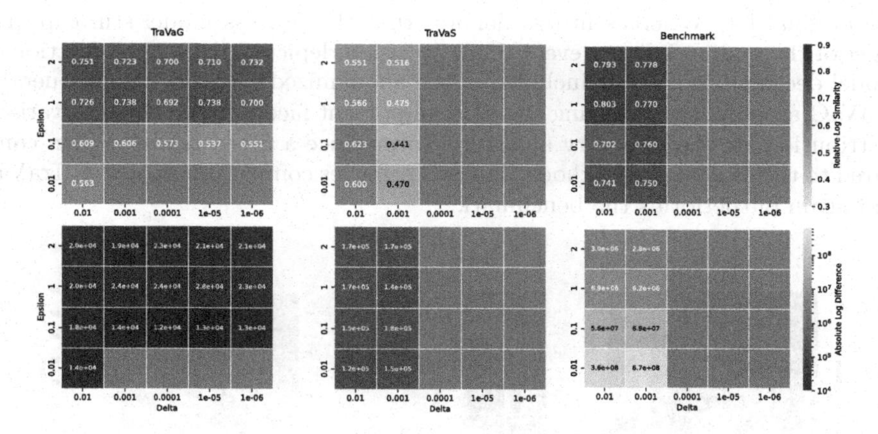

Fig. 3. The *relative log similarity* and *absolute log difference* results of anonymized Sepsis logs generated by TraVaG, TraVaS, and the benchmark method. Each value represents the mean of 100 generations for TraVaG and 10 algorithm runs for TraVaS and the benchmark method.

The data utility results for the Sepsis log are presented in Fig. 3. With only 1050 instances at 846 variants (features), this dataset is even smaller and thus more difficult to train for TraVaG than BPIC13. As a result, we observe similar, but more pronounced behavior of relative log similarity and absolute log difference metrics compared to Fig. 2. An extreme example are the results at $\epsilon = 0.01, \delta < 10^{-2}$, where the introduced gradient noise turned out as too intense for the generative model to converge under the given training data size.For the remaining privacy settings, TraVaG again outperforms its competitors at the absolute log statistics while the relative log similarity performs slightly better than TraVaS and at the same order as the benchmark results for $\epsilon > 0.1$.

5.3 Process Discovery Analysis

Figure 4 illustrates the result utility analysis of TraVaG, TraVaS, and the benchmark on BPIC13. As discussed in Subsect. 5.2, TraVaG successfully manages to produce results for $\delta < 10^{-3}$ where the other methods are not applicable. Except for the three outliers at $\epsilon = 0.1$, both fitness and precision show a stable distribution without considerable dependence on the different privacy parameters. In accordance with Fig. 2, we thus conclude that the absolute log difference provides a better proxy for process-discovery-based performance of TraVaG than relative log similarity. Similarly, the strong scores on both metrics demonstrate a sufficient replay behavior between the model obtained from an anonymized log and the original log. Whereas fitness denotes that the process model still captures most of the real underlying event data, precision depicts only a small fraction of model decisions, not being included in the anonymized event log. Consequently, TraVaG accomplishes learning the most important facets of the BPIC13 variant distribution for the discovery algorithm to produce a fitted model. When compared to the alternative methods, TraVaG achieves comparable scores as TraVaS and again outperforms the benchmark.

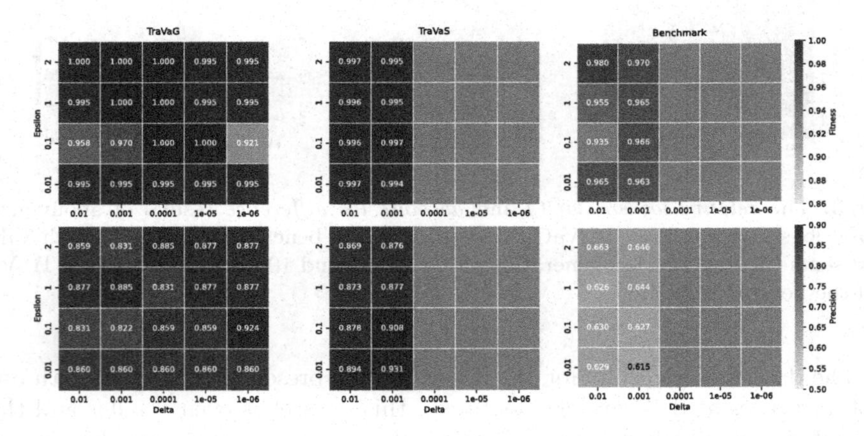

Fig. 4. The *fitness* and *precision* results of anonymized BPIC13 event logs generated by TraVaG, TraVaS, and the benchmark method. Each value represents the mean of 100 generations for TraVaG and 10 algorithm runs for TraVaS and the benchmark method.

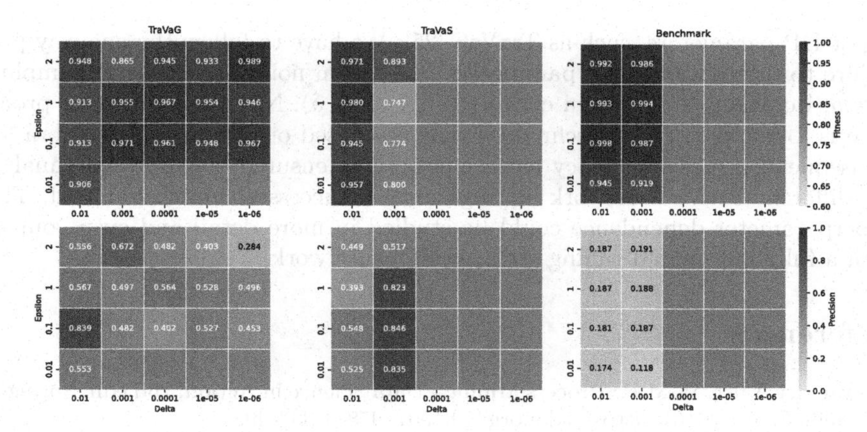

Fig. 5. The *fitness* and *precision* results of anonymized Sepsis event logs generated using TraVaG, TraVaS, and the benchmark method. Each value represents the mean of 100 generations for TraVaG and 10 algorithm runs for TraVaS and the benchmark method.

The result utility evaluation of the high trace-unique Sepsis log is presented in Fig. 5. With respect to fitness, TraVaG shows similar values as TraVaS but a slight under-performance compared to the benchmark method. The main cause for this observation again refers to the infrequent variants and the small log size. While TraVaS maintains a strong δ-related threshold and TraVaG copes with the limited training data, the benchmark method introduces many artificial variants but tends to match the frequent traces. As a result, the discovered process models are able to replay most of the original behavior in contrast to TraVaG and TraVaS results. According to the aforementioned explanation, precision reflects an inverted trend. Here, the larger models of the benchmark method contain many possible decision paths that are nonexistent in the underlying event log. For TraVaS and TraVaG, we thus achieve more precise anonymized process models.

6 Conclusion

TraVaG has shown that training a differentially private combination of autoencoders and GANs to synthesize anonymized event data from an underlying original variant distribution outperforms current state-of-the-art selection-based variant anonymization techniques and prefix-based approaches. Particularly, for strong privacy at the low δ range. Moreover, TraVaG has the unique advantages of outstanding resource-efficient execution, the absence of distorting noise thresholds, a general acceptance of continuous data streams, and no fake variant generation. In combination, these characteristics allow TraVaG to efficiently operate with infrequent variant data in the low δ regime without real competitors. Nevertheless, we note that the framework comprises a more complex training procedure and privacy budget accounting than approaches that directly digest DP parameters, such as TraVaS [25]. We have to follow the one-way procedure to first obtain RDP parameters (ϵ, α) from noise multiplier Φ, sampling rate q, iterations T and then convert (ϵ, α) to (ϵ, δ). Note that a similar procedure is followed by other techniques that are based on RDP, such as Libra [8]. Consequently, specific privacy levels can only be ensured by repeatedly analyzing different TraVaG network settings until a successful match is found. This hyperparameter dependence could be studied in more detail and even coupled with a fully automated tuning strategy in future work.

References

1. van der Aalst, W.M.P.: Process Mining - Data Science in Action, 2nd edn. Springer, Heidelberg (2016). https://doi.org/10.1007/978-3-662-49851-4
2. Abadi, M., et al.: Deep learning with differential privacy. In: Proceedings of the 2016 ACM SIGSAC Conference on Computer and Communications Security, Vienna, Austria, 24–28 October 2016, pp. 308–318. ACM (2016)
3. Ács, G., Melis, L., Castelluccia, C., Cristofaro, E.D.: Differentially private mixture of generative neural networks. IEEE Trans. Knowl. Data Eng. **31**(6), 1109–1121 (2019)
4. Chen, Q., et al.: Differentially private data generative models. CoRR abs/1812.02274 (2018)
5. Cohen, A., Nissim, K.: Towards formalizing the GDPR's notion of singling out. Proc. Natl. Acad. Sci. USA **117**(15), 8344–8352 (2020)
6. van Dongen, B.F., Weber, B., Ferreira, D.R., Weerdt, J.D.: BPI challenge 2013. In: Proceedings of the 3rd Business Process Intelligence Challenge (2013)
7. Dwork, C.: Differential privacy: a survey of results. In: Agrawal, M., Du, D., Duan, Z., Li, A. (eds.) TAMC 2008. LNCS, vol. 4978, pp. 1–19. Springer, Heidelberg (2008). https://doi.org/10.1007/978-3-540-79228-4_1
8. Elkoumy, G., Dumas, M.: Libra: high-utility anonymization of event logs for process mining via subsampling. In: 4th International Conference on Process Mining, ICPM. IEEE (2022)
9. Elkoumy, G., Pankova, A., Dumas, M.: Mine me but don't single me out: differentially private event logs for process mining. In: 3rd International Conference on Process Mining, ICPM 2021, pp. 80–87. IEEE (2021)
10. EU: EU General Data Protection. OJ L **119**(1) (2016)
11. Fahrenkrog-Petersen, S.A., Kabierski, M., Rösel, F., van der Aa, H., Weidlich, M.: Sacofa: semantics-aware control-flow anonymization for process mining. In: 3rd International Conference on Process Mining, ICPM 2021, Eindhoven, The Netherlands, 31 October–4 November 2021, pp. 72–79. IEEE (2021)
12. Frigerio, L., de Oliveira, A.S., Gomez, L., Duverger, P.: Differentially private generative adversarial networks for time series, continuous, and discrete open data. In: Dhillon, G., Karlsson, F., Hedström, K., Zúquete, A. (eds.) SEC 2019. IAICT, vol. 562, pp. 151–164. Springer, Cham (2019). https://doi.org/10.1007/978-3-030-22312-0_11
13. Goodfellow, I.J., et al.: Generative adversarial networks. Commun. ACM **63**(11), 139–144 (2020)
14. Gulrajani, I., Ahmed, F., Arjovsky, M., Dumoulin, V., Courville, A.C.: Improved training of Wasserstein GANs. In: Advances in Neural Information Processing Systems 30: Annual Conference on Neural Information Processing Systems (2017)

15. Kingma, D.P., Welling, M.: Auto-encoding variational bayes. In: 2nd International Conference on Learning Representations, Conference Track Proceedings (2014)
16. Leemans, S.J.J., Fahland, D., van der Aalst, W.M.P.: Discovering block-structured process models from incomplete event logs. In: Ciardo, G., Kindler, E. (eds.) PETRI NETS 2014. LNCS, vol. 8489, pp. 91–110. Springer, Cham (2014). https://doi.org/10.1007/978-3-319-07734-5_6
17. Li, K., Yang, S., Sullivan, T.M., Burd, R.S., Marsic, I.: Generating privacy-preserving process data with deep generative models. CoRR abs/2203.07949 (2022)
18. Liashchynskyi, P., Liashchynskyi, P.: Grid search, random search, genetic algorithm: a big comparison for NAS. CoRR abs/1912.06059 (2019)
19. Lu, Y., Chen, Q., Poon, S.K.: A deep learning approach for repairing missing activity labels in event logs for process mining. Information **13**(5), 234 (2022)
20. Mannhardt, F.: Sepsis cases (2016). https://doi.org/10.4121/UUID:915D2BFB-7E84-49AD-A286-DC35F063A460
21. Mannhardt, F., Koschmider, A., Baracaldo, N., Weidlich, M., Michael, J.: Privacy-preserving process mining - differential privacy for event logs. Bus. Inf. Syst. Eng. **61**(5), 595–614 (2019)
22. McMahan, H.B., Andrew, G.: A general approach to adding differential privacy to iterative training procedures. CoRR abs/1812.06210 (2018)
23. Mironov, I.: Rényi differential privacy. In: 30th IEEE Computer Security Foundations Symposium, CSF 2017, pp. 263–275. IEEE Computer Society (2017)
24. Rafiei, M., van der Aalst, W.M.P.: Towards quantifying privacy in process mining. In: Leemans, S., Leopold, H. (eds.) ICPM 2020. LNBIP, vol. 406, pp. 385–397. Springer, Cham (2021). https://doi.org/10.1007/978-3-030-72693-5_29
25. Rafiei, M., Wangelik, F., van der Aalst, W.M.P.: TraVaS: differentially private trace variant selection for process mining. In: Montali, M., Senderovich, A., Weidlich, M. (eds.) ICPM 2022. LNBIP, vol. 468. Springer, Cham (2022). https://doi.org/10.1007/978-3-031-27815-0_9
26. Tang, J., Korolova, A., Bai, X., Wang, X., Wang, X.: Privacy loss in apple's implementation of differential privacy on macos 10.12. CoRR abs/1709.02753 (2017)
27. Tantipongpipat, U.T., Waites, C., Boob, D., Siva, A.A., Cummings, R.: Differentially private synthetic mixed-type data generation for unsupervised learning. Intell. Decis. Technol. **15**(4), 779–807 (2021)

User Interface and Experience

When Dashboard's Content Becomes a Barrier - Exploring the Effects of Cognitive Overloads on BI Adoption

Corentin Burnay, Sarah Bouraga, and Mathieu Lega(✉)

NaDI University of Namur, Rue de Bruxelles 61, 5000 Namur, Belgium
mathieu.lega@unamur.be

Abstract. Decision makers in organizations strive to improve the quality of their decisions. One way to improve that process is to objectify the decisions with facts. Big data, business analytics, business intelligence, and more generally data-driven Decision Support Systems (data-driven DSS) intend to achieve this. Organizations invest massively in the development of data-driven DSS and expect them to be adopted and to effectively support decision makers. This raises many technical and methodological challenges, especially regarding the design of dashboards, which can be seen as the visible tip of the data-driven DSS iceberg and which play a major role in the adoption of the entire system. This paper advances early empirical research conducted on one possible root cause for data-driven DSS dashboard adoption or rejection, namely the dashboard content. We study the effect of dashboards over- and underloading on traditional Technology Adoption Models, and try to uncover the trade-offs to which data-driven DSS interface designers are confronted when creating new dashboards. The result is a Dashboard Adoption Model, enriching the seminal TAM model with new content-oriented variables to support the design of more supportive data-driven DSS dashboards.

Keywords: Decision Support Systems · Information Overloads · Dashboards Adoption · Decision Making · Structural Equation Modeling

1 Introduction

Decision makers in organizations strive to improve the quality of their decisions. One way to improve the process is to objectify the decisions with facts and avoid judgments, with the inevitable biases they can lead to [18]. Doing this implies to feed decision makers with information about the organization and its environment and hence to reduce the perverse effects of uncertainty [25]. Big data, business analytics or business intelligence are some examples of fields which intend to achieve this. Decision Support Systems (simply DSS hereafter) is used in this paper to refer to the set of tools, softwares and techniques used in an organization to provide support to decision makers in an integrated and timely manner. DSS have been recognized for long as an important success factor to guide decisions and actions [19].

S. Nurcan et al. (Eds.): RCIS 2023, LNBIP 476, pp. 435–451, 2023.
https://doi.org/10.1007/978-3-031-33080-3_26

Companies spend more and more resources to implement DSS; in 2020, revenue from DSS and analytics software is expected to amount to 22.8 billion U.S. dollars in 2020 [17]. As the amount of resources invested in this type of system increases, norms for implementation and return on investment become more stringent. Organizations invest resources in the development of DSS and expect it to be adopted and to actually support decision makers. This raises many technical and methodological challenges since DSS are multi-faceted and complex by nature [8]. In this paper, we focus on the part of DSS, and more particularly data-driven DSS [29], that is directly exposed to decision makers, namely dashboards.

Dashboards are interactive interfaces that help visualize and analyse performance metrics from the organization [8]. They typically include a set of indicators, graphs and tables together with interactive features to provide decision makers a consistent yet flexible representation of an organization [11]. In practice, high quality data-driven DSS architectures may be rejected by decision makers simply because the dashboards - representing the tip of the data-driven DSS iceberg - are not properly designed and aligned with business needs. This risk is real; pitfalls during the design and implementation of dashboards are numerous. One can cite, for instance, the risks of facing poor information contextualisation, a lack of exploration and interaction capacities, a reduced information accessibility or even an excessive cognitive load generated by information overloads [2]. This paper focuses specifically on the later risk and its impact on data-driven DSS dashboard adoption.

The risk of overloads for managers is not new, with first contributions to this question around the 70's [30]. In this paper, we understand information overload in a classical way as the situation where "the amount of information actually integrated into the decision begins to decline. Beyond this point, the individual's decisions reflect a lesser utilization of the available information" [13,32]. The impact of this extreme information exposure on the use and adoption of data-driven DSS systems has received little attention from the research community (we deepen this claim in the next section). In this paper, we therefore propose to extend well established models of technology adoption [10,21] in the specific context of data-driven DSS dashboards as a way to better understand the problem of information overload. The main hypothesis we want to investigate is that the over- or underloading of a dashboard influences its perceived quality, which in turn reduces the chances of this dashboard to be actually used by decision makers.

The rest of the paper is organized as follows. In Sect. 2, we position our contribution in the field of Information Management, cover important baselines for a proper understanding of our proposition and introduce our research hypotheses. In Sect. 3, we discuss the methodological aspects of the study. Section 4 presents our results, and Sect. 5 proposes a discussion about the results of our study and the possible future works. Section 6 concludes and examines the limitations of the paper.

2 Theoretical Background

2.1 Dashboard Content and Cognitive Load

In this work, we focus our experiment on Business Intelligence (BI) dashboards but we believe the results are transposable to any data-driven DSS interface [8]. The success of a dashboard is inextricably linked to its content and the way it relates to the strategy of an organization. There is little agreement over how exactly a dashboard should look like and what it should do [38]. There have been various attempts to guide dashboard designers (be it a technical or a business person) in the definition of a dashboard content [7,23] or in the way to structure, organize and represent that content [4,15,28]. These methods however tend to encourage more and more content in dashboards, with the risk to increase the *cognitive load* – the amount of working memory resource the brain needs to process a task – of the decision maker, to a point that it cannot be processed by managers in reasonable delays.

Informational content – the content that carries meaningful information about a given context – influences cognitive load. This has been observed in various contexts such as communication, information retrieval and analysis, consumer choice and of course managerial decision making [13]. It also applies in dashboards; too much informational content in a dashboard increases cognitive load and likely decreases dashboard usefulness. The opposite, however, is also true; providing too little information to a decision maker is not likely to be of any help. This translates in an inverted U-shaped curve between supplied information and accuracy of decision [3].

Non-informational content – the content displayed in the dashboard that does not carry particular information about the organization itself, but helps the decision maker to better understand the informational content [14] – also influences cognitive load. It includes things like titles, descriptions or glossary, firm's logo and other related visuals, arrows and color areas, color coding, etc. Such non-informational content also influences cognitive load and may even become in some cases visual distractors [12].

Cognitive load in systems' interfaces has been broadly covered in Human-Computer Interface (HCI) literature [22,37]. Existing HCI contributions however remain generic to any kind of interface and therefore overlook important aspects of data-driven DSS like the support it intends to provide in complex decision-making processes, the dashboard's features like drill-down/pivot/slice & dice or the problem of information visualisation, which all influence the cognitive load of a dashboard user. Some research has been conducted in DSS literature about adoption, but none of them investigates the root cause of adoption/rejection of DSS [33]. To the best of our knowledge, no research has been conducted so far on the influence of informational and non-informational content on the chances of success of BI dashboards, and more generally of data-driven DSS. This paper therefore intends to answer the following research question: Do informational and non-informational content influence the chance of adoption of a data-driven DSS dashboard? If so, how?

2.2 Hedonic Dimension of Dashboards

Dashboards constitute a kind of technology that is particular in several regards. Firstly, they intend to display serious (and relatively tedious) content in a more flexible, exciting and pleasing layout. The aim is to offer great flexibility to users, who can interact smoothly with data and visuals. It is also not uncommon to have strong non-functional requirements on dashboards such as "nice-looking", "eye catching", "thoughts generating". Secondly, they are designed to hide to the end-users the full complexity of the data-driven DSS architecture, leaving only the responsibility to the decision makers to make a final business decision. Finally, they intend to quicken the decision process and reduce the impact of data quality issues on decision-making processes. Put all together, these specific aspects bring us to consider data-driven DSS dashboards as hedonic systems. An hedonic information system finds its value in the fun that a user experiences when using the system, as opposed to a purely utilitarian system where value is entirely instrumental, originating for instance from functionalities offered by the system to increase performance [9].

To answer our research question, we want to relate informational and non-informational content to a well established model that fulfills three important conditions. First of all, the basis model should be applicable to dashboards. Secondly, as our focus is on decision makers, the model should be user-centric. Thirdly, the model should include an hedonic dimension, as discussed above.

The Technology Acceptance Model (TAM) fulfills our first two conditions. The model has been proposed to predict the intention to use IT in the workplace [10]. TAM identifies two variables explaining the behavioral intention to use a technology, namely perceived usefulness and perceived ease of use. Perceived usefulness refers to the degree to which a person believes that using a particular system would enhance his or her job performance, while perceived ease of use refers to the degree to which a person believes that using a particular system would be free of effort. In TAM, it is also claimed that perceived ease of use predicts the perceived usefulness of the technology. Data-driven DSS dashboards fit perfectly with the definition of Technology, as understood in the TAM model.

TAM is a seminal model on which many contributions have been built [21,24,34]. Notably, Van der Heijden proposed the Hedonistic Information System Theory (HIST) where it is argued that the nature of the system plays an important role in user acceptance [21]. The author adds to TAM one additional predictor to the behavioral intention to use a technology: perceived enjoyment, referring to the extent to which the activity of using the computer is perceived to be enjoyable in its own right. This meets our third condition, making the HIST our theoretical starting point, as depicted in Fig. 1.

In comparison, the UTAUT model does not include the hedonic dimension [34] and the UTAUT 2 model is consumer-centric [35], these two models are therefore harder to mobilize in our context.

2.3 Cognitive Theory of Multimedia Learning

Data-driven DSS dashboards are, as mentioned earlier, inexorably linked to their content. Also, the dominant paradigm in data-driven DSS is that dashboards are designed to avoid information overload. It follows that analyzing the types of content displayed on - and by extension the loads of - data-driven DSS dashboards is called for. Our theory proposes a fine-grained examination of dashboards loads by drawing on the concepts of informational content and non-informational content introduced earlier and Mayer's Cognitive Theory of Multimedia Learning [26]. The author distinguished three types of loads: essential load, representational load and incidental load.

Firstly, the load resulting from essential processing aims to make sense of the material displayed on a screen. In a data-driven DSS dashboard, it relates to the business information that the user will exploit in the decision-making process. This type of cognitive load relates directly to the Informational Content introduced earlier, so that we refer to this as the Informational Load. We expect that the more relevant - and clear - the information on a data-driven DSS dashboard, the more the dashboard will be perceived as useful by a decision maker.

Secondly, the Representational Load relates to the processing necessary for a user to compare information and draw conclusions. For instance, a gauge has a low representational load, while a pivot table or a combined chart is more demanding in terms of processing. As it has no direct correspondence with the contents presented above, we keep the name Representational Load. We expect that the higher the Informational load, the higher the representational load for the user. A case in point is the following: multiple pieces of information distributed across the dashboard will require the user to retain information. Further, this leads us to believe that a high representational load will impair the ease-of-use of a dashboard. Indeed, if a user has to retain a significant amount of information when using a BI dashboard, she will unlikely perceive the tool as easy-to-use.

Finally, the load resulting from incidental processing refers to the non essential aspects of the material displayed on a screen. The non essential characteristic of an element follows from, either the ability a user has to derive this element from essential data (for instance, color coding or a KPI displayed in green); or the element's uselessness in decision making (for instance, a logo or a title). Due to its relation with non-informational content, we call this third type of load the Non-Informational Load. Although a non-informational load is unavailing when it comes to decision making, we claim it is valuable in the overall user experience for two reasons. Firstly, the non-informational load can help reduce the representational processing of the dashboard, for instance through color coding. Secondly, the non-informational load can enhance the enjoyment the user perceives when using the dashboards (for instance by providing the user with a logo, a slogan, or export buttons).

Based on the elements discussed in this section, we develop a set of hypotheses for dashboard adoption. They are presented in Table 1. Their combination forms our Dashboard Adoption Model (DAM), as depicted in Fig. 1.

Table 1. Hypotheses for DAM model

Id	Hypotheses - P. = Perceived
H_1	P. ease-of-use of a dashboard predicts the P. usefulness of the dashboard
H_2	P. usefulness of a dashboard predicts the intention to use the dashboard
H_3	P. ease of use of a dashboard predicts the intention to use the dashboard
H_4	P. enjoyment of a dashboard predicts P. ease of use of the dashboard
H_5	P. ease of use of a dashboard predicts the P. enjoyment of the dashboard
H_6	P. informational load predicts P. usefulness of a dashboard
H_7	P. informational load predicts P. representational load of a dashboard
H_8	P. representational load predicts P. ease-of-use of a dashboard
H_9	P. non-informational load predicts P. representational load of a dashboard
H_{10}	P. non-informational load predicts P. enjoyment of a dashboard

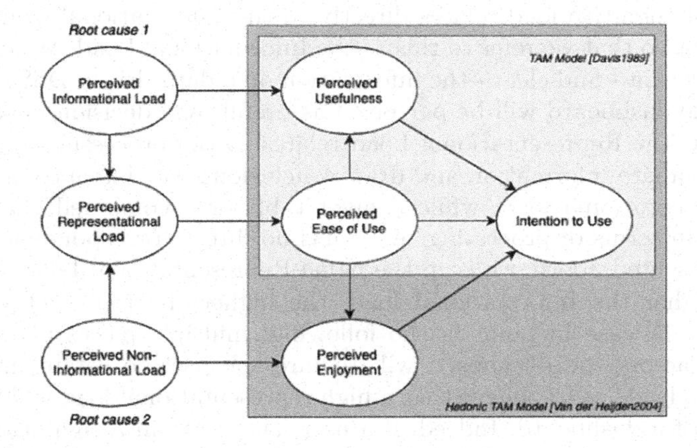

Fig. 1. Imbrication of TAM, Hedonic TAM and DAM

3 Research Methodology

3.1 Experimental Settings

To answer our research questions, we conducted a survey experiment [20]. The latter implies to randomly assign participants to at least two distinct experimental conditions, and to analyse the impact of this random assignment on some variable of interest using a set of indirect questions, under the form of a survey. This approach is particularly useful in contexts where direct questions cannot be submitted to respondents solely based on their own practice or experience, and causal relationships must be investigated [1]. In the present case, a 2×2 experimental design was adopted along two dimensions [5]; low and high Informational Loads and low and high Non-Informational Loads of a dashboard. The variable of interest is the intention of the respondent to adopt the dashboard.

The experiment is organized in two parts. Part 1 describes the process to respondents; they are asked to advise a fictitious company (i.e., AdventureWork)

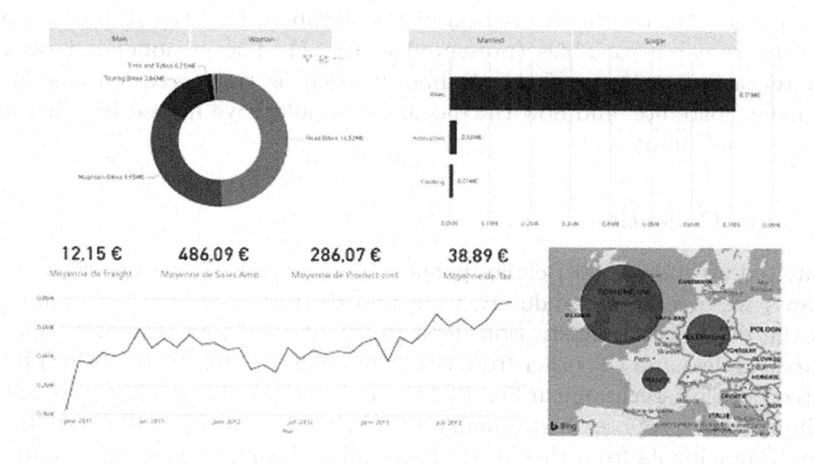

Fig. 2. Dashboard with Low Informational and Non-Informational Load

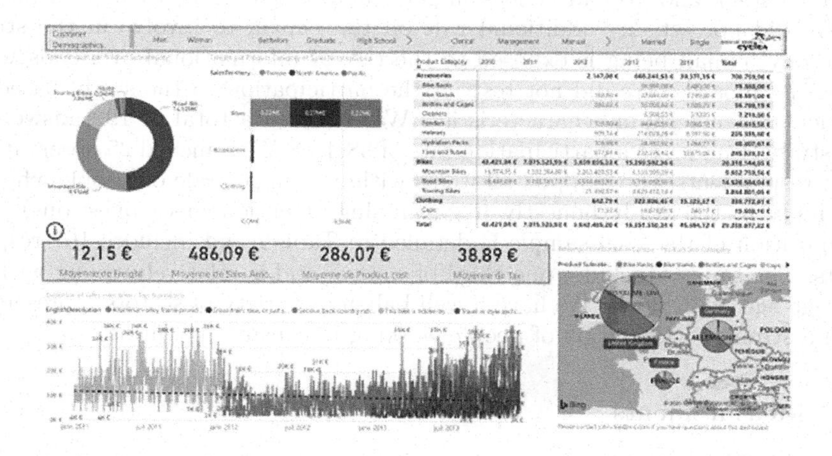

Fig. 3. Dashboard with High Informational and Non-Informational Load

about which products it should keep in its commercial offer, and which should be removed. The respondent is then invited to open a dashboard containing various details about the products of the company, and to interact with it as long as needed. The dashboard was implemented using Microsoft Power-BI and Microsoft sample database "AdventureWorks". Power-BI is used because it is a leading vendor on the BI market, is intuitive to the respondent and inexpensive to the experimenters. Each respondent was assigned randomly to one of the 24 different dashboards we created (4×6 dashboards, for each combination of low/high informational and non-informational load, in line with the 2×2 experimental design). Two examples of dashboards are depicted in Fig. 2 and Fig. 3.

Part 2 focuses on the evaluation of the dashboard by the respondent using our seven research variables (presented in Fig. 1). The respondent is asked to evaluate various aspects of the dashboard such as the perceived ease of use, enjoyment, load, etc. and how the dashboard would have helped him/her giving advice to the company.

3.2 Data Collection

We invited a total of 500 Belgian people to participate in our experiment. Participants were selected randomly from two distinct databases of alumni; one from the business administration department of the University of Namur (200 selected entries) and the other from the competence center TechnofuturTIC (300 selected people). TechnofuturTIC (TTIC) is a non-profit organization located in Belgium which proposes high quality training on advanced ICT topics to companies. Participants from these two groups all had completed a higher education program (University) and/or had significant professional experience (Competence Center), and all had at least a good understanding of what a dashboard is. We did not set any additional rule on the person to involve in the study, since any human being is exposed to the risk of informational overloads when using dashboards. We did not restrict the participation to the study based on particular demographic characteristics. We collected a total of 192 answers to our study (leading to a participation rate of 38.4%). We removed 25 observations from respondents who skipped all items within a single scale or people who did not reply to the study honestly (e.g., missing attention-check questions). The composition of the final sample is detailed in Table 2 and includes 167 respondents. The sample includes one third of women and is relatively young with an average age of 35 years. We have a well balanced variety of sectors represented, with a strong participation of people working in consultancy firms.

3.3 Measurements

Each variable used in our experiment was evaluated using multi-item scales presented in Appendix A. Variables "Intention to use" (ITU), "Ease of use" (EOU), "Usefulness" (USE) are classical 5 points likert-scales adapted directly from existing scales described in the Hedonic-TAM model [21]. Minor wording or phrasing adaptations were made to fit with the specific context of data-driven DSS dashboards. Variable "Enjoyment" (ENJ) is a semantic scale adapted from the same Hedonic-TAM model [21]. Variables "Representational Load" (RL), "Informational Load" (IL) and "Non-Informational Load" (NIL) were measured using scales of our own, directly derived or adapted from the literature discussed in Sect. 2.4.

Table 2. Sample by Gender, Age and Sector of Activity

Variable	Value	Number	Proportion
Gender	Male	113	67.7%
	Female	54	32.3%
Age	20 or younger	3	1.8%
	20–30	70	41.9%
	30–40	37	22.1%
	40–50	29	17.4%
	50 and more	28	16.8%
Sector	Consulting	41	24.5%
	Industry & Engineering	26	15.5%
	Finance & Banking	22	13.2%
	Teaching	16	9.6%
	Public sector	13	7.8%
	Tech companies	11	6.6%
	Freelance	11	6.6%
	Transport	11	6.6%
	Retail	9	5.4%
	HealthCare	7	4.2%

4 Results

To test our model, we applied Structural Equation Modeling (SEM) technique using a Maximum likelihood estimator on the data presented in Sect. 3. All our computations have been made using the R package "Lavaan" [31].

4.1 Constructs Quality

The seven variables of our DAM model are latent variables, meaning that they are not observable directly. The scales introduced in Sect. 3.3 were therefore necessary to measure those latent variables. The present section discusses the quality of these different scales from four different perspectives; Cronbach alpha's (CA), Composite Reliability (CR), Average Variance Extracted (AVE) and Discriminant Analysis. The different quality metrics are reported in Table 3 and discussed in the next paragraphs. The target values reported in Table 3 are extracted from [36].

The analysis of construct validity is threefold. First, we look for the internal consistency of our constructs. To do this, we rely on Cronbach's alpha and the Composite Reliability indicators. As reported in Table 3, the Cronbach's alpha are all above the 0.7 threshold and higher than the 0.8 threshold suggested by Nunnally & Bernstein for applied research [27]. The Composite Reliability, nearly all above 0.8, are higher than the 0.6 threshold suggested by Fornell and Larcker [16]. We conclude Internal consistency is good.

Second, we need to ensure that our scales have sufficient convergent validity, meaning that different items within the same scale are actually measuring the same thing. For a construct to have acceptable convergent validity, it is recommended to have an Average Variance Extracted (AVE) above 0.5. From Table 3,

Table 3. Analysis of Construct Quality

Variable	Code	CA	CR	AVE	Loadings
Target Value	-	> 0,7	> 0,6	> 0,5	
Represent. load	RL	0.836	0.835	0.562	0.818, 0.718, 0.805, 0.643
Informat. load	IL	0.825	0.824	0.542	0.722, 0.734, 0.728, 0.755
Non-informat. load	NIL	0.828	0.825	0.545	0.762, 0.768, 0.776, 0.632
Ease of use	EOU	0.904	0.834	0.700	0.852, 0.890, 0.869, 0.736
Usefulness	USE	0.837	0.904	0.566	0.809, 0.615, 0.766, 0.788
Enjoyment	ENJ	0.878	0.877	0.646	0.826, 0.754, 0.855, 0.767
Intention to use	ITU	0.861	0.786	0.753	0.885, 0.851

we see that this is the case for all our variables without exception. Beside, convergent validity is also controlled using the lambda scores (i.e. factor loadings) extracted from the model, which are also all above the threshold of 0.5 and all except one above the 0.7 recommendation of Wong [36]. We conclude convergent validity is good.

Thirdly, we need to ensure that our scales which are designed to measure different things are actually unrelated. We do this using the criterion proposed by Fornell and Larcker [16], suggesting that discriminant validity is good if the square root of the AVE value of each construct (reported in Table 3) is larger than the correlations of this construct with all other constructs. We report the discriminant reliability test in Table 4, and observe that the diagonal of the table (in bold) is always larger than the correlation coefficients. We conclude that the discriminant validity of our constructs is good.

Table 4. Discriminant Validity Criterion from Formell and Larcker [16]

	RL	IL	NIL	EOU	USE	ENJ	ITU
RL	**0.750**						
IL	0.41	**0.736**					
NIL	0.52	0.45	**0.738**				
EOU	0.58	0.31	0.47	**0.837**			
USE	0.55	0.44	0.59	0.42	**0.752**		
ENJ	0.50	0.49	0.40	0.38	0.44	**0.804**	
ITU	0.17	0.07	0.14	0.23	0.16	0.16	**0.868**

4.2 Structural Model

We report the structural part of the estimated model in Fig. 4. The values reported on paths are standardized coefficients. We follow the guidelines from Brown [6] to assess the overall fit of our model. The RMSEA value of the model is equal to 0.051, CFI = 0.959, TLI = 0.953 and the Chi-squared/df ratio is 1.42. All together, these metrics suggest an important fit of our model with the data collected. We observe from Fig. 4 that H1 and H2 from the TAM model are confirmed in our study; we conclude that Perceived Ease of Use of dashboards predicts Perceived Usefulness (H1: validated - 0.503, p-value = 0.000) and Perceived Usefulness of dashboards predicts Behavioral Intention to Use (H2: validated, 0.625, p-value = 0.000). We do not find evidence that supports H3 according to which Perceived Ease of Use of dashboard predicts Behavioral Intention to Use (H3: rejected - 0.057, p-value = 0.653). We develop possible explanations of this in the next Section.

We find support for the Hedonic TAM Model applied to data-driven DSS dashboards; the Perceived Ease of Use of a dashboards predicts the Perceived Enjoyment (H4: validated - 0.350, p-value = 0.000), and the Perceived Enjoyment predicts the Intention to Use the Dashboard (H5: validated - 0.576, p-value = 0.000). Finally, we find support for our dashboard load model; the Perceived Representational Load is predicted by the Perceived Informational Load (H6 - validated - 0.357, p-value = 0.024) and to a larger extent by the Perceived Non-Informational Load (H7 - validated - 0.732, p-value = 0.000). We also find

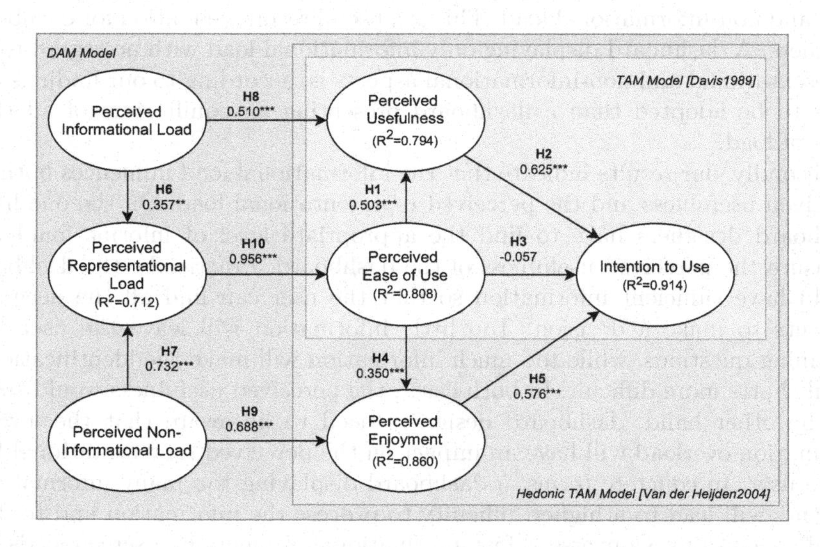

Fig. 4. Structural Equation Model with Path Coefficients, Significance and Residuals

support for our imbrication of informational overload with the TAM/Hedonic TAM; the Perceived Informational Load predicts the Perceived Usefulness (H8 - validated - 0.510, p-value = 0.000), the Perceived Non-Informational Load predicts the Perceived Enjoyment (H9 - validated - 0.688, p-value = 0.000) and the Perceived Representational Load strongly predicts the Perceived Ease of Use (H10 - validated - 0.956, p-value = 0.000).

Finally, we observe that the R-squared, measuring the proportion of variance of the latent response that was explained by the predictors, are all above 0.7. While the R-squared are especially large within the TAM/Hedonic TAM models, they remain more than satisfactory in our Dashboard Adoption Model. We see that the variance of the Perceived Representational Load for instance is explained at 71.2% by the predictors Perceived Informational Load and Perceived Non-Information Load, supporting the goodness of fit of our model.

5 Discussion and Future Works

We can draw several conclusions from our model and subsequent SEM analysis. Firstly, we find support for the importance of loads in dashboards, in that the degree of load will influence the perceived quality of the dashboard. Our study shows that a proper balance regarding the three types of loads lead to a better (perceived) quality of dashboards. BI dashboard designers should not focus only on a particular type of load - for instance the informational load - but instead should account for all three, namely informational load, representational load, and non-informational load. This is a useful recommendation for dashboard designers. A dashboard displaying only informational load with no regard to the representational and non-informational aspects is, according to our findings, less likely to be adopted than a dashboard presenting an equilibrium of all three types of load.

Secondly, our results indicate that the informational load influences both the perceived usefulness and the perceived representational load. On the one hand, dashboard designers have to find the appropriate level of informational load to ensure the perceived usefulness of the dashboard. Practically, the dashboard should have sufficient information so that the user can find all the necessary elements to make a decision. Too little information will leave the user with remaining questions, while too much information will make the identification of useful charts more difficult. In both cases, the perceived usefulness would suffer. On the other hand, dashboard designers need to be aware that the level of information overload will have an impact on the perceived representational load of the user. In concrete terms, a dashboard displaying too many informational elements will lead to a higher difficulty to process the information and to draw useful conclusions. Conversely, few informational elements in a screen leads to a light representational processing for the decision maker.

Thirdly, our findings suggest that the representational load of a dashboard influences the perceived ease of use. Dashboard designers should pay attention to the choice and the amount of charts or tables displayed on the screen. Choosing graphical elements heavy on representational load (for instance, pivot tables or combine charts) will have an impact on the user's perceived ease of use. Given that data-driven DSS aim to minimize the effort put in by users in the decision-making process, BI dashboards should be designed so that to avoid a too large representational load.

Finally, our study demonstrates that the non-informational load influences the representational load and the perceived enjoyment for decision makers. A dashboard exhibiting an appropriate amount of non-informational elements (for instance, color coding or a title) will be considered of a better quality than another dashboard displaying no such element. The reasons could be twofold. On the one hand, the dashboard will lead to a smooth representational processing - color coding can help the user to draw conclusions more efficiently. On the other hand, if the user identifies a known logo or if she can play with several features present on the dashboard, she will likely have more fun using the tool.

A possibly surprising result is that we cannot corroborate the relationship between Perceived Ease of Use and Intention to use. A possible explanation for this result lies in the nature of data-oriented systems and data-driven DSS in general. The sole ease of use of such systems cannot, alone, justify their adoption by the user. An effect exists, but is indirect through the Usefulness of the dashboard. What drives decision makers to use a given dashboard or not cannot be based on the fact that the dashboard is easy to use. Rather, the dashboard's usefulness will determine the intention of decision makers to use it. The Hedonic TAM Model's relationship between Perceived Enjoyment and Intention to Use is confirmed here. This was expected, since some of the characteristics of dashboards are their interactivity and their playfulness; as such, they fit perfectly in the definition of Hedonic Information Systems.

Based on our results, several future works may be identified. First of all, it would be interesting to investigate a way to measure automatically the different loads of a given dashboard. Indeed, this would allow dashboards designers to be guided while trying to find an equilibrium for the different loads. However, this is quite challenging due to the fact that it necessitates quantifying cognitive load, which is a subjective concept. Then, dashboards being used in a lot of different domains, a possibility would be to study if there are differences in terms of dashboards adoption, and more globally data-driven DSS adoption, between these domains. This could enrich the DAM model presented here and complete our understanding of this phenomenon. Moreover, it could allow us to discover some impactful elements for dashboard adoption that are specific to some domains. Finally, the informational load having to be controlled, it would be interesting to investigate a way to select the data to include in a dashboard. Developing a selection criterion and a related measure could help dashboards designers in their work by guiding the selection process of data to include in their dashboards. Again, the difficulty of quantifying a rather subjective phenomenon may arise.

6 Conclusion and Limitations

Using a SEM analysis, we explored the effect of dashboard load on the usefulness, ease of use and enjoyment perceptions. The findings show that if a BI dashboard designer effectively adopts the fine-grained approach to the notion of load, it would then be beneficial for the perceived quality of the dashboard. By highlighting the antecedents of the perceived usefulness, ease of use and enjoyment of BI dashboards - and more generally of data-driven DSS outputs - we provided a basis for a theoretical model of BI dashboard adoption.

Although our findings provide useful contributions and open the way for future works, we acknowledge some limitations. Firstly, the experiment was organized in a way where the participants only conducted a single dashboard analysis task. While we carefully manipulated the loading of our 24 dashboards, it would be interesting to replicate the study and to expose a single participant to various types of dashboards.

Secondly, we collected data only from participants residing in Belgium. The findings might reflect the characteristics and the culture of the respondents. Related to this issue, we did not investigate other factors that may impact the perceived quality of the BI dashboard including the goals pursued when using the dashboard, the familiarity with BI dashboards, or the cultural background.

Thirdly, for our experiment, we designed ourselves the dashboard the participant had to analyze. While these "artificial" dashboards allowed us to control the load of each dashboard, it would be interesting to replicate the study with the dashboard actually used by the decision maker.

Finally, we aimed to study the adoption of data-driven DSS in general. In the category of data-driven DSS, we find many data-oriented systems such as Big data, business analytics and business intelligence. In this work, we focused on BI dashboards but it would be ideal to replicate the study using dashboards or reports generated from business analytics or big data solutions.

Appendix

A Multi-Item Scales for Each Variable

Variable	Code	Items
Dashboard Informational Load	IL1	The dashboard ventilates the products information across enough axes of analysis
	IL2	The dashboard displays adequate quantity of indicators (measures or numbers) and with enough details about the products of AdventureWorks company
	IL3	The dashboard emphasizes sufficiently trends, cycles or other patterns in the data about the products of AdventureWorks company
	IL4	The dashboard offers well adapted interactions with AdventureWorks data
Dashboard Representational Load	RL1	The dashboard makes information about AdventureWorks products easily accessible
	RL2	The dashboard makes it easy to compare and relate different products information
	RL3	The dashboard is clearly organized and not too scattered
	RL4	The dashboard displays a reasonable quantity of information that I could easily memorize
Dashboard Non-Informational Load	NIL1	The dashboard contains a reasonable number of images, logos, arrows and other static visuals
	NIL2	The dashboard contains a reasonable amount of titles, info text and other free text zones
	NIL3	The dashboard proposes a relevant use of colors and mix of 2D/3D visuals
	NIL4	The dashboard shows a reasonable amount of buttons, URL to other websites, etc
Dashboard Perceived Usefulness	USE1	The dashboard helps me decide more quickly and more easily which product AdventureWorks should keep selling
	USE2	The dashboard is more adapted than interviews or verbal discussion to decide which product to recommend to AdventureWorks
	USE3	After reading the dashboard, I feel I am better informed about AdventureWorks products
	USE4	The dashboard is important for me to produce advice for AdventureWorks
Dashboard Perceived Ease of Use	EOU1	The interaction with the dashboard is clear and understandable
	EOU2	The dashboard is easy to use and I find information easily
	EOU3	I find it easy to extract the information I need from the dashboard
	EOU4	I would share the dashboard even with someone who has no IT background
Dashboard Perceived Enjoyment	ENJ1	Botched vs. neat
	ENJ2	Exciting vs. dull
	ENJ3	Pleasant vs. unpleasant
	ENJ4	Interesting vs. boring
Behavioral intention to use the Dashboard	ITU1	In the near future, I predict I'll use the dashboard again to complete my consulting mission for AdventureWorks
	ITU2	If hired by Adventure Works company, I would keep working with the present dashboard, and would not ask for another dashboard

References

1. Auspurg, K., Hinz, T.: Factorial Survey Experiments, vol. 175. Sage Publications, Thousand Oaks (2014)
2. Bačić, D., Fadlalla, A.: Business information visualization intellectual contributions: an integrative framework of visualization capabilities and dimensions of visual intelligence. Decis. Support Syst. **89**, 77–86 (2016)
3. Bollen, D., Knijnenburg, B.P., Willemsen, M.C., Graus, M.: Understanding choice overload in recommender systems. In: Proceedings of the Fourth ACM Conference on Recommender Systems, pp. 63–70 (2010)
4. Brath, R., Peters, M.: Dashboard design: Why design is important. DM Direct **85**, 1011285-1 (2004)
5. Broota, K.D.: Experimental design in behavioural research. New Age International (1989)
6. Brown, T.A.: Confirmatory Factor Analysis for Applied Research. Guilford Publications, New York (2015)
7. Burnay, C., Jureta, I.J., Linden, I., Faulkner, S.: A framework for the operationalization of monitoring in business intelligence requirements engineering. Softw. Syst. Model. **15**(2), 531–552 (2016)
8. Chen, H., Chiang, R.H., Storey, V.C.: Business intelligence and analytics: from big data to big impact. MIS Q. 1165–1188 (2012)
9. Chesney, T.: An acceptance model for useful and fun information systems. Hum. Technol. Interdisc. J. Hum. ICT Environ. (2006)
10. Davis, F.D.: Perceived usefulness, perceived ease of use, and user acceptance of information technology. MIS Q. 319–340 (1989)
11. Eckerson, W.W.: Performance Dashboards: Measuring, Monitoring, and Managing Your Business. Wiley, Hoboken (2010)
12. Emami, Z., Chau, T.: The effects of visual distractors on cognitive load in a motor imagery brain-computer interface. Behav. Brain Res. **378**, 112240 (2020)
13. Eppler, M.J., Mengis, J.: The concept of information overload-a review of literature from organization science, accounting, marketing, MIS, and related disciplines (2004). Kommunikationsmanagement im Wandel 271–305 (2008)
14. Few, S.: Intelligent dashboard design. Inf. Manage. **15**(9), 12 (2005)
15. Few, S.: Information Dashboard Design: The Effective Visual Communication of Data, vol. 2. O'Reilly, Sebastopol (2006)
16. Fornell, C., Larcker, D.F.: Evaluating structural equation models with unobservable variables and measurement error. J. Mark. Res. **18**(1), 39–50 (1981)
17. Gartner: Worldwide business intelligence and analytics software market revenue from 2010 to 2020 (in billion U.S. dollars) (2019). https://www.statista.com/statistics/294653/enterprise-software-revenue-worldwide/
18. Gigerenzer, G., Gaissmaier, W.: Heuristic decision making. Annu. Rev. Psychol. **62**, 451–482 (2011)
19. Glazer, R.: Measuring the value of information: the information-intensive organization. IBM Syst. J. **32**(1), 99–110 (1993)
20. Groenendyk, E.W.: Diana C. Mutz. Population-based survey experiments. Princeton, NJ: Princeton University Press. 2011. 177 pp. 49.50(cloth). 24.95 (2012)
21. Van der Heijden, H.: User acceptance of hedonic information systems. MIS Q. 695–704 (2004)
22. Hollender, N., Hofmann, C., Deneke, M., Schmitz, B.: Integrating cognitive load theory and concepts of human-computer interaction. Comput. Hum. Behav. **26**(6), 1278–1288 (2010)

23. Horkoff, J., et al.: Strategic business modeling: representation and reasoning. Softw. Syst. Model. **13**(3), 1015–1041 (2014)
24. Hsu, C.L., Lu, H.P.: Why do people play on-line games? An extended tam with social influences and flow experience. Inf. Manag. **41**(7), 853–868 (2004)
25. Kahneman, D., Slovic, S.P., Slovic, P., Tversky, A.: Judgment Under Uncertainty: Heuristics and Biases. Cambridge University Press, Cambridge (1982)
26. Mayer, R.E.: Cognitive theory of multimedia learning. In: The Cambridge Handbook of Multimedia Learning, vol. 41, pp. 31–48 (2005)
27. Nunnally, J.C.: The assessment of reliability. Psychometric Theory (1994)
28. Palpanas, T., Chowdhary, P., Mihaila, G., Pinel, F.: Integrated model-driven dashboard development. Inf. Syst. Front. **9**(2–3), 195–208 (2007)
29. Power, D.J.: Decision Support Systems: Concepts and Resources for Managers. Greenwood Publishing Group, Westport (2002)
30. Rappaport, A.: Management misinformation systems-another perspective. Manag. Sci. B133–B136 (1968)
31. Rosseel, Y.: Lavaan: an R package for structural equation modeling and more. Version 0.5-12 (BETA). J. Stat. Softw. **48**(2), 1–36 (2012)
32. Schroder, H.M., Driver, M.J., Streufert, S.: Human Information Processing: Individuals and Groups Functioning in Complex Social Situations. Holt, Rinehart and Winston, New York (1967)
33. Shibl, R., Lawley, M., Debuse, J.: Factors influencing decision support system acceptance. Decis. Support Syst. **54**(2), 953–961 (2013)
34. Venkatesh, V., Morris, M.G., Davis, G.B., Davis, F.D.: User acceptance of information technology: toward a unified view. MIS Q. 425–478 (2003)
35. Venkatesh, V., Thong, J.Y., Xu, X.: Consumer acceptance and use of information technology: extending the unified theory of acceptance and use of technology. MIS Q. 157–178 (2012)
36. Wong, K.K.K.: Partial least squares structural equation modeling (PLS-SEM) techniques using SmartPLS. Mark. Bull. **24**(1), 1–32 (2013)
37. Xie, J., et al.: The role of visual noise in influencing mental load and fatigue in a steady-state motion visual evoked potential-based brain-computer interface. Sensors **17**(8), 1873 (2017)
38. Yigitbasioglu, O.M., Velcu, O.: A review of dashboards in performance management: implications for design and research. Int. J. Account. Inf. Syst. **13**(1), 41–59 (2012)

The Effect of Visual Information Complexity on Urban Mobility Intention and Behavior

Thomas Chambon[1](✉), Ulysse Soulat[2](✉), Jeanne Lallement[2], and Jean-Loup Guillaume[1](✉)

[1] Laboratoire Informatique, Image et Interaction (L3i), La Rochelle University, 23 Avenue Albert Einstein, 17000 La Rochelle, France
{thomas.chambon1,jean-loup.guillaume}@univ-lr.fr

[2] Laboratoire Usages du Numérique pour le Développement Durable (NUDD), La Rochelle University, 39 rue de Vaux De Foletier, 17000 La Rochelle, France
aymeric-ulysse.soulat@univ-lr.fr

Abstract. Encouraging soft mobility practices is a central issue for the ecological transition. Green information systems and more specifically self-tracking applications are tools that can be used to raise awareness and changing behavior. Based on the theoretical framework of visual complexity, this paper examines how the level of visual complexity of a mobile application influences users' urban mobility intentions and behaviors. We conducted two experimental studies. The first one investigated how the visual complexity of homepages affects mobility intentions with an application to measure one's carbon footprint in a situational setting. The first result of our research is that moderate information visual complexity positively influences the acceptability of a mobile application as well as mobility intentions. A second experimental research is divided into two parts, firstly, participants responded to our questionnaire, secondly, in a longitudinal approach, 51 subjects used the application over a 3-month period. The conceptual framework was tested using regression analyses. We find that intention to change behavior influences responsible urban mobility behavior. However, our experiment shows that the visual complexity of information does not have a significant influence on behavior. We then propose theoretical implications.

Keywords: Green Information System · responsible behavior · urban mobility · visual complexity · self-tracking

1 Introduction

Despite a significant decrease in greenhouse gas emissions from transport, the latest IPCC report of 2022 warns that global warming in Europe will increase more than the world average [19]. The report also insists on the effectiveness of adopting gasoline-free urban mobility choices. Therefore, individuals need to be

S. Nurcan et al. (Eds.): RCIS 2023, LNBIP 476, pp. 452–466, 2023.
https://doi.org/10.1007/978-3-031-33080-3_27

made aware of the environmental impact of their mobility so that they can, if necessary, change their behavior and reduce their impact on the environment. Despite the growing organizational and research efforts on transforming behavioral intentions into virtuous behaviors, much remains to be discovered about bridging this 'green gap' [37] that calls for work by researchers in sustainable behavior [14].

Information framing is an effective approach used in many research studies [24]. It considers that the way information is presented to the public influences the choices people make about how to process the information. In particular, the visual complexity of a piece of information impacts consumers' processing and final decisions [29]. The visual complexity of a stimulus refers to the amount of elements as well as the level of detail present. It can be defined as the diversity of visual elements [12]. Most research on this topic is based on Berlyne's theory of visual complexity [4] and the literature tells us that a visual stimulus of moderate visual complexity (as opposed to simple or complex) is more effective on purchase intentions [12,42]. This is supported by several subsequent studies [11,28].

Our research is particularly interested in the effect of technology use, adoption and visual complexity on behavior. We combine behavioral analysis from a computer science and a management science perspectives. The association of the two disciplines appeared to be an asset for the implementation of our experiments. We adopt a quantitative, experimental design combining a questionnaire and a longitudinal mobile app experiment with 189 participants over three months to test the potential impact of information complexity on people's behavior and intent. The first study presented is a situational setting with a questionnaire and the second is a two-stage study with the use of a mobile application on a longitudinal format with a questionnaire. The research presented here aims to provide empirical evidence of environmental awareness and behavior change through visual complexity.

Contributions:

- We observed that moderate visual complexity has a more significant effect on the intention of responsible transportation behavior than higher or lower visual complexity home pages.
- We find that moderate visual complexity does not have a more significant effect on responsible mobility behavior than higher or lower visual complexity home pages.

The theoretical aspects of behavior change incentives, self-monitoring, and the impact of visual complexity will be developed in Sect. 2. We will then present our hypotheses and the protocol of the two studies in Sect. 3. The results of impact of our application on people's intention and behavior will be exposed in Sect. 4. Finally, we summarize our work and present future directions in Sect. 5.

2 Materials and Methods

2.1 A Lever for Behavioral Change, the Green IS and Self-tracking

To encourage sustainable behavior change, the use of Green Information Systems (IS) has emerged. The roles played by IS can be classified into three types: automate, inform (inform up and inform down) and transform, even if the majority of IS can only fulfill one of these three roles in practice [35]. Beyond the specific roles, it is the objectives of user-centered green IS that remain to be defined, as this area of research is still under-explored [6]. The most common application areas of these IS are oriented towards health and education. Studies are generally conducted with mobile devices (i.e., pervasive technologies) in these domains with feedback to lose weight or reduce alcohol consumption [26] or with the use of microlearning [7]. These studies show the effectiveness of these methods on user behavior.

In the context of our research - sustainable mobility applications - the most employed role is to inform [6] through the use of tools and/or feedback systems [15]. Self-tracking technologies allow people to access, monitor and analyze personal data in several areas [9] and for many individuals it is now a common practice to assess their performance, step count, diet, sleeping and more [21]. In the literature, studies about the adoption of self-tracking technology are mostly based on the unified theory of technology acceptance (UTAUT) [39]. The work of Ajana and al. [1] demonstrates that greater adoption of a self-tracking application for sports practices positively influences individuals' intentions. But, although these digital solutions are being developed to encourage responsible behavior, there is little research on their real impact. We know that tracking personal data can lead individuals to perform in a task 'for oneself' [27]. Still, the effect of information in changing and maintaining green mobility behavior raises many questions.

Compared to self-tracking applications with an essentially individual dimension, Gabrielli et al. explored the impact of a mobile application offering different transport options [17]. They show a slight increase in sustainable transport choices over a one-month period. Tulusan and al. provide a mobile application based on eco-feedback to reduce employees fuel consumption during driving [38]. The impact of the feedback is evaluated and revealed that drivers using the app save 3.23% of fuel. While this impact may seem minimal for the 25 participants of the study, it would make a notable difference on a larger population. Still with the objective of more responsible mobility, the measurement of the impact on individual mobility of a more environmentally friendly transport offer is also interesting [16]. The results of this early-stage study on 13 volunteers show the engagement in the application but not a change of behavior.

According to [1], individuals, who grant a higher acceptability of the application, should be more inclined to have stronger behavioral intentions towards the application's goal. In this paper, we predict that app acceptability mediates the effect of information complexity on more responsible behavioral intentions.

It may be noted that in the context of urban mobility, the results of some studies based on feedback are mixed. We can therefore legitimately ask the question of the impact of the information transmitted to the user. Here, we add the criterion of visual complexity of a digital tool to understand whether or not it can impact sustainable mobility intentions and behaviors. To our knowledge, no research has investigated visual complexity in an environmental setup.

2.2 The Effects of Visual Complexity of Information on Individuals

Berlyne is a pioneer in the research field of the perception and the behavior in response to the stimuli, with his book "Conflict, arousal, and curiosity" [4]. He defines four tools into stimuli selection namely Novelty, Uncertainty, Conflict and Complexity. He considers that complexity is "the most impalpable of the four elusive concepts" and he enumerates some properties of complexity. First, the number of elements that increases the complexity. Second, complexity can be reduced by having identical elements with specific conditions, such as locations, or even with a common property. The complexity varies inversely with the degree of response to multiple elements as a unit. Indeed, the closer the elements are to each other, the more an individual will consider these close elements as a whole (i.e., a unit), the less complexity he perceives.

To determine whether a stimulus is efficient, it is important to define the features that focus attention. This is the goal of Treisman and Gelade in 1980, who proposed the theory of feature integration [36]. This theory defines color, orientation, spatial frequency, brightness and direction of motion as attractors. Other works in the field of visual attention research has led to a better understanding of the mechanisms of perception and attention. Computational models have made it possible to determine the salience of a scene from the various elementary feature maps by extracting the activity level of each element that composes it [23]. This prediction allows to know where the attention is focused. Subsequently, discriminative properties such as contrast, color, edge orientation and movement have been defined to increase the visual salience of a scene [20].

Work on visual complexity illustrates negative effects of high visual complexity on consumers' emotional reactions [12], on information processing fluency showing that individuals with limited acquisition capacity are more sensitive to low visual complexity information [29]. In website design, low visual complexity leads to experience higher levels of trust and more positive judgments of website aesthetics [42]. Conversely, some studies have shown that consumer satisfaction is higher when the visual complexity of a web page is high [31]. Other researchers found that advertising with high visual complexity increases consumers' attention to the communication and generates a positive attitude [32]. A significant body of research, based on Berlyne's theory, concludes that moderate visual complexity has a more significant effect on purchase intentions [12,40]. According to the work of Geissler et al. [18], web pages with moderate visual complexity provoke a more favorable response in the user, what has been confirmed by other studies [12,41].

2.3 Experiments and Hypothesis

In an attempt to determine the impact of the information provided on mobility intentions and behaviors, we propose two experiments measuring the effects of the visual complexity of CO_2 information. To achieve this goal we have created our own application to measure this variable on intentions and behaviors. More precisely, we conducted two complementary experimental studies, the first study is a situational setting experiment exploring the effect of visual complexity on behavioral intentions and the second study is in real condition experiment through a mobile application. For each setting we formulate several closely related hypothesis.

Situational Setting Experiment:

HS1: An application with moderate visual complexity will have greater acceptability than one with low or high complexity. All of the above-mentioned literature leads us to consider that the level of complexity could determine the impact of information on the acceptability of a self-tracking application and then on behavior.

HS2: An application with moderate visual complexity elicits more responsible behavior intentions. In addition to the effectiveness of moderate complexity on the acceptability of a technology, we assume that this level of complexity will also elicit more favorable behavioral intentions depending on the purpose of the application. In our case, this is the intention to use means of transport that emits less CO_2 than the car.

HS3: High acceptability of the application has a positive effect on intentions for more responsible behavior. If information visual complexity produces this effect, acceptability should in turn influence behavioral intentions towards more responsibility. This hypothesis is supported in particular by research examining the effectiveness of self-tracking tools on attitudes and behavioral intentions.

Real Condition Experiment:

HR1: An application with moderate visual complexity leads to more responsible behavior. Our theoretical framework allows us to assume that a level of visual complexity on a digital device influences urban mobility behavior. We then assume that an application with moderate complexity generates less CO_2 emitting urban mobility behaviors.

HR2: Intentions to behave more responsibly generate more responsible behavior. Although the literature has addressed this problem between behavioral intention and actual behavior, the gap persists [2]. Our hypothesis assumes that intentions to emit less CO_2, are transformed into more responsible behavior, thanks to the application.

3 Experiment

3.1 Protocol

Our hypotheses were tested in two experiments with a mobile application with three levels of complexity by varying the characteristics, quantity and arrangement of elements. More specifically, three homepages (see Fig. 1) have been created by taking into account the criteria described by Wang et al. [41], namely quantity of elements, variety of elements, asymmetry of elements, irregularity of elements. These elements are also shared in the visual attention community as intensity (or intensity contrast, or luminance contrast), color, and orientation [5]. After developing the different stimuli, we checked the validity of the manipulation. Our concern was to check the perception of the level of visual complexity prior to conducting our experiments. We pre-tested the perceived visual complexity of the stimuli by asking 74 participants to rate the quantity, variety and irregularity of the stimuli on 9-items using a 7-point scale (1 = "Strongly disagree", 7 = "Strongly agree"), adapted from Lee, Hwangbo and Ji [25]. A descriptive analysis and ANOVA were conducted. The results of this pretest confirm significant differences ($F = 28.959$, $p = 0.000$) between the three complexity levels. The high complexity homepage is well perceived as more complex (mean = 4.359, SD = 0.6675) than the moderate complexity homepage (mean = 3.340, SD = 0.7398) and the low complexity homepage (mean = 2.270, SD = 1.0123), thus confirming the validity of the manipulation.

3.2 Situational Setting Experiment Methodology

For the first experiment, 362 people (39% between 15 and 20 years old, 46% between 21 and 25 years old, 15% between 26 and 35 years old, 49.5% of women), recruited in several French cities, took part in this study. Recruitment was carried out using the snowball method on a voluntary basis.

After randomly viewing one of the three versions (i.e., homescreens), participants responded to a questionnaire in order to rate their acceptability of the application as a function of the complexity of the stimulus using 21 items and 7 dimensions based on the UTAUT model (Perceived Usefulness, Perceived Ease of Use, Social Influence, Facilitating Conditions, Hedonic Motivations, Habit, Intention to Use) using a 7-point scale (1 = "Strongly Disagree", 7 = "Strongly Agree") [39], averaged in a single measure.

They also rated their intention to use transport that emits less CO_2 than the car on a 4-item, 7-point scale (1 = "Strongly disagree", 7 = "Strongly agree") [8], averaged into one measure. We told subjects that they had to answer the questions by taking into account the visual (with one of the three levels of complexity) before each construct.

3.3 Real Condition Experiment Methodology

The second experiment follows a two-stage design. First, participants were asked to fill in a questionnaire before installing the "Eco-Mobility" application.

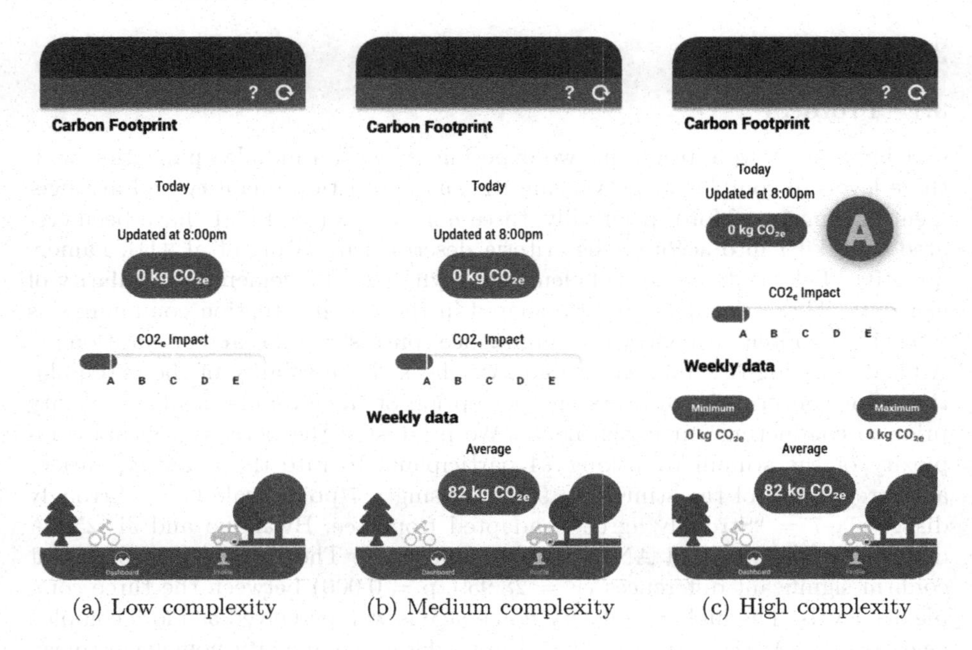

(a) Low complexity (b) Medium complexity (c) High complexity

Fig. 1. Homepage with three different levels of complexity (low, medium and high for left to right). *The visuals have been translated into English for comprehension purposes even though the application has been designed in French.*

Volunteers were then randomly assigned to three groups of visual complexity when they registered on the application.

Front End of Eco-Mobility: The front end of the Eco-Mobility application (named dashboard) is designed with the results of our preliminary study. In an effort to avoid certain biases and to isolate our information complexity variable, we have decided to remove some features usually present in self-tracking applications such as a detailed profile for the user (i.e., long-term history, consumption evolution, etc.) or a function to share one's consumption with other users or a wider audience. The social aspect of this kind of approach has been discarded in order to improve the measurement of the impact of the information on behavior.

The application dashboard includes common elements at the three levels of visual complexity, see Fig. 1. In the upper right part of the screen, two functionalities are presented: the user's manual which explains the meaning of the data displayed in the different indicators (question mark symbol) and a data update functionality to know the CO_2 consumption in real time (the refresh symbol). In the lower part of the screen, the users find two icons (i.e., Dashboard and Profile) to navigate in our application. The first allow displaying the different data on his CO_2 consumption in line with assigned visual complexity group. The second gives access to the privacy functionalities such as the disconnect button and the possibility to download its own data in a readable format (JSON). This page has also a "push up" function to send the journey data to the server.

In the central part of the screen, all users receive two information on their daily CO_2 consumption: first, a numeric value presented in kilograms of CO_2 emitted during the day and, second, a CO_2 impact on a scale from A to E. The different levels of consumption represent respectively a range between 0 kg emitted up to 10 kg with a change of letter every 2,5 kg of CO_2 consumed, and above 10 kg for level E. This indicator is based on the idea of the Nutri-Score, also known as the 5-Color Nutrition Label or 5-CNL [22] and has already proven its effectiveness in consumption behavior [13,30]. Below the CO_2 impact, users in the moderate and complex group also have their average weekly consumption. Finally, in the complex group, users have two additional pieces of information between the CO_2 impact indicator and the average weekly consumption: the minimum and the maximum consumption values over the week. We developed a specific display according to the user's complexity group. We also removed gamification or sharing features to avoid social bias.

Back End of Eco-Mobility: The back-end of Eco-Mobility is designed with several requirements using the "privacy by design" concept. First of all, the geolocation, although permanent, can be easily disconnected by the user and, in this case, the trips will not be tracked. In order to respect privacy, the collection of geolocation data is triggered two minutes after the user's movement is detected. For the most accurate calculation and therefore the most relevant information for the user, the application had to be able to detect as many means of transport as possible and also a combination of several of them. The detection phase of a means of transportation is done using the accuracy of the GPS points and the time associated with each point. These treatments are also used to detect the routes of each user in order to calculate the associated CO_2 consumption. First, the algorithm eliminates all the low accuracy points, then eliminates the identical points. Then, it considers all the points of the last five minutes and the last 10 GPS points associated with the user. It chooses the set of points containing the oldest point. If all the points in this set are within 100 m of the last point found, then the path is considered complete. This processing also allows to take into account the weaknesses or possible errors of the GPS signal of the phone.

The application also needed to be able to use different reference values in the calculation of CO_2 emissions for each mode of transport. All these prerequisites allow us to avoid asking the user to perform many actions, such as filling out a questionnaire about their trip or transportation modes, which can be tedious. From the point of view of scientific protocol, not asking the user for additional information does not introduce any bias in our experiment. Indeed, as the application is not declarative, the participants cannot modify the result of their consumption. We added some features, such as the addition of group data and the timestamp of the analysis pipeline to indicate to the user when his CO_2 consumption is updated. We have also secured the data exchange between the client and the server with the HTTPS protocol.

Development of Eco-Mobility: For the implementation of these ideas, we chose a hybrid framework (Apache Cordova [10]) to save development time in order to deploy eco-mobility on Android and IOS platforms. After initial inconclusive tests, with restricted automatic geolocation that activates when the user is in specific areas to preserve privacy, we selected an already proven development base, the E-mission project [34]. E-mission is an open source mobility platform developed at UC Berkeley that has already proven itself with several deployments around the world [33]. Our application is based on the E-mission architecture: the server is a python web application, the data is stored in a MongoDB database, the client is a Cordova application for Android and IOS. We have completely redesigned the client, with the integration of our protocol. We removed all third-party services (e.g., Google services), developed a single view (dashboard) with all data and stored all user data in our own MongoDB instance.

Deployment of Eco-Mobility: To facilitate the user's adoption, we have deployed Eco-Mobility on AppStore and Play Store. Google having refused us the access to its store, we took in hand the complete deployment for the Android platform. We have provided a APK file in our own cloud. This version is compatible with Android 6 to Android 12. For IOS devices, Eco-Mobility is compatible with IOS 8 to IOS 13.

Conduct of the Experiment: After a short presentation of the experiment, the future users were invited to use the application regularly (2-weekly reminders were sent to users) and committed to visualize their CO_2 emission at least once a day. The Eco-mobility application transmitted to the user the CO_2 impact of their travels. We focused on students from the University of La Rochelle who were personally approached in the university's classrooms in March and April 2022. In total, 186 participants installed the application out of 1000 requests, a penetration rate of (18%). Of the 186 individuals, 51 used the application for a period of at least 9 days. Our study focused on these 51 participants, concentrating our analysis on the first 9 days of use for reasons of equity. The data collected by the Eco-mobility application is stored on the university's server and analyzed. The reduced sample of 51 participants varied in terms of gender (57% female), age (38 between 15 and 20 years old, 12 between 21 and 25 years old and 1 above 26 years old) and academic year. We randomly assigned the three levels of visual application complexity to the subjects. The distribution is as follows, 22 subjects in the simple group, 14 subjects in the moderate group and 16 subjects in the complex group. Respondents first rated their intention to use transport modes that emit less CO_2 than the car. All constructs were measured by 7-point Likert scales. A score is assigned to each participant for each three-day period and an overall score for the whole nine days. An average is calculated according to the main mode of transport recorded by the application (1 = walking or cycling, 2 = bus or train, 3 = car). The score is then compared to the questionnaire data.

4 Results

4.1 Situational Setting Experiment

Our experiments were analyzed in order to validate or refute our hypotheses through statistical analysis. For this we used the SPSS software. We analyzed the effects of an independent variable on a dependent variable by the analyses of variance (ANOVA). We relied on classical statistical tests of analysis of variance such as the average difference, standard error, F-test, R-test, R^2-test and the P-value. As shown in Table 1, the average values tell us that the groups are from the same population and thus demonstrate the influence between the variables. We evaluated the differences between each homepage in order to test the greater effectiveness of a moderated homepage as we assumed. This tool also allows us to perform statistical analyses of mediation. Mediation effects reflect the influence of one variable on another through a third variable. We will now present the results of each hypothesis.

HS1. An ANOVA test is performed (acceptability, $p = 0.001$). The results show that the page with moderately complex stimuli generates a higher acceptability of the application (mean = 4.60, SE = 0.110) compared to stimuli of low (mean = 3.99, SE = 0.125) or high (mean = 4.18, SE = 0.100) complexity. This is confirmed by a Tukey significant difference test: Complex vs. Simple = 0.184, $p = 0.468$; Moderate vs. Simple = 0.609, $p = 0.001$; Moderate vs. Complex = 0.426, $p = 0.016$ (see Table 1, upper part). Thus, Hypothesis **HS1** is confirmed, stimuli of moderate complexity generate higher acceptability of the application, while stimuli of high or low complexity generate lower acceptability of the application.

HS2. A 1-factor ANOVA test is performed (transport intention, $p = 0.080$). The results show that moderately complex stimuli does not significantly generate, in a direct way, a higher intention of responsible urban mobility behavior with the application (mean = 5.33, SE = 0.181) compared to stimuli of low (mean = 4.71, SE = 0.216) or high (mean = 4.93, SE = 0.155) complexity. This is confirmed by a Tukey significant difference test (Complex vs. simple = 0.221, $p = 0.659$; Moderate vs. simple = 0.615, $p = 0.071$; Moderate vs. complex = 0.394, $p = 0.255$). Nevertheless, the Pearson correlation showed a significant effect ($p = 0.027$). Thus, Hypothesis **HS2** is partly confirmed, there is no direct relationship.

HS3. To test this hypothesis, we performed a linear regression. We found that the regression of the acceptability of the application on the intention to take more responsible transport gave a positive result (SE = 0.078, $p = 0.000$) (see Table 1, bottom part) thus validating our third hypothesis. Acceptability has a significant effect on the behavioral intention to take more responsible transport. In other words, the participants perceived the application of moderate complexity (compared to high or low complexity) as more acceptable which influences the intention to use more sustainable transport than a car (see Table 1).

After having conducted the first study on behavioral intentions, it seemed appropriate to observe the effects on behavior under real conditions.

Table 1. Results of the situational setting experiment.

Acceptability of the application

Complexity	Av. diff.	SE	P-value
Moderate vs. complex	0.426	0.153	0.016
Moderate vs. simple	0.609	0.171	0.001
Complex vs. simple	0.184	0.156	0.468
Intergroup	F	P	
	6.840	0.001	

Intention to use means of transport that emit less CO_2 than the car

Summary of the model	R	R^2	F	P-value
	0.358	0.148	6.689	0.000
Independent variable	B	SE	T	P-value
Acceptability	0.619	0.078	7.918	0.000

4.2 Real Condition Experiment

In order to assess the effect of visual complexity on urban mobility behavior, the overall score for the 9 days of the study is taken into account. **HR1** hypothesis is tested by a 1-factor ANOVA analysis considering each level of complexity. We use the same statistical tools as in the previous study. As shown in Table 2, the values show the influence between the variables. In order to test the greater effectiveness of a moderate home page, as we had assumed, we evaluated the differences between each home page. The results show a non-significant effect of complexity on urban mobility behavior (behavior, $p = 0.252$). According to the mobility score, we find that participants with moderate complexity application visuals do not have more responsible urban mobility behaviors ($mean = 1.900$, $SD = 0.5148$) than participants with low ($mean = 2.092$, $SD = 0.6262$) or high ($mean = 2.286$, $SD = 0.6075$) complexity application visuals (see Table 2, upper part). A Tukey significant difference test is then performed to confirm these results. No difference is observed between the groups: Simple vs. moderate = 0.1917, $p = 0.621$; Simple vs. complex = −0.194, $p = 0.599$; Moderate vs. complex = −0.3857, $p = 0.222$ (see Table 2, middle part). Our hypothesis **HR1** is therefore rejected: the moderate complexity visual did not lead participants who used it to behave more responsibly than participants who used an application with a lower or higher complexity visual. We know that complexity has an impact on behavioral intentions [41] and that it affects information processing positively or negatively [29]. From the literature we know that moderate visual complexity positively influences intentions, our study then shows that in our context, framing does not work on responsible behavior.

The **HR2** hypothesis is tested by regression analysis. The effect of behavioral intention on more responsible urban mobility behavior is significant and positive ($B = -0.118$, $SE = 0.042$, $p = 0.007$) thus validating our hypothesis (see Table 2, bottom part). The study shows that behavioral intentions are antecedents of urban mobility behaviors. However, the influence of urban mobility behaviors should be checked before the experiment. We controlled for the influence of transport behaviors before the use of the application. The analysis of the control variable shows a non-significant effect of transport habits before the experimentation both on the effect of technology adoption on behavioral intentions (SE = 0.4058, p = 0.2147), the effect of technology adoption on behavior (SE = 0.1403, p = 0.1484) and the effect of behavioral intentions on behavior (SE = 0.0734, p = 0.4528), thus confirming our hypothesis **HR2**.

Table 2. Results of the real condition experiment.

Descriptive statistics on the effect of complexity on behaviour

Dependent variable	Simple	Moderate	Complex
Behaviour (M)	2.092	1.900	2.286
Standard deviation (σ)	0.6262	0.5148	0.6075

Effect of visual complexity on urban mobility behaviour

Complexity	Av. diff.	SE	P-value
Simple vs. Moderate	0.1917	0.2049	0.621
Simple vs. Complex	−0.1940	0.2001	0.599
Moderate vs. complex	−0.3857	0.2292	0.222
Intergroup	F	P	
	1.417	0.252	

Effect of behavioural intention on urban mobility behaviour

Summary of the model	R	R^2	F	P-value
	0.371	0.138	7.826	0.007
Independent variable	B	SE	T	P
Behavioural intention	−0.118	0.042	−2.798	0.007

5 Conclusion and Perspectives

By examining the effect of information complexity on intention and behavior, we contribute to the growing body of visual complexity literature that seeks to understand the effects of information complexity on individuals [3,41]. We were interested in conducting two studies, one in the laboratory and one in real-world conditions, to account for the changing effects of visual complexity under different conditions. The goal of our research is to provide elements for organizations offering responsible urban mobility programs. The results of our study demonstrate that while intentions reveal responsible behavior, the green gap dilemma is not reduced by visual complexity. Therefore, designers of mobile application should not only think about the framing of the information presented, but also consider factors that may encourage users to continue using the app, such as gamification or features for increasing interactivity between users or sharing on social networks [6]. In fact, the aspect of acceptance of the media on which the information is transmitted allows to reinforce the intention and the responsible behavior. Taking into account environmental variables (e.g., weather, public transport networks, terrain topography...) may also be relevant to explain the difference between intention and behavior, with a negative impact on bicycle use for example [15].

This study is not free of research limitations. The experimentation does not allow us to know whether the repetition of information influences user's processing during the use of the device. Which, in turn, through habituation, could reduce the effectiveness of the visual complexity. We assume that presenting the visuals before asking about technology acceptability could contribute to the effect of the visual complexity on intention and behavior. Indeed, we surveyed our participants before they saw the homepages and used the application. It would also be interesting to check the following order of the variables (Complexity; Acceptability; Intention; Behaviour). This may help explain the difference in effectiveness between visual complexity on intentions (situational study) and visual complexity on behavior (real conditions study). Another limitation is that our study mobilizes users from a single city and a specific age group. In summary, we highlight the importance of visual complexity in the intention as well as in the behaviour of responsible urban mobility, underlining the need to take into account other parameters such as the acceptability of the application as well as environmental variables that have not been taken into account in our explanation of the phenomenon. Based on this assumption, we want to integrate features with components such as social comparison and individualization of information into a second version of the mobile application. We also want to extend the scheme to a larger and heterogeneous public.

References

1. Ajana, B.: Personal metrics: users' experiences and perceptions of self-tracking practices and data. Soc. Sci. Inf. **59**(4), 654–678 (2020)
2. Ajzen, I.: The theory of planned behavior. Organ. Behav. Hum. Decis. Process. **50**(2), 179–211 (1991)
3. Althuizen, N.: Revisiting Berlyne's inverted u-shape relationship between complexity and liking: the role of effort, arousal, and status in the appreciation of product design aesthetics. Psychol. Mark. **38**(3), 481–503 (2021)
4. Berlyne, D.E.: Conflict, Arousal, and Curiosity. McGraw-Hill Book Company, New York (1960)
5. Borji, A., Itti, L.: State-of-the-art in visual attention modeling. IEEE Trans. Pattern Anal. Mach. Intell. **20**, 1395–1398 (2010)
6. Brauer, B., Ebermann, C., Hildebrandt, B., Remane, G., Kolbe, L.: Green by app: the contribution of mobile applications to environmental sustainability. In: Proceedings of 20th Pacific Asia Conference on Information Systems (PACIS 2016) (2016)

7. Bruck, P.A., Motiwalla, L., Foerster, F.: Mobile learning with micro-content: a framework and evaluation. In: BLED 2012 Proceedings (2012)
8. Carrus, G., Passafaro, P., Bonnes, M.: Emotions, habits and rational choices in ecological behaviours: the case of recycling and use of public transportation. J. Environ. Psychol. **28**(1), 51–62 (2008)
9. Constantiou, I., Mukkamala, A., Sjöklint, M., Trier, M.: Engaging with self-tracking applications: how do users respond to their performance data? Eur. J. Inf. Syst. 1–21 (2022)
10. Apache cordova framework. https://cordova.apache.org/
11. Day, H.: Evaluations of subjective complexity, pleasingness and interestingness for a series of random polygons varying in complexity. Percept. Psychophys. **2**, 281–286 (1967)
12. Deng, L., Poole, M.S.: Aesthetic design of e-commerce web pages - webpage complexity, order and preference. Electron. Commer. Res. Appl. **11**(4), 420–440 (2012)
13. Dubois, P., et al.: Effects of front-of-pack labels on the nutritional quality of supermarket food purchases: evidence from a large-scale randomized controlled trial. J. Acad. Mark. Sci. **49**, 119–138 (2021)
14. ElHaffar, G., Durif, F., Dubé, L.: Towards closing the attitude-intention-behavior gap in green consumption: a narrative review of the literature and an overview of future research directions. J. Clean. Prod. **275**, 122556 (2020)
15. Flüchter, K., Wortmann, F., Fleisch, E.: Digital commuting: the effect of social normative feedback on e-bike commuting - evidence from a field study. In: ECIS 2014 Proceedings - 22nd European Conference on Information Systems (2014)
16. Froehlich, J., et al.: Ubigreen. In: Proceedings of the SIGCHI Conference on Human Factors in Computing Systems, pp. 1043–1052. ACM (2009)
17. Gabrielli, S., et al.: Designing motivational features for sustainable urban mobility. In: CHI 2013 Extended Abstracts on Human Factors in Computing Systems, pp. 1461–1466. Association for Computing Machinery (2013)
18. Geissler, G.L., Zinkhan, G.M., Watson, R.T.: The influence of home page complexity on consumer attention, attitudes, and purchase intent. J. Advert. **35**, 69–80 (2006)
19. Intergovernmental Panel on Climate Change: Climate change 2022: Impacts, adaptation and vulnerability. Contribution of working group ii to the sixth assessment report of the intergovernmental panel on climate change. In: Pörtner, H.O., et al. (eds.) Climate Change 2022: Impacts, Adaptation and Vulnerability. Contribution of Working Group II to the Sixth Assessment Report of the Intergovernmental Panel on Climate Change, pp. 2273–2318. Cambridge University Press, Cambridge and New York (2022). https://doi.org/10.1017/9781009325844.022.2273
20. Itti, L., Koch, C.: A saliency-based search mechanism for overt and covert shifts of visual attention. Vis. Res. **40**, 1489–1506 (2000)
21. Jin, D., Halvari, H., Maehle, N., Olafsen, A.H.: Self-tracking behaviour in physical activity: a systematic review of drivers and outcomes of fitness tracking. Behav. Inf. Technol. **41**(2), 242–261 (2022)
22. Julia, C., Hercberg, S.: Nutri-score: evidence of the effectiveness of the French front-of-pack nutrition label. Ernahrungs Umschau **64**, 181–187 (2017)
23. Koch, C., Ullman, S.: Shifts in selective visual attention: towards the underlying neural circuitry. Hum. Neurobiol. **4**, 219–227 (1985)
24. Kusmanoff, A.M., Fidler, F., Gordon, A., Garrard, G.E., Bekessy, S.A.: Five lessons to guide more effective biodiversity conservation message framing. Conserv. Biol. **34**(5), 1131–1141 (2020)

25. Lee, S.C., Hwangbo, H., Ji, Y.G.: Perceived visual complexity of in-vehicle information display and its effects on glance behavior and preferences. Int. J. Hum.-Comput. Interact. **32**(8), 654–664 (2016). https://doi.org/10.1080/10447318.2016.1184546
26. Lehto, T., Oinas-Kukkonen, H.: Examining the persuasive potential of web-based health behavior change support systems. AIS Trans. Hum.-Comput. Interact. **7**, 126–140 (2015)
27. Lim, J.S., Noh, G.Y.: Effects of gain-versus loss-framed performance feedback on the use of fitness apps: mediating role of exercise self-efficacy and outcome expectations of exercise. Comput. Hum. Behav. **77**, 249–257 (2017)
28. Morrison, B.J., Dainoff, M.J.: Advertisement complexity and looking time. J. Mark. Res. **9**, 396–400 (1972)
29. Mosteller, J., Donthu, N., Eroglu, S.: The fluent online shopping experience. J. Bus. Res. **67**(11), 2486–2493 (2014)
30. Nabec, L., Guichard, N., Hémar-Nicolas, V., Durieux, F.: The role of nutri-score front-of-pack labels on children's food products in informing parents: an analysis of the branding effect. Decis. Mark. **106**(2), 11–30 (2022)
31. Palmer, J.W.: Web site usability, design, and performance metrics. Inf. Syst. Res. **13**(2), 151–167 (2002)
32. Pieters, R., Wedel, M., Batra, R.: The stopping power of advertising: measures and effects of visual complexity. J. Mark. **74**(5), 48–60 (2010)
33. Shankari, K., Bouzaghrane, M.A., Maurer, S.M., Waddell, P., Culler, D.E., Katz, R.H.: E-mission deployment. https://www.nrel.gov/transportation/openpath.htmlutm_medium=print&utm_source=transportation&utm_campaign=openpath
34. Shankari, K., Bouzaghrane, M.A., Maurer, S.M., Waddell, P., Culler, D.E., Katz, R.H.: e-mission: an open-source, smartphone platform for collecting human travel data. Transp. Res. Rec. **2672**(42), 1–12 (2018)
35. Standing, C., Jackson, P., Chen, A.J., Boudreau, M.C., Watson, R.T.: Information systems and ecological sustainability. J. Syst. Inf. Technol. **10**, 186–201 (2008)
36. Treisman, A., Gelade, G.: A feature-integration theory of attention. Cogn. Psychol. **12**, 97–136 (1980)
37. Trudel, R.: Sustainable consumer behavior. Consum. Psychol. Rev. **2**(1), 85–96 (2019)
38. Tulusan, J., Staake, T., Fleisch, E.: Providing eco-driving feedback to corporate car drivers: what impact does a smartphone application have on their fuel efficiency? In: Proceedings of the 2012 ACM Conference on Ubiquitous Computing, UbiComp 2012, pp. 212–215. Association for Computing Machinery, New York (2012)
39. Venkatesh, V., Thong, J.Y.L., Xu, X.: Consumer acceptance and use of information technology: extending the unified theory of acceptance and use of technology. MIS Q. **36**(1), 157–178 (2012)
40. Wang, H.F., Lin, C.H.: An investigation into visual complexity and aesthetic preference to facilitate the creation of more appropriate learning analytics systems for children. Comput. Hum. Behav. **92**, 706–715 (2019)
41. Wang, Q., Ma, D., Chen, H., Ye, X., Xu, Q.: Effects of background complexity on consumer visual processing: an eye-tracking study. J. Bus. Res. **111**, 270–280 (2020)
42. Wu, K., Vassileva, J., Zhao, Y., Noorian, Z., Waldner, W., Adaji, I.: Complexity or simplicity? Designing product pictures for advertising in online marketplaces. J. Retail. Consum. Serv. **28**, 17–27 (2016)

Interoperability of Open Science Metadata: What About the Reality?

Vincent-Nam Dang[1,2](✉), Nathalie Aussenac-Gilles[1], Imen Megdiche[1,3], and Franck Ravat[1,2]

[1] IRIT, CNRS (UMR 5505), Université de Toulouse, Toulouse, France
vincent-nam.dang@irit.fr
[2] Université Toulouse 1 Capitole, Toulouse, France
[3] INU Champollion, Albi, France

Abstract. Open Science aims at sharing results and data widely between different research domains. Interoperability is one of the keys to enable the exchange and crossing of data between different research communities. In this paper, we assess the state of the interoperability of Open Science datasets from various communities. The diversity of metadata schemata of these datasets from different sources does not allow for native interoperability, highlighting the need for matching tools. The question is whether current metadata schema matching tools are sufficiently efficient to achieve interoperability between existing datasets. In our study, we first define our vision of interoperability by transversally considering the technical and semantic aspects when dealing with metadata schemata coming from various domains. We then evaluate the interoperability of some datasets from the medical domain and Earth system study domain using acknowledged matching tools. We evaluate the performance of the tools, then we investigate the correlation between various metrics characterizing the schemata and the performance related to their mapping. This paper leads to identify complementary ways to improve dataset interoperability: (1) to adapt mapping algorithms to the issues raised by metadata schema matching; (2) to adapt metadata schemata, for instance by sharing a core vocabulary and/or reusing existing standards; (3) to combine various trends in a more complex interoperability approach that would also make available and operational the (RDA) crosswalks between schemata and that would promote good practices in metadata labeling and documentation.

Keywords: Information System · Interoperability · Data Integration · Semantic Metadata Management · Open science

1 Introduction

Open science pushes further the objectives of scientific research by making more accessible and transparent the knowledge shared and developed by a collaborative network of researchers [27]. This opening up of science offers many advantages [24], in particular the possibility of working on new research questions and answering

S. Nurcan et al. (Eds.): RCIS 2023, LNBIP 476, pp. 467–482, 2023.
https://doi.org/10.1007/978-3-031-33080-3_28

them through collaborations between researchers. To benefit from this interdisciplinary and open collaboration, one of the challenges is to enable exchange between domains and communities. Practically, reaching this aim requires making it possible to retrieve data from where it is stored and to enable its use in the researcher's working environment. To do this, it is necessary to go through the metadata on these data to find out which data meets the research needs. This problem is defined as "data interoperability problem and particularly the metadata interoperability". To address this issue, the RDA[1] metadata working group provides crosswalks between several models and Schema.org[2]. However, not all Open Science data are represented by these schemata and these crosswalks are not enough to interoperate the various Open Science data. A very interesting direction to investigate is the advances in knowledge and schema-based matching tools. These tools can automatically generate alignments between the schemata of metadata. Matching tools have the potential to handle a wide variety of schemata and thus allow a multitude of data sources to be brought together in a single working environment. Nevertheless, it seems that there is a gap between the approved performance of these matching tools in the evaluation campaigns such as OAEI[3] and their real performance when challenged on real data coming from the Open Science.

In this paper, we go through the question to demystify the reality about the interoperation of Open Science metadata based on actual advances in the domain schema matching. We first propose in Sect. 3, an in-depth context-independent definition of interoperability and its application in Open Science. In Sect. 4 and 5, we report the results of our evaluation. We define the context of the experiment, the metadata schema used and the performance metrics used to evaluate the matching tools. Then we present the explanatory hypotheses and the metrics for testing the veracity of these hypotheses. In part 6, we propose three tracks for solving the problems raised by the results.

2 Related Work

2.1 Metadata Models

Metadata models or schemata are used to define the list of metadata to describe a dataset. Depending on the context, metadata may be used to enable the storage, archiving, exchange and publication of data, to facilitate their discovery, search and understanding [31]. Metadata schemata can be considered as metadata templates that may include a very different list of metadata. We can explore the literature on proposed models according to the domain[4], either inter-domain (Dublin Core[5], MARC[6], EAD[7], ...) or domain-specific (Life sciences with Darwin Core[8]

[1] www.rd-alliance.org/groups/research-metadata-schemas-wg.
[2] schema.org.
[3] oaei.ontologymatching.org/.
[4] guides.lib.utexas.edu/metadata-basics/standards.
[5] www.dublincore.org/.
[6] www.loc.gov/marc/.
[7] www.loc.gov/ead/EAD3taglib/index.html.
[8] rs.tdwg.org/dwc.htm.

or Social and Behavioral Sciences with FoAF[9]); according to their use case like the metadata models for e-Government project metadata [1] or according to the organization setting up these standards like the OGC[10] or the ISO[11]. Among these available models, we find standards that take up other standards, notably Dublin Core [29], which is used in many other models, or models defined for a specific case, such as ISO 19115, designed for metadata for geographic data and services. These standards are used to set up models for use cases, as in the DATA-TERRA[12] project, where the AERIS[13] data cluster has a specific model announced as being compatible with ISO 19115[14], or the ODATIS[15] cluster, which implements ISO 19115 in its models and some of the SeaDataNet vocabulary[16].

2.2 Schema and Ontology Matching

Establishing links between similar concepts and similar features of several metadata schemas is necessary to interoperate them. To achieve this linkage, one approach is to set up a mapping map between the models. In the literature [22], solutions may be based on the semantics inside the model, on the structure of the model or a combination of both. In particular, we find solutions based on n-grams [17], on external linguistic databases [6] and on intermediate representations [3, 10, 19]. Matching solutions can be generic, and apply to any schema [17] or they can be specialised for processing data from a specific domain [9] or specialised to a particular type of data [16]. Many solutions have been developed, necessitating the need to compare them to a common base. OAEI (Ontology Alignment Evaluation Initiative) is setting up platforms [12] and common evaluation campaigns [7]. This corresponds to a standardization of the evaluation of the solution, in particular by a selection of datasets [2]. However, this selection of evaluation datasets does not allow us to evaluate the solutions on the totality of the use cases or domains that can be encountered in Open Science or real life.

3 Interoperability and Open Science

3.1 Interoperability Definitions in the Literature

Interoperability is a concept present in all domains as soon as data or information exchange is concerned. There is a multitude of definitions in the literature:

- P. Wegner [28] defines the concept of interoperability for software components. His work is network-oriented and suggests 2 major mechanisms of interoperability: interface standardization and interface bridging.

[9] www.foaf-project.org/.
[10] www.ogc.org/docs/is.
[11] www.iso.org.
[12] www.data-terra.org/.
[13] www.aeris-data.fr/.
[14] www.iso.org/standard/32579.html.
[15] www.odatis-ocean.fr/.
[16] www.seadatanet.org/.

- M. D. Wilkinson [30] defines the FAIR criteria, including 3-point interoperability centred around metadata.
- S. Heiler [11] defines semantic interoperability as the objective to find meaning in the exchanges made.
- A. Tolk [25] proposes a theoretical model of interoperability in 7 layers in the context of Modeling and Visualization Systems.
- M. Nilsson *et al.* [20] describe the design choices in creating applications for various types of interoperability anchored in the Dublin Core model.
- M. Noura *et al.* [21] define interoperability in the Internet of Things (IoT), the taxonomy as well as the Open Challenges and solutions.
- H. Van Der Veer *et al.* [26] approach interoperability through 4 dimensions (technical, syntactic, semantic and organizational) in the context of telecommunication technologies and the process and criteria for validating a standard in ETSI.
- M. L. Zeng [32] proposes an exploration of interoperability in the world of Knowledge Organizations (KO) with 4 interoperability issues: system, syntactic, structural and semantic.

The exploration of different works on interoperability has brought out many interesting points. Most works are anchored in a particular domain or context without generically taking the problem of interoperability. This contextualisation of interoperability creates differences in its definition. There are definitions in layers, stacks, independent groups or without distinctions of different groups. In the layered models, we find a large variety of layer decomposition [11,20,21,25,26,28]: syntactic, semantic, technical, organizational, vocabulary-related, structural, conceptual, communication, transport, document, etc. Several approaches can use the same terms while including different mechanisms needs or aspects which makes it difficult to understand interoperability not related to a domain, objective or concept. Although there are commonalities in most proposals, there is still no complete consensus on interoperability. S. Y. Diallo *et al.* [4] have described this problem, citing the need for a "formal theory of interoperability". They propose 5 objectives that should be addressed by a formal theory of interoperability:

- Objective 1 (O1): A formal theory of interoperability should meet the necessary and sufficient requirements for interoperability (Information exchange and Usability of information).
- Objective 2 (O2): A theory of interoperability should be capable of formally defining data, information, useful information, and context.
- Objective 3 (O3): A theory of interoperability should be capable of explaining the duality of interoperability.
- Objective 4 (O4): A theory of interoperability must be able to explain interoperability without falling into an infinite recursion.
- Objective 5 (O5): A theory of interoperability should be able to explain what interoperability is as a whole.

Table 1. Comparison of interoperability proposals on the objectives of a formal theory of interoperability

	Objective 1	Objective 2	Objective 3	Objective 4	Objective 5
P. Wegner [28]	++	−	?	+	+
M. D. Wilkinson [30]	?	−	?	−	−
A. Tolk [25]	+	+	?	− −	++
S. Heiler [11]	+	−	?	+	−
M. Nilsson et al. [20]	−	+	?	+	−
M. Noura et al. [21]	+	+	+	−	++
H. Van Der Veer et al. [26]	+	−	?	−	+
M.L. Zeng [32]	+	−	?	?	+

++ *: Objective fully met* + *: Some points to address the objective* − *: Points in contradiction with the objective* − − *: Contradiction with the objective* ? *: No evidence to assess*

According to these criteria, we draw in Table 1 a comparison of the different definitions quoted above. As we can notice, none of the definitions checks all the objectives. If we consider for example objective O3, the only definition that provides keys to address it is the one proposed by M. Noura *et alii* [21]. It defines in detail the different systems and goes in depth into this distinction allowing us to have a vision of the difference between intra-system and inter-system interoperability. Apart from the proposal by A. Tolk [25] and M. Noura *et alii* [21], definitions do not go in depth of the interoperability and focus on the solutions proposed in the related domain or address a subpart of interoperability. The definition of usefulness in exchanges is often taken into account but not well defined. The proposal of M. Nilsson *et alii* [20] allows us to approach a definition of utility, but the proposal focuses only on the interoperability of vocabularies with Dublin Core. The definition by M. Noura *et alii* [21] is one of the best to handle the concept of interoperability but is too anchored in the IoT domain to allow it to be used as a generic basis. The model proposed by A. Tolk [25] is probably the most appropriate and is very rich. However, it defines interoperability in 7 layers of interoperability, but it leaves a blur on the definition. Jacobsen *et alii* [13] have proposed an interpretation of the FAIR principles that complements them. However, this proposal does not address all the issues found in other domains, such as IoT.

3.2 Our Definition: Context-Independent Interoperability

We have observed that the contextualization of interoperability includes criteria and needs that may differ in the different steps to achieve interoperability. To allow an unambiguous understanding of interoperability, it is necessary to define this concept independently of the context. This definition does not question the problems encountered in each context, whether it be the interoperability of data management systems, data interoperability or syntactic or semantic interoperability. We define interoperability as *the ability of 2 entities to work cooperatively through an exchange of information to achieve an objective* (O1). The content of the communication can be divided into two groups of exchanged objects: (1) the data to be consumed to achieve the objective of the communication (which means the data to be reused), and (2) the information useful for the consumption of this data (which means the metadata). The utility in a communication is any data or information needed to realize a mechanism or the initial need of the communication. ISO 11179 standard defines metadata as "descriptive data about an object", allowing to define (2) as the part of the metadata necessary to achieve the purpose of the communication (communication context) (O2). The use of the transmitted data is the objective of interoperability. Depending on what this use is (consumption, visualisation, crossing, storage, archiving, etc.), the requirements will change.

Interoperability concerns communicating entities, whatever the scale. Thus, it is possible to deal with *intra-system interoperability*, e.g. by observing the interoperability between blocks composing the system having different functions, as well as *inter-system interoperability*, e.g. by observing the interoperability between systems having similar functions but with different domains and/or operating mechanisms (O3). The steps of interoperability are the same but the context is different for intra-system and inter-system interoperability.

Interoperability is defined according to the purpose of the communication and the environment in which the communication takes place. Depending on the complexity of the task, it is necessary to implement several steps. The OSI model[17] allows the definition of the network protocol stack, which defines the mechanisms in place for network communication. In essence, the objective of the OSI model (system interconnection) has much in common with the definition of interoperability. We define interoperability in a multi-layer model (O4, O5), inspired by the OSI model (see Fig. 1).

- Layer 1 - Exchange layer: The data can transit from entity A (sender) to entity B (receiver). Communication between the two systems is possible and the two systems can listen to each other. The receiver receives a mass of data quantum defined as raw data.
- Layer 2 - Segmentation layer: The information quantum can be segmented into separate information segments. These segments can be stored in the receiving system as a packet with a beginning and an end. The data at this stage is readable.
- Layer 3 - Parsing layer: The data segments can be parsed and read in the original format of the data, associating to the segments a structure allowing to read the content. The segments are formatted.
- Layer 4 - Metadata layer: The formatted data is described by minimal technical metadata allowing the internal structure of the formatted data to be understood, without understanding the content. The data can be used.
- Layer 5 - Management layer: The data is described by general metadata (e.g. user, date, name, id, etc.). This allows data to be used in the context in which it was created or originally used. At this stage, the data can be valued.

[17] https://www.iso.org/ics/35.100/x/.

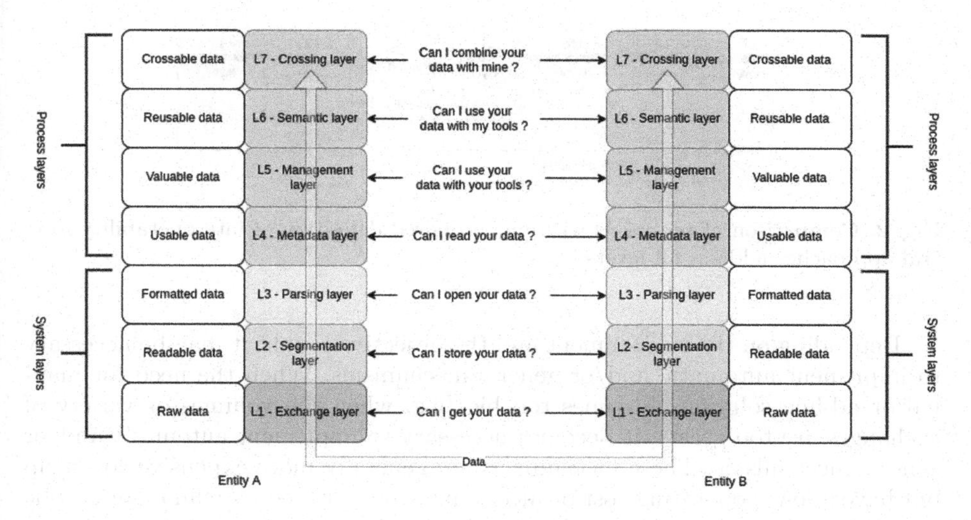

Fig. 1. Interoperability model

- Layer 6 - Semantic layer: The data is described by semantic metadata allowing the content of the data to be understood. It is possible to adapt the tools to the data and to use the data in a context different from the initial context of the data. The data is reusable.
- Layer 7 - Crossing layer: The metadata can be crossed with metadata from other data that the receiving system may have. Links can be set between data, allowing the received data to be sent to the knowledge repository of the receiving system. The data is crossable.

To meet the needs of each layer, the communicating entities must implement several mechanisms and associated metadata required for the implementation of the mechanisms. The lower layers, known as "system layers", are linked to technical communication mechanisms and are fairly low-level. The high layers, called "process layers", are layers related to the data use and the creation of value on this data. Metadata can be found in all layers, from information about the characteristics of a data quantum to recognise it (layer 1), to create links between concepts in metadata to cross and combine data (layer 7). Each layer requires the presence of metadata related to the purpose of the layer as well as the mechanisms needed to process those metadata.

There are 2 types of approaches to design layer mechanisms: standardisation and implementation of gateways. Standardisation consists of defining a standardized mechanism used by every entity to eliminate differences between entities. This can avoid the need to send metadata. Standards create native interoperability between entities. The implementation of a gateway is necessary when native interoperability is not present. The purpose of the gateway is to set up translation or information transformation operations to allow the passage from one frame of reference to another.

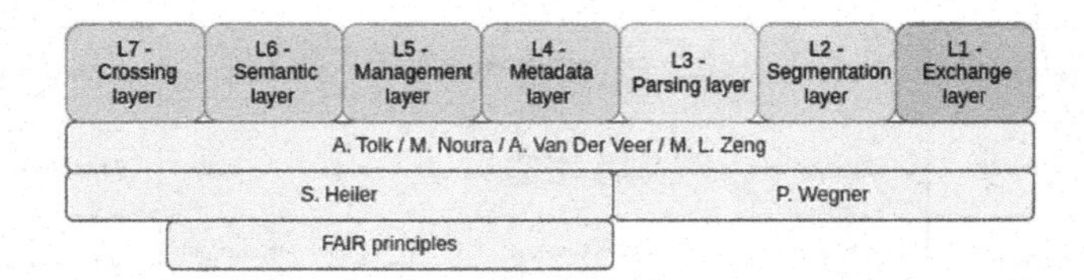

Fig. 2. Comparison of proposals with our proposed definition of interoperability: several approaches address all layers

Depending on the environment and the objective sought it may be necessary to implement automatic and/or generic mechanisms. When the need for manual or ad-hoc solutions becomes too big (e.g. when the volumes or variety of exchanges are too great), it becomes necessary to implement automatic and/or generic mechanisms. These mechanisms are generally more expensive to set up but have a lower operating cost or meet a need not covered by manual or ad-hoc mechanisms.

The proposals found in the literature address all the layers (see Fig. 2). However, even if we find the whole set of layers in 4 proposals [21,25,26,32], none of them fully meets all the criteria for a formal theory of interoperability.

The FAIR principles [30] only approach the semantic aspect of metadata. We find the idea of the need to cross metadata but it is only approached through the idea of references (i.e. "I3: (meta)data include qualified references to other (meta)data"). This does not define what these references are, how they are managed and their precise purpose. In general, the FAIR principles do not go deeply enough into interoperability. Proposals have been made[18] to provide concrete solutions. However, this does not address the need for definitions of interoperability.

Interoperability in Open Science. In the context of Open Science, the aim is to successfully exchange and cross data to enable researchers to answer new research questions. This data comes in datasets available in different open research data management systems. The data management system interoperability in Open Science consists in implementing all the layers up to layer 7.

Layers 1 to 3 are supported by the use of IT tools in particular managed by the TCP/IP stack. Layer 4 and 5 are generally managed by metadata management including the type of data, allowing to reproduce the operations performed when creating these datasets.

Layer 6 requires answering the question "how to describe metadata so that the content of the data is understandable". A single standard shared by the whole Open Science community does not seem to be a realistic solution. There are many different representations of metadata, even within the same domain [30]. A wide variety of metadata models have emerged that do not allow native interoperability between metadata.

In layer 7, one way is to create links between the metadata models and to allow the retrieval of metadata that represent the same concept. Links can be created manually through crosswalks to retrieve different representations of

[18] www.fairsfair.eu.

concepts. Automation is needed in Open Science to manage the great diversity of metadata models. Matching tools are one way of automating the exploitation of these crosswalks.

4 Schema Matching Tools Evaluation on Real Open Science Data

To observe the state of metadata interoperability in Open Science, we wish to observe the performance of tools that address the needs of the last layer of data management platforms interoperability: metadata model matching tools. These tools are evaluated with good performance in controlled contexts such as OAEI, but in the context of real Open Science data, it is necessary to evaluate their performance. In this section, we first present our experimentation setup then we report the results and their analysis.

4.1 Experimentation Setup

Open Science Metadata Models: To avoid biases, we select our metadata models from several domains. We have selected 2 models in 2 defined domains with similar needs.

- The first pair of models contains the metadata schema from the ODATIS platform[19], oceanographic data, and the AERIS platform[20], atmospheric data. These two platforms are data hubs of the DATA-TERRA data repository, holding Earth system study data.
- The second pair of metadata models comes from the medical domain with the use of the C-CDA metadata model and the FHIR metadata model. These two metadata models are widely used in the healthcare domain. They aim to store the same information and they are developed by the same organism HL7. FHIR is announced as a replacement for C-CDA, allowing for a simpler implementation and these two models are announced as interoperable. Tools for transforming C-CDA documents into FHIR documents are available[21].

We also evaluated inter-domain interoperability by conducting our experiments on the AERIS-FHIR pair.

The metadata schemata are not available directly for download, so we had to download instantiations of these models and extract the schema. Some instantiations do not contain the entire model. We, therefore, merged the schema of several instances to ensure that we had the closest model to the theoretical model. We retrieved instantiations of AERIS/ODATIS directly from the platform, of C-CDA from a collection of example instantiations[22] and of FHIR from the official documentation[23].

[19] www.odatis-ocean.fr/.
[20] www.aeris-data.fr/.
[21] gitlab.com/smilecdr-public/cda2r4.
[22] github.com/jmandel/sample_ccdas.
[23] www.hl7.org/fhir/index.html.

To work on the metadata models, we decided to transform these models into trees. Modelling in tree form makes it possible to be sufficiently generic to be able to manage any type (format or paradigm) of metadata. A metadata is defined by a set of labels. This set of labels creates a path, from the root to the metadata. We retrieve the paths as a concatenation of the labels, similar to the method applied in COMA [5] (for example, the unit of measurement ("uom") of the measured physical quantity ("parameterSet") is concatenated into "parameterSet.uom" in AERIS metadata model).

Ground Truth Mappings: Like most open science metadata schemata, there is no reference alignment that we can consider as ground truth. Hence, we have constructed three ground truth alignments.

- For the ODATIS/AERIS and AERIS/FHIR pair, we had to perform a ground truth manually
- For the FHIR/C-CDA pair, we used a tool to transform C-CDA documents into FHIR documents[24]. The matches were extracted automatically based on the similar values between the C-CDA document and its FHIR version.

The Schema Matching Tools Evaluated: To evaluate the tools in a global way, we have selected tools representative of the two main categories of tools: structural and semantic.

- For the structural solution, we have chosen COMA [5] through the Valentine library [14]. COMA is a robust solution that demonstrated very good results on different tracks of the OAEI campaign [18]. It is also described in the validation of the Valentine tools as the best approach for dataset discovery and data integration solution [14]. No pre-processing is required when using COMA.
- For the semantic solution, we have chosen a pre-trained word-embedding model from Gensim library [23] that implements a Word2Vec technique on a generalist corpus (the Wikipedia corpus). This class of tools is reused in other matching solutions [9,10] showing performances up to 0.9 of Macro-F1 measure [10].

We perform several pre-processing operations on the labels to retrieve contained words: a Camel-case splitting operation, a statistical splitting, and removing stop-words from the English language. Then words are transformed into vectors. To perform matching with word embedding models, we use two distances between the associated vectors contained in the labels of the paths (from the root to the leaf). For each path in one model, we keep only the path with the smallest distance in the other model. The first distance is the Euclidean distance. We apply it to the sum of the vectors. We base our approach on the additivity property of vector as word-embedding models [8]. Then we approximate the

[24] gitlab.com/smilecdr-public/cda2r4.

concepts defined as the semantic sum of the terms used in the model path. The second distance used is the Word mover's distance [15] for a path. The objective is to try to capture the interactions between the used terms according to their position in the path by approaching the path as a corpus.

Performance Metrics: To evaluate the quality of matching of the selected solutions, we use three metrics:

$$Precision = \frac{TP}{TP+FP} Recall = \frac{TP}{TP+FN} F_1 - score = \frac{2 * Precision * Recall}{Precision + Recall}$$

with TP the correct matches, FP the incorrect matches, FN the matches that should have been made.

4.2 Evaluation Results and Observations

Matching Results. We performed 631 matching scenarios[25], based on 3 pairs of models (ODATIS/AERIS; FHIR/C-CDA; AERIS/FHIR). Each scenario is a combination of a pair of models, a metadata model matching tool and a maximum size of sub-path selected in the model starting from the leaf. This sub-selection allows us to simulate changes in the models and to observe the evolution of tool performance according to the model characteristics.

The result are summarised in Table 2. It can be seen that the performance results of the evaluated tools are not sufficient to meet the needs of layer 7, whatever the objectives. The best F1-score is 0.21 for COMA on the ODATIS/AERIS pair. It is neither possible to set up an automatic data crossing system for a non-human user nor to set up a recommendation system for a human user. The C-CDA/FHIR model pair is the pair expected to have the best interoperability. These 2 models are defined to keep the same type of document, in the same domain and made by the same entity (HL7). However, the results for this pair are less good whatever the tool. The format of the two models is different (XML for C-CDA and JSON FHIR) and there are different modelling choices: some metadata in one model are described with a specific path in the model while these same metadata are described in the metadata content in the other model.

Surprising results are observed for the AERIS/FHIR pair. The two models are not defined to be interoperable and perform better than the FHIR/C-CDA pair with much higher recall. We explain these results by the much smaller size of the ground truth. The ground truth represents only 7.7% of the size of the smallest model compared to 44.7% for the FHIR/C-CDA pair and 48.6% for the ODATIS/AERIS pair. Matches are either defined by the same terms and on unambiguous and easily alignable concepts such as identifiers and both models are initially in the same format (JSON). Recall, therefore, carries a much greater weight in these results without indicating that these methods are suitable.

---
[25] github.com/vincentnam/OS_data_interop_RCIS_2023.

Table 2. Results of schema matching

		Model pair	Precision	Recall	F1-score
COMA		ODATIS/AERIS	0.00–**0.27**	0.00–0.2	0.00–**0.21**
		FHIR/C-CDA	0.00–0.20	0.00–0.04	0.00–0.06
		AERIS/FHIR	0.05–0.10	0.25–0.42	0.08–0.16
Word-embedding	Euclidian distance	ODATIS/AERIS	0.00–0.09	0.00–0.09	0.00–0.07
		FHIR/C-CDA	>0–0.05	>0–0.05	>0–0.04
		AERIS/FHIR	0.01–0.05	0.08–**0.50**	0.01–0.10
	Word mover distance	ODATIS/AERIS	0.00–0.15	0.00–0.26	0.00–0.19
		FHIR/C-CDA	>0–0.09	0.01–0.09	0.01–0.07
		AERIS/FHIR	0.02–0.05	0.25–**0.50**	0.04–0.10

The observed performances do not allow to meet the needs of layer 7 of the interoperability of data management systems. We could propose manual matchings but the scalability would remain a problem and an unmet challenge in the Open Science. As the implementation of interoperability of data management systems is not solved, the interoperability of metadata and data is not possible. Indeed, additional semantic aspects, such as meaning preservation between actors/tools, need to be added to reach this broader objective.

Result Analysis and Comment: We propose two hypotheses and a set of associated metrics that could explain the lack of efficiency of the matching solutions on real open science data.

- **The first hypothesis is the size of the models.** To evaluate its relevance, we define a first group of 5 metrics "Size metrics" that we can extract from the schema: (1) Quantity of metadata stored: the number of leaves in the tree, (2) Number of intermediate nodes: the number of nodes that are not leaves, (3) Tree order: the total number of nodes (leaf and non-leaf), (4) Vocabulary size: the number of different words in the set of node labels and (5) Path size: minimum and maximum path lengths from the root to a leaf in the model.
- **The second hypothesis is the domain of models.** The vocabulary used in the medical domain may be more specific than in the study of the Earth system. To assess the relevance of the vocabulary specificity hypothesis, we choose to observe "Complexity metrics" thanks to the word frequency from the word count in the English Wikipedia corpus[26]. For the metrics, we define the average of the word frequencies of the vocabulary and this same average weighted by the frequency of occurrence of the words in the vocabulary.

In Table 3, Spearman correlation between schema intrinsic metrics and performance metrics are calculated for models in each pair (i.e. 1 cell of the table represents [(ODATIS/FHIR/AERIS) Spearman correlation value with each metric/(AERIS/C-CDA/FHIR) Spearman correlation value]). For example, the

[26] github.com/IlyaSemenov/wikipedia-word-frequency.

Table 3. Spearman correlation between performance metrics and schema metrics

Spearman correlation	Precision	Recall	F1-score
	Size metric		
Amount of metadata stored	−0.34/−0.28	−0.41/−0.46	**−0.45/−0.44**
Intermediate node number	−0.27/−0.34	−0.31/−0.52	**−0.32/−0.49**
Order	−0.31/−0.33	−0.36/−0.51	**−0.38/−0.49**
Voc length	−0.35/−0.20	−0.44/−0.03	**−0.46/−0.15**
Min path size	0.21/−0.31	0.20/−0.58	0.28/−0.52
Max path size	−0.13/−0.26	−0.06/−0.45	**−0.08/−0.42**
	Complexity metric		
Average of term frequencies	0.16/−0.14	−0.19/−0.18	0.00/−0.22
Weighted average of term frequencies	0.13/−0.18	−0.24/−0.22	−0.03/−0.25

first cell (−0.34/−0.28) represents the Spearman correlation between the amount of metadata stored in each model and the precision. To get a global view of the results, we focus our analysis on **F1-score**.

All correlation values have P-values < 0.05 except the weighted average of term frequencies.

AMOUNT OF STORED METADATA/INTERMEDIATE NODE NUMBER/TREE ORDER/VOCABULARY SIZE: The correlation values (resp. (−0.45/−0.44), (−0.32/−0.49), (−0.38/−0.49) and (−0.46/−0.15)) show a *significant negative impact*.

MINIMUM PATH SIZE: Correlations values are the opposite for the schemata of each pair. *No conclusion* can be drawn on the impact of this metric.

MAXIMUM PATH SIZE: The correlation values are negative. However, the correlation of model 1 is much lower than the correlation value of model 2. The results show *a negative trend* of the maximum path size on the performance of the solutions.

SPECIFICITY METRICS: *No clear trend* emerges from these results.

A significant cause of performance degradation of matching tools is observed with increasing model size (the *amount of stored metadata*, the *global amount of nodes* in the tree and the *amount of different words and the depth of concepts* used to define these labels. The domain impacts the interoperability of the models but we cannot conclude on its impact on the performance of the solutions.

5 Lessons Learned and Future Directions

To improve interoperability in the Open Science field, we propose the following three directions:

Metadata Schema: The first direction suggests to work on the metadata schema. Results show that the size of the metadata schema influences the performance of matching solutions. Reducing the size of models is a simple solution to improve interoperability. However, this solution is not relevant for already designed models or for some datasets. Some metadata models already have a minimal set of metadata to achieve the desired objective. Another approach is to separate technical, general and domain-specific metadata for their interoperation and matching. This breakdown depends on the metadata implementation (i.e. can be done through a label on each metadata, new metadata or through the implementation of meta-metadata). This separation reduces the size of the vocabulary and model and increases the similarity of the metadata to be matched, resulting in improved performance of interoperability tools.

Matching Solutions: The second direction suggests to better explore the matching solutions. We observed in our results that almost only general and technical metadata are correctly matched (such as users related to the dataset, date of creation, codes, identifiers, etc.). The tools we have selected are designed to be generic. In the literature, we find matching solutions specific to a type of data [16] or a particular domain [9]. These tools show high performance on data with specific constraints. The use of these 2 groups of tools in cooperation could improve global performance.

Data and Metadata Management System: The third direction to explore is on data and metadata management systems. In contrast to the other two directions, here interoperability must be approached through all layers of interoperability and not only on the process layers. It means integrating the variety of metadata models and data-crossing solutions (matching solutions, crosswalks, metadata transformations, etc.). Solutions, such as crosswalks, have scaling issues. This becomes an issue in Open Science because of large volume and variety of datasets. Decentralisation provides a solution to this variety and volume. It allows for a balance of loads between the different actors and the different data and metadata management systems. Each actor processes and manages its data locally, reducing the problem of solution scalability. However, it is necessary to set up intercommunication protocols between actors. A decentralised architecture in the form of federations of metadata management systems can help to achieve this. It is necessary to take into account existing solutions and new solutions without too many constraints.

6 Conclusion

The exploration of the literature on interoperability has shown a lack of completeness on the concept of interoperability. We have proposed a context-independent and in-depth definition of the concept of interoperability in 7 layers that can serve as a basis for a formal theory of interoperability. Applied to Open Science, this proposal highlights the needs to enable data exchange. A potential solution to achieve the goal of data crossing in Open Science is the use of metadata schema matching tools. The evaluation of matching tools proposed in the literature has shown a lack of efficiency of these tools on the metadata of datasets found in Open Science. This shows that these tools are not the solution to the whole Open Science data crossing problem at the moment. But they can be used as one brick in the overall solution. We have proposed 3 directions to improve interoperability and move towards this goal of data circulation. The next step of our work will focus on exploring the solution of decentralised data management. Through a decentralised approach, avenues to overcome the problems of scaling manual interoperability mechanisms and taking into account the great diversity of metadata schema, solutions and sources of data and metadata management are possible. Through this approach, it is possible to approach all the directions proposed to move towards data interoperability in Open Science.

References

1. Alasem, A.: An overview of e-government metadata standards and initiatives based on Dublin core. Electron. J. e-Gov. **7**(1), 1–10 (2009)
2. Auer, S., Bizer, C., Kobilarov, G., Lehmann, J., Cyganiak, R., Ives, Z.: DBpedia: a nucleus for a web of open data. In: Aberer, K., et al. (eds.) ASWC/ISWC -2007. LNCS, vol. 4825, pp. 722–735. Springer, Heidelberg (2007). https://doi.org/10.1007/978-3-540-76298-0_52
3. Bengio, Y., Ducharme, R., Vincent, P.: A neural probabilistic language model. In: Advances in Neural Information Processing Systems, vol. 13 (2000)
4. Diallo, S.Y., et al.: Understanding interoperability. In: Proceedings of the 2011 Emerging M&S Applications in Industry and Academia Symposium, pp. 84–91 (2011)
5. Do, H.H., Rahm, E.: Coma-a system for flexible combination of schema matching approaches. In: VLDB 2002: Proceedings of the 28th International Conference on Very Large Databases, pp. 610–621. Elsevier (2002)
6. Embley, D.W., Jackman, D., Xu, L.: Multifaceted exploitation of metadata for attribute match discovery in information integration. In: Workshop on Information Integration on the Web, pp. 110–117 (2001)
7. Fallatah, O., et al.: A gold standard dataset for large knowledge graphs matching. In: Ontology Matching 2020: Proceedings of the 15th International Workshop on Ontology Matching co-located with the 19th International Semantic Web Conference (ISWC 2020), vol. 2788, pp. 24–35. CEUR Workshop Proceedings (2020)
8. Gittens, A., Achlioptas, D., Mahoney, M.W.: Skip-gram- zipf+ uniform= vector additivity. In: Proceedings of the 55th Annual Meeting of the Association for Computational Linguistics (Volume 1: Long Papers), pp. 69–76 (2017)
9. Gonçalves, R.S., Kamdar, M.R., Musen, M.A.: Aligning biomedical metadata with ontologies using clustering and embeddings. In: Hitzler, P., et al. (eds.) ESWC 2019. LNCS, vol. 11503, pp. 146–161. Springer, Cham (2019). https://doi.org/10.1007/978-3-030-21348-0_10
10. He, Y., Chen, J., Antonyrajah, D., Horrocks, I.: BERTMap: a BERT-based ontology alignment system. In: Proceedings of the AAAI Conference on Artificial Intelligence, vol. 36, no. 5, pp. 5684–5691 (2022). https://doi.org/10.1609/aaai.v36i5.20510
11. Heiler, S.: Semantic interoperability. ACM Comput. Surv. (CSUR) **27**(2), 271–273 (1995)

12. Hertling, S., Portisch, J., Paulheim, H.: MELT - matching EvaLuation toolkit. In: Acosta, M., Cudré-Mauroux, P., Maleshkova, M., Pellegrini, T., Sack, H., Sure-Vetter, Y. (eds.) SEMANTiCS 2019. LNCS, vol. 11702, pp. 231–245. Springer, Cham (2019). https://doi.org/10.1007/978-3-030-33220-4_17
13. Jacobsen, A., et al.: Fair principles: interpretations and implementation considerations (2020)
14. Koutras, C., et al.: Valentine: evaluating matching techniques for dataset discovery. In: 2021 IEEE 37th International Conference on Data Engineering (ICDE), pp. 468–479. IEEE (2021)
15. Kusner, M., et al.: From word embeddings to document distances. In: International Conference on Machine Learning, pp. 957–966. PMLR (2015)
16. Li, Z., et al.: Temporal knowledge graph reasoning based on evolutional representation learning. In: Proceedings of the 44th International ACM SIGIR Conference on Research and Development in Information Retrieval, pp. 408–417 (2021)
17. Madhavan, J., Bernstein, P.A., Rahm, E.: Generic schema matching with cupid. In: VLDB, vol. 1, pp. 49–58 (2001)
18. Massmann, S., Engmann, D., Rahm, E.: COMA++: results for the ontology alignment contest OAEI 2006. Ontol. Matching **225** (2006)
19. Mikolov, T., Chen, K., Corrado, G., Dean, J.: Efficient estimation of word representations in vector space. arXiv preprint arXiv:1301.3781 (2013)
20. Nilsson, M., Baker, T., Johnston, P.: Interoperability levels for Dublin core metadata. Technical report, Dublin Core Metadata Initiative (2008). https://www.dublincore.org/specifications/dublin-core/interoperability-levels/
21. Noura, M., et al.: Interoperability in internet of things: taxonomies and open challenges. Mob. Netw. Appl. **24**(3), 796–809 (2019)
22. Rahm, E., Bernstein, P.A.: A survey of approaches to automatic schema matching. VLDB J. **10**(4), 334–350 (2001)
23. Rehurek, R., Sojka, P.: Software framework for topic modelling with large corpora. In: Proceedings of the LREC 2010 Workshop on New Challenges for NLP Frameworks. Citeseer (2010)
24. National Academies of Sciences, Engineering, and Medicine: Open science by design: Realizing a vision for 21st century research. National Academies Press, Washington DC (2018). https://doi.org/10.17226/25116. https://nap.nationalacademies.org/catalog/25116/open-science-by-design-realizing-a-vision-for-21st-century
25. Tolk, A., Diallo, S.Y., Turnitsa, C.D.: Applying the levels of conceptual interoperability model in support of integratability, interoperability, and composability for system-of-systems engineering. J. Syst. Cybern. Inform. **5**(5) (2007)
26. Van Der Veer, H., Wiles, A.: Achieving technical interoperability. European Telecommunications Standards Institute (2008)
27. Vicente-Saez, R., Martinez-Fuentes, C.: Open science now: a systematic literature review for an integrated definition. J. Bus. Res. **88**, 428–436 (2018)
28. Wegner, P.: Interoperability. ACM Comput. Surv. (CSUR) **28**(1), 285–287 (1996)
29. Weibel, S., Kunze, J., Lagoze, C., Wolf, M.: Dublin core metadata for resource discovery. Technical report, IETF Request for Comments (1998). http://www.ietf.org/rfc/rfc2413.txt
30. Wilkinson, M.D., et al.: The fair guiding principles for scientific data management and stewardship. Sci. Data **3**(1), 1–9 (2016)
31. Willis, C., Greenberg, J., White, H.: Analysis and synthesis of metadata goals for scientific data. J. Am. Soc. Inform. Sci. Technol. **63**(8), 1505–1520 (2012)
32. Zeng, M.L.: Interoperability. KO. Knowl. Organ. **46**(2), 122–146 (2019)

Forum Papers

Online-Notes System: Real-Time Speech Recognition and Translation of Lectures

Tjaša Jelovšek(✉), Marko Bajec, Iztok Lebar Bajec, Kaja Gantar, and Slavko Žitnik

Faculty of Computer and Information Science, University of Ljubljana, Ljubljana, Slovenia
tjasa.jelovsek@fri.uni-lj.si

Abstract. Student mobility gives students the opportunity to visit different universities across the world. Not all courses are offered in English or in other languages foreign students might understand, so they often face problems with following the lectures. To resolve these problems, we propose Online Notes (ON), which is a real-time speech recognition and translation system. The system is trained using existing course materials. During lectures, the lecturer wears a microphone, while the students can follow the lecture by using the ON system. After the lecture, the professor can edit and update the transcripts and translations, and students have the option to listen to the lecture and read the materials. We have conducted a series of one-time tests and currently, we are in the middle of two whole-semester pilot tests at the University of Ljubljana. In the tests, a speech recognition accuracy of up to 87.4% was achieved. Preliminary results have shown that the tool is especially useful for students who either do not understand the language of the course or understand it to a limited extent. Additionally, the transcripts of the lectures have shown to be useful for creating additional learning materials.

Keywords: speech recognition · machine translation · real-time lecture translation

1 Introduction

The use of real-time speech recognition and translation has been increasing in recent years. Platforms, such as Skype and Zoom, already offer language interpretation to their users, which enables users to overcome any language barriers during meetings, courses, etc. Such technologies are useful in many scenarios, including in the educational sector. In 2021, there were 1,275 exchange students at the University of Ljubljana [1]. Since not all courses are available in English, these students are faced with limited options. In addition, there is a considerable number of full-time students whose first language is not Slovene, mainly from the Balkan region. These students often understand Slovene to a limited extent and are in the process of learning Slovene. Such students may benefit both from

S. Nurcan et al. (Eds.): RCIS 2023, LNBIP 476, pp. 485–492, 2023.
https://doi.org/10.1007/978-3-031-33080-3_29

the transcription as well as from the translations, as they can learn the language through written text as well as through comparison of the text with the English translation.

Online Notes is a system that offers real-time transcriptions and translations during and after lectures. It is designed to help students overcome the language barrier and follow the lectures in real time. We believe the system also benefits students with sensory impairments, particularly those who are hard of hearing. We have started to collaborate with experts to make the system as useful as possible for students with sensory impairments. Additionally, the transcriptions and translations can be useful learning material for students, as well as for creating additional learning materials for the lecturers.

The paper is organized as follows: in the next chapter, we briefly present similar systems. In Sect. 3, we present the system architecture, its main components, and setup, and in Sect. 4, the experiments are presented.

2 Related Work

Many universities in non-English speaking countries face similar problems with not being able to provide courses to foreign students, or only providing a limited number of English courses, and hence, not being attractive for foreign students due to the language barrier.

Similar systems exist and have been tested in other university settings already. To the best of our knowledge, these are the most recent systems with similar use-cases. A very similar system is the Karlsruhe Institute of Technology (KIT) Lecture Translation System [2,4] which offers transcriptions and machine translations of German lectures in the form of subtitles. What was observed in their evaluation is that most students who used the system found it helpful, especially the transcript, but among negative aspects, the latency of the lecture translation was reported, as well as the fact that it is hard to follow the slides and the lecture translation at the same time [4]. One of the participants in the study also pointed out that the system was particularly useful for students that are learning German.

Shadiev and Huang [6] investigated the use of speech recognition and machine translation in the case of foreign lecturers that speak in English to students to whom English is a foreign language. They investigated the cognitive load, attention, and meditation when listening to a lecture in a foreign language, and established that students who followed the lectures with the machine translations in their first language had the lowest cognitive load and highest satisfaction level in comparison to students who followed the lecture with English speech recognition only and students who had no speech technology support at all.

3 System Architecture and Setup

The system is implemented as a web application and is available for lecturers and students. The system architecture can be seen in Fig. 1. The process starts with sound recording on the frontend (1). Frontend is implemented in the React[1] framework. Frontend then sends data to the backend (2) through a Websocket. Backend is implemented in Java using the Spring[2] framework. Backend then sends the audio to the speech recognition model (3) through the gRCP[3] protocol. The speech recognition module is based on the Kaldi toolkit [5]. Once the speech recognition component returns the raw recognized text to the backend, the text is sent into the punctuator (4) through a REST API. The punctuator returns the punctuated text which is then sent to the translator (5), also through a REST API. Both texts are displayed on the web page (1). All data is stored in a Postgres database (6). Active directory is used for user authentication (7).

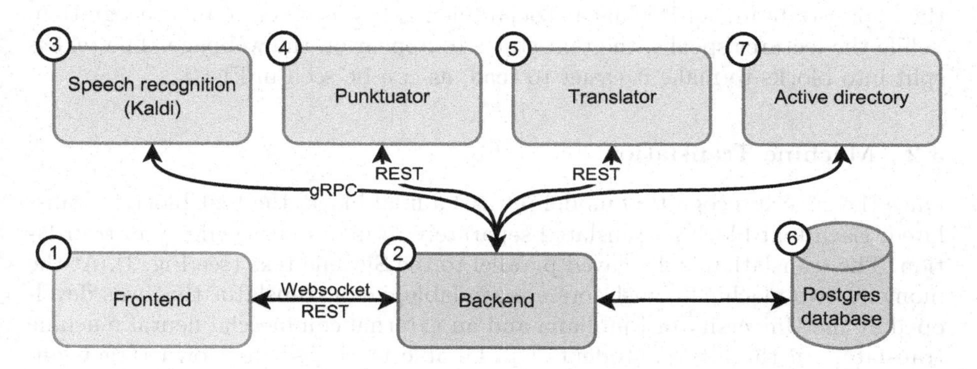

Fig. 1. System Architecture

3.1 Speech Recognition

The speech recognition model is based on the Kaldi toolkit [5] which uses Weighted Finite-State Transducers (WFST) for training and decoding. The first step in the speech recognition pipeline is waveform-reading and feature extraction from the audio which is presented in a form of a Mel spectrogram. This data is then processed by the acoustic model, which returns a probability matrix of subword units over time. This probability matrix is then decoded together with a language model, which returns text without any punctuation. The text is then sent to the punctuator (a separate service), which inserts punctuation into the text.

At the moment, two separate speech recognition models are used within the ON system, namely a model for social science lectures and a model for technical science lectures. As in the Kaldi framework, the acoustic model and the language model can be built independently, the two speech recognition models share a common acoustic model. The base for the language model is a text

[1] https://reactjs.org/.
[2] https://spring.io/.
[3] https://grpc.io/.

corpus. The content of the lectures is usually highly specialized, therefore it is essential that the language model is adjusted to the lecture. To do so, lecturers are asked to send us any audio or video recording of their lectures from previous years. These recordings are automatically transcribed and manually corrected. In rare occurrences where no material is available, other materials related to the topic are used, i.e. articles, journals, theses, etc. After the subcorpus of a lecture is prepared, it is added into one of the two main corpora, i.e. technical or social science corpus. So far, 4 subcorpora have been added to the social sciences model, and 8 subcorpora to the technical model. Once the text corpus is compiled, we prepare a lexicon of tokens with their corresponding pronunciations that match the chosen subword unit while training. During the live session, the appropriate model (technical/social sciences) is used for speech recognition. While the lecturer speaks, the text starts to appear on the webpage. The text is split into blocks to make it easier to read, as can be seen in Fig. 3.

3.2 Machine Translation

Once the speech recognition model returns a final block, the text block is translated. Each text block is translated separately to minimize the delay in translation. The translation is displayed parallel to the Slovene text (see Fig. 3). At the moment, two machine translators are available, i.e. a translator that was developed at the University of Ljubljana and an external commercial neural machine translator. In the future, students will be able to choose their own target language when using an external machine translator.

3.3 Setup

The system is implemented as a web application and is quite simple to use (see Fig. 2). Before the lecture, the lecturer signs in with a PIN number, turns on the microphone, and opens the streaming webpage. Once everything is set, they start recording. As soon as they start streaming, they can see the output of speech recognition in Slovene. Students log in using their university identity through Active directory and select the lecture they wish to follow.

During the lecture, the students can access the system through the webpage on their own devices (laptops, smartphones, or tablets). Once the session is live, they can read the Slovene transcription and/or the English translation of the transcripts that are displayed side-by-side. They cannot, however, listen to the audio during the live session. The Slovene transcription is written out as the lecturer speaks, so the students can also see the interims, and is split into blocks of texts to make the content easier to follow. Once the Slovene transcription is considered final, the text block is translated. The speaker can pause the system at any time during the lecture. One of the features that is being explored is including an external speech synthesizer to generate audio of the translation, which would give foreign students the option to listen to the translation instead of reading it, and thus enable them to focus on the lecturer and the presentation.

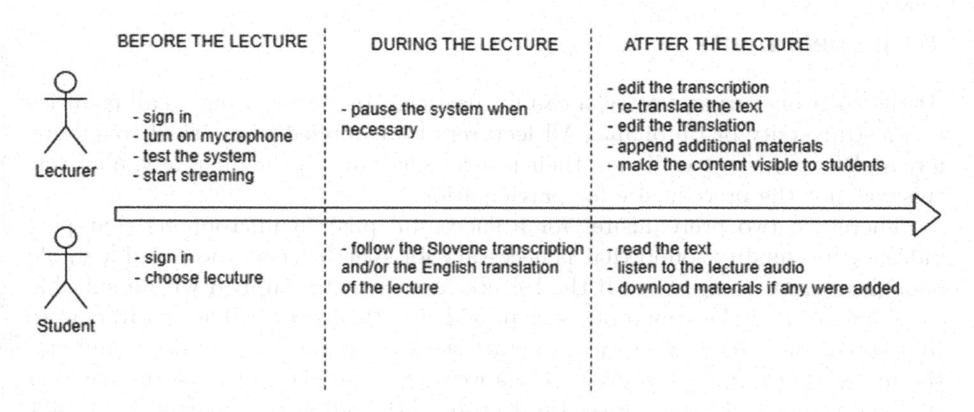

Fig. 2. Funcionalities by user role

Once the lecture is finished, the lecturer stops the stream and the live session is no longer available. After the live session, the speaker can view the content (the Slovene transcription and the English translation), and they can listen to the audio of the lecture. They can edit both the Slovene transcription and the English translation, or re-translate the text. They can publish (i.e. make it visible for students inside the system) the text once they are satisfied with it. They can also append additional materials, such as PDF files, PowerPoint presentations, etc. After the lecture has been published, the students can see both texts side by side, and listen to the audio of the lecture, as seen in Fig. 3. They can also download the additional materials if any were uploaded by the lecturer.

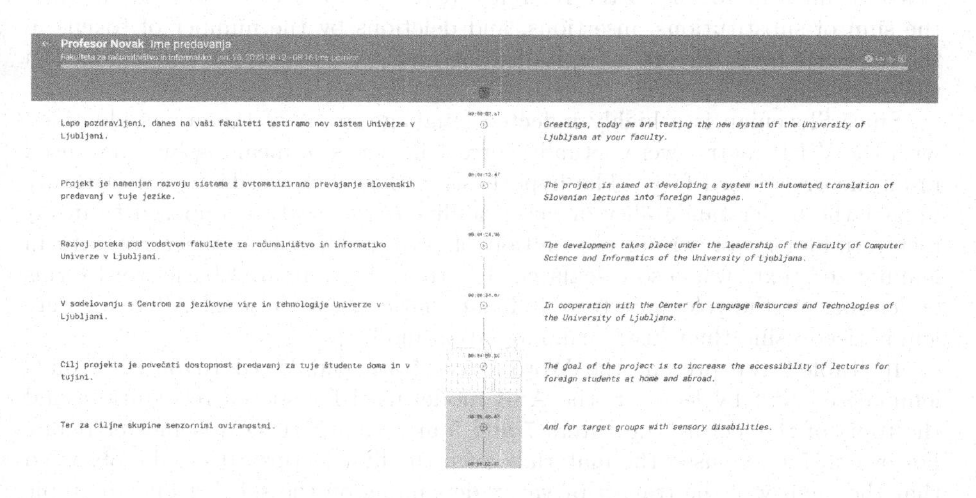

Fig. 3. Display of transcript and translation after the session.

4 Experiments

At the start of every semester, a call for participants was sent out to all faculties at the University of Ljubljana. All lecturers that applied were asked to submit any audio or video recording of their previous lectures, if they had any, however, this was not the prerequisite for participation.

There are two prerequisites for a successful pilot: a microphone that provides a good quality sound that is necessary for speech recognition, and a stable enough internet connection. If the lecture room is not equipped with a suitable microphone, a wireless microphone is provided to the lecturer. The lecturers were told how to use the system (how to start streaming, pause the system, and end the live session). No other instructions were given to the lecturers on changing the way they speak or organize the lecture, although some reported later that they adapted the structure of the lecture. At the beginning of each lecture, the system was briefly presented and students were given the option to use it on their own devices. Students log in with their university ID through Active directory (see Fig. 1).

So far, 12 tests have been carried out, encompassing 8 faculties at the University of Ljubljana. Of that, two are whole semester tests. In this section, we present the results of automatic speech recognition evaluation. For the evaluation of speech recognition, we used the Word Error Rate (WER) metric, which calculates the ratio of errors in the automatic transcription in relation to reference texts, that were made manually. We differentiate between three types of errors, i.e. insertions (tokens that are added in the automatic transcript and are not present in the reference text), deletions (tokens that are omitted from the transcript), and substitutions (different tokens that appear in the transcript in place of different tokens in the reference text). WER is calculated by dividing the sum of substitutions, insertions, and deletions by the number of tokens in the reference texts. Errors in capitalization and punctuation are not included in the evaluation.

Since Slovenian is a highly inflected language, some of the errors identified with the WER metric were actually correct in terms of meaning but were used in an incorrect word form. The hypothesis is that such words do not critically impact the understandability or acceptability of the text, as opposed to incorrect terminology or wrong recognition of words. Therefore, the error rate in lemmatized texts was also calculated. For that, we compared basic word forms or lemmas in the reference text and automatic transcriptions. All texts were lemmatized using the Classla lemmatization model [3].

In Table 1, we present the Word Error Rate and Word Error Rate on a lemmatized text by lecturer, the ASR model used for speech recognition, and the topic of the lecture. Lecturers 7 and 9 are participating in semester pilots. For evaluation, we used the materials from the first lecture. It can be observed that the quality of the transcript varies depending on the speaker and the topic of the lecture, with the lowest WER of 12.7%, and the highest WER of 31.3%. As can be seen from Table 1, an average drop of 5.5% can be observed when calculating the error rate on lemmatized texts, and the highest drop of 12.8%

in the case of Lecturer 12, which again points to the possible explanation that the quality of speech recognition is perhaps lower due to tokens that are not included in the vocabulary, while a different word form of the same lemma is.

The main factors that affect the quality of speech recognition are the following. It is necessary that the sound quality is as good as possible. This also includes background noise or speaking in the background. Perhaps the most important factor in the quality of the output text is the quantity (and quality) of the materials received for building the language model for speech recognition. If the text corpus that is used for updating the language model is not significant enough, terminology may be missing, hence, it cannot be recognized, as in the case of Lecturers 11 and 12, for example, where only a limited amount of material was received. The quality of speech recognition is slightly lower if the lecturer uses non-standard words, and even more so if they use words in a foreign language (most common are words in English).

We observe best results when the speech is clear and well structured, without prolonged pauses, and the language model (i.e. the materials received for building the language model) contains all necessary terminology. We believe that speech recognition will improve further with the size of the corpus, and by including additional support for non-standard pronunciations in the vocabulary.

Table 1. Error rate by meaning by lecture

Lecturer	ASR Model	Field	WER	Lemmatized WER
Lecturer 1	Social	Social Sciences	0.126	0.092
Lecturer 2	Technical	Chemistry	0.127	0.091
Lecturer 3	Social	Law	0.141	0.094
Lecturer 4	Technical	Statistics	0.143	0.095
Lecturer 5	Technical	Economics	0.163	0.110
Lecturer 6	Technical	Computer Science	0.177	0.127
Lecturer 7	Technical	Computer Science	0.181	0.137
Lecturer 8	Technical	Multimedia	0.190	0.136
Lecturer 9	Technical	Statistics	0.195	0.145
Lecturer 10	Social	Social Sciences	0.199	0.141
Lecturer 11	Technical	Veterinary medicine	0.214	0.152
Lecturer 12	Social	Sociology of economics	0.313	0.186

5 Conclusion

The Online Notes system offers real-time speech recognition and machine translation into English in a university setting. This enables exchange students, or other students that understand Slovene to a limited extent or not at all to attend the lectures in Slovene. It also enables students with sensory impairments to follow the lectures. The quality of speech recognition varies depending on the field and the speaker, but it is mostly affected by the size of the language model and the materials that were used to update the language model. In the tests where only a limited amount of materials were included in the language model, the word error rate is much higher (up to 31.3%, while the best WER in our tests was 12.6%). Word error rate was also calculated on the lemmatized text where an average drop of 5.5% can be observed in WER, with the best WER of 9.1% and the highest of 18.6%. The best results are achieved when the lecturer speaks clearly and the speech and well structured, without prolonged pauses. It is also essential that the language model contains all the necessary terminology. The language models are periodically updated, i.e. the corpora are extended and new terminology is added. We can already observe progress in the accuracy of speech recognition in optimal conditions.

Acknowledgements. The research presented in this paper received funding from the University of Ljubljana, and partial support from the research project titled Basic Research for the Development of Spoken Language Resources and Speech Technologies for the Slovenian Language (J7-4642) funded by the Slovenian Research Agency.

References

1. Univerza v številkah. https://www.uni-lj.si/o_univerzi_v_ljubljani/univerza_v_stevilkah/. Accessed 02 July 2023
2. Dessloch, F., et al.: KIT lecture translator: multilingual speech translation with one-shot learning. In: Proceedings of the 27th International Conference on Computational Linguistics: System Demonstrations, pp. 89–93. Association for Computational Linguistics, Santa Fe, New Mexico, August 2018. https://aclanthology.org/C18-2020
3. Ljubešić, N.: The CLASSLA-StanfordNLP model for lemmatisation of standard croatian 1.2 (2020). http://hdl.handle.net/11356/1357. slovenian language resource repository CLARIN.SI
4. Müller, M., Fünfer, S., Stüker, S., Waibel, A.: Evaluation of the KIT lecture translation system. In: Proceedings of the Tenth International Conference on Language Resources and Evaluation (LREC'16). European Language Resources Association (ELRA), Portorož, Slovenia, May 2016
5. Povey, D., et al.: The Kaldi speech recognition toolkit. In: IEEE 2011 Workshop on Automatic Speech Recognition and Understanding. IEEE Signal Processing Society, December 2011. IEEE Catalog No.: CFP11SRW-USB
6. Shadiev, R., Huang, Y.M.: Investigating student attention, meditation, cognitive load, and satisfaction during lectures in a foreign language supported by speech-enabled language translation. Comput. Assist. Lang. Learn. **33**(3), 301–326 (2020)

Temporal Relation Extraction from Clinical Texts Using Knowledge Graphs

Timotej Knez(✉) and Slavko Žitnik

Faculty of Computer and Information Science, University of Ljubljana, Ljubljana, Slovenia
{timotej.knez,slavko.zitnik}@fri.uni-lj.si

Abstract. An integral task for many natural language processing applications is the extraction of the narrative process described in a document. For understanding such processes we need to recognize the mentioned events and their temporal component. With this information, we can understand the sequence of events i.e. construct a timeline. The main task dealing with the temporal component of events is temporal relation extraction. The goal of temporal relation extraction is to determine how the times of two events are related to one another. For example, such relation would tell us whether one event happened before or after another one. In this paper, we propose a novel architecture for a temporal relation extraction model combining text information with information captured in the form of a temporal event graph. We present our initial results on the domain of clinical documents. Using a temporal event graph with only correct relations, the model achieves F1 score of 83.6% which is higher than any of our state-of-the-art baseline models. This shows the promise of our proposed approach.

Keywords: natural language processing · temporal relation extraction · common knowledge injection · clinical text · knowledge graph

1 Introduction

Extraction of temporal information about events in a document is important as it allows us to understand the presented narrative. The most common temporal information extracted about events are their temporal relations. The task of temporal relation extraction is to identify temporal relations between events described in a text document. Such relations tell us whether the first event occurred before or after the second event. We propose a new architecture for temporal relation extraction that combines information from a text document with additional information captured in the form of a knowledge graph.

In our work, we focus on temporal relation extraction from medical documents, as the automatic understanding of the sequence of events could enable many further applications. For example, we could use such a method for finding patterns

S. Nurcan et al. (Eds.): RCIS 2023, LNBIP 476, pp. 493–500, 2023.
https://doi.org/10.1007/978-3-031-33080-3_30

in symptoms and treatment progression. Such sequences are also useful in tasks like clinical dead-end prediction and disease progression detection. We train and test our model on medical discharge summaries contained in the i2b2 2012 corpus [10]. We recognise the before, after, and overlap relations that are captured in the corpus. We recognise the relations as a three-class classification problem.

The rest of this paper is organised as follows. In Sect. 2 we present existing models for temporal relation extraction as well as some multi-modal approaches used in other similar domains. In Sect. 3 we present the proposed architecture of our model. After that in Sect. 4 we present our preliminary results that show the promise of using a bimodal architecture for temporal relation extraction. Finally, in Sect. 5 we summarise our research and present the future directions of our work.

2 Related Work

Most of the models proposed in recent years in the area of temporal relation extraction are based on deep neural network architectures. The two main architectures used for the task are the use of long short-term memory (LSTM) layers and the use of pretrained language models. The models based on the LSTM architecture work by passing the initial embeddings of the tokens in a document through the LSTM network and using the state of the network at each token as its contextualized embedding. Tourille et al. [11] and Cheng and Miyao [2] both presented a model that extracts the relations using a feed-forward network on the combination of embeddings from the first and second event. The embeddings are computed using an LSTM network, while the difference between the approaches is that Cheng and Miyao use the sequence of tokens based on the dependency paths in a sentence instead of using consecutive words in a sentence.

Recently the best-performing models for temporal relation extraction are based on the pretrained language models. The most commonly used such model is Bidirectional Encoder Representations from Transformers (BERT). The model presented by Lin et al. [4] works by first marking the events with special tokens and passing the text through a BERT network. In addition to the contextualized token embeddings, the BERT network also computes an embedding that represents the entire sentence. The final classification gets performed using a multi-layer perceptron over the sentence embedding. Zhou et al. [13] improve the results, by using a similar network architecture and adding soft logic regularization, that ensures consistency between predicted relations. This approach looks at the probabilities of each relation and constructs consistent combinations that get the highest combined probability. Lin et al. [5] also perform temporal relation extraction using an architecture based on a BERT pretrained language model. They improve results over previous approaches by using a new training strategy to train a new version of a BERT model called EntityBERT. EntityBERT is trained by masking entity mentions in medical documents. Compared to more general approaches, where we mask random tokens, this focuses the training on terms that are important for tasks such as temporal relation extraction. In our research, we use the EntityBERT language model to recognise relations based on text as Lin et al. [5] showed good results using it. The main difference between the model used by Lin et al. and the text part of our model is that our model uses event embeddings to classify temporal relations, while Lin et al. use sentence embeddings.

2.1 Improving Relation Extraction Using Additional Knowledge

The main goal of our research is to use general knowledge in the process of temporal relation extraction to improve accuracy. The possibility of improving temporal relation extraction has been already demonstrated in some existing research. Ning et al. [8] introduce general knowledge to the model by including statistics about the relations from a large corpus. They developed a statistical resource that contains probabilities for each relation between common events in the news domain. They generated the statistical resource based on the news articles from New York Times over a 20-year span. In their later study [7] they show that the introduction of the statistical resource improves temporal relation extraction results by 3%. A limitation of such an approach is that it only uses global information and ignores the already extracted relations from the current document. We believe that information about other relations from the same document is very important. We propose an alternative approach where we store all of the general information in a knowledge graph. We then generate event embeddings based on the structure of the knowledge graph and use them for extracting relations.

2.2 Combining Text Data with Knowledge Graphs

Introducing a knowledge graph as a source of additional information for a model analyzing text is challenging as analyzing text requires a different architecture than analyzing a knowledge graph. Some existing models already combine information from a knowledge graph with information from text for certain domains [3, 6, 12]. Recently such approaches were used on domains similar to temporal relation extraction. Lin et al. [6] created a model combining a relational knowledge graph with text for extracting relations between diseases in medical documents. They use a SciBERT model for encoding text, while the knowledge graph gets encoded using a heterogeneous graph attention network. The model computes the final classification by combining the embeddings from the knowledge graph with the ones from the text. A similar idea is also explored by Yasunaga et al. [12], who proposed a general pretrained language model named DRAGON that combines information from text with information from a knowledge graph. Unlike the previous approach, DRAGON combines the information from both modalities while creating the token and graph node embeddings. By integrating the information from both sources earlier in the model, the model can learn more complex ways of combining the information. On the other hand, by computing two separate sets of embedding and only combining them at the end, the model proposed by Lin et al. is able to use only a single modality when necessary. For example, if the diseases are not present in the knowledge graph, the model can predict the relation solely based on text.

3 Model Architecture

In order to improve temporal relation extraction, we introduce additional common knowledge into the classification process. The knowledge is presented in the form of an event graph that contains already extracted relations between events. The event graph is a graph structure, where events are represented as nodes and the temporal relations are represented as directed edges in the graph. We store each event under its UMLS [1] identifier which enables linking the events in the text to the ones represented in the graph. The idea of using existing relations as additional knowledge is that they capture statistics about which relations are common between certain event types. The existing relations also enable the model to learn rules that hold between neighbouring relations, like for example transitivity.

We present a multi-modal architecture for temporal relation extraction presented in Fig. 1. Our proposed architecture first constructs two sets of event embeddings. First, the embeddings of the first and second events are computed based on the text input using a pretrained language model. Second, the model computes another set of relations based on the knowledge graph input. The model then produces a classification based on each event embedding pair and finally combines both classifications into a single relation prediction. Our architecture is similar to the one presented by Lin et al. [6]; however, we use a different architecture for encoding the graph input which is optimized for recognising temporal relations between events. We also use EntityBERT for encoding text compared to SciBERT which they use as EntityBERT was specifically tuned for tasks concerning medical entities and events. An example of how a relation is extracted is shown in Fig. 2.

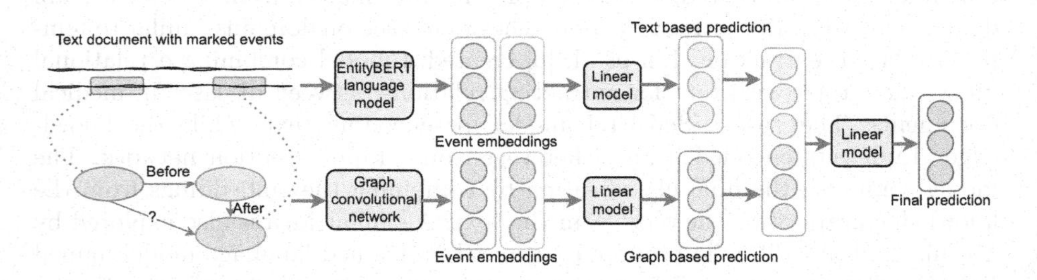

Fig. 1. A diagram showing the model architecture.

A large advantage of such architecture is that we can also perform classification using only one of the input types when only a single one is available. For example, if we only have text without the knowledge graph, our model is still capable of classifying temporal relations.

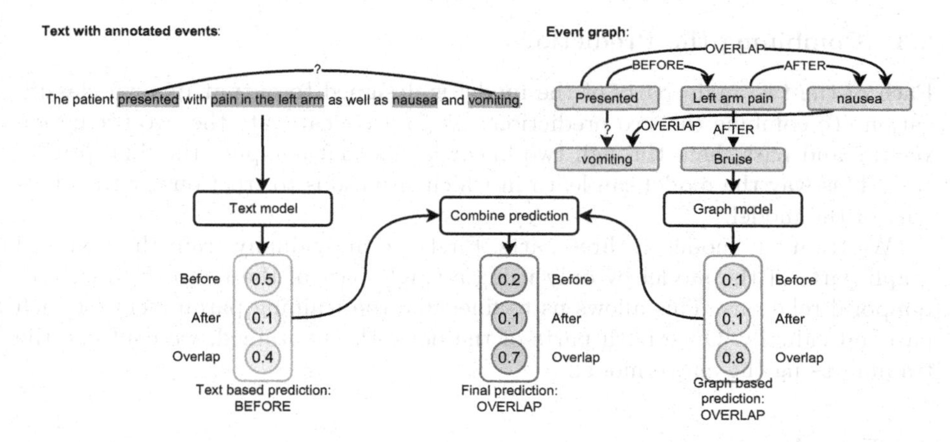

Fig. 2. An example of how recognition of a relation is performed. We are recognising the relation between events presented and vomiting.

3.1 Text Embeddings

The part of the model responsible for extracting information from text is structured in the following way. In the text, we first mark the beginning and end of each event using special tokens. After that, we compute token embeddings using the EntityBert pretrained model [5]. The model was trained on a general task focused on recognising entities and events in the medical literature. We compute the event embeddings by averaging the embeddings of all tokens corresponding to an event. This gives us a single embedding for each event that contains information about the event and its role in the text. We combine the embeddings of both events by concatenating the two vectors to create a single vector. We pass the resulting vector through a linear neural layer which classifies the vector into one of the three temporal relation types. We found that using this approach achieves better results on our dataset than the process described by Lin et al. [5].

3.2 Graph Embeddings

The second part of the model is responsible for classifying relations based on the knowledge graph. To achieve this we first use a graph neural network to compute event embeddings. The observed knowledge graph contains events and already extracted temporal relations around the events we are exploring. We initialize the embedding of each event using the GloVe embeddings [9]. We then apply three layers of graph convolution that aggregate event embeddings across temporal relations. By doing so each event embedding contains information about its related events and the relevant relations. The embeddings of the observed events get concatenated and passed through two linear layers that perform the classification.

3.3 Combining the Predictions

Each of the two main parts of the model is designed to output its own classification. To combine the two predictions, we first concatenate the two prediction vectors and pass them through two linear layers that compute the final prediction. This way the model can learn in which situations to trust one or the other part of the model.

We train the model in three parts. First, we individually train the text and graph parts of the model by only using a single part of the network to extract temporal relations. This allows us to fine-tune the training parameters for each part individually. Once both parts of the network are trained, we continue the training using the entire model.

4 Results

We performed some preliminary tests of the proposed model on the i2b2 2012 [10] dataset. For training, we used part of the i2b2 2012 training split, while we used the rest of the data as validation. For testing the performance of the model we used the i2b2 2012 test split. We performed the preliminary evaluation on three scenarios. In the first, the temporal relations are predicted based only on the text modality. This configuration shows how the model performs if no additional knowledge is available. Second, we test the performance of the model using only the knowledge graph as an input. This result tells us how well the relations can be predicted as a knowledge graph completion task. Finally, we test the entire model with both modalities used at the same time. This shows us how well both predictions can be combined to form a single more reliable prediction. All of the results are presented in Table 1. As a baseline, we use the two state-of-the-art models presented by Zhou et al. [13]. The CTRL model is based on a BERT model with a single linear layer for classification. CTRL-PG is the same model with the addition of soft logic regularization.

Table 1. Temporal relation extraction results achieved on the i2b2 2012 test set using the best-case scenario event graph as a source of additional knowledge.

Testing scenario	F1 score
Text only model	72.9%
Graph only model	78.4%
Combined predictions	**83.6%**
State-of-the-art baseline models	
CTRL [13] (BERT based model)	78.6%
CTRL-PG [13] (Soft logic regularization)	80.2%

4.1 Constructing the Event Graph for Temporal Relation Extraction

The goal of using the event graph in the process of temporal relation extraction is that we formulate relation extraction as a knowledge graph completion task. For relations extracted in this way, the most important events and relations are the ones extracted from the same document. As a proof of concept, we used event graphs constructed from all relations in the document except for the relation we are extracting. This kind of knowledge graph simulates the best-case scenario, where we already have access to all other relations in a document, but is unrealistic for practical use. In practice, we could gather the relations by first predicting them using the text model and then constructing the event graph. This would mean, that some of the relations in the event graph would contain errors likely lowering the results.

4.2 Discussion

In Table 1 we see that the performance of the model significantly improved (11% improvement) with the addition of the event graph compared to only using the text. However, it has to be noted that to achieve this result, the model had access to an event graph containing all but the target relations from the document, which is not realistic for real-world use. In most real-life applications, the model would not have access to any relations from the observed document. To enable such use-case, we could first build the event graph using the text model and then perform the relation extraction using the constructed graph. The achieved results show promise for future research in the area of using such temporal event graphs to aid temporal relation extraction.

5 Conclusion

We present a new model architecture for temporal relation extraction. The model uses a pretrained language model to predict temporal relations from text. In addition to the information from the text, it also uses information about other relations captured in a knowledge graph. Using preliminary tests, we have shown that under ideal conditions the proposed architecture significantly outperforms existing models. This result shows that the proposed approach has the potential to achieve new state-of-the-art results on the task of temporal relation extraction. The current limitation of the proposed approach is that it relies on a knowledge graph to contain correct relations between events. In real-world scenarios, the relations would likely contain errors, as they would come from previously extracted information. In our future research, we will improve the real-world usability of the proposed approach and evaluate it on multiple scenarios.

Acknowledgements. This work has been financially supported by the Slovenian Research Agency in the young researchers grant.

References

1. Bodenreider, O.: The unified medical language system (UMLS): integrating biomedical terminology. Nucleic Acids Res. **32**(Database-Issue), 267–270 (2004)
2. Cheng, F., Miyao, Y.: Classifying temporal relations by bidirectional LSTM over dependency paths. In: Proceedings of the 55th Annual Meeting of the Association for Computational Linguistics (Volume 2: Short Papers), pp. 1–6 (2017)
3. Koloski, B., Perdih, T.S., Robnik-Sikonja, M., Pollak, S., Skrlj, B.: Knowledge graph informed fake news classification via heterogeneous representation ensembles. Neurocomputing **496**, 208–226 (2022)
4. Lin, C., Miller, T., Dligach, D., Bethard, S., Savova, G.: A BERT-based universal model for both within-and cross-sentence clinical temporal relation extraction. In: Proceedings of the 2nd Clinical Natural Language Processing Workshop, pp. 65–71 (2019)
5. Lin, C., Miller, T., Dligach, D., Bethard, S., Savova, G.: EntityBERT: entity-centric masking strategy for model pretraining for the clinical domain. Association for Computational Linguistics (ACL) (2021)
6. Lin, Y., Lu, K., Yu, S., Cai, T., Zitnik, M.: Multimodal learning on graphs for disease relation extraction. arXiv preprint: arXiv:2203.08893 (2022)
7. Ning, Q., Subramanian, S., Roth, D.: An improved neural baseline for temporal relation extraction. In: Proceedings of the 2019 Conference on Empirical Methods in Natural Language Processing and the 9th International Joint Conference on Natural Language Processing (EMNLP-IJCNLP), pp. 6203–6209 (2019)
8. Ning, Q., Wu, H., Peng, H., Roth, D.: Improving temporal relation extraction with a globally acquired statistical resource. In: Proceedings of the 2018 Conference of the North American Chapter of the Association for Computational Linguistics: Human Language Technologies, Volume 1 (Long Papers), pp. 841–851 (2018)
9. Pennington, J., Socher, R., Manning, C.D.: Glove: global vectors for word representation. In: Proceedings of the 2014 Conference on Empirical Methods in Natural Language Processing (EMNLP), pp. 1532–1543 (2014)
10. Sun, W., Rumshisky, A., Uzuner, O.: Evaluating temporal relations in clinical text: 2012 i2b2 challenge. J. Am. Med. Inform. Assoc. **20**(5), 806–813 (2013)
11. Tourille, J., Ferret, O., Neveol, A., Tannier, X.: Neural architecture for temporal relation extraction: A Bi-LSTM approach for detecting narrative containers. In: Proceedings of the 55th Annual Meeting of the Association for Computational Linguistics (Volume 2: Short Papers), pp. 224–230 (2017)
12. Yasunaga, M., et al.: Deep bidirectional language-knowledge graph pretraining. arXiv preprint: arXiv:2210.09338 (2022)
13. Zhou, Y., et al.: Clinical temporal relation extraction with probabilistic soft logic regularization and global inference. In: Proceedings of the AAAI Conference on Artificial Intelligence (AAAI2021) (2021)

Domain TILEs: Test Informed Learning with Examples from the Testing Domain

Niels Doorn[1(✉)], Tanja Vos[1,2], Beatriz Marín[2], Christoph Bockisch[3], Steffen Dick[3], and Erik Barendsen[1]

[1] Open Universiteit, Heerlen, The Netherlands
niels.doorn@ou.nl
[2] Universitat Politècnica de València, Valencia, Spain
[3] Philipps-Universität, Marburg, Germany

Abstract. Test Informed Learning with Examples (TILE) helps educators to add testing to their programming courses early, easily and in a subtle way. Currently, TILE describes how to add informed examples of testing to test runs, test cases, and messages. In this paper, we extend TILE by incorporating information from the testing domain itself into the examples. Our non-conclusive results from a survey with 300 participants indicate that using TILE, results in students creating more tests that cover more parts of the code.

Keywords: Software Testing · Repository · Example-based learning

1 Introduction

Software testing is a critical to assess whether the program works as intended. Despite the importance, testing is often taught late in the computer science curricula [1,3,7,15]. Students typically encounter software testing after learning programming, leading to a lack of appreciation for its necessity as prior programming exercises were completed without testing.

The use of worked examples is a common approach in programming education as it effectively demonstrates concepts and improves cognitive skills acquisition [16]. Test Informed Learning with Examples (TILE) [9] promotes test awareness in programming courses through the use of exercises with testing-related examples. The goal of these examples is to provide students with a better understanding of software testing concepts, to illustrate how testing works in real-world scenarios, and to develop software testing skills. Currently, TILE encourages adding informed testing examples to test runs, cases, and messages. This paper presents a new variant of TILE: the *domain* TILE that employs exercises that utilise information from the testing domain.

The paper is structured as follows. First, we describe some relevant related works. Secondly, we present the existing types of TILEs. Then, we introduce the domain TILEs, followed by the results of a study to understand the effects on test-awareness of students. Finally, our conclusions are presented.

S. Nurcan et al. (Eds.): RCIS 2023, LNBIP 476, pp. 501–508, 2023.
https://doi.org/10.1007/978-3-031-33080-3_31

2 Related Work

Several studies [4,5,7,13] emphasise the importance of teaching testing techniques at the outset of a computer science education, ideally in introductory programming courses. It has been widely recognised that worked examples play a crucial role in building a foundational understanding of testing knowledge [2,14].

One approach to promoting early exposure to testing is Test-Driven Development (TDD) [8,12]. In TDD, students follow an iterative process in which they must define test cases, that at the first instance should fail, before writing just enough code to make those test cases pass. Although the name might suggest that the focus of TDD is on testing, it can be seen as a software design philosophy [6]. Despite these benefits, research has shown that students may struggle to create effective test cases due to a lack of programming experience and testing knowledge [11,17].

The TILE approach [9] focuses on helping students develop conceptual knowledge about testing while simultaneously learning programming concepts. TILE is distinct from TDD as it aims to foster a deeper understanding of testing principles without the need for initial programming knowledge. Moreover, it is independent from any development approach. Another difference is the type of testing addressed with TDD and TILE. TDD has a strong focus on unit and regression testing. With TILE, in principle all aspects of testing can be taught, in particular with domain TILEs.

3 Test Informed Learning with Examples (TILE)

TILE currently uses three ways of using examples that revolve around testing in exercises: **test run** TILEs, **test cases** TILEs, and **test message** TILEs.

Test Run. TILEs consist of slightly changing the words we use when explaining what happens when we run a program. For example, imagine we are explaining a computer program that solves a given problem or does some calculations. Let us look at the following commonly used type of instruction: *let us run this program, the user gives input and the results are shown on the screen.* – and compare it with an instruction using a 'Test run TILE': – *let us test this program by running it and entering test input data and checking the resulting output on the screen.* Although this is a very subtle difference in telling the students what we are going to do, it has the benefit of making them test aware by introducing them to a form of testing: test your program by running it with test input data to check whether the resulting output is as expected.

Test Case. TILEs use concrete examples of possible test cases to exercises to nudge students into testing their code. These example test cases should not just include the 'happy path' tests, but should also include corner cases and other less happy tests. This can be done in various ways. One way is to add examples test cases, or of test executions, such that students can use and repeat these for their own program. Another way is to ask questions to force students to think about good test cases for the exercise.

Test Message. TILEs contain subliminal messages about importance of testing. For example a MadLib with a predefined text about the importance of testing, or an example input file with an explanation about a testing subject.

4 Test Domain TILEs

When teaching students certain concepts, examples are given using a familiar domain. The concept of inheritance is for example commonly explained using domains as shapes, animals, cars, fruits, etc. Using commonly known relations, we can make the inheritance relationship of a hierarchy clearer for students. In a later stage, when the concept is more consolidated and the domain does not impact the educational concepts being taught, instructors can replace these examples with those from the testing domain. This way, besides practising the intended concept, students also learn from the domain. For this reason, we extend the TILE approach with "Test Domain TILEs". Let us look at some examples.

- *Writing a simple expert system* is a well-known exercise in introductory programming. In Fig. 1 we can see two versions of such an exercise. The left-hand version of this exercise is part of a Python course offered at the bachelor's degree level by one of the authors. It uses fruits as its domain. We could transform this into the domain of testing, as shown in the version on the right-hand side. This way, students learn about testing and programming at the same time.
- *Format strings to buy beers or to estimate testing time.* Let us compare two other similar exercises. By using the testing domain as is shown in Fig. 2, we can get students acquainted with the names of different testing activities and make them aware that these can make up 50% of development time.
- *Searching through files* and teaching file manipulation use sample files that students must open, process, and then close again. For example, in the open PY4E book[1], Fig. 3 shows a sample of a file is given that records e-mail activity from a development team[2]. To understand the process of searching through a file, the book explains a common pattern to read through a file and only process lines which meet a particular condition, for example, lines that start with the prefix "From:", we can use the Python string method `startswith` to select only those lines with that prefix. To TILE this exercise to the testing domain, we could give the students a sample file that contains the results from a test execution. For example, the one in Fig. 3 (bottom). This way, the exercise still consists of reading a file, but now we only print out lines which started with the prefix "testcase =:" because those contain the identifiers of the test cases that failed. To make sure the students test there solution, we can give them various text files that come from different tests as possible test cases. Again, students learn something about testing while practising with files.
- *Using dictionaries to make histograms* is a typical exercise in Python programming. Students are asked to traverse some data and create a dictionary

[1] https://www.py4e.com.

[2] The complete file with the mail interactions is available from https://edu.nl/vvtw8.

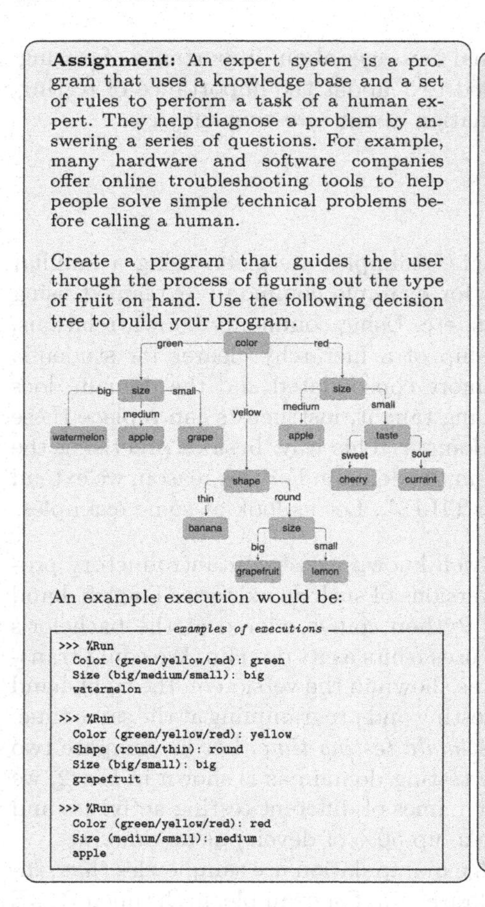

Assignment: An expert system is a program that uses a knowledge base and a set of rules to perform a task of a human expert. They help diagnose a problem by answering a series of questions. For example, many hardware and software companies offer online troubleshooting tools to help people solve simple technical problems before calling a human.

Create a program that guides the user through the process of figuring out the type of fruit on hand. Use the following decision tree to build your program.

An example execution would be:

examples of executions

```
>>> %Run
 Color (green/yellow/red): green
 Size (big/medium/small): big
 watermelon

>>> %Run
 Color (green/yellow/red): yellow
 Shape (round/thin): round
 Size (big/small): big
 grapefruit

>>> %Run
 Color (green/yellow/red): red
 Size (medium/small): medium
 apple
```

Assignment: An expert system is a program that uses a knowledge base and a set of rules to perform a task of a human expert. They help diagnose a problem by answering a series of questions. For example, many hardware and software companies offer online troubleshooting tools to help people solve simple technical problems before calling a human.

Now imagine Ana wrote a program that sorts a list of numbers. Evidently, this needs to be tested. Create a program that guides her through the process of deciding whether she has tested sufficiently in the following way:

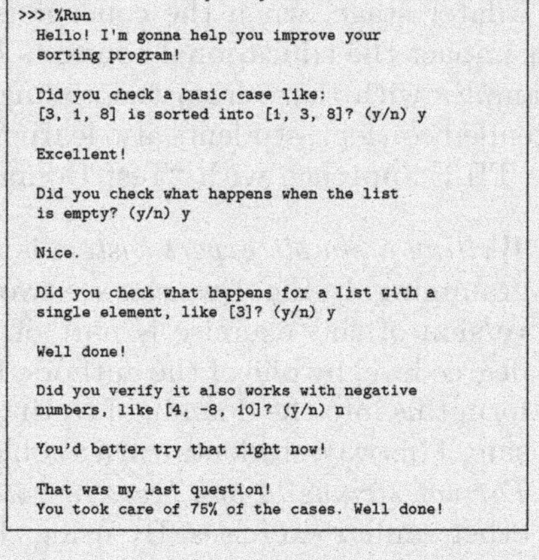

examples of executions

```
>>> %Run
 Hello! I'm gonna help you improve your
 sorting program!

 Did you check a basic case like:
 [3, 1, 8] is sorted into [1, 3, 8]? (y/n) y

 Excellent!

 Did you check what happens when the list
 is empty? (y/n) y

 Nice.

 Did you check what happens for a list with a
 single element, like [3]? (y/n) y

 Well done!

 Did you verify it also works with negative
 numbers, like [4, -8, 10]? (y/n) n

 You'd better try that right now!

 That was my last question!
 You took care of 75% of the cases. Well done!
```

Try to extend the program with possible cases that test other important things (at least two are useful to add).

Fig. 1. Two similar exercises: one on the left-hand side uses fruits as a domain, the one on the right-hand side uses testing as a domain.

that contains the frequencies of some element in the data. To TILE exercises of this type, the data that needs to be traversed can be related to testing. For example, in pytest, a fixed set of result types are possible for running automated test: PASSED, FAILED, SKIPPED, XFAIL, XPASS or ERROR. We could give the students a file containing the results of a test suite run with pytest and ask them to make a histogram that shows the frequencies of each type of test result.

The objective of the exercises stays the same, practising with dictionaries and making histograms. Whether this is about counting first-names or about test results is irrelevant for the learning objective. However, when we use testing as the domain, students also learn about testing. Being creative, we can think of a myriad of possibilities to introduce testing in subtle ways using these type of TILEs.

Assignment: We want to format a shopping list for a party with the following products: beer (2 €), wine (3 €) and chips (1.15 €). You have to ask the user for the quantities they want to buy (we assume that they are always < 100 units). The program must return the shopping list of the purchase formatted as in the following example execution:

example of execution

```
>>> %Run
  How much beer? 99
  How much wine? 23
  How many bags of chips? 1
  -------------------------
  Total purchase
  -------------------------
  Beer        99   198.00
  Wine        23    69.00
  Chips       01     1.15
                   ------
          Total    268.15
```

Assignment: We want to format a summary of the time spend in different phases of a testing process using the historical data from our company indicating that, from the whole time dedicated to a software project, 30% is spent on unit testing, 5% on system testing and 15% on acceptance tests. Your program should ask the user for the amount of time (in hours) that was spent on the software project, and then it must return the summary formatted as in the following example execution:

example of execution

```
>>> %Run
  How many hours for the whole project? 1455
  --------------------------------------
  Summary of the testing activities
  --------------------------------------
  unit testing            436 hours 30 min
  system testing           72 hours 45 min
  acceptance testing      218 hours 15 min
                          ----------------
               Total      727 hours 30 min
```

Fig. 2. Two similar exercises: one on the left-hand side uses buying beer as the domain and one on the right-hand side uses time spent on testing as the domain.

part of an example of an e-mail interaction

```
From stephen.marquard@uct.ac.za Sat Jan 5 09:14:16 2008
    Return-Path: <postmaster@collab.sakaiproject.org>
Date: Sat, 5 Jan 2008 09:12:18 -0500
To: source@collab.sakaiproject.org
Subject: [sakai] svn commit: r39772 -content/branches/
Details: http://source.sakaiproject.org/viewsvn/?view=rev&rev=39772 ...
```

example of pytest output

```
==================================== FAILURES ===================================
____________________ test_union[4-input13-input23-output3] _____________________
testcase = 4, input1 = [1, 1], input2 = [], expected_output = [1]
    @pytest.mark.parametrize("testcase, input1, input2, expected_output",[
    .....
    ])
    def test_union(testcase, input1, input2, expected_output):
E       AssertionError: caso 4
E       assert [1, 1] == [1]
union_test.py:23: AssertionError
____________________ test_union[7-input16-input26-output6] _____________________
testcase = 7, input1 = [1, 1, 2, 2, 3, 3], input2 = [], expected_output = [1, 2, 3]
    @pytest.mark.parametrize("testcase, input1, input2, expected_output",[
    .....
    ])
    def test_union(testcase, input1, input2, expected_output):
E       AssertionError: caso 7
E       assert [1, 1, 2, 2, 3, 3] == [1, 2, 3]
union_test.py:23: AssertionError
========================= 2 failed, 6 passed in 0.07s ==========================
```

Fig. 3. Sample files that can be used for an exercise

5 Using TILE — A Report on Our Experience

We used TILE in two runs of a first-semester course on *Object-Oriented Programming* (OOP) for the computer science bachelor at the *Philipps University of Marburg*. This 15-week course consists of two lectures per week and 12 mandatory, graded exercises that lead up to a written final exam. The final exam includes a diagnostic task on software quality and testing.

Assignment: The Fibonacci-sequence begins with the numbers: $\{1, 1, 2, 3, 5, 8, \ldots\}$. By means of the following recurrence relation, it is possible to compute the Fibonacci-number at the position $n \in \mathbb{N}$:

$$F_n = \begin{cases} 1 & n = 1 \\ 1 & n = 2 \\ F_{n-1} + F_{n-2} & otherwise \end{cases} \qquad P_n = \begin{cases} 3 & n = 0 \\ 0 & n = 1 \\ 2 & n = 2 \\ P_{n-2} + P_{n-3} & otherwise \end{cases}$$

Given the following method, which implements this recurrence relation. Assume that this method is part of a reusable library. It could, e.g., be part of the class `java.lang.Math`.

code snippet 2020

```
int fib(int n){
  if(n==1){ return 1; }
  if(n==2){ return 1; }
  return fib(n-1) + fib(n-2);
}
```

code snippet 2021

```
public static int perrin(int n) {
  if (n == 0) { return 3; }
  else if (n == 1) { return 0; }
  else if (n == 2) { return 2; }
  else {
    return perrin(n-2) + perrin(n-3);
  }
}
```

a) The quality of the given implementation of the method `fib/perrin` has some flaws. Name one of these flaws and outline how to improve the code. If possible, use examples.[a]

b) Write a meaningful number of meaningful tests for the method `fib/perrin` and specify whether the implementation passes or fails each test. Write tests as JUnit-Tests. **Hint**: You can assume that all required classes are on the Classpath and all required imports and static imports are specified.

[a] Possible answers include: missing JavaDoc comments, inefficiency due redundant computation of the same Fibonacci-Number in the course of the recursion, violating the style guide or missing parameter validation.

Fig. 4. The assignments added to the exams

The assignments in the exams of both runs had identical wording but utilised different code snippets, as illustrated in Fig. 4.

Although the participants in both runs met the same prerequisites and had a similar background, we improved the course. While this may have affected the study's validity, we still compared the results to the previous cohort. The code snippets in both exams use a recursive implementation of a tree-based algorithm. The first code snippet was the Fibonacci sequence, while the second was the Perrin sequence, constructed in a similar manner as the previous. Due to the difference in termination cases, the Perrin sequence allows for the easier creation of more test cases compared to the Fibonacci sequence. Figure 5 depicts the result of our findings between the two runs. We examined the answers of the students using the four criteria shown in Fig. 5 and counted the occurrences within their answers.

We also noticed that far fewer students forgot to annotate their tests with `@Test`. That number fell by nearly 30%. Another improvement could be seen in that more students tested for corner cases. Although a small improvement, that number fell from nearly 7% to about 1%. Although these are preliminary

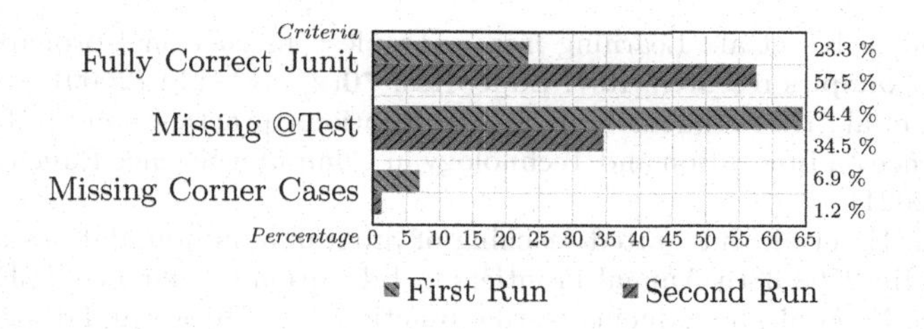

Fig. 5. Correctness of unit tests of both exams. The bar in blue depicts the results *using* the TILE approach, whereas the bar in orange depicts the run *without* using the TILE approach. (Color figure online)

results, and we have more data to be evaluated, we did already see an overall improvement of the students' performance in the area of software testing.

6 Conclusion

TILE is an approach to improve students' test awareness in programming courses by using examples related to testing. This paper presents a new type of TILE in which we use a context related to software testing. We call these exercises' domain TILEs. We hypothesise that using a testing domain as a context will improve the student's testing awareness.

We have created an open repository of exercises on GitHub [10] to share assignments that are adapted or developed using the TILE approach.

We have performed an exploratory empirical evaluation to analyse the test-awareness produced in students when TILE is used. We compared two different runs of the same programming course, one run without using TILE and one run with using TILE. Results show that students of the course *with* TILE wrote more test cases of better quality and with more coverage than the students of the course *without* TILE.

Our exploratory study shows preliminary results that TILE can be effective to produce more test-awareness in students in introductory programming courses. Nevertheless, we are conscious that more research is needed to evaluate TILE. A follow-up experiment is planned to gain knowledge of the effectiveness of TILE.

Acknowledgements. The work leading to this paper has received funding from the EU Erasmus+ projects European innovation alliance for testing education (101055874-ENACTEST-ERASMUS-EDU-2021-PI-ALL-INNO) and Quality-focused Programming Education (2020-1-NL01-KA203-064626).

References

1. Aniche, M., et al.: Pragmatic software testing education. In: Proceedings of the 50th ACM Technical Symposium on Computer Science Education. SIGCSE 2019, Minneapolis, MN, USA. Association for Computing Machinery (2019)

2. Atkinson, R.K., et al.: Learning from examples: instructional principles from the worked examples research. Rev. Educ. Res. **70**(2), 181–214 (2000)
3. Bai, G., et al.: How students unit test: perceptions, practices, and pitfalls. In: 26th Conference on Innovation and Technology in Computer Science Education ITiCSE. ACM (2021)
4. Barbosa, E., et al.: Integrated teaching of programming foundations and software testing. In: 2008 38th Annual Frontiers in Education Conference. IEEE (2008)
5. Barbosa, E., et al.: Introducing testing practices into objects and design course. In: 16th Conference on Software Engineering Education and Training, 2003. (CSEE&T 2003). IEEE (2003)
6. Beck, K.: Test-Driven Development: By Example. Addison-Wesley Professional, Boston (2003)
7. Desai, C., et al.: A survey of evidence for test-driven development in academia. ACM SIGCSE Bull. **40**(2), 97–101 (2008)
8. Desai, C., et al.: Implications of integrating test-driven development into CS1/CS2 curricula. ACM SIGCSE Bull. **41**(1), 148–152 (2009)
9. Doorn, N., et al.: Set the right example when teaching programming: test informed learning with examples (TILE). In: 16th IEEE International Conference on Software Testing, Verification and Validation (ICST) (2023)
10. Doorn, N., et al.: Test informed learning with examples assignments repository. https://tile-repository.github.io/
11. Isomöttönen, V., Lappalainen, V.: CSI with games and an emphasis on TDD and unit testing: piling a trend upon a trend. ACM Inroads **3**(3), 62–68 (2012)
12. Janzen, D.S., Saiedian, H.: Test-driven learning. intrinsic integration of testing into the CS/SE curriculum. In: SIGCSE 2006 (2006)
13. Janzen, D., Saiedian, H.: Test-driven learning in early programming courses. In: 39th SIGCSE Technical Symposium on Computer Science Education (2008)
14. Renkl, A.: Toward an instructionally oriented theory of examplebased learning. Cogn. Sci. **38**(1), 1–37 (2014)
15. Tuomi, P., Multisilta, J., Saarikoski, P., Suominen, J.: Coding skills as a success factor for a society. Educ. Inf. Technol. **23**(1), 419–434 (2017). https://doi.org/10.1007/s10639-017-9611-4
16. van Merrienboer, J.: Training Complex Cognitive Skills: A Four-component Instructional Design Model for Technical Training. Educational Technology Publications, Englewood Cliffs (1997)
17. Whalley, J.L., Philpott, A.: A unit testing approach to building novice programmers' skills and confidence (2011)

Using GUI Change Detection for Delta Testing

Fernando Pastor Ricós[1(✉)], Rick Neeft[2], Beatriz Marín[1], Tanja E. J. Vos[1,2], and Pekka Aho[2]

[1] Universitat Politècnica de València, València, Spain
fpastor@pros.upv.es
[2] Open Universiteit, Heerlen, The Netherlands

Abstract. Current software development processes in the industry are designed to respond to rapid modification or changes in software features. Delta testing is a technique used to check that the identified changes are deliberate and neither compromise existing functionality nor result in introducing new defects. This paper proposes a technique for delta testing at the Graphical User Interface (GUI) level. We employ scriptless testing and state-model inference to automatically detect and visualize GUI changes between different versions of the same application. Our proposed offline change detection algorithm compares two existing GUI state models to detect changes. We present a proof of concept experiment with the open-source application Notepad++, which allows automatic inference and highlights GUI changes. The results show that our technique is a valuable amplification of scriptless testing tools for delta testing.

Keywords: Delta testing · Scriptless testing · State-model inference · GUI change detection

1 Introduction

Agile methods [1,17] have evolved into an efficient software development process that emphasizes collaboration between Development and Operations (DevOps) teams and leverages Continuous Integration (CI) tools to automate the build and test stages [6]. This shift towards rapid-release deployment leads to many software versions and needs a comprehensive delta testing approach to maintain the high quality of the software product.

Software updates and new versions often involve multiple changes to enhance features or fix issues. Delta testing checks that the changes are intentional and do not compromise existing functionality or result in introducing new defects. In this paper, we concentrate on delta testing at the software's Graphical User Interface (GUI), the primary end-users entry point to interact with applications.

Existing tools perform change detection at the code level [15], compare GUI changes to repair test scripts [4], or are restricted to comparing data differences between individual documents [11]. However, none of these tools intend to highlight how GUI transitions evolve. Therefore, we aim to develop a novel tool that

S. Nurcan et al. (Eds.): RCIS 2023, LNBIP 476, pp. 509–517, 2023.
https://doi.org/10.1007/978-3-031-33080-3_32

helps enrich GUI change detection of delta testing to safeguard against software failures that might negatively affect the end-user experience.

Our solution employs scriptless GUI testing and state-model inference to automatically detect and visualize GUI changes when testing incremental releases of the Software Under Test (SUT). In order to do that, we present an offline change detection algorithm that compares previously inferred GUI state models to detect how the GUI evolves. This helps to determine whether the detected changes are intentional modifications or unintended failures.

We use the scriptless testing tool TESTAR [22] to infer different state models from different versions of a SUT. Subsequently, we apply the new change detection mechanism to compare these state models. The results show that the highlighted GUI changes correspond to the updates made in the SUT.

As a proof of concept, we will use two different versions of the open-source application Notepad++, and check whether our solution reports the changes found in the Notepad++ repository changelog.

The rest of this paper is structured as follows. Section 2 presents the related work and background. Section 3 introduces the GUI state model inference of the TESTAR tool. Section 4 describes the Notepad++ application used for the proof of concept evaluation. Section 5 explains the change detection solution. Finally, Sect. 6 summarizes the conclusion and future work.

2 Related Work

2.1 State of the Art in GUI Change Detection

For Android systems, tools, such as CAT [19] and ATUA [15], detect changes in the code and use them to improve the testing effectiveness by prioritizing the interactions with GUI elements that invoke the changed code. To do that, they identify changes in the source code of the SUT and map them to possible interactions on the GUI to generate test cases. CAT uses a JSON file to manually indicate changed methods, while ATUA compares source code versions to identify changes. Rather than analyzing changes in the code, our approach utilizes information from the GUI to identify differences between versions of the SUT.

ATOM [10] and CHATEM [4] are Android testing tools that aim to maintain test scripts as the GUI changes. They compare delta Event Sequence Models (ESMs) of two SUT versions to compute alternative event paths with intermediate states to repair broken test paths affected by GUI changes. The construction of these ESMs requires manual effort to build or evaluate the model behavior. By contrast, we do not try to calculate alternative paths to repair model transitions but highlight how these GUI transitions change. Furthermore, we try to reduce manual effort by using GUI state models that are inferred automatically.

For web systems, there are change detection tools [11] developed to monitor and notify changes on web pages. These tools use crawling techniques [9] to traverse over the graph-hyperlink structure of a web and save the HTML data in unstructured documents. Changes are detected by comparing different document versions to detect and highlight HTML data changes (i.e., text, images, videos, etc.). These tools [8,21] detect changes on individual HTML pages but do not analyze how the GUI evolves. Using state model information allows highlighting the GUI transitions that occur when the end-user interacts with the SUT.

To summarize the current state of the art, Android code-change and GUI-change analysis tools do not aim to highlight GUI changes, web change detection tools focus their comparison algorithms on web document data without considering GUI state model transitions, and there is a lack of active research tools for detecting GUI changes for desktop applications. For this reason, we consider it necessary to evaluate the active tools capable of inferring a GUI desktop, web, and mobile state models and build a change detection mechanism on top of these models. Accordingly, our contribution aims to employ state models inferred from the GUI to highlight to software engineers how the GUI evolves over different versions of the SUT. This enables us to provide valuable insights on changes that may have occurred in the GUI, which can be useful in guiding development efforts and ensuring consistent user experience across different software versions.

2.2 Background of Model-Inference GUI Tools

Augusto [13] and Autoblacktest [12] are automated testing tools capable of inferring GUI event flow graphs. However, these tools are not actively maintained and are restricted to desktop applications.

APE tool [7] uses the runtime information observed during testing to build and refine a GUI model gradually. AMOGA [20] is a strategy that uses a crawling algorithm to generate UI behavior models. Although these tools have recent studies about building models, their system is restricted to android applications.

GUITAR [16] is a model-based testing framework that supports a plugin-based architecture. This allows the tool to be extended with multiple plugins to obtain the hierarchy of the GUI elements and build GUI event model graphs for desktop, web, and mobile systems. Nevertheless, the tool has not been actively maintained or evaluated in the industry in recent years.

TESTAR [22] is an active scriptless testing open-source tool that integrates multiple plugins to connect and test different types of systems. Windows Automation API, Java Access Bridge, Selenium WebDriver, and Appium allow TESTAR to test desktop, web, and mobile applications through the GUI. The tool has been evaluated in the industry [3,18] and uses an inference technique to automatically infer a GUI state model at run-time when testing a SUT [14]. For these reasons, we decided to use the state model information of the TESTAR tool as a basis to research and develop a novel GUI change detection software.

3 GUI State Model Inference with TESTAR

TESTAR connects with the SUT and obtains the state s of the GUI. The state consists of a hierarchical set of widgets w, known as a *widget tree*, that are present in the GUI state s. All widgets contain a set of properties that indicates the position, title, role, etc., of each GUI element. TESTAR generates two types of GUI state identifiers based on the properties of widgets. The first is the concrete state, which includes all the properties of all widgets in the widget tree. The second is the abstract state, which can be configured to include only a subset of widget properties. With TESTAR, testers can select which properties to include in the abstract state. This feature enables greater flexibility and customization.

Consequently, also actions have two types of identifiers. The *concrete action identifier* takes the concrete state identifier and includes the coordinates of the interacted widget, the role of the action (i.e., click or type), and the specific text if the action includes typing. And the *abstract action identifier* takes the abstract state identifier and includes the hierarchical index of the interacted widget and the role of the action without considering the typed text.

When TESTAR selects and executes an action a in a state s, the GUI will reach a new state s'. This transition $s \rightarrow a \rightarrow s'$ is stored in a graph database where the state model is created. Transitions are stored in two main model layers: abstract and concrete. Because the abstract identifiers come from a subset of widgets and their properties, abstract and concrete states are mapped in a surjective manner. Each concrete state is associated with a unique abstract state, and an abstract state can be composed of multiple concrete states.

When TESTAR finishes the testing process and the GUI state model is inferred, it is possible to apply offline oracles [5]. The GUI change detection follows this same offline underlying idea. After TESTAR finishes the run-time testing process, the GUI change detection software will use the inferred GUI state model information to detect and highlight the GUI changes.

4 SUT for Proof of Concept Evaluation

Notepad++[1] is a popular open-source code editor that supports multiple programming and natural languages. The GitHub repository contains around 18.000 stars, 2.000 open issues, 7.000 closed issues, and 3.000 closed pull requests. The source code has over 5.000 daily evolving commits from over 300 contributors.

The repository contains a change-log section that documents the bug fixes and new features of each Notepad++ version release[2]. The incremental release from Version 8.4.5 (2022-09-07) to version 8.4.6 (2022-09-29) contains various GUI changes in the menu items and settings panels. This proof of concept evaluation aims to infer a partial TESTAR state model as accurately as possible for the menu options and settings configuration panels of these two Notepad++ releases and validate that the change detection tool can detect GUI changes.

5 Change Detection Solution

Our ChangeDetection[3] solution compares different state models that were automatically generated by TESTAR on different versions of the software. Let us consider having two state models, SM_{old} and SM_{new}, for two different versions of the

[1] https://notepad-plus-plus.org/.
[2] https://github.com/notepad-plus-plus/notepad-plus-plus/wiki/Changes.
[3] https://github.com/TESTARtool/ChangeDetection.NET.

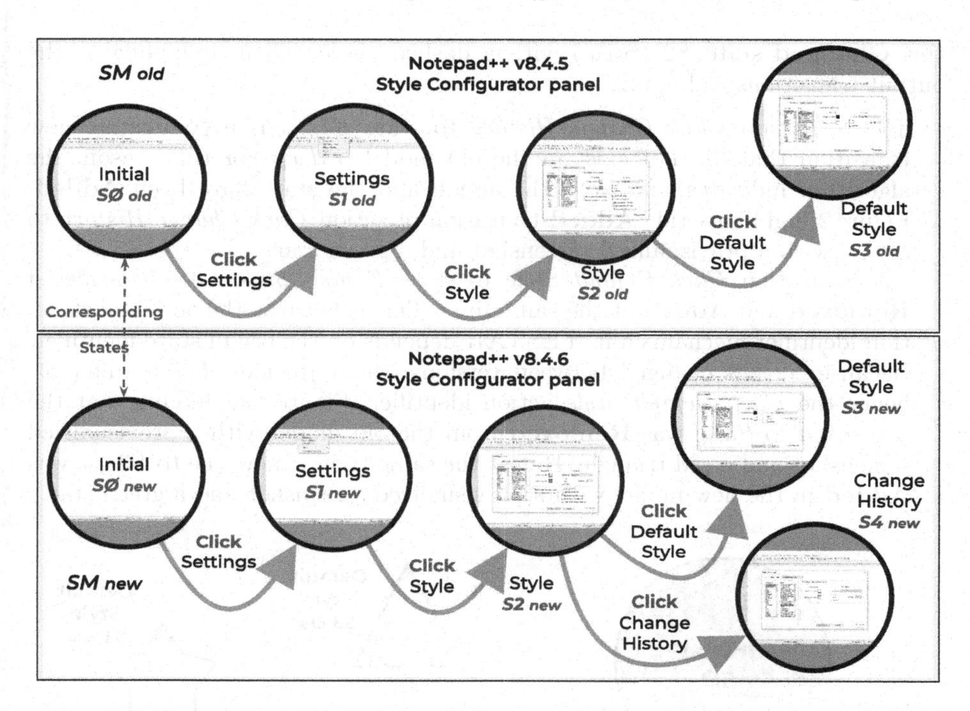

Fig. 1. Notepad++ style configurator panel models from v8.4.6 and v8.4.5

SUT. The fundamental idea of the underlying change detection algorithm is to take transitions $s \rightarrow a \rightarrow s'$ simultaneously in both SM_{old} and SM_{new} while comparing the properties of the destination states s'. This comparison automatically transits, detects, and marks the differences between the properties that were changed, added, or removed. Next, we use a merged graph technique [2] that helps to visualize the matching and changed states, as well as the transitions that were added in the new model SM_{new} and removed from the old model SM_{old}.

Figure 1 shows partial state models of two incremental versions of Notepad++. The algorithm starts with the *Initial States* ($S\emptyset_{old}$ and $S\emptyset_{new}$) and creates an association called *corresponding states* between $S\emptyset_{old}$ and $S\emptyset_{new}$. Then, the algorithm iterates over the existing actions a from $S\emptyset_{new}$. Since action *Click Settings* also exists in $S\emptyset_{old}$, the algorithm transits to the destination state *Settings* ($S1$) and updates the corresponding states association between $S1_{old}$ and $S1_{new}$. Because there are no changes in the corresponding states $S1_{old}$ and $S1_{new}$, the algorithm continues.

In the next iteration, the algorithm continues, for example, with action *Click Style* from $S1_{new}$. In both models, a transition to destination state *Style* ($S2$) is made, updating the corresponding states association between $S2_{old}$ and $S2_{new}$. At this point, the algorithm detects a change in the corresponding states $S2_{old}$ and $S2_{new}$, i.e., a widget list element *Change History margin* has been added to the panel, which provokes the GUI of $S2_{new}$ to be changed. Figure 2 highlights this **Changed** state $S2$ (*Style*) with a dashed circle. Then, it highlights the output transitions as follows:

- The transition *Click Change History* to state $S4_{new}$ is a completely new transition that did not exist in the old model SM_{old}. For this reason, the algorithm indicates that both the action and the state have been **Added**. Figure 2 highlights this **Added** transition of action *Click Change History* to $S4_{new}$ with a previsualized screenshot and a green star.
- The transition *Click Default Style* to state *Default Style* $S3$ is considered **Removed** and **Added** at the same time. This is because the action abstraction identifier mechanism in TESTAR depends on the origin state identifier. Because $S2$ has changed between versions, the state identifier is different, hence the *Click Default Style* action identifier. Figure 2 highlights that the transition to $S3_{old}$ was **Removed** from the old model with a previsualized screenshot and a red triangle. But at the same time, a new one to $S3_{new}$ was **Added** in the new model with a previsualized screenshot and a green star.

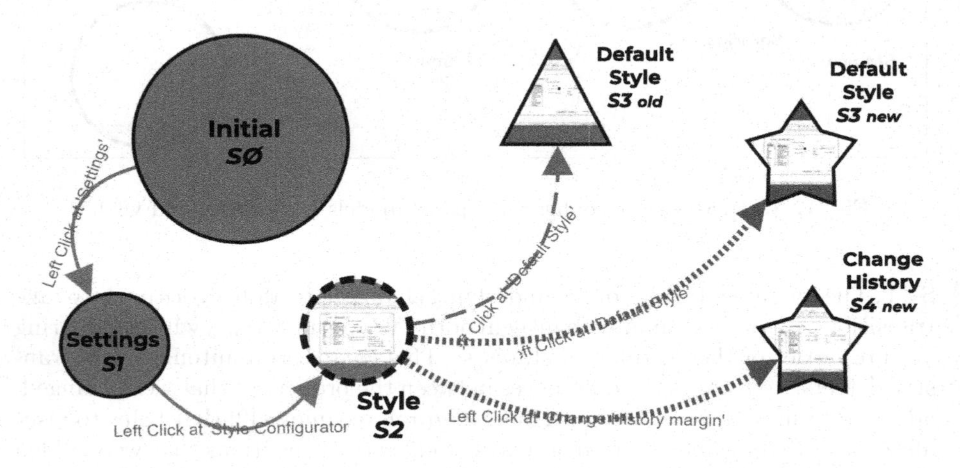

Fig. 2. Merged graph of Notepad++ comparing v8.4.6 with v8.4.5

The GUI change detection tool can detect various states containing specific changes as well as the added and removed actions and states when comparing two Notepad++ versions[4]. Figure 3 shows an overview of the model comparison between version 8.4.6 with 8.4.5. The actions and states can be selected to visualize the screenshot and inspect the element properties. For example, the *Edit → Copy to Clipboard Menu* state is selected to visualize that the titles of the menu elements have changed.

To detect SUT changes, the algorithm relies on the abstract properties of the compared models. If a major SUT release implies the GUI to change widget properties that affect the model's abstraction strategy, the change detection tool may detect false positive results. Therefore, in a DevOps process, it will be necessary to maintain an adequate abstraction strategy in the tool that infers the models.

[4] https://doi.org/10.5281/zenodo.7762904.

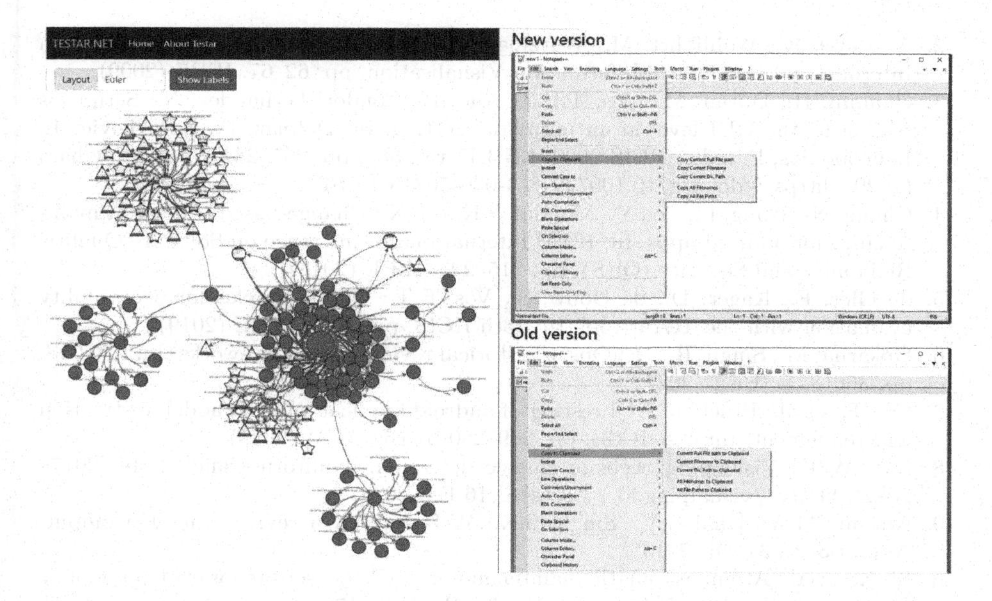

Fig. 3. Comparison of Notepad++ v8.4.6 with v8.4.5 with the change detection tool

6 Conclusions and Future Work

Delta testing is essential in rapid development cycles because it helps to evaluate that the software changes are intentional and do not compromise the SUT functionality. GUI change detection for delta testing can help software engineers to validate that no unintentional changes affect the end-user experience. This paper describes a change detection algorithm that uses automatically inferred state models to detect and highlight changes at the GUI level. We present a proof of concept using different versions of Notepad++.

Future work includes extended empirical evaluations involving more delta versions of Notepad++ and web and android SUTs, which will also help to measure the performance of the solution in a DevOps environment. Furthermore, we plan to research the impact and maintenance requirements of the abstraction strategy within the change detection software algorithm and the use of change detection at GUI in learning testing courses to obtain usability feedback.

This research has been funded by the project European innovation alliance for testing education (ENACTEST), ERASMUS+ number 101055874, 2022-2025.

References

1. Al-Saqqa, S., Sawalha, S., AbdelNabi, H.: Agile software development: methodologies and trends. Int. J. Inter. Mob. Technol. **14**(11) (2020)

2. Andrews, K., Wohlfahrt, M., Wurzinger, G.: Visual graph comparison. In: 13th International Conference Information Visualisation, pp. 62–67. IEEE (2009)
3. Chahim, H., Duran, M., Vos, T.E.J., Aho, P., Condori Fernandez, N.: Scriptless testing at the GUI level in an industrial setting. In: Dalpiaz, F., Zdravkovic, J., Loucopoulos, P. (eds.) RCIS 2020. LNBIP, vol. 385, pp. 267–284. Springer, Cham (2020). https://doi.org/10.1007/978-3-030-50316-1_16
4. Chang, N., Wang, L., Pei, Y., Mondal, S.K., Li, X.: Change-based test script maintenance for android apps. In: IEEE International Conference on Software Quality, Reliability and Security (QRS), pp. 215–225. IEEE (2018)
5. de Gier, F., Kager, D., de Gouw, S., Vos, E.T.: Offline oracles for accessibility evaluation with the Testar tool. In: 13th RCIS, pp. 1–12. IEEE (2019)
6. Gokarna, M., Singh, R.: DevOps: a historical review and future works. In: ICCCIS, pp. 366–371. IEEE (2021)
7. Gu, T., et al.: Practical GUI testing of android applications via model abstraction and refinement. In: 41st ICSE, pp. 269–280. IEEE/ACM (2019)
8. Inc., W.T.: Visualping: website change detection, monitoring and alerts (2017–2022). https://visualping.io/. Accessed 16 Feb 2023
9. Kausar, M.A., Dhaka, V., Singh, S.K.: Web crawler: a review. Int. J. Comput. Appl. **63**(2), 31–36 (2013)
10. Li, X., et al.: Atom: automatic maintenance of GUI test scripts for evolving mobile applications. In: 10th ICST, pp. 161–171. IEEE (2017)
11. Mallawaarachchi, V., Meegahapola, L., Madhushanka, R., Heshan, E., Meedeniya, D., Jayarathna, S.: Change detection and notification of web pages: a survey. ACM Comput. Surv. (CSUR) **53**(1), 1–35 (2020)
12. Mariani, L., Pezze, M., Riganelli, O., Santoro, M.: Autoblacktest: automatic black-box testing of interactive applications. In: 5th ICST, pp. 81–90. IEEE (2012)
13. Mariani, L., Pezzè, M., Zuddas, D.: Augusto: exploiting popular functionalities for the generation of semantic GUI tests with oracles. In: ICSE, pp. 280–290 (2018)
14. Mulders, A., Rodriguez-Valdes, O., Pastor Ricós, F., Aho, P., Marín, B., Vos, T.E.J.: State model inference through the GUI using run-time test generation. In: Guizzardi, R., Ralyte, J., Franch, X. (eds.) Research Challenges in Information Science. RCIS 2022. Lecture Notes in Business Information Processing, vol. 446, pp. 546–563. Springer, Cham (2022). https://doi.org/10.1007/978-3-031-05760-1_32
15. Ngo, C.D., Pastore, F., Briand, L.: Automated, cost-effective, and update-driven app testing. ACM TOSEM **31**(4), 1–51 (2022)
16. Nguyen, B.N., Robbins, B., Banerjee, I., Memon, A.: Guitar: an innovative tool for automated testing of GUI-driven software. ASE **21**(1), 65–105 (2014)
17. Pantiuchina, J., Mondini, M., Khanna, D., Wang, X., Abrahamsson, P.: Are software startups applying agile practices? The state of the practice from a large survey. In: Baumeister, H., Lichter, H., Riebisch, M. (eds.) XP 2017. LNBIP, vol. 283, pp. 167–183. Springer, Cham (2017). https://doi.org/10.1007/978-3-319-57633-6_11
18. Ricós, F.P., Aho, P., Vos, T., Boigues, I.T., Blasco, E.C., Martínez, H.M.: Deploying TESTAR to enable remote testing in an industrial CI pipeline: a case-based evaluation. In: Margaria, T., Steffen, B. (eds.) ISoLA 2020. LNCS, vol. 12476, pp. 543–557. Springer, Cham (2020). https://doi.org/10.1007/978-3-030-61362-4_31
19. Peng, C., Rajan, A., Cai, T.: Cat: Change-focused android GUI testing. In: ICSME, pp. 460–470. IEEE (2021)
20. Salihu, I.A., Ibrahim, R., Ahmed, B.S., Zamli, K.Z., Usman, A.: AMOGA: a static-dynamic model generation strategy for mobile apps testing. IEEE Access **7**, 17158–17173 (2019)

21. s.r.o., W.: Wachete: monitor web changes (2014–2022). https://www.wachete.com/. Accessed 16 Feb 2023
22. Vos, T.E.J., Aho, P., Pastor Ricos, F., Rodriguez Valdes, O., Mulders, A.: TESTAR-scriptless testing through graphical user interface. STVR **31**(3), e1771 (2021)

Enterprise Modeling for Machine Learning: Case-Based Analysis and Initial Framework Proposal

Dominik Bork[1](✉), Panagiotis Papapetrou[2], and Jelena Zdravkovic[2]

[1] Business Informatics Group, TU Wien, Vienna, Austria
dominik.bork@tuwien.ac.at

[2] Department of Computer and Systems Sciences, Stockholm University, Stockholm, Sweden
{panagiotis,jelenaz}@dsv.su.se

Abstract. Artificial Intelligence (AI) continuously paves its way into even the most traditional business domains. This particularly applies to data-driven AI, like machine learning (ML). Several data-driven approaches like CRISP-DM and KKD exist that help develop and engineer new ML-enhanced solutions. A new breed of approaches, often called canvas-driven or visual ideation approaches, extend the scope by a perspective on the business value an ML-enhanced solution shall enable. In this paper, we reflect on two recent ML projects. We show that the data-driven and canvas-driven approaches cover only some necessary information for developing and operating ML-enhanced solutions. Consequently, we propose to put ML into an enterprise context for which we sketch a first framework and spark the role enterprise modeling can play.

Keywords: Enterprise modeling · Conceptual modeling · Artificial intelligence · Machine learning · Model-driven engineering

1 Introduction

Artificial Intelligence (AI) and Machine Learning (ML) are continuously paving their way into even the most traditional business domains. Many products and services nowadays entail data-driven components, often realized by an ML model. This new breed of products and services require adjustments to the development and operations practices in enterprises as existing methods are either purely (software) product-focused (like Scrum, Waterfall, and Business Model Canvas) - i.e., not incorporating the ML part - or from the opposite direction, ML focused (like CRISP-DM, KDD, and MLOps), thereby lacking focus on the enterprise context within the ML solution needs to be integrated into, e.g., the business domain, business processes, resources, etc. What is essential to consider is that such ML-based solutions do most often not run in isolation [4]. In contrast, our experience - which we will report on later- shows that ML-based solutions must be integrated into an enterprise context. Similarly to, e.g., the business and IT alignment, enterprises need to ensure that an ML solution is aligned with its enterprise context [2].

S. Nurcan et al. (Eds.): RCIS 2023, LNBIP 476, pp. 518–525, 2023.
https://doi.org/10.1007/978-3-031-33080-3_33

The paper at hand stresses the need to account for such an extended context when designing, implementing, monitoring, and deploying ML-enhanced solutions in an enterprise. We report on two recent cases where the authors were involved and use these cases to identify relevant context dimensions that sketch a vision of a comprehensive framework. This framework aims to put ML into an enterprise context. We equip this framework with exemplary questions that aim to engage business people in the AI/ML discussion, which is, based on the cases we present in the following, dominated by data scientists and focused on data aspects of an ML solution, ignoring to a great extent the enterprise context.

In the remainder of this paper, we first introduce state-of-the-art methods for developing and operating ML-enhanced solutions. Based on two exemplary ML project cases, we derive a multi-dimensional conceptual framework that we envision as capable of addressing challenges adhering to such ML projects from the enterprise context perspective.

2 Background

In the following, we provide an overview of existing approaches that support enterprises in realizing ML projects.

Data-Driven Approaches. The *CRoss Industry Standard Process for Data Mining (CRISP-DM)*, describes how the data science research process is currently implemented. The model comprises six sequential steps. The first, referred to as business understanding, aims to understand and exemplify the objectives and requirements of the project. The second step focuses on understanding the data that is available for the problem at hand, verifying data quality, exploring and visualizing the data variables, and eventually determining whether the data is appropriate for addressing the objectives and requirements. The third step concerns data processing and preparation (e.g., missing values, normalization, integration and transformation). Step four determines which algorithms should be selected and applied for solving the defined problem, deciding how to split the data into training, validation, and testing, and evaluating whether the model solves the particular task(s) at hand effectively. The fifth step focuses on the evaluation of the model with regard to meeting the defined objectives and requirements. Finally, during step six a deployment plan is defined, including monitoring and maintenance of the deployed model while in operation, final reporting of the project and the results with directions for future improvements.

The *Knowledge Discovery in Databases (KDD)* process overlaps with CRISP-DM. The first step is to develop an understanding of the underlying application domain and the existing domain knowledge, identify the available data, and set the requirements from a customer's perspective. The next step is to select a data subset that is more relevant and suitable to the problem solution. The third step is about data preprocessing, removing missing values, and dealing with data complexity. Next, the appropriate features and their representations are chosen and defined (e.g., using feature selection and dimensionality reduction). The fifth step is modeling, i.e., selecting and applying a set of appropriate ML/data mining techniques. Next, model evaluation and analysis of the findings is performed, coupled with the previous step. The appropriate parameters are tuned and the model outperforming the other candidate models in terms of chosen performance metrics is selected. Moreover, the results, extracted patterns and rules are visualized. The final step concerns the exploitation of the extracted knowledge (and model) by either integrating it into the current knowledge base or domain knowledge in general or deploying the model to a software system.

Machine learning operations (MLOps) is an approach for streamlining ML software application life cycle management with the main principles inherited from DevOps in software engineering. MLOps aims for a higher software quality, release frequency, and user customization, by integrating and automating the tasks of development and operations, and by moving between them continuously. MLOps supports continuous integration and testing of ML models. MLOps is a collaborative approach, comprising different roles such as data scientists and architects, DevOps engineers, and traditional software engineers. By following MLOps practices, these roles increase the pace and synergy of development and production, monitoring, validation, and governance of ML models. The approach also enables high scalability where a number of ML models can be managed and monitored for continuous delivery and deployment. It also provides reproducibility of ML pipelines, thereby enabling more tightly-coupled collaboration across data teams, efficiency in model regulatory scrutiny and drift-check through transparency, and compliance with organizational policies.

Canvas-Driven Approaches. The previously introduced data-driven approaches primarily target data scientists which is also reflected in the focus on the data-related aspects of a ML project. To account for the needs of business people (i.e., domain experts) and the characteristics of the business that aims to develop and use a new ML solution, several canvas-based approaches have been proposed recently. All of these approaches follow the general paradigm of design canvases as pioneered by the Business Movel Canvas. These canvas-driven approaches all entail a view on the data (e.g., [3]) of a ML project but also accommodate other aspects like the value that is aimed to be delivered by a ML solution [7,9], the affected business processes [6], the heterogeneous stakeholders [4,5], and regulative aspects [9]. These approaches are not yet matured and industry-proven compared to the data-driven ones.

3 ML in an Enterprise Context: Two Cases

Case 1: Explainable Machine Learning for Healthcare

This case concerns a 5-year ongoing collaborative research project between two universities, one research institute, and medical practitioners as stakeholders from two hospitals in Sweden. The first objective of the project is to develop explainable ML models and workflows that exploit the complexity of medical data sources and produce useful and insightful predictions and treatment recommendations for patients suffering from cardiovascular conditions as well as patients suffering from adverse drug effects. Secondly, the project aims to leverage explainable and responsible AI principles when used for decision making in healthcare, and demonstrate the benefits and pitfalls of AI-assisted diagnostics. Moreover, the third goal is to develop a software prototype that integrates several ML methods and provides a diagnosis and a treatment recommendation for an ongoing patient visit alongside a medically valid and trustworthy rationale behind these recommendations.

In terms of its implementation, the project follows a standard ML workflow, starting with the exploration of the available data sources and their relevance to the project objectives. More concretely a database of 1.3 million patients has been used, with one of the two hospitals being the data owner. The dataset comprises over 180 million timestamped hospital events, including diagnoses, medications, lab tests, medical procedures, and clinical text. The dataset can be readily used to train ML models, while having been de-anonymized in order to preserve the privacy and integrity of the individual patients. At a first stage, an exploratory analysis of the data variables was performed to assess data quality in terms of missingness, uncertainty, and any other potential errors. Next, the elicitation of the requirements was done by consulting with some of the stakeholders, including a small number of medical practitioners (doctors and specialists in cardiology and clinical pharmacology), patients, and lawyers.

Several ML models have been explored, such as random forests and logistic regressors, and new models have been designed, including deep learning-based architectures, such as RNNs and CNNs. The models were trained and validated using common ML model evaluation procedures, such as cross validation and repeated holdout. Standard predictive performance metrics were used, such as precision, recall, and AUC. At the same time, model explainability was assessed both qualitatively and quantitatively. Finally, and in accordance to the project's objectives, the compliance of the ML models to the existing legislative rules of Sweden was assessed and enforced during design and evaluation.

Case 2: Simulation of Pandemics Using Machine Learning
This was a 1-year collaborative project between university researchers and a public agency in Sweden. The project had two main objectives: To use reinforcement learning (RL) for policy recommendation during an ongoing pandemic (with COVID-19 as a case), and to update and improve the existing simulator used by the agency so as to include policy recommendations for contact reduction of the pandemic spread using ML. Following a similar workflow, the data sources available for this project were explored and assessed in terms of quality, validity, and relevance. Real data from the COVID-19 pandemic spread was available both at a national as well as international level. The data variables included various epidemic spread indicators, mitigation measures taken, as well as people's sentiment as quantified by Twitter posts. The epidemic data was partly public and partly owned by the agency as it concerned the epidemic spread in Sweden in particular.

Several contrasts to the first case emerged. First, policy and decision makers were strongly present in the development of both the RL-based ML models as well as during their integration to the existing software of the health agency. Hence, all constraints related to the feasibility and potential societal implications of the designed model and the contact reduction measures proposed by the model were taken into consideration. Second, methods and software had to be updated several times due to *context drift* caused by the mutation of the virus and the availability of vaccinations against the virus.

The ML methods used in this project were mostly restricted to RL techniques as they were highly suitable for the particular problem formulation. The learned policies and proposed mitigation measures were thoroughly assessed both quantitatively and qualitatively. On one hand, quantitative metrics such as model convergence, stability, and accuracy were used. On the other hand, the policies were assessed by a team of policy makers in terms of feasibility, before being integrated into the existing epidemic simulator software tool. Finally, the agency was continuously in the loop, protecting the RL agent from taking infeasible and potentially unlawful decisions. The RL model was never employed and used in practice, but it was only used for retrospective analysis while the pandemic was ongoing. In that respect, any unlawful or unreasonable recommendation would not effectively impact the population.

Lessons Learned
While the workflow followed by both cases is consistent with CRISP-DM and KDD, the two cases differ in terms of implementation. Firstly, the omission of some hospital stakeholders (e.g., nurses, specialists in other pathologies except for cardiology and clinical pharmacology) and the hospital leadership from the process resulted in several deficiencies during the development of the software prototype. On one hand, omitting nurses may neglect inaccuracies related to the content of the electronic health records, such as delayed registry of the patients' blood tests and erroneous or incomplete diagnosis codes. On the other hand, missing the hospital's leadership team resulted in inadequate information concerning software adoption and integration at the hospital.

Furthermore, the extracted rules and reasoning used by the ML models had to be aligned with the medical guidelines and the hospital decision-making processes. However, they have not been taken into full consideration, since the focus was, as is often the case, mostly on the quantitative side of model performance. These processes would have been detrimental for the design of the software prototype, as well as on the underlying mechanisms and rules that the ML models employ during training. For example, in the case of treating heart failure, the national guidelines in Sweden recommend a particular line of medication unless some other underlying condition is present. Such rules are easily integrated into the ML models, e.g., by means of constraints during model training and validation. Nonetheless, patient prioritization may differ between hospital units as they are primarily based on demand, underlying costs, or availability of specialists and personnel. Such constraints are harder to be integrated and require thorough consultation with the hospital leadership and decision-making team.

Table 1. Mapping ML cases and related approaches to an enterprise context

	Enterprise context dimension						
	Business	**Data**	**Process**	**Stakeholders**	**Technology**	**Regulation**	**Legislation**
Case: Issues faced in which dimensions							
Case#1:	◐	●	○	◐	◐	●	●
Case#2:	◐	●	○	◐	◐	◐	●
Data-driven approaches: Coverage of which dimensions							
CRISP-DM	●	●	○	○	◐	○	○
KDD	●	●	○	○	◐	○	○
MLOps	◐	●	○	◐	●	◐	◐
Canvas-driven approaches: Coverage of which dimensions							
Data Collection Map [3]	○	●	○	○	○	○	○
Enterprise AI Canvas [4]	◐	●	○	◐	○	○	○
Data Innovation Board [5]	◐	●	○	●	○	○	○
Machine Learning Canvas [6]	◐	●	◐	○	○	○	○
Data Science Canvas [7]	●	●	○	○	●	○	○
Prescriptive Modeling Canvas [9]	●	●	◐	◐	○	◐	○

○ = Not applicable; ◐ = Partially applicable; ● = Fully applicable

With regard to the second case, the methods and software used experience context drift, which had to be taken into consideration during the development and implementation. Finally, while in the first case law experts were part of the project both as researchers and stakeholders, in this case law experts were not included in the training and development of the RL agent. This implies that some recommended pandemic mitigation policies were not thoroughly assessed in terms of their actual feasibility.

4 A Framework for ML Projects in Enterprise Context

We now elicit the relevant dimensions when developing enterprise-wide ML solutions and combine them toward a vision for a framework. By analyzing the two presented cases and the primary existing ML development methods, we collected the following dimensions: *Business*, describing business value(s) for organizations and business goals to be achieved; *Data*, that is relevant for the ML models; *Process*, the workflows of business activities related to a specific concern; *Stakeholder*, people (internal or external to the enterprise) with a particular interest or role to the development or use of the ML solution; *Technology*, encompasses all technological frameworks, algorithms, and tools used in the ML solution development and operation; *Regulation*, refers to enterprise policies; and *Legislation*, concerns laws, directives, or decisions of a relevant governing body (e.g., state or municipality).

Upon a detailed analysis of the support of each case and the existing works, we have concluded their outcomes as presented in Table 1. Regarding the two cases, the results show that: the Business dimension was supported in terms of goals (objectives) for the project, while business values were not considered; both

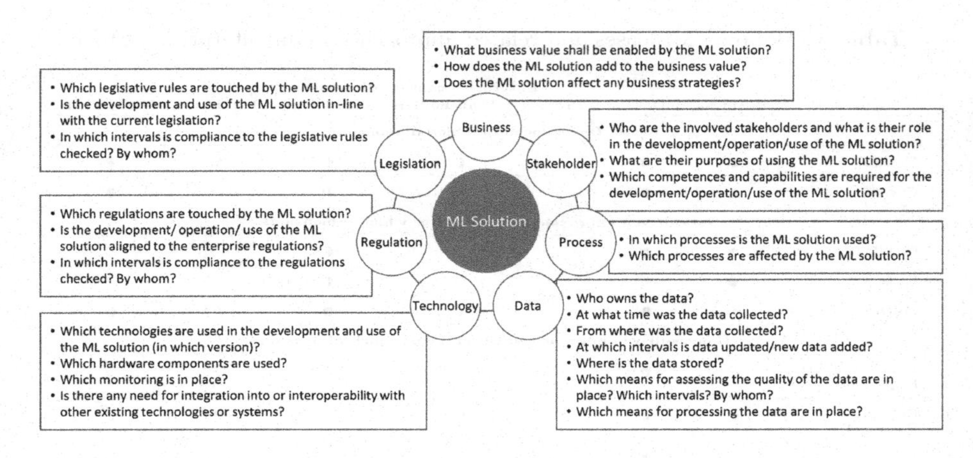

Fig. 1. Enterprise context of AI/ML solutions

cases covered the Data dimension, as well as the algorithmic part of Technology, while deployment environments where not exercised; understanding of the existing Processes was neglected which negatively influenced the quality of the ML models; the Regulation and Legislation dimensions were reasonably considered, confirming thus a maturity of these aspects about the use of ML in enterprises.

Regarding the data-driven approaches, the analysis has shown full support for the Data and Technology dimensions; Stakeholders are considered only in MLOps, yet with the focus on the development team roles; the Business is well-recognized in KDD and CRISP-DM, i.e., both values and objectives, while in MLOps only the latter are considered. The Process dimension needs to be addressed, thus showing low support for aligning the software systems and its enterprise. In contrast, the Regulation and Legislation dimensions are not considered in the first two and, to an extent, are guided for addressing in the MLOps specification. Regarding the canvas-driven approaches, a recent systematic review [11] also confirms our observations of a need for more integration of the enterprise context. The authors analyzed a total of 25 ML canvases. They concluded that many canvases focus on the Data and Technology dimensions, and a few on the Business and (partly) the Process dimensions. At the same time, more consideration of the Regulation and Legislation dimensions must be considered.

Figure 1 shows our vision for a framework comprising seven dimensions, each equipped with exemplary questions supporting collecting relevant information about an ML project. These questions are based on the lessons learned from the two presented cases and shall aim to operationalize our framework. The aim is twofold: first and foremost, these questions shall help mitigate many of the issues faced in the two presented cases; second, they shall enable business people and domain experts to engage in the ML discussion already during the development stage and not only at the time they face issues when using the ML solution. Enterprise models can play an influential role here by providing richer specifications of individual dimensions and representing an integrated view of the different dimensions (cf. [10]). This endeavor fits nicely within the prospective future of combining enterprise modeling and AI [1,8].

5 Conclusive Discussion

In this paper, we reflected on the lessons learned from two recent ML projects to derive a set of dimensions forming a vision for a framework that puts ML-enhanced solutions into an enterprise context. Based on the limitations of existing approaches, we spark the role enterprise modeling can play in providing a more holistic description of ML-enhanced solutions. In our future work, we aim to validate and revise our framework with more cases and to formalize the presented dimensions using well-defined requirements of a modeling method. Ultimately, we aim to propose a model-driven method for guiding the development of ML-enhanced solutions encompassing the enterprise context. Such a method might be informed by the data-driven and canvas-driven approaches and follow a structured process model.

References

1. Bork, D., Ali, S.J., Roelens, B.: Conceptual modeling and artificial intelligence: a systematic mapping study. CoRR abs/2303.06758 (2023). https://doi.org/10.48550/arXiv.2303.06758
2. Haller, K.: Managing AI in the Enterprise. Springer, Berlin (2022)
3. Kayser, L., Mueller, R.M., Kronsbein, T.: Data collection map: a canvas for shared data awareness in data-driven innovation projects. In: Pre-ICIS Symposium on Inspiring Mindset for Innovation with Business Analytics and Data Science (2019)
4. Kerzel, U.: Enterprise AI canvas integrating artificial intelligence into business. Appl. Artif. Intell. **35**(1), 1–12 (2021)
5. Kronsbein, T., Müller, R.M.: Data thinking: a canvas for data-driven ideation workshops. In: Bui, T. (ed.) 52nd Hawaii International Conference on System Sciences, HICSS 2019, pp. 1–10. ScholarSpace (2019)
6. Marin, I.: Data science and development team remote communication: the use of the machine learning canvas. In: Calefato, F., Tell, P., Dubey, A. (eds.) 14th International Conference on Global Software Engineering, pp. 18–21 (2019)
7. Neifer, T., Lawo, D., Esau, M.: Data science canvas: evaluation of a tool to manage data science projects. In: 54th Hawaii International Conference on System Sciences, HICSS 2021, pp. 1–10. ScholarSpace (2021)
8. Rittelmeyer, J.D., Sandkuhl, K.: Features of AI solutions and their use in AI context modeling. In: Modellierung 2022 - Workshop Proceedings, pp. 18–29. GI (2022)
9. Shteingart, H., Oostra, G., Levinkron, O., Parush, N., Shabat, G., Aronovich, D.: Machine learning prescriptive canvas for optimizing business outcomes. CoRR abs/2206.10333 (2022). https://doi.org/10.48550/arXiv.2206.10333
10. Takeuchi, H., Ito, Y., Yamamoto, S.: Method for constructing machine learning project canvas based on enterprise architecture modeling. In: International Conference on Knowledge-Based and Intelligent Information & Engineering Systems (2022)
11. Thiée, L.W.: A systematic literature review of machine learning canvases. In: für Informatik, G. (ed.) 51. Jahrestagung der Gesellschaft für Informatik. LNI, vol. P-314, pp. 1221–1235. Gesellschaft für Informatik, Bonn (2021)

Towards Creating a Secure Framework for Building Mirror World Applications

Panos Mazarakis(✉)

Privacy Engineering and Social Informatics Laboratory, Department of Cultural Technology and Communication, University of the Aegean, University Hill, 81100 Mytilene, Greece
pmazarakis@aegean.gr

Abstract. The growing interest in Virtual Reality and Augmented Reality technologies has led to the development of the concept of Mirror Worlds, the Metaverse, and the AR Cloud. Different companies are creating their own version of the Metaverse and Mirror Worlds serve as a foundation for these experiences. As these technologies continue to grow, there is a need for the design of a systematic way for designing and implementing applications in virtual spaces. There are emerging technologies and standards that could form the basis of such a framework, however, privacy concerns like data privacy and location tracking need to be addressed while designing any potential framework or solution.

Keywords: Mirror Worlds · Augmented Reality · AR Cloud · Privacy

1 Introduction

In recent years, there has been increasing interest in the development of Virtual Reality (VR) and Augmented Reality (AR) technologies that have the potential to transform the way we experience and interact with the world. Mirror Worlds, the AR Cloud, and the Metaverse are related concepts in this field that have attracted significant attention in recent years and are part of the broader trend toward creating more realistic and accurate digital representations of the real world.

The Metaverse as a term has seen a significant surge recently (see Fig. 1), after the re-branding of Facebook to Meta in October 2021 [1]. Other companies and organizations are building their own proprietary versions of the Metaverse, such as Niantic building *Lightship* [2], an AR Cloud platform. Mirror Worlds, on the other hand, expanding on the original concept presented by Gelernter [3], can be seen as a building block for such virtual worlds, serving as a foundational layer that enables the creation of immersive and interactive AR experiences.

With the increasing popularity of such technologies, the need for a common framework for building applications in these virtual spaces is becoming increasingly important. The field is still in its early stages and there are many different approaches being taken. However, there are some emerging standards and technologies that could serve as a basis for a common framework for building Mirror Worlds in the future.

S. Nurcan et al. (Eds.): RCIS 2023, LNBIP 476, pp. 526–533, 2023.
https://doi.org/10.1007/978-3-031-33080-3_34

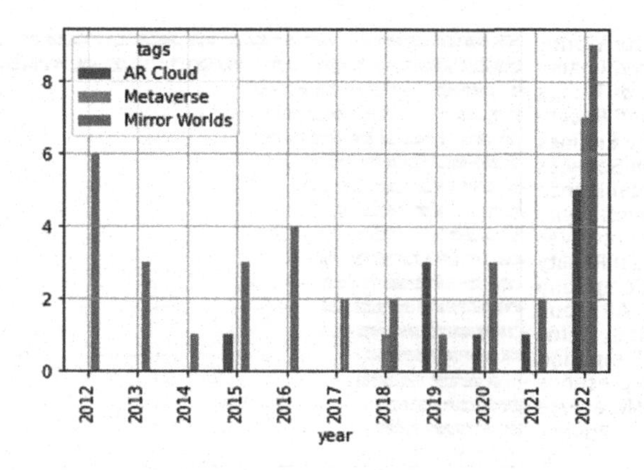

Fig. 1. The increasing popularity of the term "Metaverse" in the past year.

Although the development of these technologies presents new opportunities, there are also several potential privacy risks, that are often not taken into account when designing such systems. These stem from the underlying technologies supporting such systems and include location tracking, data privacy, surveillance, and targeted advertising. The extent of these risks will depend on how the technology is developed and used, as well as the laws and regulations in place to protect personal data and privacy, so they should be taken into account when designing a common framework for Mirror Worlds.

2 Methodology

To compile this paper, a series of distinct steps were taken. The keywords "ar cloud" and "mirror worlds" were used across two different citation databases: Scopus [4] and Web of Science [5], and the search was focused to research efforts conducted during the last ten years. The term "Metaverse" was not used in the search, as the increasing popularity of the term produced many irrelevant results in the context of AR application frameworks. Despite of this, the term "Metaverse" was an emerging keyword in the resulting papers. This search resulted in a total of 199 papers, which after de-duplication came down to 153 papers. The next step was to exclude not accessible papers, short papers and irrelevant results based on the title and abstract. This process resulted in 51 relevant papers. After reading these, reoccurring keywords relevant to Mirror Worlds and the AR cloud were identified. The last step followed was to group the most common keywords to tags (see Fig. 2).

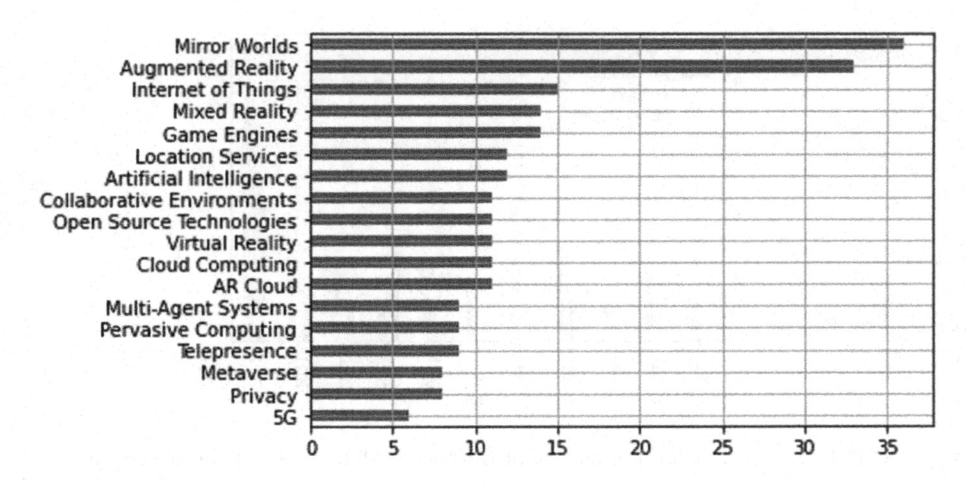

Fig. 2. Number of papers belonging to each of the most common tags.

3 Challenges in Creating a Common Framework for Mirror Worlds

Mirror worlds are systems that bring together research aspects from various fields, being laboratories in which to explore together inter-disciplinary aspects [6]. In order to create a framework for building such applications, we must first identify the building blocks most commonly found in relevant works.

3.1 Building Blocks

Augmented and Mixed Reality serve as a digital layer to augment the physical world with situated information and services that can be perceived by wearable devices, such as smart glasses [7], therefore they are considered an integral part of building such applications. Frameworks for the development of AR applications have been readily available in the past years, for example ARKit [8] for *iOS* devices and ARCore [9] for *Android* devices. Although these SDKs provide tools for "anchoring" entities to the real world, there is no solid concept of a Mirror World provided by them.

To bridge the gap between the physical and virtual world, the Internet of Things plays a central role. Connected sensors and devices make it possible to keep a continuous coupling between the two [7], by feeding real-world data to the Mirror World. These devices can also become active participants to our lives, by acting autonomously on our behalf [10]. These sensor networks can scale from rooms or houses to entire cities. Therefore the Internet of Things is an enabling technology for Mirror Worlds.

Game engines, such as *Unity* [11] are often used as a mid-level representation layer between the virtual and real world [12]. Apart from visualization purposes, the game engine's environment can be used to simulate the behavior of virtual agents, such as robots [13], before applying their actions to the real world. These engines also provide many of the tools necessary for building digital twins, like real-time 3D representations, networking, and multi-platform support for AR and VR. However, in order to keep the virtual world in-sync with the real world, and consistent across multiple devices, communication with external servers is needed.

In order to maintain consistent real-world AR applications across multiple users, a shared global coordinate system is necessary [14]. This allows users to be positioned in a unified system, allowing 3D virtual content to be seamlessly displayed on different devices. By sharing the local map created by one user with others, a collaborative AR experience, such as a Mirror World, can be provided [15]. In current AR headsets, the location of the device is determined utilizing Simultaneous Localization and Mapping (SLAM). The goal of SLAM is to determine the position of the user and construct a map of the surrounding environment simultaneously. SLAM has been a major area of focus in computer vision and has seen considerable advancements in the past decades. Commercial AR systems, such as ARKit and ARCore provide highly accurate positioning through their implementation of SLAM, which fuses multiple inputs, including cameras, odometry, depth, and inertial sensors, in real-time. Other techniques, such as orientation using magnetic field measurements [16], have also been used.

Multi-agent systems (MAS) can also be considered a key aspect of Mirror Worlds. The agents in a Mirror World can be either human users, usually joining the virtual world in the form of avatars [17], or software agents powered by Artificial Intelligence (AI). These virtual agents can perceive the Mirror World and act autonomously according to its changes. As such, Mirror Worlds constitute cooperative environments, where humans and AI agents can communicate, collaborate, and negotiate with each other.

Furthermore, Mirror Worlds are inherently distributed virtual environments [18] that require inter-connectivity in order to remain accessible to everyone. Cloud computing is the main enabling technology for this feature, as it can provide the infrastructure necessary to store and process large amounts of data. It can also scale dynamically, according to the amount of users and virtual agents connected. To achieve integration with the cloud, it is necessary to implement common APIs [14] that are open-source and standardized. A common way to achieve this is using web-based AR platforms to take advantage of the inherent characteristics of web services, including their lightweight design, and cross-platform compatibility [19].

3.2 Challenges

Mirror worlds are designed to simulate real-world environments, and therefore, real-time constraints play a crucial role in ensuring the realism and accuracy of these applications. For example, when creating a virtual mirror of a city, events such as traffic flow, pedestrian movement, and weather conditions need to be monitored. If these events are not accurately updated in real time, the virtual environment would lack the ability to provide a useful simulation. For this reason, these frameworks need to be robust enough to support changes in the environment and, at the same time, flexible enough to dynamically add elements [20]. Additionally, it is important to enable real-time interactions between the various agents. For example, if a user is engaging in a virtual conversation with a virtual agent, the timing and duration of their responses must be done in real-time in order to ensure a convincing and immersive interaction.

Another significant challenge for creating Mirror Worlds is working with current-generation hardware. Current mobile hardware has limited performance in real-world AR applications. The current AR systems on mobile devices lack the capability to store persistent data across multiple devices and user sessions [16], and to reduce energy consumption, it is a common practice for the mobile device transfers all computation-heavy tasks to the cloud [21].

3.3 Privacy Implications

The development of Mirror Worlds, AR cloud and the Metaverse has the potential to greatly impact society and individuals. However, there are also potential privacy and security risks associated with these technologies that must be addressed and managed.

Mirror worlds inherit the privacy risks of the underlying technologies that enable them. AR applications require the collection and processing of data from the user's surroundings, such as camera data, or from the user itself, such as eye tracking and movement. This, in combination with the location data necessary, could be a potential source of privacy risks for the users, especially if the individuals substitute their smartphones with devices they wear all the time [17]. Research has also demonstrated that people value the idea of virtual cities being filled with current content from social media platforms [22], so the privacy risks from these platforms are also inherited by Mirror Worlds.

The integration of cloud computing and the Internet of Things also presents several privacy risks that must be considered, including data collection and storage, personal preferences and usage patterns. This data is often stored in the cloud, where it can be vulnerable to unauthorized access and cyber attacks. In addition, the companies owning such cloud infrastructure, collect and process personal information as a primary business model, not only to improve their products, but also to steer user behaviour towards the desired outcomes of the companies [17].

4 Discussion

In recent years, the fields of Mirror Worlds, AR cloud and the Metaverse have gained significant attention in academic research. Despite this, there remains a gap in the existing bibliography regarding the interplay between these areas, particularly with respect to privacy (see Fig. 3).

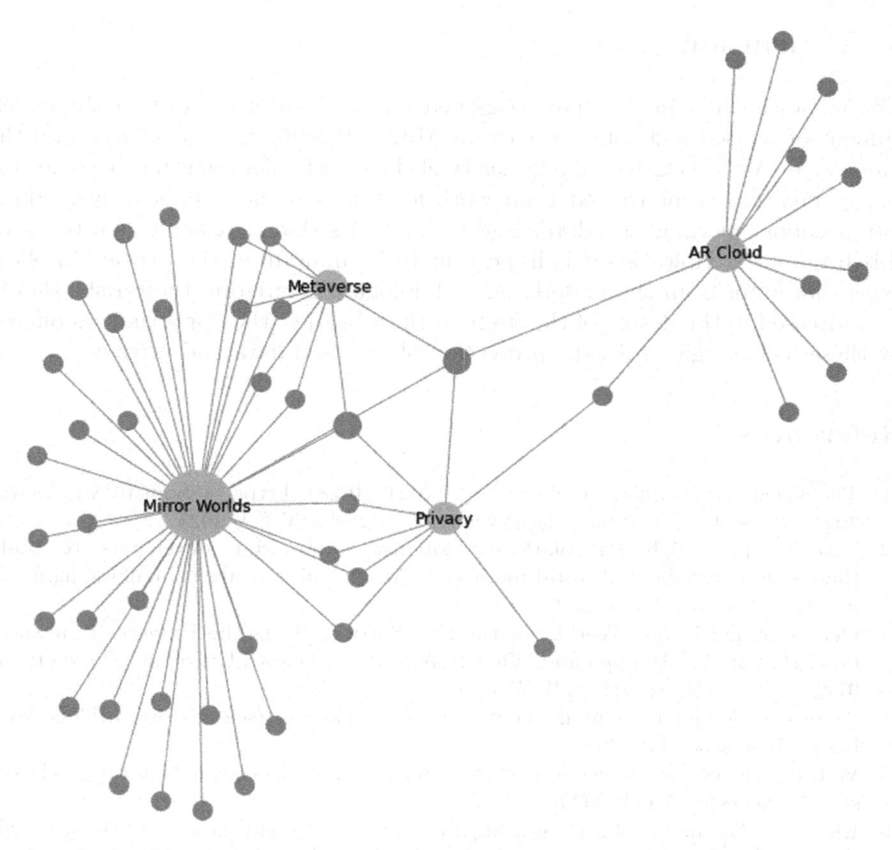

Fig. 3. Network graph for papers in the categories: *Mirror Worlds, AR Cloud, Metaverse and Privacy.* Shown in red are papers with connections to 3 or more categories.

As the field of Mirror Worlds continues to grow, it is becoming evident that there is a need for a common framework for building applications in this space. It is clear that reality will be augmented in a dynamic and context-dependent way, rather than a static way [7], supported by software that scales to the number of simultaneously connected entities. Using the multi-agent system approach seems to be compatible with these requirements, in contrast to the traditional client/server paradigm [20]. Additionally, taking advantage of emerging technologies such as open source AR cloud platforms, Multi-access Edge Computing (MEC) and 5G networks [15,19], it is now easier to overcome the real-time constrain of Mirror Worlds.

However, it is important to consider the privacy risks associated with these virtual environments, and incorporate privacy-by-design principles into such framework. These should include encryption, access control, and auditing mechanisms, to ensure that user's personal information remains protected.

5 Conclusion

The advancements in AR have triggered an increase in interest in the development of related technologies, such as Mirror Worlds, the AR Cloud, and the Metaverse. With the growing popularity of these technologies, it has become crucial to create a standardized framework for building such applications, taking into account emerging standards and technologies that may serve as a basis for this framework. However, it is important to be mindful of the privacy implications that stem from the underlying technologies. Therefore, these risks should be addressed in the design of the framework to balance the opportunities offered by these technologies with the protection of personal data and privacy.

References

1. The Facebook company is now meta (2021–10-28). https://about.fb.com/news/2021/10/facebook-company-is-now-meta/. Accessed 6 Feb 2023
2. Niantic opens lightship platform globally, empowering developers to build their visions for the real-world metaverse. https://nianticlabs.com/news/lightship launch/. Accessed 6 Feb 2023
3. Gelernter, D.: Mirror Worlds: or the Day Software Puts the Universe in a Shoebox...How It Will Happen and What It Will Mean. Oxford University Press (1993–01-28). google-Books-ID: jh2U379fq18C
4. Scopus - document search. https://www.scopus.com/search/form.uri?display=basic. Accessed 7 Feb 2023
5. Web of science core collection. https://www.webofscience.com/wos/woscc/basic-search. Accessed 7 Feb 2023
6. Ricci, A., Tummolini, L., Piunti, M., Boissier, O., Castelfranchi, C.: Mirror worlds as agent societies situated in mixed reality environments. In: Ghose, A., Oren, N., Telang, P., Thangarajah, J. (eds.) COIN 2014. LNCS (LNAI), vol. 9372, pp. 197–212. Springer, Cham (2015). https://doi.org/10.1007/978-3-319-25420-3_13
7. Ricci, A., Piunti, M., Tummolini, L., Castelfranchi, C.: The mirror world: preparing for mixed-reality living. IEEE Pervasive Comput. **14**(2), 60–63 (2015)
8. ARKit. https://developer.apple.com/augmented-reality/. Accessed 7 Feb 2023
9. Build new augmented reality experiences that seamlessly blend the digital and physical worlds — ARCore. https://developers.google.com/ar. Accessed 7 Feb 2023
10. Ehlers, M., Woodgate, P., Annoni, A., Schade, S.: Advancing digital earth: beyond the next generation. Int. J. Digit. Earth **7**(1), 3–16 (2014)
11. Unity real-time development platform — 3d, 2d VR & AR engine. https://unity.com/. Accessed 8 Feb 2023
12. Eckstein, B., Lugrin, B.: Augmented reasoning in the mirror world. In: Proceedings of the ACM Symposium on Virtual Reality Software and Technology, VRST. 02–04-November-2016, pp. 313–314 (2016)
13. McLin, J., Hoang, N., Deneke, W., McDowell, P.: A mirror world-based robot control system. In: Proceedings - 2017 2nd International Conference on Information Systems Engineering, ICISE 2017. January 2017, pp. 74–77 (2017)
14. Jackson, J., Vogt, M., Sörös, G., Salazar, M., Fedorenko, S.: Demo: the first open AR cloud testbed. In: Proceedings - 2021 IEEE International Symposium on Mixed and Augmented Reality Adjunct, ISMAR-Adjunct 2021, pp. 495–496 (2021)

15. Duong, N.D., Cutullic, C., Henaff, J.M., Royan, J.: AR cloud: towards collaborative augmented reality at a large-scale. In: Proceedings - 2022 IEEE International Symposium on Mixed and Augmented Reality Adjunct, ISMAR-Adjunct 2022, pp. 733–738 (2022)
16. Rajagopal, N., Miller, J., Reghu Kumar, K., Luong, A., Rowe, A.: Demo abstract: welcome to my world: Demystifying multi-user AR with the cloud. In: Proceedings - 17th ACM/IEEE International Conference on Information Processing in Sensor Networks, IPSN 2018, pp. 146–147 (2018)
17. Harborth, D.: Human autonomy in the era of augmented reality-a roadmap for future work. Information **13**(6), 289 (2022)
18. Croatti, A., Ricci, A.: From virtual worlds to mirror worlds: a model and platform for building agent-based eXtended realities. In: Bassiliades, N., Chalkiadakis, G., de Jonge, D. (eds.) EUMAS/AT -2020. LNCS (LNAI), vol. 12520, pp. 459–474. Springer, Cham (2020). https://doi.org/10.1007/978-3-030-66412-1_29
19. Park, G., Kim, R., Song, H.: Collaborative virtual 3D object modeling for mobile augmented reality streaming services over 5G networks. IEEE Trans. Mob. Comput. (2022)
20. Rincon, J., Poza-Lujan, J.L., Julian, V., Posadas-Yagüe, J.L., Carrascosa, C.: Extending MAM5 meta-model and JaCalIV e framework to integrate smart devices from real environments. PLoS ONE **11**(2), e0149665 (2016)
21. Chung, J.M., Park, Y.S., Park, J.H., Cho, H.: Adaptive cloud offloading of augmented reality applications on smart devices for minimum energy consumption. KSII Trans. Internet Inf. Syst. **9**(8), 3099–3111 (2015)
22. Kukka, H., Pakanen, M., Badri, M., Ojala, T.: Immersive street-level social media in the 3D virtual city: anticipated user experience and conceptual development. In: Proceedings of the ACM Conference on Computer Supported Cooperative Work, CSCW, pp. 2422–2435 (2017)

Towards a Secure and Privacy Compliant Framework for Educational Data Mining

Polydorou Eleni(✉)

Privacy Engineering and Social Informatics Laboratory, Department of Cultural Technology and Communication, University of the Aegean, University Hill, 81100 Mytilene, Greece
epolidorou@aegean.gr

Abstract. Digital education technology has become an essential component of the educational process and has greatly impacted the learning experience in schools and higher education. Especially, in recent years, there has been a surge in the adoption of educational applications and the use of e-learning platforms resulting in a vast amount of educational data being generated. Using Learning Analytics and Educational Data Mining can assist in extracting valuable information from educational data, which can enhance the learning experience for both learners and educators. However, there is a growing need to address privacy concerns and ensure data security, and thus research focus has shifted to the development of privacy frameworks and principles to ensure data privacy and security. This paper highlights the significance of data protection and privacy regulations, such as GDPR, in utilizing data mining methods in educational environments aiming on establishing the foundations for the creation of a privacy-focused framework for data processing within the educational platform.

Keywords: Learning Analytics · Educational Data Mining · Privacy · Educational Applications

1 Introduction

Digital education technology is considered an irreplaceable part of educational practice and has fundamentally transformed the learning experience of schools and higher education [1]. Especially, in recent years due to the COVID-19 pandemic there has been a significant increase in the use of educational applications. This integration has produced a vast amount of educational data that are generated through various learning resources such as the emerging Massive Online Open Courses (MOOCs), e-learning platforms, and learning management systems (LMS) like Moodle, Edmondo, and Canvas [1,2]. Many higher education institutions have developed educational programs with distance and blended learning that benefits accessibility to several online resources.

The education sector is always striving to improve learning outcomes and student performance. This requires the evaluation of the efficiency and impact of

S. Nurcan et al. (Eds.): RCIS 2023, LNBIP 476, pp. 534–541, 2023.
https://doi.org/10.1007/978-3-031-33080-3_35

teaching techniques and methods. Through the use and application of Learning Analytics (LA) and Educational Data Mining (EDM), a large amount of information is seamlessly captured regarding students' engagement, thinking patterns, cognitive load, enjoyment, and learning progress. Rather than merely document and store that information, educators are given the opportunity to interpret and evaluate students' needs and competencies. Moreover, they can redirect learners' educational trajectories into a more flexible and adaptable learning process [3]. Students, on the other hand, receive personalized feedback that gives them the necessary guidance to improve their performance and effectively optimize their overall knowledge accumulation.

Although the results obtained from data mining can greatly aid the decision-making process [4] many researchers are concerned about the issues of privacy and data security. This is because the possibility of students' data being leaked poses a threat to all parties involved in the process [5–7]. Students' data protection requires meticulous consideration of respective data protection and usage policies, implementation of appropriate security methods, and compliance with privacy laws and regulations. Moreover, in order to implement any technological methods for data processing the students must be aware of the potential consequences that may occur from using these systems both in real-time and in the long term. Educational stakeholders must be informed of the functionalities of learning analytics and educational data mining such as the data collection and processing, the privacy measures, and the validity of the generated results [8].

The purpose of this paper is to highlight the significance of data protection and privacy regulations in utilizing data mining methods in educational environments, which are defined in Sect. 2. Following with Sect. 3, the value of educational data in enhancing the learning experience and providing valuable insights for educational stakeholders is showcased. This paper also examines the need for the creation of a framework that combines features from existing frameworks but with a stronger focus on legal, organizational, and technical privacy requirements, allowing for processing to start within the educational platform without jeopardizing personal data outside of the system's boundaries. Current privacy frameworks are presented in Sect. 4, and in Sect. 5 a framework layout is proposed, in the form of a plugin, that has the potential to be added to already established LMS. Finally, conclusions are drawn in Sect. 6.

2 Privacy Regulations Regarding Educational Data

The definition of privacy derives from different theoretical research approaches and is consolidated in its final form with the binding regulations decided within each country [9]. In terms of law and regulations, privacy is acknowledged and protected as a fundamental human right and is safeguarded through legal agreements, the most prominent being the Universal Declaration of Human Rights (UDHR) and the International Covenant on Civil and Political Rights (ICCPR), Article 12 and Article 17 respectively [10]. According to Kalloniatis et al. [11], privacy as a concept is multi-dimensional and comes in many forms depending on the type of information each individual wants to keep private.

Educational platforms and educational technology, in general, are tightly interconnected with student's everyday life from an early age and from the very first stages of their educational journey [5]. This has resulted in the creation of large repositories of student data gathered from these sources that include among others: interaction data between educators and students like level of participation, navigation patterns, answers in interactive exercises, and communication through e-learning platforms (messages, comments, etc.), demographic data like age, gender, enrollment, attendance as well as administrative data [12]. For this reason, the issue of privacy is a major concern when learning analytics and educational data mining techniques are used as the personal data being collected and analyzed are of personal and sensitive nature. Personal information about students, such as their name, address, contact details, marks, and progress reports, is considered personal data. Student-sensitive data includes biometric data (e.g. fingerprints, photos), health (e.g. allergies), religious beliefs (e.g. a student's opting out of religion class), or dietary requirements (that might hint at their religious beliefs or health status) [13].

General Data Protection Regulation (GDPR) is a privacy regulation of the European Union (EU) that sets out specific principles for the handling of personal data. Key principles of the GDPR regarding user data, demonstrated in Article 6, are: i) lawfulness, fairly, and transparency, the data are processed accordingly to the data subject, ii) purpose limitation, the collected data are processed for specific and legitimate purposes and only for those purposes iii) data minimization, the use of personal data should be as minimum as possible for the intended use, iv) accuracy, personal data must be kept up to date while inaccurate data must be rectified or erased altogether, v) storage limitation, personal data must retain a specific form in order for the subjects to be identified and processed only for as long as it is needed for the purposes for which it was collected, vi) integrity and confidentiality, data must be processed in a secure way and protected from potential damage with the use of technical methods. The controller of the data must be compliant with all principles aforementioned (accountability). Furthermore, GDPR outlines specific rights that data subjects have regarding the use of their personal data and gives them control over how their data is processed: i) right of access by the data subject, ii) right to rectification iii) Right to erasure ('right to be forgotten'), iv) right to restriction of processing v) right to data portability, v) right to object and vi) right no to be subject to automated individual decision-making, including profiling. More importantly, when the learner's age is under the age of 16 the controller should confirm that consent is given or authorized by the holder of parental responsibility [14].

An important step in utilizing LA and data mining techniques effectively is to carefully study the GDPR's regulations on processing data in a fair and lawful manner, and examine whether the GDPR guidelines encourage or hinder the use of these techniques. For instance, Amo et al. [1], raised an issue regarding data ownership and digital traces left behind by learners. If a computer model is created using a set of data records from a particular system, there should be a way for a student to decline the inclusion of their data records in such a model. In addition, it is currently difficult for educators to sufficiently manage the GDPR's requirements while using LA and data mining tools.

3 Learning Analytics and Educational Data Mining

Data mining is the process of discovering and extracting useful information stored in large databases ultimately shaping it into information that can be easily perceived by humans [3]. It has incorporated many techniques from other domain fields like statistical analysis, artificial intelligence (machine learning), information retrieval, data warehouse, pattern recognition, algorithms, and high-performance computing. The use of data mining techniques has begun to expand significantly [15] as it is recognized that it can provide valuable insights and transform the field of education into a technologically advanced environment with multiple dynamics. Decision makers use the data-driven models provided by data mining techniques to achieve their objectives for improving the efficiency and quality of learning processes [15].

LA and EDM are two data-centric fields that have emerged in recent decades and utilize machine learning in educational research. Lemay et al. did an extensive review and reported that although these two fields are considered interchangeable, due to their shared themes, some researchers maintain distinct boundaries that differentiate them [16]. Romero et al. [12], state that there is a significant overlap between these areas both in terms of the objectives and the methods used by researchers. Baker et al. [17], support that there is considerable overlap between the two terms regarding the benefits provided to the learners and education in general but they also highlight key differences. For instance, EDM researchers are interested in the automated discovery of information within educational data, while LA researchers are more interested in the human aspect of data discovery.

LA is the process of collecting, analyzing, and reporting data about learners' behavior, that are generated through various digital learning environments. As a field, it combines education, psychology, computer science, and data science. LA provides valuable insights into how students learn, what education strategies and methods are effective as well as how online learning platforms can be improved [6]. EDM is used to uncover multiple patterns and relationships that can enable decision-making about teaching and learning utilizing statistical methods to educational data, data mining, and machine learning. Moreover, the use of data mining reveals valuable insights that are often difficult to uncover through alternative methods [18]. LA and EDM share, among others, a common interest in the learning outcomes of students from a learning process.

The ability to capture so much information about the learning experience holds great potential for increasing our understanding of the educational process [19] nevertheless it is imperative to address the privacy and ethical concerns associated with the use of data in education and to ensure that users' data is being handled responsibly. Therefore, it is vital for decision-makers to clarify to the learners the reasons for using analytics, who has access to the data, and the anonymization process of user's data [20].

4 Privacy Frameworks

In 2016, Steiner et al. [21], proposed LEA's framework in order to establish a set of principles relating to privacy and data protection for LEA's BOX project. They conducted a thorough analysis of previous frameworks and combined different aspects including ethical considerations obtained from discussions, national and European regulations, and existing guidelines and approaches. The eight fundamental requirements are data privacy, purpose and data ownership, consent, transparency and trust, access and control, accountability and assessment, data quality, and data management and security. This framework has a significant impact on learning analytics on a large scale and it may be adopted as a foundational basis for other learning analytics projects. Although this research acknowledged the significance of privacy policies and safeguarding personal information leading to creating the LEA's project, there is a lack of definite execution in terms of privacy measures.

In the same year, Drachsler et al. [8], developed a practical framework/checklist in order to ensure that educational data for LA are employed in a manner that is both acceptable and complies with regulations. The initials of the framework's name represent eight action points. DELICATE: Determination (Why do we want to apply Learning Analytics?), Explain (We have to be open about our intentions and objectives), Legitimate (Why are we allowed to have data?), Involve (We should Involve all data subjects and stakeholders), Consent (We should make a contract with the data subjects), Anonymize (Make the individual not retrievable), Technical (Procedures to guarantee privacy), and External (If we work with external partners). As a result, the DELICATE framework suggests that it is important for LA projects to be approached with a value-sensitive design process, where ethical and privacy considerations are given equal weight to functional requirements. This approach helps ensure that LA can deliver valuable insights for students and educators while protecting the privacy and security of students' data.

In 2020, Amo et al. [1], presented the LEDA framework(Local Educational Data Analytics) to perform data analysis locally within the classroom without transferring it and relying on cloud computing to process it. The LEDA framework addresses the privacy and security requirements for The New Learning Context (NCA), a new educational framework developed at La Salle institutions, in Barcelona, Spain. They synthesize a set of guidelines to serve as a basis for the secure and ethical analysis of educational data, personal data, and metadata regardless of the individual's educational role. The principles are Legality, Transparency, Information, and Expiration, Data Control, Anonymous Transactions, Responsibility in the Code, Interoperability, and Local First. LEDA's emphasis on prioritizing local data processing is effective in safeguarding the privacy of students during synchronous learning. However, it may be challenging to implement these same principles in an asynchronous setting such as an online course or distance learning. Additionally, it is unclear how to export the data for further processing once the lesson has concluded.

5 Discussion

The frameworks that were examined provide useful guidance for maintaining privacy in EDM and LA processes. However, there is a lack of clear instructions on how to put these requirements into practice. Compliance with GDPR is a cumbersome process for the stakeholders of learning platforms. This process could be integrated into existing learning platforms (such as Moodle) and such be made easier to maintain. For example, a plugin that integrates into an existing learning platform could offer a simpler interface for both students and educators to manage their privacy preferences and affect the data mining process accordingly. Another gap that rises from existing frameworks is how to handle the already processed data in the case of students modifying their privacy preferences. The plugin could automate the process of re-analyzing the data each time the user's preferences are modified, thus maintaining compliance with GDPR.

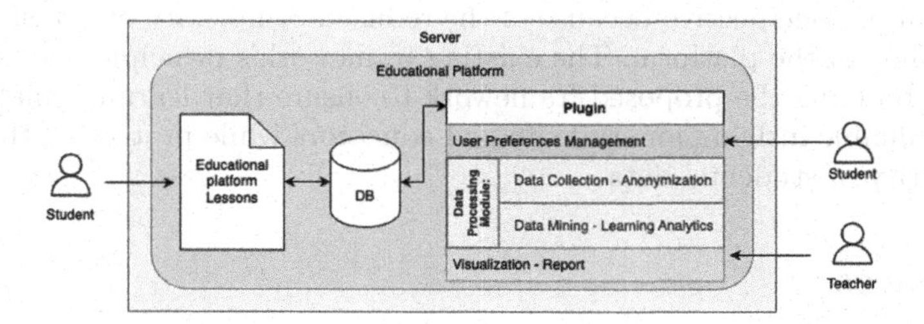

Fig. 1. Proposed framework layout.

Based on the above a proposed framework could be as the following (see Fig. 1). The proposed framework could be integrated into an existing LMS platform in the form of a plugin. As such, the data processing can be handled internally within the application without requiring communication with external APIs. A UI could be provided to the students to allow them to adjust their user preferences at any time, without requiring technical expertise. This UI would be designed to be user-friendly and intuitive, presenting the privacy policy in a clear and easy-to-understand way. Furthermore, if the privacy policy is modified, the plugin would handle the user's notification process. One of the key components of the framework would be a data processing module, responsible for collecting, cleaning, and transforming data. Upon the completion of the data collection and mining processes, the processed data are stored in the existing LMS database and are presented to the stakeholders (i.e., teachers) using visualization techniques, minimizing the need to export results from the application.

6 Conclusion and Future Work

The increasing digitization of educational environments produces an extensive amount of information that can be leveraged to improve the learning process. LA and EDM play a crucial role in effectively extracting valuable insights from different digital educational environments [20]. At the same time, it is important to recognize that data collected and processed are an asset that requires sustained protection through data protection policies and regulations. Educational institutions should devise plans and policies for learning analytics, clarify the reasons for using it, and create protocols for anonymization. GDPR introduced the 'Privacy by Design' approach which means that privacy policies are incorporated into the system architecture instead of being added later in the process [14].

This study presented various privacy frameworks establishing data protection, security principles, and applications. At the same time, it was observed that the data analysis usually happens outside the platform to be done locally or on different trusted servers. In light of this, we propose the development of a framework that incorporates GDPR and security policies and immediately processes the learner's data within the digital educational platform in the form of a plugin. With this type of solution, we can also achieve a better user experience from the educator's perspective because of the reduced complexity of transferring the data outside of the platform. The existing framework's principles can serve as a basis for building the proposed framework to ensure that learning analytics can deliver valuable insights for students and educators while protecting the privacy and security of students' data.

References

1. Amo, D., Torres, R., Canaleta, X., Herrero-Martín, J., Rodríguez-Merino, C., Fonseca, D.: Seven principles to foster privacy and security in educational tools: Local Educational Data Analytics. In: ACM International Conference Proceeding Series, pp. 730–737 (2020)
2. UNESCO: minding the data: protecting learners' privacy and security. UNESCO (2022). https://unesdoc.unesco.org/ark:/48223/pf0000381494
3. Vidakis, N., Barianos, A., Trampas, A., Papadakis, S., Kalogiannakis, M., Vassilakis, K.: In-game raw data collection and visualization in the context of the "ThimelEdu" educational game. Commun. Comput. Inf. Sci. **1220**, 629–646 (2020)
4. Peña-Ayala, A.: Educational data mining: a survey and a data mining-based analysis of recent works. Exp. Syst. Appl. **41**(4, Part 1), 1432–1462 (2014), https://www.sciencedirect.com/science/article/pii/S0957417413006635
5. Pardo, A., Siemens, G.: Ethical and privacy principles for learning analytics: Ethical and privacy principles. Br. J. Educ. Technol. **45**(3), 438–450 (2014), https://onlinelibrary.wiley.com/doi/10.1111/bjet.12152
6. Marshall, R., Pardo, A., Smith, D., Watson, T.: Implementing next generation privacy and ethics research in education technology. Br. J. Edu. Technol. **53**(4), 737–755 (2022)

7. Slade, S., Prinsloo, P.: Learning analytics: ethical issues and dilemmas. Am. Behav. Sci. **57**(10), 1510–1529 (2013)
8. Drachsler, H., Greller, W.: Privacy and analytics - it's a DELICATE issue. a checklist for trusted learning analytics. In: LAK 2016: Proceedings of the Sixth International Conference on Learning Analytics & Knowledge (April 2016)
9. Viberg, O., Mutimukwe, C., Grönlund, A.: Privacy in LA research: understanding the field to improve the practice. J. Learn. Anal. **9**(3), 169–182 (2022)
10. The OECD privacy framework, organisation for economic co-operation and development. https://www.oecd.org/sti/ieconomy/oecd_privacy_framework.pdf. Accessed 14 Feb 2023
11. Kalloniatis, C., Kavakli, E., Gritzalis, S.: Addressing privacy requirements in system design: the PriS method. Requir. Eng. **13**, 241–255 (2008)
12. Romero, C., Ventura, S.: Educational data mining and learning analytics: an updated survey. WIREs Data Min. Knowl. Discov. **10**(3), e1355 (2020), https://onlinelibrary.wiley.com/doi/abs/10.1002/widm.1355, _eprint: https://onlinelibrary.wiley.com/doi/pdf/10.1002/widm.1355
13. A brief guide to GDPR for schools and teachers. https://www.schooleducationgateway.eu/en/pub/resources/tutorials/brief-gdpr-guide-for-schools.htm
14. Regulation (EU) 2016/679 of the European Parliament and of the Council of 27 April 2016 on the protection of natural persons with regard to the processing of personal data and on the free movement of such data, and repealing Directive 95/46/EC (General Data Protection Regulation) (Text with EEA relevance) (April 2016). http://data.europa.eu/eli/reg/2016/679/oj/eng, Legislative Body: EP, CONSIL
15. Aldowah, H., Al-Samarraie, H., Fauzy, W.M.: Educational data mining and learning analytics for 21st century higher education: a review and synthesis. Telemat. Inform. **37**, 13–49 (April 2019), https://linkinghub.elsevier.com/retrieve/pii/S0736585318304234
16. Lemay, D.J., Baek, C., Doleck, T.: Comparison of learning analytics and educational data mining: a topic modeling approach. Comput. Educ. Artif Intell. **2**, 100016 (2021). https://linkinghub.elsevier.com/retrieve/pii/S2666920X21000102
17. Baker, R., Siemens, G.: Educational data mining and learning analytics. In: The Cambridge Handbook of the Learning Sciences, pp. 253–272. Cambridge University Press (January 2014)
18. Romero, C., Ventura, S.: Educational data mining: a review of the state of the art. Syst. Man Cybernet. Part C Appl. Rev. IEEE Trans. **40**, 601–618 (2010)
19. Gasevic, D., Dawson, S., Jovanovic, J.: Ethics and privacy as enablers of learning analytics. J. Learn. Anal. **3**(1), 1–4 (2016), https://learning-analytics.info/index.php/JLA/article/view/4956, number: 1
20. Tsai, Y.S., Whitelock-Wainwright, A., Gašević, D.: The privacy paradox and its implications for learning analytics. In: Proceedings of the Tenth International Conference on Learning Analytics & Knowledge, pp. 230–239. ACM, Frankfurt Germany (Mar 2020), https://dl.acm.org/doi/10.1145/3375462.3375536
21. Steiner, C.M., Kickmeier-Rust, M.D., Albert, D.: LEA in private: a privacy and data protection framework for a learning analytics toolbox. J. Learn. Anal. **3**(1), 66–90 (2016), https://learning-analytics.info/index.php/JLA/article/view/4588, number: 1

Internet of Cloud (IoC): The Need of Raising Privacy and Security Awareness

Asimina Tsouplaki(✉)

Privacy Engineering and Social Informatics Laboratory, Department of Cultural Technology and Communication, University of the Aegean, University Hill, 81100 Mytilene, Greece
atsouplaki@aegean.gr

Abstract. In our fast-paced society, due to the interconnection of millions of devices on the Internet, a great amount of complex data is shared, received, and managed by facilitating communication among users and devices globally. As a result, users from different educational backgrounds and professions are exposed to cyber threats and risks. Current megatrends such as Industry 4.0, Internet of Things (IoT), Cloud Computing, Metaverse, and 6G technology introduce a new era of user experience through strong connectivity, and costless service with high adaptability, however, it raises major data privacy and security concerns. These trends can ameliorate people's lives; however, they are characterized as "attractive vulnerable targets" for cyber-attacks with high intent to harm and disrupt people's daily lives. This paper focuses on two main pillars: the privacy and security fundamental challenges of the synergy between the Internet of Things (IoT) and the Cloud which creates the Internet of Cloud (IoC) and the need of raising awareness to prevent violation of user's data on the IoC. The scope of this paper is to indicate the importance of privacy and security concerns on IoC and to explain why privacy awareness training should be a priority.

Keywords: Internet of Cloud · Internet of Things · Security · Privacy · Cybersecurity and training awareness · Cloud Computing

1 Introduction

Nowadays, data play an integral and vital role in our daily lives and in our economies. According to ENISA, considering only the EU 27 area, the value of data to the economy predicted for 2025 will be €829 billion, up from €301 billion (2.4% of EU GDP) in 2018. Over the last twenty years, we have experienced a steady increase in the amount of data being generated, processed, and transmitted to different parties to create new value [1]. According to Gartner [2], data sharing is a business necessity as it can empower digital transformation and innovation. However, all these create vulnerable data privacy and security ecosystems, through the Internet of Things (IoT) devices and Cloud Computing, for general users and for stakeholders such as policymakers, Data Protection Officers (DPOs), developers, designers, etc. It's worth mentioning that the introduction and integration of Industry 4.0 devices, platforms, and frameworks to existing systems

S. Nurcan et al. (Eds.): RCIS 2023, LNBIP 476, pp. 542–550, 2023.
https://doi.org/10.1007/978-3-031-33080-3_36

come with the issue of interoperability. Securing interconnectivity among complex types of devices such as the Internet of Things (IoT) which are fully related and dependent on Cloud services, arises data privacy and cybersecurity issues [3]. In our contemporary society, 5G technology also generates important security concerns, due to the continuous and wireless connection of a tremendous number of IoT devices [4]. Moreover, 6G wireless networks improve 5G by increasing reliability, enhancing connectivity, the speed of data transmission, and the increase of the available bandwidth [5]. All these benefits arise from 6G networks accelerate the integration of IoT and Cloud into the Internet of Cloud (IoC) which created a new series of privacy challenges [6]. Apart from the current known challenges, the Internet of Cloud (IoC) reveals a new set of cyber threats and risks for all users. Simultaneously, the metaverse is expected to be the next major evolution of the Internet. It will bring a great array of benefits to human life, society, and the economy; however, security and privacy concerns are inevitable. User information has always been a very important and sensitive concern related to cyber issues in modern society [7]. According to Cisco Privacy Survey in 2022, 46% of consumers believe they cannot sufficiently protect their data, with the biggest cause being a lack of understanding of the nature, purpose, and use of their data by companies. Privacy laws continue to be viewed very positively by consumers around the world, but awareness of these laws and the protections they afford remain low. A very controversial key statistic resulting from this survey is that most respondents (54%) said they are willing to share their anonymized personal data to help improve AI products and decision-making. They believe that the potential benefits outweigh the risk, assuming proper anonymization and de-identification techniques are employed [8]. From the above, it is obvious that the level of users' unawareness of data privacy issues remains high. Hence, proper training is essential to be cultivated for all types of users. According to NIST Computer Security Handbook, awareness stimulates and motivates those being trained to care about security and to remind them of important security practices. In addition, the purpose of training is the upskilling of individuals so as to perform their daily digital activities more securely. Training can address many levels, from basic security practices to more advanced or specialized skills. Security education is more in-depth than security training and is targeted at security professionals [9]. The scope of this paper is to present the key privacy and security challenges derived from the Internet of cloud (IoC) and to mention the importance of raising awareness to prevent malicious attacks on users' data. More specifically in Sect. 2, we will present the two interrelated technologies (IoT and Cloud) and then we will give an overview of the Internet of Cloud (IoC). In Sect. 3, we will state the privacy issues that arise from the two technologies based on ENISA's Reports and other sources. Then, in Sect. 4, we will place an emphasis on special training and the value of raising awareness. Finally, in conclusion, we will indicate the next steps for our future work.

2 Internet of Cloud (IoC)

In recent years, we live a technological evolution, with the use of billions of IoT interconnected devices that share and store all types of data (i.e., health, personal, governmental) through Cloud Servers globally, such as: Amazon Web Services (AWS), Microsoft

Azure, Google Cloud Platform (GCP), Alibaba Cloud, Oracle Cloud. The symbiosis seems smooth and obvious because Cloud has attributes that directly benefit the IoT and enables its continued growth, nonetheless, this technological partnership has brought new security challenges, apart from the existing privacy concerns [6]. In this section, we will present the IoT as well as the Cloud components and then we will examine both technologies in combination, to approach some of their most significant privacy and security challenges. The below figure depicts the rapid growth of the Internet of Things (IoT) technologies is revolutionizing how communities live and interact by enabling IoT devices (e.g., mobiles, sensors) to work as interconnected systems of digital and personal information (Fig. 1).

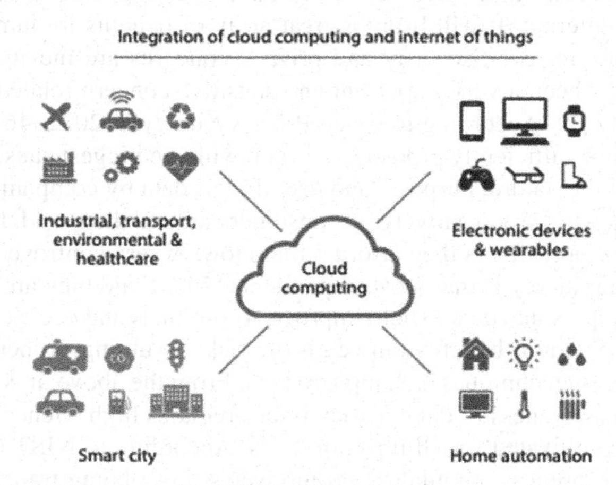

Fig. 1. Correlation between IoT and Cloud [13]

2.1 Cloud Computing

According to NIST, Cloud computing is a model for enabling ubiquitous, convenient, on-demand network access to a shared pool of configurable computing resources (e.g., networks, servers, storage, applications, and services) that can be rapidly provisioned and released with minimal management effort or service provider interaction. This cloud model is composed of five essential characteristics, three service models, and four deployment models [10]. Cloud computing gained much attention due to numerous advantages that offer especially flexibility and elasticity to customers through the existence of pay-as-you-use services. In contrast, many drawbacks do exist that make cloud environments vulnerable to various threats depending on the service model used [11]. In addition, Cloud computing consists of complex technologies using different types of services which leads to extra growth of IoT devices. This combination of IoT and the Cloud created future opportunities regarding the exploitation of smart devices [6].

2.2 Internet of Things (IoT)

Kevin Ashton invented the term 'IoT' back in 1999 for promoting the Radio Frequency Identification (RFID) concept, which includes embedded sensors and actuators. However, the original idea was introduced in the 1960s. During that period, the idea was called pervasive computing or embedded Internet [12]. The Internet of Things is an emerging technology that consists of and connects different kinds of devices: smart appliances, smart cars, smartphones, sensors, surveillance systems, etc. All these devices can communicate, exchange, share, collect and analyze heterogeneous data. The popularity of IoT or the Internet of Things has increased rapidly, as these technologies are used for various purposes, including communication, health, transportation, education, and business development. The heterogeneous nature of the IoT technology ecosystem is characterized as critical and as a result, security plays a vital and essential role that must be taken into consideration [13]. Additionally, this poses a significant risk as lack of security makes users skeptical about using IoT devices [14]. Consequently, IoT devices are vulnerable to security attacks, ultimately causing enormous financial and reputational losses. The fast multiplication of IoT devices in correlation with the lack of data security, transforming these devices into a "mecca" for malicious activities [15].

2.3 Combined Technologies

The synergy between the Cloud and the IoT creates the Internet of Cloud (IoC). As a result, IoC depicts the need for a combination of these two technologies due to the large volume of data as well as the use of billions of interconnected new-generation devices. The adoption of Cloud services from IoT offers many advantages to users such as limitless storage with low cost and easy access remotely via an internet connection. However, the complexity of this technological fusion arises the existing security challenges and underpins new privacy and security concerns both for general users and for sector stakeholders. Moreover, when two or more parties decide to share their data, they become part of a larger data ecosystem where they can take advantage of the combined data that enables the discovery of new information or trends relating to individuals, groups of individuals, or to society as a whole [6]. According to ENISA, many Europeans still fail to take basic cybersecurity measures: many say they care a lot about their personal data, but then give them away for free on social networks. The cybersecurity sector has been at the forefront due to the greater exposure of citizens to multiple cyber threats. [16]. In the third section, we will mention and enumerate the most significant privacy and security issues on IoC.

3 Security and Privacy Challenges

As reported on the ENISA Threat Landscape 2022, cybersecurity attacks continued to increase during the last two years, not only in numbers but also in terms of their societal and economic impact. Due to the volatile geopolitical situation and the Russia-Ukraine war, we face a new era of international cyber warfare and hacktivism that will anticipate having more serious and convoluted consequences. Among the top cybersecurity threats that the ENISA report summarizes, three of them are spotted because they are reflected in the scope of this paper. These are the: a) threats against data, b) threats against availability: Denial of Service, and c) Threats against availability: Internet threats [17]. Moreover, the pandemic has more than confirmed the importance of being prepared for the digital decade as well as the need to continually improve cyber-resilience, particularly for those who operate essential services such as healthcare and energy [18]. Subsequently, from ENISA's report key findings as well as from further bibliography research, it was detected that a research gap still exists, regarding the data privacy and security issues arising from both IoT and Cloud technologies as well as from their combination which may have new, more complex, worse, and more impactful privacy and security challenges [17]. In the below section, we will give an overview of IoC privacy and security challenges derived from the correlation and the interconnection of IoT and Cloud ecosystems.

3.1 Internet of Cloud (IoC) Privacy and Security Challenges

ENISA in its short report regarding the secure convergence of Cloud and IoT, characterizes the IoT as a cyber-physical ecosystem of interconnected devices, which facilitates intelligent decision-making by generating a great amount of data. The cloud is a vital part of this ecosystem that manages the storage and flow of this infinite amount of data. With IoT devices proliferation, the Cloud evolved in such a way to accommodate the extended needs of IoT ecosystem by providing new features regarding the storage and the process of data. Among these features are device virtualization, business intelligence tools, machine learning, command and control (C&C), processing to perform complex analytics, and Application Programming Interfaces (APIs). As a result of these new, complex, and extended features, IoT and Cloud are in full technological alignment which also raises new risks and multidimensional privacy challenges that have not been addressed totally yet [19]. By taking into consideration all the above, IoT can be described as a multiple-domain ecosystem and each domain has its own trust, privacy, and security challenges. Some of these challenges are: 1. User privacy and data protection in process of communication and data exchange among devices, 2. Authentication and Identity Management (this is achieved by combining both authentication techniques and identity management techniques), 3. Policy integration for establishing secure communication between user and IoT device, 4. Authorization and access control (access control is concerned with the process of controlling resources' access and authorization can be achieved upon the use of access control), 5. End–to–End Security (ensures both sides secure communication which is hidden from anyone and it is not possible to modify the transmitting data), 6. Attack-resistant security solutions (diverse devices in IoT have different characteristics and are prone to attacks) [20]. The vulnerable structural characteristics of IoC remain a major concern of regulatory bodies. Hence, the introduction of the European General Data Protection Regulation (GDPR) enables users to control how their data is accessed and processed, requiring consent from users before any data manipulation is carried out on their personal data by smart devices or cloud-hosted services. Moreover, GDPR has been extended to also include IoT environments, to give users the right to control their data and restrict how such data is shared and processed. [21]. Several privacy challenges of Cloud are presented below: 1) Privacy data security (data and privacy disclosures, threats on access right management, data destruction difficulties),

2. Authentication and access control (Cloud computing involves massive resources; the management complexity of access control and identity authentication expands dangerously), 3. Virtualization security (attacks among virtual machines cannot be completely avoided), 4. Multi-tenant and cross-domain sharing (multi-user security is in jeopardy because service authorization and access control are more complex), 5. System security vulnerability (due to the complexity of a cloud computing system, many service providers have different management and service levels, so security vulnerabilities will increase the cyber threats), 6. Insider Threat (International information leakage of service providers often overcomes security policies and causes security issues on Cloud), 7. Wrong or illegal use of Cloud service (causes security problems to users, service providers, and third parties), 8. Service availability (usual Cloud service unavailability and denial of service attacks have become significant privacy issues) [21]. We can conclude that some of the recognizable security challenges are overlapped in both technologies. According to ENISA, three main security challenges of the Internet of Cloud are identified and presented in the below table in correlation with security recommendations: 1. Connectivity (interactions and communications among endpoints, gateways, and Cloud), 2. Analysis (processing, filtering, and aggregation of the data coming from the IoT devices in different levels of the IoT ecosystem), 3. Integration (features that enable real-time bidirectional flow of data (eg. Cloud APIs and remote command and control (C&C) of IoT devices through the Cloud) [19] (Table 1)

Table 1. IoC security challenges and security takeaways identified by ENISA [19]

IoT security challenges	Security challenges	Security takeaways
1. Connectivity	-Heterogeneous protocols for communication	-Device virtualization to bring homogeneity
	-Insecure data flow from the Edge to the Cloud	-Secure communications, security stream analysis and security of data at rest
2. Analysis	-Real-time processing at the edge overshadows security	Physical and cyber security in edge devices
	-Impact of Cloud decentralisation on security	
3. Integration	- Security depends on the vertical that Cloud is serving	Addition of security elements to IoT environment Adoption of baseline security measures
	-Security relies much on the implementation from IoT developers	Automated, secure software updates
	-Outdated devices	End-to-end security, through the whole environment

Therefore, the combination of these two technology environments creates new and more challenging privacy threats that we face now and in the near future, so they need further multi-scope and in-depth research.

4 Training and Awareness

It is obvious from all the above mentioned that as the technology evolution is growing with extreme growth, however, the percentage of unaware users, regarding data security and privacy, it remains high. The need for targeted, and well-structured training and raising awareness programs is inevitable, significant, and mandatory. According to NIST, related training programs should be user-friendly and designed with useful information regarding the current technological and cyber threats to address security issues that directly affect the users. The goal is to improve basic security practices, not to make everyone literate in all the jargon or philosophy of security [9]. ENISA's Report for Cyber awareness identifies and presents five key objectives for a successful design of raising cybersecurity awareness programs and promoting cybersecurity education and culture for all types of users. These objectives are: 1. Awareness (generate awareness about the impact of different types of attacks), 2. Information (provide detailed information on how to react in the event of phishing and ransomware attacks). 3. Engagement (Prompt the target audience to act and eventually spread the word on what they learned from you), 4. Promotion (Promote the safer use of the internet for end users and promote existing cybersecurity recommendations and best practices to prevent cyberattacks), 5. Empowerment (provide users with resources to protect themselves online and prevent attacks [23]. Make people become 'human firewalls' by empowering them to play their part in preventing attacks [24].

5 Conclusion

It is understandable, from all the above, that the exploitation, the use, and a combination of disruptive technologies, with complex data, such as the Internet of Things (IoT), AI, and the Cloud cause numerous privacy and security issues, and as a result, users should face multiple complexities of cyber, economical, and societal treats, as a collective system. GDPR Compliance, to be fully implemented, requires strong training and awareness background for all users. Moreover, the new emerging era of the Internet of Cloud (IoC) brings more challenging cyber threats. Hence, there is an imperative need for customized training and awareness programs for the different types of users and stakeholders. For in-depth scientific contribution, our future related work will focus on 1. Further analysis and comparison of existing security training tools and frameworks that ensure privacy in the IoC, 2. Mapping and analyzing the key privacy threats, and providing a recommended list of ways to deal with cyber threats which will ensure IoC security, 3. Develop and design a methodology of a training guide for users' familiarization, practice and awareness of the public on privacy issues in IoC, 4. Verification of this methodology and 5. Apply to a real environment so as to help people become cyber-responsible and ethical users.

References

1. Drogkaris, P.: Engineering Personal Data Sharing, (ENISA). ISBN:978-92-9204-602-6 (2023). https://doi.org/10.2824/36813
2. Gartner Research: Data Sharing Is a Business Necessity to Accelerate Digital Business (2021)
3. Malatras, A., Skouloudi, Ch., Koukounas, A.: Industry 4.0 Cybersecurity: Challenges & Recommendations (ENISA). ISBN: 978-92-9204-293-6 (2019). https://doi.org/10.2824/143986
4. Sicari, S., Rizzardi, A., Coen-Porisini, A.: 5G In the internet of things era: an overview on security and privacy challenges. Comput. Netw. **179**, 107345 (2020). https://doi.org/10.1016/j.comnet.2020.107345
5. Nguyen, T., et al.: Privacy-aware blockchain innovation for 6G: challenges and opportunities (IEEE), In: 2020 2nd 6G Wireless Summit. https://doi.org/10.1109/6GSUMMIT49458.2020.9083832 (2020)
6. Cook, A., Robinson, M., Ferrag, M. A.: Internet of Cloud: Security and Privacy issues (2017), http://arxiv.org/abs/1711.00525
7. Zhao, R., Zhang, Y., Zhu, Y., Lan, R., Hua, Z.: Metaverse: Security and Privacy Concerns (IEEE). arXiv. http://arxiv.org/abs/2203.03854 (2022)
8. Cisco 2022 Consumer Privacy Survey. https://www.cisco.com/c/dam/en_us/about/doing_business/trust-center/docs/cisco-consumer-privacy-survey-2022.pdf
9. Guttman B., Roback E.: An Introduction to Computer Security: The NIST Handbook 1995)
10. Mell, P., Grance, T.: The NIST Definition of Cloud Computing (2011)
11. Simou, S., Kalloniatis, C., Mouratidis, H., Gritzalis, S.: A meta-model for assisting a cloud forensics process. In: Lambrinoudakis, C., Gabillon, A. (eds.) CRiSIS 2015. LNCS, vol. 9572, pp. 177–187. Springer, Cham (2016). https://doi.org/10.1007/978-3-319-31811-0_11
12. Al-Sarawi, S., Abdullah, R., Al Hawari, A.: Internet of Things market analysis forecasts. 2020–2030 (IEEE),INSPEC Accession Number: 20023664 (2020), https://doi.org/10.1109/WorldS450073.2020.9210375
13. Nivedita, M., Sharnil, P.: Internet of Things Applications, security challenges, attacks, intrusion detection, and future visions: a systematic review IEEE Access 8, 59353–59377 (2021). https://doi.org/10.1109/ACCESS.2021.3073408
14. Pan, Y., Naixue, X.: Data security and privacy protection for cloud storage: a survey. IEEE Access **8** (2020). https://doi.org/10.1109/ACCESS.2020.3009876
15. European Union Agency for Cybersecurity (ENISA): Guidelines for Securing the Internet of things (2020). ISBN: 978-92-9204-411-4. https://doi.org/10.2824/314452
16. European Union Agency for Cybersecurity (ENISA): Cybersecurity Education Initiatives in the EU Member States (2022). https://doi.org/10.2824/486119
17. European Union Agency for Cybersecurity (ENISA): Threat Landscape (2022). https://www.enisa.europa.eu/publications/enisa-threat-landscape-2022
18. Negreino, M.: The NIS2 Directive, a High Common Level of Cybersecurity in the EU, European Parliament (2022)
19. Skouloudi, Ch., Fernaddezn, G.: Towards secure convergence of cloud and IoT. In: ENISA (2018)
20. Abomhara, M., Køien, G.: Security and privacy in the Internet of Things: current status and open issues. In: International Conference on Privacy and Security in Mobile Systems (PRISMS), pp. 1–8 (2014)
21. Barati, M., Rana, O.: GDPR Compliance Verification in Internet of Things, (IEEE) (2020). https://doi.org/10.1109/ACCESS.2020.3005509
22. Sun, P.: Security, and privacy protection in cloud computing: discussion and challenges. J. Netw. Comput. Appl. **160**, 102642 (2020). https://doi.org/10.1016/j.jnca.2020.102642

23. Karale, A.: The challenges of IoT addressing security, ethics, privacy, and laws, Internet of Things **15** (2021). https://doi.org/10.1016/j.iot.2021.100420
24. European Union Agency for Cybersecurity (ENISA): Raising Awareness of Cybersecurity: A Key Element of National Cybersecurity Strategies. Publications Office (2020). https://doi.org/10.2824/363629

Comparative Study of Unsupervised Keyword Extraction Methods for Job Recommendation in an Industrial Environment

Bissan Audeh[1(✉)], Maia Sutter[2], and Christine Largeron[2]

[1] INASOFT Company, 2507 avenue de l'Europe, 69140 Rillieux-La-Pape, France
bissan.audeh@inasoft.fr
[2] UJM-Saint-Etienne, CNRS, Institut d'Optique Graduate School, Laboratoire Hubert Curien UMR 5516, 42023 Saint-Etienne, France
maia.sutter@etu.univ-st-etienne.fr, chistine.largeron@univ-st-etienne.fr

Abstract. Automatic keyword extraction has important applications in various fields such as information retrieval, text mining and automatic text summarization. Different models of keyword extraction exist in the literature. In most cases, these models are designed for English-language documents, including scientific journals, news articles, or web pages. In this work, we evaluate state-of-the-art unsupervised approaches for extracting keywords from French-language Curricula Vitae (CVs) and job offers. The goal is to use these keywords to match a candidate and a job offer as part of a job recommendation system. Our evaluation showed that statistical baselines obtain good results with an interesting processing time in an industrial context. It also allowed us to highlight, on the one hand, biases related to pre-trained word embedding models on corpora of a different nature than CVs and job offers, and on the other hand, the difficulties of annotation within the framework of job search platforms.

Keywords: Information extraction · Keyword extraction · Job recommendation system

1 Introduction

Matching a job offer and the CV of a job seeker is a challenging task that requires advanced natural language processing, text analysis and information retrieval. In traditional hiring systems, employers create job offers describing the functions associated to the proposed position and the required profile. Job seekers navigate through these offers and apply to the ones that interest them, or they send a spontaneous candidacy hoping to be contacted for the opening of a pertinent position. Most hiring systems keep job offers and candidates profiles in a structured storage, often a database, that enables executing job seekers' queries to retrieve corresponding job offers. The difficulty in this configuration is that the candidate needs to know suitable keywords when searching for job offers.

S. Nurcan et al. (Eds.): RCIS 2023, LNBIP 476, pp. 551–558, 2023.
https://doi.org/10.1007/978-3-031-33080-3_37

In many cases, candidates are not familiar with the vocabulary used by employers when creating the job propositions, which causes them to miss pertinent job offers just because they did not use the matching keywords in their queries. Proposing job offers to candidates based on semantic matching can increase hiring possibilities and facilitate job retrieval. Assigning keywords to job offers and CVs in such context is essential, as it facilitates indexing and matching and allows for the realization of statistical analysis and classification, which can be a powerful tool for business intelligence and decision making.

Keywords are simple or composed terms that represent essential information about a text. In the context of hiring systems, manually assigning keywords by job seekers and recruiters is not always possible or not systematically applied. In such contexts, automatic keyword extraction can be very helpful. We focus in this paper on the use of unsupervised approaches for keyword extraction. We aim to evaluate to what extent keywords extracted with these approaches can represent text content in French even with no training and no external knowledge in a context of job recommendation. For this, we established an evaluation protocol that includes the construction of a gold standard from real CVs and job offers, on which we evaluated six unsupervised methods for keyword extraction.

We start this paper by a presentation of existing unsupervised keyword extraction models in Sect. 2. Section 3 details our experimental protocol including the choice of corpus and models. Section 4 describes the results of the compared keyword extraction methods followed by a discussion in Sect. 5. Finally, conclusion and perspectives are proposed in Sect. 6.

2 Unsupervised Keyword Extraction Models

The most simple unsupervised model for keyword extraction is TF-IDF. This statistical model computes the product of the term frequency (TF) and the inverse document frequency (IDF) for each term in a document and selects as keywords the terms with the higher score. Candidate keywords in this model are n-grams that fall within a given range, such as [1..2] as frequently used in practice. Because of the use of IDF, this model requires computations at the corpus level. Another statistical model often used as a baseline is RAKE [9], which stands for "Rapid Automatic Keyword Extraction". It considers word co-occurrence to calculate the score for the candidate keywords. The model builds a frequency distribution and a co-occurrence graph, then goes through the candidate list to calculate the rank/score of each candidate. Campos et al. [2] introduced YAKE!, a method that implements a maximum similarity above which terms are removed to keep only keywords that differ from each other.

TextRank [6] is based on Google's PageRank algorithm for the purposes of keyword extraction. It constructs a graph, with tokens as nodes and edges between nodes that appear next to each other in the text, using a window of predefined size. Once all the edges have been created, nodes without any connecting edges or neighbors are removed and the remaining nodes are ranked using the PageRank algorithm. Among the other models that were developed based on PageRank, we can mention RaKUn [13] that introduces meta vertices and builds a graph that can maintain sequential information, which means that context is not lost. Sung et al. [11] uses a Hierarchical Semantic Network (HSN) and a multiple centrality network to extract keywords that are representative of a document and its topics.

Another family of unsupervised approaches is based on embedding representations which encode textual data in a vector space that can then be used alongside distance/similarity measures in order to score potential words. EmbedRank [1] is one of these interesting embedding approaches that levels embeddings and distance metrics to calculate the keyness of the candidate keywords based on their embedding's distance from the document embedding. KeyBERT [3] is another embedding approach based on EmbedRank. It aims to maximize the distance between a candidate keyword and other keywords and minimizing the distance between a candidate keyword and the document. Both of these distances are calculated with cosine distance. Papagiannopoulou et al. [7] introduced Reference Vector Algorithm (RVA), an embedding model that learns local embeddings rather than using a pre-trained model in order to capture the context of the specific document being treated. There exist several methods that combine different approaches. For instance, TF-IDFISW [12] uses a TF-IDF model in addition to a set of synonyms, trained from a Word2vec model. Mahdi et al. [5] combine BERT embeddings, a version of TF-IDF based on class, and the topic modeling method Latent Dirichlet Allocation. Recently, Jingxia Ma [4] proposed a hierarchical clustering method that combines embedding vectors with cluster analysis to identify and select keywords based on positional importance which are representative of important ideas or concepts in the document.

3 Methodology

In the state of the art, the models that were the most used as baselines were TF-IDF, TextRank, and RAKE. Thus, we used these approaches for our comparative study. In addition to these baselines, we evaluated two versions of the KeyBERT model. The first, (Keybert) uses the term/document matrix to identify candidates' terms. The second (Keybert+) uses a language model to identify combined terms as candidates instead of individual terms [10]. Similarly, as TF-IDF also uses term/document matrix, we added to our experiments a sixth approach (TF-IDF+) based on the detection of combined terms for TF-IDF instead of individual terms. We implemented TF-IDF and TF-IDF+, but an implementation was already available for all the other approaches evaluated and presented in Table 1. For combined-terms identification in TF-IDF+ and Keybert+, we used KeyVectorizer implementation[1] that needs a language model and an identification pattern to detect a composed terms. As language model we used the French model "fr_core_news_lg" of Spacy[2] and as identification pattern we chose "name followed by 0 or more adjectives".

[1] KeyphraseVectorizers: https://github.com/TimSchopf/KeyphraseVectorizers.

[2] French model Spacy: https://spacy.io/models/fr.

3.1 Corpus and Gold Standard Construction Preparation

For this work, 818 CVs and 858 job offers were provided by Inasoft company after anonymization and removing personal data. This data comes from a panel of active companies specialized in distribution sector.

To measure to what extent keywords extracted by the evaluated models are representative, we evaluated each model m based on the list of keywords Lm_d it produces for each document d. This was achieved by comparing Lm_d to the gold standard list Lv_d of the document d using precision, recall and F-score. To generate the list Lv_d, a manual annotation of a subset of our corpus was needed. This subset was selected randomly from our initial corpus after size-based stratification to respect the initial distribution of the data. To facilitate the manual annotation, for each document d, we used the union of a maximum of top 20 keywords extracted by each evaluated model as a starting base B_d. For each evaluated document d, the annotation protocol consisted in manually selecting the valid keywords from the starting base B_d. The annotators were three, one specialised in recruitment systems, an information retrieval researcher and a master student in machine learning. Annotators had also the possibility of adding additional keywords that do not figure in B_d. Unfortunately, due to time constraints, we were able to assign only 12, 316 confirmed keywords that corresponded to 57 CVs and 29 job offers.

For our protocol, a keyword is defined as a word that is "useful" in representing a profile of a person if the document is a CV, or the main characteristics of a position if the document is a job offer. Keywords considered as valid should also be well-formed and coherent. This removes keywords such as "anglais espagnol," which is badly formed; "commerciales baccalaureat," which is incoherent; or "journée," which is simply not useful. Because this extraction is being done in the context of a future job recommender system, names of locations and companies are considered useful, as long as they are complete.

4 Results

Table 1 and Table 2 present for CVs and job offers respectively, the average (arithmetic mean) of the precision *Avg.Prec*, recall *Avg.Recall*, and F-Score for each model over all the documents at 5 and 15 extracted keywords.

The low F-score results can partially be explained by the bias related to annotation, as annotations were done by three separate annotators. Such bias could be overcome as we will discuss in Sect. 5. Nevertheless, our annotation seams correct since these scores are of the same order as those reported frequently in the literature with automatic keyword extraction.

In our industrial context, keyword extraction should be done in real-time. Thus, execution time is an important evaluation criteria. Model timings were calculated for all 1,676 documents in initial base of the two data sources and the average is presented in Tables 3 and 4, along with the longest and shortest time by model.

Table 1. Evaluation results on CVs at 5 and 15 keywords extracted. At each level of number of keywords, the best score overall is marked in bold.

	Model	*Avg.Prec*	*Avg.Recall*	F-Score
@5	TF-IDF	0.232	0.052	0.092
	TF-IDF+	0.249	0.069	0.108
	TextRank	0.186	0.043	0.070
	RAKE	**0.295**	**0.070**	**0.112**
	KeyBERT	0.133	0.034	0.054
	KeyBERT+	0.221	0.063	0.098
@15	TF-IDF	0.177	0.124	0.146
	TF-IDF+	0.214	0.162	0.185
	TextRank	0.158	0.093	0.117
	RAKE	**0.300**	**0.203**	**0.242**
	KeyBERT	0.148	0.117	0.131
	KeyBERT+	0.205	0.159	0.179

Table 2. Evaluation results on job offers at 5 and 15 keywords extracted. At each level of number of keywords, the best score overall is marked in bold.

	Model	*Avg.Prec*	*Avg.Recall*	F-Score
@5	TF-IDF	**0.193**	0.060	0.092
	TF-IDF+	0.172	0.057	0.085
	TextRank	**0.193**	**0.072**	**0.105**
	RAKE	0.166	0.063	0.091
	KeyBERT	0.041	0.014	0.021
	KeyBERT+	0.090	0.032	0.047
@15	TF-IDF	0.172	0.183	0.177
	TF-IDF+	**0.205**	**0.222**	**0.213**
	TextRank	0.133	0.098	0.122
	RAKE	0.186	0.217	0.200
	KeyBERT	0.053	0.058	0.055
	KeyBERT+	0.115	0.124	0.119

Table 3. Timing results by model on CVs, in seconds. The fastest average time is in bold.

Model	Average Time	St Dev	Max Time	Min Time
TF-IDF	0.0834	0.0181	0.3871	0.0700
TF-IDF+	0.0132	0.0025	0.0274	0.0090
TextRank	0.0244	0.0200	0.1540	0.0020
RAKE	**0.0018**	0.0009	0.0090	0.0
KeyBERT	1.3718	0.2861	4.3494	0.9517
KeyBERT+	2.0873	0.2095	3.6025	1.7957

Table 4. Timing results by model on offers, in seconds. The fastest average time is in bold.

Model	Average Time	St Dev	Max Time	Min Time
TF-IDF	0.0089	0.0150	0.2991	0.0050
TF-IDF+	0.0021	0.0005	0.0060	0.0009
TextRank	0.0076	0.0029	0.0453	0.0040
RAKE	**0.0008**	0.0004	0.0021	0.0
KeyBERT	1.2578	0.1379	2.1478	1.0543
KeyBERT+	2.0939	0.1996	3.4176	1.8722

We can see a large difference in the speed of various models. RAKE lives up to the "rapid" in its name, while KeyBERT and KeyBERT+ rarely, if ever, run in under one second. The others are all decently fast, but TF-IDF and TextRank have the worst maximum times of the three. This places TF-IDF+ in second place for speed, which makes sense given that the vectorizer it uses reduces the possible candidates, allowing it to run much faster.

5 Discussion

Our results showed that among the tested approaches, RAKE and TF-IDF+ models, outperformed the others in speed and accuracy. It should be noted that the other models were used with their default parameters with no optimization. In addition, embedding-based approaches (Keybert and Keybert+) were used with pre-trained embeddings. Additional experiments including fine-tuning these models for our context could bring interesting results. However, in real production environments, the gain in accuracy should be balanced against the cost of training these models, and their processing time which is higher than simpler approaches such as RAKE.

One major difficulty when evaluating unsupervised methods on real data is annotation. Although this difficulty does not concern the final usage of these approaches, it is time consuming at evaluation time, and its subjectivity makes the results difficult to interpret. One possible solution is having multiple annotators working on the same document so that the annotation becomes a consensus, rather than the opinion of a single annotator. The proper annotation of a dataset could be a project in and of itself, in order to annotate enough data, include present keywords that are not found by the models, and integrate the addition of absent keywords. Although our annotation protocol enabled the manual addition of keywords to a document or a CV, annotators mainly selected valid keywords from the list that represents the union of top 20 extracted keywords per model. Another consideration is that CV texts were mainly extracted from PDF documents. This extraction can lead to errors in detecting keywords as it might omit some text from the original document, or introduce misspelled words due to OCR.

While annotating, we noticed that the unigrams that the models extracted were often false positives, but that the bigrams seemed to give more interesting information. This brings up the question of whether the window length for candidate keywords should be expanded to [1..3]. This would allow for the capture of keywords such as "assistante ressources humaines" instead of just "assistante ressources" or "ressources humaines," as the models currently output when only going as far as bigrams. Unigrams in particular lack context, but still remain useful in certain situations, such as with the languages someone lists in their CV or that a job offer might require.

Because evaluated approaches are originally developed and tested with text in English, they don't always take into account punctuation differences that may be common in another languages and give odd results when not considered, such as the use of parentheses in French to add the "e" for the feminine form of a noun or adjective that takes that form, like "communicant(e)". These types of situations need to be taken into account, either to maintain the punctuation or to remove it completely and put the word into either its feminine or masculine form. There are also issues with words that are connected by a dash or that have an apostrophe, such as "e-commerce". In this case, the dash is removed and the two parts are separated, which becomes a problem when the keywords are post-processed and anything under three characters is removed, as "e-commerce" becomes just "commerce".

6 Conclusion

Automatic keyword extraction is a field that has seen a lot of new development in recent years. From a lack of extraction of absent keywords to difficulty taking context into account, the various proposed models have different approaches and solutions. However, much of the work and literature remains focused on English text, especially from long documents such as academic journal articles. This can be a problem when models are needed for domain-specific texts, such as CVs and job offers. The lack of related work on French text in such context prevents us from selecting an appropriate method for keyword extraction. In this work, we tested six models on real anonymized CVs and job offers to see which model would then be chosen to be integrated into an existing recruitment software. Our experiments shows that RAKE and TF-IDF+ models generally outperform the others in speed and accuracy. While TF-IDF+ needs statistics at corpus level to calculate keyword scores, RAKE could be a more suitable choice to take into account in production. In addition, the competitive execution time of this approach is suitable for industrial environment where a resource manager expects real-time responses in the hiring process.

In future work, it would be interesting to investigate keyword extraction methods that go beyond unsupervised or supervised into self-supervision. While not a keyword extraction model, the concept of relational reasoning [8] could presumably be applied to keyword extraction tasks, particularly in cases where annotated data is either minimal or not available. Finally, if this study sheds light on the options for extracting keywords from CVs and job offers, it is only one step in the development of a job referral system that requires not only improvement on this step, as our results show, but also to go further, in particular by integrating semantics into the classification of extracted keywords for the suggestion of relevant job offers for a candidate CV.

Acknowledgement. We would like to thank Servan Cazenave, director of Inasoft, for his support for this project, his contribution on domain knowledge, and his help in annotation.

References

1. Bennani-Smires, K., Musat, C., Hossmann, A., Baeriswyl, M., Jaggi, M.: Simple unsupervised keyphrase extraction using sentence embeddings. arXiv preprint arXiv:1801.04470 (2018)
2. Campos, R., Mangaravite, V., Pasquali, A., Jorge, A., Nunes, C., Jatowt, A.: Yake! keyword extraction from single documents using multiple local features. Inf. Sci. **509**, 257–289 (2020)
3. Grootendorst, M.: Keybert: Minimal keyword extraction with bert (2020). https://doi.org/10.5281/zenodo.4461265
4. Ma, J.: Research on keyword extraction algorithm in English text based on cluster analysis. Comput. Intell. Neurosci. **2022** (2022)
5. Mahdi, H.F., Dagli, R., Mustufa, A., Nanivadekar, S.: Job descriptions keyword extraction using attention based deep learning models with BERT. In: 2021 3rd International Congress on Human-Computer Interaction, Optimization and Robotic Applications (HORA), pp. 1–6. IEEE (2021)
6. Mihalcea, R., Tarau, P.: Textrank: bringing order into text. In: Proceedings of the 2004 Conference on Empirical Methods in Natural Language Processing (2004)
7. Papagiannopoulou, E., Tsoumakas, G.: Local word vectors guiding keyphrase extraction. Inf. Process. Manag. **54** (2018)
8. Patacchiola, M., Storkey, A.J.: Self-supervised relational reasoning for representation learning. Adv. Neural. Inf. Process. Syst. **33**, 4003–4014 (2020)
9. Rose, S., Engel, D., Cramer, N., Cowley, W.: Automatic keyword extraction from individual documents. In: Text Mining: Applications and Theory, pp. 1–20 (2010)
10. Schopf, T., Klimek, S., Matthes, F.: Patternrank: Leveraging pretrained language models and part of speech for unsupervised keyphrase extraction. In: Proceedings of the 14th International Joint Conference on Knowledge Discovery, Knowledge Engineering and Knowledge Management - KDIR. INSTICC, SciTePress (2022)
11. yeon Sung, Y., Kim, S.B.: Topical keyphrase extraction with hierarchical semantic networks. Decis. Supp. Syst. **128** (2020)
12. Zhang, Z., Wu, Z.: Improved TF-IDF algorithm combined with multiple factors. In: 2021 3rd International Conference on Applied Machine Learning (ICAML), pp. 492–495. IEEE (2021)
13. Škrlj, B., Repar, A., Pollak, S.: *RaKUn*: *Ra*nk-based *K*eyword extraction via *Un*supervised learning and meta vertex aggregation. In: Martín-Vide, C., Purver, M., Pollak, S. (eds.) SLSP 2019. LNCS (LNAI), vol. 11816, pp. 311–323. Springer, Cham (2019). https://doi.org/10.1007/978-3-030-31372-2_26

A Meta-model for Digital Business Ecosystem Design

Chen Hsi Tsai(✉), Jelena Zdravkovic, and Janis Stirna

Department of Computer and Systems Sciences, Stockholm University, 7003, SE-16407 Kista, Sweden
{chenhsi.tsai,jelenaz,js}@dsv.su.se

Abstract. The Digital Business Ecosystem (DBE) theory has evolved to facilitate the functioning of open business networks by adopting the ecosystem paradigm from nature in a shared digital environment. While it enables exhibiting diverse interests, it also places high demands on managing the DBE's resilience. The current research lacks support for how a DBE-based business model can be integrated with its supporting information system's data structure to enable the design and monitoring of resilience for indicating the needed adaptations caused by changes in DBE's actors, their engagement or performance balance. We propose and instantiate a meta-model that describes the DBE's entities relevant to its design using the CIVIS ecosystem. The meta-model provides a foundation for a modelling language for the management of the resilience of DBE.

Keywords: Enterprise Modelling · Digital Business Ecosystem · Meta-modelling

1 Introduction

Information systems (IS) have evolved from supporting single organisations or well-defined networks of organisations to supporting networks of partners to deliver inter-linked services. The Digital Business Ecosystem (DBE) theory has emerged as an approach for addressing such highly dynamic, open, self-organising, and resilient networks [1]. A DBE is defined as: "a socio-technical environment, enabled by shared digital platforms and ICTs, where loosely-coupled interdependent organisations and individuals in an economic community deliver, consume, or exchange resources and co-evolve their capabilities and roles [2]." DBE is designed to make the most of resources and capabilities for its various independent actors, even if they have different business intentions. Given the number of actors involved, it is crucial to capture, integrate, and analyse a large amount of information related to the DBE and its actors [3, 4]. Design and management of DBEs require integration of views on the business analysis, organisation design, IS engineering, as well as system operation and management.

Analysis of the state of the art of DBE analysis and design in literature [5] concludes that despite some existing proposals for modelling, many aspects relevant to organising a DBE and ensuring its sustainability are insufficiently addressed or unsolved. From

S. Nurcan et al. (Eds.): RCIS 2023, LNBIP 476, pp. 559–567, 2023.
https://doi.org/10.1007/978-3-031-33080-3_38

the point of view of Enterprise Modelling (EM), the more traditional way of modelling enterprises or constellations of enterprises assumes a small and manageable number of organisations and accessible information about an actor's business design. This might not be the case considering the loose coupling of actors in a DBE. DBEs need to know their operations' status at runtime and execute corrective actions based on performance or business context. The existing modelling methods have been developed and used in settings of a single company or a well-defined network of companies. In these settings, dealing with data needed to calculate application context and performance KPIs is easier because much of the data is local, foreseeable, or can be agreed to be shared. These challenges highlight the scarcity of existing efforts on EM methods for modelling of DBEs in their deployment and operations phases, which is needed to support resilience management. To this end, the goal of this study is to propose a meta-model of DBE to be used as a definition basis for a modelling language for DBE design and management. To demonstrate the preliminary feasibility of the proposal, a case of the CIVIS [6] ecosystem – a network of collaborating universities has been used.

The rest of the paper is structured as follows. Section 2 outlines the research design. Section 3 elaborated on the proposal of the DBE meta-model and its instantiations using the case. A discussion and concluding remarks are presented in Sect. 4.

2 Research Approach

This study is part of a design science research [7] project. In previous studies [2, 5], we have attained the objectives of explicating the problem and eliciting and analysing requirements. Concerning the design and development of the artefact, a DBE modelling method, its outline in terms of method components (chunks) and usage intentions has been constructed in [8]. This study presents (1) the next step, the creation of the meta-model for the envisioned method based on the processes of the method in [8], and (2) a validation of the meta-model as an early demonstration activity using a case study.

The case study of CIVIS [6] is a European university alliance of over ten higher education (HE) institutions. It should enable richer interactions and co-creation of knowledge and skills with citizens, schools, business companies and enterprises, and social and cultural associations by following traditions and welcoming transformations. Each CIVIS member university contributes to the ecosystem and promotes European values such as inclusiveness, gender equality, non-discrimination and social equity.

CIVIS aims to increase access to quality education and create opportunities for the exchange of students and staff among the member universities as an inter-university campus where students, academics, researchers and staff collaborate within their institution of origin. The main activities of the alliance include, among others, the development of physical and online courses and education programs and student and teacher mobility.

3 Proposed Meta-model and Instantiation

The proposed meta-model for a DBE is divided into three aspects - actor & role, goal & performance, and fitness & qualification, presented in Sects. 4.1, 4.2, and 4.3. Some of the concepts were defined in previous work [2]. The definitions and more detailed descriptions of the resilience of a DBE and the four pillars of resilience, diversity, efficiency, adaptability, and cohesion, can be found in [2]. The DBE roles and their corresponding responsibilities are significant for the design of a DBE and its resilience concerning the continuous operation of the DBE (c.f. [2]).

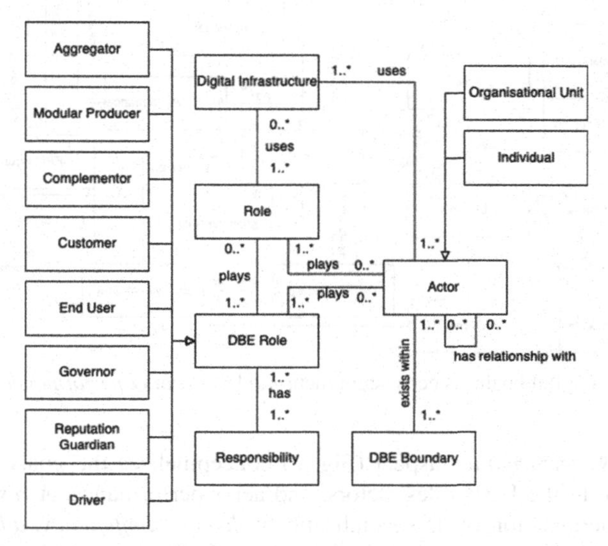

Fig. 1. Digital business ecosystem meta-model – *actor & role* aspect

The *actor & role* aspect (Fig. 1) conceptualises the actors within the boundary of a DBE as a digital environment where they play different (business) roles and DBE roles and utilise digital infrastructures. *A DBE boundary* is the digital border of the DBE - it determines which actors take part in the DBE. *An actor* is an abstract generalisation of *an individual* or an organisational actor (*organisational unit*). Actors can have relationships with each other, such as collaborating or competing relationships. An individual is a specific person who participates in a DBE. An organisational unit represents any organisational structure, such as an enterprise, a department, or a subsidiary. An actor may play different *roles*, e.g., customer or supplier. In some business contexts, it is essential to identify the roles without knowing the actor playing them. This may also depend on at what stage of development a DBE is at the time of modelling. For example, a university is a role of significance in a DBE. However, it might not be necessary to identify all universities participating in a DBE when designing it. This can be done later once the DBE is operational. *A digital infrastructure* is any technical infrastructure used by a specific actor or shared by several actors to distribute digital representations, such as software applications, services, or descriptions of skills in a DBE. *A DBE role* is a generalisation of the eight significant DBE roles: Driver, Aggregator, Modular Producer, Complementor, Customer, End User, Governor, and Reputation Guardian (c.f. [2]). A DBE role can be played by one or many actors and roles within the same DBE. Each DBE role has its corresponding responsibilities, which are DBE specific, as presented in [2].

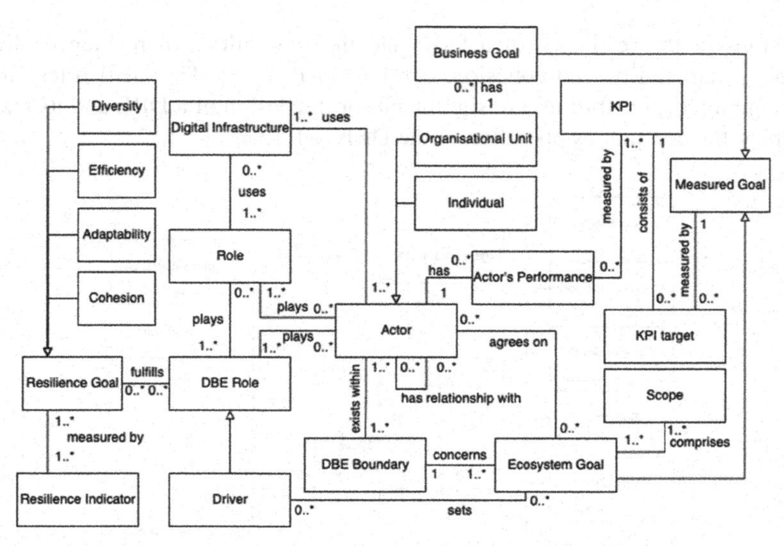

Fig. 2. Digital business ecosystem meta-model – *goal & performance* aspect

The *goal & performance* aspect (Fig. 2) conceptualises the goals of a DBE and how they relate to the DBE roles, actors, and actor performance. *A resilience goal* is an abstract generalisation of its specialisations: *diversity*, *efficiency*, *adaptability*, and *cohesion* (c.f. [2]). Any of the resilience goals are fulfilled to a different extent by the DBE roles. A resilience goal can be measured by resilience indicators which quantify to what extent the resilience goal has been fulfilled. Various resilience indicators are used during the design, deployment, and runtime stages of a DBE. *An ecosystem goal* is a goal that concerns a DBE (boundary) as a whole. It is a common goal shared among the actors in the DBE, commonly initiated by Driver. *A scope* captures high-level goals or mission statements of a DBE and consists of several ecosystem goals. *A business goal* is a desired state of an organisational unit, which expresses what is to be achieved or avoided concerning the business or state of business affairs of the organisational unit. *A measured goal* is an abstract generalisation of an ecosystem or business goal. It denotes how an ecosystem goal or a business goal can be measured by several *key performance indicator targets* and their related *key performance indicators (KPIs). A key performance indicator (KPI)* is a description of a measurable property used to evaluate performance and the progress of fulfilling a goal. *A KPI target* is used to express a target value, i.e., a minimum requirement, in terms of a quantifiable figure or a qualitative description, that is used together with a KPI to assess the achievement of a measured goal. A KPI instance can consist of several instances of KPI targets if the same KPI is used for several *measured goals.* An instance of a KPI target can be used to assess exact one instance of an ecosystem or business goal, whereas an instance of an ecosystem or business goal can be measured by several instances of KPI targets and their related KPIs. An *actor's performance* concerns the runtime performance and is calculated and measured by the relevant KPIs. An instantiation of *the goal & performance* aspect is given in Fig. 4.

The *fitness & qualification* aspect (Fig. 3) conceptualises the fitness & qualification of the actors and the assets they provide into a DBE under the context of design, deployment, and runtime phases.

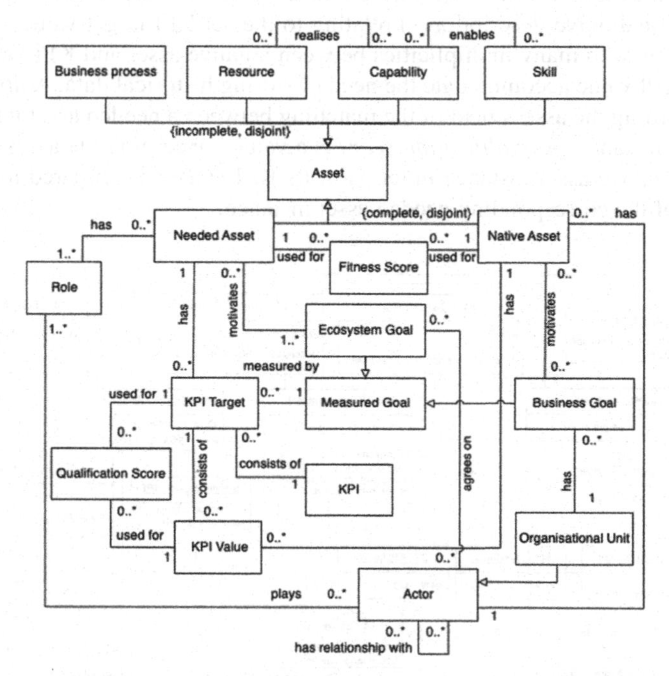

Fig. 3. Digital business ecosystem meta-model – *fitness & qualification* aspect

An asset is an entity of value and use in the DBE. Asset has two generalisation sets: the first indicates that an asset may be classified as a business process, a resource, a capability, a skill, or some other types of assets as specified by the constraints (incomplete and disjoint); the second set defines the possible asset specialisation as a needed asset and native asset where the set is constrained to be complete and disjoint. *A business process* is a collection of activities that, based on a set of rules, consumes input and produces output, such as information and/or material. *A resource* can be a product, a service, knowledge, or money, which may be enabled by capabilities. *A capability* is an ability and capacity that enable the achievement of goals in a certain context. *A skill* is an ability to perform a function acquired or learnt with practice, which may enable capabilities. *A needed asset* is an asset considered necessary in a DBE as it is motivated by one or more ecosystem goals and should be owned by one or more roles in the DBE. It can have KPI target(s), meaning that the needed asset has minimum required value(s) in terms of the specific KPI(s) and may later be assessed for its qualification. The zero to many multiplicities indicate that each instance of a needed asset can have no KPI target, especially during the design stage, or it can have several different KPI targets of different KPIs or several different KPI targets of the same KPI as historical data. *A native asset* is an asset owned by an actor. It may be motivated by the business goal(s) of an organisational unit. Note that an actor in a DBE may, for specific reasons, choose not to share information about its business goals or native assets. Hence, the zero to many multiplicities are set on several associations to accommodate this possibility. A native asset may have *KPI value*(s), which indicates its runtime value(s) in quantifiable figure(s) or qualitative description(s) relating to the set KPI target value(s) of specific KPI(s). The zero to many multiplicities between a native asset and KPI value and KPI target and KPI value accommodate the need of storing historical data. A *fitness score* is a value regarding the assessment of the matching between a needed asset instance and a native asset instance. A *qualification score* is a value concerning the assessment of the quality of a native asset instance in terms of its KPI value as compared to one related KPI target of the corresponding needed asset instance.

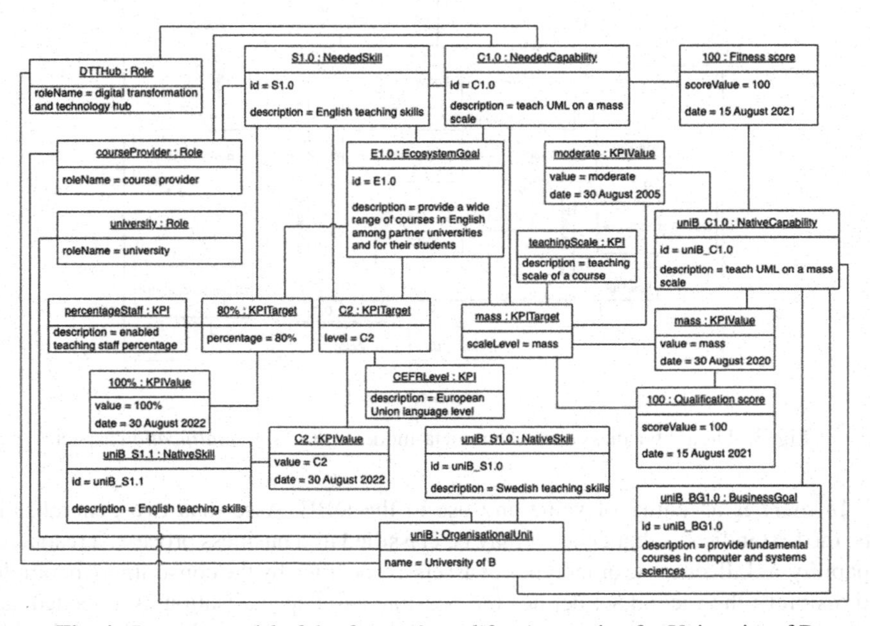

Fig. 4. Instance model of the *fitness & qualification* section for University of B

An instantiation of the *goal & performance* and the *fitness & qualification* aspects of the meta-model is in Fig. 4 illustrates the responsibility and performance of the actor, University B. The ecosystem goal is *E1.0-provide a wide range of courses in English among partner universities and for their students.* Since it is a goal set by the Driver of the CIVIS DBE, the ecosystem goal instance is not connected to the actor instances existing in the instance model. This ecosystem goal motivates an instance of a needed capability (needed asset with the asset type capability) *C1.0-teach UML on a mass scale* and an instance of needed skill *S1.0-English teaching skills.* The needed skill enables the needed capability. They are needed assets owned by the role instances – *digital transformation and technology (DTT) hub* and *course provider* in the CIVIS DBE.

The needed capability has a KPI target (instance) and related KPI (instance), which is the minimum required teaching scale of a course; in this case, a UML course at a mass scale. The needed skill has two different sets of KPI targets and related KPIs: the minimum required level of the English language based on the *Common European Framework of Reference for Languages (CEFR) level* should be at a C2 level, and the minimum required *percentage of teaching staff* who has C2 level English is 80%. The KPI targets and related KPIs are used to measure the ecosystem goal *E1.0.*

In Fig. 4, the example actor instance is the University of B as an organisational unit. University of B plays three roles – *digital transformation and technology hub*, *course provider*, and *university*. It owns the instance of native skill (a native asset with asset type skill) *uniB_S1.0-Swedish teaching skills*. This instance is not associated with business goals, KPI values, or fitness scores. The reasons are: (1) the actor did not share any business goals related to this native skill instance, and (2) the example used in this instance model concerns English teaching skills for the DBE and not Swedish. This illustrates the possibility of an instance of a native asset not being further analysed because it is, at this point, irrelevant to the DBE. University of B has a business goal *uniB_BG1.0- provide fundamental courses in computer and systems sciences*, which motivates its native capability *uniB_C1.0-teach UML on a mass scale*. This capability instance is used for assessing the matching with the instance of needed capability *C1.0* on the 15th of August 2021, resulting in an instance of fitness score with a score value of 100 out of 100. This means that the native capability and the needed capability are an exact match. The native capability *uniB_C1.0* has two stored historical data concerning KPI value instances – a moderate value recorded on the 30th of August 2005 and a mass value recorded on the 30th of August 2020. The latest data from the KPI value instance is used for assessment on the same date (15th of August 2021) to evaluate the quality of this native capability as compared to the related KPI target of the corresponding needed capability *C1.0*. The assessment results in an instance of qualification score with value of 100 out of 100, meaning that the quality of the native capability fulfils the target quality of the needed capability.

The native capability *uniB_C1.0* is enabled by its native skill *uniB_S1.1- English teaching skills* which has two historical data concerning two different sets of KPI value and related KPI instances - a 100% value for the KPI enabled teaching staff percentage and a C2 value for the KPI CEFR level both recorded on the 30th of August 2022. As these KPI values are recorded after the last assessment date (15th of August 2021), no qualification score has been assigned (no qualification score instance has been created).

4 Discussion and Conclusions

In a previous analysis in [5], meta-models and language constructs used for modelling DBEs were investigated to see if they encompassed the five essential elements of a DBE, *actor, role, capability, relationship,* and *digital component*, based on its definition. The findings suggested that none of the analysed studies included all five of these essential DBE elements. We have attempted to encompass all five essential DBE elements for our proposed meta-model. They are represented in the *actor & role* aspect of the meta-model as actor, role, DBE role, relationship (as associations among actors), and digital infrastructure. The element capability was included in the *fitness & qualification* aspect of the meta-model as a type of asset. The meta-model, with its three aspects, *actor & role*, *goal & performance*, and *fitness & qualification*, is also intended to describe, capture, and analyse a DBE during its deployment and operation phases. Understanding these phases is more difficult as compared to the design phase. The dynamics, i.e., how the actors behave, interact, and react under different circumstances, occurring in a DBE during runtime contribute to the complexity of comprehending the relations among entities in the DBE.

Preliminarily validated through demonstration in a real-life DBE, the CIVIS case, the meta-model demonstrates its potential for supporting the resilience of DBEs and their runtime contexts. The examples in the instance model (Fig. 4) show the dynamics as the actors behaved by providing different information to the DBE at different time points when the assessment of fitness and qualification was conducted at a specific time point during the operation phase. The model-based approach better supports these dynamics by providing valuable insights concerning the related entities and historical data (i.e., KPI values of a KPI on different dates concerning native capabilities and goals of various actors). These examples in Fig. 4 also highlight that regular assessment is needed for runtime monitoring and decision-making. The modelling effort with the proposed DBE meta-model could support these activities (i.e., continuous collection of data, runtime monitoring, decision support, etc.) necessary in the deployment and operation (runtime) phases. Together with the essential DBE elements (actor, role, capability, relationship, and digital component), the newly identified elements in the meta-model, especially the DBE roles, the resilience goals, the specialisations of needed and native assets, the KPI target and value, and the fitness and qualification scores in the *fitness & qualification* aspect, support a DBE's resilience. Using models based on these elements, tools can be developed to support DBE resilience.

Future research should focus on the validation of the meta-model with other real-life DBE cases to improve its validity and generalisability as well as tool development.

References

1. Moore, J.F.: Predators and prey: a new ecology of competition. Harv. Bus. Rev. **71**(3), 75–86 (1993)
2. Tsai, C. H.: A method for designing resilient digital business ecosystems. Licentiate dissertation, Department of Computer and Systems Sciences, Stockholm University (2023)
3. Kampars, J., Zdravkovic, J., Stirna, J., Grabis, J.: Extending organisational capabilities with open data to support sustainable and dynamic business ecosystems. Softw. Syst. Model. **19**(2), 371–398 (2020)
4. Basole, R.C., Russell, M.G., Huhtamäki, J., Rubens, N., Still, K., Park, H.: Understanding business ecosystem dynamics: a data-driven approach. ACM Trans. Manag. Inf. Syst. **6**(2), 1–32 (2015)
5. Tsai, C.H., Zdravkovic, J., Stirna, J.: Modeling digital business ecosystems: a systematic literature review. Complex Syst. Informat. Model. Q. **30**, 1–30 (2022)
6. CIVIS homepage, https://civis.eu/en. Accessed 10 Feb 2023

7. Hevner, A., March, S.T., Park, J., Ram, S.: Design science in information systems research. Manag. Inf. Syst. Q. **28**(1), 75–105 (2004)
8. Tsai, C. H., Zdravkovic, J., Söder, F.: A method for digital business ecosystem design: situational method engineering in an action research project. Softw. Syst. Model. **22**, 573–598 (2022)

Supporting Students in Team-Based Software Development Projects: An Exploratory Study

Carles Farré(✉), Xavier Franch, Marc Oriol, and Alexandra Volkova

Universitat Politècnica de Catalunya (UPC), Barcelona, Catalonia, Spain
{carles.farre,xavier.franch,marc.oriol,
alejandra.volkova}@upc.edu

Abstract. Team-based software development projects (TBSDP) are a useful instrument to expose students to teamwork in an industry-like working context. However, TBSDP exposes students to a number of challenges. This paper has a twofold objective. First, understand the practices and challenges that students face in TBDSP in an Agile context. Second, investigate whether the use of a teaching domain-specific information system, a learning dashboard, could help them in improving these practices and facing these challenges. We conducted a multi-instrument exploratory study at the Polytechnical University of Catalunya. We gathered information about the progress of 39 students organised in 6 teams during one semester by mining two software repositories and conducting a number of questionnaires and interviews related both to working practices and to students' perception of learning dashboard adoption. Results show that many students do not follow adequate practices for some TBSDP activities. On the other hand, metrics informing about the use of code repositories and task management were generally well understood and perceived as potentially useful by students when shown in a learning dashboard. We conclude that the adoption of learning dashboards is a viable approach to improve student practices in TBSDP, but it needs to be carefully considered which metrics provide the most value to face the identified challenges.

Keywords: Team-based Software Development Project · Learning Dashboard

1 Introduction

Learning Dashboards are a type of information system in the education domain designed to support students and instructors in their learning/teaching activities [11]. In this paper, we focus on their use in courses that comprise a software development project executed by student teams, which we call *Team-Based Software Development Project* (TBSDP). TBSDPs expose students to industry-like teamwork and project management challenges but sometimes fail to meet students' expectations [7].

This paper reports a multi-instrument empirical study with two goals: 1) to identify student practices during various activities in agile TBSDPs, and 2) to explore the potential benefits of a learning dashboard in enhancing these practices. For this later purpose, we have customised the Q-Rapids dashboard [8], designed to be used in a continuous

S. Nurcan et al. (Eds.): RCIS 2023, LNBIP 476, pp. 568–576, 2023.
https://doi.org/10.1007/978-3-031-33080-3_39

improvement agile life cycle, into a learning dashboard called *Q-Rapids$_{LD}$*. Considering the two goals, it can be said that our study provides a systematic approach to gathering the requirements for an effective TBSDP learning dashboard.

We have introduced Q-Rapids$_{LD}$ in *Software Engineering Project* (SEP), a compulsory 6-ECTS, 15-week-long course taught at the Polytechnical University of Catalunya (UPC). SEP is designed to reproduce, as far as possible, the agile project development of a software system in a professional environment.

2 Research Method

We designed a multi-instrument study combining quantitative and qualitative data to answer the two research questions in Table 1. **RQ1** aims to identify poorly performed development practices among students working on agile software projects in the context of SEP and to understand the needs that a learning dashboard for the SEP course should address. RQ1 is divided into subquestions related to the indicators presented in Sect. 2.1. **RQ2** evaluates Q-Rapids$_{LD}$ in the SEP teaching context to inform us whether the students perceive this customized tool as beneficial. RQ2 is divided into three research subquestions related to understandability and perceived usefulness of the metrics computed by the dashboard and the visualisation capabilities offered.

Table 1. Research questions of our study

RQ1. To what extent do students in SEP follow established good practices in TBSDPs? Subquestions: What are the practices that students follow for:
RQ1.1. eliciting and managing tasks in the project's backlog? **RQ1.2**. completing the information related to user stories and tasks in the project backlog? **RQ1.3**. developing software and managing the code repository?
RQ2. What is the students' perception on the possible use of Q-Rapids$_{LD}$ in the SEP course?
RQ2.1. Do students understand the proposed metrics and the way they are computed? **RQ2.2**. Do students find these metrics useful to monitor their own progress over time? **RQ2.3**. Which visualisation strategies do students find most informative?

2.1 Data Collection

For RQ1, we used a mix of qualitative and quantitative instruments to collect data, while RQ2 is based entirely on qualitative data. The data for this study was gathered during the Fall 2022 academic term from 6 teams comprising 39 students. Their software projects were developed in three iterations or sprints.

Questionnaires in RQ1. We created two questionnaires to investigate agile development practices among students in the SEP course. One questionnaire is intended for team response, while the other is for individual students. Both questionnaires cover the same three dimensions as the three research sub-questions but have different questions and answer options. Each question has four answer choices, ranked in decreasing order of established good practices, and includes an open text field for students to provide additional information. Questions and answers are available in our shared dataset [3].

These questionnaires were handed over to students during the review meetings at the end of each sprint in hard copy form. They were anonymous for those who answered individually (although the team's name was always included). Students were informed that the responses would not affect their evaluation.

Mining Repositories in RQ1. Q-Rapids$_{LD}$ retrieves data from the Taiga project management tool and the GitHub code repository service. From these data, Q-Rapids$_{LD}$ computes several *metrics* and aggregates them into *strategic indicators* to provide an overview of the software project's development status (see Table 2). Metrics generated by Q-Rapids$_{LD}$ are available in our dataset [3].

Table 2. Metrics and indicators defined in Q-Rapids$_{LD}$ for the SEP course

Indicator	Metric		Data source
	Code	Description	
Information Completeness	AC	Number of user stories that include acceptance criteria	Taiga
	USP	Number of user stories that follow the definition pattern	Taiga
	TwEE	Number of tasks with estimated effort information	Taiga
	CTwEE	Number of closed tasks with actual effort information	Taiga
	DiTEE	Deviation in task effort estimation (estimated vs. actual)	Taiga
Backlog Management	TT	Number of tasks assigned per student	Taiga
	CT	Number of closed tasks per student	Taiga
	UT	Number of unassigned tasks	Taiga
	TSD	Standard deviation of TT in the team	Taiga
Repository Contribution	CTR	Number of commits that refer to task id	Both
	TC	Number of commits per student	GitHub
	ML	Number of modified lines per student	GitHub
	CSD	Standard deviation of TC in the team	GitHub

Questionnaires and Interviews in RQ2. We granted access to Q-Rapids$_{LD}$ to only three teams and did not mandate its usage. We designed a questionnaire for students using Q-Rapids$_{LD}$ to gather their perceptions on (i) the understandability of the thirteen metrics offered by the dashboard; (ii) the usefulness of the said metrics; (iii) the clarity of the six types of visualisations offered by the dashboard. Students could also provide open-ended responses. The questions and answers are in our dataset [3]. The questionnaires were available online to students in the third sprint review meeting. Questionnaires were not anonymous since we wanted to perform a follow-up interview with some selected students (again, we informed students that answers would not affect evaluation). In this respect, we selected one student per team, based on the responses they gave to the questionnaire.

2.2 Threats to Validity

Internal. Issues in mining software repositories arose from our infrastructure and cloud deployment, causing occasional data collection disruptions. The most significant occurred twice in December 2022, affecting all teams. These disruptions were promptly detected and addressed.

Construct. To reduce bias, instructors not involved in this paper's design and analysis participated in the study, and students were encouraged to provide candid, anonymous (when possible) responses to the questionnaires. However, as Hundhausen et al. suggest [5], students might misrepresent their contributions. To mitigate this, we incorporated objective metrics from Q-Rapids$_{LD}$.

Conclusion. The study involved six student teams and three sprints, limiting the findings. We analyzed results per student (39 total) to mitigate the low team count but advise cautious interpretation of the observations. Further research is needed for definitive conclusions.

External. The study was conducted in one semester with a specific, agile, course profile and covered six distinct projects. However, to generalize results, further research in other contexts is necessary.

3 RQ1: Students' Practices and Challenges in TBSDPs

This section reports the questionnaire results concerning students' agile development practices, complemented by Q-Rapids$_{LD}$ metrics. The response rates were:

- Team questionnaire: 94.4% (17 responses from 6 teams across three sprints).
- Student questionnaire: 82.9% (97 responses from 39 students across three sprints).

Figure 1 displays team (**T1-T6**) and individual (**I1-I4**) question results for each sprint (**S1-S3**), with frequencies for answers (**R1-R4**), where **R1** is best and **R4** worst. We analyze these results by topic below.

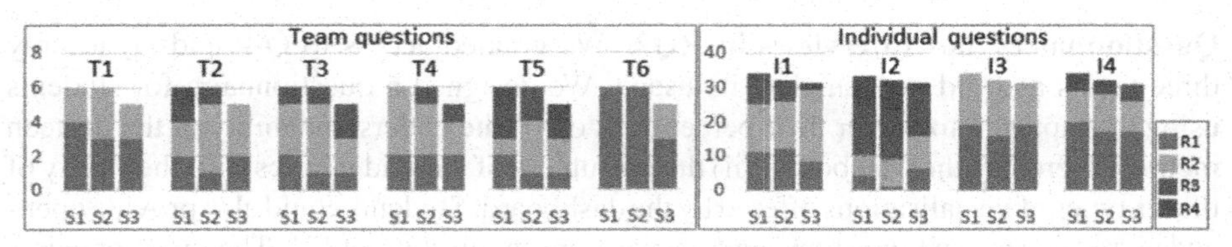

Fig. 1. Results of the questionnaires regarding the practices followed by SEP students.

RQ1.1. What are the practices that students follow for eliciting and managing the tasks in the project's backlog?

T1: How have the teams defined and distributed tasks? Teams defined and distributed tasks by common agreement (**R1**) or mostly by agreement except for those identified later during the sprint, defined and assigned by individual team members (**R2**). However, Q-Rapids$_{LD}$, data reveals uneven task distribution, with high standard deviations (TSD) and varying task percentages (TT) among team members.

T2: How have the teams tracked the backlog progress? Backlog tracking improved over time, with all response types present in Sprint 1 but consistently whole-team tracking (**R1** or **R2**) by Sprint 3. Q-Rapids$_{LD}$'s metrics "Fulfillment of Tasks" (computed as the average of team's CT) and UT have, in all teams, short intervals of fluctuation, typically around sprint deadlines, combined with longer steadiness intervals.

I1: How often and how have students used Taiga? Taiga usage increased as the course progressed, with the percentage of students checking their tasks and others' tasks (**R1** or **R2**) growing from 73.0% in Sprint 1 to 93.5% in Sprint 3.

RQ1.2. What are the practices that students follow for completing the information related to the user stories and tasks in the project's backlog?

T3: How have the teams managed user stories' acceptance criteria? All teams included user story acceptance criteria, but most completed them as the sprint progressed (**R2**). Q-Rapids$_{LD}$'s AC metric showed varied results, with some teams scoring high and others having a flat zero because they did not apply the required format.

T4: How have the teams estimated the development effort of user stories and tasks? Most teams estimated development effort jointly (**R1** or **R2**). Q-Rapids$_{LD}$ shows that only one team consistently had high TwEE values, with other teams estimating effort occasionally or never.

I2: When students completed a task in Taiga, did they record the time actually needed? Students' task completion and effort recording in Taiga varied. The percentage of students reporting effort upon task completion was low, while those who never included effort decreased over time.

T5: How have the teams validated the development effort of user stories and tasks? Few teams consistently recorded and analyzed the final effort of tasks for future iterations (**R1**). Q-RapidsLD's CTwEE metric fluctuated for some teams and remained at zero for others. DiTEE metric results also varied among teams.

RQ1.3. What are the practices that students follow for developing the software and managing the code repository?

T6: How have the teams managed the code repository? In the first two sprints, all teams reported using GitFlow with separate branches for each task (**R1**). In the third sprint, 40% used GitFlow with separate branches for each developer (**R2**).

I3: How have the students programmed in relation to tasks? Most students programmed one task at a time (**R1**) or multiple tasks in parallel (**R2**), with all contributing to the project's coding activities.

I4: How and when have you done commits? Over half of the students regularly committed to task-specific branches (**R1**), while others committed regularly without referencing tasks (**R2**) or infrequently (**R3**). Some students relied on peers for commits (**R4**). However, the Q-Rapids$_{LD}$'s CTR metric shows that only one of the teams referenced Taiga task identifiers in their commit messages. Q-Rapids$_{LD}$'s CSD revealed high levels of uneven contribution among students, but it decreased over time for most teams.

4 RQ2: Students' Perception on Learning Dashboard Benefits

We report the Q-Rapids$_{LD}$ utility perception questionnaire results. Of the students with dashboard access, 73.7% (14 out of 19) completed the questionnaire, resulting in 13 valid responses (68.4%) after removing one low-quality answer.

RQ2.1. Students' understanding of metrics is categorized into 5 groups (Fig. 2(a)):

- **Group 1**. Code repository use: 77% fully understood GitHub-related metrics.
- **Group 2**. Task management: 69–77% fully understood the three related metrics.
- **Group 3**. Effort estimation: 54–62% fully understood the two related metrics.
- **Group 4**. Artefact quality: 38–54% fully understood three metrics (acceptance criteria application, use of user story pattern, commit-task relation), with no more than 16% not understanding them.
- **Group 5**. Standard deviation: Full understanding of three deviation metrics (on task assignment, on effort estimation, and on commit frequency), did not exceed 38%.

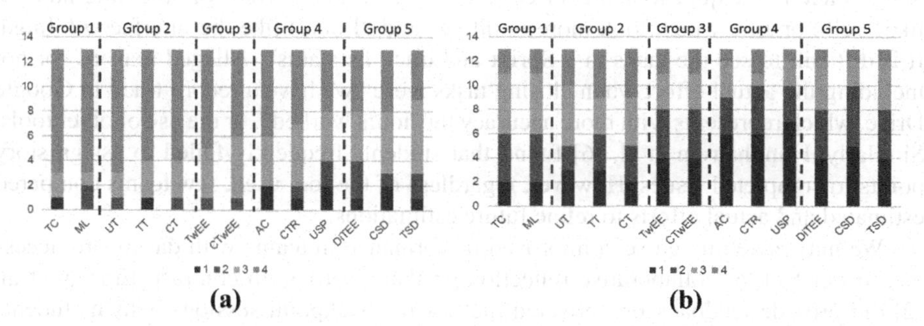

Fig. 2. (a, left) Understandability of the Q-Rapids$_{LD}$ metrics (1:low; 4: high); (b, right) Perceived usefulness of the Q-Rapids$_{LD}$ metrics (1: low; 4: high).

RQ2.2. Figure 2(b) shows the low perceived usefulness of Q-Rapids$_{LD}$ metrics, with no metric deemed very useful by over 31% of students. Group 1 and 2 metrics received over 50% positive responses, but Group 2 was slightly better regarded. Negative perceptions prevailed for other groups. Group 5 showed high variability (23%–38%), and Group 4 and 5 metrics were seen as potentially most useful by only one respondent.

RQ2.3. Gauge-based visualizations were the clearest (92% approval), while the table-based view was least preferred (54% approval). The other three had similar results, but none were considered clear in all situations.

Additionally, we interviewed three students (*Stu1*, *Stu2*, *Stu3*) from different dashboard-using teams and identified the following feedback:

- **Difficulty of use:** *Stu1*'s team found Q-Rapids$_{LD}$ interesting but difficult to integrate into their way of working, while *Stu2* recommended instructors introduce the dashboard alongside the other tools they use, GitHub and Taiga.
- **Integration with other tools**. All students recommended further integration of Q-Rapids$_{LD}$ with Taiga and GitHub.
- **New metrics**. *Stu1* suggested evolving the metric TT by considering the estimated effort of the tasks. *Stu2* proposed a new metric measuring how well the team members go together. *Stu3* requested GitHub metrics to identify bad practices.
- **Better explanations**. *Stu1* advised adding legends to dashboard values and suggested including mitigation actions for underperforming metrics.

5 Discussion

RQ1 results reveal that students face more challenges managing project tasks and ensuring information completeness than managing code repositories. However, Q-Rapids$_{LD}$ metrics show a clear disparity in contributions to the code repositories among team members, which aligns with the results reported in previous studies [4, 6, 9].

Although tasks were defined and assigned jointly, Q-Rapids$_{LD}$ data suggest uneven distribution and increased activity near Sprint deadlines. This is consistent with our findings that not all team members track project status in Taiga, and most teams do not record the final effort to improve future estimations. Students who did not regularly use Taiga stated in the questionnaire's free text, "I prefer the interface of other alternatives, like Trello or Jira, which look more intuitive" and "I don't like the interface of Taiga. It is difficult to see the tasks in a Sprint and track its status". Alleged reasons for not including the actual effort when closing tasks were "we have a record track in Google Drive, which represents with more accuracy the hours worked", or the use of other tools. Similarly, Hundhausen et al. [6] found that students frequently failed to assign story points to completed issues. However, regardless of the tool used, few teams compared estimated and actual efforts to refine future estimations.

We may have missed seen improved performance in teams with dashboard access due to our lack of collaborative reflection on their metrics. In contrast, Eraslan et al. [2] did provide feedback on collected metrics in checkpoint sessions, which students appreciated for helping them to improve their practices and identify issues earlier.

RQ2's analysis brings some interesting remarks, even considering its exploratory nature. First, the categorization of metrics into five groups clearly illustrates how easy and useful each of them are. The first two groups, code repository use and task management, are the easiest to understand and the most useful, as they are closely related to the everyday practices of the team. Future metrics could address other frequent activities, such as continuous integration practices. However, as Group 4 (artifact quality) reveals, this is not always the case; while understandable, these metrics were deemed less useful, perhaps due to their focus on documentation rather than coding or project management. The SEP course should emphasize the significance of documentation.

Finally, the low consideration of metrics evaluating individual student contributions within teams calls for discussion and understanding. Students can gain insights from comparing their performance to peers for self-improvement and timely identification of team issues. Employing gamification could effectively enhance engagement in this context. Data from Q-Rapids$_{LD}$ shows that teams do not assign a balanced number of tasks to each member. The dashboard could help teams recognize this and reflect on its impact on project management. Metrics may require minor adjustments, such as considering task quantity and estimated effort, as one student suggested during interviews.

In conclusion, the Learning Dashboard presented in this paper has the potential to facilitate a process of self-reflection, as proposed by Sedelmaier and Landes [10], by enabling teams to monitor their progress and identify corrective actions to improve their performance. However, using metrics to assess students and make them aware of it carries the risk of students altering their work methods to optimize the values for these metrics [1]. Therefore, the dashboard cannot be seen by no means as a panacea.

6 Conclusions

This paper has addressed the potential use of a learning dashboard as a tool for team-based software development project management. The study is exploratory and represents the first step towards offering this technology to students as a regular project management tool alongside others, such as SonarCloud, that have similar aims. To advance toward the generalization of results, we plan to replicate the study in a different course with smaller teams and less code. We also aim to expand the set of metrics, improve the visualization capabilities of Q-Rapids$_{LD}$, and activate additional techniques offered in the original Q-Rapids dashboard, such as predictions and alerts.

Acknowledgements. This paper has been funded by MCIN/AEI/10.13039/501100011033 and NextGenerationEU/PRTR under project PDC2021-121195-I00. We would like to thank the instructors of the SEP course at UPC, Silverio Martínez-Fernández and Jordi Piguillem, for their collaboration and support during the study.

References

1. Campbell, D.T.: Assessing the impact of planned social change. Eval. Program Plann. **2**(1), 67–90 (1979)
2. Eraslan, S., et al.: Integrating GitLab metrics into coursework consultation sessions in a software engineering course. J. Syst. Softw. **167**, 110613 (2020)
3. Farré, C., Franch, X., Oriol, M., Volkova, A.: Supporting Students in Team-Based Software Development Projects: An Exploratory Study. https://doi.org/10.5281/zenodo.7622293

4. Hamer, S., Quesada-López, C., Martínez, A., Jenkins, M.: Measuring students' contributions in software development projects using git metrics. In: CLEI 2020, pp. 531–540 (2020)
5. Hundhausen, C., Carter, A., Conrad, P., Tariq, A., Adesope, O.: Evaluating commit, issue and product quality in team software development projects. In: SIGCSE 2021, pp. 108–114 (2021)
6. Hundhausen, C.D., Conrad, P.T., Carter, A.S., Adesope, O.: Assessing individual contributions to software engineering projects: a replication study. Comput. Sci. Educ. **32**(3), 335–354 (2022)
7. Iacob, C., Daily, S.: Exploring the gap between the student expectations and the reality of teamwork in undergraduate software engineering group projects. J. Syst. Softw. **157**, 110393 (2019)
8. López, L., et al.: Q-rapids tool prototype: supporting decision-makers in managing quality in rapid software development. In: CAiSE-Forum 2018, pp. 200–208 (2018)
9. dos Santos, A.P., et al.: Mining undergraduate students' code repositories: insights from interdisciplinary software projects. In: Meirelles, P., Nelson, M.A., Rocha, C. (eds.) WBMA 2019. CCIS, vol. 1106, pp. 61–75. Springer, Cham (2019). https://doi.org/10.1007/978-3-030-36701-5_5
10. Sedelmaier, Y., Landes, D.: Practicing soft skills in software engineering: a project-based didactical approach. In: Computer Systems and Software Engineering: Concepts, Methodologies, Tools, and Applications, pp. 232–252. IGI Global (2018)
11. Verbert, K., et al.: Learning dashboards: an overview and future research opportunities. Pers. Ubiquitous Comput. **18**, 1499–1514 (2014)

Ontology of Product Provenance for Value Networks

Lohanna Saraiva[1(✉)], Patricio Silva[1], Angelica Castro[1], Claudia Ribeiro[2], and Joao Moreira[3]

[1] Universidade Federal Rural do Semi-Árido (UFERSA), Mossoró, RN, Brazil
lohanna.saraiva@alunos.ufersa.edu.br,
{patricio.alencar,angelica}@ufersa.edu.br

[2] Instituto Federal de Educação, Ciência e Tecnologia do Rio Grande do Norte (IFRN), Natal, RN, Brazil

[3] Semantics, Cybersecurity & Services (SCS), University of Twente, Enschede, The Netherlands
j.luizrebelomoreira@utwente.nl

Abstract. The economic issues surrounding value networks have received much attention from the research community in business modeling. However, other aspects can also influence the success of a network. One of them is sharing subjacent information that has value for the actors involved and fulfills a consumer's business need, such as product provenance. Thus, it is also essential to address these aspects in business modeling. Considering that provenance is a significant value to be explored, this work proposes an ontology for modeling value networks with an indication of provenance, focused on geographical indications. The ontology allows configuring models that show different ways of sharing information to assist business analysts in making strategic decisions about their value propositions for consumers. The Design Science Methodology and an Ontology Engineering methodology (SAMOD) guided the design of the ontology. Technical Action Research (TAR) supported the validation of the ontology in a Brazilian biotechnology company providing organic beverages. Expert opinion helped evaluate the utility of the ontology according to the support to decision-making provided by its derived models. This research also provides new insights into the importance of considering provenance information essential for modeling business networks, especially in economic and environmental sustainability cases.

Keywords: Ontology · Value Networks · Provenance · Technical Action Research

1 Introduction

A value network comprehends an organization of actors performing value transactions to fill a business need [8]. Each part of a value creation system is key to the network's success. Thus, it is necessary to understand its structure and analyze the ideas that contribute to its development [9]. One of the techniques used

S. Nurcan et al. (Eds.): RCIS 2023, LNBIP 476, pp. 577–584, 2023.
https://doi.org/10.1007/978-3-031-33080-3_40

for planning strategies that generate value for the network actors is the Value Network Modeling [3]. The starting point for designing these models is a consumer's need [4], which is constantly changing and always determines the value of what is offered [13]. Business needs become increasingly complex over time, and organizations need to be aware of them to offer better value propositions to stand out in the market environment.

There is a current and urgent demand for product provenance information as a factor of quality differentiation [6]. In 2021, the Brazilian National Supporting Office for SMEs (SEBRAE)[1] launched the *SEBRAE Origens*[2] portal to disseminate Geographical Indications (GI) from Brazil. Sharing product provenance information can generate opportunities for an organization and improve its relationship with the consumer [12]. In this sense, leveraging business network models with explicit provenance information could help represent more efficient strategies to create value and define better proposals for final consumers.

This paper proposes an approach for enriching value networks with provenance information. As a methodological strategy to achieve this goal, we adopted the Design Science Research methodology [14]. According to the context and the recommendations of the methodology, we addressed the following research question: *How to characterize product provenance information in value network models?* This question was subdivided into more specific ones: (1) *How to define product provenance in value networks?* (2) *How to build value network models with indications of product provenance?* (3) *Could modeling value networks with provenance assist companies in making decisions about their value propositions?*

To address questions (1) and (2) above, we propose an ontology to assist companies in designing business network models with indications of product provenance information. There are different requirements associated with a product that can be interpreted as provenance, e.g., traceability. This work focuses on realizing the concept of product provenance as Geographical Indication (GI). The proposed ontology was designed in OntoUML and then translated to the Web Ontology Language (OWL) under the guidance of the Simplified Agile Methodology for Ontology Development (SAMOD) [10]. The ontology evaluation process proposed by Gómez-Pérez [2], which includes verification and validation steps, helped evaluate the ontology. In the second step, Technical Action Research (TAR) helped explore a real business case of a Brazilian biotechnology company specialized in producing organic beverages. Business expert opinion helped evaluate the ontology's usefulness and acceptance.

The organization of the remaining of this paper follows. Section 2 describes the proposed ontology for modeling value networks with product provenance. Section 3 reports on the validation and evaluation of ontology in the case study and threats to validity and limitations. Last, a summary of research contributions and a plan for future research closes this work in Sect. 4.

[1] *Serviço Brasileiro de Apoio às Micro e Pequenas Empresas* in portuguese.
[2] Available in: https://www.sebrae.com.br/sites/PortalSebrae/origens.

2 Value Network with Provenance Ontology (VNPO)

This section describes the ontology proposed for modeling value networks with provenance. The concepts and properties described in the ontology extend the e^3*value* framework ontology [3] with concepts from the SVNO [11] and REA model [7]. The Food Supply Chain literature also identified roles of actors who are part of the value creation systems with Geographical Indication (GI), their activities, and types of products. The code of the ontology is available online[3].

The Value Network with Provenance Ontology (VNPO) was designed in OntoUML [5] and manually translated into OWL-DL with the support of the Protégé[4] tool. The reason why we translated the ontology into OWL is that in this language it is possible to instantiate and test the ontology competence questions with SPARQL. The development followed the engineering process recommended by SAMOD [10], which consisted of three steps: (1) requirements collection and model development; (2) merge of the current model with the final model; and (3) refactoring. At the end of each stage, the ontology underwent consistency, correctness, and completeness criteria tests. The company participating in the case study provided a dataset to support the ontology verification.

2.1 Concepts - Business Need View

The main aspects of the VNPO are the concepts related to the provenance of products, e.g., *provenance*, *geographical indication*, *declared provenance*, and *provenance assessment*. Besides those, the idea of the *provenance path* will represent the composition of several pieces of information about the provenance of an end product. The OntoUML model in Fig. 1 shows how these concepts are connected. The idea of business need takes a central role of importance.

The structure of the concepts mentioned above derives from the system of objective and subjective values found in SVNO [11]. The *provenance* is an accurate value characterizing a value object (*raw* or *end product*). At the same time, the *provenance assessment* is a subjective value, as consumers use it to evaluate a product or service before purchasing it. Similar to what is proposed by Reis et al. [11], in which the desired object and its value (objective or subjective) satisfies the consumer's business need, an *end product* and its *provenance* (*geographical indication*, *declared provenance*, or *provenance path*) could also meet a consumer's *business need*. Here, *provenance* is associated with value objects, composes a business need, and consists of a particular value object.

In VNPO, *provenance* is represented by a defined class equivalent to *geographical indication* or *declared provenance*. In principle, the class *provenance* would have only the subclass *geographical indication*, but considering that a company might create provenance models in which products are certified by a GI, the concept of *declared provenance* was also specified to describe an uncertified provenance of a product. The idea of *provenance assessment* represents the

[3] Available in: https://github.com/lohannaaires/VNPO/tree/main/OWL.

[4] Available in: https://protege.stanford.edu/.

subjective value of the origin of a product in the view of a consumer and can be used to assist others in the purchase decision. Each assessment can receive a value from the *declared value* class, representing consumer satisfaction levels.

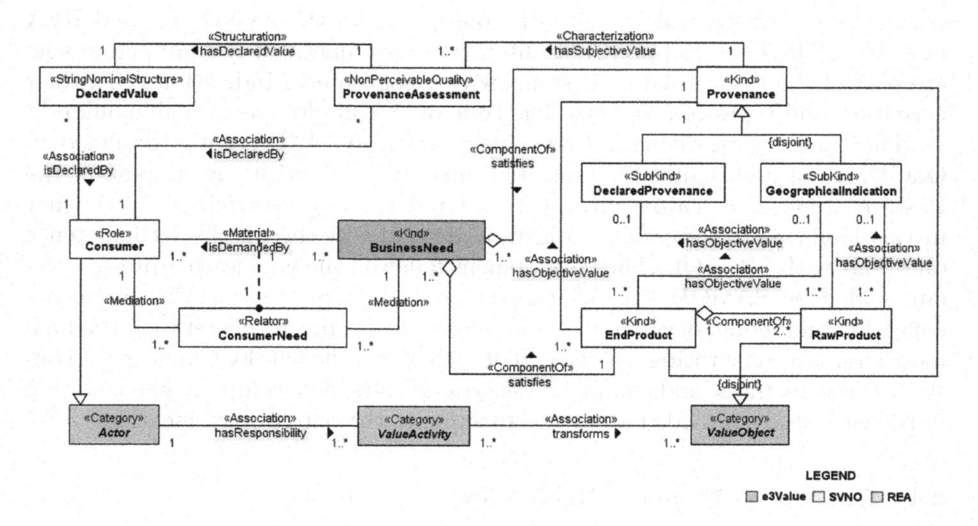

Fig. 1. Business Need Model

2.2 Concepts - Organizational View

In addition to the concepts already presented, to configure a value network with provenance, it is also necessary to consider the organizational policy that includes the roles of existing actors, their business activities, the valuables exchanged between them, and how to share provenance information within the value network. The ontology defined these elements as follows (Fig. 2).

Actors exchange objects of economic value to one another. In this work, they are local producers and their associations, companies (e.g., processors, distributors, and suppliers), consumers, the government, and any other institution part of the GI system [12]. In VNPO, therefore, the role of the *consumer* belongs to the final consumer of the network, i.e., the one whose business need must be satisfied. The *provider* role represents producers, associations, and other raw materials suppliers. Still, processors and distributors can assume the *agent* role.

A *value activity* is the core competency task performed by an *actor* [4] and transforms one or more economic value objects into aggregated value. In VNPO, there are four types of *value objects*: *raw product*, *end product*, *provenance*, and *counter object*. The value activities that act on them are *receipt*, *payment*, *production*, *sale*, and *purchase*.

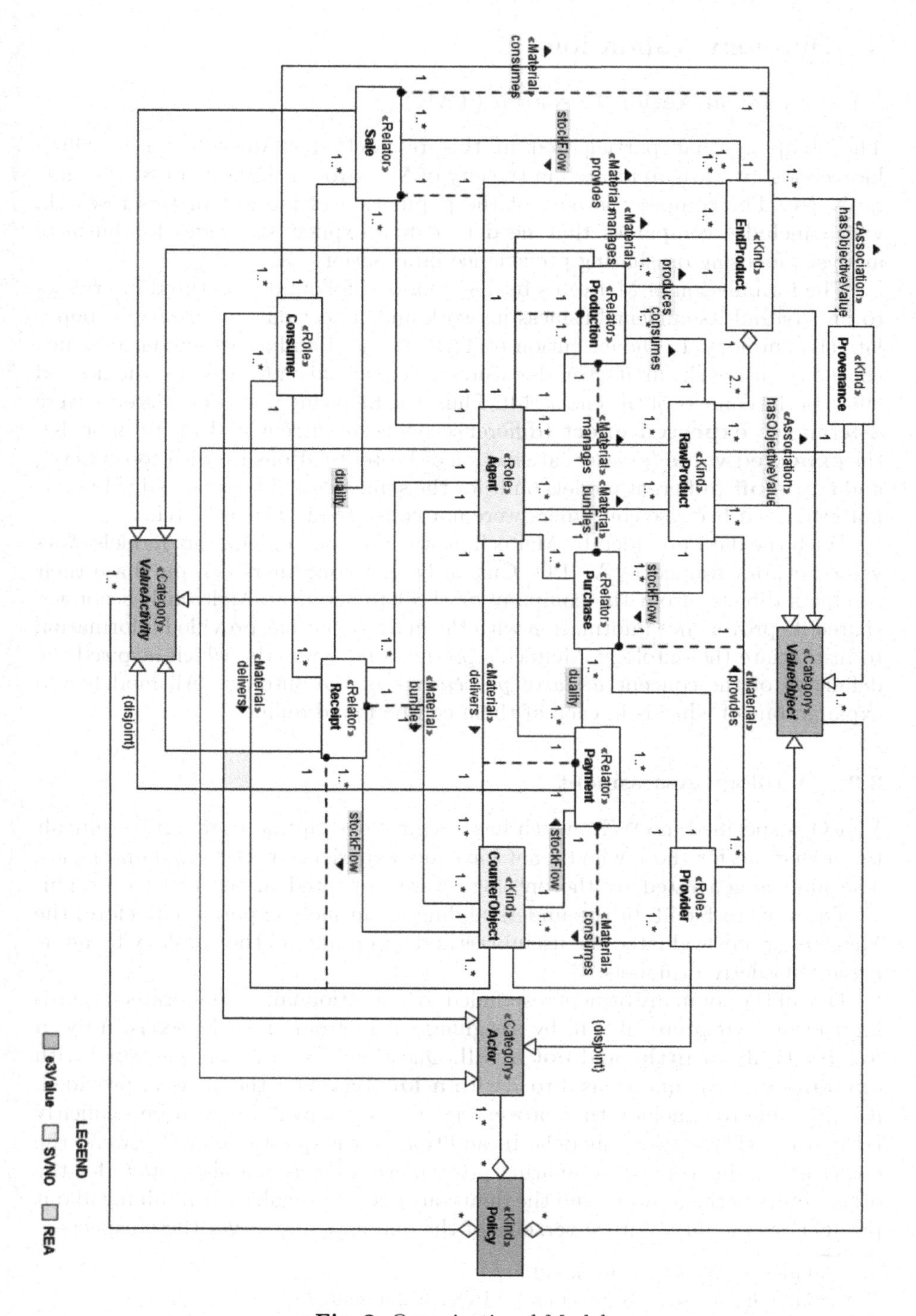

Fig. 2. Organizational Model

3 Ontology Validation

3.1 Technical Action Research (TAR)

The company that participated in this research was Meltech[5], a Brazilian biotechnology start-up located in the city of Mossoró, in the state of Rio Grande do Norte. The company is part of the population of interest of this research, which includes companies that need to define explicit strategies for business analysis, focusing on sharing provenance information.

The formulation of Meltech's business models followed structured interviews to understand its current business network and collect data necessary to populate the ontology. In the execution of TAR, we used abductive inference, which consists of using the artifact under study as a mechanism to produce the desired effect in the context of the client [14]. Thus, the following points of inference were determined: **expected effect** (difference between current and future models), the **expected value** (greater satisfaction or better relationship with consumers), and **trade-off** (different models built for the same case). The expected value was not evaluated because consumers were not consulted in this research.

With the data provided by Meltech, it was possible to define six models of its value network framed by VNPO. Currently, end consumers can purchase their products directly from the company or through retailers. Meltech has not yet shared its provenance information with the market, but has provided information to instantiate the ontology. Meltech's products have no GIs, which inspired the definition of the concept *declared provenance* of the ontology. All models and explanations of what is in each of them can be found online[6].

3.2 Ontology Assessment

VNPO is specified in OWL, which has a logical description that can be difficult to understand for those who do not have any experience with formal ontologies. The models generated by the ontology were presented in the e^3*value* graphical notation to facilitate the understanding of business experts. Therefore, the Meltech owners evaluated the usefulness and acceptance of the ontology by interpreting its derived models.

One of the company owners responded to a questionnaire[7], elaborated according to the TAR protocol [14], by assigning values from an scale: **extremely**, **a lot**, **partially**, **a little**, and **not at all**. Questions 1 and 2 were answered with **extremely**, while questions 3 to 7 with **a lot**. Based on the answers provided, it is possible to conclude that provenance is a significant value to be explicitly treated in value network models. In addition to the questionnaire answers, the owner also pointed out that sharing provenance information also counts for the other actors of the network and the final consumer, especially when talking about product traceability. Future versions of the ontology may cover these aspects.

[5] Available in: https://meltech.ind.br/.

[6] Available in: https://github.com/VNPO/e3value-models.

[7] Available in: https://github.com/VNPO/assessment/questionnaire.md.

3.3 Research Validity Threats and Limitations

One of the threats to this research's validity is the sharing of declared provenance information, addressed in the ontology and presented in the models generated for the Meltech case. Sharing this information works better for markets with already built trusted channels. In trusted markets, third-party mediation and certification bureaucracy can cause companies not to share their data. On the other hand, companies that do not have trust relationships would benefit from certifications seals, among others. If the company still wishes to adopt the models with declared provenance, it is crucial to conduct a network trust analysis initially. Meltech's case models were constructed without this preliminary analysis because the ontology is limited to modeling of provenance sharing. Another threat is that the models generated are only valid for a company that develops organic products. It is necessary to apply the ontology in other practical and observational cases. The main limitation of the work is the approach used, which assumes that the person responsible for configuring the value network models needs to have prior knowledge of the OWL language and the e^3*value* framework to perform the manual transformation of the models for a graphical view.

By further analyzing and discussing the ontology, we identified that it might need improvements to adequate its semantic validity and clarity. For example, *Value Activity* is itself an *Event*, and *Value Object* might be a *roleMixin* since it is an anti-rigid and relational dependent class that aggregates individuals with different identity principles. Furthermore, a *Need* seems to be a mode (a desire) inhering in a *Consumer*, as well as *Business Need. Policy*, in general, is a *Relator* between the *Policy Maker* and the *Community* that the *Policy* is directed. It can also be represented as a normative description accepted by a collective agent (e.g., organization, society, social group). Finally, the concept of provenance adopted here is grounded on relation of historical dependence, which might bring consequences such as the absence of particular qualities that would support the relationship and a deeper analysis of its nature [1].

4 Conclusion

To conclude, this study discusses the explicit inclusion of product provenance information as an essential element for filling consumers' needs in value network models. Consumers have increasingly demanded transparency from companies about their products, which requires opening competitive advantage data. Sharing this kind of data is a sensitive issue for value networks since there is a boundary on sensitive data exchange, i.e., between what may and may not be revealed by a company and its business partners. Business relationships may not be transparent, or competing value networks might take advantage of sensitive data sharing. Thus, companies could benefit from analyzing the disclosure of data provenance from a business strategy point of view. With this perspective, in this paper we propose an ontology that allows the configuration of value network models with organizational arrangements for sharing provenance information.

For future work, we highlight the need of: (1) apply the ontology in other cases to identify new organizational structures and consolidate the current models; (2) develop a GUI to simplify the construction of the models and the presentation of the inserted data; (3) evaluate the perception of consumers regarding sharing provenance information; (4) add certifying actors and mediators to the ontology to indicate accountancy or auditing of product provenance; and (5) extend the ontology with concepts that broaden the meaning of provenance presented, such as information related to the traceability of products.

References

1. Fonseca, C.M., Porello, D., Guizzardi, G., Almeida, J.P.A., Guarino, N.: Relations in ontology-driven conceptual modeling. In: Laender, A.H.F., Pernici, B., Lim, E.-P., de Oliveira, J.P.M. (eds.) ER 2019. LNCS, vol. 11788, pp. 28–42. Springer, Cham (2019). https://doi.org/10.1007/978-3-030-33223-5_4
2. Gómez-Pérez, A.: Ontology evaluation. In: Staab, S., Studer, R. (eds.) Handbook on Ontologies. International Handbooks on Information Systems, pp. 251–273. Springer, Heidelberg (2004). https://doi.org/10.1007/978-3-540-24750-0_13
3. Gordijn, J.: E-business value modelling using the E3-value ontology. In: Value Creation from E-business Models, pp. 98–127. Elsevier (2004)
4. Gordijn, J., Wieringa, R.: E^3value User Guide. Designing Your Ecosystem in a Digital World, The Value Engineers (2021)
5. Guizzardi, G.: Ontological foundations for structural conceptual models. No. 25, Telematica Instituut Fundamental Research Series (2005). ISBN 90-75176-81-3
6. Herschel, M., Diestelkämper, R., Ben Lahmar, H.: A survey on provenance: what for? what form? what from? VLDB J. **26**(6), 881–906 (2017)
7. McCarthy, W.E.: The rea accounting model: A generalized framework for accounting systems in a shared data environment. Account. Rev. 554–578 (1982)
8. Normann, R., Ramirez, R.: From value chain to value constellation: designing interactive strategy. Harv. Bus. Rev. **71**(4), 65–77 (1993)
9. Peppard, J., Rylander, A.: From value chain to value network: insights for mobile operators. Eur. Manag. J. **24**(2–3), 128–141 (2006)
10. Peroni, S.: A simplified agile methodology for ontology development. In: Dragoni, M., Poveda-Villalón, M., Jimenez-Ruiz, E. (eds.) OWLED/ORE -2016. LNCS, vol. 10161, pp. 55–69. Springer, Cham (2017). https://doi.org/10.1007/978-3-319-54627-8_5
11. Reis, J.D.S., Silva, P.D.A., Castro, A.F.D.: Ontologia para configuração semi-automática de redes de valor. iSys-Braz. J. Inf. Syst. **13**(1), 77–113 (2020)
12. Vandecandelaere, E., et al.: Linking people, places and products. FAO (2009)
13. Vargo, S.L., Lusch, R.F., Akaka, M.A., He, Y.: Service-Dominant Logic, vol. 3. Routledge, Abingdon (2020)
14. Wieringa, R.J.: Design Science Methodology for Information Systems and Software Engineering. Springer, Heidelberg (2014). https://doi.org/10.1007/978-3-662-43839-8

A Data Value Matrix: Linking FAIR Data with Business Models

Ben Hellmanzik(✉) and Kurt Sandkuhl

University of Rostock, Rostock, Germany
{ben.hellmanzik,kurt.sandkuhl}@uni-rostock.de

Abstract. Data is the raw material of digitization, but its economic use and potential economic benefits are not always clear. Therefore, we would like to show that the processing of data, especially according to FAIR principles, plays an enormous role in enabling business models and improving existing business models. This is being tested within the EU funded project Marispace-X, part of the Gaia-X initiative. The maritime domain in particular currently still suffers from a lack of digitization: while as much data is being collected as ever before, this data is often kept in silos and hardly reused or even shared. This work therefore involved linking the FAIR principles to a data value chain, which together with business model dimensions form a data value matrix. This application of this matrix was carried out together with practice partners using the example of maritime data processing in the use case "offshore wind" and can be used and adapted as an analysis tool for data-driven business models.

Keywords: Data value chain · FAIR · dataspace · business model

1 Introduction

The Economist's headline "The world's most valuable resource is no longer oil, but data"[1] has motivated research into the value of data, business models for exploiting data, the data economy, data ecosystems, data-driven products and services, and other related topics even further. Increasing digitalization, new technologies for capturing data, the growing availability of computing power and big data or artificial intelligence techniques to make sense out of data supported the Economist's statement of data turning into a commodity. One of the downsides of this development is that many data owners nowadays are concerned about losing control of who uses their data for what purposes. In this context, approaches for ensuring sovereignty gain importance, like the attempts to established large-scale infrastructure for FAIR dataspaces. However, preserving data sovereignty should not be equivalent to preventing economic activities or data-driven services. Data in many application areas with clear sovereignty

[1] https://www.economist.com/leaders/2017/05/06/the-worlds-most-valuable-resource-is-no-longer-oil-but-data.

S. Nurcan et al. (Eds.): RCIS 2023, LNBIP 476, pp. 585–592, 2023.
https://doi.org/10.1007/978-3-031-33080-3_41

requirements are very valuable in the transformation to a sustainable low-carbon industry and society. For these application areas, new business models have to be developed that acknowledge the FAIR use of data but also allow for value creation and propositions based on the data. The intention of our work is to contribute to this field by supporting enterprises in discovering business model potential.

The aim of our work is to provide an instrument for companies collecting, processing, managing or providing data to analyse the effects of shifting to FAIR data on their business model. This instrument is supposed to support decision making about positioning the enterprise on the market, the requested investments in (technical and staff resources) and new services.

Among the essential elements of business models are the value offering made to target groups and the value creation required for this (cf. Sect. 3). When developing new business models, identification and analysis of these aspects are core challenges. For business models related to existing industry or domain developments, knowledge of these mechanisms can be an inspiration or even blueprint (see, e.g., Schallmo [4]). However, if such related areas do not exist or are not known, other approaches for identification and analysis are required.

The contributions of this paper are (1) an innovative dataspace as example case for business model development, and (2) an instrument for the analysis of change needs in business models when adapting them to FAIR principles.

The paper is structured as follows: Sect. 2 introduces the research methodology used in the paper. Section 3 discusses relevant background from business models. Section 4 proposes the business model analysis matrix as an instrument for identifying the required changes in a business model when adapting a FAIR data value chain. In Sect. 5, the analysis matrix is applied in the application example of a maritime dataspace. Section 6 summarizes the findings and gives an outlook to future work.

2 Research Methodology

This paper is part of a research project aiming at developing new business models for data-driven services in the context of dataspaces following the FAIR principles, and for implementing these business models in organizations, including the required adaptations of IT infrastructures, organisational structures and processes. The project follows the paradigm of design science research (DSR) [2] and this paper concerns first steps towards designing the envisioned artefact, a business model prototype and methodical/technical support for implementing it in organizations. More concretely, we focus in this paper on how to determine change needs in exising business models and propose the instrument of an analysis matrix that will form part of the envisioned methodical support. The research question for this paper is: How to support enterprises offering data-focused business models to adapt to FAIR principles? The research approach used is a combination of argumentative-deductive work and descriptive case study. The resulting analysis matrix is subject to an initial validation by applying it in a use case. The motivation for the use case is that we need to explore the nature and phenomenon of business model adaptation to FAIR principles in real-world environments, which is possible in case studies.

3 Business Models

Business models have been an important element of economic behavior for many years and received significantly growing attention in research with the beginning of the Internet economy. In general, the business model of an enterprise describes the essential elements that create and deliver a value proposition for the customers, including the economic model and underlying logic, the key assets and core competences [3]. Zott and Amit identified three major fields of research in business model developments [6]: (1) business models for electronic business contexts and the use of IT in organizations, (2) strategic business models for competitive advantage, value creation and organizational performance, and (3) business models in innovation and technology management. The first two fields are considered relevant for our research as they affect value creation processes and positioning in the market. Analysis and design of business models can be supported by approaches dividing business models into several perspectives. Partial business models, as defined by Wirtz [5] or business model dimensions proposed by Schallmo [4] can support these tasks. The perspectives covered in both approaches are: financial perspective, customer perspective, value creation perspective, partner and supplier perspective, value offer and service perspective.

4 Analysis Matrix

As a tool for analysing the impact of the FAIR principles on business models, a matrix was developed that focuses on the value creation dimension and can be used in interactive analysis sessions. The horizontal axis of the matrix covers the value creation dimension of business models, i.e. the individual value-creating steps in data-based business models are broken down here. Specifically, a data value chain adapted to the FAIR principles is used here, that was developed in previous work and can be seen in Fig. 1, where the differences to a"classical" value chain are highlighted in dark blue. The previous article is still in the publishing process, but can later be found under the title"Towards a FAIR-ready Data Value Chain for Dataspaces".

The idea behind the FAIR-aware data value chain is that conventional data value chains [1] contain generic elements transferable to dataspaces but implementation of FAIR principles requires services or techniques altering the established view of elements in data value chains. Combination of both is expected to create the nucleus for a future business model.

The vertical axis of the matrix defines the aspects of each value step that are needed to analyse FAIR impacts. These aspects are:

- Description is a rough account and definition of the actual activities as performed in this step in the enterprise, i.e., the value creation process.

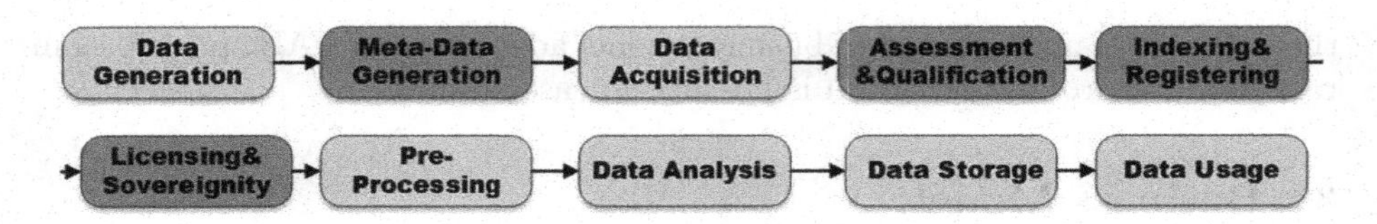

Fig. 1. Changes made to the Data Value Chain to make it FAIR-ready

- Contribution to value creation corresponds to the service offering, with a particular focus on the value proposition offered.
- Within the matrix, the model of service creation is once again linked to a resource model: this is particularly useful in digital business models, as the steps of data resources and their processing can be clearly represented here. The corresponding aspects are the input and output dimensions.
- Trend corresponds to a part of the strategic component of a business model, in which current developments in the corresponding industry are reflected. This is of particular interest in highly dynamic and innovative business fields, as the positioning of the business fields is of special importance here.
- The strategy component also includes the aspect of the actors in the matrix, in which existing actors in the market are identified, who, for example, offer competing products or can act as partners within the framework of the business model.
- Standards, architectures and challenges also contribute to the analysis of the business environment: Explicitly writing down the challenges can serve to clearly define the company's mission. Standards and existing architectures help to better define the service delivery process.

4.1 Procedure for Matrix Use

The procedure consists of the following steps, which usually are performed sequentially. Not all steps are mandatory; the optional ones are marked accordingly:

- The preparation step is dedicated to investigating the suitability of the analysis matrix for the case at hand and to define the scope. For deciding on suitability, the main question is if the company under consideration has a substantial part of the value creation in the field of data processing and finishing. If the data value chain dimension of the matrix is applicable, this usually is the case. Scoping sets the focus unit or product line relevant for analysis. For a company with several data-intensive products or digital information products, this step is essential to increase the significance of the analysis results. The result of this step is a decision on the scope of the analysis. This phase is optional if the company has a clear data processing focus and only one information product line.

- The population step aims at capturing the current situation in an enterprise in the relevant cells of the matrix. For this purpose, the value chain (horizontal axis) has to be considered phase after phase. For each phase, the core question is if the enterprise performs corresponding value creation steps and, if so, what the situation is for the different aspects (vertical axis). The result of this step is the matrix populated with the current situation.
- Opportunity analysis is a step that systematically analyses the potential to extend or change the current business model. Most important aspects to consider in the matrix are
  - the value creation process (aspect "description"): can the process be automated by changed resource use, extended to neighbouring steps, or substituted by selecting less cost-intensive services of partners?
  - the service offering (aspect "contribution"): can the offer be bundled with other offers, split into different parts or has it to be changed?
  - the used resources (aspect "resource"): does a change in resources or alteration of inputs and outputs (including the required processes changes) offer gains in efficiency or effectiveness?
- The strategy definition phase is optional and required, if the opportunity analysis step shows various options to develop the business model, for example, by extending the currently performed value creation steps to neighbouring phases or by improving the automation of these steps by new kinds of resources. In this case, the strategic goals of the company have to be clarified and the opportunities have to be put into relation to strategic goals: are the opportunities supporting the goals or contradicting them? Should opportunities substitute or change existing goals? What problems exist in the enterprise that might delay or hinder the exploitation of opportunities? The result of this step is a priority list of what business model changes to do.
- The roadmap planning is dedicated to identify the different steps or projects to be performed to make the envisioned changes in the business model. The result is an implementation plan.

5 Use Case: Maritime Dataspace MARISPACE-X

For an initial evaluation of the business model analysis matrix, this section uses the case of an ecosystem related to the maritime dataspace MARISPACE-X. Starting from a brief introduction of the use case in Sect. 5.1, Sect. 5.2 discusses the use of the analysis matrix with an enterprise that is part of the MARISPACE-X ecosystem and focuses on data-driven services.

5.1 Case Description

Among the application cases considered in the Marispace-X project is the case of Offshore Wind Farms: During the life cycle of an offshore wind farm, immense amounts of data are generated by the turbines themselves, but also by project planning and asset monitoring. Intelligent management and analysis of data in the context of Federated Services has to support the overall lifecycle of the wind farm, from the planning phase through to decommissioning. In this context, the services and collaboration options developed within MARISPACE-X enable a significant increase in performance, the optimization of asset handling as well as the possibility of efficient multiple use of data. As this has a concrete impact on the competitiveness of the offshore wind farm operators, the targeted cost reduction can have a direct impact on the EEG (Law for renewable energies in germany) surcharge financed by the taxpayer and thus on electricity prices. The targeted digitalization leap will also lead to more efficient processes and thus also contributes to accelerating the energy transition.

5.2 Analysis Matrix Use in MARISPACE-X

The analysis matrix was applied in a small company from Northern Germany that acts as a data provider and data service provider in the MARISPACE-X ecosystem. The focus of the services is in supporting offshore wind park planning; the data available are subsea information about the geological formation, current, and detected metal objects in the soil.

Starting point for the matrix use was a scoping workshop that at the same time also included an analysis of the suitability of the matrix. Participants were the CEO of the company, the head of the business unit data and services, and two researchers involved in the matrix development. The company representatives introduced the existing business services. The researchers presented the FAIR-aware data value chain. The company representatives expressed an interest in using the data value chain and it was decided to conduct a modelling workshop.

In a second step, a full day modelling workshop was conducted. This workshop basically was structured by walking through the different value creation steps and filling the different rows of the matrix for each step. We used a visualization of the matrix in the Miro tool (for documentation) in combination with a physical whiteboard (for discussion). For some value creation steps, there was a discussion about the naming of these phases. All in all, the matrix proved applicable.

The populated matrix (see Fig. 2) showed a focus of the company on services in data acquisition, indexing and pre-processing plus basic data storage. This means that the FAIR data value chain is not completely addressed, as for example, registry and licensing are missing and meta-data generation can be extended. Thus, in the opportunity analysis, the primary focus was on the value creation perspective and on investigating the gaps in the value chain. Resource use and service opportunities were considered by the company as secondary aspects.

In case of ambiguities within the Data Value Matrix, colour coding was applied: Blue stands for different views of the respective Data Value Chain Matrix, green for unknowns.

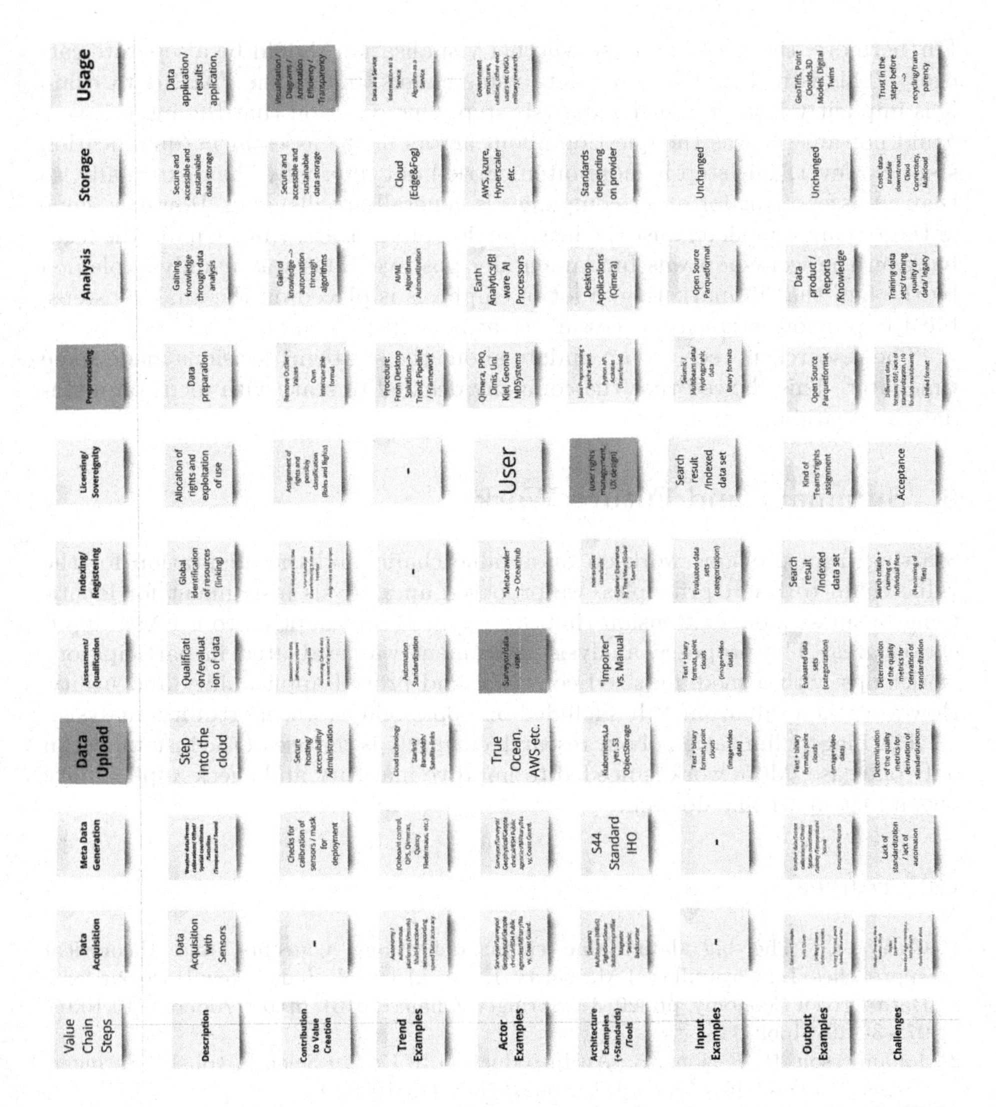

Fig. 2. Property profile of the diverse library compared to the compound pool. (Color figure online)

The biggest differences in theory and practice concerned the part "data acquisition", which was understood/used in the company as "data storage". Another challenge was the definition of preprocessing with the partners involved and the question of whether preprocessing and analysis should be considered separately.

Furthermore, the question arose whether visualisation should be a separate category. This is included in some data value chains, but in the FAIR data chain it is implicitly included in the analysis step. One question that the interviewees could not answer was the question about actors in the assessment/qualification step. However, this step is more often carried out internally than externally, so that an assessment for other companies is generally challenging. However, none of these points speak against the introduction of our Data Value Chain; the feedback on the workshop was fundamentally positive. This was mainly explained by the fact that in marketing, a lot of emphasis is placed on the analysis steps, but the previous steps are necessary to achieve good results.

The researchers received an update about the strategic decisions made some time later. This showed that the company decided to start with focus on metadata generation.

6 Summary and Future Work

Starting from previous work on data value chains and the adaptation for the value chain to FAIR principles, we proposed an analysis instrument for identifying business model extension that are required for adapting to a FAIR application area. The proposed analysis instrument was evaluated in participatory workshops with a medium-sized company and proved applicability and utility. However, the evaluation only included one possibility for opportunity analysis.

The biggest limitation of our research currently is the use of the instrument in only one case. More work is needed to improve maturity and check applicability beyond the maritime domain.

References

1. Curry, E.: The big data value chain: definitions, concepts, and theoretical approaches. In: Cavanillas, J.M., Curry, E., Wahlster, W. (eds.) New Horizons for a Data-Driven Economy, pp. 29–37. Springer, Cham (2016). https://doi.org/10.1007/978-3-319-21569-3_3
2. Johannesson, P., Perjons, E.: An Introduction to Design Science, vol. 10. Springer, Cham (2014). https://doi.org/10.1007/978-3-319-10632-8
3. Osterwalder, A., Pigneur, Y.: Business Model Generation: A Handbook for Visionaries, Game Changers, and Challengers, vol. 1. Wiley, Hoboken (2010)
4. Schallmo, D., Williams, C.A., Boardman, L.: Digital transformation of business models–best practice, enablers, and roadmap. In: Digital Disruptive Innovation, pp. 119–138. World Scientific (2020)
5. Wirtz, B.W., Pistoia, A., Ullrich, S., Göttel, V.: Business models: origin, development and future research perspectives. Long Range Plan. **49**(1), 36–54 (2016)
6. Zott, C., Amit, R.: Business model design: an activity system perspective. Long Range Plan. **43**(2–3), 216–226 (2010)

An Information Privacy Competency Model for Online Consumers

Aikaterini Soumelidou(✉) and Thanos Papaioannou(✉)

Ionian University, Corfu, Greece
{ksoumelidou,thanospapa}@ionio.gr

Abstract. E-commerce has taken a prominent role in our everyday life. However, the increasing disclosure of personal information as a prerequisite to join in such services in combination with consumers' low privacy knowledge, has raised serious concerns regarding the protection of information privacy. To address this issue, the current study proposes a novel framework for the design of information privacy competency model for online consumers, incorporating Protection Motivation Theory and Big Personality Theory. Additionally, synthesizes the results into an indicative privacy competency model for online consumers. The results of this paper can act as a guide for the development of privacy awareness and training programs.

Keywords: Information privacy · Competency model · Competencies

1 Introduction

E-commerce has taken a prominent place in our everyday life, as it offers plenty of advantages, such as faster selling procedure or abolishment of location limitations [1]. Nevertheless, services which e-commerce employs, such as product viewing, recommendations or retrieval, which however is based on automatic collection of personal data, have raised serious concerns regarding consumers' online privacy [2].

In this changing landscape, several researchers have dedicated their work in variously studying consumers' information privacy, such the study of privacy paradox phenomenon [3] or the proposal of ways that firms should adopt to provide more secure environments for their customers [4].

However, while much of the early privacy research relating to consumers' online information disclosure has been focused on providing insights into the antecedents of their privacy concerns and attitudes [5], little attention has been given to the competencies that a consumer should hold to make informed choices against information privacy risks. Research in the information privacy competency domain can significantly contribute not only to the deeper understanding of users' privacy behavior but also to the optimal design of privacy awareness programs [6]. Nonetheless, this is an under-investigated area, for both generic Internet use, as well as targeted Internet behaviors (e.g., online shopping). To fill this gap, our research investigates the following research question "What are the competencies that an online consumer should hold to protect own information privacy

S. Nurcan et al. (Eds.): RCIS 2023, LNBIP 476, pp. 593–602, 2023.
https://doi.org/10.1007/978-3-031-33080-3_42

against relevant risks, while performing online shopping activities?". To do so, we first compose the framework for the design of an information privacy competency model for consumers through the combination of the conceptual analysis of privacy competencies implied in the literature along with the two main theories that shed light on the way that people act in order to protect themselves against privacy risks, namely Protection Motivation Theory and Big Personality theory. Further, we propose an extension of the Iceberg Competency Model for privacy behavior, by classifying the privacy knowledge competencies into declarative and procedural, and the privacy skill competencies into skills for preventive and confronting behavior. Following our synthesized framework, we develop an indicative information privacy competency model for online consumers, comprising the information privacy protection competencies that consumers should hold in every phase of the purchasing cycle, as proposed by the European Commission [7].

2 Literature Review

2.1 Competency

The concept of competency has been widely studied in many areas such as information systems, psychology, education, human resources and is related with an individual's superior performance to a task. Since its emergence, in 1973, when McClelland introduced competency as a critical differentiator of performance [8], researchers have proposed different definitions for it. However, a widely used definition of competency is that it is a measurable human capability which is necessary for effective performance and includes three elements namely knowledge, skill and attitudes [9].

2.2 Competency Model

A competency model is a set of competencies which defines primary behaviors that are necessary for excellent performance in a particular role and is usually organized into tiers of competencies [10]. Such models are valuable tools to human resources management [11]. Spencer and Spencer [12] introduce the Iceberg Model which represents the set of competencies as an iceberg; knowledge and skills are "visible" competencies on the top of the iceberg, whereas motives and traits are "hidden" at the bottom of the iceberg as they are deeper and central to personality.

There are few competency models in the Information System (IS) literature and moreover, most of them refer to IS professionals but limited research on the development of competency models for IS end-users [6]. Additionally, little attention has been paid to the investigation of users' competencies that lead to more effective privacy protection. Specifically, majority of the existing models are generic and include privacy as one component; whilst a specialized privacy competency model is absent. Table 1 below summarizes privacy competencies implied in the literature.

3 Methodology

Aiming to define the main competencies of the privacy competency model for online consumers, we primarily relied on the Information Privacy Competency Model for Citizens [6] which is adapted to the Iceberg Model. Following the above approach, our framework represents knowledge and skills as visible competencies on the top of the iceberg, whereas Social Role, Traits, Self-Image and Motives as hidden ones.

Table 1. Privacy competencies implied in the literature.

Privacy competencies	References
Privacy protection; Understanding of online tracking and digital footprint	[13]
Privacy literacy training	[14]
Knowledge of privacy risks and privacy laws	[15]
Knowledge about the recipients of their data; Metacognitive accuracy	[16]
Planning of security and privacy systems; Developing policies	[17]
Identifying privacy issues; Reviewing privacy policies;	[18]
Knowledge; Skills; Attitudes; Values	[6]

Starting from the top of the Iceberg Model, in order to define knowledge, we adopted the double definition given by [19], which we adapted as follows; The "privacy declarative knowledge" which means that user knows about the existence of privacy risks and is familiar with the existence of an appropriate countermeasure and the "privacy procedural knowledge" which means that the user knows how to operate the privacy-control.

The next type of visible competency is skills. In the information privacy context, we adopted the term "digital privacy skills" as introduced in [20], referring to a subset of digital skills that allow users to apply strategies for individual online privacy regulation and data protection, such as ability to install and customize safeguards [6].

Additionally, literature indicates that effective privacy protection requires the combination of attitudes and low-risk behaviors on behalf of users [21]. For this reason, we extended the visible part of the Iceberg model by adding the competency of "behavior", as adopted by [20]. Thus, we argue that behavior is a competency that a consumer should hold when in online purchases and it consists of two elements; Privacy Preventive behavior which includes actions that consumer performs in order to mitigate the risk of being exposed to privacy violations and Privacy Confronting behavior which includes actions that consumer performs while facing a privacy risk.

The remainder of our competency model adopted the hidden competencies as introduced in [6]. Social role refers to attitudes and values projected to others. Literature implies that privacy concerns determine attitudes towards the disclosure of personal data online, namely privacy attitudes [22]. In the e-commerce context, privacy concerns have negative influence on information disclosure behavior [23].

The next type of competencies is values, which are defined as people's decisions to prevent others from obtaining given information about oneself [24]. Anonymity, which is referred to the condition of being unknown to others, is considered to be one of the most appreciate values of online consumers, as it can effectively contribute to the protection of information privacy [25]. Moreover, anonymity comes with the values of freedom and fear-free living, meaning that, when users adopt anonymity tools, they feel free, in terms that they protect their online privacy [26].

The next competency type refers to personality traits that have direct impact on users' privacy concerns. Literature implies that self-disclosure intention is directly connected to the Big Personality Theory [27] which demonstrates that there are five traits used to describe the human personality. Conscientiousness, which is defined as a person's ability to be self-disciplined, is negatively related to the disclosure personal information [27]. Extraversion, which refers to people who seek to create interpersonal relationships, leads to low level of privacy concerns [28]. Agreeableness, that indicates an individual's degree of trust in others and by others is negatively related to privacy concerns [28]. Openness, which refers to the degree someone looks for new experiences, leads to less cautious personal data disclosure [28]. Neuroticism, which refers to a person who is susceptible to psychological distress and other distressing emotions, is negatively related to information disclosure [29].

Self-image, which is the next competency in the hidden part of our competency model, indicates a person's sense of identity and worth [30] and includes the concept of self-esteem. Self-esteem refers to an individuals' sense of self-worth and is an important competence that consumers should hold as users with low level of self-esteem tend to disclose more personal information looking for others' approval [31].

The last competency of the iceberg model are motives [6]. According to the revised Protection Motivation theory (PMT), protective behavior is motivated by; perceived vulnerability, namely the extent of ones' belief that health condition may happen to them and positively affects users' privacy concerns [32]. Perceived severity, namely an individual's understanding of the potential harm of the behavior to themselves or to others, which positively affect users' privacy concerns and effectively motivate users to the adoption of protective measures [32]. Response efficacy, which is the belief on the effectiveness of the recommended coping response in reducing risk and positively affects privacy protective behaviors [33]. Self-efficacy, which refers to individuals' belief that they are capable of successfully performing protective behaviors and positively affects privacy protective behavior [32]. And finally, response costs, which refers to the perceived costs of adopting preventive response including money, time, or personal effort and negatively affects protecting behaviors [33]. Furthermore, self-representation is another motive which refers to the establishment of an image of oneself [34]. According to the literature, this motive is leading factor which urge users to disclose their personal information a view to raise their popularity [35]. Table 2 summarizes the elements of competencies in our proposed framework for the development of Information Privacy Competency Model for Consumers.

4 Results

To identify the main competencies that consumers should hold for online shopping, we first needed to clarify which actions are involved in online purchases. For that purpose, we relied on the Digital Competence Framework for Consumers of European Commission [7] which defines general competences for consumers in the digital marketplace. Specifically, the purchase of a product or service constitutes a "purchasing cycle" which is divided into three phases; the Pre-purchase phase which includes actions taken before the purchase, such as browsing and searching. The "Purchase" phase which includes actions related to purchasing, such as buying or selling goods and services. And finally, the "Post-purchase" phase, which comprises actions taken after the purchase, such as sharing information with others [7].

Based on this separation, we thoroughly identified every competence that is required in all three phases of the purchasing cycle of online shopping with respect to information privacy. As a result, we argue that consumers should hold specific competencies in order to address information privacy issues that appear in each phase of the purchasing cycle. Table 3 bellow represents indicative competencies that consumers should hold in the three phases of the purchasing cycle and depicts a part of our proposed Information Privacy Competency Model for Consumers (see Table 2).

Table 2. The proposed framework of an Information Privacy Competency Model for Consumers

Element of competency	Description
Knowledge	Privacy Declarative knowledge; Pr. Procedural knowledge
Digital privacy skills	Ability to e.g., read privacy policy, exercise own rights
Behavior	Privacy Preventive behavior; Privacy Confronting behavior
Social Role	Privacy concerns
Values	Anonymity; freedom; fear-free online
Self-image	Self esteem
Traits	Consciousness, neuroticism, and moreover, low level of agreeableness, openness and extraversion
Motives	Perceived severity, perceived vulnerability, response efficacy, self-efficacy, low response costs, low self-representation

Table 3. Indicative consumers' competencies in the three phases of the purchasing cycle

Actions	Recognize/evaluate commercial communication and advertisement	Interact in the digital marketplace to buy and sell	Share information in the digital marketplace
Pr.Competencies	**Pre - Purchase Competencies**	**Purchase Competencies**	**Post – Purchase Competencies**
Declarative knowledge	Consumers know that websites may use customized functions to explicitly save and store their information for later use	Consumers know that privacy terms of use/ privacy policies/ privacy rights may vary across countries	Consumers know that, when sharing reviews and experiences, specific personal information is disclosed
Procedural knowledge	Consumers know how to install blocking tools in order to protect themselves against online advertising, such as MyTracking	Consumers know each time how to find, read and understand the privacy policy	Consumers know how to act in order to protect their personal information such as anonymously sharing
Pr. Preventive behavior	Actions against online advertising (e.g., installation of ad blocking tools, such as Privacy Budger or MyTracking)	Installation of privacy preference tools (Privacy Bird), reading of privacy policies	Installation of privacy preference tools (Privacy Bird)
Pr.Confronting behavior	Communication with the data protection officer of the online shop or filing a complaint with the national data protection authority	Exercise of privacy rights	Actions that consumer performs while facing a privacy risk, such as exercise of privacy rights
Skills	Ability to distinguish online advertising techniques, install safeguards against online advertising (e.g. Adblock Plus, cookie settings), exercise their own privacy rights	Ability to read and understand the privacy policy of each website, install browser add-on services (Privacy Badge)/ privacy preference tools	Ability to use anonymous sharing orders as well as adopt tools such as Tor Browser or InPrivate browsing
Social Role	High privacy concerns about cookies on behalf of online shops or search engines. High confidence; Consumers are critical towards the promotion of goods on behalf of services providers and they act accordingly	High privacy concerns about the collection of information on behalf of different vendors. Anonymity; Use of tools such as Tor Browser	High privacy concerns. Anonymity; Consumers value anonymity as valuable regarding the protection of their information privacy when sharing reviews

(*continued*)

Table 3. *(continued)*

Actions	Recognize/evaluate commercial communication and advertisement	Interact in the digital marketplace to buy and sell	Share information in the digital marketplace
Traits	Conscientiousness; Self-disciplined consumers will both focus on real needs and distinguish online advertising techniques. Low Agreeableness; Consumers should express higher distrust regarding online advertising techniques	Low Agreeableness; Higher distrust in service providers. Low openness; Consumers should restrict the number of online vendors they visit	Low Agreeableness; Consumers should express higher distrust about the collection of their data when sharing reviews
Self - image	High self – esteem; Consumers should hold a high level of self-esteem, so as to be more protective regarding their online privacy in all three phases of the purchasing cycle		
Motives	Perceived severity; Implicit understanding of the potential privacy risks of online advertising techniques. Perceived vulnerability; Belief that privacy risks derive from advertising techniques. Response efficacy; Belief on the effectiveness of protection tools against advertising techniques. Self-efficacy; Belief on the ability to use privacy protection tools. Low response costs of adopting preventing response such as the time of reading privacy policy	Perceived severity; Implicit understanding of the potential privacy risks e.g. different terms of privacy. Perceived vulnerability; Belief that privacy risks exist when interacting to the global marketplace. Response efficacy; Belief on the effectiveness of privacy protection tools such as InPrivate browsing. Low response costs of adopting preventing response such as the time of adjusting privacy settings	Perceived severity; Implicit understandimg of the potential privacy risks when visiting websites in order to find or share reviews. Perceived vulnerability; Belief on the existence of privacy risks (e.g. digital footprint). Response efficacy; Belief on the effectiveness of privacy protection tools. Low response costs of adopting preventing response. Low self-representation level; Careful regarding the sharing of personal information as a means to create more likeable self -image in sharing reviews communities

5 Conclusions

This paper presents, to the best of our knowledge, the first attempt for the development of privacy competency model for consumers. The study was conducted conceptually by examining competency theories as well as by incorporating widely known personality theories. Based on this exploratory analysis, the paper presents a novel framework for the development of privacy competency models and moreover, develops an indicative information privacy competency model for online consumers.

This paper provides important theoretical as well as practical implications. Specifically, the systematic literature review results into a novel and detailed framework for the design of competency model and moreover, sheds light on researchers' efforts to investigate privacy in the e-commerce field. Additionally, this framework can act as a guide for online service providers to make transparent the way that personal data are processed. Finally, this paper is also beneficial not only for policy makers and educators who conduct educational interventions for citizens, but also for consumers for the protection of their privacy. Nevertheless, a main limitation of the study is the lack of empirical research, which will allow the evaluation of the proposed model.

References

1. Niranjanamurthy, M., Kavyashree, N., Jagannath, S., Chahar, D.: Analysis of e-commerce and m-commerce: advantages, limitations and security issues. Int. J. Adv. Res. Comput. Commun. Eng. **2**(6), 2360–2370 (2013)
2. Wu, Z., Shen, S., Zhou, H., Li, H., Lu, C., Zou, Do.: An effective approach for the protection of user commodity viewing privacy in e-commerce website. Knowl. Based Syst. **220**, 106952 (2021). https://doi.org/10.1016/j.knosys.2021.106952
3. Willems, J., Schmid, J. M., Vanderelst, D., Vogel, D., Ebinger, F.: AI-driven public services and the privacy paradox: do citizens really care about their privacy? Public Manage. Rev. (2022)
4. Bird, D., Neeman, Z.: What should a firm know? Protecting consumers' privacy rents. Am. Econom. J. Microecon. **14**(4), 257–295 (2022)
5. Kang, J., Lan, J., Yan, H., Li, W., Shi, X.: Antecedents of information sensitivity and willingness to provide. Mark. Intell. Plan. **40**(6), 787–803 (2022)
6. Tsohou, A.: Towards an information privacy and personal data protection competency model for citizens. In: FischerHübner, S., Lambrinoudakis, C., Kotsis, G., Tjoa, A.M., Khalil, I. (eds.) Trust, Privacy and Security in Digital Business. LNCS, vol. 12927, pp. 112–125. Springer, Cham (2021). https://doi.org/10.1007/978-3-030-86586-3_8
7. Brečko, B., Ferrari, A., edited by Vuorikari R., Punie Y.: The Digital Competence Framework for Consumers. Joint Research Centre Science for Policy Report, EUR28133
8. McClelland, D.C.: Testing for competence rather than for "intelligence." Am. Psychol. **28**(1), 1–14 (1973). https://doi.org/10.1037/h0034092
9. Holtkamp, P., Pawlowski, J.M.: A competence-based view on the global soft-ware development process. J. Univ. Comput. Sci. **21**(11), 1385–1404 (2015)
10. Kansal, J., Singhal, S.: Development of a competency model for enhancing the organizational effectiveness in a knowledge-based organisation. Int. J. Indian Cult. Bus. Manage. **16**(3), 287–301 (2018)
11. Boyatzis, R.E.: The Competent Manager: A Model for Effective Performance. John Wiley & Sons, New York (1982)
12. Spencer L.M., Spencer S.M.: Competence at Work: Models for Superior Performance, John Wiley & Sons, Inc., New York (1993)
13. Lauricella, A.R., Herdzina, J., Robb, M.: Early childhood educators' teaching of digital citizenship competencies. Comput. Educ. **158**, 103989 (2020)
14. Desimpelaere, L., Hudders, L., Van de Sompel, D.: Knowledge as a strategy for privacy protection: how a privacy literacy training affects children's online disclosure behavior. Comput. Human Behav. **110**, 106382 (2020). https://doi.org/10.1016/j.chb.2020.106382

15. James, L.T., Wallace, L., Warkentin, M., Kim, B., Collignon, S.: Exposing others' information on online social networks (OSNs): perceived shared risk, its determinants, and its influence on OSN privacy control use. Inf. Manage. **54** (2017)
16. Moll, R., Pieschl, S., Bromme, R.: Competent or clueless? Users' knowledge and misconceptions about their online privacy management? Comput. Human Behav. **41**, 212–219 (2014)
17. Mekovec, R., Oreški, D.: Competencies for professionals in the fields of privacy and security. In: Proceedings of the International Conference on Privacy-friendly and Trustworthy Technology for Society – COST Action CA19121 - Network on Privacy-Aware Audio- and Video-Based Applications for Active and Assisted Living (2022)
18. Schueller, S.M., Armstrong, C.M., Neary, M., Ciulla, R.P.: An introduction to core competencies for the use of mobile apps in cognitive and behavioral practice. Cogn. Behav. Pract. **29**(1), 69–80 (2022). https://doi.org/10.1016/j.cbpra.2020.11.002
19. Bitton, R., Finkelshtein, A., Sidi, L., Puzis, R., Rokach, L., Shabtai, A.: Taxonomy of mobile users' security awareness. Comput. Secur. **73**, 266–293 (2018)
20. Trepte, et al.: Do people know about privacy and data protection strategies? Towards the "Online Privacy Literacy Scale" (OPLIS). In: Gutwirth, S., Leenes, R., de Hert, P. (eds.), Reforming European data protection law. Springer, Dordrecht, Netherlands (2015)
21. Soumelidou A., Tsohou, A.: Towards the creation of a profile of the information privacy aware user through a systematic literature review of information privacy awareness. Telem. Inform. **61** (2021)
22. Li, Y.: Empirical studies on online information privacy concerns: literature review and an integrative framework. Commun. Assoc. Inf. Syst. **28**, 28 (2011). https://doi.org/10.17705/1CAIS.02828
23. Sun, Y., Fang, S., Hwang, Y.: Investigating privacy and information disclosure behavior in social electronic commerce. Sustainability **11**(12), 3311 (2019). https://doi.org/10.3390/su11123311
24. Stone, E.F., Gueutal, H.G., Gardner, D.G., McClure, S.: A field experiment comparing information-privacy values, beliefs, and attitudes across several types of organizations. J. Appl. Psychol. **68**(3), 459–468 (1983)
25. LapidotLefler, N., Barak, A.: Effects of anonymity, invisibility, and lack of eye-contact ontoxic online disinhibition. Comput. Hum. Behav. **28**(2), 434–443 (2012)
26. Skalkos, A., Tsohou, A., Karyda, M., Kokolakis, S.: Identifying the values associated with users' behavior towards anonymity tools through means-end analysis. Comput. Human Behav. Rep. **2**, 100034 (2020). https://doi.org/10.1016/j.chbr.2020.100034
27. Erdem, Ö.: Why do consumers behave differently in personal information disclosure and self-disclosure? The role of personality traits and privacy concern. Alphanum. J. **6**(2), 257–276 (2018). https://doi.org/10.17093/alphanumeric.460158
28. Junglas, A.I., Johnson, N.A., Spitzmüller, C.: Personality traits and concern for privacy: an empirical study in the context of location-based services. Eur. J. Inf. Syst. **17**(4), 387–402 (2008). https://doi.org/10.1057/ejis.2008.29
29. Skrinjaric, B., Budak, J., Žokalj, M.: The effect of personality traits on online privacy concern. Ekonomski Pregled. **69**, 106–130 (2018). https://doi.org/10.32910/ep.69.2.2
30. İsmail, N., Tekke, M.: Rediscovering Rogers's self theory and personality. J. Educ. Health Commun. Psychol. **4**, 2088–3129 (2015)
31. Stone, C.B., et al.: Why do people share memories online? An examination of the motives and characteristics of social media users. Memory **30**(4), 450–464 (2022)
32. Adhikari, K., Panda R.K.: Users' Information Privacy Concerns and Privacy Protection Behaviors in Social Networks. J. Global Market. (2018)

33. Sedek, M., Ahmad, R., Othman, N.F.: Motivational factors in privacy protection behaviour model for social networking. MATEC Web Conf. **150**, 05014 (2018). https://doi.org/10.1051/matecconf/201815005014
34. Kokolakis, S.: Privacy attitudes and privacy behaviour: a review of current research on the privacy paradox phenomenon. Comput. Secur. **64**, 122–134 (2015)
35. Felim, P., Dimyati, D., Shihab, M.: ASKfm: motives of self-disclosure to anonymous questions. Jurnal Komunikasi **13**(1), 93–108 (2018)

DECENT: A Domain Specific Language to Design Governance Decisions

Fadime Kaya[1(✉)], Francisco Perez[2], Joris Dekker[3], and Jaap Gordijn[1]

[1] Vrije Universiteit, Amsterdam, The Netherlands
{f.kaya,j.gordijn}@vu.nl
[2] Universidad Rey Juan Carlos, Madrid, Spain
francisco.perez@urjc.es
[3] ABN AMRO Commercial Bank, Amsterdam, The Netherlands
joris.dekker@nl.abnamro.com

Abstract. Decentralized ecosystems, such as the Bitcoin, claim to be decentralized to avoid power concentrations such as commercial banks. This is perhaps true for their operations, but often not the case for their governance, which is about deciding the rules of monitoring and controlling protocols. In previous work, we have developed DECENT a domain specific language (DSL) to conceptualize the domain of decentralized governance design. In this paper, we focus on deriving governance design decisions based on the DECENT language. We do so by taking the case of Fractional Reserve Banking (FRB), which is about governance rules for commercial banks to create and destroy money. As many banks are licensed to do FRB, under the control of national central banks and the European Central Bank (ECB), this is already a case of a decentralized ecosystem. The governance design decisions are developed in close cooperation with our case study partner, a commercial bank.

Keywords: Decentralized Governance Design · Blockchain Governance · Fractional Reserve Banking · Money Creation · Design Decision

1 Introduction

Blockchain governance design gained traction as a societal and research topic, mainly as the result of the emergence of the crypto-currencies. For example considering the Bitcoin, who actually may create money, under which conditions, and how this done, is an example of a governance question. In case of the crypto's many governance design decisions are ultimately reflected in the protocols of these crypto's. Individuals consider the decentralized nature (e.g. bypassing traditional banks) of crypto-currencies as an important motivation to invest in it. However, although the crypto-currency philosophy promotes values as democracy and openness, there are multiple governance design and decisions problems with crypto-currencies platforms [2, 4, 7]. First, Bitcoin claims to be highly decentralized (no centralized actor involved in transaction processing), however its governance processes are organized to a large extent rather centrally [2,

S. Nurcan et al. (Eds.): RCIS 2023, LNBIP 476, pp. 603–610, 2023.
https://doi.org/10.1007/978-3-031-33080-3_43

4]. Secondly, the exchange of crypto-currencies are facilitated by highly centralized exchange platforms, and this has lead to issues such as fraud, money laundering, and tax avoidance [1, 2].

A method to create the euro currency is by means of Fractional Reserve Banking (FRB). This is a governed system in which commercial banks can create loans based on saving accounts of others, and these banks are allowed to *multiply* the deposited savings with a certain factor, such that banks can loan more money than they actually have on the savings account. If a loan is paid off, the inverse happens and the earlier created money is 'burned'. As commercial banks are licensed to do FRB, the traditional money system is already decentralized to a certain extent. In [5] it is argued that a decentralized ecosystem can only succeed if the governance is decentralized and more importantly, the governance structure is organized through a design approach. That is exactly our *research goal*, as we developed DECENT, which is a domain specific language to develop governance design decisions for blockchain ecosystems [3]. In this paper, we are interested if the DECENT language can be applied to derive governance design decisions. We do so by partnering with a Dutch leading commercial bank to design and develop the governance design decisions for Fractional Reserve Banking. The contribution of this paper is to demonstrate that governance design decisions can be conceptualized by using DECENT language. We do that by deriving and presenting the governance design decisions from a business and process model perspective. These are all positioned and mapped with the DECENT concepts. This paper is structured as follows. Section 2 explains our research approach, Sect. 3 we present the DECENT Governance Design Decisions, and in Sect. 4 we conclude the paper.

2 Research Approach

For this research paper we are interested in developing governance design decision by employing DECENT. Therefore, we have formulated the following research question: Can we conceptualize governance design decisions by using DECENT? We employ a single-case study experiment within the design science context as we are designing governance decisions by using the DECENT language. We do this in a real-world context with the commercial bank ABN AMRO. The topic of the case study will be Fractional Research Banking (FRB), which is a financial system to create and destroy money by commercial and central banks.

3 Case Study: Fractional Reserve Banking in DECENT

3.1 Fractional Reserve Banking in DECENT Meta Language

We employ a single case mechanism experiment on the topic of Fractional Reserve Banking (FRB). Our goal in this paper is to assess the usability of DECENT as a language to representing, explaining, and analyzing governance design decisions in an ecosystem. In several highly interactive sessions with the commercial bank, we elicited the governance constructs for FRB by using the DECENT meta model. What follows now is an expression and description of FRB in terms of the DECENT meta model language. To emphasize the use of the relevant DECENT decentralized governance concept as clear as possible for the case, we use *Italic words* that represents and refers to the DECENT meta model concepts. A debitor, creditor, commercial banks, the European and National Central Banks and the Basel Committee are all instances of *Party*. A *Party* can be either an *Actor* or a *Group*. There is a clear distinction between these two; a debitor is an example of a single *actor* who wants to deposit money, whereas the Basel Committee is a *group* of *actor* s, amongst others the national banks. Each *party* plays one or more *role*s towards a *governance construct*. In FRB, the *Group* Basel committee, in collaboration with *Group* European Central Banks *sets* how much money can be created to *stimulate* the economy with the total amount of loans. As a *group* in their 'setting' role, they use a *decision making* procedure, for example a majority vote. FRB is steered and *decided* through *goals* (Fig. 1).

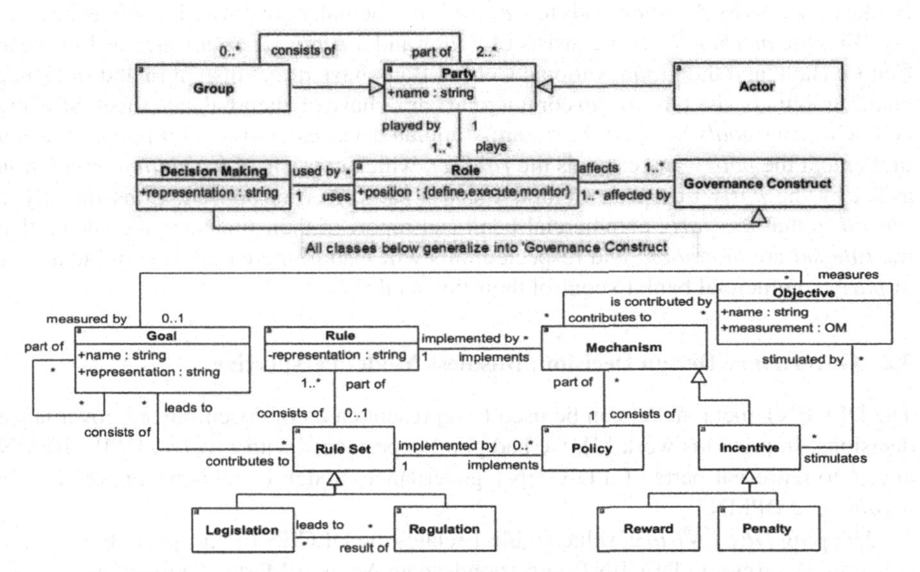

Fig. 1. DECENT Meta Model for Decentralized Governance Design

The *group* Basel Committee has the *goal* to maintain resilient banking systems through world-wide regulation. The *goal* of the *group* European Central Banks (consisting of the National Central Banks) is to maintain trust in the euro as a currency. Also they have the *goal* to regulate and provide financial stability to mitigate bank stress.

The debitor *actor* has the goal to increase capital through receiving interest; the creditor's *goal* is attract capital in return for interest paid.

The amount of money that can be created using FRB is determined through a *rule-set* that takes the shape of *legislation* and *regulation*. Also *Legislation* and *regulation* contributes to *rule-set*.

The *rule-set* is implemented and set by a *policy*. Each *party* has their *policy* in which the agreed *governance constructs* are represented. Actual execution of the *policy* defines how much loans *legally* can be created based on the amount of capital a *party* Commercial

Table 1. DECENT meta model in e^3*value*

DECENT Concept	Representation in e^3*value*
Actor	Actor or Market Segment
Group	Group (extension of e^3*value* in [6])
Goal, Objective	Customer Need
Mechanism, Policy	Value Activity, Value Transfer, Value Object
Reward, Penalty	Value Transfer, Value Object
Rule set, Rule	Dependency Path

Bank has access to. A *policy* leads to *mechanism* to actually govern and *regulate* the *rule-set*. With the *mechanism* that consists of *policy* and *incentive* a *group* such as European Central Bank and the *group* National Central Bank have direct insight in and influence on the amount of risks the *group* commercial banks have on their balance sheet. Steering and achieving *goals* of every *party* are stimulated via *incentives*. If a *party* does not implement the *policy* and exceeds the *rule-set*, which consists of *legislation, regulation* as set by the *party* European Central Bank, a *penalty* is introduced. Consequently, a *reward* is that the *party* commercial bank can report in their financial statements that the *rule-set* are *monitored* and respected, this will lead to increased trust in the ability of *party* commercial bank to control their financial risks.

3.2 Governance Design Decision: Business Model Perspective

The DECENT meta model can be used to represent cases of decentralized governance decisions. In previous work [3], we had good experiences with e^3*value*, UML, BPMN and *i** to represent parts of a DECENT governance model. For this paper we employ e^3*value* and BPMN.

Mapping DECENT to e^3value. Table 1 relates the DECENT concepts to the e^3*value* concepts. An Actor in DECENT corresponds to an Actor or Market Segment in e^3*value*. Actors are the same because in DECENT and e^3*value*, both refer to actors who take decisions independently. A Market Segment in e^3*value* is a set of Actors, of which one is randomly selected to exchange objects of economic value with. Hence a Market Segment can be considered as a DECENT Actor. There is no corresponding concept of the Group concept in DECENT with e^3*value*, but in [6], we have proposed an extension to e^3*value*, with the notion of Group, that has the same meaning as in DECENT.

There is a multiplier effect, since the creditor can puts its money on a deposit account (E 90,000.-) and the same dependency path repeats. This process is continuously executed, and the total amount of money creates is then $1/R$. This multiplier effect is not shown in the e^3*value* model. The reverse situation happens if the creditor pays off its loan; this is not shown in the e^3*value* model. In e^3*value*, the notion of customer need represents that an actor or market segment (*Actor*) has a need. Table 2 presents the governance design decisions, as expressed by the e^3*value* model in Fig. 2.

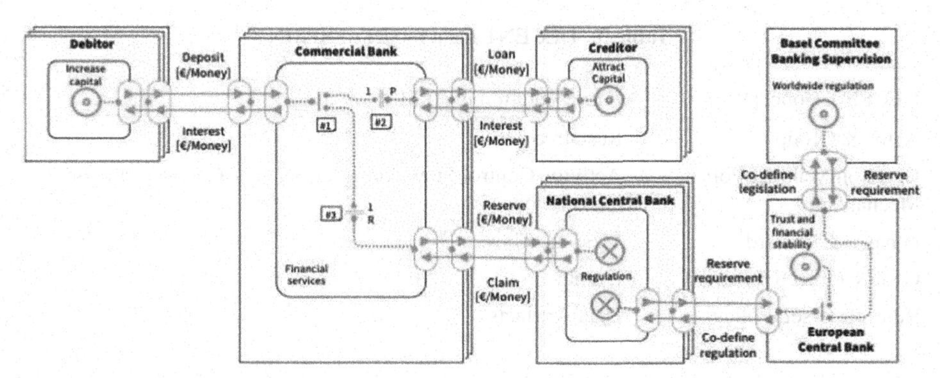

Fig. 2. DECENT Governance e^3*value* FRB Model

Table 2. Governance Design Decision: Business model Perspective

Representation in e^3*value*	FRB Governance Design Decision
Actor Market Segment Customer Need	Debitor: Increase capital CB: Earn profit through interest on loans. Creditor: Attract capital NCB: Controlling inflation through monetary policy. ECB: Assert trust & financial stability in € currency. BC: Worldwide regulation to achieve uniformity in capital markets
Value Activity, Transfer, Object	Reserve requirement to control inflation Financial stability in euro currency achieved through reserve requirement and worldwide alignment
Dependency Path	Formula Fraction (*R*) reserve requirement: [1] R Formula Fraction *(P)* to create money: 1 − R Total amount of loans is regulated by NCB through fraction *R* requirement Remaining *P* a CB is allowed to create money CB stores deposits and create loans

3.3 Governance Design Decisions: Process Model Perspective

Mapping DECENT to BPMN. Table 3 relates the DECENT meta model concepts, and their relationships to the relevant construct(s) in BPMN. In BPMN, the way to model Actors is via Resource Pools. There is no direct corresponding construct for the Group, but often the Resource Pool is used too, with the addition that many of these Resource Pools of the same kind may exist. BPMN is in particular very strong in modeling the DECENT Policy and Mechanism constructs in depth. In fact, we consider the BPMN language as a tool to specify policies and mechanisms in a much more fine grained way.

Table 3. DECENT meta model in BPMN

DECENT Concept	Representation in BPMN
Actor & Group	Resource pool
Decision Making, Policy, Mechanism	Activity, Control Flow constructs such as Gateway, and Sequence flow
Penalty & Reward	Message Flow
Goal & Objective	Event
Rule, Rule-set, Policy	Data Artifacts

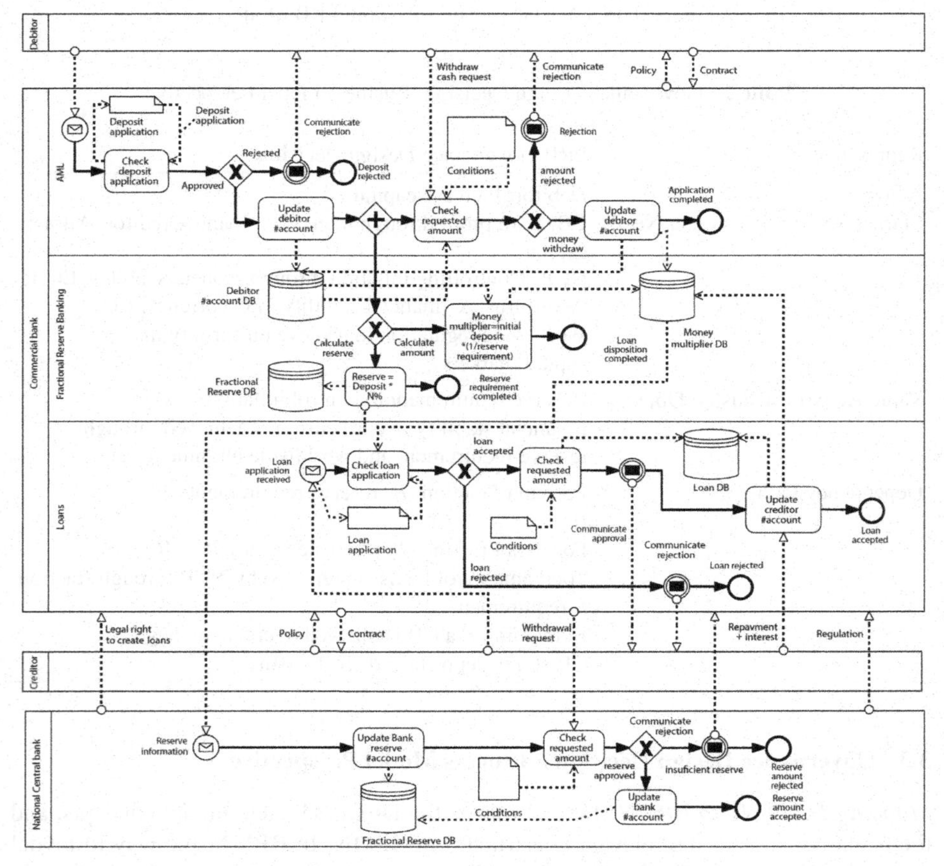

Fig. 3. DECENT Governance BPMN FRB model

The BPMN notion of activity, plus all concepts to express a control flow, such as Gateways correspond to the DECENT Policy and Mechanism concepts. In case Decision

Making in DECENT can be expressed as a (collection of) activity it can be modelled in BPMN as well. Penalties and Incentives will result in Message.

Table 4. Governance Design Decision: Process model Perspective.

Representation in BPMN	Governance Design Decision
Resource Pool	Debitor, Commercial Bank (CB), Creditor, National Central Bank (NCB)
Activity, Control flow, Sequence, Gateway	Deposit check for anti money laundering by CB Money inflow from Debitor: Deposit Money outflow from from CB to Debitor: Interest on deposit. Deposit claim increase/decrease at debitor account Update FRB database with cash inflow/cash withdrawal based on deposit and loans in and outflow Money outflow to Creditor at CB: Create loan and interest. Money inflow from Creditor at CB: Interest + repayment loan. Money outflow from CB to NCB: Reserve requirement Calculate reserve requirement by NCB and CB: Deposit*R%. Calculate money multiplier by CB: Deposit*($\frac{1}{R}$) Update reserve requirement account at NCB
Message Flow	Acceptance and/ or rejection of money deposit by CB Acceptance and/ or rejection of reserve requirement by NCB. Debitor and CB exchange terms of contract for deposit Creditor and CB exchange terms of contract for loan. NCB and CB exchange terms of regulation Legal right to create loans by NCB
Event	Objective measurable by quantifiable formulas
Data Artifacts	Debitor and Creditor Database at CB Money Multiplier Database at CB Fractional Reserve Database at CB and NCB. Loan and Deposit Database at CB Deposit and loan application sheet via Debitor at CB. Condition sheet for Deposit and Loan provided by CB Condition sheet for Reserve requirement by NCB

Flows between Actors, hence they are related. Rules and Rule-sets (DECENT) may require BPMN Data Stores, to make them persistent.

Figure 3 presents an BPMN model of the case of FRB, based on the description in Sect. 3.1. With this DECENT Governance BPMN model we have modeled and explain how Fractional Reserve Banking is executed. Concepts from the DECENT model are shown in *italic* again. Benefit of BPMN is that the power structure is made explicitly, which *party* executes which *role* and how these processes of money creation are *defined* and *monitored*. A *role* is to *define* their need to deposit cash and a need to obtain an asset by requesting a loan. Both these processes for requesting a loan and depositing cash are facilitated and *executed* by the *group* Commercial Bank. The amounts available for being able to create loans, the *group* Commercial Bank is dependent on *actor* Debitor and on the *group* European Central Bank. The total amounts a commercial bank has in savings, is multiplied by the *rule-set* which consists of *legislation* and *regulation* that *defines* how much money can be created. All these financial transactions are executed via *group* commercial bank. An important *role* in FRB is anti-money laundering (AML). This *role* of AML is *executed* by the *group* commercial bank and *monitored* by National Central Bank. Table 4 presents the governance design decisions, as expressed by the BPMN model in Fig. 3.

4 Conclusion

We presented the notion of decentralized governance design, and by following the DECENT language we were able to derive governance design decisions. We used the DECENT language first to describe the case at hand and consequently we presented the governance design decisions from a business and a process model perspective. We have used DECENT in the domains of peer-to-peer energy trading, and in CBDC; first with a national central bank and now with a commercial bank. The outcomes of this study will be further used and discussed in the EU setting in the digital euro context. We are planning to extend this research to design how money creation and destruction will function in a digital euro mechanism.

Acknowledgement. We thank the Market Infrastructure Group (ABN AMRO), Jesse Donkervliet (VUA), Massivizing Computer Systems Group (VUA), Knowledge Engineering Group (Vienna), and Paul Ellsiepen (VUA) for the support. Declaration. This research is not funded by ABN AMRO.

References

1. Corbet, S. (ed.): Understanding Cryptocurrency Fraud: The Challenges and Headwinds to Regulate Digital Currencies. De Gruyter (2021). https://doi.org/10.1515/9783110718485
2. Fritsch, R., Mu¨ller, M., Wattenhofer, R.: Analyzing voting power in decentralized governance: Who controls daos? arXiv preprint arXiv:2204.01176 (2022)
3. Kaya, F., Gordijn, J.: Decent: An ontology for decentralized governance in the renewable energy sector. In: 2021 IEEE 23rd Conference on Business Informatics (2021)
4. Mariem, S.B., Casas, P., Romiti, M., Donnet, B., Stu¨tz, R., Haslhofer, B.: All that glitters is not bitcoin–unveiling the centralized nature of the btc (ip) network. In: IEEE Network Operations and Management Symposium (2020)
5. Pocher, N., Zichichi, M.: Towards cbdc-based machine-to-machine payments in consumer IoT. In: Proceedings of the 37th ACM/SIGAPP, pp. 308–315 (2022)
6. Sarkar, R., Gordijn, J.: Modeling communities in e3 value. In: Workshop On Value Modeling and Business Ontologies (2018)
7. Ziolkowski, R., Miscione, G., Schwabe, G.: Decision problems in blockchain governance: old wine in new bottles or walking in someone else's shoes? J. Manag. Inf. Syst. **2**, 316–348 (2020)

Doctoral Consortium Papers

Business User-Oriented Recommender System of Data

Sarah Pinon(✉)

NaDI, Namur Digital Institute, University of Namur, Namur, Belgium
sarah.pinon@unamur.be

Abstract. Companies nowadays are increasingly dependent on data. In an environment that is more dynamic than ever, they are looking for tools to leverage those data and obtain valuable information in a rapid and flexible way. One way to achieve this is by using Data-Driven Decision Support Systems (Data-Driven DSS). In this project, I focus on one such type of DSS, namely the Self-Service Business Intelligence (SSBI) Systems. These systems are designed specifically to avoid the involvement of the IT department when creating business reports by empowering businesspeople in the production of their own reports, thereby reducing the time-to-release of a given report and improving the responsiveness of companies. Business decision-makers, when developing their own reports, however face barriers. These challenges are related to the current self-service features that are not sufficiently adapted to their business needs and their lack of technical knowledge. The objective of my project is to build a framework based on Artificial Intelligence (AI) techniques such as Natural Language Processing techniques, Semantic and Recommender Systems to solve one of the main challenges faced by businesspeople, namely: the data picking within technical and large current databases. These AI systems offer a number of benefits that are strongly linked to the problems encountered by business users in the data picking process. This paper introduces the three main research questions of my thesis and positions them in the current literature. It then elaborates on the different theoretical, methodological and empirical contributions I plan to advance as part of my project.

Keywords: Self-Service Business Intelligence · Business User · Data · Recommender System · Semantic Systems · Natural Language Processing

1 Context

The current success of companies depends on their ability to use the available data [6]. In fact, companies that make decisions based on data are more productive than companies that rely on intuition [21]. Extracting useful knowledge from data to solve business problems is becoming a priority for companies in almost every industry [21]. However, this task is increasingly complex as the volume

S. Nurcan et al. (Eds.): RCIS 2023, LNBIP 476, pp. 613–621, 2023.
https://doi.org/10.1007/978-3-031-33080-3_44

of data continues to grow. Companies are therefore investing in Decision Support Systems (DSS) to manage this amount of information [6]. These systems are nowadays a business necessity but also an opportunity to gain a significant competitive advantage [20].

A Business Intelligence (BI) is one such tool that assists decision makers in their process by transforming data into information and knowledge [29]. Indeed, BI systems collect data, store it and produce knowledge with analytical tools to present complex information to decision makers [16]. Several actors are involved in the BI process. These actors can be divided into two groups. The first are IT professionals or experienced BI users. They have the technical skills to efficiently set up and run BI systems. The second group consists of decision-makers, called Business Users. They do not have technical skills and use BI tools to support their decision making [2,14].

Over the last few years, BI had to adapt to a new type of requirement from companies that face an increasingly dynamic environment, and which as a result need to be more and more reactive in their decision-making [14,31]. To meet this new requirement, a new version of BI tools is being developed, called Self-Service Business Intelligence (SSBI) [13,14].

SSBI systems aim to provide reports and dashboards to decision makers much faster. To achieve this, these systems allow decision makers to build reports and dashboards themselves. In this way, all the necessary interactions between IT experts and business users to define the report requirements are eliminated and a considerable amount of time is saved [30,31].

SSBI tools are increasingly used in companies [31]. Therefore, there is a growing interest in their efficiency in the literature. Nowadays, it turns out that SSBI tools are not as easy to use as they promise [14]. This issue comes from the many challenges faced by Business Users throughout the BI process, from the access to data to the interpretation of reports produced [13,14].

In my research project, I focus on one of the challenges faced by Business Users: finding relevant data fields within the current large and technical databases. There is a lot of evidence showing the difficulties business people face when dealing with databases, such as exploring them, locating specific data or identifying data fields for a particular need [2,13,14,25]. These challenges imply three negative effects. First, the adoption of SSBI tools is hindered, even though it is critical in today's dynamic business environment [14]. Second, these obstacles imply an under-exploitation of the available data, which remains not understood by the user. Third, there is a high risk of wrong data selection, which can compromise the quality of decision outputs [13,14].

This paper is organized as follows: Sect. 2 explains the research objectives of the project by developing the state of the problem and the research questions. Section 3 develops the research approach by explaining for each research question the objective, methods used, related work, current achievement and preliminary results. Section 4 concludes the paper.

2 Research Objectives

2.1 Problem Statement

The challenges faced by Business Users when dealing with current databases are mainly related to two of their characteristics:

1. **The amount of data fields**: In the era of Big Data, the volume of available data is continuously increasing [28]. This quantitative growth comes from technological advances in digital sensors, communication, computation, and storage [8]. While collecting data is becoming easier, identifying relevant information within that data is more and more complex. Indeed, several data fields do not provide any interesting information or do not concern all the Business Users of the company [14]. This task of identification requires special skills and technologically-adapted tools [21,28]. Business Users have limited IT skills and therefore have difficulties to perform the task properly [14].
2. **Technical jargon**: Databases frequently use highly technical jargon. The terminologies employed are not necessarily familiar to Business Users. Indeed, the labels of the data can, for example, correspond to names of codes or can have several definitions according to the type of Business User [14]. This jargon implies a lack of understanding of the data meaning for the users [24].

In addition to the complexity of today's databases, Business Users do not always have a clear vision of what data they actually need [14]. There are many situations where the decision maker does not know from which angle to analyze a company: Is it the marketing actions that need to be reviewed? Is it the performance of the employees that needs to be checked? Is the cost of products too high? The volume and technicality of today's databases will further obscure their fuzzy vision of what they are looking for.

The ultimate goal of my project is to help Business Users with these challenges by providing recommendations about which data from a large data repository should be used by the decision maker, based on question submits with his/her own words and without any knowledge of the underlying naming of the database, tables or columns. Moreover, when producing these recommendations, I take care to adopt a broad perspective on all possible data, opening the eyes of Business Users to the field of possibilities. In this way, I optimize the value generated by the data for decision makers.

To achieve this, I propose a solution mobilizing different Artificial Intelligence (AI) techniques, namely: Natural Language Processing (NLP) techniques, Semantic Systems (e.i. Ontologies) and Recommender Systems (RSs). These different techniques offer indeed many advantages to help users to face these challenges.

NLP techniques aim to understand and manipulate the natural languages of human beings. These techniques are, for example, used within user interfaces to translate requests expressed in natural language into requests expressed in language understandable by the system [5]. These techniques are more and more exploited within SSBI tools in order to allow Business Users to explore their data in a more natural way and without requiring the help of IT experts [1,18].

Ontologies are semantic modeling systems. They offer a formal semantic representation of the system's structure [9,17]. By using Ontologies within SSBI, the business semantics used by the Business Users can be integrated to, for example, define the different data of the company's Data Warehouse (DWH) [4,23]. This business jargon simplifies and makes more intuitive the interaction between the Business User and the system. One can indeed speak of business-oriented interaction rather than technology-centered interaction [4].

Regarding the RS, its objective is to generate meaningful recommendations to users for items or products that might interest them [15]. In the context of SSBI, the RS can allow to detect the Business Users's interests and thus to characterize the type of interaction desired with the SSBI tool [7]. These systems can then guide and support the Business User in his/her process of transforming data into information [26].

2.2 Research Questions

The previous large objective brings me to the definition of a number of research questions:

- **RQ1**: How to help Business Users to pick relevant data within a voluminous and technical DWH through the Business Glossary and a Semantic RS ?
- **RQ2**: What are the barriers to adoption of business decision makers using SSBI, and how does our implemented Data RS help mitigate those barriers ?
- **RQ3**: How can we account for the user profile in the Data RS to offer personalized data suggestions ?

3 Research Approach

This PhD research is currently composed of three ongoing studies corresponding to the three research questions. Each study has its own methodology and contribution:

1. Data RS for SSBI tools;
2. Empirical study on Data RS;
3. Business User-centric RS of Data.

The three studies together follow the methodology called Design Science Research methodology [10,19]. This methodology aims to create new innovative artifacts that solve real-world problems. This approach develops and evaluates new solutions [10,19]. My research project follows this approach through these three studies. Indeed, the first research consists in developing my solution answering a problem identified in the literature. This step corresponds to the rigor cycle of the methodology. Then, the empirical study of this research consisting in demonstrating the solution relevance for the companies and their employees refers to the relevance cycle of the methodology. Finally, the last study of the project corresponds to an improvement of the system involving all the cycles of the methodology, namely the existing literature and a field study of the users of the existing tools.

3.1 Study 1 - Data RS for SSBI Tools

Objective. The first step of my research project is to support Business Users to pick the right data, among technical and voluminous databases, based on a request expressed in semi-natural language. Concretely, my data picking support system will allow these users to express their data needs with their own business words without necessarily referring to the technical jargon of the database. Compared to what is proposed in the literature, my system will exploit the Business Glossary associated with the database, develop a Data RS and offer a great transparency and flexibility.

Methods. To achieve the objective of this paper, I exploit RSs. These systems help people to cope with information overload [22] which is completely aligned with the problem my system is facing, namely data overload. Moreover, my RS is based on an ontological representation of the DWH schema integrating the associated business semantics via the Business Glossary. Ontology-based RSs allow to deal with most of the conventional RS problems such as cold-start problem, rating sparsity and overspecialization [27]. As for the Business Glossary, it allows to add the business context to the available data [12]. To provide the user with data recommendations, my system exploits a NLP technique and a set of business rules to link the terms used by the user in his/her request to the numerous technical data fields of the DWH.

Related Work. Systems to make databases more easily usable by Business Users have already been developed in the literature. Some papers develop, indeed, Natural Language Interface for Databases (NLIDB). These systems transform data queries expressed in natural or semi-natural language into complex languages such as SQL [1,18]. Other works exploit Semantic Systems to add a semantic layer to databases [23,25]. To the best of my knowledge, no Data RS has been already developed.

Current Achievement and Preliminary Results. This first step of my research project is already implemented and the associated paper is being written.

3.2 Study 2 - Empirical Study on Data RS

Objective. To address the second research question of my project: "What are the barriers to adoption of business decision makers using SSBI, and how does our implemented Data RS help mitigate those barriers?", I will study my problem of interest in a real-world setting and develop my RS developed in the first study within several industries in order to identify its added value. The companies used as use cases will be a large consortium of hospitals thanks to a collaboration with the research center CETIC ("Centre d'Excellence en Technologies de l'Information et de la Communication") and the international company TotalEnergies.

Methods. First, a qualitative study will be conducted with the business decision-makers of the two collaborators mentioned above in order to identify and confirm the challenges they face with large and technical databases. Second, after developing our RS on the data of our two partners, we will do another qualitative study to demonstrate the added value of our system and identify its limitations.

Related Work. To the best of my knowledge, no study has been done specifically on the limitations encountered by Business Users in relation to current databases. Moreover, as this study serves as a use case for the system we developed in the first article, the contribution is completely new to the literature.

Current Achievement and Preliminary Results. This phase of the project has not yet been implemented. Only collaborative initiatives with businesses have been launched.

3.3 Study 3 - Business User-Centric Data RS

Objective. In this study, I will try to answer the third question of my research: "How can we account for the user profile in the Data RS to offer personalized data suggestions?". This paper aims to offer an advanced version of our tool developed in the first study. Indeed, I want to integrate the user's profile (e.g. his/her data preferences, historical interactions) as well as additional functionalities (e.g. feedback on recommendations). These different features will allow me to refine my recommendations.

Methods. To improve the RS, I will exploit the weighting of ontologies to, for example, represent the user's historical interactions with certain data on the DWH ontology. The weights are then exploited by the RS to recommend the most relevant fields, as is done by the RS based on the case-based approach [3]. In addition to the ontology weighting, this study will identify the different data needs of different Business User profiles in order to help the RS. To do this, a qualitative study will be conducted with different Business Users from different companies. I will identify their data needs and analyze the differences between the different users according to their profile. This study will be inspired by what has been done by [11] for the audience of visualization technologies.

Related Work. As previously explained, no Data RS has been, to my knowledge, developed in the current literature. Consequently, no Data RS considering the user's profile and my additional features exists. Concerning the user profile, some studies have been conducted to characterize, not the Business Users in the context of Data Picking within DWH, but, for example, the audience of visualization technologies [11].

Current Achievement and Preliminary Results. During the first months of research, an exploration of the different existing RSs was carried out. This allowed me to understand how they work, their advantages and disadvantages. I also analyzed different studies studying the categorization of BI users in terms of visualization tools for example.

4 Conclusion

This PhD research proposes a complete system to support business decision-makers in the process of data picking within large and technical databases. The proposed system is completely adapted to the business world by considering the business semantic and the user's profile. I suggest integrating AI techniques and in particular, NLP techniques, Semantic Systems and RSs, to achieve this goal. The research project will be divided into three studies. These studies will provide methodological contributions to the existing literature to address the problem identified and developed in this paper. Fundamental contributions will also be presented in these different studies by the definitions of several ontologies and the integration methods of these knowledge models.

Acknowledgements. I would like to thank the supervisors of this project, Dr. Corentin Burnay and Dr. Isabelle Linden, for their review and support on this paper. Moreover, this research was financed by the Walloon Region in the scope of the ARIAC project.

References

1. Affolter, K., Stockinger, K., Bernstein, A.: A comparative survey of recent natural language interfaces for databases. VLDB J. **28**(5), 793–819 (2019). https://doi.org/10.1007/s00778-019-00567-8
2. Alpar, P., Schulz, M.: Self-service business intelligence. Bus. Inf. Syst. Eng. **58**(2), 151–155 (2016)
3. Bridge, D., Göker, M.H., McGinty, L., Smyth, B.: Case-based recommender systems. Knowl. Eng. Rev. **20**(3), 315–320 (2005)
4. Cao, L., Zhang, C., Liu, J.: Ontology-based integration of business intelligence. Web Intell. Agent Syst. Int. J. **4**(3), 313–325 (2006)
5. Chowdhary, K.: Natural language processing. Fundam. Artif. Intell. 603–649 (2020)
6. Cody, W.F., Kreulen, J.T., Krishna, V., Spangler, W.S.: The integration of business intelligence and knowledge management. IBM Syst. J. **41**(4), 697–713 (2002)
7. Drushku, K., Aligon, J., Labroche, N., Marcel, P., Peralta, V., Dumant, B.: User interests clustering in business intelligence interactions. In: Dubois, E., Pohl, K. (eds.) CAiSE 2017. LNCS, vol. 10253, pp. 144–158. Springer, Cham (2017). https://doi.org/10.1007/978-3-319-59536-8_10
8. Emani, C.K., Cullot, N., Nicolle, C.: Understandable big data: a survey. Comput. Sci. Rev. **17**, 70–81 (2015)

9. Guarino, N., Oberle, D., Staab, S.: What is an ontology? In: Staab, S., Studer, R. (eds.) Handbook on Ontologies. IHIS, pp. 1–17. Springer, Heidelberg (2009). https://doi.org/10.1007/978-3-540-92673-3_0
10. Hevner, A.R., March, S.T., Park, J., Ram, S.: Design science in information systems research. Manag. Inf. Syst. Q. **28**(1), 6 (2008)
11. Kerren, A., Stasko, J., Fekete, J.D., North, C.: Information Visualization: Human-Centered Issues and Perspectives, vol. 4950. Springer, Heidelberg (2008). https://doi.org/10.1007/978-3-540-70956-5
12. Kotsis, G., et al.: Database and Expert Systems Applications - DEXA 2021. Springer, Cham (2021). https://doi.org/10.1007/978-3-030-86472-9
13. Lennerholt, C., van Laere, J., Söderström, E.: Implementation challenges of self service business intelligence: a literature review. In: 51st Hawaii International Conference on System Sciences, Hilton Waikoloa Village, Hawaii, USA, 3–6 January 2018, vol. 51, pp. 5055–5063. IEEE Computer Society (2018)
14. Lennerholt, C., Van Laere, J., Söderström, E.: User-related challenges of self-service business intelligence. Inf. Syst. Manag. **38**(4), 309–323 (2021)
15. Melville, P., Sindhwani, V.: Recommender systems. Encyclopedia Mach. Learn. **1**, 829–838 (2010)
16. Negash, S., Gray, P.: Business intelligence. In: Burstein, F., Holsapple, C. (eds.) Handbook on Decision Support Systems, vol. 2, pp. 175–193. Springer, Heidelberg (2008). https://doi.org/10.1007/978-3-540-48716-6_9
17. Obrst, L.: Ontologies for semantically interoperable systems. In: Proceedings of the Twelfth International Conference on Information and Knowledge Management, pp. 366–369 (2003)
18. Őzcan, F., Quamar, A., Sen, J., Lei, C., Efthymiou, V.: State of the art and open challenges in natural language interfaces to data. In: Proceedings of the 2020 ACM SIGMOD International Conference on Management of Data, pp. 2629–2636 (2020)
19. Peffers, K., et al.: Design science research process: a model for producing and presenting information systems research. arXiv preprint arXiv:2006.02763 (2020)
20. Power, D.J.: Decision Support Systems: Concepts and Resources for Managers. Greenwood Publishing Group, Westport (2002)
21. Provost, F., Fawcett, T.: Data science and its relationship to big data and data-driven decision making. Big Data **1**(1), 51–59 (2013)
22. Resnick, P., Varian, H.R.: Recommender systems. Commun. ACM **40**(3), 56–58 (1997)
23. Sell, D., et al.: SBI: a semantic framework to support business intelligence. In: Proceedings of the First International Workshop on Ontology-Supported Business Intelligence, pp. 1–11 (2008)
24. Smuts, M., Scholtz, B., Calitz, A.: Design guidelines for business intelligence tools for novice users. In: Proceedings of the 2015 Annual Research Conference on South African Institute of Computer Scientists and Information Technologists, pp. 1–15 (2015)
25. Spahn, M., Kleb, J., Grimm, S., Scheidl, S.: Supporting business intelligence by providing ontology-based end-user information self-service. In: Proceedings of the First International Workshop on Ontology-Supported Business Intelligence, pp. 1–12 (2008)
26. Sulaiman, S., Gómez, J.M.: Recommendation-based business intelligence architecture to empower self service business users. In: Multikonferenz Wirtschaftsinformatik, pp. 1–12 (2018)

27. Tarus, J.K., Niu, Z., Mustafa, G.: Knowledge-based recommendation: a review of ontology-based recommender systems for e-learning. Artif. Intell. Rev. **50**(1), 21–48 (2018)
28. Tsai, C.-W., Lai, C.-F., Chao, H.-C., Vasilakos, A.V.: Big data analytics: a survey. J. Big Data **2**(1), 1–32 (2015). https://doi.org/10.1186/s40537-015-0030-3
29. Vercellis, C.: Business Intelligence: Data Mining and Optimization for Decision Making. Wiley, Hoboken (2011)
30. Weiler, S., Matt, C., Hess, T.: Understanding user uncertainty during the implementation of self-service business intelligence: a thematic analysis (2019)
31. Yu, E., Lapouchnian, A., Deng, S.: Adapting to uncertain and evolving enterprise requirements: the case of business-driven business intelligence. In: IEEE 7th International Conference on Research Challenges in Information Science (RCIS), pp. 1–12. IEEE (2013)

Secure Infrastructure for Cyber-Physical Ranges

Vyron Kampourakis(✉)

Department of Information Security and Communication Technology, Norwegian University of Science and Technology, 2802 Gjøvik, Norway
vyron.kampourakis@ntnu.no

Abstract. Industrial systems (IS), including critical ones, swiftly move towards integrating elements of modern Information Technology (IT) into their formerly air-gapped Operational Technology (OT) architectures. And, naturally, the more such systems become interconnected, the more alluring they pose to attackers. Concurrently, the twenty-four-seven availability of these systems renders it harder for defenders to promptly apply contemporary security controls. In this context, cyber ranges have emerged as a proper complementary solution for better comprehending and subsequently tackling the relevant risks without endangering the operation of the real systems. This work aspires to contribute a reference architecture for designing and developing cross-sector critical infrastructure (CI) cyber-physical ranges and security testbeds. A second key goal is to demonstrate the soundness of the proposed reference architecture through the implementation and evaluation of a number of cyber range instances specifically tailored for CIs of interest, including manufacturing, energy, and healthcare.

Keywords: Industry 4.0 · cyber-physical system · reference architecture · cyber-physical range · security testbed · risk assessment

1 Introduction

With the penetration of modern IT devices and networks into the traditional and old-fashioned OT-oriented Industrial Systems (IS), the industry has moved towards its fourth revolution era. Precisely, moving from siloed CI environments to fully interconnected and remotely accessible systems, made possible a plethora of functionalities towards enabling technological advances across all such areas, including healthcare, smart manufacturing, energy supply, and so on. In this respect, Cyber-Physical Systems (CPS) introduce innovative applications that impact the worldwide economy and sustainable development.

At the same time, interconnectivity which goes hand-in-hand with Industry 4.0 and relevant CPS, must, among others, depend on reliable and secure communications, either machine-to-human or machine-to-machine. That is, networked systems and infrastructures, even the most barricaded ones, present vulnerabilities and may be exploited at any moment by a variety of threat actors. Even more,

S. Nurcan et al. (Eds.): RCIS 2023, LNBIP 476, pp. 622–631, 2023.
https://doi.org/10.1007/978-3-031-33080-3_45

such complex system-of-systems may expose a large attack surface, which is cumbersome to control in the context of an information security risk management process based on ISO/IEC 27005. Therefore, the protection of any type of critical IT or OT system along with the communication links between them is of utmost importance. Such provision is key to the successful implementation and growth of Industry 4.0. Put simply, as governments, companies, and individuals have become critically reliant on Information and Communication Technology (ICT), cybersecurity is anymore a sine qua non for guaranteeing the continuous delivery of critical services like water supply, healthcare, and electricity.

Altogether, cyber threats pertinent to critical IT and OT systems should be nowadays considered an omnipresent and imminent danger that modern CI and Industry 4.0, in general, have to face. On top of that, the Internet of Things (IoT) and 5G and beyond cellular networks are gradually forming a tightly connected union of systems. This means that certain processes in smart cities, smart buildings, and transport systems are already remotely controlled over the Internet. This means that solely concentrating on threats and vulnerabilities in specific sectors is anymore inefficient; instead, a cross-sector cyber thinking and defense approach is required. This must be considered a high priority for the cybersecurity community because of the ramifications that a cyberattack can inflict on an IS and especially a CI. Scrutinizing the security status of CI systems while in operation, say, through red teaming, is not always suggested or even feasible; a CI system provides vital services and even a single oversight may be catastrophic. Instead, CI cybersecurity assessment can be performed through a "testbed", which emulates the real-life system in part or in full. Ideally, such a testded would comprise a Digital Twin (DT) of its physical twin, on top of which a cyber range can be built. In this respect, the practical application and assessment of hands-on cybersecurity, including training, can be done through the cyber range.

The ***contribution*** of this work is twofold. First off, to provide a reference architecture based on which a cyber-physical range (CPR) can be built. This would establish a common reference point for future research in this area. Based on this reference architecture, a second goal is to develop and assess CPR targeted to certain CI. This will provide further insight both to the scientific community and the industry regarding the existing threats and respective countermeasures.

The remainder of this work is structured as follows. The next section discusses the related work. Sections 3 and 4 provide details on CPR and the need for a generic reference architecture, respectively. The objectives and research questions regarding this work are given in Sect. 5. Section 6 outlines the proposed research method. The expected results are summarized in Sect. 7, while the last section concludes Sect. 8.

2 Related Work

The work in [1] detailed a variety of security aspects in terms of industrial CPS focusing on vulnerabilities, attacks, and CPS components. They conducted a decade-wide survey, resulting in a comparative analysis among the solutions identified in the literature; Driven by this analysis, they pinpointed a plethora of security issues that are omnipresent in industrial CPS settings, categorizing them based on a number of criteria.

The authors in [8] reviewed industrial cyber-physical systems (ICPS) in terms of cybersecurity. They meticulously examined the ICPS attack surface to comprehend threats, challenges, and countermeasures. Specifically, they first detailed the ICPS architecture characteristics by defining its components, with an emphasis on the OT counterpart of an ICPS. Moreover, they inspected used communication technologies and protocols, concentrating on their advantages and disadvantages, using the domain of application as a benchmark. They carried on with an adaptive attack taxonomy, evaluating also real-life ICPS cyber incidents.

The contribution in [9] surveyed several pertinent works, indicating the need for securing the CPS. First, they concentrated on specific attacks linked with CPS. Second, they provided an extensive overview of the various security scenarios, emphasizing the need to confront such threats. From this standpoint, the authors conclude that CPS testbeds are instrumental in developing proper security controls. Even though the authors devoted a significant portion of their paper to CPS testbeds, the main focus of this review is on the different attacks against CPS.

The study in [14] reviewed cyber ranges and security testbeds. The authors also proposed a taxonomy for cyber ranges with a focus on architectures and attack scenarios, including also capabilities, roles, tools, and evaluation criteria. They concluded that the derived taxonomy can serve as a cornerstone for future initiatives on this topic.

The authors in [7] conducted an extensive survey regarding the already deployed cyber-physical testbeds on five different major sectors. They also exhibited a comparison among these testbeds to recognize analogies. The pinpointed similarities led to the suggestion of a more generic reference architecture, which could be used for designing future CPR.

A survey of Industrial Control System (ICS) testbeds is presented in [6]. Particularly, the authors elaborated on 30 ICS testbeds destined for scientific research. They identified the needs these testbeds were built to address, say, vulnerability analysis, education, and defense mechanisms. A closer look was taken at the nature of the testbeds' hardware components, namely physical, virtual, or emulated, as well as the way the requirements for the ICS systems, which are being simulated, are met.

Based on the above analysis, it becomes clear that apart from individual security-oriented testbeds, which are custom-built to meet the needs of a certain system, either CPS or CI, so far, no generally admitted reference architecture exists. Actually, an initial rudimentary proposal has been given in [7], but still, the reference architecture proposed by the authors is rather primitive. The current work aspires to fill this gap by designing and subsequently assessing a more generic and adaptable reference architecture, which will serve as a common ground for the development, testing, and analysis of CI-oriented cyber testbeds.

3 Cyber-Physical Ranges

Defense mechanisms and countermeasures generally referred to as "controls" with reference to ISO/IEC 27005 and 27032, can be devised after scrutinizing the CPS against attacks to observe its behavior and revise and adjust the system appropriately. However, it is quite cumbersome and, in some cases, infeasible to test an actual CPS, especially if it is part of a CI. In such a case, the ramifications could be catastrophic [9,10].

Instead, a widely adopted solution refers to CPR and DT, generally used for simulating and emulating in whole or in part, say, a real CI system. Such a facility allows for observing, testing, and adjusting the virtual system for improving the operation of its real-world counterpart in a safe manner, as also pinpointed by the National Institute of Standards and Technology (NIST) in [11]. From a cybersecurity viewpoint, this improvement refers to enhancing the security posture of the real-life system, mostly through using or developing the appropriate controls and applying system hardening techniques. This endeavor should follow a proper information security risk management process as detailed in ISO/IEC 27005.

With reference to Fig. 1, a CPR is fed with information stemming from the physical twin, e.g., the relevant threats and vulnerabilities, which are then used by evaluators (trainers) to define, deploy and maintain specific cybersecurity scenarios and challenges. Next, the latter are used by challengers to perform self- or team-based assessment scenarios. The lower level of a CPR comprises the real-life or virtualized ICT, OT, or mixed infrastructure, which is maintained by the administrators; this infrastructure may be represented by the DT, if any. The insight gained from this interplay between the twins and the corresponding CPR serves as feedback towards ameliorating the cybersecurity posture of both twins, especially the physical one.

Under this mindset, a DT can produce at least one CPR with as many security challenges as desired, each of them oriented in the same or similar CI sector of application. Apart from cybersecurity testing by means of offensive or defensive mechanisms, a cyber-physical testbed can also serve as an education platform. This enables the interested parties to learn in a hands-on way the underlying system, evaluate their cyber capability and capacity, test and become familiarized with new procedures and controls, and many more.

So far, diverse variations of such cyber testbeds have been proposed in the literature [4,7,9]. With reference to Sect. 2, one can differentiate between a complete physical representation of a real-life system, a fully simulated real-life system with the usage of software tools, and hybrid implementations. The first two options are problematic for different reasons, namely the first is a stiff and expensive solution that requires multiple resources and funds, while the second may lead to inaccurate deductions due to the sometimes loose or inaccurate emulation of the real-life system by means of software and the lack of interaction between them. In this regard, hybrid cyber testbeds are generally considered the superior option for modeling a system and especially a CI, which comprises both IT and OT physical and virtual components. Overall, the progression from the classic

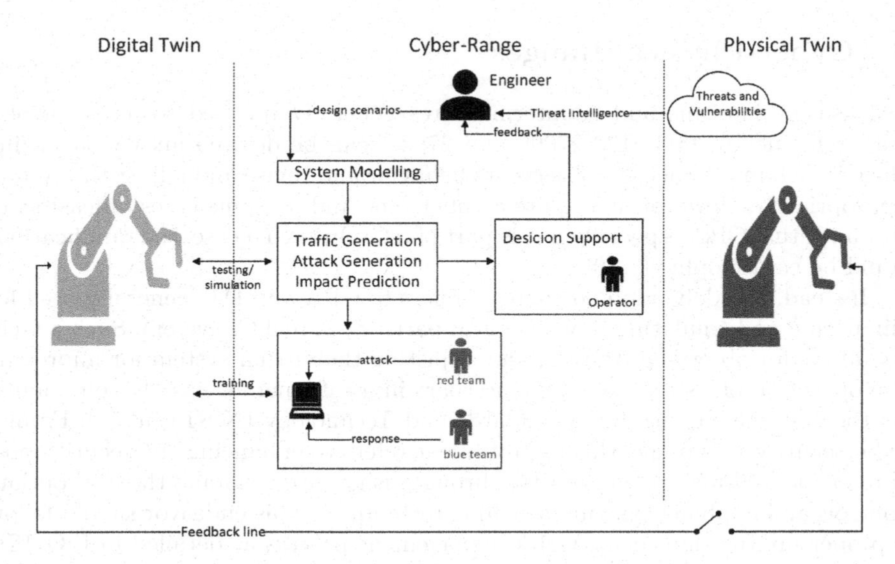

Fig. 1. Interplay between digital and physical twins and a respective cyber-physical range (adapted from [2])

cyber ranges toward CPR sounds like the fittest solution for modeling, testing, and safeguarding a CI, which naturally embraces CPS.

4 Reference Architecture for Cyber-Physical Ranges

The innate heterogeneity that characterizes the nature of CPSs is an inherent problem of such systems, which has not yet received adequate attention in the literature. Particularly, so far, only a limited number of works attempt to address this sparsity and propose more generic and comprehensive solutions, rather than developing individual cyber testbeds for specific case scenarios [5,7,14]. To this end, a reference architecture that overlaps the commonalities of CPS met in a number of CIs could provide a mold for future testbeds focusing on cyber-physical systems security.

In other words, a reference architecture can be used to explain the relations among the different components of a large and complex system. The typical use of a reference architecture is to divide the functionality of a system into sub-layers and define how each layer communicates (or serves) its adjacent ones, if any. Precisely, a reference architecture could be seen as the base layer on top of which lies the corresponding backbone infrastructure, either virtualized or mixed, which in turn can be exploited by a CPR, as also illustrated in Fig. 2.

To better conceptualize the difference among the terms "reference architecture", "backbone infrastructure", and "cyber-physical range" as considered in the context of this work and shown in Fig. 2, we provide their definition below.

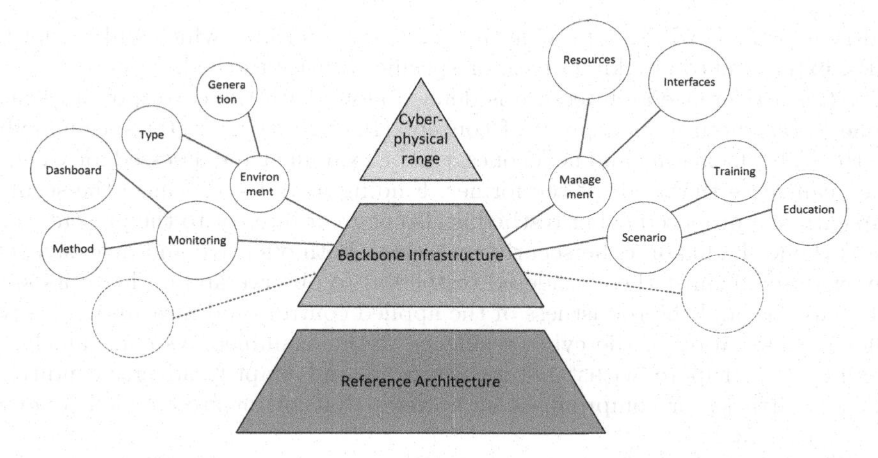

Fig. 2. A layered, coarse-grained association between the terms "reference architecture", "backbone infrastructure", and "cyber-physical range".

- **Reference architecture**: It is a layered abstract representation of the key components, functions, and modules (simulated or emulated) existing in a generic CPR. Apart from the distribution of the aforementioned components along the different layers, a reference architecture also describes possible interrelations among them.
- **Backbone infrastructure**: It is defined as the group of the components, functions, and modules that an IS or CI sector-oriented CPR integrates. Examples of such modules are depicted in Fig. 2.
- **Cyber-physical range**: It is the actual implementation of a given backbone infrastructure.

Putting it another way, following a bottom-up approach, the backbone infrastructure of a CPR should be designed and developed based on a reference architecture. This provides a common ground (translated to a certain degree of uniformity) for designing, building, and comparing CPR across diverse IS/CI sectors.

5 Objectives and Research Questions

The focus of this work is on designing, building, and evaluating CPRs that are applicable to a number of IS/CI sectors. This overall objective includes the following intermediate key objectives.

Given that a CPR can have application to numerous sector-specific CIs, including energy, manufacturing, or healthcare an initial important task is to devise and define a generic architecture upon which individual cyber testbeds of several compatible sectors can be built on. Such an overarching architecture, also referred as to reference architecture, along with the underlying backbone infrastructure, will bring together and possibly conflate similar characteristics of existing CPR. It will also provide the necessary interfaces which will render the CPR extensible to any kind of sector-specific security testbed.

After a reference architecture is defined along with the corresponding backbone infrastructure, as a proof of concept (PoC), various sector-specific cyber testbeds, both existing and novel ones, can be examined and assessed for whether they can serve the needs of the former. Finding ways to scrutinize the security capacity of a proposed cyber testbed is also of great interest to the present work. To this end, hands-on cybersecurity practices, both offensive and defensive, can be exercised against the considered testbed(s) to observe and evaluate its security capacity and the robustness of the applied controls and measures. This will offer an estimation of the cyber resilience of the examined system, namely, its "ability to anticipate, withstand, recover from, and adapt to adverse conditions, stresses, attacks, or compromises on systems that use or are enabled by cyber resources" [13].

With reference to Fig. 2, the research questions relevant to the current work can be summarized as follows. Recall that this study focuses on the top part of the depicted pyramid, namely CPR.

- ***RQ1***: Which is the state of the art regarding CPS and IS/CI cybersecurity testbeds? With reference to Sect. 3, this research axis will consider both CPR and security-focused testbeds created solely for research purposes. Considering the investigated security testbeds, which are the baseline operational and functional requirements for defining a reference architecture focused on IS and CI?
- ***RQ2***: Given the state of the art and the identified requirements in RQ1, which are the key components for the development of a generic reference architecture destined for IS and CI? Prominent CI sectors here include energy, manufacturing, and health.
- ***RQ3***: In the context of a risk assessment process, which are the appropriate cybersecurity assessment methods and schemes to efficiently assess the cyber testbed's security capacity? This will also include the creation of proper training scenarios and challenges, both from an offensive and defensive standpoint.
- ***RQ4***: Based on the results of RQ3, which is the appropriate risk treatment for the potential risks identified?

6 Research Method

The objectives of the work at hand can be fulfilled through a step-based, problem-centered approach as summarized in Table 1. As the applicability of the respective architectures and systems are of major significance, we opt to follow the well-known Design Science Research Methodology (DSRM) methodology [12].

With reference to Table 1, the interviews with the industrial partners will be performed so that the state of the art as given in the relevant literature can be placed vis-à-vis practices already applied by the respective market players. In

Table 1. Methods applied according to the RQs.

RQ1	RQ2	RQ3	RQ4
Systematic literature review	Systematic literature review	Vulnerability identification and analysis	Modeling of controls
	Interviews with industrial partners	Penetration testing	Simulation
	Modeling of the reference architecture	Compliance testing	
		Simulation	
		Development of training material	

this regard, further clarification regarding the gap analysis will be provided, so that the use cases can be fully mapped and potentially addressed. Experimental data from existing analogous systems may also be useful during the design and development steps included in Table 1. The latter will facilitate the comparison between the virtual and the physical components that will be used. Last but not least, risk assessment will be performed according to the ISO/IEC 27005 and 27032 standards, and especially the ISA/IEC 62443 [3] one, which addresses cybersecurity for OT in automation and control systems.

7 Expected Results

The anticipated results regarding this work are presented below in accordance with the objectives and research questions mentioned in Sect. 5. To explicitly define the expected results, an analogy to the five first steps of DSRM [12] is followed.

Identify the Problem and Define Objectives of a Solution: Examine the current literature with an eye towards the shaping of a generic reference architecture on top of which a CPR can be hosted and operated. Therefore, the literature survey associated with RQ1 will not concentrate on specific IS/CI sectors of applications, but on any industrial testbed that can provide relevant input. This effort is also key to identifying the differences between testbeds proposed by the scientific community and analogous systems which are actually used in the CI domain. The feedback received from the industrial partners will be valuable in the achievement of this goal.

Design: Based on the outcomes of the identification and definition of the objectives given in Sect. 5, a subsequent goal is to provide a modular reference architecture upon which future CPR can be built on. The proposed model should not be IS/CI sector-specific, rather it should be abstract enough to meet its CI sector-agnostic nature.

Development and Demonstration: By exploiting the results of the suggested reference architecture, the aim here is to adjust the proposed reference architecture along with the underlying backbone infrastructure to accommodate sector-specific CPS with a high degree of fidelity. The way the derived reference architecture will properly fit into a particular backbone infrastructure is also a significant concern here.

Evaluation: Given the fact that at least one cyber testbed will be developed according to the reference architecture, the results of this phase revolve around the assessment of each cyber testbed's security capacity. This is done by selecting the appropriate cyber assessment methods, applying pertinent attack scenarios, and, if necessary, crafting efficient simulation mechanisms. The security evaluation of each testbed will be examined from both an offensive and defensive perspective, that is, from both a red and blue team viewpoint. This phase also involves the creation of proper training scenarios and challenges for training and educational purposes. Based on the outcomes of the preceding phase, fitting security controls can be proposed. This typically involves a risk treatment process.

Altogether, the first three phases concentrate on designing, developing, and demonstrating a reference architecture suitable for hosting CPR of specific IS/CI sectors. Their fulfillment will lead to the first milestone of this work, namely a reference architecture – generic to a certain point – which can provide a common ground for the development and deployment of a CPR. The second milestone, associated with the last two phases, has to do with the scrutiny of the security of CPS that exist in an IS/CI through the proposed CPR.

8 Conclusions

Despite the evolution of IS towards highly interconnected and remotely accessible systems, the literature lacks a fully featured reference architecture based on which CPR and security testbeds can be built on. The integration of OT systems with modern IT ones immensely broadened the attack surface and put it within the grasp of a plethora of parties, both malevolent externals and insiders. Security testbeds and CPR in general can aid in better comprehending such risks without jeopardizing or disturbing the operation of the real system.

Under this context, the establishment of an overarching reference architecture for designing and developing CPR or security testbeds is of paramount importance. This work aspires to contribute to this goal with an eye toward facilitating future initiatives and providing homogeneity among different CPRs and security testbeds. Moreover, such a reference architecture is expected to offer a common ground for multiple critical sectors, instead of just emphasizing a single sector of application. Another key objective of this study is to create and assess targeted CPRs based on the proposed reference architecture. These testbeds, along with the associated educational material, will facilitate the proper training and skill development of diverse groups of stakeholders, including security professionals and facility and risk managers from the industry.

Acknowledgements. This Ph.D. is carried out under the supervision of Profs. Vasileios Gkioulos and Sokratis Katsikas of Norwegian University of Science and Technology, Gjøvik, Norway and Dr. Habtamu Abie of the Norwegian Computing Centre.

References

1. Agrawal, N., Kumar, R.: Security perspective analysis of industrial cyber physical systems: a decade-wide survey. ISA Trans. **130**, 10–24 (2022)
2. Bécue, A., et al.: 1-securing the industry 4.0 with cyber-ranges and digital twins. In: Proceedings of the 2018 14th IEEE International Workshop on Factory Communication Systems (WFCS), Imperia, Italy, pp. 13–15 (2018)
3. International Electrotechnical Commission: IEC 62443 security for industrial automation and control systems standard (2018)
4. Geng, Y., Wang, Y., Liu, W., Wei, Q., Liu, K., Wu, H.: A survey of industrial control system testbeds. In: IOP Conference Series: Materials Science and Engineering, vol. 569, p. 042030. IOP Publishing (2019)
5. Habib, M.K., Chimsom, C., et al.: CPS: role, characteristics, architectures and future potentials. Procedia Comput. Sci. **200**, 1347–1358 (2022)
6. Holm, H., Karresand, M., Vidström, A., Westring, E.: A survey of industrial control system testbeds. In: Buchegger, S., Dam, M. (eds.) NordSec 2015. LNCS, vol. 9417, pp. 11–26. Springer, Cham (2015). https://doi.org/10.1007/978-3-319-26502-5_2
7. Kavallieratos, G., Katsikas, S.K., Gkioulos, V.: Towards a cyber-physical range. In: Proceedings of the 5th Workshop on Cyber-Physical System Security, pp. 25–34 (2019)
8. Kayan, H., Nunes, M., Rana, O., Burnap, P., Perera, C.: Cybersecurity of industrial cyber-physical systems: a review. ACM Comput. Surv. **54**(11s), 1–35 (2022)
9. Lydia, M., Kumar, P.E.G., Selvakumar, A.I.: Securing the cyber-physical system: a review. Cyber-Phys. Syst. 1–31 (2022)
10. Makrakis, G.M., Kolias, C., Kambourakis, G., Rieger, C., Benjamin, J.: Industrial and critical infrastructure security: technical analysis of real-life security incidents. IEEE Access **9**, 165295–165325 (2021)
11. NIST: Cyber ranges. https://www.nist.gov/system/files/documents/2018/02/13/cyber_ranges.pdf. Accessed 27 Mar 2023
12. Peffers, K., Tuunanen, T., Rothenberger, M.A., Chatterjee, S.: A design science research methodology for information systems research. J. Manag. Inf. Syst. **24**(3), 45–77 (2007)
13. Ross, R., Victoria, P., Gary, G., Ryan, W., Richard, G., Deborah, B.: Enhanced security requirements for protecting controlled unclassified information (2021). https://doi.org/10.6028/NIST.SP.800-172
14. Yamin, M.M., Katt, B., Gkioulos, V.: Cyber ranges and security testbeds: scenarios, functions, tools and architecture. Comput. Secur. **88**, 101636 (2020)

Guidelines for Developers and Recommendations for Users to Mitigate Phishing Attacks: An Interdisciplinary Research Approach

Javara Allah Bukhsh(✉)

University of Twente, Enschede, The Netherlands
j.allahbukhsh@utwente.nl

Abstract. Phishing attacks are common these days. If successful, these attacks cause psychological, emotional, and financial damage to the victims. Such damages may have a long-term impact. The overall objective of this Ph.D. research is to contribute to mitigating phishing victimization risks by exploring phishing prevalence, user-related risk factors, and vulnerable target groups and by designing (1) guidelines for social website developers focused on internet user vulnerabilities and (2) recommendations for users to avoid such attacks. The Ph.D. research acknowledges that phishing attacks are technical in nature, while the impact is financial and psychological. Therefore, an interdisciplinary research approach focusing on empirical research methods from social sciences (i.e., focus groups and surveys) and computer science (i.e., data-driven techniques such as machine learning) is adopted for the research. In particular, we aim to use a machine learning model for data analytics and quantitative and qualitative research design for psychological analysis. The research outcome of this Ph.D. work is expected to provide recommendations for internet users and organizations developing social-media-based software systems through more phishing aware development practices.

Keywords: Phishing · Repeat phishing · User perspective · Risk factors · Vulnerability · Guidelines · Recommendations · Empirical research methods

1 Introduction

Information and Communication Technology (ICT) is ubiquitous in life today as people increasingly rely on multiple ICT components (e.g., mobile devices, personal computers, and intelligent appliances) in day-to-day life. Specifically, by July 2022, the number of internet users has reached 5.3 billion [1]. However, the widespread digital communication poses several risks and has important implications, including common cyber threats, namely social engineering attacks (phishing), ransomware, and mobile security attacks. Due to the rising number of cyber-attacks, computer privacy and cyber security have become a global concern [2, 3]. As per Statista, a market and consumer data provider [4], in the 2nd quarter of 2022, 5.18 million data records were exposed worldwide.

S. Nurcan et al. (Eds.): RCIS 2023, LNBIP 476, pp. 632–640, 2023.
https://doi.org/10.1007/978-3-031-33080-3_46

The Ph.D. research in this paper is concerned with one of the prominent types of cybercrime, namely *phishing*, which compromises personal information, including banking and credit card details, passwords, and individual files. Lastdrager [5] defined *phishing* as 'a scalable act of deception whereby impersonation is used to obtain information from a target". Each year an increasing number of phishing attacks is reported. According to the Anti-Phishing Working Group (APWG), more than one million phishing cases were reported in the first three months of 2020, which was the highest number of attacks in one quarter until now. It was followed by 384,291 cases reported in March 2022, which was the highest number of attacks in one month thus far [6].

Despite the concerted efforts of government institutions and private organizations to limit phishing victimization, overcoming phishing attacks is still considered to be extremely challenging because of the rapidly evolving technological capabilities available to attackers, and the types of the attacks themselves, e.g. email, social media, mobile phone. So far, many studies have been published on phishing detection and its countermeasures [7, 8]. These mostly focused on technical aspects of overcoming phishing attacks. Unlike prior publications and leveraging the author's background in psychology, this PhD work is interested in the human factors that play a significant role in successful phishing attempts. Our PhD research interest is motivated by the observation of Abroshan et al. [9] that human behavior is one of the most important factors determining the success rate of phishing attacks. Moreover, our research is also motivated by the observation [10] that people can quickly disclose confidential information even when they are being warned or nudged by an awareness campaign. While many national surveys and studies have been conducted on phishing prevalence, on its causes, and on the countermeasures to reduce their success rate [3, 8, 11, 12], a steep rise in phishing attacks is still being reported each year.

The present PhD research initiative is set out to help both public and private institutions tackle and reduce the impact of phishing attacks. To this end, there are two significant areas that we aims to work on:

First, exploring and understanding of phishing prevalence, risk factors, and vulnerable target groups. Developing more profound knowledge of victims will enable us to propose and design robust countermeasures. Furthermore, this knowledge would serve as foundation to create recommendations to make internet users more aware about their risky behavior, i.e., sharing credentials with strangers and its consequences in future.

Second, developing specific guidelines for 'attentive' software systems that manage the user's attention for risky behavior, based on known risk factors and phishing techniques. Such guidelines are expected to be helpful to software developers while designing software in the best interest of users' security and privacy.

The context of this Ph.D. work includes both user victimization and *repeat* victimization (i.e., becoming victim more than once) due to phishing. We deliberately include repeat victimization, because on one side scholars acknowledge its importance [13], while on another side, it is an under-researched phenomenon [14]. As per Milani et al. [15], in 2018, ten percent of repeated data breach events were reported. Moreover, Wittebrood & Nieuwbeerta [16] indicated that previous victimization and routine activity increase the chance of repeat victimization. While the literature on cybercrime has mainly focused on cyber victimization generally [17, 18], little attention has been given so far to studying the phenomena of repeat phishing.

This doctoral paper is structured as follows: Sect. 2 provides background on (repeat) phishing and its causes. It summarizes literature on phishing prevalence and its socio-demographic vulnerabilities. Section 3 presents the motivation of this Ph.D. project. Section 4 formulates the research goal and identifies research questions to support this goal. Section 5 describes (i) the interdisciplinary research method to answer the research questions and (ii) the research design that will be implemented to achieve the results. Section 6 discusses findings that have been obtained so far and sheds light on work implemented these days. Finally, Sect. 7 summarizes our progress and our immediate next steps and plans in the long run.

2 Background

Users' insufficient awareness, advances in phishing email technology, and human errors are prominent reasons behind successful phishing attacks [9]. Existing empirical studies [19, 20] indicate the following user vulnerability factors, among others, for phishing attacks: user age and gender, level of education, duration of internet usage, dispositional (e.g., individual aspects) and situational (e.g., environment and others) aspects, phishing awareness and victim personality factors. Furthermore, studies also highlighted personality factors, e.g., those included in the Big Five Personality Theory, behind successful phishing [21, 22]. These authors reported that narcissistic, female users are more frequently tricked through phishing attacks and possess a higher level of conscientiousness, than male users. Moreover, a few people are targeted more often by phishing attacks, significantly if they have fallen victim to a phishing attack in the past. Although there is a consensus among scholars that a few internet users are at risk of repeat exploitation by offenders, little so far has been done to consolidate the published knowledge on the prevalence of repeat phishing and its risk factors among individuals and organizations. The current Ph.D. research intends to bridge this gap of knowledge. To this end, we expect our Ph.D. work to bring two contributions: the first is a framework for understanding the human factors involved in victimization and repeat victimization due to phishing, and the second is to design (1) guidelines for developers to help them design software systems in such a way that leads to users avoiding victimization due to phishing, and (2) recommendations for users to prevent (repeat) victimization due to phishing.

3 Research Goal and Questions

This Ph.D. project is meant to add up to the collective efforts of scholars working towards protecting users against phishing attacks. In line with this, the PhD project's goal is twofold: (1) using acquired knowledge on user vulnerabilities that takes socio-demographic and cultural differences into account, provide recommendations for increasing privacy awareness to people; and (2) based on the knowledge acquired on user vulnerabilities and phishing techniques, provide guidelines for developers of ssoftware systems that manage the user's attention for risky behavior and recommendations for users to avoid victimization. To achieve our research goal, we designed the following research questions (RQs) for this Ph.D. work:

RQ1: What are the prevalence, sociodemographic correlates, and risk factors of users' vulnerabilities toward phishing, according to published literature?
RQ2: What is the prevalence of repeat phishing victimization in relation to sociodemographic and the users' vulnerabilities, according to publicly available data sources?
RQ3: Are there cultural differences in security and privacy awareness across countries?
RQ4: What guidelines and recommendations can be designed to minimize the vulnerability of internet users to phishing attacks, based on a combination of the knowledge acquired in answering RQ1, RQ2 and RQ3, insights obtained from a focus group discussion with phishing victims, and a model developed using the previous results and tested against real-world phishing attacks?
RQ5: To what extent are these guidelines usable and useful in practice?

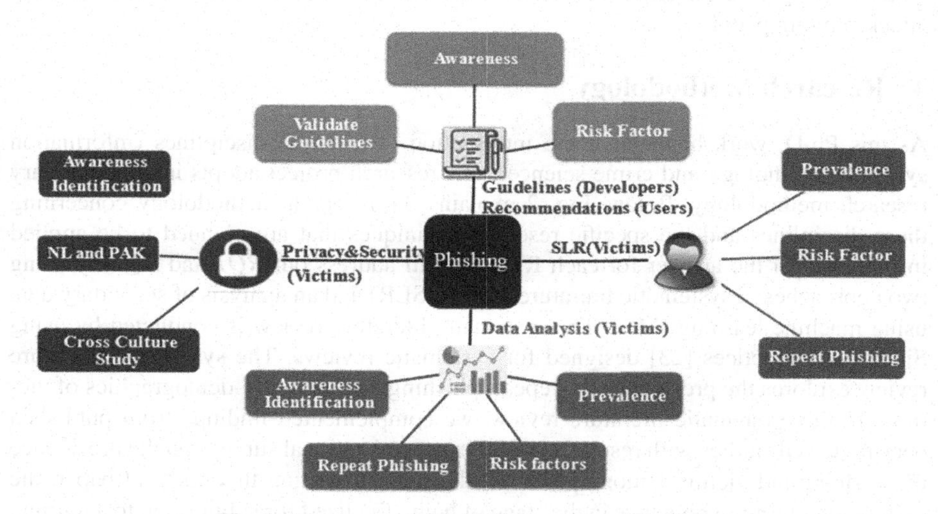

Fig. 1. The scope of this PhD research: a Mind Map

Figure 1. Shows a mind map that is grounded on our RQs and puts together the inputs and outputs of this Ph.D. work and the research activities. As Fig. 1 indicates, phishing will be explored from different perspectives and by employing different research techniques. The area in green, following the line labeled Systematic literature review **SLR (Victims)** means that the phishing victimization phenomenon will be examined in order to know the prevalence and risk factors of phishing and repeat phishing. In another perspective labeled as **Data Analysis (Victims)**, in orange, phishing will be investigated through data analysis of secondary data belonging to a more significant population, i.e., the Dutch population. It is planned that at the end of the study, we will be able to know the figures of prevalence and sociodemographic vulnerability of phishing and repeat phishing. It will also help us understand the Dutch population's awareness of the privacy and security of their confidential data.

In the **Privacy & Security (Victims),** a cross-cultural study will be conducted in two countries, i.e., the Netherlands and Pakistan, that will give us approximate figures of privacy awareness about both countries. Furthermore, in the area labeled **Guidelines (Developers) Recommendations (Users)**, in blue color, we indicate that based on the results of the empirical work to be done until that point (i.e. the studies that the author will do to answer RQ1, RQ2 and RQ3), we plan to create two artefacts. First of all, a set of guidelines will be designed for software developers that highlight the risk factors of phishing victims. The intention behind these guidelines is to be considered by developers for implementation while designing social media websites, in order to minimize user vulnerability to phishing attacks. Moreover, a case study will be designed with software professionals to validate the usefulness of proposed guidelines. Second, a set of recommendations will be developed for users that emphasize the victimization risk factors and help users to avoid phishing attacks in the future. In addition, these recommendations will be shared with educational and professional institutions in the Netherlands and Pakistan to specific audiences as part of actions to increase the awareness of phishing attacks among people.

4 Research Methodology

As this Ph.D. work happens at the intersection of multiple disciplines (information systems, psychology and crime science), this research project adopts interdisciplinary research methodology. Below, Fig. 2 explains the research methodology concerning these disciplines and the specific research techniques that are planned to be applied in order to get the answer for each RQ. We will address our *RQ1* and *RQ2* by using two approaches: a systematic literature review (SLR) and an analysis of secondary data using machine learning (ML). The systematic literature review is conducted by using Siddaway's practices [23] designed for systematic reviews. The systematic literature review explores the prevalence of (repeat) phishing and the socio-demographics of victims. In this systematic literature review, we complemented findings from published peer-reviewed studies with results reported in several national surveys on the prevalence of phishing and victim demographics. At the time of writing this doctoral paper, the systematic literature review is in the stage of being finalized for submission to a journal. Through the systematic literature review, we learned that there are no exact figures in the literature about the prevalence of (repeat) phishing. As we didn't find any concrete answer to the vulnerable demographics of phishing victims, we plan to continue with a data-driven approach. We will analyze secondary data using ML methods to predict prevalence and user vulnerability concerning users' demographic and particular risky behavior.

For *RQ3*, we will design a quantitative study using the method of survey research [24] to uncover the privacy awareness and sensitivity toward user confidential information. Moreover, we will perform a cross-cultural study between two countries, Pakistan (PAK) and the Netherlands (NL), to account for possible (cultural) differences in phishing victimization. (Note that in the leftmost side of Fig. 1, NL and PAK indicate the two countries.)

For *RQ4*, we plan two pieces of research that build upon each other. First, we will conduct qualitative research [24] in which focus groups will be used to gather information from phishing victims. Our focus will be to discover why users become victimized through phishing. What are the commonly observed behaviors that become the reason for their victimization? Moreover, what could be the psychological tricks and information that, if people know before attacks, even if they are aware of cybercrime, will help them to avoid such attacks? Our plan is to analyze the results by using content analysis [24].

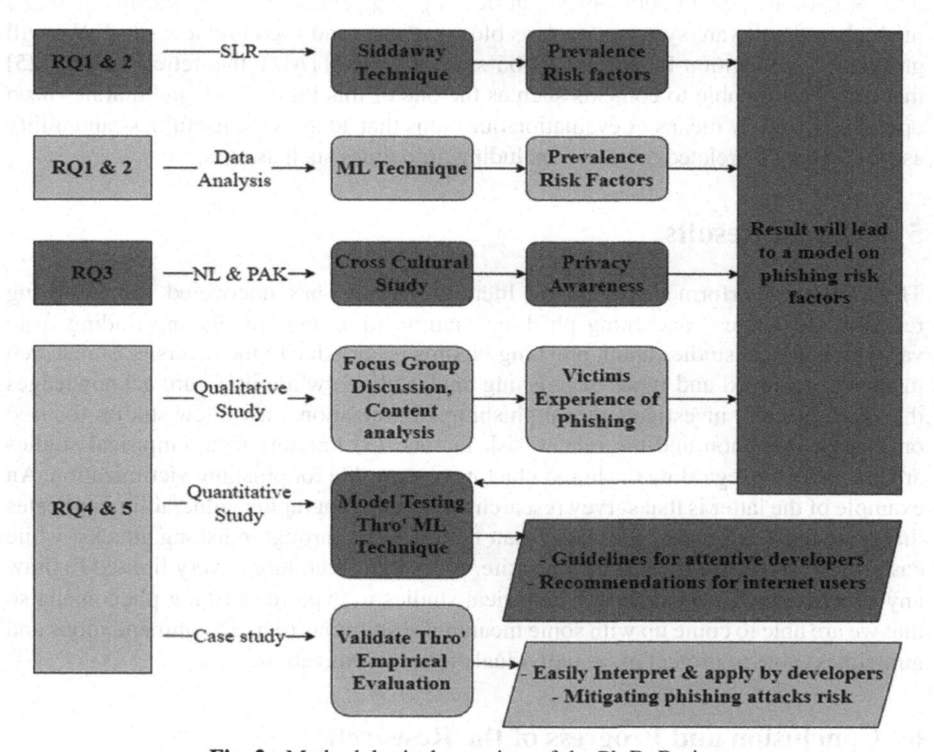

Fig. 2. Methodological overview of the Ph.D. Project

Second, based on the previous study's results, we will make a model using risky behaviors and information to avoid attacks and test that model on random internet users using a survey method. The data of this study will be analyzed by using machine learning techniques.

Our second study results will serve two purposes: (1) we will make guidelines (e.g., use of password strength indicators and use brief terms of services) for software developers to guide their software development processes that account for users' vulnerabilities. These guidelines are expected to be helpful while designing software for internet users. (2) we will make recommendations for internet users to avoid victimization and will share these with public and private institutions interested in and responsible for creating and maintaining users' awareness to avoid phishing victimization. For example, the school boards. We expect that the research to be done in order to answer RQ4, will provide foundation for these organizations to come up with educational measures and policies that are helpful for users when dealing with phishing attacks.

Finally, for answering *RQ5*, we plan to design empirical evaluation research process with software practitioners that will help us understand the extent to which the proposed guidelines are useful and usable. For example, we will evaluate how easy developers can interpret and apply the guidelines, and how effective the developed software is in mitigating phishing attack risks. For this, we will do a perception-based evaluation with practitioners from companies that develop, e.g., social media platforms or social media based software systems (such as blogging sites and social review sites). We will ground our perception-based evaluation study on the UTAUT theoretical model [25] that has been suitable to contexts such as the one of this Ph.D. work and that has been operationalized by means of evaluation questions that address the usefulness and utility aspects of any IT-related artefacts, including guidelines such as ours.

5 Current Results

Thus far, our performed systematic literature review has uncovered the following research challenges concerning phishing victimization: (1) aggregating finding from various empirical studies about phishing victims is hard due to the diversity of research methods employed and types of phishing analyzed; (2) while literature acknowledges the urgent need to investigate repeat phishing victimization, only a few studies focused on this phenomenon and the related risk factors; (3) findings from empirical studies are inconclusive regarding the human factors responsible for phishing victimization. An example of the latter is that survey research on sociodemographic vulnerability indicates that male users are more victimized than female users through phishing attacks, while case study research indicates the opposite. As current literature is very limited to draw any conclusions, we plan further empirical studies to explore phishing phenomena so that we are able to come up with some meaningful solutions (e.g., recommendations and guidelines) able to protect more individuals from victimization.

6 Conclusion and Progress of the Research

The phenomenon of phishing victimization and repeat victimization is only partly understood as it has been researched until now in a fragmentary way, either from technical standpoint or holistically from cybercrime standpoint, while the risks due to human factors evaded the scholar's attention. To the best of our knowledge, this PhD research is one of the first initiatives that addresses this gap and creates a model for understanding the phishing victimization risks due to human factors as well as proposes guidelines for developers to help design software systems that reduce or prevent phishing victimization of users. Unlike existing works, this Ph.D. research takes the perspective of individual users and their contexts. Until now, we completed a systematic literature review on the prevalence of repeat victimization and its social demographics. Currently, we are working towards getting authorized access to secondary data from a large public organization in the Netherlands, in order to measure the prevalence of phishing and to analyze the sociodemographic vulnerability of phishing victimization among the Dutch population. It is a specialized government institution that keeps the records of millions of Dutch citizens about phishing victimization.

In parallel, we are thinking over strategies to collect phishing victimization data from Twitter. The purpose of the Twitter data is to check people's vulnerability based on their social demographic. We plan to take the social profile of people who claim phishing victimization. The data acquisition task through Twitter is in progress, with 50% completion at the time of writing this paper.

Acknowledgement. This Ph.D. work is carried out under the supervision of Dr. Marten van Sinderen and Dr. Maya Daneva of the University of Twente, the Netherlands.

References

1. Statista: Internet and social media users in the world 2022, Statista (2022). https://www.statista.com/statistics/617136/digital-population-worldwide/. Accessed 30 September 2022
2. Cyber Security Breaches Survey 2020. GOV.UK (2020). https://www.gov.uk/government/statistics/cyber-security-breaches-survey-2020/cyber-security-breaches-survey-2020. Accessed 14 Apr 2021
3. Proofpoint: State of the Phish Report: Attack Rates Rise, Account Compromise Soars. Proofpoint (2019). https://www.proofpoint.com/us/corporate-blog/post/2019-state-phish-report-attack-rates-rise-account-compromise-soars. Accessed 01 Oct 2022
4. Statista: Data records breached worldwide 2022, Statista (2022). https://www.statista.com/statistics/1307426/number-of-data-breaches-worldwide/. Accessed 30 Sep 2022
5. Lastdrager, E.E.H.: Achieving a consensual definition of phishing based on a systematic review of the literature. Crime Sci. **3**(1), 1 (2014). https://doi.org/10.1186/s40163-014-0009-y
6. APWG:APWG | APWG 1Q 2022: Phishing Reaches Record High; APWG Observes One Million Attacks Within the Quarter – For the First Time – in the First Quarter of 2022 (2022). https://apwg.org/apwg-1q-2022-phishing-reaches-record-high-apwg-observes-one-million-attacks-within-the-quarter-for-the-first-time-in-the-first-quarter-of-2022/. Accessed 01 Oct 2022
7. Aleroud, A., Zhou, L.: Phishing environments, techniques, and countermeasures: a survey. Comput. Secur. **68**, 160–196 (2017).https://doi.org/10.1016/j.cose.2017.04.006
8. Huang, H., Zhong, S., Tan, J.: Browser-side countermeasures for deceptive phishing attack. In: 5th International Conference on Information Assurance and Security, IAS 2009, September 2009, pp. 352–355 (2009). https://doi.org/10.1109/IAS.2009.12
9. Abroshan, H., Devos, J., Poels, G., Laermans, E.: Phishing attacks root causes. In: Cuppens, N., Cuppens, F., Lanet, J.-L., Legay, A., Garcia-Alfaro, J. (eds.) Risks and Security of Internet and Systems. LNCS, vol. 10694, pp. 187–202. Springer, Cham (2018). https://doi.org/10.1007/978-3-319-76687-4_13
10. Junger, M., Montoya, L., Overink, F.-J.: Priming and warnings are not effective to prevent social engineering attacks. Comput. Hum. Behav. **66**, 75–87 (2017). https://doi.org/10.1016/j.chb.2016.09.012
11. Garera, S., Provos, N., Chew, M., Rubin, A.D.: A Framework for Detection and Measurement of Phishing Attacks, pp. 1–8 (2007)
12. Hutchings, A., Hayes, H.: Routine activity theory and phishing victimisation: who gets caught in the 'Net'? Current Issues Crim. Justice **20**(3), 433–452 (2018). https://doi.org/10.1080/10345329.2009.12035821
13. Canham, M., Posey, C., Strickland, D., Constantino, M.: Phishing for long tails: examining organizational repeat clickers and protective stewards. SAGE Open **11**(1), 215824402199065 (2021). https://doi.org/10.1177/2158244021990656

14. Correia, S.G.: Patterns of online repeat victimisation and implications for crime prevention. In: 2020 APWG Symposium on Electronic Crime Research (eCrime), November 2020, pp. 1–11 (2020). https://doi.org/10.1109/eCrime51433.2020.9493258
15. Milani, R., Caneppele, S., Burkhardt, C.: Exposure to cyber victimization: results from a Swiss survey. Deviant Behav. 1–13 (2020).https://doi.org/10.1080/01639625.2020.1806453
16. Wittebrood, K., Nieuwbeerta, P.: Criminal victimization during one's life course: the effects of previous victimization and patterns of routine activities. J. Res. Crime Delinq. **37**(1), 91–122 (2000). https://doi.org/10.1177/0022427800037001004
17. Brown, C.F., Demaray, M.K., Secord, S.M.: Cyber victimization in middle school and relations to social emotional outcomes. Comput. Hum. Behav. **35**, 12–21 (2014). https://doi.org/10.1016/j.chb.2014.02.014
18. Whitty, M.T.: Predicting susceptibility to cyber-fraud victimhood. J. Financ. Crime **26**(1), 277–292 (2019). https://doi.org/10.1108/JFC-10-2017-0095
19. Darwish, A., Zarka, A.E., Aloul, F.: Towards understanding phishing victims' profile. In: 2012 International Conference on Computer Systems and Industrial Informatics, December 2012, pp. 1–5 (2012). https://doi.org/10.1109/ICCSII.2012.6454454
20. Parsons, K., Butavicius, M., Delfabbro, P., Lillie, M.: Predicting susceptibility to social influence in phishing emails. Int. J. Hum. Comput. Stud. **128**, 17–26 (2019). https://doi.org/10.1016/j.ijhcs.2019.02.007
21. Curtis, S.R., Rajivan, P., Jones, D.N., Gonzalez, C.: Phishing attempts among the dark triad: patterns of attack and vulnerability. Comput. Hum. Behav. **87**, 174–182 (2018). https://doi.org/10.1016/j.chb.2018.05.037
22. Halevi, T., Memon, N., Nov, O.: Spear-phishing in the wild: a real-world study of personality, phishing self-efficacy and vulnerability to spear-phishing attacks. SSRN Electron. J. (2015).https://doi.org/10.2139/ssrn.2544742
23. Siddaway, A.P., Wood, A.M., Hedges, L.V.: How to do a systematic review: a best practice guide for conducting and reporting narrative reviews, meta-analyses, and meta-syntheses. Ann. Rev. Psychol. **70**(1), 747–770 (2019). https://doi.org/10.1146/annurev-psych-010418-102803
24. Oakley, J.G.: Access. In: Waging Cyber War, pp. 101–114. Apress, Berkeley (2019). https://doi.org/10.1007/978-1-4842-4950-5_8
25. Venkatesh, V., Morris, M.G., Davis, G.B., Davis, F.D.: User acceptance of information technology: toward a unified view. MIS Q. **27**(3), 425–478 (2003). https://doi.org/10.2307/30036540

Leveraging Exogeneous Data for the Predictive Monitoring of IT Service Management Processes

Marc C. Hennig(✉)

University of Applied Sciences Munich, Lothstr. 64, 80335 Munich, Germany
mhennig@hm.edu

Abstract. Accurate prediction of process execution time in IT Service Management (ITSM) is essential for service providers to meet service-level agreements (SLA). However, traditional pre dictive process monitoring methods struggle with processes delivering complex process artifacts, where event log data is insufficient to understand the flow of instances. To overcome this challenge, exogenous predictive process monitoring is proposed, utilizing exogenous data sources available in IT organizations to improve the accuracy of ITSM process predictions. This approach leverages a wide range of exogenous data sources, such as the service knowledge management system, to enhance the predictions and decision-making process. The resulting planning and decision support system, incorporating exogenous data, improves SLA compliance through better resource allocation and decisions throughout the ITSM process instance lifecycles.

Keywords: IT Service Management · ITSM · Service-Level Agreement · SLA · Predictive Process Monitoring · Exogenous Data · Process Context · AITSM · AIOps

1 Introduction

IT Service Management (ITSM) provides IT services to internal and external clients through a process-based operating model for managing and controlling IT organizations [1, 2]. Clients' demands are met by services that are bound by contractual commitments in the form of service level agreements (SLA). SLAs define the financial, legal, and quality constraints of service delivery [2]. Ensuring that the provision of services complies with SLAs is the goal of the IT organization and a major indicator of its service quality and resilience.

To achieve this goal ITSM employs several processes such as Incident, Problem, Change, and Configuration Management [1]. Their proper monitoring and control are crucial to meet SLAs. In particular, the duration of process instances must be predicted to detect process instances that will cause a violation of SLAs. Process instances causing SLA violations trigger countermeasures such as functional or hierarchical escalation. These countermeasures may include the relocation of resources and the rescheduling of planned activities.

S. Nurcan et al. (Eds.): RCIS 2023, LNBIP 476, pp. 641–650, 2023.
https://doi.org/10.1007/978-3-031-33080-3_47

The need to detect process instances that will cause SLA violation implies that the organization's performance in achieving SLA compliance and hence the successful application of ITSM is primarily influenced by its capabilities of effective predictive process monitoring [3]. However, as part of a service domain delivering complex process artifacts, processes in ITSM, are difficult to assess and predict [3]. These difficulties obstruct operational and strategic optimizations, and, as a result, adherence to SLAs. This is caused by the ITSM processes whose process artifacts [4] are complex and heterogenous due to their dependence on the dynamic socio-technological environment [5]. Therefore, the event logs are frequently insufficient to predict the following course of actions and the durations of process instances. Additionally, the planning is hampered by competing short-term goals [6] and resources [7] in different ITSM processes such as the incident and change management processes.

This Ph.D. project aims to improve the predictive monitoring of processes in ITSM by taking into account available process context information from exogenous [8, 9] data. A holistic view of ITSM processes incorporating predictive and prescriptive process monitoring is developed using domain-specific contextual i.e., exogenous, data sources to then enable the accurate prognosis for improved planning and execution. The primary objectives are reducing the risk of failing to comply with SLAs and supporting operational and strategic decision-making across the ITSM process life cycles.

2 Related Work

In process mining and predictive and prescriptive process monitoring a considerable amount of research has been done. The following sections will reflect on the current state-of-the-art and highlight the current work's gaps.

2.1 Predictive Monitoring of ITSM Processes

The terms exogenous data [8, 9] and context [10–13] are used mostly interchangeably, referring to data that is beyond the common notion of process instance-specific attributes commonly annotated in the event log [14]. The inclusion of exogenous data from the processes' context has been done in process mining [8, 9] and predictive process monitoring [10–13, 15, 16] before but despite the previous efforts, ITSM event logs and some other ITSM-related data sources have not been analyzed jointly. Specifically, only the graph-like [17] Configuration Management Databases (CMDB) [18] were analyzed [19–21] while other constituents of the Service Knowledge Management System (SKMS) were mostly neglected.

ITSM event logs were covered in the BPI challenges 2013 and 2014 and other publications. However, the full range of contextual information usually available in ITSM processes via the SKMS [18] is not present in publicly available data sets and has not yet been included in a process monitoring model for ITSM. Previous efforts have tried to model relations of process instances executing in parallel [22, 23] to capture organizational workloads and dependencies. However, this has not yet been extended to the ITSM service operation and transition processes [18], such as Incident, Problem, and Change Management and their mutual impacts on the process contexts.

Additionally, an approach to operationalizing predictive insights in the organization's work is missing [24]. Hence, the planning and decision-making have not been systematically studied as an optimization problem of process instance criticality, resource allocation, and other prompted escalations that might be necessary to ensure SLA compliance with additional predictive knowledge.

2.2 AITSM and AIOps

Mostly in the industry, two related approaches are emerging, aiming to support the successful provision of IT services using predictive analytics. On the one hand, AI-driven ITSM (AITSM) [24] is the automation and support of ITSM processes using machine learning (ML). AITSM takes a strategic view of ITSM and its role in ITSM. On the other hand, artificial intelligence for IT operations (AIOps) [25–27] is a data-driven approach to provide operators with the information required to handle complex systems efficiently. AIOps primarily uses telemetry data and focuses on system operation.

AITSM and AIOps have only been covered sparsely so far in the scientific literature. Despite the overlaps between predictive and prescriptive process monitoring, AIOps, and AITSM, comprehensive studies on the intersection and potential use cases within the fields are lacking. This is interesting as the outcomes usually supplied by process monitoring solutions [28] are valuable for these fields. Furthermore, interventions may be obtained using prescriptive process monitoring [29] to proactively optimize ongoing instances in these fields.

3 Research Objectives

As illustrated in the preceding sections, there are gaps in the existing research in forecasting process durations and impending SLA violations in ITSM processes. The context of the ITSM process instances, as delineated by exogenous data, is not fully considered, resulting in insufficient predictions. Ascertained predictions are also difficult to apply in real decision situations on running process instances, lacking a support and planning system that considers the trade-offs between different possible interventions. The research will thus be structured into two subprojects focusing on data analytics and their organizational application, as displayed in Fig. 1.

RQ1: How can exogenous data be leveraged to improve predictive process monitoring?

The first subproject is concerned with the analysis and preprocessing of exogenous data and their impact on predictive performance and model choice in predictive process monitoring. To achieve this, the relevant data sources must be assessed, the data collected and extracted first. The CMDB as the core repository of employed hard- and software, as well as the relevant human resources and their interdependencies in the SKMS [18] might be an interesting starting point. Alongside the CMDB, other parts of the SKMS might be leveraged to improve the performance of prediction models for contextual exogenous data. Building upon the extracted data, predictive monitoring models can be used to provide insights about the process instance progressions.

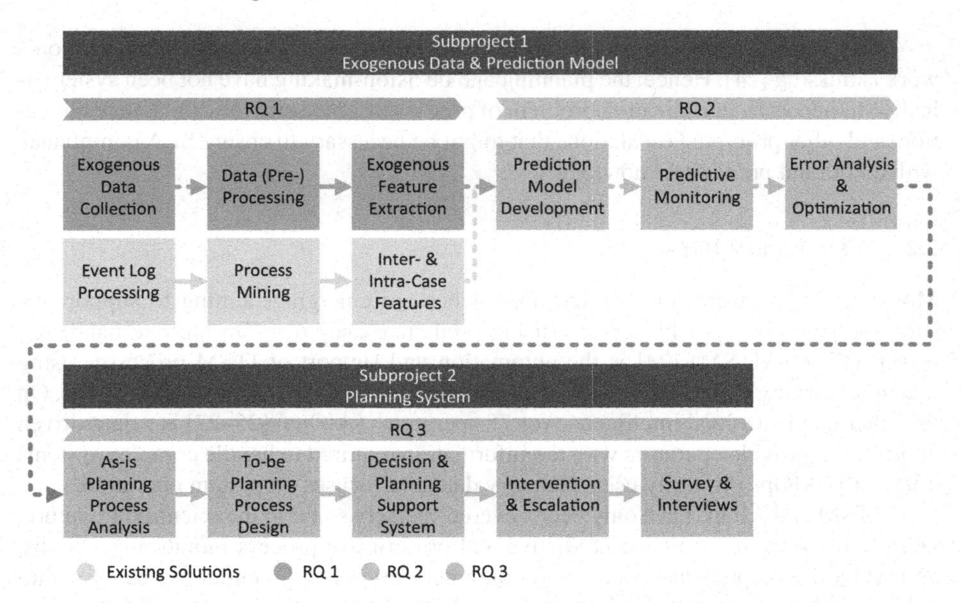

Fig. 1. Research context and envisaged research objectives in the first subproject

RQ2: How can predictions and interventions on the event log and exogenous ITSM data be derived to assure SLA adherence, and what domain-specific adjustments to ML architectures are necessary or encouraged?

Based on the prepared data, it has to be figured out, whether and how event logs and exogenous data can be utilized to enhance predictive performance compared to existing baseline solutions. The objective is to identify models that can analyze sequential data, such as that found in event logs, as well as static and dynamic exogenous data extracted from the SKMS. The models should enable the early detection of SLA violations such that proactive interventions may be implemented, and associated risks can be mitigated through improved planning. The following activities and instance completion times are typically predicted in predictive process monitoring [28] and can be used as the starting point for additional research in the second subproject on the predicted variables. As RQ1 and RQ2 are focused on data analytics, they are summarized in the first subproject.

RQ3: How can predictive analytics improve the planning processes and how can the decision-making in the IT organization be systematically improved therewith?

Extending on the predictions, the second subproject continues the research with the integration of the models into a planning and decision support system. This system allows for actionable implementation in the IT organization and accommodates changes in decision-making, service provision, and value generation enabled by the application of predictive analytics in ITSM. It will be looked into how the predictions derived might be used to improve the planning procedures of the IT organization. Therefore, it must be identified how the current ITSM processes are impacted by the predictions and how functional and hierarchical escalations might be applied systematically in an IT organization. In particular, this pertains to the planning and decision-making procedures and how identifying the critical process instances influences the priority and execution of countermeasures. The interest lies in the impact on resource allocation and alterations in the activity flow in particular.

The artifacts generated during the research shall be combined in the second subproject into an architecture that uses predictive technologies to allow for improved planning in ITSM processes. This architecture shall ultimately result in the form of a decision and planning support system, which allows for informed decisions on escalations. In detail, this system should answer the question of how and when to intervene in process instances that can be considered critical concerning SLA adherence.

4 Method

As common in information systems research, this Ph.D. project will follow the design science research approach as outlined by Johannesson and Perjons [30]. Due to the work's strong reliance on data science and machine learning on new data, for the development of these parts specifically, Huyen's [31] iterative process will be used.

Extending on the research questions, the initial problems will be detailed using literature analyses of the corresponding area in the problem explication phase [30]. To supplement in academic literature, it will be researched how challenges in ITSM process and service monitoring are currently addressed in practice by leading ITSM software platforms using market analyses and industry publications.

Based upon the challenges, requirements for the artifacts are defined in cooperation with the project partner and the prerequisites of the domain extracted from the literature. As a medium-sized software and consulting enterprise with international customers focused on mapping complex and heterogeneous IT infrastructures, the project partner can provide data for analysis as well as insights into problems commonly encountered in the daily work and customer base.

The defined requirements serve as the formal base of evaluation for demonstrating goal attainment after the development and consist of functional and non-functional parts (Fig. 2).

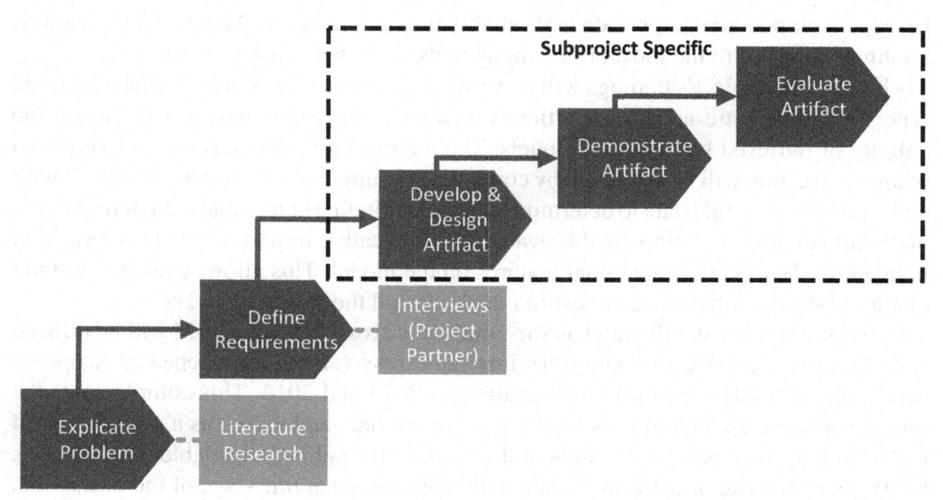

Fig. 2. Design science phases and employed research strategies [30]

An overview of the artifacts is covered in the following sections, detailing the strategies employed during the development, demo, and evaluation phases in both subprojects.

4.1 Exogenous Data and Prediction Model

To better comprehend the data sources available in the context of ITSM, an exploratory case study is conducted for the development of this artifact. The case study is chosen as a research strategy since it is well-suited for cooperating with an industry project partner and the explorative nature of collecting and analyzing real-world data. The study will be carried out across various IT services, with the main focus being on service operations, to ensure that there is an adequate amount of data available. The ITSM service operation processes [18] are chosen since they are customer-facing, create dynamic, often high, organizational workload [32] and are directly associated with service delivery and thus of high relevance for SLA adherence. In addition, some data sets, especially in Incident and Problem Management, are publicly available and can serve as a reference.

The study follows the steps outlined in Fig. 3. First, the event logs and exogenous data from other relevant sources, like the SKMS, are systematically collected from the systems. The viability of the use-case and further utility in predictive models is then assessed with an exploratory data analysis including statistical methods and graph analysis. Further preprocessing, transformation, and error-handling techniques are also applied to each data source during this step. This preprocessed data will be used to determine how relevant features can be extracted. The specific focus in this step lies in the representation and modeling of organizational and processual states as well as other attributes from the exogenous data and the event log.

The model types suitable for including exogenous data from the previously extracted features are selected and developed based on the extracted data. The developed models are used to provide predictive monitoring insights. During this step, suitable statistical and machine learning methods used in the field [28] capable of including exogenous data as well as potential novel models shall be identified. Identified methods are then compared in their potential to provide high-quality predictions on ITSM data. Optimizations are finally applied to the most promising models to further improve the results.

Finally, the project's findings will be verified for internal consistency. Since multiple types of artifacts are used, a bipartite evaluation technique is required to ensure the validity of retrieved features and models. The metrics and characteristics derived from feature extraction will be evaluated by computing common ML quality metrics [33] with and without contextual data to determine whether including contextual data improves the prediction quality. To complete this evaluation, the feature importance [31] is measured to quantify the effect of individual features on the model. This allows evaluation of the quality of different feature composition methods and the feature relevance.

To assure model stability and quality, multifold cross-validation [33] will be utilized in each step of the model evaluation. Training and cross-validation are repeated on openly accessible data sets such as the BPI challenges 2013 and 2014. This complements the previous feature extraction tests on the proprietary data and serves as a comparison of model quality with baseline models in the field. Since publicly available data sets lack SKMS data, the use of exogenous data will be eschewed at this stage of the evaluation.

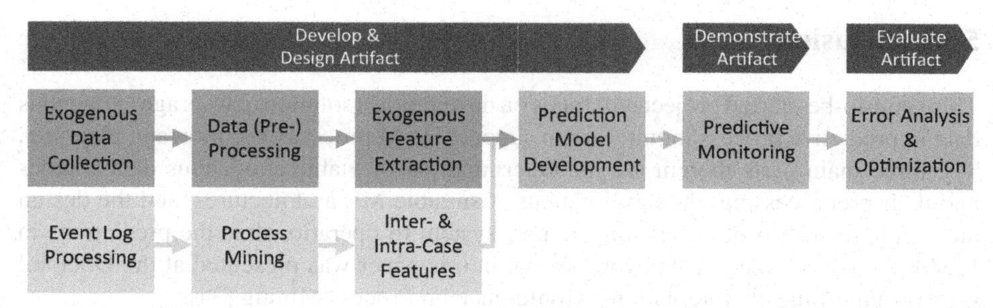

Fig. 3. Proceeding and methods for the prediction solution following Huyen [31]

The result of this subproject will be a systematic assessment of the different data sources available in ITSM. For each data source, an evaluation based on the effectiveness of the model, relevant features and their extraction is to be derived. Besides, a benchmark of different model architectures capable of effectively working extracted features in comparison to baseline models on different data sets shall be created.

4.2 Decision Support System

Building on the predictions of process instance progressions from the extracted data, it is elicited how the results can be used to improve planning and decision-making in the IT organization. Hence it is investigated how the results can be optimally applied to improve the ITSM processes. To analyze this, the challenges and tangible solutions to them are identified in an analysis of ITSM planning processes to understand how planning and decision-making therein are affected. Extending on the analysis results, a decision support system for ITSM processes using the predictions shall be conceptualized and developed extending on the previous case study. Specifically, the focus lies on how escalations can be organized and planned to minimize negative impacts on other non-critical process instances.

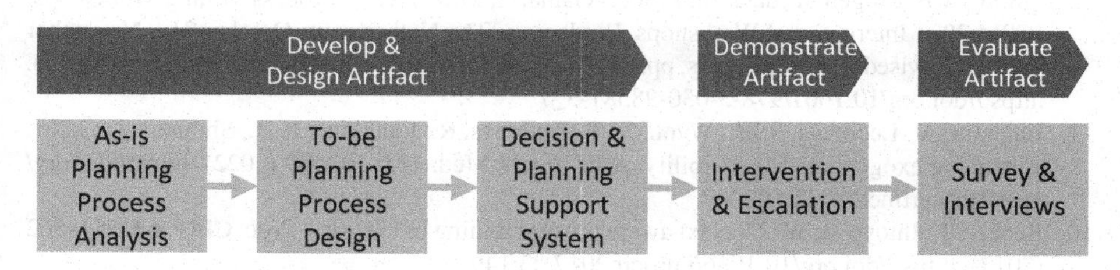

Fig. 4. Proceeding and methods for the decision support system

The decision support system is then validated in a survey of ITSM practitioners and decision-makers to determine where and how the system can be used. Additionally, potential future research direction might be uncovered during this step. Interviews are conducted to gain additional qualitative insights into the field and solution and complement the survey results. The planned research process is displayed in Fig. 4.

5 Conclusion

This soon-to-be-started project will focus on developing a solution to leverage exogenous data in predictive process monitoring to support ITSM process execution and planning. The three main parts therein are the determination of viable exogenous data sources and their preprocessing, the development of suitable ML architectures, and the design and implementation of a planning support system to operationalize the predictions in ITSM. An earlier state of the work shown in this paper was presented at the Doctoral Consortium of the 4th International Conference on Process Mining [34].

References

1. Galup, S.D., Dattero, R., Quan, J.J., Conger, S.: An overview of IT service management. Commun. ACM. **52**, 124–127 (2009). https://doi.org/10.1145/1506409.1506439
2. Bardhan, I.R., Demirkan, H., Kannan, P.K., Kauffman, R.J., Sougstad, R.: An interdisciplinary perspective on it services management and service science. JMIS. **26**, 13–64 (2010). https://doi.org/10.2753/MIS0742-1222260402
3. Serrano, J., Faustino, J., Adriano, D., Pereira, R., da Silva, M.: An IT service management literature review: challenges, benefits, opportunities and implementation practices. Information **12**, 111 (2021). https://doi.org/10.3390/info12030111
4. Dumas, M., La Rosa, M., Mendling, J., Reijers, H.A.: Fundamentals of Business Process Management. Springer, Heidelberg (2018). https://doi.org/10.1007/978-3-662-56509-4
5. Müller, S.D., de Lichtenberg, C.G.: The culture of ITIL: values and implementation challenges. ISM. **35**, 49–61 (2018). https://doi.org/10.1080/10580530.2017.1416946
6. Guven, S., Murthy, K.: Understanding the role of change in incident prevention. In: 2016 12th International Conference on Network and Service Management (CNSM), pp. 268–271. IEEE Press, Montreal, QC, Canada (2016). https://doi.org/10.1109/CNSM.2016.7818430
7. Zia, L., Diao, Y., Rosu, D., Ward, C., Bhattacharya, K.: Optimizing Change Request Scheduling in IT Service Management. In: IEEE International Conference on Services Computing, pp. 41–48. IEEE Press, Honolulu, HI, USA (2008). https://doi.org/10.1109/SCC.2008.144.
8. Banham, A., Leemans, S.J.J., Wynn, M.T., Andrews, R.: xPM: a framework for process mining with exogenous data. In: Munoz-Gama, J., Lu, X. (eds.) Process Mining Workshops: ICPM 2021 International Workshops, Eindhoven, The Netherlands, October 31 – November 4, 2021, Revised Selected Papers, pp. 85–97. Springer International Publishing, Cham (2022). https://doi.org/10.1007/978-3-030-98581-3_7
9. Banham, A., Leemans, S.J.J., Wynn, M.T., Andrews, R., Laupland, K.B., Shinners, L.: xPM: enhancing exogenous data visibility. Artif. Intell. Med. **133**, 102409 (2022). https://doi.org/10.1016/j.artmed.2022.102409
10. Becker, T., Intoyoad, W.: Context aware process mining in logistics. Proc. CIRP. **63**, 557–562 (2017). https://doi.org/10.1016/j.procir.2017.03.149
11. Ogunbiyi, N., Basukoski, A., Chaussalet, T.: Incorporating spatial context into remaining-time predictive process monitoring. In: Proceedings of the 36th ACM Symposium on Applied Computing, pp. 535–542. ACM, Virtual (2021). https://doi.org/10.1145/3412841.3441933.
12. Yeshchenko, A., Durier, F., Revoredo, K., Mendling, J., Santoro, F.: Context-aware predictive process monitoring: the impact of news sentiment. In: Panetto, H., Debruyne, C., Proper, H.A., Ardagna, C.A., Roman, D., Meersman, R. (eds.) On the Move to Meaningful Internet Systems. OTM 2018 Conferences. LNCS, vol. 11229, pp. 586–603. Springer, Cham (2018). https://doi.org/10.1007/978-3-030-02610-3_33

13. Gunnarsson, B.R., vanden Broucke, S.K.L.M., De Weerdt, J.: Predictive process monitoring in operational logistics: a case study in aviation. In: DiFrancescomarino, C., Dijkman, R., Zdun, U. (eds.) Business Process Management Workshops. LNBIP, vol. 362, pp. 250–262. Springer, Cham (2019). https://doi.org/10.1007/978-3-030-37453-2_21
14. van der Aalst, W.: Process Mining. Springer, Heidelberg (2016). https://doi.org/10.1007/978-3-662-49851-4
15. Brunk, J., Stierle, M., Papke, L., Revoredo, K., Matzner, M., Becker, J.: Cause vs. effect in context-sensitive prediction of business process instances. Inf. Syst. **95**, 101635 (2021). https://doi.org/10.1016/j.is.2020.101635
16. Marquez-Chamorro, A.E., Resinas, M., Ruiz-Cortes, A.: Predictive monitoring of business processes: a survey. IEEE Trans. Serv. Comput. **11**, 962–977 (2018). https://doi.org/10.1109/TSC.2017.2772256
17. Stiefel, S., Möstl, C., Bär, F., Schmidt, R., Möhring, M.: Graph-datenbanken als grundlage des configuration managements – Eine Untersuchung am Beispiel von Neo4. J. HMD. **53**, 470–485 (2016). https://doi.org/10.1365/s40702-016-0241-x
18. Long, J.: ITIL Version 3 at a Glance - Information Quick Reference. Springer, Boston, MA, USA (2008). https://doi.org/10.1007/978-0-387-77393-3.
19. Sarnovsky, M., Surma, J.: Predictive models for support of incident management process in IT service management. AEI. **18**, 57–62 (2018). https://doi.org/10.15546/aeei-2018-0009.
20. Anchuri, P., Zaki, M.J., Barkol, O., Bergman, R., Felder, Y., Golan, S., Sityon, A.: Graph mining for discovering infrastructure patterns in configuration management databases. Knowl. Inf. Syst. **33**, 491–522 (2012). https://doi.org/10.1007/s10115-012-0528-3
21. Li, H., Zhan, Z.: Business-driven automatic IT change management based on machine learning. In: 2012 IEEE Network Operations and Management Symposium. pp. 1374–1377. IEEE, Maui, HI, USA (2012). https://doi.org/10.1109/NOMS.2012.6212078.
22. Senderovich, A., Di Francescomarino, C., Ghidini, C., Jorbina, K., Maggi, F.M.: Intra and inter-case features in predictive process monitoring: a tale of two dimensions. In: Carmona, J., Engels, G., Kumar, A. (eds.) Business Process Management, pp. 306–323. Springer International Publishing, Cham (2017). https://doi.org/10.1007/978-3-319-65000-5_18
23. Pourbafrani, M., Kar, S., Kaiser, S., van der Aalst, W.M.P.: Remaining time prediction for processes with inter-case dynamics. In: Munoz-Gama, J., Lu, X. (eds.) Process Mining Workshops: ICPM 2021 International Workshops, Eindhoven, The Netherlands, October 31 – November 4, 2021, Revised Selected Papers, pp. 140–153. Springer International Publishing, Cham (2022). https://doi.org/10.1007/978-3-030-98581-3_11
24. Mao, H., Zhang, T., Tang, Q.: Research framework for determining how artificial intelligence enables information technology service management for business model resilience. Sustainability. **13**, 11496 (2021). https://doi.org/10.3390/su132011496
25. Dang, Y., Lin, Q., Huang, P.: AIOps: real-world challenges and research innovations. In: IEEE/ACM 41st International Conference on Software Engineering: Companion Proceedings. pp. 4–5. IEEE Press, Montreal, QC, Canada (2019). https://doi.org/10.1109/ICSE-Companion.2019.00023
26. Notaro, P., Cardoso, J., Gerndt, M.: A systematic mapping study in AIOps. In: Hacid, H., et al. (eds.) Service-Oriented Computing – ICSOC 2020 Workshops: AIOps, CFTIC, STRAPS, AI-PA, AI-IOTS, and Satellite Events, Dubai, United Arab Emirates, December 14–17, 2020, Proceedings, pp. 110–123. Springer International Publishing, Cham (2021). https://doi.org/10.1007/978-3-030-76352-7_15
27. Sabharwal, N., Bhardwaj, G.: Hands-on AIOps: Best Practices Guide to Implementing AIOps. Apress, Berkeley (2022). https://doi.org/10.1007/978-1-4842-8267-0

28. Di Francescomarino, C., Ghidini, C., Maggi, F.M., Milani, F.: Predictive process monitoring methods: which one suits me best? In: Weske, M., Montali, M., Weber, I., vom Brocke, J. (eds.) Business Process Management: 16th International Conference, BPM 2018, Sydney, NSW, Australia, September 9–14, 2018, Proceedings, pp. 462–479. Springer International Publishing, Cham (2018). https://doi.org/10.1007/978-3-319-98648-7_27
29. Kubrak, K., Milani, F., Nolte, A., Dumas, M.: Prescriptive process monitoring: Quo vadis? PeerJ. Comput. Sci. **8**, e1097 (2022). https://doi.org/10.7717/peerj-cs.1097
30. Johannesson, P., Perjons, E.: An Introduction to Design Science. Springer International Publishing, Cham (2021). https://doi.org/10.1007/978-3-030-78132-3
31. Huyen, C.: Designing machine learning systems: an iterative process for production-ready applications. O'Reilly, Sebastopol, CA, USA (2022)
32. Diao, Y., Heching, A.: Workload management in dynamic IT service delivery organizations. In: Bartolini, C., Gaspary, L.P. (eds.) Integrated Management of Systems, Services, Processes and People in IT, pp. 123–137. Springer, Heidelberg (2009). https://doi.org/10.1007/978-3-642-04989-7_10
33. Raschka, S.: Model Evaluation, Model Selection, and Algorithm Selection in Machine Learning (2020). http://arxiv.org/abs/1811.12808
34. Hennig, M.C.: Transformer for predictive and prescriptive process monitoring in IT service management (extended abstract). In: Hassani, M., Koschmider, A., Comuzzi, M., Maggi, F.M., and Pufahl, L. (eds.) Proceedings of the ICPM Doctoral Consortium and Demo Track 2022, pp. 22–26. CEUR-WS.org, Bolzano, Italy (2022)

Predicting Privacy Decisions in Mobile Applications and Raising Users' Privacy Awareness

Rena Lavranou(✉)

Department of Informatics, Ionian University, Corfu, Greece
lavranou@ionio.gr

Abstract. Smartphones and mobile applications are now an integral part of our daily lives. Managing mobile privacy is a challenging task, that overwhelms users with numerous and complex privacy decisions related to permission settings. Many users do not have sufficient knowledge and understanding of how applications use their personal data, and others do not spend enough time configuring these settings. Various approaches to assist users by automating their decisions have been proposed in the literature. This paper presents a literature review of the current state of knowledge in the area of permission settings and the solutions proposed by different researchers, focusing on the use of machine learning techniques. Machine learning can address the challenges of mobile privacy management by learning users' preferences and predicting their decisions based on a relatively small number of factors. We then describe our future research plans to reduce the user's burden in configuring application permissions, to increase their awareness, and to protect their privacy.

Keywords: Information Privacy · Application Permissions · Users' Privacy Decisions Prediction · Machine Learning

1 Introduction

Smartphones are nowadays essential personal devices to support our activities in almost every aspect of our lives [5]. Nevertheless, in addition to the capabilities they offer, they collect significant amounts of users' personal data [21], endangering their privacy. Many mobile applications have access to users' personal and sensitive data, not only to support and provide their basic functionality (e.g., navigation), but also for marketing and advertising purposes [11] and other purposes. The access rights to most data, such as those related to phone contacts, location, camera, photos, microphone, etc. are granted by the users of the devices, through the applications' permission requests (app permission requests). Regulations such as the GDPR (General Data Protection Regulation) require users to have control over the collection and use of their personal data, such as securing informed consent [20].

Considering the number of applications that users typically install on their mobile phones, it is obvious that they are faced with numerous and complex decisions regarding

S. Nurcan et al. (Eds.): RCIS 2023, LNBIP 476, pp. 651–660, 2023.
https://doi.org/10.1007/978-3-031-33080-3_48

their information privacy, the number of which may become unmanageable [21] by them. Studies on user-defined security controls on mobile devices have shown that the average user must make over a hundred permission decisions per device [13]. While Raber & Krueger talk about over 470 permission decisions for an average user with 95 installed applications and 5 permissions on average for each of them [19]. Therefore, it is no surprise that most users do not spend time configuring many of these settings [21]. Only a few people are aware of permission settings or make changes to them. Moreover, according to earlier works, many users are unaware of the number or the context of permissions they have granted to mobile applications [2], or at least feel uncomfortable with the permissions they granted to their apps [7].

Similarly, other studies show that few users actually read the requested application permissions and even fewer understand them correctly [7, 8]. Another major problem is the familiarity of users with the permissions' requests and the careless clicks on them [6]. Studies have pointed out that most users do not pay attention to system messages related to permission requests [7]. Obstructive factors in reading application permissions are the extensive or difficult to understand texts, the permissions' acceptance without thinking and the fact that the users think the permissions are insignificant or a waste of time [9].

On the other hand, users are often surprised and concerned when they learn about data access practices and the frequency, the extent of the collection and use of sensitive data through mobile apps [8, 11], and the sharing of data with third parties [1]. Researchers have also highlighted the difficulty of users in describing the potential harm that can be caused by collecting and sharing their personal information by applications [8]. Other studies indicate that many users may not be able to assess the security risks associated with granting permissions [19], especially if we also consider apps that require more permissions than they need to run [6]. Thus, many users do not have sufficient knowledge and understanding of how applications use their personal data, so they are vulnerable [15], without knowledge of future implications [4].

In summary, users seem to increasingly not have the capacity to make privacy decisions due to their time constraints, motivation and their cognitive decision-making skills. As the relevant literature shows, managing information privacy is an increasingly challenging task for the average internet user [21]. Users still do not seem to be sufficiently equipped and well prepared to make well-considered, self-regulated and informed privacy decisions [8] when downloading and installing new software on their phones.

Various ways of assisting users in their privacy decisions have been suggested in the literature. [19] refers to a better visualization of permissions and the associated privacy risks, allowing the user to decide on each permission. Biswas et al. highlight the need to alleviate the user burden from complex permission decisions [4]. There have been many approaches to overcome this user burden by automating the configuration of privacy settings [19].

Many researchers suggest several mechanisms for predicting users' privacy decisions, using machine learning techniques, based on a relatively small number of factors, such as previous privacy decisions [13] or answers to privacy-related questions [12–14].

Software agents can then use these machine learning models to give users personalized privacy recommendations, helping them to better control their privacy [10], while reducing the number of decisions that users must make themselves.

Taking into account this research problem, we aim to first study users' behavior and attitudes towards privacy and then to investigate how we can effectively learn users' preferences and predict their decisions to accept or reject permission requests with significant accuracy. Our main objective is to use deep or machine learning techniques to provide support in the decision-making process, adapted to each user, in an effort to alleviate their burden and at the same time strengthen the protection of their privacy in the various applications and in general on the internet.

Our research aims to answer most of the following questions:

- Can we accurately predict the users' privacy preferences using deep (or machine) learning techniques?
- What factors significantly influence the user's final decision to accept or reject application permissions?
- Is there a correlation between personality or privacy attitude and application permission decisions?
- Is the role of contextual information important in decision-making? And if so, which parts of contextual information?

This paper is structured as follows: Sect. 2 presents the literature review, which describes the current knowledge in the field of permission settings, as well as the solutions proposed by several researchers. In Sect. 3 we describe our plans for the research methodology we intend to follow.

2 Literature Review

In this section we describe the current state of in the area of permission settings, as well as the solutions proposed by various researchers. First, the evolution of permissions systems in the Android operating system is mentioned, highlighting their strengths and weaknesses. Then, we refer to privacy management solutions found in the literature, mainly using machine learning techniques, in order to reduce the users' burden and encourage them to protect their privacy more. We also refer to users' privacy profiles, as an approach to personalized privacy, that has been used by several researchers. The section concludes with the factors that have been studied in various studies in an attempt to predict users' privacy decisions.

2.1 Android Permission System

Android protects privacy-critical device functionality, with permissions being one of the key points of the Android security mechanism [1]. Android applications must request permissions in order to access sensitive user data (such as contacts, call logs, and SMS) and certain system functions (such as camera, microphone, and Internet). If the application requires a dangerous permission that could potentially affect the user's privacy or the normal operation of the device, the user is asked to grant this permission. The way Android asks users to grant dangerous permissions has changed over the years [1].

The Android operating system has so far used two different permission models: ask-on-install (AOI) and ask-on-first-use (AOFU) to give users control over permissions [22]. Prior to Android 6.0, all dangerous permissions had to be granted during the installation time (AOI), which limited user flexibility [4]. Then, users could either accept all requested permissions to be able to install the application, or deny them (all) and not install it [22]. In other words, smartphone users did not have the ability to choose the permissions that each app would have access to [19].

Android version 6, released in 2015, introduced a major change to the permission system, adopting the runtime permission model. From Android 6 onwards, like iOS, the user is asked to grant permission the first time an app tries to use a sensitive resource. This is the ask-on-first-use (AOFU) model. The user's response to this permission request is carried over to all future requests by that app for that permission, until the user denies it via the phone's settings, which they usually don't do or, in newer versions of Android (since Android 11), until it automatically reverts to the denied state after a few months of not using the app [16].

Although AOFU offers an improvement over the AOI model in that it provides users with contextual cues as to why an application requires a protected resource, it does not take into account that user preferences may change in different contexts [22]. AOFU has been characterized as insufficient to meet users' privacy wishes and needs, according to the contextual integrity framework, as well as inflexible, because it does not take into account the context surrounding the data streams [17].

A natural extension of AOFU is ask on every use, i.e. every access requires user input. In theory, this would be optimal, as the user would be able to consider the context and then make decisions on a case-by-case basis [22]. However, this approach is not feasible in practice. Research has shown that applications request access to permission-protected resources with high frequency: on an average smartphone, about once every 15 s [23]. Such a high frequency of requests could overwhelm the user with privacy controls that are often useless, not only risking user habituation, but also rendering the device inoperable [22].

2.2 Privacy Management with Machine Learning

There are many approaches in the literature that attempt to address the privacy management challenges related to application permissions. Automated management of privacy decisions is necessitated by numerous accesses to sensitive resources that applications make - hundreds a day [16]. Recent research on permission models has focused on using machine learning [12–14, 24] to automatically predict users' permission preferences.

The first attempt to predict users' permission settings for mobile apps was made by Liu et al. [14]. Liu et al. successfully trained a machine learning prediction using the settings of 4.8 million users of the LBE Privacy Guard app in China [14]. In a later work, Liu et al. attempted to model users' privacy preferences in a Personalized Privacy Assistant (PPA) [13]. They proposed and evaluated a PPA capable of helping users with permission settings. Again, they relied on machine learning, using the purpose of each permission. They also used privacy nudges, where users are prompted to review and change their settings, aiming to increase user awareness regarding the use of resources. For their paper, the authors in [13] relied on App Ops, a permission manager, introduced in Android 4.3, but removed in version 4.4.2 [22].

The same permission manager was also studied by Almuhimedi et al. [2]. For the first time, AppOps gave users the ability to modify, selectively grant or restrict permissions for installed apps. Almuhimedi et al. evaluated the benefits of AppOps and sending privacy nudges to users in an effort to complement and increase the effectiveness of Android's permission manager [2]. Their goal was to increase users' awareness of the data collected by their apps and empower them to more effectively control their privacy through privacy nudges.

Olejnik et al. developed SmarPer, an advanced user permission decision prediction engine for Android, based on contextual information and machine learning methods and mimics user decisions based on decision patterns [18]. Olejnik et al. implemented a partial version of SmarPer with a fully manual mode, compatible with Android 4.0.3 to 5.1.1, where users make decisions at runtime. However, no automated decisions or learning from user behavior is provided [18]. Lin et al. present the design and evaluation of a new privacy summary interface that highlights of mobile apps behavior that does not meet crowd expectations [11]. That is, it points out the points where mobile apps violate people's expectations, highlighting users' common misconceptions and the purpose of accessing resources, thus increasing user privacy awareness.

Tsai et al. proposed the design of a new, fully functional privacy feedback interface for the Android platform, for mobile privacy management [22]. More specifically, they designed a new permission manager, TurtleGuard, which helps users to vary their privacy preferences based on some selected contextual circumstances. According to its creators, it is the first context-aware permission manager for third-party applications on Android [22].

Building on the previous work of Wijesekera et al. [24] and Tsai et al. [22], Wijesekera et al. [25] implemented and evaluated the usability of a novel mobile privacy management system on Android. Specifically, they implemented the first contextual-aware permission system that performs dynamic permission denial, an advance over previous work that only performed offline learning or did not set permissions in real time [25]. Using contextual signals, they created a classifier that predicts users' privacy preferences in different scenarios. They trained an offline classifier, with machine learning, using participants' responses to runtime prompts. The machine learning model runs entirely on the device and uses infrequent user prompts to retrain and improve its accuracy over time.

To address the long-standing challenges in mobile privacy management, Wijesekera et al. proposed the use of machine learning to dynamically manage app permissions [24]. Wijesekera et al. investigated permissions in Android in several studies. They modified the Android operating system and upgraded a custom version of Android 5.1.1 (pre-Marshmallow) on participants' devices to collect data on privacy-related behaviors and frequency of app access to protected resources regulated by permissions. In [23] Wijesekera et al. investigate whether participants expected applications to access these resources and whether permissions should have been granted. In their next work [24], Wijesekera et al. investigate the factors that influence users' privacy decisions and design a classifier to predict users' permission decisions, achieving a four-fold reduction in error rate compared to current systems.

Smullen et al. [21] present a quantitative assessment of how machine learning can contribute to mitigating the trade-off between accuracy and user burden, between effectiveness and efficiency, when configuring permission for Android apps. In their paper they compared models with permission settings that consider the purpose of permissions against models that do not. They also used machine learning to assign users to privacy profiles and infer a set of permissions for each user based on these profiles [21]. The results of Smullen et al. show that it is indeed possible to capture users' privacy preferences more accurately, while reducing their burden [21].

A new strategy that simplifies the process of granting permissions to apps was attempted by Raber & Krueger [19]. Specifically, they trained a privacy wizard that automatically sets individual app permissions based on machine learning. Based on users' personalities and privacy attitudes, they used machine learning techniques to predict the mobile app permission settings of Android OS 6.0 users. Mendes et al. [16] use machine learning approaches to automate privacy decisions, combining privacy profiles, context-awareness and user expectations. They analyze the context dependency of privacy, by assessing which features are actually relevant for privacy decisions, and then they use these findings to develop an automated, personalized and context-aware permission manager to predict user preferences in relation to permission requests.

2.3 Users' Privacy Profiles

Another approach to personalized privacy is the creation and assignment of privacy profiles, i.e. a set of predefined rules defined according to users' preferences [12, 14]. Such work, such as Lin et al. [12], Liu et al. [13, 14], showed that while people's privacy preferences vary, a small number of privacy profiles can predict people's decisions to allow app permissions, deny, or be prompted for app permissions with a high level of accuracy [14]. Additionally, these profiles can be assigned through a small number of questions, thus reducing the amount of input required from users [12, 13].

Lin et al. use clustering techniques to identify privacy profiles [12], taking into account purpose information and users' self-reported willingness to potentially grant access, which emerged in a scenario-based online study. Smullen et al. also use machine learning to assign users to privacy profiles and then use these profiles to infer a set of permissions for each user [21]. Biswas et al. use the user's profile, combined with their opinion about the purpose of using an application to selectively assign permissions [4].

Andriotis et al. propose the use of heat maps, as a visualization scheme, to represent users' privacy profiles [3]. Mendes et al. differ from other works in that they consider and evaluate the impact of contextual features and user expectations in combination with privacy profiles [16]. Thus, they go beyond the traditional privacy profiles that are created only with the category of the requested application and the requested permission [14], in order to integrate context-awareness into personalization [16].

2.4 Predictive Factors

There are many studies in the literature that try to predict users' privacy decisions about application permissions, each based on different factors. The first attempts to predict users' privacy decisions, such as the work by Liu et al. [14], used only the application category, the permission type and the user ID. In their subsequent work [13], Liu et al. found that important predictors for accepting or not accepting permissions are application category and permission type, while information such as purpose or privacy concerns are not determinant for users' decisions. In contrast, according to Olejnik et al. [18], contextual information is important.

Lin et al. developed a recommendation algorithm that predicts whether users would share private information with a mobile application based on the type of app, the type of private information and the purpose of using the app [12]. Users' opinions about the purpose of using each app were used and by Biswas et al. [4]. In [4] they developed SDroid (Secured anDroid), a tool that evaluates requested permissions and allows users to selectively grant permissions by receiving a single opinion from the user about the purpose of using an app. This work took into account the user's profile and thus suggested the best compatible permission decision.

Andriotis et al. found that users' privacy settings seem to be directly influenced by application functionality [3]. Similarly, Smullen et al. revealed that individuals' privacy preferences are strongly influenced by the purpose for which permissions are requested [21]. The results of Lin et al. [11] move in the same direction, suggesting that both the purpose of using sensitive resources and the users' expectation about different apps significantly influence the users' subjective feelings and trust decisions. On the other hand, Raber & Krueger found a significant correlation between user personality, privacy attitudes and application permission settings [19].

Wijesekera et al. noted that users' decisions are mainly driven by two influencing factors: privacy concerns for the specific type of data and understanding of the relevance of a permission request to the functionality of the requested application [23]. They also showed that the visibility of the requested application and the frequency with which requests appear are two important factors used by users when making privacy decisions in mobile applications [23].

In a subsequent paper, Wijesekera et al. [24] confirm the importance of application visibility as a determinant of user decisions, while adding the importance of the current foreground application. The analysis by Mendes et al. shows that contextual features such as the visibility of the requesting application, the user location, and the network status are important contextual cues that partially explain the user's decision to grant or deny permission [16]. Furthermore, they find that the category of the requesting application and the requested permission moderately encode the context, as the user uses different applications in different contexts.

3 Research Methodology

Our research focuses on analyzing mobile users' application permissions and making predictions about them. For this reason, we will try to collect real-world application permission settings from our research participants. We are in the process of developing an application that will extract users' permissions data directly from their devices. In this way we will collect real world data about users' privacy decisions on their smartphones. At the same time, we intend to collect self-reported data from users about their privacy concerns and knowledge, demographic data, and data about their privacy intentions.

Our main aim is to have a sample as less biased as possible, so that it can better represent a large part of the population, both in terms of age and technical expertise. In contrast to the samples used in other works, such as [16], which were biased towards young adults with technical expertise, we will aim for as diverse a population as possible to better validate our findings. For example, we intend to conduct our research initially with students from the department of Informatics, but also from other departments at the university, in order to limit the technical expertise bias. Students could also be selected from different semesters, before and after taking information privacy courses. At a later stage, the research could be extended to other environments outside the university to ensure a wider age range of participants.

Next, our goal is to analyze the data using more sophisticated techniques than those that have been mainly used in research so far. Using more sophisticated learning techniques, such as deep learning, instead of machine learning, could potentially further improve the accuracy of predictions about user behavior in relation to application permissions. We would also like to infer the factors that significantly influence users' decisions about permission settings, so that we can make the most accurate predictions about their behavior. Finally, our research could be completed with an exit survey, possibly including interviews with users about their application permission settings, their understanding of them and their choices.

In the future, our findings could be used to create a sophisticated permissions manager, that makes decisions which will align with users' preferences as much as possible, without tiring them, balancing efficiency with effectiveness. Through all the above, we aim to alleviate the user's burden in making decisions about application permissions and to raise their awareness, with the ultimate goal of increasing their privacy protection. Our research is at the early stage of planning. It has not yet proceeded with the collection and analysis of the relevant data.

Our methodology is not yet mature, so in terms of data availability and more generally the open science principles to be followed, these will be prepared accordingly, depending on the research strategy that will eventually be implemented. The aim is to have all data and tools used as open and accessible as possible to benefit the research community and make our findings reproduceable and extendable. Finally, it should be noted that this work has not been presented at other doctoral consortia or conferences.

4 Conclusion

Users of mobile device are confronted with a multitude of applications and configurations of privacy and security settings on a daily basis. Literature shows that managing information privacy is an increasingly difficult task for the average internet user [21]. At the same time, users still do not seem to be sufficiently equipped and well prepared to make well-considered, self-regulated and informed privacy decisions [8]. Therefore, automated solutions are needed in order to assist users.

As highlighted in the literature, automating users' decisions with machine learning is an important solution that strikes a balance between accurately applying users' privacy preferences and not overburdening them with too many decisions, while reducing habituation [24]. Our own effort is moving in this direction, trying to apply more sophisticated techniques, in order to make more accurate predictions about users' behavior on mobile permission settings. Our ultimate goal is to raise their privacy awareness and protect their privacy.

Aknowledgement. This Ph.D. is carried out under the supervision of Aggeliki Tsohou of the Ionian University, GREECE.

References

1. Alecakir, H., Can, B., Sen, S.: Attention: there is an inconsistency between android permissions and application metadata! Int. J. Inf. Secur. **20**(6), 797–815 (2021). https://doi.org/10.1007/s10207-020-00536-1
2. Almuhimedi, H., et al.: Your location has been shared 5,398 times! A field study on mobile app privacy nudging. In: Proceedings of the 33rd annual ACM Conference on Human Factors in Computing Systems, pp. 787–796 (2015)
3. Andriotis, P., Li, S., Spyridopoulos, T., Stringhini, G.: A comparative study of android users' privacy preferences under the runtime permission model. In: Tryfonas, T. (ed.) Human Aspects of Information Security, Privacy and Trust. LNCS, vol. 10292, pp. 604–622. Springer, Cham (2017). https://doi.org/10.1007/978-3-319-58460-7_42
4. Biswas, S., Haipeng, W., Rashid, J.: Android permissions management at app installing. Int. J. Secur. Appl. **10**(3), 223–232 (2016)
5. Cao, H., Lin, M.: Mining smartphone data for app usage prediction and recommendations: a survey. Pervasive Mob. Comput. **37**, 1–22 (2017)
6. Chia, P.H., Yamamoto, Y., Asokan, N.: Is this app safe? A large scale study on application permissions and risk signals. In: Proceedings of the 21st International Conference on World Wide Web, pp. 311–320 (2012)
7. Felt, A.P., Ha, E., Egelman, S., Haney, A., Chin, E., Wagner, D.: Android permissions: user attention, comprehension, and behavior. In: Proceedings of the eighth symposium on usable privacy and security, pp. 1–14 (2012)
8. Kelley, P.G., Consolvo, S., Cranor, L.F., Jung, J., Sadeh, N., Wetherall, D.: A conundrum of permissions: installing applications on an android smartphone. In: Blyth, J., Sven Dietrich, L., Camp, J. (eds.) Financial Cryptography and Data Security, pp. 68–79. Springer, Heidelberg (2012). https://doi.org/10.1007/978-3-642-34638-5_6
9. Kusyanti, A., Catherina, H.P.A.: An empirical study of app permissions: a user protection motivation behaviour. Int. J. Adv. Comput. Sci. Appl. **9**(11), 106–111 (2018)
10. Lee, H., Kobsa, A.: Privacy preference modeling and prediction in a simulated campuswide IoT environment. In: 2017 IEEE International Conference on Pervasive Computing and Communications (PerCom), pp. 276–285. IEEE (2017)
11. Lin, J., Amini, S., Hong, J.I., Sadeh, N., Lindqvist, J., Zhang, J.: Expectation and purpose: understanding users' mental models of mobile app privacy through crowdsourcing. In: Proceedings of the 2012 ACM Conference on ubiquitous computing (pp. 501–510)
12. Lin, J., Liu, B., Sadeh, N., Hong, J.I.: Modeling users' mobile app privacy preferences: Restoring usability in a sea of permission settings (2014)

13. Liu, B.: Follow my recommendations: A personalized privacy assistant for mobile app permissions. In: 12th symposium on usable privacy and security (SOUPS 2016), pp. 27–41 (2016)
14. Liu, B., Lin, J., Sadeh, N.: Reconciling mobile app privacy and usability on smartphones: could user privacy profiles help?. In: Proceedings of the 23rd International Conference on World Wide Web, pp. 201–212 (2014)
15. Lutaaya, M.: Rethinking App permissions on iOS. In: Extended Abstracts of the 2018 CHI Conference on Human Factors in Computing Systems, pp. 1–6, April 2018
16. Mendes, R., Cunha, M., Vilela, J.P., Beresford, A.R.: Enhancing user privacy in mobile devices through prediction of privacy preferences. In: Atluri, V., Di Pietro, R., Jensen, C.D., Meng, W. (eds.) Computer Security – ESORICS 2022, pp. 153–172. Springer International Publishing, Cham (2022). https://doi.org/10.1007/978-3-031-17140-6_8
17. Nissenbaum, H.: Privacy as contextual integrity. Wash. L. Rev. **79**, 119 (2004)
18. Olejnik, K., et al.: Smarper: context-aware and automatic runtime-permissions for mobile devices. In: 2017 IEEE Symposium on Security and Privacy (SP), pp. 1058–1076. IEEE
19. Raber, F., Krueger, A.: Towards understanding the influence of personality on mobile app permission settings. In: Bernhaupt, R., Dalvi, G., Joshi, A., Balkrishan, D.K., ONeill, J., Winckler, Marco (eds.) Human-Computer Interaction – INTERACT 2017. LNCS, vol. 10516, pp. 62–82. Springer, Cham (2017). https://doi.org/10.1007/978-3-319-68059-0_4
20. Regulation, P.: Regulation (EU) 2016/679 of the European parliament and of the council. Regulation (EU) **679**, 2016 (2016)
21. Smullen, D., Feng, Y., Zhang, S., Sadeh, N.M.: The best of both worlds: mitigating trade-offs between accuracy and user burden in capturing mobile app privacy preferences. Proc. Priv. Enhanc. Technol. **2020**(1), 195–215 (2020)
22. Tsai, L., et al.: Turtle guard: helping android users apply contextual privacy preferences. In: Symposium on Usable Privacy and Security (SOUPS), vol. 2017 (2017)
23. Wijesekera, P., Baokar, A., Hosseini, A., Egelman, S., Wagner, D., Beznosov, K.: Android permissions remystified: a field study on contextual integrity. In: 24th USENIX Security Symposium (USENIX Security 2015), pp. 499–514 (2015)
24. Wijesekera, P., et al.: The feasibility of dynamically granted permissions: Aligning mobile privacy with user preferences. In: 2017 IEEE Symposium on Security and Privacy (SP), pp. 1077–1093. IEEE, May 2017
25. Wijesekera, P., et al.: Contextualizing privacy decisions for better prediction (and protection). In: Proceedings of the 2018 CHI Conference on Human Factors in Computing Systems, pp. 1–13 (2018)

Information Overload: Coping Mechanisms and Tools Impact

Philippe Aussu(✉)

University of Paris-Saclay, University of Evry, IMT-BS, LITEM, 91025 Evry, France
philippe.aussu@universite-paris-saclay.fr

Abstract. The issue of information overload is becoming increasingly prevalent in society and the workplace, with the widespread use of digital technologies and big data. However, this phenomenon has led to negative effects at the organizational level, such as time loss, decreased efficiency, and poor employee health, among others. In academic literature, information overload is part of a new research trend in information systems management (MSI), known as the "dark side of IT," which focuses on studying the negative effects of organizational use of ICT to propose solutions. Thus, our research objective explores the role of software tools and their features as coping strategies in response to information overload. Specifically, we want to study the features and uses of software tools used by managers to reduce their information overload. The literature on information overload identifies its determinants, consequences, and mechanisms for alleviation. Research on alleviation mechanisms, particularly on the role of tools, is still in its early stages and deserves greater attention. This research is important because it addresses the call from the information systems research community for the urgent need to promote healthy management of the interaction between humans and technologies and its implications for workplace health.

Keywords: Information Overload · Coping Strategies · Use of software tools

1 Introduction

With ubiquitous and intrusive digital technology, information overload has become a major problem in the world of work and society. Initially considered a problem mainly affecting researchers and academics, information overload has gradually spread to businesses and professions with high information consumption, such as medicine, for example [1]. Information overload is not a recent phenomenon, as it emerged in the 1990s through many reports presenting its negative effects such as time loss, de-creased efficiency, and poor health [1]. It has accelerated with the advent of information and communication technologies (ICT) and big data [2–5]. Since the end of the 20th century, information overload has affected all aspects of society, including education, government, domestic life, leisure, and citizens [1].

In academic literature, information overload is a theme that falls under a new research trend in Information Systems Management (ISM) commonly referred to as the

S. Nurcan et al. (Eds.): RCIS 2023, LNBIP 476, pp. 661–669, 2023.
https://doi.org/10.1007/978-3-031-33080-3_49

"Dark Side of IT." This new research trend focuses on studying the negative ef-fects of the organizational use of ICTs [6]. The same literature teaches us that infor-mation overload is presented as a much-discussed concept since there is no common-ly accepted definition [1]. Bertram Gross [7], an American specialist in social scienc-es, is a researcher to whom the concept of information overload is attributed. The same author proposed an initial definition of information overload by stating that it refers to a state in which information inputs into a system exceed processing capabilities. However, researchers agree that information overload refers to a situation in which the quantity of relevant, useful, and available information becomes a problem rather than an aid [7, 8]. In academic literature, different types of overloads have been presented that are closely associated with information overload, such as cognitive overload [9–11], communication overload [12–14], knowledge overload [15], information fatigue syndrome [16], information anxiety [8, 17], and information obesity [18]. Although researchers agree that information overload refers to a situation in which the quantity of relevant, useful, and available information becomes a problem rather than an aid [7, 8], information overload is presented as a much-discussed concept since there is no commonly accepted definition [1]. The various terms referring to information overload mentioned earlier demonstrate a lack of con-sensus on the definition of information overload [19]. Therefore, recently, Belabbes et al. [20] attempted to propose a definition of information overload through a conceptual analysis using Rodgers' approach (1970), stating that "Information overload is a state of cognitive overload caused by an excessive amount of information that exceeds an individual's capacity to process, resulting in negative emotional and cognitive consequences" [20].

Moreover, the main themes addressed in the literature related to information overload concern the factors that determine it [1, 20, 21], its consequences at the individual level [20, 22, 23], and the organizational level [1, 20, 21], and finally the mecha-nisms for mitigating information overload [1, 24, 25].

As part of this research work, we are interested in attenuation mechanisms and want to examine the potential role of software tools and their functionalities in the emergence and implementation of these mechanisms. Indeed, information overload is correlated with the ability to process information, and this processing necessarily involves software tools with increasingly sophisticated functionalities, which users appropriate to varying degrees. We will focus on the issues of managers using soft-ware tools and how they contribute to reducing information overload.

We will start with a presentation of the problem we are trying to solve, followed by a review of the existing literature on the subject. We will then present the methodology and objectives of our research, as well as the theoretical context in which our study is situated. Finally, we will conclude by presenting the expected results of our research.

2 The Problem

The issue of information overload is a growing topic of interest in information systems research, as it represents a diffuse organizational problem that affects all professions. Although numerous studies have sought to define the concept of overload, present its determinants, and highlight its impact on the individual and in professional contexts, research on mitigation mechanisms, particularly on the role of tools, is still in its infancy and deserves greater attention. The importance and originality of this research lie in its exploration of the role of tools as a mechanism for attenuating information overload in an organizational context that is still largely unexplored in the information systems (IS) literature. It is in this context that this proposed thesis topic is situated, contributing to the general call to the research community on the urgent need to promote healthy management of the interaction between humans and technologies and its implications for workplace health [26].

3 State of the Art

Eppler, Martin, and Mengis [21] were the first researchers in the field of Information Systems to highlight the causes of information overload in management discipline, which are related to personal factors, information characteristics, task and process parameters, organizational design, and information technology. Previous research has shown that the use of email in a professional context is a source of information overload [21, 27, 28]. Furthermore, Kalika et al. [28] have demonstrated that com-munication channels overlap with each other without really blending, thus creating a "mille-feuille" effect. According to these authors, this mille-feuille effect is a source of information overload [28]. Recently, Bawden and Robinson [1], in the context of their study, identified the causes of information overload grouped into four themes: too much information; diversity, complexity, and novelty of information; ubiquitous and pushed information; and personal factors and individual differences. Belabbes et al. [20] refer to triggers that designate the elements leading to information overload such as an individual's cognitive state, poorly defined information needs, characteristics of the information, information environment, or the environment in which an individual interacts with information.

Many authors have enumerated the consequences of information overload. For instance, Eppler and Mengis [21] provide a detailed list of consequences observed in management disciplines up to the early years of the millennium, which are categorized as follows: limited strategies for information search and retrieval, arbitrary analysis and organization of information, suboptimal decision-making, and personal distress. More recently, Bawden and Robinson [1] proposed summarizing the consequences of overload into three categories: health effects, inefficiency, disinformation, and fake news. Belabbes et al. [20] identified on the one hand, internal consequences, which refer to the impacts of information overload on individuals such as low creativity [11, 14, 29, 30], attention problems [31], poor learning and acquisition of skills [32], poor well-being [19, 33], lower confidence [34], and increased demands on their working memory [18, 19].

There are three significant orientations in recent research on adaptation strategies for dealing with information overload: human-centered, information process-centered and technology-centered. The human-centered perspective focuses on decision-maker behavior or emotional or physical effects (e.g., stress reduction). This perspective draws on research on IS [35–37]. The information process-centered perspective relies on countermeasures to deal with the complexity and volume of information [38, 39]. The technology-centered perspective involves using technical countermeasures such as filtering agents, search protocols, and visualization. This view has slightly increased in recent research [40]. For example, several academic studies have shown that filtering is an effective strategy for reducing information overload [24, 25, 41–43]. In addition, information architecture and technical solutions such as interactive dashboards for presenting filtered information [25, 44, 45] and "intelligent agents" [46] can also help reduce overload [47]. However, authors such as Badwen and Robinson [1] argue that the use of new tools based on artificial intelligence (AI) can also reduce information overload, although this remains to be proven. However, we can see that the literature in IS is still in its infancy when it comes to adaptation strategies, particularly the role of tools and their usage in reducing information over-load. This is why we have chosen, as part of our doctoral research, to study the functionalities and uses of tools in reducing information overload.

4 Research Objectives and Methodology

Information overload is a pervasive problem that is difficult to characterize precisely and persists in organizations. This research focuses on the role of tools and more specifically their use in mitigating information overload, articulated around three main questions:

1. What are the features and uses of the tool that reduce information overload?
2. How do these features and uses reduce or help control information overload?
3. Are there general principles that can be identified in these features or uses that contribute to the reduction of information overload?

Several empirical studies will be conducted during this research project to identify tools and their functionalities that reduce the information overload of managers. Indeed, we want to study four software tools specific to different application domains such as medicine, academic research, support services, and email management for top executives to deduce general principles of functionalities and uses that can reduce information overload. These domains are strongly affected by information overload. Moreover, the results can also be generalized to other domains, as they highlight the benefits of using tools to manage information overload. Finally, this study provides a solid foundation for further studying the role of tools in reducing information overload in these and other domains.

Firstly, we plan to conduct a controlled experiment in a specific domain using a particular software tool to highlight causal relationships. [48]. Indeed, controlled experimentation is a data collection method that can be used to study the effect of software tools on reducing information overload. This method involves dividing participants into two groups, one group using the software tool and the other group not using it, and then comparing the results between the two groups to assess the effect of the software tool's functionalities and usages in reducing information overload. The use of controlled experiments is limited by several factors, such as the generalization of results, the effect of the social environment, selection bias, the Hawthorne effect, costs and required time, and the influence of the experiment [41]. Therefore, it is important to consider these limitations and complement the results with other methods.

To complement the results of our experiment, we plan to conduct case studies to gather quantitative and qualitative data. The purpose of these case studies is to help us analyze, understand, and explain a complex phenomenon in its natural context [49, 50]. In fact, in the context of this research, a case study could be conducted on an organization that uses a software tool in a business context. The study will involve conducting interviews with user managers to identify the features and uses of this software tool that help reduce information overload. The advantage of case studies is that they allow for a detailed and in-depth analysis of real situations. However, the results of a case study may be limited to the specific situation studied and cannot be generalized to other contexts.

5 Theoretical Background

These analyses and explanations will be based on two relevant theoretical frame-works. First, the Technology Acceptance Model [51], proposes that technology adoption and usage are influenced by two main factors: perceived usefulness and perceived ease of use. This theory helps us understand the factors that influence the perceived effectiveness of software tools in reducing information overload. In other words, through this theory, it would be possible to understand how user managers perceive the usefulness and ease of use of software tools in managing information overload.

Secondly, the Lazarus and Folkman model [52], represents the most influential model in the literature regarding coping strategies. These authors have studied how individuals adapted to stressful situations [53, 54]. According to these authors, the adaptation process is broken down into three sub-processes; individuals evaluate the consequences of the stressful event in terms of threats and opportunities and then a second evaluation of the level of control they must deal with it [55]. The authors distinguish two types of coping strategies: problem-focused coping strategies and emotion-focused coping strategies. Problem-focused coping strategies involve targeting the suppression or attenuation of the stressful event. Emotion-focused coping strategies, on the other hand, aim to minimize the emotions associated with stressful situations [52]. In the context of information overload, user managers of software tools may be faced with a large amount of information to process, which can be stressful and difficult to manage. Software tools can help managers cope with this stress by providing them with means of managing and sorting information more effectively. Specifically, how the features and usage of these software tools can be used as coping strategies to reduce stress and improve the psychological well-being and performance of managers.

6 Expected Tangible Results

As part of this research project, we conducted an initial qualitative study with five manager users of a ticket management tool within a French customer relationship management service provider company. The results showed that colors on the interface, settings of new features, and individual coping strategies are effective in reducing information overload. The results also showed that these features allow for quick access to information, facilitate employee responsiveness, and enhance the productivity of manager users.

As part of a future study in academic research, we plan to investigate the features and uses of AI software tools that would not only help reduce researchers' information overload but also accelerate research by improving the quality of results obtained. It is believed that these AI software tools can assist researchers or students in collecting, analyzing, and interpreting data more quickly and accurately, as well as writing reports and articles more efficiently.

As part of this research project, academic contributions will take the form of conceptual, descriptive, and explanatory models related to the use of tools that contribute to reducing information overload. From a managerial perspective, this research is part of the general problem of the Dark Side of IT with the aim of proposing solutions to overcome information overload in an organizational context. The objective is to contribute to existing knowledge in information systems on coping strategies, highlighting the role of features and uses of software tools in reducing information overload.

7 Conclusion

This study contributes to the existing knowledge in information systems regarding coping strategies, highlighting the role of software tool functionalities and usage in reducing information overload in organizational contexts. It falls within the management of information systems (MIS) research stream commonly referred to as the "Dark Side of IT". By analyzing the role of tools and providing practical recommendations, this research will help managers better understand the role of tools in overcoming information overload and thus improving the effectiveness and well-being of employees.

References

1. Bawden, D., Robinson, L.: Information overload: an introduction. In: Bawden, D., Robinson, L. (eds.) Oxford Research Encyclopedia of Politics. Oxford University Press (2020). https://doi.org/10.1093/acrefore/9780190228637.013.1360
2. McAfee, A., Brynjolfsson, E.: Big data: The management revolution. Harv Bus Rev. 90, (2012)
3. Floridi, L.: Big data and information quality. In: Floridi, L., Illari, P. (eds.) The Philosophy of Information Quality. SL, vol. 358, pp. 303–315. Springer, Cham (2014). https://doi.org/10.1007/978-3-319-07121-3_15
4. Gupta, D., Rani, R.: A study of big data evolution and research challenges. J Inf Sci. **45**, 322–340 (2019). https://doi.org/10.1177/0165551518789880
5. Merendino, A., et al.: Big data, big decisions: the impact of big data on board level decision-making. J. Bus. Res. **93**, 67–78 (2018). https://doi.org/10.1016/j.jbusres.2018.08.029
6. Tarafdar, M., Gupta, A., Turel, O.: The dark side of information technology use: the dark side of information technology use. Inf. Syst. J. **23**(3), 269–275 (2013). https://doi.org/10.1111/isj.12015
7. Gross, B.M.: The Managing of Organizations: The Administrative Struggle. Free Press of Glencoe, New York (1964)
8. Bawden, D., Robinson, L.: The dark side of information: Overload, anxiety and other paradoxes and pathologies. J. Inf. Sci. **35**(2), 180–191 (2009). https://doi.org/10.1177/0165551508095781
9. Junco, R.: In-class multitasking and academic performance. Comput. Human Behav. **28**(6), 2236–2243 (2012). https://doi.org/10.1016/j.chb.2012.06.031

10. Mayer, R.E., Moreno, R.: Nine ways to reduce cognitive load in multimedia learning. Educ. Psychol. **38**(1), 43–52 (2003). https://doi.org/10.1207/S15326985EP3801_6
11. Sabeeh, Z.A., Ismail, Z.: Effects of information overload on productivity in enterprises: A literature review. In: 2013 International Conference on Re-search and Innovation in Information Systems (ICRIIS), pp. 210–214. IEEE (2013)
12. Marques, R.P.F., Batista, J.C.L. (eds.): Information and Communication Overload in the Digital Age. IGI Global (2017)
13. KarrWisniewski, P., Lu, Y.: When more is too much: operationalizing technology overload and exploring its impact on knowledge worker productivity. Comput. Human Behav. **26**(5), 1061–1072 (2010). https://doi.org/10.1016/j.chb.2010.03.008
14. Virkus, S., Mandre, S., Pals, E.: Information overload in a disciplinary context. In: Kurbanoglu, S., Boustany, J., Špiranec, S., Grassian, E., Mizrachi, D., Roy, L. (eds.) Information Literacy in the Workplace, pp. 615–624. Springer International Publishing, Cham (2018). https://doi.org/10.1007/978-3-319-74334-9_63
15. Al-Shamsi, M.: Addressing the physicians' shortage in developing countries by accelerating and reforming the medical education: Is it possible? J Adv Med Educ Prof. **5** (2017)
16. Kabachinski, J.: Coping with information fatigue syndrome, (2004)
17. Hartog, P.: A generation of information anxiety : refinements and recommendations. The Christian librarian : J. Assoc. Chiristian Librarians. 60, (2017)
18. Lauri, L., Virkus, S.: Information overload of academic staff in higher education institutions in Estonia. In: Kurbanoglu, S., et al. (eds.) Information Literacy in Everyday Life, pp. 347–356. Springer International Publishing, Cham (2019). https://doi.org/10.1007/978-3-030-13472-3_33
19. Jackson, T.W., Farzaneh, P.: Theory-based model of factors affecting infor-mation overload. Int J Inf Manage. 32, (2012). https://doi.org/10.1016/j.ijinfomgt.2012.04.006
20. Belabbes, MAmine, Ruthven, I., Moshfeghi, Yar, Pennington, Diane R.: Information overload: a concept analysis. J. Document. **79**(1), 144–159 (2022). https://doi.org/10.1108/JD-06-2021-0118
21. Eppler, M.J., Mengis, J.: The concept of information overload: a review of literature from organization science, accounting, marketing, MIS, and related disciplines, (2004)
22. Aral, S., Brynjolfsson, E., van Alstyne, M.: Information, Technology, and Information Worker Productivity. Inf. Syst. Res. **23**, 849–867 (2012). https://doi.org/10.1287/isre.1110.0408
23. Cameron, A.-F., Webster, J.: Multicommunicating: juggling multiple conversations in the workplace. Inf. Syst. Res. **24**, 352–371 (2013). https://doi.org/10.1287/isre.1120.0446
24. Feng, Y., Agosto, D.E.: The experience of mobile information overload: Struggling between needs and constraints. Information Research. 22, (2017)
25. Saxena, D., Lamest, M.: Information overload and coping strategies in the big data context: evidence from the hospitality sector. J. Inf. Sci. **44**(3), 287–297 (2018). https://doi.org/10.1177/0165551517693712
26. Kefi, H., Kalika, M., Saidani, N.: Dépendance au courrier électronique: effets sur le technostress et la surcharge informationnelle et répercussions sur la per-formance. Syst. Inf. Manage. **26**, 1–45 (2021). https://doi.org/10.3917/SIM.211.0045
27. MaryLizGrisé, R., Gallupe, B.: Information overload: addressing the productivity paradox in face-to-face electronic meetings. J. Manage. Inf. Syst. **16**(3), 157–185 (1999). https://doi.org/10.1080/07421222.1999.11518260
28. Kalika, M., Charki, N.B., Isaac, H.: La théorie du millefeuille et l'usage des tic dans l'entreprise. Revue française de gestion **33**(172), 117–129 (2007). https://doi.org/10.3166/rfg.172.117-129
29. Roetzel, P.G., Fehrenbacher, D.D.: On the role of information overload in information systems (IS) success: Empirical evidence from decision support systems. In: 40th International Conference on Information Systems, ICIS 2019 (2019)

30. Strother, J.B., Ulijn, J.M., Fazal, Z.: Information Overload: An International Challenge for Professional Engineers and Technical Communicators. (2012)
31. Koltay, T.: Information overload in a data-intensive world. In: Advanced Information and Knowledge Processing (2017)
32. Green, Alyssa: Information overload in healthcare management: how the READ Portal is helping healthcare managers. J. Canad. Health Libr. Assoc. **32**(3), 173–176 (2011). https://doi.org/10.5596/c11-041
33. Tan, W.K., Kuo, P.C.: The consequences of online information overload confusion in tourism. Inf. Res. **24**, (2019)
34. Furner, C.P., Zinko, R.A.: The influence of information overload on the development of trust and purchase intention based on online product reviews in a mobile vs. web environment: an empirical investigation. Electron. Mark. **27**(3), 211–224 (2016). https://doi.org/10.1007/s12525-016-0233-2
35. DArcy, J., Gupta, A., Tarafdar, M., Turel, O.: Reflecting on the "Dark side" of information technology use. Commun. Assoc. Inf. Syst. **35**, 109–118 (2014). https://doi.org/10.17705/1CAIS.03505
36. Lee, A.R., Son, S.-M., Kim, K.K.: Information and communication technology overload and social networking service fatigue: a stress perspective. Comput. Human Behav. **55**, 51–61 (2016). https://doi.org/10.1016/j.chb.2015.08.011
37. Plotnick, L., Turoff, M., van den Eede, G.: Reexamining threat rigidity: implications for design. In: Proceedings of the 42nd Annual Hawaii International Conference on System Sciences, HICSS (2009)
38. Lee, B.-K., Lee, W.-N.: The effect of information overload on consumer choice quality in an on-line environment. Psychol. Mark. **21**, 159–183 (2004). https://doi.org/10.1002/mar.20000
39. Sumecki, D., Chipulu, M., Ojiako, U.: Email overload: exploring the moderating role of the perception of email as a 'business critical' tool. Int. J. Inf. Manage. **31**, 407–414 (2011). https://doi.org/10.1016/j.ijinfomgt.2010.12.008
40. Koroleva, K., Bolufe-Röhler, A.: Reducing information overload: Design and evaluation of filtering & ranking algorithms for social networking sites. In: ECIS 2012 - Proceedings of the 20th European Conference on Information Systems (2012)
41. Jones, S.L., Kelly, R.: Dealing With information overload in multifaceted personal informatics systems. Human–Comput. Interact. **33**(1), 1–48 (2018). https://doi.org/10.1080/07370024.2017.1302334
42. Savolainen, R.: Filtering and withdrawing: strategies for coping with information overload in everyday contexts. J. Inf. Sci. **33**(5), 611–621 (2007). https://doi.org/10.1177/0165551506077418
43. Shachaf, O., Aharony, N., Baruchson, S.: The effects of information over-load on reference librarians. Libr. Inf. Sci. Res. **38**, 301–307 (2016). https://doi.org/10.1016/j.lisr.2016.11.005
44. Davis, N.: Information overload, reloaded. Bull. Am. Soc. Inf. Sci. Technol. **37**(5), 45–49 (2011). https://doi.org/10.1002/bult.2011.1720370513
45. Koltay, T.: Information overload, information architecture and digital literacy. Bull. Am. Soc. Inf. Sci. Technol. **38**(1), 33–35 (2011). https://doi.org/10.1002/bult.2011.1720380111
46. Edmunds, A., Morris, A.: The problem of information overload in business organisations: a review of the literature. Int. J. Inf. Manage. **20**(1), 17–28 (2000). https://doi.org/10.1016/S0268-4012(99)00051-1
47. Klerings, I., Weinhandl, A.S., Thaler, K.J.: Information overload in healthcare: too much of a good thing? Zeitschrift für Evidenz, Fortbildung und Qualität im Gesundheitswesen **109**(4–5), 285–290 (2015). https://doi.org/10.1016/j.zefq.2015.06.005
48. Zelkowitz, M.V., Wallace, D.R.: Experimental models for validating technology. Computer **31**(5), 23–31 (1998). https://doi.org/10.1109/2.675630

49. Yin, R.K.: Case study research: design and methods, Applied Social Research Methods Series. (2009)
50. Walsham, G.: Interpretive case studies in IS research: nature and method. Euro. J. Inf. Syst. **4**(2), 74–81 (1995). https://doi.org/10.1057/ejis.1995.9
51. Venkatesh, M., Davis, D.: User acceptance of information technology: Toward a unified view. MIS Q **27**(3), 425 (2003). https://doi.org/10.2307/30036540
52. Lazarus, R.S., Folkman, S.: Stress, Appraisal, and Coping - Richard S. Laza-rus, PhD, Susan Folkman, Ph.D. (1984)
53. Folkman, S., Lazarus, R.S.: An analysis of coping in a middle-aged community sample. J. Health Soc. Behav. **21**(3), 219 (1980). https://doi.org/10.2307/2136617
54. Folkman, S., Lazarus, R.S.: If it changes it must be a process: study of emotion and coping during three stages of a college examination. J. Personal. Soc. Psychol. **48**(1), 150–170 (1985). https://doi.org/10.1037/0022-3514.48.1.150
55. Beaudry, P.: Understanding user responses to information technology: a coping model of user adaptation. MIS Q. **29**(3), 493 (2005). https://doi.org/10.2307/25148693

Tutorials

Blockchain Technology in Information Science: How Blockchains Enable the Connected World

Pooyan Kazemian(✉)

Department of Operations, Weatherhead School of Management, Case Western Reserve University, Cleveland, OH 44106, USA
pooyan.kazemian@case.edu

Abstract. Blockchain is a distributed database (ledger) that is shared among the nodes of a computer network. Network participants follow a protocol (called the consensus mechanism) to agree on which data to add to the ledger. Blockchain makes records of any digital asset transparent and unchangeable without relying on a third-party intermediary.

Blockchain technology has the potential to revolutionize many aspects of information science, enabling a more connected and transparent world. Blockchain allows for the permanent and immutable recording of data and transactions. This makes it possible to exchange anything of value, whether that is a physical item or something less tangible, in a transparent and tamper-proof way. Blockchain can be used to track orders, payments, accounts, production, and much more, enabling more efficient and secure information exchange.

We discuss the fundamentals of blockchain and how this innovative technology can disrupt the field of information science and enable a more connected world.

Keywords: Blockchain · Information science · Connected world

1 Blockchain Technology

Blockchain is a type of distributed ledger or database that stores data in sequential blocks. Network participants maintain a copy of the ledger and jointly agree on adding new blocks of data to the ledger according to a consensus mechanism. These blocks of data are linked together by including the cryptographic hash of the previous block into the current block. Doing so makes blockchain a tamper-proof and immutable ledger because manipulating a block of data after it is added to the blockchain would change the cryptographic hash of that block, which subsequently alters the cryptographic hash of the next block, making the entire chain invalid.

Bitcoin was the first blockchain developed in 2009 by an anonymous person (or group of individuals) under the pseudonym Satoshi Nakamoto. The Bitcoin white paper introduced a peer-to-peer electronic cash system that is decentralized (no single entity controls it) and trustless (no need to trust anybody for the system to work) that operates on a public blockchain. Bitcoin uses the proof-of-work consensus mechanism

S. Nurcan et al. (Eds.): RCIS 2023, LNBIP 476, pp. 673–674, 2023.
https://doi.org/10.1007/978-3-031-33080-3

to secure the network, in which the probability of being selected as the next block producer is tied to the amount of computational power each participant contributes to the network. This ensures that the network remains decentralized and secure without the need for a central intermediary. Most other blockchains, including Ethereum, use a proof-of-stake consensus mechanism, in which the probability of being selected as the next block producer is tied to the number of coins a participant stakes. Proof-of-stake offers more security, fosters decentralization, and significantly reduces the network's energy consumption [1].

2 Blockchain's Role in Information Science and Supply Chain Management

In addition to financial applications, Blockchain technology has the potential to revolutionize information systems and supply chains by addressing fundamental problems related to inefficiency, opacity, and fraud. Blockchain technology can play a vital role in connecting supply chain participants and facilitating the flow of information in a transparent and secure way. For example, blockchain can enhance the traceability of products in a complex supply chain, which is often lacking in traditional supply chains due to data silos. Blockchain achieves so by connecting multiple tiers of a supply chain and allowing for data to be shared with relevant parties in a secure, tamper-proof, and transparent way. Blockchain-based traceability also makes it possible to quickly identify the source of a problem and selectively remove defective products, instead of having to recall the entire product line. This, along with eliminating the need for paper records and carrying documents, can lead to notable gains in speed and efficiency and result in cost savings [2]. Moreover, blockchain technology can enhance product quality and reduce counterfeiting in supply chains by providing a secure end-to-end tracking system. By using blockchain, companies can ensure quality checks at every level, from production to delivery, and identify the provenance or proof-of-origin of a product, making it difficult for counterfeit goods to be introduced across the supply chain [3].

References

1. Buterin, V., Schneider, N.: Proof of Stake: The Making of Ethereum and the Philosophy of Blockchains, 1st edn. Seven Stories Press, New York (2022)
2. Kshetri, N.: Blockchain and Supply Chain Management, 1st edn. Elsevier, Amsterdam (2021)
3. Waller, M.A., Van Hoek, R., Davletshin, M., Fugate, B.: Integrating Blockchain into Supply Chain Management: A Toolkit for Practical Implementation, 1st edn. Kogan Page, London (2019)

Comparing Products Using Similarity Matching

Mike Mannion[1] and Hermann Kaindl[2(✉)]

[1] Glasgow Caledonian University, Glasgow G4 0BA, UK
m.a.g.mannion@gcu.ac.uk
[2] Institute of Computer Technology, TU Wien, Vienna, Austria
hermann.kaindl@tuwien.ac.at

Abstract. The volume, variety and velocity of products in software-intensive systems product lines is increasing. Product comparison is difficult when each product has hundreds of features. Reasons for product comparison include (i) concern for sustainability reasons whether to build a new product or not (ii) evaluating how products differ for strategic positioning reasons (iii) gauging if a product line needs to be reorganized (iv) assessing if a product falls within legislative and regulatory boundaries. We will describe a product comparison approach using similarity matching. A product configured from a product line feature model is represented as a weighted binary string. The similarity between products is compared using a binary string metric. The allocation of feature weights is contested. We will describe one weight allocation method based on a feature's position in the feature model. We will discuss the benefits and limitations of this method using a mobile phone example.

Keywords: Similarity metrics · Feature models

1 Background of Attendees

This tutorial is aimed at product managers, product line engineers, requirements engineers, project managers, software engineers and PhD students.

2 Added Value of Tutorial

The added value of this tutorial is:

- to provide an opportunity to reflect on how significant societal and behavioral trends are affecting what software we build and how we build it
- to draw attention to the role that software reuse can play in the context of societal concerns about financial and environmental sustainability
- to encourage a wider search for software and computing research inspiration given that product comparison techniques and binary string metrics have been deployed in many other fields e.g. biology, ecology.

S. Nurcan et al. (Eds.): RCIS 2023, LNBIP 476, pp. 675–677, 2023.
https://doi.org/10.1007/978-3-031-33080-3

3 Tutorial Plan

Content	Interaction with Attendees
Part 1: Trends in Software Product Line Development Software Product Line Development; Demand Trends – Personalization; Supply Trends – Customization; Matching Supply & Demand – Product Comparisons	• Attendees describe their background and motivation • Invite attendees to take out their mobile phones and begin with an interactive discussion about what features they wanted, what products and features they compared before purchase, what were ignored, how they did the comparison, what were the difficulties of comparison
Part 2: Reuse Approaches – Feature Model-Based Development Feature Models; Product Derivation; Selection from a Product Line Model of Features; Product Verification: Constraint-based Satisfaction, Examples; Tools	• Make point that questions are encouraged throughout session • Attendees invited to contribute with any relevant experience to trying to reuse software • Summary of section presented to audience and attendees invited to confirm understanding
Part 3: Product Comparison using Similarity Matching Techniques Product and Feature Comparisons; Similarity in Domain Engineering & Application Engineering; The Role of Feature-Similarity Models; A Feature-Similarity Model Construction Process	• Make point that questions are encouraged throughout session • Attendees invited to contribute other ideas for doing product and feature comparisons • Summary of section presented to audience and attendees invited to confirm understanding
Part 4: Using Binary Strings for Feature Similarity Matching Representing Product Configurations with Binary Strings; Binary String Metrics; Positive vs. Negative Matches; Weighted Binary Strings; Allocating Weights, Examples; Open-ended Research Questions	• Make point that questions are encouraged throughout session • Invite discussion on benefits of positive and negative matches • Summary of section presented to audience and attendees invited to confirm understanding
Part 5: Summary Lessons Learned, Future Challenges	• Summary of section presented to audience and attendees invited to confirm understanding

References

1. Mannion, M., Kaindl, H.: Using binary strings for comparing products from software-intensive systems product lines. In: Proceedings of the 23rd International Conference on Enterprise Information Systems (ICEIS 2021), pp. 257–266 (2021)
2. Kaindl, H., Mannion, M.: A feature-similarity model for product line engineering. In: Schaefer, I., Stamelos, I. (eds.) ICSR 2015. LNCS, vol. 8919, pp. 34–41. Springer, Cham (2014). https://doi.org/10.1007/978-3-319-14130-5_3

3. Mannion, M., Kaindl, H., Using similarity metrics for mining variability from software repositories. In: Proceedings of 18th the International Conference on Software Product Line Engineering: Companion Volume for Workshops, Demonstrations and Tools, vol. 2, Florence, Italy, pp. 32–35 (2014). 2nd Workshop on Reverse Variability Engineering
4. Kaindl, H., Śmiałek, M., Nowakowski, W.: Case-based reuse with partial requirements specifications. In: Proceedings of 18th IEEE International Requirements Engineering Conference (RE 2010), Sydney, NSW, Australia, pp. 399–400 (2010)

Getting Started with Scriptless Test Automation Through the Graphical User Interface, a Hands-on Tutorial

Olivia Rodriguez Valdés[2(✉)], Beatriz Marín[1], Tanja E. J. Vos[1,2], Fernando Pastor Ricós[1], and Lianne V. Hufkens[2]

[1] Universitat Politècnica de València, València, Spain
{bmarin,tvos,fPastor}@dsic.upv.es
[2] Open Universiteit, Heerlen, The Netherlands
{olivia.rodriguezvaldes,tanja.vos,lianne.hufkens}@ou.nl

Abstract. Automated testing of the graphical user interface (GUI) has traditionally been done using scripted methods. However, these can often lead to expensive maintenance issues. Additionally, script-based testing typically only covers happy-path user scenarios, while neglecting other scenarios that are crucial for testing overall system robustness. Therefore, there is a need to complement the scripted approach with another automated method for GUI testing: script*less* testing.

Script*less* test automation through GUI is a potential solution where tests are generated during execution based on the observed state of the system under test (SUT), rather than using pre-defined scripts. TESTAR (https://testar.org/) [1] is an open-source tool that implements a scriptless approach that consists of repeatedly doing the following: (1) Obtain the state of the SUT, (2) Check the state-based test oracles, (3) Derive the possible actions that the end-user could do, (4) Select and execute one of these actions, (5) Wait for the GUI to update its state.

This tutorial is aimed at a diverse audience including students, researchers, professors, and practitioners who are interested in GUI testing. In this tutorial, attendees will obtain a solid understanding of GUI testing, as well as practical hands-on experience using the testing tool TESTAR.

1 Learning Objectives and Added Value

The goal of this tutorial is to educate attendees on GUI testing, including its benefits and drawbacks. The tutorial starts with a general introduction that helps to understand the importance of software testing, the main characteristics of GUI testing, the scriptless testing technique, and the TESTAR tool.

In addition to theoretical knowledge, the tutorial will provide a hands-on learning experience with the specific GUI testing tool TESTAR. Attendees will learn how to run the tool and identify failures in the generated reports.

By the end of the tutorial, attendees will have an understanding of GUI testing, as well as practical experience using a scriptless testing tool to improve

S. Nurcan et al. (Eds.): RCIS 2023, LNBIP 476, pp. 678–679, 2023.
https://doi.org/10.1007/978-3-031-33080-3

the reliability and quality of their software. Moreover, attendees can use the knowledge of the hands-on tutorial to use TESTAR in their own systems in order to improve and complement their current testing processes. In addition, Professors can use the tool and the instructional material of the tutorial to teach scriptless GUI testing in their courses.

2 History of the Tutorial

Prior to the COVID-19 pandemic in 2019, the TESTAR team conducted multiple tutorials and workshops in a "TESTAR tour" to promote the significance of scriptless testing techniques and the capabilities of the TESTAR tool. These events proved to be fruitful, as they resulted in attendees adopting the tool or collaborations in European research projects. A list of the tutorials performed is presented below:

- 2017, A-TEST, Paderborn (DE)
- 2018, TNO, Groningen (NL)
- 2018, InnSpire company, Utrecht (NL)
- 2018, JISBD, Sevilla (ES)
- 2018, Software Testing course, Quitto (EC)
- 2019, TestNet, Nieuwegein (NL)
- 2019, Newspark BV company, Nieuwegein (NL)
- 2019, Xebia company, Amsterdam (NL)
- 2019, ING bank, Amsterdam (NL)
- 2019, Ministry of Justice and Security, Gouda (NL)

Over the past four years, the TESTAR tool has undergone active development and enhancement, as it has been used to collaborate on innovative research projects. These updates have introduced new techniques, enabling the testing of systems beyond desktop applications, such as web, android, and virtual reality (VR) systems, while also automatically inferring state models.

Given these new software testing capabilities, which go beyond the state-of-the-art, our objective is to resume dissemination tutorials to showcase these capabilities to interested stakeholders.

Acknowledgements. This work was funded by ENACTEST - European innovation alliance for testing education, ERASMUS+ Project 101055874, 2022–2025 (enactest-project.eu).

Reference

1. Vos, T.E.J., Aho, P., Pastor Ricos, F., Rodriguez-Valdes, O., Mulders, A.: Testar-scriptless testing through graphical user interface. Softw. Test. Verification Reliab. **31**(3), e1771 (2021)

How to Develop and Realize Conceptual Models? The Bee-Up Research & Education Support Tool

Wilfrid Utz[1(✉)], Patrik Burzynski[1], and Robert Andrei Buchmann[2]

[1] OMiLAB NPO, Lützowufer 1, 10785 Berlin, Germany
{wilfrid.utz,patrik.burzynski}@omilab.org
[2] Faculty of Economics and Business Administration, Babeş-Bolyai University, Str. T. Mihali 58-60, 400591 Cluj-Napoca, Romania
robert.buchmann@econ.ubbcluj.ro

Abstract. This tutorial focuses on the Bee-Up multi-language modeling tool and its key features that expand the value of conceptual models beyond their basic function as diagrammatic documentation and communication support. Bee-Up supports modeling with several established languages, starting from an initial selection on which its acronym is based - BPMN, EPC, ER, UML, Petri Nets – and expanding with each version to meet diversifying requirements. The goal of the tutorial is to highlight how Bee-Up goes beyond the diagramming use case, to help generate model value - through model analysis (e.g. model queries, simulation), transformation (e.g. to RDF graphs), execution, and integration with external systems (e.g. cyber-physical devices and interfaces). A selection of these is demonstrated during the tutorial, while also covering aspects of what is under the hood of Bee-Up's model processing capabilities.

Keywords: Conceptual modeling education · Model value · Metamodeling

1 Motivation

The tutorial[1] presents Bee-Up[2] - a conceptual modeling tool designed to meet specific needs of educators and researchers, allowing them to exploit the dual nature of models - as human-oriented diagrammatic representations and as knowledge structures that can be processed by machines. Initially, the tool provided integrated support for a selection of established languages [1] - BPMN, EPC, ER, UML, and Petri Nets - and later evolved by expanding this design space (e.g. with DMN, flowcharts), by enriching semantics (with data annotations and machine-readable links between semantically related diagrams) and by expanding functionality to highlight the value of models.

From an educational perspective, Bee-Up is a treatment for a previously discussed "design problem" of conceptual modeling education [2]. Its integration-oriented and

[1] A video teaser of the tutorial is available at https://www.omilab.org/bee-up/tutorial/.

[2] The tool can be obtained at https://bee-up.omilab.org/.

S. Nurcan et al. (Eds.): RCIS 2023, LNBIP 476, pp. 680–682, 2023.
https://doi.org/10.1007/978-3-031-33080-3

iterative nature aims to facilitate the positioning of conceptual modeling as a standalone discipline rather than as auxiliary chapters scattered between discipline-specific use cases (e.g. from business process management, systems design). From a scientific perspective, the tutorial aims to showcase the tool as a typical result of the Agile Modeling Method Engineering framework (AMME) - which enables semantic and functional agility in modeling methods and has recently become the methodological core of OMiLAB's Digital Innovation Environment [3], the experimentation installation used by the OMiLAB Community of Practice having Bee-Up as one of its central toolkits.

2 Tutorial Topics and Tool Capabilities

The tutorial assumes familiarity with the languages supported by Bee-Up. On this foundation, it employs exemplary models to demonstrate a diversity of features expanding model value, implemented with the help of the ADOxx metamodeling platform, available to educators and researchers.[3] The key capabilities demonstrated are (a) language-specific simulations (e.g. the Petri Nets token game); (b) language-independent model queries (e.g. process queries, queries over linked models of different types); (c) model transformation/code generation capabilities - both language-specific (SQL statements out of ER diagrams) and generic (models of any type semantically enriched and converted into RDF graphs to enable semantic queries and reasoning cf. patterns discussed in [4]); and (d) model-driven execution of cyber-physical behavior for robotic devices benefitting from model-IoT integration. Bee-Up evolves through the iterative application of AMME and based on feedback from its user community or hands-on interactions during tutorials and the NEMO Summer School Series.[4] Tutorial participants, educators, and design-oriented researchers are encouraged to adopt or adapt Bee-Up for their own purposes and model-driven environments, and thus to join the OMiLAB Network of Nodes[5] and Community of Practice.

References

1. Karagiannis, D., Buchmann, R.A., Burzynski, P., Reimer, U., Walch, M.: Fundamental conceptual modeling languages in OMiLAB. In: Karagiannis, D., Mayr, H., Mylopoulos, J. (eds.) Domain-Specific Conceptual Modeling, pp. 3–30. Springer, Cham (2016). https://doi.org/10.1007/978-3-319-39417-6_1
2. Buchmann, R.A., Ghiran, A.M., Doeller, V., Karagiannis, D.: Conceptual modeling education as a design problem. Complex Syst. Inf. Model. Q. 21–33 (2019) https://csimq-journals.rtu.lv/article/view/csimq.2019–21.02

[3] The ADOxx metamodeling platform, https://www.adoxx.org/live/home.

[4] NEMO Summer School Series, https://nemo.omilab.org/.

[5] The OMiLAB Network of Nodes, https://www.omilab.org/nodes/.

3. Karagiannis, D., Buchmann, R.A., Utz, W.: The OMiLAB digital innovation environment: Agile conceptual models to bridge business value with digital and physical twins for product-service systems development. Comput. Ind. **138**, 103631 (2022)
4. Karagiannis, D., Buchmann, R.A.: A proposal for deploying hybrid knowledge bases: the ADOxx-to-GraphDB interoperability case. In: Proceedings of the 51st Hawaii International Conference on System Sciences (2018). http://hdl.handle.net/10125/50399

Author Index

S. Nurcan et al. (Eds.): RCIS 2023, LNBIP 476, pp. 683–685, 2023.
https://doi.org/10.1007/978-3-031-33080-3

Printed in the United States
by Baker & Taylor Publisher Services

Printed in the United States
by Baker & Taylor Publisher Services